UJF NATO ASI

Les Houches

Session LXXVIII

2002

Accretion discs, jets and high energy phenomena in astrophysics

Disques d'accrétion, jets et phénomènes de haute énergie en astrophysique

Conférenciers

Abraham Achterberg
Venya Berezinsky
Vasily Beskin
Omer Blaes
Ulrich Briel
Nuria Calvet
Max Camenzind
John Carr
Annalisa Celotti
Bozena Czerny
Frédéric Daigne
Wolfram Freudling
Lee Hartmann
Jean Heyvaerts
Cyril Lachaud
Conor Masterson
Thierry Montmerle
Ralph Pudritz
Rashid Sunyaev

École d'été de Physique des Houches

Session LXXVIII, 29 July-23 August 2002

NATO Advanced Study Institute

Euro Summer School

École thématique du CNRS

Accretion discs, jets and high energy phenomena in astrophysics

Disques d'accrétion, jets et phénomènes de haute énergie en astrophysique

Edited by

Vassily Beskin, Gilles Henri, François Menard, Guy Pelletier and Jean Dalibard

EDP SCIENCES

Les Ulis, Paris, Cambridge

Springer

Berlin, Heidelberg, New York, Barcelona, Hong Kong, London Milan, Paris, Tokyo

Published in cooperation with the NATO Scientific Affair Division

ISSN 0924-8099 print edition
ISSN 1610-3459 online edition

ISBN 3-540-20171-8 Springer-Verlag Berlin Heidelberg New York
ISBN 2-86883-702-6 EDP Sciences Les Ulis

École de Physique des Houches

Service inter-universitaire commun
à l'Université Joseph Fourier de Grenoble
et à l'Institut National Polytechnique de Grenoble

Subventionné par le Ministère de la Jeunesse,
de l'Éducation Nationale et de la Recherche,
le Centre National de la Recherche Scientifique,
le Commissariat à l'Énergie Atomique

Directeurs scientifiques de la session LXXVIII :
Vassily Beskin (Lebedev Institute, Moscow, Russia)
Gilles Henri (Observatoire de Grenoble, France)
François Menard (Observatoire de Grenoble, France)
Guy Pelletier (Observatoire de Grenoble, France)

The students have always been very attentive,
even during the toughest lectures!

Understanding the mysteries of INTEGRAL data processing in Geneva...

Lecturers

Abraham Achterberg, Sterrenkundig Instituut, Utrecht University & Center for High Energy Astrophysics, Amsterdam, The Netherlands

Venya Berezinsky, Laboratori Nazionali Gran Sasso, INFN, Assergi AQ, Italy

Vasily Beskin, Lebedev Physical Institute, Moscow, Russia

Omer Blaes, Physics Department, University of California, Santa Barbara, CA 93106-9530, USA

Ulrich Briel, Max-Planck-Institut für Extraterrestrische Physik, Postfach 1312, 85741 Garching, Germany

Nuria Calvet, Harvard-Smithsonian Center for Astrophysics, 60 Garden Street, Cambridge, MA 02138, USA

Max Camenzind, Landessternwarte, Königstuhl, 69117 Heidelberg, Germany

John Carr, Centre de Physique des Particules de Marseille, 163 avenue Luminy, case 907, 13288 Marseille Cedex 09, France

Annalisa Celotti, International School for Advanced Studies, SISSA/ISAS, via Beirut 2-4, 34014 Trieste, Italy

Bozena Czerny, Nicolaus Copernicus Astronomical Center, Bartycka 18, 00-716 Warsaw, Poland

Frédéric Daigne, Service d'Astrophysique, Bât. 709, CEA Saclay, 91191 Gif-sur-Yvette, France

Wolfram Freudling, European Southern Observatory, Karl Schwarzschild Strasse 2, 85748 Garching, Germany

Lee Hartmann, Harvard-Smithsonian Center for Astrophysics, 60 Garden Street, Cambridge, MA 02138, USA

Jean Heyvaerts, Observatoire Astronomique, 11 avenue de l'Université, 67000 Strasbourg, France

Cyril Lachaud, LPCC, Collège de France CNRS/IN2P3, 11 place Marcellin Berthelot, 75231 Paris Cedex 05, France

Conor Masterson, Max-Planck-Institut für Kernphysik, Saupfercheckweg 1, 69117 Heidelberg, Germany

Thierry Montmerle, Service d'Astrophysique, CEA Saclay, 91191 Gif-sur-Yvette, France

Ralph Pudritz, Department of Physics and Astronomy, McMaster University, Hamilton, ON L8S 4M1, Canada

Rashid Sunyaev, Max-Planck-Institut für Astrophysik, Karl Schwarzschild Strasse, Garching, Germany

A lively babyfoot match.
Some superluminal motions of the ball have been reported.

Yes scientists can also dance!

Participants

Meltem Akyilmaz, University of Cukurova, Faculty of Arts and Science, Physics Department, 01330 Adana, Turkey

Denis Allard, IPN Orsay, Bâtiment 100, 91405 Orsay, France

Carsten Arbeiter, Theoretische Physik IV, Ruhr-Universität Bochum, 44780 Bochum, Germany

Klaus Michael Aye, University of Durham, Department of Physics, Rochester building, Science Laboratories, Durham DH1 3LE, U.K.

Rumen Bachev, Institute of Astronomy, Tsarigradsko Shose 72, Sofia 1784, Bulgaria

Maxim Barkov, Space Research Institute (IKI), 84/32 Profsoyuznaya Str., Moscow 117997, Russia

Matthias Beilicke, Institut für Experimentalphysik, Luruper Chaussee 149, 22761 Hamburg, Germany

Michal Blazejowski, Nicolaus Copernicus Astronomical Center, Polish Academy of Sciences, ul Bartycka 18, 00-716 Warsaw, Poland

Stanislav Boldyrev, Department of Astronomy and Astrophysics, University of Chicago, 5640 S. Ellis Ave., Chicago, IL 60637, U.S.A.

Niccolò Bucciantini, Dipartimento di Astronomia e Scienza dello Spazio, Universita di Firenze, Largo E. Fermi 5, 50125 Firenze, Italy

Manojendu Choudhury, CM 34, Department of Astronomy and Astrophysics, Tata Institute of Fundamental Research, Homi Bhabha Road, Colaba, Mumbai 400005, India

Miguel de Val Borro, Stockholm Observatory, SCFAB, Roslagstullsbacken 21, SE 106 91 Stockholm, Sweden

Johann Dischler, Lund Observatory, Box 43, SE 221 00 Lund, Sweden

Stefano Gabici, Dipartimento di Astronomia e Scienza dello Spazio, Universita di Firenze, Largo E. Fermi 5, 50125 Firenze, Italy

Loïc Girard, LAPP, Chemin de Bellevue, BP. 110, 74941 Annecy le Vieux Cedex, France

Agueda Gras-Velazquez, Dublin Institute for Advanced Studies, School of Cosmic Physics, 5 Merrion Square, Dublin 2, Ireland

Minfeng Gu, Shanghai Astronomical Observatory, Chinese Academy of Sciences, 80 Nandan Road, Shanghai 200030, China

Julien Guy, LPNHE IN2P3/CNRS, Université Paris 6, Tour 33, RdC, 4 place Jussieu, 75252 Paris Cedex 05, France

Ricardo Hueso, Laboratoire Cassini, Observatoire de la Côte d'Azur, BP. 6339, 06304 Nice Cedex 04, France

Ira Verena Jung, MPI für Kernphysik, Saupfercherk weg 1, 69117 Heidelberg, Germany

Yoshiaki Kato, Astrophysics Laboratory, Department of Science and Technology, Chiba University, 1-33 Yayoi-cho, Inage-ku, Chiba 263-8522, Japan

Elena Kazakevich, Astronomical Institute of Saint Petersburg State University, Universitetsky Pr. 28, Stary Peterhoff, St. Petersburg 198504, Russia

Ralf Keil, Max-Planck-Institut für Extraterrestrische Physik, Giessenbachstrasse, 85748 Garching, Germany

Dmitry Khangoulian, Moscow State Engineering Physics Institute Dep. 32, 115409 Kashirskoe shosse 31, Moscow, Russia

Nicolas Leroy, Laboratoire Leprince Ringuet, École Polytechnique, 91128 Palaiseau Cedex, France

Hakima Manseri, Laboratoire Leprince Ringuet, École Polytechnique, 91128 Palaiseau Cedex, France

Sergio Mendoza, Instituto de Astronomia, Universidad Nacional Autonoma de Mexico, Ap. 70-264, Ciudad Universitaria, Distrito Federal CP 04510, Mexico

Bruno Merin Martin, LAEFF-INTA, PO Box 50727, 28080 Madrid, Spain

James Miller-Jones, Astrophysics Department, Oxford University, Denys Wilkinson Building, Keble Road, Oxford, OX1 3RH, U.K.

Joachim Moortgat, Department of Astrophysics, University of Nijmegen, Postbus 9010, 6500 GL Nijmegen, The Netherlands

Marek Nikolajuk, Nicolaus Copernicus Astronomical Center, Polish Academy of Sciences, ul Bartycka 18, 00-716 Warsaw, Poland

Darren O'Brien, Centro de Astrofisica da Universidade de Porto, Rua das Estrelas, 4150-762 Porto, Portugal

Chingis Omarov, Astrophysical Institute, Observatory, Almaty 480020, Kazakhstan

Hideki Ozawa, Institut d'Astrophysique de Paris, 98 bis Boulevard Arago, 75014 Paris, France

Gabriela Pavalas, Str. Ferentari No. 14, Bl. 124, Sc. 2, Et. 2, Ap. 55, Sector 5, Bucuresti, Romania

Donald Payne, School of Physics, University of Melbourne, Parkville, Vic 3010, Australia

Jochem Peitz, University of Tübingen, Department for Theoretical Astrophysics and Computational Physics, Auf der Morgenstelle 10, 72076 Tübingen, Germany

Manuel Perucho Pla, Department of Astronomy and Astrophysics, University of Valencia, C/Dr Moliner, 50 46100 Burjassot, Valencia, Spain

Frédéric Piron, Groupe d'Astroparticules de Montpellier, Bât. 11, CC 085, Université Montpellier II, Place Eugène Bataillon, 34095 Montpellier Cedex 5, France

Nathalie Playez, Laboratoire de l'Accélérateur Linéaire, Université Paris Sud, Bâtiment 200, Bp. 34, 91892 Orsay Cedex, France

Oleksiy Poludnenko, University of Rochester, Department of Physics and Astronomy, Bausch and Lomb Hall 206, Rochester, NY 14627, U.S.A.

Aldo Serenelli, Facultad de Ciencias Astronomicas y Geofisicas, Universidad Nacional de La Plata, Paseo del Bosque S/N, La Plata - B1900FWA, Argentina

Mohsen Shadmehri, Department of Physics, School of Science, Ferdowsi University, 91775 Mashhad, Iran

Anna Sidorova, Department of Physics, Volograd State University, 2-ya Prodolnaya 30, Volograd 400062, Russia

Malgorzata Sobolewska, Nicolaus Copernicus Astronomical Center, Bartycka 18, 00-716 Warsaw, Poland

Ivana Stoklasova, Charles University, Faculty of Mathematics ans Physics, Astronomical Institute, V. Holesovickach 2, 180 00 Praha 8, Czech Republic

Ladislav Subr, Astronomical Institute, Charles University, V. Holesovickach 2, 180 00 Praha 8, Czech Republic

Valeriu Tudose, Astronomical Institute of the Romanian Academy, Str. Cutitul de Argint 5, Bucharest, Romania

Dmitri Uzdensky, Institute for Theoretical Physics, Kohn Hall, University of California, Santa Barbara, CA 93106, U.S.A.

Ana Vasile, Str. Maguricea, No. 3, Bl. 3F, Sc. 2, Ap 23 Sector 1, Bucharest, Romania

Tatiana Yelenina, Keldysh Institute of Applied Mathematics of Russian Academy of Science 4, Miusskaya Sq., Moscow 125047, Russia

Preface

Astrophysics has always left an important imprint on Les Houches summer school. Thirty years after the famous session on "Black Holes" organized by Evry Schatzman (1972), where many great ideas of modern astrophysics were launched, and where many of the students became renowned astrophysicists, we felt it timely to take stock of all the impressive progresses made in the domains of compact galactic and extragalactic objects, young stellar objects and high energy astrophysics, from both the theoretical developments and observational standpoints. These topics are at a watershed and are blossoming at the forefront of new trends opened by the so-called "astroparticles" physics, which proposes new means of investigation of the universe where particles (neutrinos, cosmic rays, and also gravitational waves) serve as messengers of cosmic information. Therefore, we felt it was important, as one of the schools goals, to prepare the community for new breakthroughs.

During this school, several classes of phenomena have been studied such as X-ray binary systems, microquasars, active galactic nuclei, young stellar objects, pulsars and gamma ray bursts. They all share common features, namely: a central compact object, an accretion disk, outflows (sometimes collimated jets), and high energy radiations. These topics address important key-issues of modern astrophysics.

A series of fundamental lectures (Sect. 1) provided the audience with both the required background and the new physical processes involved in the research field. They covered Magnetohydrodynamics (Jean Heyvaerts), Relativistic Hydrodynamics (Vasily Beskin), Physics of accretion disks around Black Holes (Omer Blaes), Physics of Astrophysical Jets and Accretion-Ejection models (Ralph Pudritz). The important issue of the excitation of turbulence in magnetised accretion disks recently recieved a decisive insight, analysed by impressive numerical simulations.

The lectures devoted to high energy astrophysics (Sect. 2) opened our minds to new frontiers of Astrophysics. Gamma Ray Bursts (Frédéric Daigne) are probably the most attractive phenomena of high energy astrophysics nowadays, because they are not only the favorite events of gamma ray astronomy, but also potential candidates for future neutrino astronomy, as well as promising sources of gravitational waves. They could also be the main astronomical sources of ultra high energy cosmic rays (UHECR's), whose origin remains an enigma. New important progresses in Fermi acceleration, especially in the relativistic regime, have been derived recently (Abraham Achterberg), reviving the possibility of UHECRs generation by some astrophysical objects; this is the so-called "bottom up" scenario to solve the enigm of UHECR origin (note that, because of an unexpected impediment to Abraham Achterberg's coming, the lecture was given by Guy Pelletier). The "top down" scenario, on the other hand, involves new physics, beyond the standard model of particle physics (Venya Berezinski).

The investigation of Active Galactic Nuclei (Sect. 3) has benefited from a new impulse provided by new high energy observatories and multiwavelength campaigns from radio to gamma rays, including very crucial observations in the X-ray range, allowing a deeper understanding of their nature. Together with X-ray binary systems (Rachid Sunyaev, not reported in the book), they are one of the main manifestations of Black Hole environments (Max Camenzind), giving the opportunity of testing the General Relativity Theory. The accretion that takes place in these environments is particularly rich and has been studied in detail for many years (Bozena Czerny). These objects are also powerful sources of high energy emission, originating close to the central Black Hole at the beginning of the jets (Annalisa Celotti). Intermediate mass Black Holes would now complete the bestiary but they were not, at the time of the school, in a stage of clear identification. Pulsars were also presented as interesting high energy objects with their specific nebulae (not reported in the book). Strange Quark Stars were not yet discovered at the time of the school. If they really exist, they will be at the origin of a high energy emission that will have to be compared with neutron stars and stellar black holes.

At first sight, it may look strange to have included lectures on young stellar objects (YSO's) (Sect. 4) in a school discussing, for the large part, high energy astrophysics. Rapid inspection quickly reveals however that the similarities are numerous. Furthermore, their proximity, *i.e.*, the solar neighbourhood, guarantees that their accretion disks are well resolved and well studied (Lee Hartmann), with promises of one day looking directly at its interface with the central engine when large interferometers become operational. The powerful outflows and jets they drive are also reminiscent of jets in other classes of objects. Accurate diagnostics of accretion can therefore be obtained by studying them (Nuria Calvet). YSO's also emit X-rays abundantly and their production mechanisms and impact on the immediate circumstellar environment are now being studied with unprecedented details with the new facilities available (Thierry Montmerle).

In that context, comparing and cross fertilising both fields of astrophysics, that of compact objects and that of young stellar objects, held the promises of a very exciting and fruitful summer school for a large community of astrophysicists.

At the time of the school, several important instrumental projects were nearly completed and the first observation campaigns are expected rapidly, especially the Pierre Auger Observatory (Cyril Lachaud) and Gamma Ray Observatories (Conor Masterson) such as HESS and VERITAS. Seminars have been also given on Neutrino Observatories and Gravitational Waves Observatories (they are not reported in the book).

We thank all contributors for the effort they have made in preparing these lecture notes, which often cover much more material than the lectures themselves. We strongly hope that this set of lectures will, in the tradition of many earlier volumes of the Les Houches series, provide a useful introduction and serve as a reference for several years for researchers and students in this very open and rapidly evolving field.

People from Chamonix valley are used to enjoying conferences on astrophysics topics, given by lecturers of Les Houches school. Gilles Henri gave a very successful public conference on "Black Holes" at Les Houches village.

Acknowledgments

The 78th Les Houches summer school and the present volume have been made possible by the financial support of the following institutions, whose contribution is gratefully acknowledged:

- the "High level scientific conference" program of the Research Directorate of the European Commission under grant HPCF-CT-2001-00018.
- the "Advanced Scientific Institute" program of the Scientific and Environmental Affairs division of NATO, under grant ASI-978104.
- the "Lifelong learning" program and the "Groupement de recherche Phénomènes Cosmiques des Hautes Énergies" of the Centre National de la Recherche Scientifique (France).
- the Université Joseph Fourier, the French Ministry of Research and the Commissariat à l'Énergie Atomique, through their constant support to the Physics School.

The permanent staff of the School, especially Brigitte Rousset and Isabelle Lelièvre, have been of invaluable assistance at every stage of the preparation and development of the school, and we would like to thank them warmly on behalf of all students and lecturers. We also thank the former director François David for his help during the early stages of preparation, and the school scientific council for encouraging us in preparing a session on this topic.

Guy PELLETIER
Gilles HENRI
François MENARD
Vasily BESKIN
Jean DALIBARD

CONTENTS

Section I: The physics of accretion and ejection

Course 1. Accretion and Ejection-Related MHD

by J. Heyvaerts *3*

Section II: High energy astrophysics

Course 7. Cosmic Rays and Particle Acceleration at Astrophysical Shocks

by A. Achterberg *313*

Section III: Black holes and AGN's

Course 8. The Black Hole Environments

by M. Camenzind *405*

Course 9. Accretion Around Active Galactic Nuclei

by B. Czerny *461*

Section IV: Pre-main sequence objects

Section V: New instrumentation

The “grolle”, a typical object in Savoie. It is filled with a mixture of coffee, alcohol and sugar. Not to be used before a lecture!

Section I: The physics of accretion and ejection

COURSE 1

ACCRETION AND EJECTION-RELATED MHD

J. HEYVAERTS

Observatoire de Strasbourg,
11 rue de l'Université,
67000 Strasbourg,
France

Contents

ACCRETION AND EJECTION-RELATED MHD

J. Heyvaerts

Abstract

This lecture is an introduction to MHD. Relevant equations, both in the classical and special-relativistic regimes are derived. The magnetic field evolution is considered both in the perfect-MHD limit and when weak resistivity is present, giving rise to reconnection flows. A short section gives a flavour of dynamo theory. Examples of simple stationnary flows and equilibria are then presented. Stationnary, axisymmetric, rotating perfect-MHD winds and jets are discussed in some more detail. Their asymptotic structure is described. The last sections deal with small motions about an equilibrium and stability. These issues are illustrated by a few classical examples. The last section discusses linear aspects of the magneto-rotationnal instability.

1 Introduction

Accretion-ejection structures in AGN's and binary systems consist most often of ionized, sometimes hot, gases [26], which are electrically conducting, even when weakly ionized. Electric currents are easily generated in these plasmas by the $(\boldsymbol{v} \times \boldsymbol{B})$ electromotive field. The magnetic field $\boldsymbol{B}$ partly ows its existence to these *in situ* currents which exert on the plasma a Lorentz force density $\boldsymbol{j} \times \boldsymbol{B}$ which bears on its motion. The fluid velocity, the current and the magnetic field are intimately coupled by this interaction. Magneto-Hydro-Dynamics (MHD) is an extension of Hydro-Dynamics (HD) to electrically conducting media. It provides equations which describe the plasma motion under the effect of all forces, including electro-magnetic ones. It also provides equations for the associated evolution of the electromagnetic field. This series of lectures is an introduction to the rather vast subject of MHD. The applications described are motivated by accretion-ejection physics. The reader is assumed to have some knowledge of electromagnetism, basic university physics and vector calculus. The following section is a quick presentation of HD and MHD. Pedagogically oriented presentations can also be found in [12] and [18].

2 The MHD model

MHD visualizes the medium as a single fluid described by "MHD fields", such as the mass density of the fluid $\rho(\boldsymbol{r},t)$, its local velocity $\boldsymbol{v}(\boldsymbol{r},t)$, its local pressure $P(\boldsymbol{r},t)$ or its local temperature $T(\boldsymbol{r},t)$. The electromagnetic field is described by the local values of the magnetic and electric fields, $\boldsymbol{B}(\boldsymbol{r},t)$ and $\boldsymbol{E}(\boldsymbol{r},t)$. These fields are supported by electric current and charge densities, $\boldsymbol{j}(\boldsymbol{r},t)$ and $\rho_{\mathrm{e}}(\boldsymbol{r},t)$. This is an Eulerian description, because $\boldsymbol{r}$ represents a fixed position in the chosen reference frame, hereafter "the lab frame". The velocity $\boldsymbol{v}(\boldsymbol{r},t)$ is the velocity of the fluid that happens to be present at $\boldsymbol{r}$ at time t. A fluid, especially when under-dense, may not be "hydrodynamical" in character. The condition for an hydrodynamical description to be valid is that the mean-free-path of any particle in the medium be shorter than any relevant gradient scale.

2.1 Following the motion

An observer riding a fluid element observes time-dependent local physical quantities. such as density, temperature or velocity of this parcel of fluid. Let $g(\boldsymbol{r},t)$ be the Eulerian field representing such a quantity and $g_{\mathrm{l}}(t)$ be its time variation as seen by an observer following the fluid element which is present at $\boldsymbol{r}$ at some particular time t. Let $\boldsymbol{r}_{\mathrm{l}}(t)$ be its motion. From the derivative chain rule,

$$\frac{\mathrm{d}g_{\mathrm{l}}}{\mathrm{d}t}=\frac{\mathrm{d}}{\mathrm{d}t}\left(g(\boldsymbol{r}_{\mathrm{l}}(t),t)\right)=\frac{\partial g}{\partial t}+\boldsymbol{v}\cdot\boldsymbol{\nabla}g. \tag{2.1}$$

2.2 Conservation equations in general

Any scalar extensive quantity, G, is characterized by its Eulerian density field, $g(\boldsymbol{r},t)$, its flux $\boldsymbol{\Phi}(\boldsymbol{r},t)$ and its volumic creation rate $s(\boldsymbol{r},t)$ which satisfy the conservation equation

$$\frac{\partial g}{\partial t}+\mathrm{div}\boldsymbol{\Phi}=s. \tag{2.2}$$

This equation expresses the fact that the change of G in any small volume V is due to creation or destruction inside V and to the flux entering or leaving through the boundary of V. If the extensive quantity, $\boldsymbol{G}$, is vectorial, such as momentum for example, its density and its volumic creation rate are also vectorial. Its flux is a second order tensor of Euclidean space. In a cartesian frame $\mathcal{R}$, the ab component of this flux tensor is the a-component of the flux vector associated with the b-component of $\boldsymbol{G}$. The corresponding conservation equation can be written, in geometrical notation, as:

$$\frac{\partial \boldsymbol{g}}{\partial t}+\mathrm{div}\overline{\overline{\Phi}}=\boldsymbol{s}. \tag{2.3}$$

2.3 Conservation of mass and charge

In non-relativistic physics mass can neither be created nor destroyed. Therefore its volumic creation rate vanishes. The mass flux is $\rho \boldsymbol{v}$. The mass conservation equation is:

$$\frac{\partial \rho}{\partial t} + \mathrm{div}(\rho \boldsymbol{v}) = 0. \qquad (2.4)$$

Let, similarly, ρ_{e} be the charge density. The charge flux is the total electric current density $\boldsymbol{j}$. Since charge is globally conserved, its volumic creation rate vanishes and the charge conservation equation is:

$$\frac{\partial \rho_{\mathrm{e}}}{\partial t} + \mathrm{div}\boldsymbol{j} = 0. \qquad (2.5)$$

The current $\boldsymbol{j}$ is the sum of a convected charge flux, $\rho_{\mathrm{e}}\boldsymbol{v}$, which is due to charge transport by the bulk motion of the fluid, and of a microscopic flux, or conduction current, $\boldsymbol{j}_{\mathrm{cond}}$:

$$\boldsymbol{j} = \rho_{\mathrm{e}}\boldsymbol{v} + \boldsymbol{j}_{\mathrm{cond}}. \qquad (2.6)$$

2.4 Conservation of momentum

A quick way to momentum conservation equation is to write down the fundamental law of mechanics for a small fluid parcel, following its motion. The neighbouring parcels of fluid exert pressure and viscosity forces at its boundary. The net force exerted by the environment through a surface element $\mathrm{d}\boldsymbol{S}_{\mathrm{out}}$ of this boundary can be written as:

$$\mathrm{d}\boldsymbol{F} = -P\mathrm{d}\boldsymbol{S}_{\mathrm{out}} + \overline{\overline{\sigma}} \cdot \mathrm{d}\boldsymbol{S}_{\mathrm{out}} \qquad (2.7)$$

where $\overline{\overline{\sigma}}$ is the viscous stress tensor and $\mathrm{d}\boldsymbol{S}_{\mathrm{out}}$ is oriented outwards. The viscous stress tensor results from the velocity gradients. In an isotropic fluid, the ab component of this tensor is:

$$\sigma_{ab} = \eta(\rho, T)\left(\nabla_a v_b + \nabla_b v_a - \frac{2}{3}\,\delta_{ab}\,\mathrm{div}\boldsymbol{v}\right) + \zeta(\rho, T)\,\delta_{ab}\,\mathrm{div}\boldsymbol{v}. \qquad (2.8)$$

The coefficients η is the shear dynamical viscosity coefficient and ζ is the bulk, or second, dynamical viscosity coefficient. The latter is usually negligible under astrophysical conditions [12]. The fluid parcel is also subject to volume forces, such as gravity or Lorentz forces. Let $\boldsymbol{f}_{\mathrm{vol}}$ be the total volume force density. The fundamental law of mechanics applied to an infinitesimal fluid element reduces to the equation of motion of hydrodynamics:

$$\rho\,(\partial \boldsymbol{v}/\partial t + (\boldsymbol{v} \cdot \boldsymbol{\nabla})\boldsymbol{v}) = -\boldsymbol{\nabla} P + \mathrm{div}\,\overline{\overline{\sigma}} + \boldsymbol{f}_{\mathrm{vol}}. \qquad (2.9)$$

This equation can be given a conservative form by using equation (2.4):

$$\partial(\rho \boldsymbol{v})/\partial t + \mathrm{div}(\rho \overline{\overline{vv}} + P\,\overline{\overline{\delta}} - \overline{\overline{\sigma}}) = \boldsymbol{f}_{\mathrm{vol}}. \tag{2.10}$$

The ab component of the tensor $\overline{\overline{vv}}$ is $v^a v^b$. Equation (2.10) is the equation of conservation for *matter* momentum. However, matter interacts with other entities, such as the electromagnetic field, exchanging momentum with them. These exchanges of momentum between the matter and the electro magnetic field are represented by the electromagnetic volume force $\boldsymbol{f}_{\mathrm{em}} = \rho_{\mathrm{e}}\boldsymbol{E} + \boldsymbol{j} \times \boldsymbol{B}$. From Maxwell equations it can be shown that:

$$\boldsymbol{f}_{\mathrm{em}} = -\frac{\partial}{\partial t}\left(\frac{\boldsymbol{E}\times\boldsymbol{B}}{\mu_0 c^2}\right) - \mathrm{div}\left(\frac{1}{2}\left(\epsilon_0 E^2 + \frac{B^2}{\mu_0}\right)\overline{\overline{\delta}} - \epsilon_0\overline{\overline{EE}} - \frac{\overline{\overline{BB}}}{\mu_0}\right). \tag{2.11}$$

Substituting this in equation (2.10) a more global equation of conservation, for the material and electromagnetic forms of momentum, is obtained:

$$\frac{\partial}{\partial t}\left(\rho\boldsymbol{v} + \epsilon_0(\boldsymbol{E}\times\boldsymbol{B})\right) + \mathrm{div}\left(\rho\overline{\overline{vv}} + P\overline{\overline{\delta}} + \overline{\overline{\psi}}_{\mathrm{em}}\right) = \boldsymbol{g} + \boldsymbol{f}_{\mathrm{vol}\neq\mathrm{em}} \tag{2.12}$$

where the total momentum flux tensor of the electromagnetic field, the so-called Maxwell stress tensor, is

$$\overline{\overline{\psi}}_{\mathrm{em}} = (\epsilon_0 E^2/2 + B^2/2\mu_0)\,\overline{\overline{\delta}} - \epsilon_0\overline{\overline{EE}} - \overline{\overline{BB}}/\mu_0. \tag{2.13}$$

The momentum density of the electromagnetic field is $\epsilon_0(\boldsymbol{E}\times\boldsymbol{B})$.

2.5 Conservation of energy

The first law of thermodynamics can be similarly written for a moving fluid parcel which is followed in its motion. Care should be taken of the fact that the work performed by the fluid element against the external world includes both pressure and viscosity forces. Denoting by U the internal energy density, by $\boldsymbol{q}$ the heat flux and by H the volumic heating rate, this leads to the conservation equation for the internal energy of matter:

$$\partial U/\partial t + \mathrm{div}\,(U\ \boldsymbol{v} + \boldsymbol{q}) = H - P\,\mathrm{div}\boldsymbol{v} + (\overline{\overline{\sigma}}\cdot\boldsymbol{\nabla})\cdot\boldsymbol{v}. \tag{2.14}$$

The volumic rate of creation on the right of equation (2.14) describes the exchanges of internal energy with other forms of energy, such as kinetic energy of organized motions or electromagnetic energy. It is of course possible to write down conservation equations for more global forms of energy. For example, the conservation equation for the kinetic energy of organized motions is obtained by dot-multiplying the equation of motion (2.9) by $\boldsymbol{v}$.

Using equation (2.4) and adding the resulting equation to equation (2.14) we obtain the conservation equation for both these forms of material energy:

$$\frac{\partial}{\partial t}\left(\frac{1}{2}\rho v^2 + U\right) + \mathrm{div}\left(\frac{1}{2}\rho v^2\ \boldsymbol{v} + (U+P)\boldsymbol{v} + \boldsymbol{q} - \boldsymbol{v}\cdot\overline{\overline{\sigma}}\right) =$$
$$\rho\boldsymbol{v}\cdot\boldsymbol{g} + \boldsymbol{v}\cdot\boldsymbol{f}_{\mathrm{vol}\neq\mathrm{em}} + \boldsymbol{v}\cdot(\rho_{\mathrm{e}}\boldsymbol{E} + \boldsymbol{j}\times\boldsymbol{B}) + H. \quad (2.15)$$

It is also possible to transform those source terms in equation (2.15) which are associated with exchanges between material and electromagnetic energy to a conservative form and to incorporate them into a conservation equation for material and electromagnetic forms of energy. The Joule heating term should be singled out of the general heating rate H. Using Maxwell's equations, it can be shown that

$$\boldsymbol{j}\cdot\boldsymbol{E} = -\frac{\partial}{\partial t}\left(\frac{\epsilon_0 E^2}{2} + \frac{B^2}{2\mu_0}\right) - \mathrm{div}\left(\frac{\boldsymbol{E}\times\boldsymbol{B}}{\mu_0}\right). \quad (2.16)$$

This changes equation (2.15) to:

$$\frac{\partial}{\partial t}\left(\frac{1}{2}\rho v^2 + U + \frac{1}{2}\epsilon_0 E^2 + \frac{1}{2}\frac{B^2}{\mu_0}\right) +$$
$$\mathrm{div}\left(\frac{1}{2}\rho v^2\boldsymbol{v} + (U+P)\boldsymbol{v} + \boldsymbol{q} - \boldsymbol{v}\cdot\overline{\overline{\sigma}} + \frac{\boldsymbol{E}\times\boldsymbol{B}}{\mu_0}\right) =$$
$$\rho\boldsymbol{v}\cdot\boldsymbol{g} + \boldsymbol{v}\cdot\boldsymbol{f}_{\mathrm{vol}\neq\mathrm{em}} + H_{\neq\mathrm{joule}}. \quad (2.17)$$

Further progress can be made in globalizing the energy conservation equation when gravity is independent of time. This assumption excludes motions of self-gravitating systems. Introducing the gravitational potential Φ_{g}, supposedly independent of time, and using equation (2.4), it is found that

$$\rho\boldsymbol{v}\cdot\boldsymbol{g} = -\frac{\partial}{\partial t}(\rho\Phi_{\mathrm{g}}) - \mathrm{div}(\rho\Phi_{\mathrm{g}}\boldsymbol{v}). \quad (2.18)$$

Inserting this in equation (2.17) an equation for internal, kinetic, gravitational and electromagnetic forms of energy is obtained. In the absence of other forces and heating this equation reduces to:

$$\frac{\partial}{\partial t}\left(\frac{1}{2}\rho v^2 + U + \frac{1}{2}\epsilon_0 E^2 + \frac{1}{2}\frac{B^2}{\mu_0} + \rho\Phi_{\mathrm{g}}\right)$$
$$+\,\mathrm{div}\left(\frac{1}{2}\rho v^2\boldsymbol{v} + (U+P)\boldsymbol{v} + \boldsymbol{q} - \boldsymbol{v}\cdot\overline{\overline{\sigma}} + \frac{\boldsymbol{E}\times\boldsymbol{B}}{\mu_0} + \rho\Phi_{\mathrm{g}}\boldsymbol{v}\right) = 0. \quad (2.19)$$

2.6 Coupling to Maxwell's equations

The spatial structure and time evolution of the electromagnetic fields, $\boldsymbol{E}$ and $\boldsymbol{B}$, are described by Maxwell's equations:

$$\mathrm{rot}\boldsymbol{E} = -\frac{\partial \boldsymbol{B}}{\partial t} \tag{2.20}$$

$$\mathrm{div}\boldsymbol{B} = 0 \tag{2.21}$$

$$\frac{\mathrm{rot}\boldsymbol{B}}{\mu_0} = \boldsymbol{j} + \epsilon_0 \frac{\partial \boldsymbol{E}}{\partial t} \tag{2.22}$$

$$\epsilon_0 \mathrm{div}\boldsymbol{E} = \rho_{\mathrm{e}}. \tag{2.23}$$

The MKSA system of units is used, μ_0 being the magnetic permeability of free space ($\mu_0 = 4\pi\ 10^{-7}$ H m^{-1}) and ϵ_0 its dielectric permittivity. The product $\epsilon_0\mu_0 = (1/c^2)$, c being the speed of light. The currents and densities which appear in equations (2.22) and (2.23) refer to all charges and currents, be they "internal" or "external" to the medium. Ohm's law in its usual form, $\boldsymbol{j} = \sigma_{\mathrm{e}}\boldsymbol{E}$, gives the current density in the fluid frame in terms of the electric field in this frame. The DC electrical conductivity of a plasma, σ_{e}, is given by

$$\sigma_{\mathrm{e}} = 1.6 \times 10^{-2}\ T^{\frac{3}{2}}\ (\ln \Lambda)^{-1}\ \mathrm{Mho\ m^{-1}}. \tag{2.24}$$

This expression is valid when the thermal motion of electrons is dominated by collisions with ions. The conductivity σ_{e} is isotropic when the electron-ion collision frequency is larger than the electron gyrofrequency. This stringent condition is not always met. When it is not, the electric conductivity may be tensorial, instead of being scalar. A pedagogically-oriented discussion can be found in [8] and [18]. Here we consider only isotropic conductivity, which is of little consequence since conductivity may often be regarded as infinite. Transforming Ohm's equation from the fluid to the lab frame, it takes a form valid for a moving fluid:

$$\boldsymbol{j}_{\mathrm{cond,\ lab}} = \sigma_{\mathrm{e}}\left(\boldsymbol{E} + \boldsymbol{v} \times \boldsymbol{B}\right) \tag{2.25}$$

The total current in the lab frame, $\boldsymbol{j}$, is the sum of $\boldsymbol{j}_{\mathrm{cond,\ lab}}$ and the convected current $\rho_{\mathrm{e}}\boldsymbol{v}$.

3 Non relativistic MHD

Considerable simplification occurs when the plasma motions and the wave phase velocities are all non-relativistic. Let T be the characteristic time over which the MHD quantities vary and L be the characteristic scale length of these variations. The ratio $V \equiv L/T$ is of order of the fluid velocity or of order of the phase velocity of MHD waves propagating in this plasma. Assuming both to be much less than the speed of light, an order of magnitude analysis of equation (2.20) reveals that the ratio of the induction electric field to the magnetic field is of order V. Unless very large external electrostatic fields are imposed on the medium, the ratio E/B should be of order V. From this it results that the displacement current can be neglected in equation (2.22), being of order V^2/c^2 compared to the quantity on its left. The convection current $\rho_e \boldsymbol{v}$ is likewise negligible compared to the conduction current. The electric force density $\rho_e \boldsymbol{E}$ is negligible to the Lorentz force density $\boldsymbol{j} \times \boldsymbol{B}$ and the electric energy density $\epsilon_0 E^2/2$ is negligible to the magnetic energy density $\mu_0 B^2/2$. Then, the electric charge density ρ_e disappears from all equations, but from the Coulomb equation (2.23). The latter is then of no use in non-relativistic MHD unless one wishes to calculate the charge density explicitly.

3.1 Non relativistic MHD equations

Let us sum up and present the set of non-relativistic MHD equations. It consists of the three conservation laws of hydrodynamics, with Lorentz force and Joule heating, of the simplified Maxwell equations (2.20)–(2.22), without the unnecessary equation (2.23), and of appropriate equations for auxiliary quantities. Disregarding radiative forces and heating, MHD is described by the following set of equations

$$\frac{\partial \rho}{\partial t} + \mathrm{div}(\rho \boldsymbol{v}) = 0 \tag{3.1}$$

$$\rho \left(\frac{\partial \boldsymbol{v}}{\partial t} + (\boldsymbol{v} \cdot \boldsymbol{\nabla}) \boldsymbol{v} \right) = \rho \boldsymbol{g} - \boldsymbol{\nabla} P + \mathrm{div} \overline{\overline{\sigma}} + \boldsymbol{j} \times \boldsymbol{B} \tag{3.2}$$

$$\begin{aligned} &\frac{\partial}{\partial t} \left(U + \frac{1}{2}\rho v^2 + \rho \Phi_{\mathrm{g}} + \frac{B^2}{2\mu_0} \right) \\ &+ \mathrm{div} \left(\frac{1}{2}\rho v^2 \boldsymbol{v} + (U+P)\boldsymbol{v} + \boldsymbol{q} - \boldsymbol{v} \cdot \overline{\overline{\sigma}} + \rho \Phi_{\mathrm{g}} \boldsymbol{v} + \frac{\boldsymbol{E} \times \boldsymbol{B}}{\mu_0} \right) = 0 \end{aligned} \tag{3.3}$$

$$\mathrm{rot} \boldsymbol{E} = -\frac{\partial \boldsymbol{B}}{\partial t} \tag{3.4}$$

$$\mathrm{div} \boldsymbol{B} = 0 \tag{3.5}$$

$$\mathrm{rot} \boldsymbol{B} = \mu_0 \boldsymbol{j} . \tag{3.6}$$

Equations (3.1)–(3.6) must be complemented with supposedly known equations of state and transport relations. Neglecting the second viscosity, these are:

$$P = P(\rho, T) \qquad\qquad U = U(\rho, T) \tag{3.7}$$

$$\boldsymbol{q} = -\chi(\rho, T)\boldsymbol{\nabla} T \qquad\qquad \boldsymbol{j} = \sigma_{\mathrm{e}}(\rho, T)(\boldsymbol{E} + \boldsymbol{v} \times \boldsymbol{B}) \tag{3.8}$$

$$\sigma_{ij} = \eta(\rho, T)\left(\nabla_i v_j + \nabla_j v_i - \frac{2}{3}\delta_{ij}\mathrm{div}\boldsymbol{v}\right) \cdot \tag{3.9}$$

The gravity field $\boldsymbol{g}$, the radiative force and volumic heating rate, if present, must be calculated independently by Newton's theory of gravitation or by the theory of radiative transfer.

3.2 Incompressible limit and Navier-Stokes equation

A simple particular case is found in the limit of an incompressible, non-conducting fluid in the absence of a gravitational field. An hydrodynamical fluid may be regarded as incompressible if its motion is subsonic. When ρ is constant in space and time, equation (3.1) reduces to

$$\mathrm{div}\ \boldsymbol{v} = 0. \tag{3.10}$$

The viscous force density simplifies to:

$$\mathrm{div}\ \overline{\overline{\sigma}} = \eta \nabla^2 \boldsymbol{v}. \tag{3.11}$$

Defining the kinematic viscosity by $\nu = \eta/\rho$, equation (3.2) takes the familiar form of the Navier-Stokes equation:

$$\frac{\partial \boldsymbol{v}}{\partial t} + (\boldsymbol{v} \cdot \boldsymbol{\nabla})\boldsymbol{v} = -\frac{\boldsymbol{\nabla} P}{\rho} + \nu \nabla^2 \boldsymbol{v}. \tag{3.12}$$

The vorticity of the flow is

$$\boldsymbol{\omega} = \mathrm{rot}\boldsymbol{v}. \tag{3.13}$$

Taking the rotational of equation (3.12) and using the vector calculus identity

$$\boldsymbol{\nabla}(v^2) = 2(\boldsymbol{v} \cdot \boldsymbol{\nabla})\boldsymbol{v} + 2\boldsymbol{v} \times \mathrm{rot}\boldsymbol{v} \tag{3.14}$$

we obtain an evolution equation for $\boldsymbol{\omega}$:

$$\frac{\partial \boldsymbol{\omega}}{\partial t} = \mathrm{rot}\,(\boldsymbol{v} \times \boldsymbol{\omega}) + \nu \Delta \boldsymbol{\omega}. \tag{3.15}$$

This equation, which is similar to equation (6.2) below, is in fact non-linear, since $\boldsymbol{\omega}$ and $\boldsymbol{v}$ are related by equation (3.13).

3.3 Magnetic pressure and tension

The Lorentz force is the sum of the gradient of a magnetic pressure and of a magnetic tension force. From equation (3.14) it can be shown that:

$$\boldsymbol{j} \times \boldsymbol{B} = \frac{\mathrm{rot}\boldsymbol{B}}{\mu_0} \times \boldsymbol{B} = -\boldsymbol{\nabla}\left(\frac{B^2}{2\mu_0}\right) + (\boldsymbol{B}\cdot\boldsymbol{\nabla})\,\frac{\boldsymbol{B}}{\mu_0}\cdot \tag{3.16}$$

The first term on the right of equation (3.16) is the gradient of the "magnetic pressure"

$$P_{\mathrm{M}} = B^2/(2\mu_0). \tag{3.17}$$

The second one is the magnetic tension force. A more sophisticated expression, which makes it more obvious that the Lorentz force is perpendicular to the magnetic field $\boldsymbol{B}$, is obtained by writing the magnetic field as $B\boldsymbol{t}$, B being its modulus and $\boldsymbol{t}$ a unit vector tangent to the field line. Then,

$$(\boldsymbol{B}\cdot\boldsymbol{\nabla})\,\boldsymbol{B} = B\boldsymbol{t}\cdot\boldsymbol{\nabla}\,(B\boldsymbol{t}) = \boldsymbol{t}\left((\boldsymbol{t}\cdot\boldsymbol{\nabla})\left(\frac{B^2}{2}\right)\right) + B^2\,((\boldsymbol{t}\cdot\boldsymbol{\nabla})\boldsymbol{t})\,\cdot \tag{3.18}$$

The Frénet formula gives $(\boldsymbol{t}\cdot\boldsymbol{\nabla})\boldsymbol{t}$ in terms of the principal normal $\boldsymbol{N}$ to the field line and its radius of curvature R_{c}:

$$(\boldsymbol{t}\cdot\boldsymbol{\nabla})\,\boldsymbol{t} \equiv \frac{\mathrm{d}\boldsymbol{t}}{\mathrm{d}s} = \frac{\boldsymbol{N}}{R_{\mathrm{c}}}\cdot \tag{3.19}$$

The Lorentz force can then be written as:

$$\boldsymbol{j}\times\boldsymbol{B} = -\left(\boldsymbol{\nabla}\frac{B^2}{2\mu_0} - \boldsymbol{t}\,\left(\boldsymbol{t}\cdot\boldsymbol{\nabla}\frac{B^2}{2\mu_0}\right)\right) + \left(\frac{B^2}{\mu_0}\right)\frac{\boldsymbol{N}}{R_{\mathrm{c}}}\cdot \tag{3.20}$$

Both terms on the right are perpendicular to $\boldsymbol{B}$. The second one is properly the "magnetic tension" force. It grows larger when the radius of curvature of the field line becomes smaller.

3.4 Alfvén speed

Just as pressure and density define the square of sound speed, the magnetic pressure and density define the square of a velocity which is characteristic of the propagation of magnetic signals in an MHD fluid. This speed, the Alfvén speed, usually denoted by v_{A} or c_{A}, plays a very important role in all MHD phenomena. It is defined by

$$v_{\mathrm{A}}^2 = \frac{B^2}{\mu_0\rho}\cdot \tag{3.21}$$

4 Special relativistic MHD

4.1 Four-vectors and tensors in minkowskian space

The reader is assumed to have some knowledge of the theory of special relativity, which introduces the concept of geometrized space-time. This brief summary is only meant to establish the notations. Any event may be labeled, in a given reference frame, by space-time coordinates x^a, with the (upper placed) index a taking the values 1, 2, 3 or 4. The fourth coordinate is associated with time. A minkowskian frame uses cartesian space coordinates and a time coordinate ct, t being the time recorded by synchronized observers. The space-time distance between two neighbouring events is expressed, in minkowskian coordinates, by

$$\mathrm{d}s^2 = c^2\mathrm{d}t^2 - \mathrm{d}x^2 - \mathrm{d}y^2 - \mathrm{d}z^2 \equiv \eta_{ab}\mathrm{d}x^a\mathrm{d}x^b. \tag{4.1}$$

The relation on the right defines the components of the metric matrix η_{ab}. It makes use of the dummy index rule, according to which any repeated index implicitly implies a summation on the values that it may take. This rule is used throughout. The time-like convention assigns a positive sign to the time part. Some results may differ by a sign when the opposite, space-like, convention is used. The coordinates x^a of an event in one particular minkowskian frame $\mathcal{R}$ are related to the coordinates $x^{a'}$ in another frame $\mathcal{R}'$ by a Lorentz transformation matrix $X_a^{a'}$ so that $x^{a'} = X_a^{a'}x^a$. The inverse matrix of $X_a^{a'}$ is denoted by $X_{a'}^a$. Space-time vectors, also called four-vectors, are represented by bold-faced letters. Their components in a given frame come in two different kinds, the contra-variant and the co-variant ones. Contravariant components, denoted by an upper placed label, like V^a, change under a change of reference frame as the coordinates x^a do. The associated covariant components, denoted by a lower placed label, like V_a, are $V_a = \eta_{am}V^m$. This relation can easily be inverted. Under a change of reference frame, the covariant components of a four-vector change according to $V_{a'} = X_{a'}^m V_m$. The partial derivatives with respect to minkowskian coordinates form the *covariant* components of a four-vector operator, ∇:

$$\nabla_a = \left(\frac{\partial}{\partial x}, \frac{\partial}{\partial y}, \frac{\partial}{\partial z}, \frac{1}{c}\frac{\partial}{\partial t}\right). \tag{4.2}$$

The scalar product of two four-vectors $\mathbf{U}$ and $\mathbf{V}$ is $\mathbf{U}\cdot\mathbf{V} = U_m V^m$. The four-divergence of a four-vector field $\mathbf{V}(\boldsymbol{r}, t)$ is simply the four-vector scalar product $\nabla\cdot\mathbf{V}$. In a given minkowskian frame, tensors of the second rank are represented by a set of (doubly contravariant) components, T^{ab}, which depend on two indices. When changing from a frame $\mathcal{R}$ to a frame $\mathcal{R}'$ these components change, by definition, as $T^{a'b'} = X_a^{a'}X_b^{b'}T^{ab}$ Tensors are four-geometrical objects. They may also be represented by mixed type, or fully covariant, components. These are obtained from the fully contravariant ones, T^{ab}. For example $T_a^b = \eta_{am} T^{mb}$. Similar operations are to be used for changing the variance of any other index.

Most usual physical quantities are components of four-geometrical objects. For example, the small interval between two successive events along the life line of a particle is a four-vector $\mathrm{d}\mathbf{M}$. The associated four-velocity $\mathbf{u}$ of this particle or fluid element is $\mathbf{u} = \mathrm{d}\mathbf{M}/\mathrm{d}s$. Denoting by γ the usual relativistic Lorentz factor, the contravariant components, u^a, of the four-velocity are given in terms of the usual velocity by $(u^a) = (\gamma \boldsymbol{v}, \gamma c)$. The four-momentum of a particle of rest mass m is $\mathbf{p} = m\mathbf{u}$. Its contravariant components, p^a are the relativistic momentum and energy (divided by c) of the particle. The current density $\boldsymbol{j}$ is associated with the charge density ρ_e in a four-current, $\mathbf{j}$, with covariant components $j^a = (\boldsymbol{j}, c\rho_\mathrm{e})$ The vector potential $\boldsymbol{A}$ and the electric potential Φ_e are associated in a four-potential, $\mathbf{A}$, with components $A^a = (\boldsymbol{A}, \Phi_\mathrm{e}/c)$ Any scalar extensive quantity $\mathcal{G}$ which is distributed in space-time is characterized by a density field $g(\mathbf{x})$, a three-vector flux $\boldsymbol{F}(\mathbf{x})$ and a volumic creation rate $s(\mathbf{x})$, which is a Lorentz scalar. The numbers $g^a = (\boldsymbol{F}, cg)$ are the components of a four-vector $\mathbf{g}$, the flux-density object associated with $\mathcal{G}$. The conservation equation in which $\boldsymbol{F}$ and g appear can be written as a four-divergence:

$$\nabla_a g^a = s. \tag{4.3}$$

An antisymmetric four-tensor, the electro-magnetic field tensor, $\mathbf{F}$, represents both the electric and the magnetic field. Its fully *covariant* components are defined by:

$$F_{ab} = \nabla_a A_b - \nabla_b A_a. \tag{4.4}$$

To within a sign, all non-vanishing components of $\mathbf{F}$ are cartesian components of either the electric or the magnetic field. Maxwell's equations can be elegantly written in terms of the tensor $\mathbf{F}$.

A most important four-tensor is the energy-momentum tensor, which is the density-flux object associated with energy-momentum. Energy and momentum are components of a four-vectorial quantity distributed in space. In a given reference frame, a volume density, g^b, a flux three-vector, $\boldsymbol{F}^b$ and a volumic rate of creation s^b are associated with each of these four components, b. They all satisfy a conservation equation. All these densities and fluxes turn out to be components of a unique four-tensorial quantity, the energy-momentum tensor $\mathbf{T}$, and the four conservation equations merge into a single four-geometrical conservation equation involving this tensor. In a given frame, the components of $\mathbf{T}$ are defined as follows. When $a = x$, y, z the component T^{ab} is the a-component of the flux vector $\boldsymbol{F}^b$. When $a = t$,

$T^{ab} = cg^b$. All four conservation equations merge in the single vectorial equation

$$\nabla_a T^{ab} = s^b. \tag{4.5}$$

The energy-momentum tensor of the electromagnetic field is:

$$T^{ab}_{\text{e.m.}} = \frac{1}{\mu_0}\left(-F^{am}F^{b}_{m} + \frac{1}{4}F^{mn}F_{mn}\right). \tag{4.6}$$

In the absence of viscosity and thermal conduction (perfect fluid), the energy momentum tensor of matter can be expressed as:

$$T^{ab}_{\text{matt.}} = \left(\rho + \frac{P}{c^2}\right)u^a u^b - P\eta^{ab}. \tag{4.7}$$

Here, ρc^2 represents the total proper energy density of matter and P the proper pressure. A "proper" quantity is the quantity measured in the instantaneous rest frame of the fluid. The proper energy density ρc^2 incorporates rest mass energy as well as other forms of matter internal energy, such as thermal internal energy. The components u^a refer to the fluid's four-velocity in the lab frame. In the presence of heat conduction and viscosity, the contribution of the viscous stress tensor and of the heat flux should be included in $\mathbf{T}_{\text{matt}}$ as well (see Chap. 22 of Ref. [19]). We shall not expand on this here, although this is of importance for discussing relativistic accretion disks.

4.2 Equations of special relativistic MHD

When the system consisting of matter and electromagnetic field does not exchange energy-momentum with any other entity, its total energy-momentum tensor obeys a perfect conservation equation, in which no volumic creation source term appears:

$$\nabla_a\left(T^{ab}_{\text{matt}} + T^{ab}_{\text{e.m.}}\right) = 0. \tag{4.8}$$

The four components of this four-vectorial equation replace the three components of the equation of motion and the energy conservation equation of non-relativistic MHD. There is no separate conservation equation for mass because mass is but one particular form of energy. A conservation equation may however be written for the baryonic number, which is a globally conserved quantity. The four-vectorial flux-density associated with baryonic number is $n_{\text{B}}\mathbf{u}$, where n_{B} is the proper baryonic number density and $\mathbf{u}$ is the fluid's four velocity. The corresponding conservation equation is $\nabla_a(n_{\text{B}}u^a) = 0$. It can be written in three-dimensional form as:

$$\frac{\partial}{\partial t}(\gamma n_{\text{B}}) + \text{div}\,(\gamma n_{\text{B}}\boldsymbol{v}) = 0. \tag{4.9}$$

The proper pressure and proper energy density ρc^2 can be expressed in terms of $n_{\rm B}$ and of the proper specific entropy per baryon, $s_{\rm B}$:

$$P = P(n_{\rm B}, s_{\rm B}) \qquad\qquad \rho c^2 = \epsilon(n_{\rm B}, s_{\rm B}). \tag{4.10}$$

For a perfect fluid, in which viscosity, heat conduction and electric resistivity are absent, $s_{\rm B}$ is constant following the motion. Ohm's law can be written as $\boldsymbol{j} = \sigma_{\rm e}\boldsymbol{E}$ in the fluid's instantaneous rest frame. This relation can be expressed in terms of four-geometrical quantities as

$$j^a - \frac{u^a u_b}{c^2} j^b = \sigma_{\rm e} F^{ab} u_b. \tag{4.11}$$

In three-dimensional form the spatial part of equation (4.11) is:

$$\boldsymbol{j} - \gamma^2\left(\rho_{\rm e}\boldsymbol{v} - \frac{(\boldsymbol{v}\cdot\boldsymbol{j})\boldsymbol{v}}{c^2}\right) = \gamma\sigma_{\rm e}\left(\boldsymbol{E} + \boldsymbol{v}\times\boldsymbol{B}\right). \tag{4.12}$$

For infinite conductivity this reduces, as in non relativistic MHD, to

$$\boldsymbol{E} + \boldsymbol{v}\times\boldsymbol{B} = 0. \tag{4.13}$$

Equation (4.8) has four components which, in usual three-dimensional form, can be written as:

$$\gamma^2\left(\rho + \frac{P}{c^2}\right)(\partial_{\rm t}\boldsymbol{v} + (\boldsymbol{v}\cdot\boldsymbol{\nabla})\boldsymbol{v}) = -\boldsymbol{\nabla}P + \rho_{\rm e}\boldsymbol{E} + \boldsymbol{j}\times\boldsymbol{B} - \frac{\boldsymbol{v}}{c^2}(\boldsymbol{E}\cdot\boldsymbol{j} + \partial_{\rm t}P) \tag{4.14}$$

$$\frac{\partial}{\partial t}\left(\gamma^2\rho c^2 + (\gamma^2 - 1)P\right) + {\rm div}\left(\gamma^2(\rho c^2 + P)\boldsymbol{v}\right) = \boldsymbol{j}\cdot\boldsymbol{E}. \tag{4.15}$$

For $P \ll \rho c^2$ and $v \ll c$, equation (4.14) reduces to the equation of motion of non-relativistivc MHD. Neglecting terms of order smaller than $(v/c)^2$ and $P/\rho c^2$, equation (4.15) reduces to:

$$\frac{\partial}{\partial t}\left(\gamma^2\rho c^2\right) + {\rm div}\left(\gamma^2(\rho c^2 + P)\boldsymbol{v}\right) = \boldsymbol{j}\cdot\boldsymbol{E}. \tag{4.16}$$

The proper energy density, ρc^2, consists of the rest mass energy density $\rho_{00}c^2$, and the thermal energy density U, which is of order P. The rest mass density in the laboratory frame is not ρ_{00} but $\rho_0 = \gamma\rho_{00}$, because of the Lorentz contraction. Therefore, to order v^2/c^2,

$$\gamma^2\rho c^2 = \gamma^2(\rho_{00}c^2 + U) = \gamma\rho_0 c^2 + \gamma^2 U \approx \rho_0 c^2 + \frac{1}{2}\rho_0 v^2 + U. \tag{4.17}$$

Separating terms of different orders, equation (4.16) can be rewritten as

$$\partial_{\rm t}(\rho_0 c^2) + \boldsymbol{\nabla}\cdot(\rho_0 c^2\boldsymbol{v}) + \partial_{\rm t}\left(\frac{\rho_0 v^2}{2} + U\right) + \boldsymbol{\nabla}\cdot\left(\frac{\rho_0 v^2}{2}\boldsymbol{v} + (U + P)\boldsymbol{v}\right) = \boldsymbol{j}\cdot\boldsymbol{E}. \tag{4.18}$$

The lowest order term gives the rest mass conservation equation, which also results from the baryonic number conservation equation. The terms of order $(v/c)^2$ reduce to the classical equation of conservation of kinetic and internal energy of matter.

To sum up, the special-relativistic MHD equations for a perfect isentropic fluid, in which the entropy per baryon is uniformly distributed in space and constant in time, consist of the relativistic hydrodynamical equations (4.9), (4.14), (4.15), the Maxwell equations (2.20)–(2.23), Ohm's law in its perfect form (4.13) and the equations of state (4.10) in which the proper entropy per baryon is to be regarded as a constant parameter. A distinctive feature of relativistic MHD is that neither the electric charge density nor the electric force density may be neglected. The full set of Maxwell equations, with the displacement current included, must be used.

5 Magneto hydrostatic equilibria

5.1 General properties of magneto hydrostatic equilibria

In an equilibrium state, the relevant equations (3.2) and (2.14) reduce to the force balance equation $-\boldsymbol{\nabla} P + \boldsymbol{j} \times \boldsymbol{B} + \rho \boldsymbol{g} = 0$ and the energy balance equation $\mathrm{div}\boldsymbol{q} = H$ which expresses the balance between heating and heat conduction. These two equilibrium equations must be supplemented by an equation of state, by some specific expression for the heating (or cooling) rate H and by a transport law giving the heat flux. Under non-relativistic conditions, the current density and the magnetic field are related by Ampere's law (3.6). In the absence of gravity the force balance equation reduces to:

$$\boldsymbol{\nabla} P = \boldsymbol{j} \times \boldsymbol{B}. \tag{5.1}$$

For such equilibria, the pressure is constant along any field line. This can be seen by dotting equation (5.1) with $\boldsymbol{B}$. Simple solutions can easily be found for planar structures, in which the magnetic field and the pressure depend on an unique cartesian coordinate, and for plama column structures in which the cylindrical polar components of the magnetic field and the pressure depend on the distance r to the axis only.

The properties of the equilibrium depend on the ratio of the forces on both sides of equation (5.1). Let L be a characteristic gradient scale, supposedly the same for the pressure and the magnetic field. The pressure gradient is of order P/L if the variations of P are of the order of P itself. Similarly, $|\boldsymbol{j}| \approx |\boldsymbol{B}|/\mu_0 L$. Then, α being the angle between the directions of $\boldsymbol{j}$ and $\boldsymbol{B}$,

$$|\boldsymbol{j} \times \boldsymbol{B}| \approx \sin\alpha \; (B^2/\mu_0 L). \tag{5.2}$$

The Lorentz force can be neglected compared to the gas pressure force whenever the ratio

$$\beta = \frac{P}{(B^2/2\mu_0)} \tag{5.3}$$

is large. Conversely, when β is small, the gas pressure is unable to balance fully developped Lorentz forces. The angle α must then be small, *i.e.* $\sin\alpha$ must be of order β. In the limit of a vanishingly small β, equation (5.1) reduces to:

$$\boldsymbol{j} \times \boldsymbol{B} = 0. \tag{5.4}$$

A magnetic structure satisfying equation (5.4) with a non vanishing current is said to be "force free". Equation (5.4) can be equivalently written as:

$$\mathrm{rot}\boldsymbol{B} = \lambda(\boldsymbol{r})\ \boldsymbol{B}. \tag{5.5}$$

The function $\lambda(\boldsymbol{r})$ is constant along field lines, which can be seen by taking the divergence of (5.5), with the result $\boldsymbol{B} \cdot \boldsymbol{\nabla}\lambda = 0$. When λ is constant in space, the force free field is said to be linear.

5.2 *Axisymmetric MHD equilibria and Grad Shafranoff equation*

Some MHD equilibria are invariant under some simple transformation. The main situations of interest are those of translational or axial symmetry. Axially symmetric structures are invariant under a rotation about an axis. We shall restrict to this case. Let us use cylindrical coordinates r, θ, z. Any scalar quantity, or any cylindrical component of a vector quantity is independent of θ. A vector quantity $\boldsymbol{V}$ has a poloidal part $\boldsymbol{V}_{\mathrm{P}}$, which is its projection on the local meridian plane and a toroidal, or azimuthal, part $\boldsymbol{V}_\theta$. Axisymmetry implies that the poloidal field $\boldsymbol{B}_{\mathrm{P}}$ is solenoidal. It then derives from a toroidal vector potential and can be written as

$$\boldsymbol{B}_{\mathrm{P}} = \mathrm{rot}\left(\frac{a(r,z)}{r}\ \boldsymbol{e}_\theta\right) = \frac{\boldsymbol{\nabla} a \times \boldsymbol{e}_\theta}{r} = -\frac{1}{r}\frac{\partial a}{\partial z}\ \boldsymbol{e}_{\mathrm{r}} + \frac{1}{r}\frac{\partial a}{\partial r}\ \boldsymbol{e}_z. \tag{5.6}$$

Taking $a(0,z) = 0$, the flux through a circle perpendicular to the axis, centered at $(0,z)$, with radius r is $\Phi_{\mathrm{m}}(r,z) = 2\pi a(r,z)$. The surfaces of constant $a(r,z)$, called magnetic surfaces, are flux tubes. The electric current is given by equation (3.6). The associated Lorentz force is:

$$\boldsymbol{j} \times \boldsymbol{B} = -\frac{1}{\mu_0 r^2}\left(\boldsymbol{\nabla}\left(\frac{r^2 B_\theta^2}{2}\right) + \Delta^* a\ \boldsymbol{\nabla} a\right) - \frac{1}{\mu_0 r^2}\left(\boldsymbol{\nabla} a \times \boldsymbol{\nabla}(rB_\theta)\right) \tag{5.7}$$

Δ^* is an operator similar to, but different from, the Laplacian, Δ:

$$\Delta^* a = r\frac{\partial}{\partial r}\frac{1}{r}\frac{\partial a}{\partial r} + \frac{\partial^2 a}{\partial z^2}\cdot \tag{5.8}$$

In the absence of gravitation, the pressure P is constant along a field line. By axisymmetry, P is also constant on the magnetic surface generated by rotating this line about the axis of symmetry. The pressure is said to be a "surface function". Its value depends only on the value, a, that $a(r,z)$ assumes on this surface:

$$P(r,z) = P(\ a(r,z)\ ) \equiv P(a). \tag{5.9}$$

Using equations (5.9), (5.1) becomes:

$$\frac{1}{r^2}\left(\boldsymbol{\nabla}\left(\frac{r^2 B_\theta^2}{2}\right) + (\Delta^* a)\ \boldsymbol{\nabla} a + \boldsymbol{\nabla} a \times \boldsymbol{\nabla}(rB_\theta)\right) + \mu_0 \frac{\mathrm{d}P}{\mathrm{d}a}\ \boldsymbol{\nabla} a. = 0. \tag{5.10}$$

By axisymmetry, $a(r,z)$ and (rB_θ) do not depend on the azimuthal variable and their gradients are poloidal. The toroidal part of equation (5.10) is

$$\boldsymbol{\nabla} a \times \boldsymbol{\nabla}(rB_\theta) = 0. \tag{5.11}$$

This indicates that $\boldsymbol{\nabla} a$ and $\boldsymbol{\nabla}(rB_\theta)$ are everywhere parrallel. Since $\boldsymbol{\nabla} a$ is along the normal to surfaces of constant $a(r,z)$ and $\boldsymbol{\nabla}(rB_\theta)$ along the normal to surfaces of constant (rB_θ), equation (5.11) shows that the surfaces of constant $a(r,z)$ and of constant (rB_θ) everywhere have the same normal. Thus surfaces of constant (rB_θ) coincide with surfaces of constant a. This means that (rB_θ) is also a surface function which may be written as:

$$rB_\theta(r,z) = \frac{J(a(r,z))}{2\pi} \equiv \frac{J(a)}{2\pi}. \tag{5.12}$$

From Ampere's law, the total current through a circle of radius r centered on the axis at z is

$$J(r,z) = 2\pi r B_\theta = J(a). \tag{5.13}$$

That J should be a surface function was to be expected, since, by dotting equation (5.1) with $\boldsymbol{j}$ it results that $\boldsymbol{j} \cdot \boldsymbol{\nabla} P = 0$, which means that the current flows on surfaces of constant P, which also are magnetic surfaces. Using equation (5.12), the poloidal component of equation (5.10) can be written as:

$$-\Delta^* a = \frac{\mathrm{d}}{\mathrm{d}a}\left(\frac{J^2(a)}{8\pi^2}\right) + \mu_0 r^2 \frac{\mathrm{d}P}{\mathrm{d}a}. \tag{5.14}$$

This is the so-called equilibrium Grad Shafranoff equation. It is a non-linear elliptic equation for $a(r,z)$. The functions $P(a)$ and $J(a)$ which appear in (5.14) must be found from other considerations than just equilibrium. Because the equation is non-linear, neither the existence nor the uniqueness of a solution for given boundary conditions is granted for any $P(a)$ and $I(a)$ [1, 11]. An axisymmetric force free field is a solution of equation (5.14) without pressure term on the right. It is easy to generalize equation (5.14) to account for the presence of an axisymmetric gravitational potential. This formulation of equilibrium condition can be extended to the relativistic rotation of pulsar magnetospheres as described in [4].

6 Magnetic field evolution

Using Maxwell's equations and Ohm's law, equations (3.4), (3.6), (3.8), we obtain an equation for the evolution of the magnetic field:

$$\frac{\partial \boldsymbol{B}}{\partial t} = \mathrm{rot}(\boldsymbol{v} \times \boldsymbol{B}) - \mathrm{rot}\left(\frac{1}{\mu_0 \sigma_\mathrm{e}} \mathrm{rot}\boldsymbol{B}\right) \cdot \tag{6.1}$$

The magnetic diffusivity $\eta_\mathrm{m} = (\mu_0 \sigma_\mathrm{e})^{-1}$ is often regarded as being constant in space, because its effect is usually either negligible or localized. Assuming this, equation (6.1) becomes:

$$\frac{\partial \boldsymbol{B}}{\partial t} = \mathrm{rot}(\boldsymbol{v} \times \boldsymbol{B}) + \eta_\mathrm{m}\, \Delta \boldsymbol{B}. \tag{6.2}$$

Equation (6.2) is similar to equation (3.15) but for the important difference that $\boldsymbol{B}$, unlike $\boldsymbol{\omega}$, is a vector field unrelated to the velocity field. Nevertheless a number of important consequences are common to equations (6.2) and (3.15) (see Sect. (6.1). The first term on the right of equation (6.2) represents the effect of the electro-motive field while the other term represents Ohmic dissipation.

6.1 The perfect MHD limit

Which one, of the electromotive term and the dissipative term, dominates in equation (6.2)? Let L be a characteristic scale of the magnetic field gradients. The ratio of these two terms is approximately

$$|\mathrm{rot}(\boldsymbol{v} \times \boldsymbol{B})| / |\eta_\mathrm{m} \Delta \boldsymbol{B}| \approx (vB/L)/(\eta_\mathrm{m} B/L^2) = vL/\eta_\mathrm{m}. \equiv R_\mathrm{m}. \tag{6.3}$$

The dimensionless number R_m is the magnetic Reynolds number. The large size of astrophysical objects and the good electrical conductibility of gaseous plasmas cause the magnetic Reynolds number of MHD flows in these objects to be usually, though not always, very large. Similarly comparing the advective and dissipative terms in equation (3.15) involves a viscous Reynolds number:

$$R_\mathrm{v} = \frac{vL}{\nu} \cdot \tag{6.4}$$

It is then appropriate to take the limit of infinite R_m and R_v since this latter number is also very large under usual astrophysical conditions. Taking the zero resitivity, zero viscosity and zero heat conduction limit defines the perfect MHD limit. In this regime the magnetic field evolves under the electromotive term alone, *i.e.* as:

$$\frac{\partial \boldsymbol{B}}{\partial t} = \mathrm{rot}(\boldsymbol{v} \times \boldsymbol{B}). \tag{6.5}$$

The electrical resistivity σ_{e}^{-1} being regarded as strictly zero, $\boldsymbol{j}/\sigma_{\mathrm{e}}$ vanishes and Ohm's law, equation (3.8), simplifies to

$$\boldsymbol{E} + \boldsymbol{v} \times \boldsymbol{B} = 0. \tag{6.6}$$

Equation (6.6) physically means that the electric field vanishes in the fluid's instantaneous rest frame: the very good plasma conductor finds at any time a state of electrostatic equilibrium in its own rest frame. Equation (6.5) follows by taking the rotational of equation (6.6).

In the perfect MHD limit, the flux through any circuit moving with the fluid remains constant in time and the fluid elements connected by a field line at some time remain so later on. These results are known as "flux freezing" and "field freezing" theorems resp.. This term refers to the fact that matter and field accompany eachother in the motion. Let us prove these theorems.

Consider at time t_1 a circuit C_1, each point of which accompanies the local fluid in its motion. At time $t_2 = t_1 + \mathrm{d}t$ a point M_1 on this circuit has moved to a position M_2 given by $\boldsymbol{OM}_2 = \boldsymbol{OM}_1 + \boldsymbol{v}(M_1, t_1)\ \mathrm{d}t = \boldsymbol{OM}_1 + \mathrm{d}\boldsymbol{M}_1$. The locus of all points M_2 defines the circuit C_2. Let S be the closed surface consisting of a surface S_1 spanning C_1, another surface S_2 spanning C_2 and the surface Σ generated by the vectors $\boldsymbol{v}(M)\mathrm{d}t$ for all M's on C_1, joining C_1 to C_2 (Fig. 1). The magnetic flux at time t_2 through S is zero. Separating the flux integral through S in three parts, this implies that

$$\int_{S_1} \boldsymbol{B}(P, t_2) \cdot \mathrm{d}\boldsymbol{S}_{\mathrm{out}} + \int_{S_2} \boldsymbol{B}(P, t_2) \cdot \mathrm{d}\boldsymbol{S}_{\mathrm{out}} + \int_{\Sigma} \boldsymbol{B}(P, t_2) \cdot \mathrm{d}\boldsymbol{S}_{\mathrm{out}} = 0. \tag{6.7}$$

The second term on the right of equation (6.7) is Φ_2, the flux at time t_2 through C_2. But the first term is not the flux Φ_1 at time t_1 through C_1 because $\boldsymbol{B}(P, t_2)$ is involved. However,

$$\boldsymbol{B}(P, t_2) = \boldsymbol{B}(P, t_1) + \frac{\partial \boldsymbol{B}}{\partial t}\mathrm{d}t = \boldsymbol{B}(P, t_1) + \mathrm{rot}(\boldsymbol{v} \times \boldsymbol{B})\mathrm{d}t. \tag{6.8}$$

In the third term on the right of equation (6.7), $\boldsymbol{B}(P, t_1)$ can be taken for $\boldsymbol{B}(P, t_2)$ because the surface Σ is infinitesimal. Substituting (6.8) in (6.7), we get:

$$\Phi_2 - \Phi_1 + \int_{\Sigma} \boldsymbol{B}(P) \cdot \mathrm{d}\boldsymbol{S}_{\mathrm{out}} + \mathrm{d}t \int_{S(C_1)} \mathrm{rot}(\boldsymbol{v} \times \boldsymbol{B}) \cdot \mathrm{d}\boldsymbol{S}_{\mathrm{out}} = 0. \tag{6.9}$$

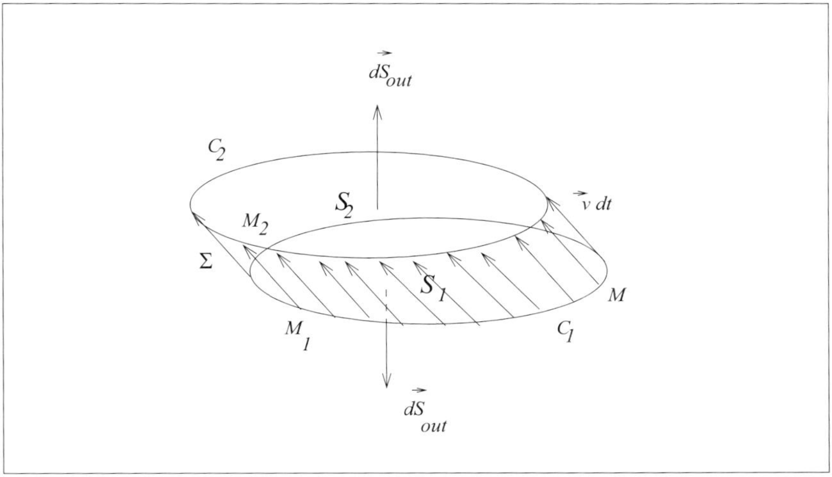

Fig. 1. A circuit moving with the conducting fluid.

Let $\mathrm{d}\boldsymbol{l}$ be the line element in the vicinity of a point M on the line C_1, oriented such that the sense of the normal vector to S_1 be the sense with respect to which the flux Φ_1 is defined. The outgoing surface element on Σ can be written as $\mathrm{d}\boldsymbol{S}_{\mathrm{out}} = \mathrm{d}\boldsymbol{l} \times \boldsymbol{v}(M)\ \mathrm{d}t$, which changes equation (6.9) into:

$$\Phi_1 - \Phi_2 = \mathrm{d}t \int_{S_1} \mathrm{rot}(\boldsymbol{v} \times \boldsymbol{B}) \cdot \mathrm{d}\boldsymbol{S}_{\mathrm{out}} + \mathrm{d}t \int_{C_1} \boldsymbol{B}(M) \cdot (\mathrm{d}\boldsymbol{l} \times \boldsymbol{v}(M)). \quad (6.10)$$

The surface integral can be transformed by Stokes theorem. The sense of the surface element $\mathrm{d}\boldsymbol{S}_{\mathrm{out}}$ to S_1 in the first term of equation (6.10) is opposite to the sense with respect to which the flux Φ_1 is defined, so that:

$$\Phi_1 - \Phi_2 = -\mathrm{d}t \int_{C_1} (\boldsymbol{v}(M) \times \boldsymbol{B}(M)) \cdot \mathrm{d}\boldsymbol{l} + \mathrm{d}t \int_{C_1} \boldsymbol{B}(M) \cdot (\mathrm{d}\boldsymbol{l} \times \boldsymbol{v}(M)). \quad (6.11)$$

The sum on the right of equation (6.11) is zero, which establishes the theorem.

The second theorem establishes the permanence of magnetic connection under perfect MHD motions. Consider two fluid elements which, at time t_1, are at points P_1 and Q_1, separated by an infinitesimal distance on a common field line. Let $\boldsymbol{P}_1\boldsymbol{Q}_1$ be $\boldsymbol{K}_1$ and $\boldsymbol{B}(P_1, t_1)$ be $\boldsymbol{B}_1$. It is assumed that $\boldsymbol{K}_1 \times \boldsymbol{B}_1 = 0$. At time $t_2 = t_1 + \mathrm{d}t$ the plasma element initially at P_1 has moved to P_2 and the one at Q_1 has moved to Q_2. Obviously $\boldsymbol{P}_1\boldsymbol{P}_2 = \boldsymbol{v}(P_1, t_1)\mathrm{d}t$ and $\boldsymbol{Q}_1\boldsymbol{Q}_2 = \boldsymbol{v}(Q_1, t_1)\mathrm{d}t$. Let $\boldsymbol{P}_2\boldsymbol{Q}_2$ be $\boldsymbol{K}_2$ and $\boldsymbol{B}(P_2, t_2)$ be $\boldsymbol{B}_2$.

A Taylor expansion gives

$$\boldsymbol{K}_2 - \boldsymbol{K}_1 = (\boldsymbol{v}(Q_1) - \boldsymbol{v}(P_1))\,\mathrm{d}t = ((\boldsymbol{K}_1 \cdot \boldsymbol{\nabla})\ \boldsymbol{v}(P))\,\mathrm{d}t \tag{6.12}$$

$\boldsymbol{B}_2$ is related to $\boldsymbol{B}_1$ by the chain rule:

$$\boldsymbol{B}_2 = \boldsymbol{B}(P_2, t_2) - \boldsymbol{B}(P_2, t_1) + \boldsymbol{B}(P_2, t_1) - \boldsymbol{B}(P_1, t_1) + \boldsymbol{B}(P_1, t_1). \tag{6.13}$$

For infinitesimal $\mathrm{d}t$, this gives:

$$\boldsymbol{B}_2 = \frac{\partial \boldsymbol{B}}{\partial t}\ \mathrm{d}t + ((\boldsymbol{v}(P_1) \cdot \boldsymbol{\nabla})\ \boldsymbol{B})\,\mathrm{d}t + \boldsymbol{B}_1. \tag{6.14}$$

The theorem claims that if $\boldsymbol{K}_1 \times \boldsymbol{B}_1$ vanishes, so does $\boldsymbol{K}_2 \times \boldsymbol{B}_2$. Using equations (6.12) and (6.14) we calculate:

$$\boldsymbol{K}_2 \times \boldsymbol{B}_2 = (\boldsymbol{K}_1 + (\boldsymbol{K}_1 \cdot \boldsymbol{\nabla})\ \boldsymbol{v}\ \mathrm{d}t)) \times \left(\boldsymbol{B}_1 + \frac{\partial \boldsymbol{B}}{\partial t}\ \mathrm{d}t + (\boldsymbol{v} \cdot \boldsymbol{\nabla})\boldsymbol{B}\ \mathrm{d}t\right). \tag{6.15}$$

For vanishing $\boldsymbol{K}_1 \times \boldsymbol{B}_1$, equation (6.15) can be transformed by using equation (6.5) and the identity

$$\mathrm{rot}(\boldsymbol{v} \times \boldsymbol{B}) = (\mathrm{div}\ \boldsymbol{B})\ \boldsymbol{v} - (\mathrm{div}\ \boldsymbol{v})\ \boldsymbol{B} + (\boldsymbol{B} \cdot \boldsymbol{\nabla})\boldsymbol{v} - (\boldsymbol{v} \cdot \boldsymbol{\nabla})\boldsymbol{B}. \tag{6.16}$$

The vectors $\boldsymbol{B}_1$ and $\boldsymbol{K}_1$ being parallel, they can be written as $\boldsymbol{B}_1 = B_1\ \boldsymbol{t}$ and $\boldsymbol{K}_1 = K_1\ \boldsymbol{t}$, where $\boldsymbol{t}$ is a common unit vector. Equation (6.15) then becomes:

$$\boldsymbol{K}_2 \times \boldsymbol{B}_2 = K_1 B_1\ \mathrm{d}t\ \ (\boldsymbol{t} \times ((\boldsymbol{t} \cdot \boldsymbol{\nabla})\boldsymbol{v}) + ((\boldsymbol{t} \cdot \boldsymbol{\nabla})\boldsymbol{v}) \times \boldsymbol{t})\,. \tag{6.17}$$

The right of equation (6.17) vanishes, which establishes the theorem.

6.2 Consequences and limits of the field freezing theorems

The field freezing theorem shows that in perfect MHD flows field lines accompany the fluid. They are thus materialized and "unbreakable" because they always link the same fluid elements. Hence, their topology cannot change. As long as perfect MHD conditions apply, fluid elements which are not initially on a common field line cannot be linked by a field line at any later time. The constraint that the field follows the fluid motion, whatever its complexity, may eventually lead to very complex magnetic structures. The electric current density, which is proportional to the rotational of the field, may develop local current concentrations.

Our estimate of the magnetic Reynolds number in Section (6.1) has been based on an *a-priori* estimate of the scale of field gradients, that is, on an a-priori estimate of the electric current density which we assumed to be of order of $B/(\mu_0 L)$, L being the global scale. But the field gradient scale really is position-dependent. It may be locally much less than the global scale L. This would happen if the electric current locally flows through unusually narrow sections, of scale ℓ, say. For small enough ℓ, the perfect MHD approximation eventually comes to a local breakdown. The realistic exception to perfect MHD is then met when the magnetic field locally develops very small scalelengths. However, such an exception is not exceptional at all, because the dynamics of perfect MHD systems may naturally develop such structures, where the flux freezing conditions are eventually violated. When so, dissipative motions develop *locally* in a flow where perfect MHD still remains valid at large. These local dissipative motions have to match a general flow which retains elsewhere its perfect MHD character. Locally, in the dissipative region, the topological constraints imposed by the freezing theorems are violated and matter can slip accross the field lines. A consequence is that fluid elements which were initially magnetically unconnected may become connected after other plasma elements to which they were initially linked happened to meet at a dissipative region. The description of such motions, which are dissipative in restricted regions of space but are otherwise perfect MHD, is the purpose of "reconnection theory", some aspects of which will be described below. Because the effect of finite resistivity is in this case only local, it is often regarded as sufficient to assume the magnetic diffusivity to be spatially constant.

6.3 Field diffusion versus field advection

To illustrate the effect of resistivity on field evolution, consider a one-dimensional field $B(x,t)\ \boldsymbol{e}_y$ in a non-moving medium of uniform magnetic diffusivity η_{m}. Equation (6.2) reduces in this case to the heat equation

$$\frac{\partial B}{\partial t} = \eta_{\mathrm{m}} \frac{\partial^2 B}{\partial x^2}. \tag{6.18}$$

The current density, given by $\mu_0 \boldsymbol{j} = (\partial B/\partial x)\boldsymbol{e}_z$, also obeys equation (6.18). If the initial current distribution is a Dirac concentration at $x = 0$, the current distribution at later times is:

$$j(x,t) = \frac{I_0}{\sqrt{4\pi\eta_{\mathrm{m}} t}}\ \mathrm{e}^{-x^2/4\eta_{\mathrm{m}} t}. \tag{6.19}$$

Note that, although resistivity is present in the system, the total current remains constant. This is an effect of Lenz's law. When, because of resistivity, the current density is reduced at some place, the magnetic field created at other places by this current changes. This variation induces an electric field which drives a Lenz current which itself is the source of some magnetic field. As a result, dissipation causes currents to diffuse rather than just to decay. Although the total electric current is conserved, magnetic energy is lost in the dissipative process. Indeed, an integration by parts shows that

$$\frac{\partial}{\partial t}\int_{-\infty}^{+\infty}\frac{B^2}{2\mu_0}\,\mathrm{d}x=\int_{-\infty}^{+\infty}\frac{B}{\mu_0}\,\eta_{\mathrm{m}}\,\frac{\partial^2 B}{\partial x^2}\,\mathrm{d}x=-\int_{-\infty}^{+\infty}\frac{j^2}{\sigma_{\mathrm{e}}}\,\mathrm{d}x. \tag{6.20}$$

The loss of magnetic energy is due to the reduction of the self inductance $\mathcal{L}$ of the current circuit when, because of the dissipation, the spatial distribution of the current density reconfigurates, broadening the current channels. The lost magnetic energy goes in this case entirely to Joule heat.

Consider now a stationnary, two-dimensional, perfect MHD motion in which $\boldsymbol{v}$ and $\boldsymbol{B}$ have only x and y cartesian components, assumed to be independent of z- and t. Such perfect MHD motions would be singular at neutral points, where $|\boldsymbol{B}|=0$. From equation (6.6), it is seen that the electric field in a perfect MHD motion is equal to $-(\boldsymbol{v}\times\boldsymbol{B})$. In this geometry $\boldsymbol{E}=E(x,y)\,\boldsymbol{e}_z$. Equation (2.20) then shows that, for a stationnary motion, $E(x,y)$ is a constant, which, in the presence of cross-field motions, does not vanish. The constancy of the electric field then implies that the velocity diverges at neutral points. What is wrong? The point is that resistive effects may not be neglected near neutral points. Indeed, the full form of Ohm's law (3.8), which can be written as

$$\boldsymbol{E}=-\boldsymbol{v}\times\boldsymbol{B}+\boldsymbol{j}/\sigma_{\mathrm{e}} \tag{6.21}$$

shows that the resistive term on the right of equation (6.21) may never be neglected at neutral points and in a small vicinity. These are priviledged regions for reconnection flows to develop.

6.4 A simple model of joint advection and diffusion

It is then appropriate to discuss the joint effect of advection and resistivity. We still consider stationnary two-dimensional motions and field distribution and assume that the flow is incompressible, with a stagnation point [27]:

$$\boldsymbol{v}(x,y)=-\frac{V_0 x}{a}\boldsymbol{e}_x+\frac{V_0 y}{a}\boldsymbol{e}_y. \tag{6.22}$$

The magnetic field is supposedly given by $\boldsymbol{B}=B(x)\,\boldsymbol{e}_y$, $B(x)$ being an odd function of x. The field lines are straight and there is a neutral sheet at $x=0$, towards which the flow converges. Ohm's law gives:

$$E=+\frac{V_0 x}{a}\,B(x)+\eta_{\mathrm{m}}\,\frac{\mathrm{d}B}{\mathrm{d}x}. \tag{6.23}$$

This equation cannot be satisfied at $x = 0$ in the absence of resistivity. For small $\eta_{\rm m}$, the resistivity can be neglected far from $x = 0$ but, by contrast, the electromotive field may be neglected in equation (6.23) for very small x. Thus,

$$B(x) \approx \frac{Ea}{V_0 x} \quad (x \text{ large}) \qquad\qquad B(x) \approx \frac{Ex}{\eta_{\rm m}} \quad (x \text{ small}). \tag{6.24}$$

The transition from one to the other regime occurs where both expressions in equation (6.24) become almost equal, at $x \approx l$, with l given by

$$\frac{l}{a} = \sqrt{\frac{\eta_{\rm m}}{V_0 a}} = R_{\rm m}^{-1/2}. \tag{6.25}$$

The smaller the resistivity, the thinner this resistive layer. For very large magnetic Reynolds numbers $R_{\rm m}$, it takes the character of a boundary layer. Equation (6.23) can be solved exactly for $B(x)$, with the result

$$B(x) = \frac{2Ea}{V_0 l}\ {\rm e}^{-(x^2/l^2)} \int_0^{(x/l)} {\rm e}^{t^2}\,{\rm d}t \tag{6.26}$$

where $l^2 = 2\eta_{\rm m} a/V_0$. Equation (6.26) is the solution for field advection and diffusion, assuming the velocity field to be given. Is it always physically possible to maintain this assumed velocity field? If not, what are the resulting constraints? In this geometry, and for a constant density ρ, the equation of motion integrates as:

$$P + \frac{B^2}{2\mu_0} + \rho\frac{v^2}{2} = C. \tag{6.27}$$

A condition for consistency is that the pressure given by equation (6.27) be positive everywhere. This is possible for very small inflow velocities only. To understand this, consider the flow in some circular region of radius $L_{\rm e}$, assuming the pressure, the magnetic field, and the inflow velocity to be known at the point $M_{\rm e}$, of coordinates $x = L_{\rm e}$, $y = 0$ where the plasma enters this region. Then consider a point where the second and third terms of (6.27) are large. The magnetic field is largest at $x = l$ and, on this line, the velocity is the largest at a distance $L_{\rm e}$ to the origin. The requirement that the pressure be positive at this point imposes a condition on the velocity $v_{\rm e}$ of the fluid entering the region at $M_{\rm e}$. This condition can eventually be expressed as a constraint on the alfvenic Mach number $M_{\rm Ae}$ at the point $M_{\rm e}$. It is found that $M_{\rm Ae}$ should be less than a quantity of order of the inverse of the alfvenic magnetic Reynolds number at $M_{\rm e}$, $R_{\rm mAe} = v_{{\rm A}_{\rm e}} L_{\rm e}/\eta_{\rm m}$. This is a very severe constraint. The reason for this is that in this flow the inflowing magnetic energy is entirely transformed into heat. The flow then proceeds at a rate controlled by the very small resistivity.

6.5 Reconnection flows

Are there more efficient reconnection flows? In this context, larger efficiency means that the inflow Alfvénic Mach number decreases less rapidly than $R_{\rm m}^{-1}$ when resistivity is reduced to zero. Finding such flows is the point of the theory of MHD reconnection [24]. The Sweet-Parker model considers a flow converging uniformly to a central dissipative sheet-like region in which reconnection occurs. The plasma is supposedly expelled out of this thin dissipative layer by the large magnetic tension force which results from the topological change of reconnected field lines, forming a two-dimensional flow, symetrical with respect to a central point O, along this layer. This model can be described by order of magnitude relations. To our knowledge, no detailed mathematical treatment has ever been worked out. It is assumed that the dissipative sheet extends over the whole length L of the system (the integral scale). Its width l is much smaller. The flow is steady and incompressible. The inflow speed on the large sides of the reconnection layer is $v_{\rm in}$ and the velocity of the plasma jetting out of it is $v_{\rm out}$. The incoming magnetic field is $B_{\rm in}$ and its amplitude at the outlet of the reconnection layer is $B_{\rm out}$. The electric field being constant, $v_{\rm in}B_{\rm in} = v_{\rm out}B_{\rm out}$. Mass conservation implies that $Lv_{\rm in} = lv_{\rm out}$. The solenoidal property of the magnetic field implies that the flux through the sides of a box, one corner of which is at O, with length L in the y-direction and l in the direction perpendicular to the sheet, vanishes, so that $LB_{\rm out} \approx lB_{\rm in}$. For stationnarity, the inflow speed $v_{\rm in}$ must balance the rate of expansion of the sheet by diffusion, so that

$$v_{\rm in} \approx \frac{\eta_{\rm m}}{l}. \tag{6.28}$$

The current density in the reconnecting layer is $j \approx 2B_{\rm in}/(\mu_0 l)$. The acceleration of the flow in the reconnection layer, $\approx v_{\rm out}/(L/v_{\rm out})$, must match the Lorentz force per unit mass $jB_{\rm out}/\rho$. From this and the soleneoidal property of the magnetic field, it results that $v_{\rm out}^2 \approx v_{\rm Ain}^2$, where $v_{\rm Ain}$ is the Alfvén speed associated with the incoming field. Summing up,

$$v_{\rm in} \approx \frac{\eta_{\rm m}}{l} \qquad Lv_{\rm in} = lv_{\rm out} \qquad v_{\rm out}^2 = B_{\rm in}^2/\mu_0\rho. \tag{6.29}$$

Eliminating $v_{\rm out}$ and l from equation (6.29) we find the inflowing alfvénic Mach number in terms the incoming alfvénic magnetic Reynolds number:

$$M_{\rm Ain} = (v_{\rm Ain}L/\eta_{\rm m})^{-1/2} \equiv R_{\rm Ain}^{-1/2}. \tag{6.30}$$

This is a much more favourable scaling than the one which applies to the purely diffusive solution of Section (6.4). $M_{\rm Ain}$ is however still too small to explain the rate of magnetic energy release in solar flares. Note that the assumption that the length L of the sheet is equal to the integral scale

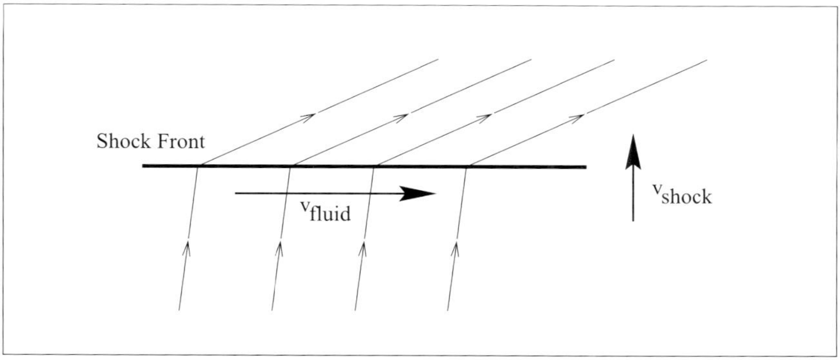

Fig. 2. Structure of a slow shock front. The front moves as indicated by the arrow. Magnetic field lines have a kink at the shock front. The tangential component of the field is reduced downstream and the plasma is accelerated tangentially.

is arbitrary. Are there reconnection flows with a dissipative region much shorter than the integral scale L? If so, how does the alfvénic Mach number scales with the incoming alfvénic magnetic Reynolds number?

The answer to this question first came from the work of Petschek [22] who stressed the fact that the incoming magnetic energy might be converted into kinetic energy instead of being dissipated into Joule heat. This would occur if the plasma is to pass through MHD shock waves at which the magnetic energy density decreases at the benefit of kinetic energy density. The type of MHD shock which achieves this is a slow, or switch-off, shock (Fig. 2). A switch-off shock propagates in an uncompressible medium at the Alfvén speed $v_{\rm An}$ associated with the field component $B_{\rm n}$ perpendicular to the shock front. In the upstream medium, the field has a tangential component which is reduced in the downstream medium. The field lines have a kink at their crossing with the front, which develops an infinite magnetic tension force density. This force impulsively accelerates the plasma in a direction tangential to the shock front at the Alfven speed associated with the upstream tangential field component. This can be vizualized by thinking that the shock front in Figure 2 advances at $v_{\rm An}$ toward the more magnetized region: a lateral plasma motion in the post shock region must be present because the magnetic field lines must remain continuuous. The shock front is a transformer of magnetic into kinetic energy at a rate that is set by the Alfvén speed $v_{\rm An}$. For the shock front to be stationnary in the laboratory frame, the upstream medium should be flowing towards the front at $v_{\rm An}$. Petschek has shown how to arrange a converging flow in such a way that it supports two pairs of stationnary slow shocks (Fig. 3). The

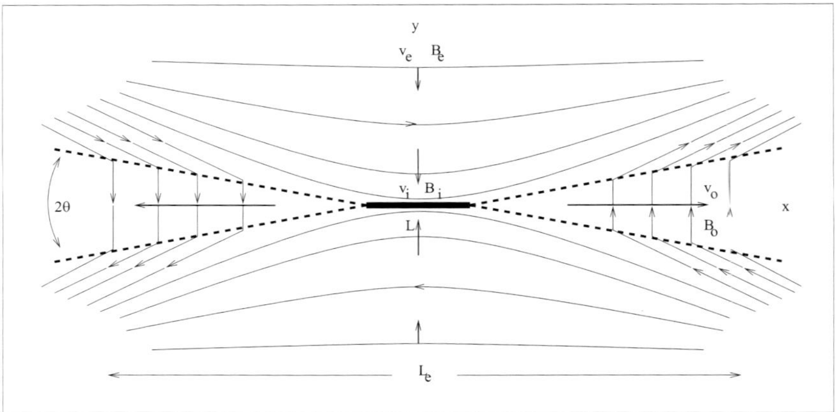

Fig. 3. Petschek reconnection flow. The flow, carrying antiparallel fields, converges to a central region bordered by four slow shock fronts (dashed lines). The field lines first hit a dissipative region (heavy line) where reconnection occurs.

shock fronts intersect at a small dissipative region. This allows the necessary topological change of the magnetic field lines. Let L be the length and l the width of this small dissipative region, which is basically a Sweet Parker sheet, although its length L is much less than the integral scale, L_{e}. The overall size of the shock fronts is assumed to be of order L_{e}. The flow in the pre-shock region must cause a slight bending of the magnetic field lines, so that they can make contact with the dissipative region first and achieve the necessary topological change before meeting the shocks. At the shocks the plasma suffers an impulsive acceleration and joins the fast flow in the inter-shock region. The magnetic structure is supposedly current-free except in the dissipative layer and on the shock fronts.

The reconnection rate is characterized by the value of the alfvénic Mach number, $M_{\mathrm{Ae}} = (v_{\mathrm{e}}/v_{\mathrm{Ae}})$ far in the upstream flow. Relations (6.29) apply to the small central dissipative region. The magnetic field at the edge of this sheet, B_{i}, slightly differs from the inflowing field, B_{e}, far from the sheet in the upstream flow and v_{i} also differs from v_{e}. Let v_{o} and B_{o} be the velocity and the magnetic field in the inter-shock region and v_{Ae} and v_{Ai} be the Alfvén speeds associated with B_{e} and B_{i}. R_{Ame} is the magnetic Reynolds number associated with v_{Ae} and L_{e}. $M_{\mathrm{Ai}} = (v_{\mathrm{i}}/v_{\mathrm{Ai}})$ is the alfvénic Mach number of the flow at the edge of the dissipative sheet. The electric field being uniform in this 2D stationnary flow, $v_{\mathrm{e}}B_{\mathrm{e}} = v_{\mathrm{i}}B_{\mathrm{i}} = v_{\mathrm{o}}B_{\mathrm{o}}$. The dynamics of the shock is such that $v_{\mathrm{o}} = v_{\mathrm{Ai}}$. The component of the field normal to the shock front is continuous, as is also, as a consequence of mass conservation and incompressibility, the normal component of the fluid velocity. From the constancy of the electric field we get:

$$\frac{B_\mathrm{i}}{B_\mathrm{e}} = \frac{v_\mathrm{e}}{v_\mathrm{i}} = \frac{M_\mathrm{Ae} v_\mathrm{Ae}}{M_\mathrm{Ai} v_\mathrm{Ai}} = \frac{M_\mathrm{Ae} B_\mathrm{e}}{M_\mathrm{Ai} B_\mathrm{i}} \qquad \rightarrow \qquad \frac{B_\mathrm{i}}{B_\mathrm{e}} = \left(\frac{M_\mathrm{Ae}}{M_\mathrm{Ai}}\right)^{1/2} \cdot \tag{6.31}$$

From equations (6.31) and (6.29) it results that

$$\frac{l}{L_\mathrm{e}} = \frac{1}{R_\mathrm{Ame}} \frac{1}{M_\mathrm{Ae}^{1/2} M_\mathrm{Ai}^{1/2}} \qquad \frac{L}{L_\mathrm{e}} = \frac{1}{R_\mathrm{Ame}} \frac{1}{M_\mathrm{Ae}^{1/2} M_\mathrm{Ai}^{3/2}} \cdot \tag{6.32}$$

To appreciate the consistency of the solution, we should calculate B_i in terms of the outer field B_e. The magnetic field in the pre-shock region is potential and its boundary values at the edge of the sheet and the shock fronts are easily found. If we neglect the small angle between both shocks, this boundary reduces to the x-axis. The normal component of the magnetic field along it is zero along the edge of the diffusive region but is non-zero on the shock fronts, which extend between $x = \pm L$ and approximately $x = \pm L_\mathrm{e}$. At larger distances, both along the x-axis and in any other direction, the field $\boldsymbol{B}$ approaches $\boldsymbol{B}_\mathrm{e}$. The normal component on the boundary, the x-axis, assumes an almost constant value B_y between $+L$ and $+L_\mathrm{e}$ and the opposite value between $-L$ and $-L_\mathrm{e}$. It vanishes elsewhere on this axis. This component B_y will be calculated below. The field which results from this boundary distribution is easily calculated: in 2D geometry, a source of flux ϕ localized at some point creates at a position $\boldsymbol{R}$ apart a field $\boldsymbol{B} = (\phi/\pi)(\boldsymbol{R}/R^2)$. From this we find, by integration, the field B_i at the origin:

$$B_\mathrm{i} = B_\mathrm{e} - \frac{2}{\pi}\, B_y \, \ln\left(\frac{L_\mathrm{e}}{L}\right) \cdot \tag{6.33}$$

To calculate B_y along the shock front in terms of B_e, the angle θ between these fronts and the x-axis, albeit small, should not be regarded as being exactly zero. Figure 3 illustrates the geometry. The component of the incident magnetic field tangent to the shock front, B_tg, is approximately $B_\mathrm{e}\cos\theta$, and the incident velocity $-v_\mathrm{e}\boldsymbol{e}_y$. Using the continuity of the normal component of the magnetic field and of the velocity as well as the constancy of the electric field we obtain the series of relations:

$$\begin{aligned} B_y &= B_\mathrm{n}\cos\theta + B_\mathrm{t}\sin\theta = B_\mathrm{o}\cos^2\theta + (v_\mathrm{o}B_\mathrm{o}/v_\mathrm{e})\sin\theta\cos\theta \\ &= B_\mathrm{o}\cos^2\theta + B_\mathrm{o}\cos^2\theta = 2B_\mathrm{o}\cos^2\theta = 2B_\mathrm{n}\cos\theta. \end{aligned} \tag{6.34}$$

For the shock to be stationnary, the inflow velocity of the upstream fluid must be the Alfvén speed v_An of the normal field component B_n. From this it results that $B_\mathrm{n} = M_\mathrm{Ae}B_\mathrm{e}$ and, from (6.34), that

$$B_y \approx 2M_\mathrm{Ae}B_\mathrm{e}. \tag{6.35}$$

Calculating (L_e/L) from (6.32) with $M_{Ai} \approx M_{Ae}$, we find from equations (6.35) and (6.33) that

$$B_i = B_e \left(1 - \frac{4}{\pi} M_{Ae} \ln \left(R_{Ame} M_{Ae}^2\right)\right). \quad (6.36)$$

Because B_i cannot be much smaller than B_e, there obviously is a limit to M_{Ae}. Considering that B_i cannot be less than approximately $B_e/2$, we may estimate this limit, M_{Ae*}, by substituting $B_i = B_e/2$ in equation (6.36). For large R_{Ame} the solution of the equation which results for M_{Ae*} is approximately

$$M_{Ae*} = \frac{\pi}{8 \ln(R_{Ame})}. \quad (6.37)$$

The reconnection rate in Petchek's flow decreases much less rapidly with R_{Ame} than the diffusive annihilation rate found in Section (6.4) or the Sweet-Parker rate derived in Section (6.5). For this reason, this flow is refered to as a "fast reconnection flow".

A branch of reconnection theory is devoted to developing more precise analytical or numerical calculations of 2D reconnection flows, taking also compressibility into account. A large class of fast reconnection models has been discovered, Petschek's flow being but one particular member of this family [24]. Extending the theory from two to three dimensions represents a challenge that will not be discussed here (see [24]). Before reconnection reaches a stationnary state, if it ever does, it must develop through a time dependent phase. An approach to this regime consists in discussing the development of reconnection from an initial perfect MHD equilibrium state as the growth of an instability allowed by finite resitivity [9]. Resistive effects play a role only locally in the vicinity of regions where $\mathrm{rot}(\boldsymbol{v} \times \boldsymbol{B})$ vanishes in the perturbation. These modes, the growth of which is allowed by finite resitivity, are called tearing modes (because their effect is to tear current sheets into separated current threads).

7 The dynamo problem

A major astrophysical MHD problem is to explain how magnetic fields observed in celestial bodies may have grown from a very small initial seed. The electromotive field $(\boldsymbol{v} \times \boldsymbol{B})$ in equation (6.2) is to be responsible for this. Could it generate organized electric currents and magnetic field against the adverse effect of Ohmic dissipation? For known plasma motions equation (6.2) is a linear equation for the magnetic field, which defines a "kinematic" dynamo problem. However the velocity field really depends on $\boldsymbol{B}$ because the Lorentz force takes part in determining the motions. Demonstrating field amplification while taking self-consistently into account this feedback constitutes the non-linear, "dynamic", dynamo problem. Assuming the velocity field to be known is an approximation that one would be tempted to regard as justified for weak field. However, in some instances, the flow would develop from an instability caused by the magnetic field itself, even when weak. In accretion disks, for example, unstable motions develop from the magnetorotational instability. In this case, there is no room left for a kinematic dynamo approach. The kinematic approach also cannot predict the amplitude of the field generated by the dynamo action because this amplitude depends on non-linear saturation mechanisms. Accounts of researches on the dynamo problem may be found in [21] and [25].

7.1 Antidynamo theorems

In the kinematic approximation some plasma flow, often a stationary one, is assumed. There is dynamo action if the linear equation (6.2) has unstable eigenmodes. This kinematic dynamo problem, however, gives rise to a number of "anti-dynamo" theorems which state that, if the flow has too much symmetry, magnetic field growth is impossible. Cowling's theorem, for example, states that no axisymmetrical flow can support dynamo action. A weaker form of it claims that no axisymmetrical field can be maintained against dissipation by an axisymmetrical, incompressible, flow. The proof is as follows. The poloidal field can be represented by its flux function $a(r, z, t)$ as in (5.6). Equation (6.2) then takes the following form for incompressible flows:

$$\frac{\partial a}{\partial t} + \boldsymbol{v} \cdot \boldsymbol{\nabla} a = \eta \Delta^* a \tag{7.1}$$

$$\frac{\partial}{\partial t}\left(\frac{B_\theta}{r}\right) + \boldsymbol{v} \cdot \boldsymbol{\nabla}\left(\frac{B_\theta}{r}\right) = \boldsymbol{B} \cdot \boldsymbol{\nabla}\left(\frac{v_\theta}{r}\right) + \frac{\eta}{r^2}\Delta^*\left(rB_\theta\right). \tag{7.2}$$

The operator Δ^* is defined by equation (5.8). Equation (7.2) has an electromotive term, its first term on the right, which represents the ω-effect, *i.e.*, the stretching of magnetic field lines by the differential rotation. However, equation (7.1) is lacking such a source term: no growth of the poloidal field is then possible. This can be formally shown by multiplying equation (7.1) by $a(r, z, t)$ and integrating over infinite space, assuming the dynamo flow to occupy only a finite volume. Integrating over infinite space, any surface term vanishes because there is now moving fluid at large distances. Taking incompressibility into account and integrating the product $a\Delta^* a$ by parts over r and z we are left, after some algebra, with

$$\frac{\partial}{\partial t}\left(\int_V \left(\frac{a^2}{2}\right)\right) = -\eta \int_V |\boldsymbol{\nabla} a|^2. \tag{7.3}$$

This shows that $a(r, z, t)$ asymptotes to zero and establishes the proposed theorem, many powerful generalizations of which exist. Thus, dynamo action should essentially be associated with 3D velocity fields, which has prompted research on dynamo action from turbulent velocity fields [29].

7.2 Kinematic turbulent dynamos

The turbulent dynamo problem considers the evolution of a magnetic field interacting with a random, turbulent, velocity field by equation (6.2). This flow generates a random field, the statistical properties of which should be determined. In its full, dynamical form, this is just the MHD turbulence problem. The kinematical version regards the statistical properties of the velocity field as known and aims at determining the resulting properties of the magnetic field. In particular, it is asked whether a large scale component would grow. For simplicity, it can be assumed here that the eddies of the velocity field are much smaller than the integral scale of the system, while in reality there is a continuuous distribution of eddy sizes, extending to the integral scale. Accepting nevertheless this assumption, the velocity and magnetic fields may be split into a large scale and a small scale part as:

$$\boldsymbol{B} = \boldsymbol{B}_0 + \boldsymbol{b} \qquad \boldsymbol{v} = \boldsymbol{v}_0 + \boldsymbol{w}. \tag{7.4}$$

The large scale fields $\boldsymbol{v}_0$ and $\boldsymbol{B}_0$ are regarded as non-random, while the small scale components, $\boldsymbol{w}$ and $\boldsymbol{b}$ are random fluctuations with zero mean value. For even more simplicity, let us assume that the large scale velocity $\boldsymbol{v}_0$ vanishes and neglect the molecular diffusivity $\eta_{\rm m}$. With these assumptions, the average of the induction equation (6.2) reduces to

$$\frac{\partial \boldsymbol{B}_0}{\partial t} = \mathrm{rot}\langle \boldsymbol{w} \times \boldsymbol{b}\rangle \cdot \tag{7.5}$$

The large scale field then evolves as a result of the average electromotive field associated with the small scale fluctuations,

$$\boldsymbol{E}^0 = \langle \boldsymbol{w} \times \boldsymbol{b}\rangle \cdot \tag{7.6}$$

In order to reduce equation (7.5) to an equation for the large scale field alone, this average electromotive field should be expressed in terms of $\boldsymbol{B}_0(\boldsymbol{r}, t)$ and of some parameters reflecting the statistical properties of the random velocity field. Consider the fluctuating part of equation (6.2), which can be obtained by inserting (7.4) in (6.2) and then substracting (7.5):

$$\frac{\partial \boldsymbol{b}}{\partial t} = \mathrm{rot}(\boldsymbol{w} \times \boldsymbol{B}_0) + \mathrm{rot}\left(\boldsymbol{w} \times \boldsymbol{b} - \langle \boldsymbol{w} \times \boldsymbol{b}\rangle\right). \tag{7.7}$$

Equation (7.7) can be solved for $\boldsymbol{b}$ in terms of $\boldsymbol{w}$ and $\boldsymbol{B}_0$ in the first order smoothing approximation. This consists in neglecting the second term, *i.e.*

the fluctuating part of $\boldsymbol{w} \times \boldsymbol{b}$ in equation (7.7). This would be justified for weak turbulence when $\boldsymbol{b}$ is small compared to $\boldsymbol{B}_0$. This is however not so in reality. Accepting this approximation nevertheless, equation (7.7) can be integrated in time to give:

$$\boldsymbol{b}(\boldsymbol{r},t) = \int_0^{\infty} \mathrm{d}\tau \ \mathrm{rot}\left(\boldsymbol{w}\ (\boldsymbol{r},t-\tau) \times \boldsymbol{B}_0(\boldsymbol{r},t-\tau)\right). \tag{7.8}$$

Expanding the rotational in equation (7.8) and substituting in equation (7.6) we obtain, for incompressible flows, the following average electromotive field:

$$\begin{aligned}\boldsymbol{E}^0 = &\int_0^{\infty} \mathrm{d}\tau \ \langle \boldsymbol{w}(\boldsymbol{r},t) \times (\ (\boldsymbol{B}_0(\boldsymbol{r},t-\tau)\cdot\boldsymbol{\nabla})\ \boldsymbol{w}(\boldsymbol{r},t-\tau)\ )\rangle \\ &- \int_0^{\infty} \mathrm{d}\tau\ \langle \boldsymbol{w}(\boldsymbol{r},t) \times (\boldsymbol{w}\ ((\ \boldsymbol{r},t-\tau)\cdot\boldsymbol{\nabla}\ )\ \boldsymbol{B}_0(\boldsymbol{r},t-\tau))\rangle. \end{aligned} \tag{7.9}$$

The averaging involves the velocity correlation at two different times, which vanishes when the delay is much larger than the correlation time τ_{corr}. Assume that the large scale field varies negligibly over such a time. Then, $\boldsymbol{B}_0(\boldsymbol{r},t)$ can be substituted for $\boldsymbol{B}_0(\boldsymbol{r},t-\tau)$ in (7.9). The calculation of $\boldsymbol{E}^0$ then becomes a matter of algebra which is best handled by working out the cartesian components of $\boldsymbol{E}_0$ in terms of the space and time velocity correlation tensor $\mathcal{C}$, with components defined by

$$\mathcal{C}_{ij}(\boldsymbol{R},\tau) = \langle w_i(\boldsymbol{r},t)\ w_j(\boldsymbol{r}+\boldsymbol{R},t+\tau)\rangle. \tag{7.10}$$

For homogeneous and isotropic turbulence, the velocity correlation tensor takes the following form, R being the modulus of the separation vector:

$$\mathcal{C}_{ij}(\boldsymbol{R},\tau) = A(R,\tau)\ \delta_{ij} + B(R,\tau)R_iR_j + C(R,\tau)\epsilon_{ijk}R_k. \tag{7.11}$$

Equation (7.9) involves averages such as $\langle w_i(\boldsymbol{r},t)\ \partial_{x_{\mathrm{m}}} w_j(\boldsymbol{r},t-\tau)\rangle$ where $\partial_{x_{\mathrm{m}}}$ is a component of the gradient with respect to $\boldsymbol{r}$. These are simply the values of $\langle w_i(\boldsymbol{r},t)\ \partial_{X_{\mathrm{m}}} w_j(\boldsymbol{r}+\boldsymbol{R},t-\tau)\rangle$ for vanishing $\boldsymbol{R}$, where $\partial_{X_{\mathrm{m}}}$ now is a component of the gradient with respect to the separation $\boldsymbol{R}$. This trick greatly simplifies the calculation of the r.h.s. of (7.9). Eventually,

$$\frac{\partial \boldsymbol{B}_0}{\partial t} = \mathrm{rot}(\alpha_{\mathrm{D}}\boldsymbol{B}_0) + \eta_*\Delta(\boldsymbol{B}_0). \tag{7.12}$$

The coefficients α_{D} and η_* are:

$$\alpha_{\mathrm{D}} = \int_0^{\infty} C(0,\tau)\ \mathrm{d}\tau \qquad\qquad \eta_* = \int_0^{\infty} A(0,\tau)\ \mathrm{d}\tau. \tag{7.13}$$

As compared to equation (6.2), which, in the absence of a velocity field would reduce to a diffusion equation, a new field-generating term, $\mathrm{rot}(\alpha_{\mathrm{D}}\boldsymbol{B}_0)$, has appeared in equation (7.12). Turbulence also causes "effective" magnetic diffusivity, represented by η_*. The physical meaning of $A(0,\tau)$ and $C(0,\tau)$ is revealed by noting that:

$$A(0,\tau) = \frac{1}{3}\langle \boldsymbol{w}(\boldsymbol{r},t)\cdot\boldsymbol{w}(\boldsymbol{r},t+\tau)\rangle \tag{7.14}$$

$$C(0,\tau) = -\frac{1}{6}\langle \boldsymbol{w}(\boldsymbol{r},t)\cdot \mathrm{rot}(\boldsymbol{w}(\boldsymbol{r},t+\tau)\rangle\cdot \tag{7.15}$$

The relation (7.15) is obtained by using (7.11) and the above mentioned trick to calculate

$$\langle \boldsymbol{w}(\boldsymbol{r},t)\cdot \mathrm{rot}\ \boldsymbol{w}(\boldsymbol{r},t+\tau)\rangle = \lim_{R=0} \epsilon_{ijk}\ \partial_{R_j}\ \mathcal{C}_{ik}(\boldsymbol{R},\tau). \tag{7.16}$$

The new α_{D} term in equation (7.12) vanishes unless the turbulence is not parity invariant and is endowed with "kinetic helicity", *i.e.* with a non vanishing average value of $\boldsymbol{w}\cdot\mathrm{rot}\boldsymbol{w}$. The turbulent magnetic diffusivity η_* is:

$$\eta_* = \frac{1}{3}\int_0^\infty \mathrm{d}\tau\langle \boldsymbol{w}(\boldsymbol{r},t)\cdot\boldsymbol{w}(\boldsymbol{r},t+\tau)\rangle\cdot \tag{7.17}$$

At $\tau = 0$, the integrand is simply $\langle w^2\rangle$. Its time dependence is described by a function $g(\tau)$ which vanishes for τ much larger than τ_{corr}. Thus

$$\eta_* = \frac{1}{3}\langle w^2\rangle \int_0^\infty g(\tau)\mathrm{d}\tau \approx \frac{1}{3}\langle w^2\rangle\tau_{\mathrm{corr}}. \tag{7.18}$$

Defining the turbulent velocity by $v_{\mathrm{turb}} = \langle w^2\rangle^{1/2}$ and the correlation length by $l_{\mathrm{corr}} = v_{\mathrm{turb}}\ \tau_{\mathrm{corr}}$, the turbulent magnetic diffusivity is written as

$$\eta_* = \frac{1}{3}\ v_{\mathrm{turb}}\ l_{\mathrm{corr}}. \tag{7.19}$$

This is an expression similar to that of a molecular diffusion coefficient, with v_{turb} in place of the thermal velocity and l_{corr} in place of mean free path.

8 Simple stationnary flows

8.1 MHD Couette flow

Accretion disks may be regarded as axially symmetric stationnary flows. An often considered example of cylindrically symmetric MHD flow is the motion of an incompressible fluid with an azimuthal velocity $\boldsymbol{v}(\boldsymbol{r}) = v_\theta(r)\boldsymbol{e}_\theta$ and an axial, homogeneous, magnetic field. This is the so-called hydromagnetic Couette flow. The stationnary form of mass conservation and momentum equation for a viscous fluid with uniform viscosity reduce, in the absence of gravity, to just the radial and azimuthal components of the equation of motion:

$$\rho \frac{v_\theta^2}{r} = \frac{\mathrm{d}P}{\mathrm{d}r} \tag{8.1}$$

$$\eta \frac{\mathrm{d}}{\mathrm{d}r}\left(\frac{\mathrm{d}v_\theta}{\mathrm{d}r} + \frac{v_\theta}{r}\right) = 0. \tag{8.2}$$

In the absence of viscosity, any $v_\theta(r)$ is a solution of equation (8.2). In the presence of viscosity, the solution of equation (8.2) is:

$$v_\theta(r) = Ar + B/r. \tag{8.3}$$

For the angular velocity to equal Ω_1 at R_1 and Ω_2 at R_2, the integration constants A and B should be:

$$A = \frac{\Omega_2 R_2^2 - \Omega_1 R_1^2}{R_2^2 - R_1^2} \qquad B = \frac{R_1^2 R_2^2 (\Omega_1 - \Omega_2)}{R_2^2 - R_1^2}. \tag{8.4}$$

8.2 Isotropic HD wind

A simple model for thermal stellar mass loss is the hydrodynamical, stationnary, spherically symmetric and isothermal flow of a perfect gas. The motion is determined by the mass and momentum conservation equations. The isothermality assumption replaces the energy equation. The known isothermal sound speed is a. Let R be the distance to the center of the star. The radial velocity is $v_\mathrm{R}(R)$, the mass density $\rho(R)$ and the gravitational field created by the star, of mass M_* and radius R_*, is $\boldsymbol{g} = -(GM_*/R^2)\boldsymbol{e}_\mathrm{R}$. Self gravitation of the coronal envelope is neglected. With $P = \rho a^2$, the relevant equations can be written as:

$$\frac{1}{R^2}\frac{\mathrm{d}}{\mathrm{d}R}\left(R^2 \rho\, v_\mathrm{R}\right) = 0 \tag{8.5}$$

$$\rho v_\mathrm{R} \frac{\mathrm{d}v_\mathrm{R}}{\mathrm{d}R} = -\frac{\mathrm{d}P}{\mathrm{d}R} - \rho \frac{GM_*}{R^2}. \tag{8.6}$$

We look for a solution between R_* and infinity. The density $\rho_* = \rho(R_*)$, is known. The velocity at R_* is unknown but should be very small. Because the interstellar medium exerts negligible pressure at large distances, it should be required that the pressure of the flow vanishes at infinity, subject to some subtle considerations to be discussed below. Equation (8.6) has a first integral, the Bernoulli constant, which expresses the conservation of the total energy per unit mass during the motion:

$$\left(\frac{v_\mathrm{R}^2}{2} + a^2 \ln \rho - \frac{GM_*}{R}\right) = K_1. \tag{8.7}$$

Equations (8.7), (8.5) and (8.6) are made dimensionless by defining

$$x = R/R_* \qquad D = \rho/\rho_* \qquad M = v_\mathrm{R}/a. \tag{8.8}$$

A dimensionless parameter, which measures the strength of the gravitational confinement, appears:

$$\lambda = \frac{GM_*}{R_* a^2}. \tag{8.9}$$

The pressure and the density can be eliminated from equation (8.6) to obtain an unique equation for the Mach number $M(x)$:

$$\left(M - \frac{1}{M}\right)\frac{\mathrm{d}M}{\mathrm{d}x} - \frac{2}{x} + \frac{\lambda}{x^2} = 0. \tag{8.10}$$

Equation (8.10) has a first integral which is just the dimensionless form of equation (8.7):

$$\mathcal{B}(M, x) \equiv \frac{M^2}{2} - \ln M - \ln x^2 - \frac{\lambda}{x} = \epsilon. \tag{8.11}$$

An equivalent form of equation (8.11) is

$$M\mathrm{e}^{-M^2/2} = K\,\frac{\mathrm{e}^{-\lambda/x}}{x^2} \tag{8.12}$$

K is a constant, the value of which should be determined by the boundary conditions. The function of M on the left of equation (8.12) reaches an unique maximum at $M = 1$ and the function of x on the right reaches an unique maximum at $x = \lambda/2$. For a given x, equation (8.12) is an equation for M which, in general, has either two or zero solutions for $M(x)$. Equation (8.10) indicates that all solutions reach an extremum at $x = \lambda/2$ and that they all have a singular derivative when M reaches unity. There is one exception however, which is when the solution happens to pass in the (x, M) plane at the point C with coordinates $x = \lambda/2$, $M = 1$. At this critical point (8.10) gives the derivative $\mathrm{d}M/\mathrm{d}x$ in the undeterminate form $\frac{0}{0}$. Using l'Hopital's rule, or expanding (8.11) to second order in the vicinity of C, it is found that two "critical" solutions pass at C, with a slope $(\mathrm{d}M/\mathrm{d}x)_\mathrm{C}$ given by

$$\left(\frac{\mathrm{d}M}{\mathrm{d}x}\right)^2_\mathrm{C} = \frac{4}{\lambda^2}. \tag{8.13}$$

The critical point C is a saddle point of the Bernoulli function $\mathcal{B}(M, x)$ of equation (8.11) which has a vanishing differential at this point. The constant K of equation (8.12) which is associated with the critical solutions is

$$K_\mathrm{c} = \frac{\lambda^2 e^{3/2}}{4}. \tag{8.14}$$

A discussion of the solutions of equation (8.12) shows that, for $K > K_C$, $M(x)$ is not defined over all the interval from $x = 1$ to infinity. For $K < K_C$, one solution is everywhere subsonic and another one is everywhere supersonic. The latter must be rejected. The former has a vanishing velocity at infinity and its pressure approaches a finite limit. It must be rejected on the ground that it cannot match the boundary condition at infinity. The only possibly acceptable solution is the accelerating critical solution, with $K = K_c$. This solution is transsonic at $x = \lambda/2$ and it has a vanishing pressure, although a non-vanishing velocity, at infinity. Its supersonic character allows to terminate it by an expanding shock. Note that this solution is not selected on the basis that it satisfies the boundary conditions at infinity but because of its transsonic character. We expand on this below. Having found the relevant value of the constant K, given by (8.14), it is a simple matter to numerically obtain the solution of equation (8.12) and, in particular, to calculate the mass loss. This is left to the reader.

Let us come back on the choice of the solution. How is it that "natural" boundary conditions at infinity have given way to an entirely different demand, namely, that the solution passes at the critical point, *i.e.* that it be transsonic? The reason is hydrodynamical causality: in an hydrodynamical fluid the physical conditions at some point and at some time exert influence at other places by transmitting signals in the form of sound waves and entropy waves. This influence is, in particular, propagated along sound characteristics which are the trajectory of localized sound signals propagating at the local instantaneous sound speed c_S with respect to the fluid. A local observer placed at a certain point at a certain time may be unaware of the conditions having existed at some other point at an earlier time if there is no path for information to travel from one to the other. In the case of supersonic flows acoustical information cannot travel in the direction opposite to the fluid velocity, whatever the direction in which a signal propagates with respect to the fluid. Figure 4 shows the sound characteristics, which are the trajectories of sound signals in the position-time plane, for a 1D transsonic flow. At at any given position and time, there are two characteristics which propagate at the velocity $\pm c_s$ with respect to the moving fluid and at the velocity $v \pm c_S$ in the lab frame, v being the local and instantaneous fluid velocity. Because of stationnarity, all the forward-propagating characteristics are deduced from eachother by a translation in time, as are all the backward propagating characteristics. An important remark to make here is that the trans-sonic point is a watershed point for backwards characteristics, which propagate in the upwind direction with respect to the fluid. In the region of subsonic flow they can propagate, in the lab frame, in the direction opposite to the flow. However, in the supersonic region, signals are blown with the wind and propagate in the lab frame only in the downstream direction. Any

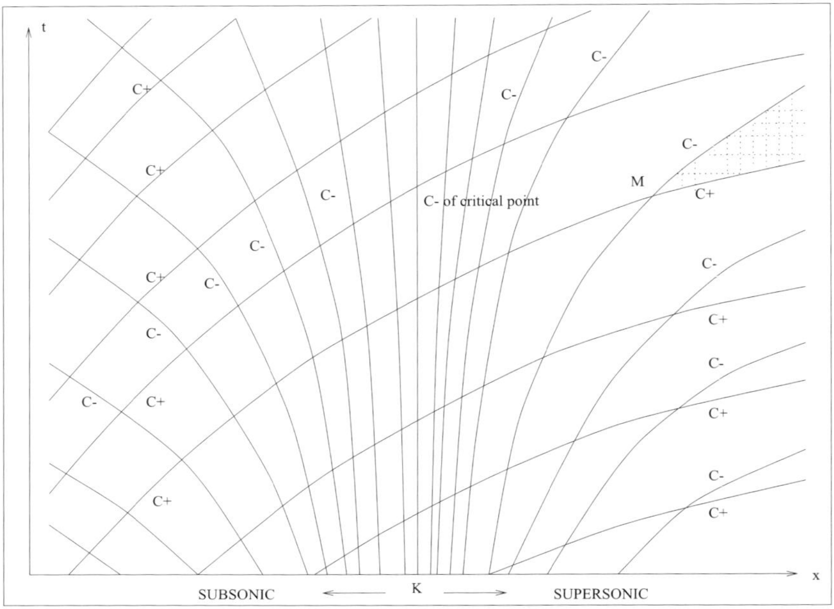

Fig. 4. Network of forward (C+) and backward (C–) characteristics in position-time plane for a one-dimensional stationnary transsonic flow. The transsonic, or critical, point is at K. The supersonic region cannot communicate with the subsonic region and therefore has no hydrodynamical influence on it. The region of influence of event M is hatched.

point in the subsonic region can send, and receive, signals to and from any other spatial point in the same region. In the supersonic region any point can only receive information from the upwind region and send information to the downwind region. This diagram makes it clear why any boundary condition at infinity cannot make itself felt. It also explains why the solution in the subsonic region is only affected by boundary conditions at the base of the flow and at the transonic point, with the boundary condition that $v = c_{\rm S}$ there. To sum up, the boundary conditions at infinity cannot, in the case of a transsonic flow, make themselves known to the subsonic region. They are replaced by the condition that the wind makes a transonic transition. For a more detailed presentation see [32] or [17].

9 Axisymmetric, rotating, stationnary, perfect MHD winds and jets

A more realistic model for a mass loss flow is that of a wind emitted by a rotating, magnetized object, which may be a star or an accretion disk. The wind, supposedly non-relativistic, axisymmetric and stationnary, consists of perfectly conducting, non viscous, plasma. Cylindrical coordinates $r\ \theta\ z$ are used. The energy balance is simplified by assuming that, for any fluid element, there is a polytropic relation (9.3) between its pressure and its density. The relevant dynamical and electrical equations are:

$$\mathrm{div}(\rho \boldsymbol{v}) = 0 \tag{9.1}$$

$$\rho(\boldsymbol{v}\cdot\boldsymbol{\nabla})\boldsymbol{v} = \rho\boldsymbol{g} - \boldsymbol{\nabla}P + +\boldsymbol{j}\times\boldsymbol{B} \tag{9.2}$$

$$(\boldsymbol{v}\cdot\boldsymbol{\nabla})\left(P/\rho^{\gamma}\right) = 0 \tag{9.3}$$

$$\mathrm{rot}\boldsymbol{E} = 0 \tag{9.4}$$

$$\mathrm{div}\boldsymbol{B} = 0 \tag{9.5}$$

$$\mathrm{rot}\boldsymbol{B} = \mu_0\boldsymbol{j} \tag{9.6}$$

$$\boldsymbol{E} + \boldsymbol{v}\times\boldsymbol{B} = 0. \tag{9.7}$$

The gravitational field derives from an axisymmetrical gravitational potential $\Phi_{\mathrm{g}}(r,z)$. Self-gravity of the wind is neglected. The poloidal magnetic field is represented as in equation (5.6). From equation (9.4) the electric field is electrostatic and given by

$$\boldsymbol{E} = -\boldsymbol{\nabla}\Phi_{\mathrm{e}}(r,z). \tag{9.8}$$

Dotting equation (9.7) with $\boldsymbol{B}$ it is found that magnetic field lines are equipotential, or, in the language of Section 5.2, that Φ_{e} is a surface function:

$$\Phi_{\mathrm{e}}(r,z) = \Phi_{\mathrm{e}}(a(r,z)) = \Phi_{\mathrm{e}}(a). \tag{9.9}$$

From $\Phi_{\mathrm{e}}(a)$ we define a rotation rate $\Omega(a)$, by no means equal to v_θ/r, by

$$\Omega(a) = \mathrm{d}\Phi_{\mathrm{e}}/\mathrm{d}a. \tag{9.10}$$

9.1 Ferraro's isorotation law

Let us write equation (9.7) by separating $\boldsymbol{v}$ and $\boldsymbol{B}$ into their poloidal and azimuthal parts, taking equation (9.9) into account. Its azimuthal part reduces to $(\boldsymbol{v}_{\mathrm{P}}\times\boldsymbol{B}_{\mathrm{P}}) = 0$, which can also be written as

$$\rho\boldsymbol{v}_{\mathrm{P}} = \alpha(r,z)\boldsymbol{B}_{\mathrm{P}}. \tag{9.11}$$

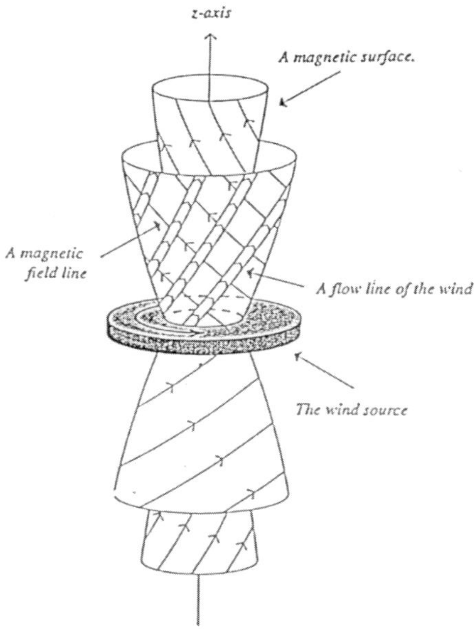

Fig. 5. General structure of an axisymmetric perfect MHD stationnary wind.

Taking the divergence of equation (9.11) it is seen, using (9.1), that α is a surface function because $\boldsymbol{B}_{\mathrm{P}} \cdot \boldsymbol{\nabla}\alpha = 0$. Thus,

$$\alpha = \alpha(a(r, z)) = \alpha(a). \tag{9.12}$$

Using (5.6) and (9.11), the projection of equation (9.7) on $\boldsymbol{\nabla}a$ gives:

$$\rho v_\theta = \rho r\Omega(a) + \alpha B_\theta. \tag{9.13}$$

Equations (9.13) and (9.11) can be summarized in Ferraro's isorotation law:

$$\rho \boldsymbol{v} = \rho r\Omega(a)\ \boldsymbol{e}_\theta + \alpha \boldsymbol{B}. \tag{9.14}$$

This law states that the flow is similar to the motion of a fluid in rotating pipes, the pipes being the field lines. The second term on the right of equation (9.14) describes the pipe-aligned motion. The pipes themselves rotate at the angular speed Ω, as the first term on the right of equation (9.14) shows. This is why Ω is often refered to as being the rotation rate of the magnetic field. This rotation is common to all field lines on a common magnetic surface. Magnetic surfaces then rotate like solid bodies. There may be however differential rotation between them, because $\Omega(a)$, may depend on a. Again, $\Omega(a)$ is not the rotation rate of the matter, (v_θ/r), unless ρ becomes infinite.

9.2 Conservation of the polytropic entropy

Equation (9.14) shows that the fluid flows on magnetic surfaces. Then equation (9.3), implies that Q is a surface function:

$$Q = Q(a(r, z)) = Q(a). \tag{9.15}$$

9.3 Conservation of specific angular momentum

The azimuthal component of equation (9.2) can be written as:

$$\rho \boldsymbol{e}_\theta \cdot ((\boldsymbol{v} \cdot \boldsymbol{\nabla})\boldsymbol{v}) = \boldsymbol{e}_\theta \cdot (\boldsymbol{j} \times \boldsymbol{B}) \,. \tag{9.16}$$

Equation (9.16) can transformed by using the identity (3.14) for $\boldsymbol{v}$ and $\boldsymbol{B}$, the identity $\mathrm{rot}(v_\theta \boldsymbol{e}_\theta) = \boldsymbol{\nabla}(rv_\theta) \times \boldsymbol{e}_\theta / r$ and the fact that $\boldsymbol{\nabla}$ has no azimuthal component. Using also equation (9.11) we get:

$$\boldsymbol{B}_{\mathrm{P}} \cdot \boldsymbol{\nabla} \left(\frac{rB_\theta}{\mu_0} - \alpha r v_\theta \right) = 0. \tag{9.17}$$

This identifies another surface function, the specific angular momentum,

$$L(a) = rv_\theta - \frac{rB_\theta}{\mu_0 \alpha} \tag{9.18}$$

$L(a)$ contains an electromagnetic contribution. The angular momentum of the matter alone is not conserved following its motion because the toroidal part of the Lorentz force exerts a torque on it. To understand how the electromagnetic term in equation (9.18) comes about, consider the flux of angular momentum. Since the gravitational force exerts no torque, the total angular momentum of the system consisting of the matter *and* the electromagnetic field together is conserved. The flux of this conserved total angular momentum, $\boldsymbol{\varphi}_{\mathrm{L}}$ must satisfy a perfect conservation law, which under stationnary conditions, can be written as $\mathrm{div}(\boldsymbol{\varphi}_{\mathrm{L}}) = 0$. The density of the angular momentum of matter is $\rho r v_\theta$ and the associated convected flux is $\rho r v_\theta \boldsymbol{v}$. In the absence of viscosity there is no diffusive flux. The amount of matter angular momentum flowing through a poloidal surface element $\mathrm{d}\boldsymbol{S}_{\mathrm{P}}$ in $\mathrm{d}t$ seconds is

$$\mathrm{d}l_{\mathrm{matt}} = \rho r v_\theta (\boldsymbol{v} \cdot \mathrm{d}\boldsymbol{S}_{\mathrm{P}}) \; \mathrm{d}t. \tag{9.19}$$

The flux of linear momentum carried by the electromagnetic field is the Maxwell stress tensor, given by equation (2.13). In the non-relativistic MHD it reduces to its magnetic part. The amount of azimuthal component of linear momentum flowing in electromagnetic form through $\mathrm{d}\boldsymbol{S}_{\mathrm{P}}$ in $\mathrm{d}t$ seconds is

$$\mathrm{d}p_{\theta\,\mathrm{em}} = \boldsymbol{e}_\theta \cdot \overline{\overline{\psi}}_{\mathrm{em}} \cdot \mathrm{d}\boldsymbol{S}_{\mathrm{P}} \; \mathrm{d}t = -\frac{B_\theta}{\mu_0} (\boldsymbol{B} \cdot \mathrm{d}\boldsymbol{S}_{\mathrm{P}}) \; \mathrm{d}t. \tag{9.20}$$

The corresponding quantity of angular momentum, $dl_{\rm em}$, is obtained by multiplying (9.20) by r. The poloidal part $\boldsymbol{\varphi}_{\rm L,P}$ of the total angular momentum flux $\boldsymbol{\varphi}_{\rm L}$ is found by adding $\mathrm{d}l_{\rm em}$ and $\mathrm{d}l_{\rm matt}$:

$$\boldsymbol{\varphi}_{\rm L,P} = \rho r v_\theta \boldsymbol{v}_{\rm P} - r B_\theta \boldsymbol{B}_{\rm P}/\mu_0. \tag{9.21}$$

This flux is aligned to the poloidal magnetic field, as is the poloidal part of the mass flux $\rho\boldsymbol{v}_{\rm P}$. Both fluxes are then conserved through the section of a flux tube. Their ratio, which is precisely L, is then also conserved.

9.4 Alfvén radius and density

From equations (9.13) and (9.18), v_θ and B_θ can be expressed as:

$$r v_\theta = r^2\Omega + \mu_0\alpha^2 \frac{L - r^2\Omega}{\mu_0\alpha^2 - \rho} \tag{9.22}$$

$$r B_\theta = \mu_0 \alpha\rho \frac{L - r^2\Omega}{\mu_0\alpha^2 - \rho}. \tag{9.23}$$

The Alfvén radius, $r_{\rm A}(a)$, and the Alfven density, $\rho_{\rm A}(a)$ are defined by

$$r_{\rm A}^2 = L/\Omega \qquad\qquad \rho_{\rm A} = \mu_0\alpha^2. \tag{9.24}$$

The point where $r = r_{\rm A}(a)$ is the Alfvén point on the poloidal field line a. The locus of all Alfvén points is the Alfvén surface. Expressions (9.22)–(9.23) can only be regular if, when $\rho = \rho_{\rm A}(a)$, $r = r_{\rm A}(a)$. It can be seen from equations (9.11) and (9.24) that, when $\rho = \rho_{\rm A}$, the poloidal velocity equals the Alfvén velocity associated with the poloidal magnetic field.

9.5 Conservation of specific energy

The acceleration term in equation (9.2) can be transformed by using equation (3.14) and by separating the poloidal and toroidal parts of vector fields. This changes equation (9.2) to:

$$\begin{aligned}\rho\boldsymbol{\nabla}\left(v^2/2\right) + \boldsymbol{\nabla}P + \rho\boldsymbol{\nabla}\Phi_{\rm g} = \rho\boldsymbol{v}_{\rm P} \times \mathrm{rot}\ \boldsymbol{v}_{\rm P} + \rho\boldsymbol{v}_\theta \times \mathrm{rot}\ \boldsymbol{v}_\theta + \\ \left(\mathrm{rot}\ \boldsymbol{B}_{\rm P} \times \boldsymbol{B}_{\rm P} + \mathrm{rot}\ \boldsymbol{B}_\theta \times \boldsymbol{B}_\theta\right)/\mu_0.\end{aligned} \tag{9.25}$$

From equations (9.3) and (9.15), we obtain, with the notation $Q' \equiv \mathrm{d}Q/\mathrm{d}a$ and similar notation for other surface functions,

$$\frac{\boldsymbol{\nabla}P}{\rho} = \frac{\gamma}{\gamma-1}\boldsymbol{\nabla}\left(Q\rho^{\gamma-1}\right) - Q'\,\frac{\rho^{\gamma-1}}{\gamma-1}\,\boldsymbol{\nabla}a. \tag{9.26}$$

It can also be shown, using equation (9.13), that

$$\boldsymbol{B}_\theta \times \mathrm{rot}\ \boldsymbol{B}_\theta = \frac{B_\theta}{r}\ \boldsymbol{\nabla}\left(rB_\theta\right) \tag{9.27}$$

$$\boldsymbol{v}_\theta \times \mathrm{rot}\ \boldsymbol{v}_\theta = \left(\Omega + \frac{\alpha B_\theta}{\rho r}\ \boldsymbol{\nabla}\left(L + \frac{r\Omega B_\theta}{\mu_0 \alpha}\right)\right)\cdot \tag{9.28}$$

After some algebra, the poloidal part of equation (9.25) can be written:

$$\frac{\mathrm{rot}\ \boldsymbol{B}_\mathrm{P} \times \boldsymbol{B}_\mathrm{P}}{\mu_0 \rho} - \mathrm{rot}\ \boldsymbol{v}_\mathrm{P} \times \boldsymbol{v}_\mathrm{P} = \boldsymbol{\nabla}\left(\frac{v^2}{2} + \frac{\gamma}{\gamma - 1} Q\rho^{\gamma-1} + \Phi_\mathrm{g} - \frac{r\Omega B_\theta}{\mu_0 \alpha}\right)$$
$$- \boldsymbol{\nabla} a \left[\frac{Q'\rho^{\gamma-1}}{\gamma - 1} + \Omega L' - \Omega' \frac{rB_\theta}{\mu_0 \alpha} + L' \frac{\alpha B_\theta}{\rho r} - \frac{\alpha'}{\alpha}\frac{B_\theta^2}{\mu_0 \rho}\right]\cdot \tag{9.29}$$

Projecting this equation on $\boldsymbol{B}_\mathrm{P}$ gives:

$$\boldsymbol{B}_\mathrm{P} \cdot \boldsymbol{\nabla}\left(\frac{v^2}{2} + \frac{\gamma}{\gamma - 1} Q\rho^{\gamma-1} + \Phi_\mathrm{g} - \frac{r\Omega B_\theta}{\mu_0 \alpha}\right) = 0. \tag{9.30}$$

This identifies yet another surface function, the total specific energy $E(a)$:

$$E(a) = \frac{v_\mathrm{P}^2 + v_\theta^2}{2} + \frac{\gamma}{\gamma - 1} Q\rho^{\gamma-1} + \Phi_\mathrm{g} - \frac{r\Omega B_\theta}{\mu_0 \alpha}\cdot \tag{9.31}$$

There is an electromagnetic contribution to this specific energy. The energy of matter alone is not conserved. This is because electromagnetic forces act on the plasma, so that energy is exchanged between matter and electromagnetic fields during the motion. The energy conservation law, which can be written here as

$$\mathrm{div}\left(\frac{1}{2}\rho v^2 \boldsymbol{v} + \frac{\gamma}{\gamma - 1} Q\rho^\gamma \boldsymbol{v} + \rho\Phi_\mathrm{g}\boldsymbol{v} + \frac{\boldsymbol{E} \times \boldsymbol{B}}{\mu_0}\right) = 0 \tag{9.32}$$

indicates that the flux of the sum of all forms of energy is conserved. Using equations (9.9) and (9.10), we find that

$$\left(\boldsymbol{E} \times \boldsymbol{B}\right)_\mathrm{P} = -r\Omega B_\theta\ \boldsymbol{B}_\mathrm{P}. \tag{9.33}$$

This shows that the poloidal part of the energy flux in equation (9.32), including the Poynting contribution, is along $\boldsymbol{B}_\mathrm{P}$, as is also the mass flux. Both fluxes are then conserved through the section of a flux tube. Their ratio, which is precisely the total specific energy E, is then also conserved.

9.6 Transfield equation

The projection of equation (9.29) perpendicular to $\boldsymbol{B}_{\mathrm{P}}$, onto $\boldsymbol{\nabla} a$, can be brought to the form of a partial differential equation for $a(r,z)$, also involving the density ρ:

$$\frac{\alpha}{\rho r}\left(\frac{\partial}{\partial r}\frac{\alpha}{\rho r}\frac{\partial a}{\partial r}+\frac{\partial}{\partial z}\frac{\alpha}{\rho r}\frac{\partial a}{\partial z}\right)-\frac{1}{\mu_0\rho r}\left(\frac{\partial}{\partial r}\frac{1}{r}\frac{\partial a}{\partial r}+\frac{\partial}{\partial z}\frac{1}{r}\frac{\partial a}{\partial z}\right)=E'-\frac{Q'\rho^{\gamma-1}}{\gamma-1}$$
$$+\frac{\alpha'}{\alpha}\frac{\mu_0\alpha^2\rho}{r^2}\left(\frac{L-r^2\Omega}{\mu_0\alpha^2-\rho}\right)^2-\frac{\rho}{r^2}\frac{(L'-r^2\Omega')(L-r^2\Omega)}{\mu_0\alpha^2-\rho}-\frac{LL'}{r^2}. \quad (9.34)$$

This is the transfield, or generalized Grad-Shafranoff, equation. Toroidal variables may be eliminated from equation (9.31) by equations (9.22)–(9.23), leading to:

$$\frac{\alpha^2|\nabla a|^2}{2\,\rho^2 r^2}+\Phi_{\mathrm{g}}+\frac{\gamma Q\rho^{\gamma-1}}{\gamma-1}-\frac{\rho\Omega(L-r^2\Omega)}{\mu_0\alpha^2-\rho}+\frac{1}{2}\left(\frac{L}{r}+\frac{\rho}{r}\frac{L-r^2\Omega}{\mu_0\alpha^2-\rho}\right)^2=E(a). \quad (9.35)$$

Equations (9.34) and (9.35) form a coupled system for $\rho(r,z)$ and $a(r,z)$ which encompasses all the physics of this perfect MHD flow, including the shaping of magnetic surfaces by the stresses exerted by the flow. For a vanishing velocity, equation (9.34) reduces to equation (5.14). To show this, it should be noted that Ω and α vanish in this limit. Careful examination shows that (α/Ω) approaches a finite limit and that L diverges as $L\approx -I/\mu_0\alpha$.

9.7 Bernoulli dynamics alone

The system (9.34)–(9.35) being formidable, much work has been devoted to solving equation (9.35) alone, or its relativistic form, the shape of magnetic surfaces being supposedly known and often taken to be conical [20, 31]. This reduces the problem to a form similar to the hydrodynamical solar wind problem, but for the fact that Lorentz forces are taken into account. As in Section (8.2), the density on a given magnetic surface a is determined as a function of r by an implicit algebraic relation

$$\mathcal{B}(\rho,r)=E(a). \quad (9.36)$$

The Bernoulli function $\mathcal{B}$ on magnetic surface a is given by equation (9.35). It depends on the first integrals α, Q, Ω and L. The functions $Q(a)$ and $\Omega(a)$ are given by the boundary conditions but E, α and L are not. Equation (9.36) can be rewritten in the form of a differential equation for the density $\rho(r)$ on the magnetic surface a

$$\frac{r}{\rho}\frac{\mathrm{d}\rho}{\mathrm{d}r}=-\frac{r(\partial\mathcal{B}/\partial r)}{\rho(\partial\mathcal{B}/\partial\rho)}. \quad (9.37)$$

For regularity, the numerator on the right of equation (9.37) must vanish when the denominator does. This is possible only when special relations between Q, Ω, α and L, the criticality conditions, are satisfied. Explicit calculation shows that

$$\rho \frac{\partial \mathcal{B}}{\partial \rho} = -\frac{\alpha^2 B_{\rm P}^2}{\rho^2} + \gamma Q \rho^{\gamma-1} + \frac{\rho \rho_{\rm A}^2}{r^2} \frac{\left(L - r^2 \Omega\right)^2}{\left(\mu_0 \alpha^2 - \rho\right)^3}. \tag{9.38}$$

This expression vanishes when the poloidal velocity of the flow equals the phase velocity of one of the magnetoacoustic modes propagating parallel to the poloidal flow. This can be seen by introducing the characteristic speeds:

$$c_{\rm S}^2 = \gamma Q \rho^{\gamma-1} \quad v_{\rm AP}^2 = \frac{B_{\rm P}^2}{\mu_0 \rho} = \frac{\rho v_{\rm P}^2}{\mu_0 \alpha^2} \quad v_{{\rm A}\theta}^2 = \frac{B_\theta^2}{\mu_0 \rho} \quad v_{\rm A}^2 = v_{\rm AP}^2 + v_{{\rm A}\theta}^2. \tag{9.39}$$

The term on th right of equation (9.38) vanishes when

$$v_{\rm P}^4 - v_{\rm P}^2 (v_{\rm A}^2 + c_{\rm S}^2) + c_{\rm S}^2 v_{\rm AP}^2 = 0. \tag{9.40}$$

It will be shown in Section (11.3) that this equation gives the phase velocity of magnetoacoustic waves propagating in the direction of the poloidal motion. There are two such waves, the slow mode and the fast mode, with phase velocities $v_{\rm s}$ and $v_{\rm f}$ resp. The slow mode speed $v_{\rm s}$ should not be mistaken for the sound speed $c_{\rm S}$. The denominator on the right of equation (9.37) vanishes when the poloidal flow velocity $v_{\rm P}$ equals $v_{\rm s}$ or $v_{\rm f}$. The flow can pass these trans-magnetosonic points regularly only if the numerator of equation (9.37) also vanishes. Thus, regular trans-magnetosonic transitions are only possible in the r–ρ plane where the differential of $\mathcal{B}(\rho, r)$ vanishes. This statement is independent of the variables used: it would apply just as well in the r–$v_{\rm P}$ plane, for example. The points where the differential of $\mathcal{B}$ vanishes are called *critical points*. The solution of equation (9.36) must pass the MHD mode speed at critical points if it is to be regular and super-fast-mode at infinity. See [17] for a more detailed account of the critical points in the Weber Davis model. The requirement that the solution passes both slow and fast critical points imposes two different conditions on Ω, α, L and Q, which allow the determination of α and L. Once these have been determined, the value of E is obtained from equation (9.36).

9.8 The Bernoulli-Transfield system

The full MHD wind problem consists in jointly solving equations (9.34)–(9.35), with unknowns $a(r, z)$ and $\rho(r, z)$, or any equivalent pair of functions. This system incorporates the fact that the geometry of the magnetic surfaces determines the flow and that, conversely, the flow shapes the magnetic surfaces. Ideally, equation (9.35) would be solved for ρ in terms of the first integrals and of $|\boldsymbol{\nabla} a|^2$ and the result substituted in equation (9.34), giving an equation for $a(r,z)$ alone. This is not possible, but very important information may be obtained by working out the expression of the highest order derivative terms of this ideal equation, which, by contrast is possible. These highest order derivative terms are all present on the left of equation (9.34), but they are not written explicitly. This is because the density ρ, which should eventually be eliminated, is still present under the derivatives in the first term on the left of (9.34). The derivatives of ρ, which may be obtained from equation (9.35), involve derivatives of $|\boldsymbol{\nabla} a|^2$, that is, second order derivative terms. Equation (9.35) should then be differentiated. Expressing the result in terms of the poloidal fluid velocity and of the speeds defined in equation (9.39), it is found that

$$\frac{\alpha^2}{2\rho r^2}\boldsymbol{\nabla}\left(|\nabla a|^2\right) - \frac{v_{\rm P}^4 - v_{\rm P}^2(v_{\rm A}^2 + c_{\rm S}^2) + c_{\rm S}^2 v_{\rm AP}^2}{v_{\rm P}^2 - v_{\rm AP}^2}\,\boldsymbol{\nabla}\rho = \boldsymbol{L}_1 \qquad (9.41)$$

where $\boldsymbol{L}_1$ involves no second order derivatives of $a(r,z)$. Equation (9.41) can then be used to sort out, among terms which are generated by the derivatives of ρ on the left of equation (9.34), those which involve second order derivatives of $a(r,z)$. This manipulation reduces equation (9.34) to the form

$$\frac{\mu_0\alpha^2-\rho}{\mu_0\rho r^2}\,\Delta a - \frac{\alpha^4}{\rho^3 r^4}\,\frac{(v_{\rm P}^2 - v_{\rm AP}^2)\,(\boldsymbol{\nabla} a\cdot\boldsymbol{\nabla}(|\nabla a|^2))}{2\,(v_{\rm P}^4 - v_{\rm P}^2(v_{\rm A}^2 + c_{\rm S}^2) + c_{\rm S}^2 v_{\rm AP}^2)} = g_{\rm A}(r,z,a,\boldsymbol{\nabla} a). \quad (9.42)$$

The term on the right, $g_{\rm A}$, involves no second order derivatives of $a(r,z)$. Δ is the Laplacian operator in cylindrical coordinates. From equation (9.39) $v_{\rm P}^2/v_{\rm AP}^2 = \rho_{\rm A}/\rho$, which turns equation (9.42) into:

$$\frac{1}{\mu_0 r^2}\left(\frac{v_{\rm P}^2}{v_{\rm AP}^2} - 1\right)\left[\Delta a - \frac{v_{\rm P}^4}{v_{\rm P}^4 - v_{\rm P}^2(v_{\rm A}^2 + c_{\rm S}^2) + c_{\rm S}^2 v_{\rm AP}^2}\,\frac{\boldsymbol{\nabla} a\cdot\boldsymbol{\nabla}(|\nabla a|^2)}{2\,|\nabla a|^2}\right] = g_{\rm A}. \quad (9.43)$$

The operator on the left of equation (9.43) is linear in the second order derivatives of $a(r,z)$. Thus, equation (9.43), though its right hand side cannot be explicitly written in terms of a and $\boldsymbol{\nabla} a$, appears to be a quasilinear partial differential equation of second order.

9.9 Alfvén regularity condition

It is important to note that all highest order derivative terms of equation (9.43) vanish when $v_{\rm P}^2 = v_{\rm AP}^2$. Equation (9.43) is therefore singular on the Alfvén surface, defined in Section (9.4). It results that on this surface, all terms on the right of equation (9.43) must vanish. Any failure to satisfy this Alfvén regularity condition is punished by obtaining a solution in which magnetic and flow surfaces have a kink at their crossing of the Alfvén surface. This kink represents a rotational discontinuity standing in the flow. The Alfvén regularity condition is expressed by the statement that

$$g_{\mathrm{A}}(r, z, a, \boldsymbol{\nabla} a) = 0 \tag{9.44}$$

at the Alfvén surface. An explicit form of it has been derived in [13].

9.10 Ellipticity or hyperbolicity of the Bernoulli-Transfield system

In a quasilinear partial differential equation for a function f of two variables x and y, the linear second order operator can be quite generally written as:

$$A\frac{\partial^2 f}{\partial x^2} + 2B\frac{\partial^2 f}{\partial x \partial y} + C\frac{\partial^2 f}{\partial y^2}\cdot \tag{9.45}$$

The coefficients A, B and C may be functions of x, y, f and its first order derivatives. The equation is hyperbolic when the operator (9.45) can be locally factored, with real p and q, as

$$A\left(\frac{\partial}{\partial x} - p\frac{\partial}{\partial y}\right)\left(\frac{\partial}{\partial x} - q\frac{\partial}{\partial y}\right)\cdot \tag{9.46}$$

The numbers p and q are the roots of equation $At^2 + 2Bt + C = 0$. One can also think of hyperbolicity as a condition for very small wavelength perturbations to propagate as waves, which implies that there should exist a vector $\boldsymbol{k}$, with real components k_x and k_y, such that

$$Ak_x^2 + 2Bk_xk_y + Ck_y^2 = 0. \tag{9.47}$$

Equation (9.47) results from equation (9.46) by replacing ∂_x and ∂_y by ik_x and ik_y resp. The condition for local hyperbolicity is that $B^2 - AC > 0$. Since the coefficients A, B, C associated with equation (9.43) are known explicitly, its elliptic or hyperbolic nature can be discussed. To write down equation (9.47) for the second order operator of equation (9.43), use cartesian coordinates x_i. By substituting wave vectors components for the partial derivatives in the second order term of equation (9.43), it is found that

$$\frac{\boldsymbol{\nabla} a\cdot \boldsymbol{\nabla}|\nabla a|^2}{2|\nabla a|^2} = \frac{\nabla_i a(\nabla_i\nabla_j a)\nabla_j a}{|\nabla a|^2} \rightarrow -\left(\frac{\boldsymbol{k}\cdot\boldsymbol{\nabla} a}{|\nabla a|}\right)\left(\frac{\boldsymbol{k}\cdot\boldsymbol{\nabla} a}{|\nabla a|}\right)a = -k_\perp^2 a. \tag{9.48}$$

The index $\perp$ indicates a component perpendicular to the magnetic field, as is $\boldsymbol{\nabla} a$. Equation (9.43) is locally hyperbolic if the equation

$$k^2 - \frac{v_{\mathrm{P}}^4 k_\perp^2}{v_{\mathrm{P}}^4 - v_{\mathrm{P}}^2(v_{\mathrm{A}}^2 + c_{\mathrm{S}}^2) + c_{\mathrm{S}}^2 v_{\mathrm{AP}}^2} = 0 \tag{9.49}$$

has a solution with real $\boldsymbol{k}$. Thus, k^2 should exceed $k_\perp^2$, which implies that

$$\frac{(v_\mathrm{A}^2 + c_\mathrm{S}^2)\left(v_\mathrm{P}^2 - \frac{c_\mathrm{S}^2 v_\mathrm{A}^2}{v_\mathrm{A}^2 + c_\mathrm{S}^2}\right)}{v_\mathrm{P}^4 - v_\mathrm{P}^2(v_\mathrm{A}^2 + c_\mathrm{S}^2) + c_\mathrm{S}^2 v_\mathrm{AP}^2} > 0. \tag{9.50}$$

It is shown in Section (11.3) that the roots of the denominator of the term on the right of equation (9.50) are the phase speeds, v_s and v_f, of the fast and slow magnetosonic modes propagating in the direction of the poloidal motion. The velocity defined by

$$v_\mathrm{cp}^2 = \frac{c_\mathrm{S}^2 v_\mathrm{A}^2}{v_\mathrm{A}^2 + c_\mathrm{S}^2} \tag{9.51}$$

is the cusp speed. It is smaller than v_AP, c_S, v_s and v_f. The hyperbolicity condition (9.50) can then be written as:

$$\left(v_\mathrm{P}^2 - v_\mathrm{cp}^2\right)\left(v_\mathrm{P}^2 - v_\mathrm{s}^2\right)(v_\mathrm{P}^2 - v_\mathrm{f}^2) > 0. \tag{9.52}$$

Thus, the coupled Transfield-Bernoulli problem is of a mixed type. It is elliptic as long as $v_\mathrm{P}^2 < v_\mathrm{cp}^2$, hyperbolic when v_P is between v_cp and v_s, elliptic again between v_s and v_f and finally hyperbolic for $v_\mathrm{P} > v_\mathrm{f}$.

9.11 An analysis of forces in axisymmetric MHD winds and jets

It is interesting to analyze the influence of the different forces exerted on an MHD wind. The equation of motion in stationnary state can be written as:

$$\rho(\boldsymbol{v} \cdot \boldsymbol{\nabla})\boldsymbol{v} = \rho \boldsymbol{g} - \boldsymbol{\nabla} P + \boldsymbol{j} \times \boldsymbol{B} + \boldsymbol{f}_\mathrm{other}. \tag{9.53}$$

The term on the left of equation (9.53) is the inertia force, which includes the centrifugal force. The first term on the right is the gravitational force, which exerts no torque nor outward acceleration. The pressure gradient force drives the solar wind. For very deep gravitational potential wells this force has however little, or only auxiliary, effect. The last term on the right of equation (9.53) would include, for example, radiative forces, the effect of which might be substancial for close-to-Eddington sources. Radiation is not expected to exert a significant torque for sub-relativistic flows. The Lorentz force may have a variety of effects. Separating poloidal and toroidal components, it may be written as:

$$\boldsymbol{j} \times \boldsymbol{B} = \boldsymbol{j}_\mathrm{P} \times \boldsymbol{B}_\mathrm{P} + \boldsymbol{j}_\theta \times \boldsymbol{B}_\mathrm{P} + \boldsymbol{j}_\mathrm{P} \times \boldsymbol{B}_\theta. \tag{9.54}$$

The first term on the right of equation (9.54) only exerts a torque. It may accelerate or decelerate the rotation of the fluid. The second term is the

Lorentz force of the poloidal magnetic field, which has no component along $\boldsymbol{B}_{\mathrm{P}}$ and therefore no effect on the field-aligned dynamics. This force only plays a role in shaping the magnetic surfaces. The third term is the Lorentz force of the toroidal field. Its value at (r, z) can be expressed, by Ampere's law, in terms of the total poloidal electric current J_{P} flowing through a circle centered on the axis passing at (r, z), *i.e.*:

$$\boldsymbol{j}_{\mathrm{P}} \times \boldsymbol{B}_{\theta} = -\frac{\mu_0}{8\pi r^2}\, \boldsymbol{\nabla} J_{\mathrm{P}}^2. \tag{9.55}$$

This force may have components both along and perpendicular to $\boldsymbol{B}_{\mathrm{P}}$. Let $\boldsymbol{e}_{\|}$ and $s_{\|}$ be the unit tangent vector and the curvilinear abcissa along a poloidal magnetic field line and $\boldsymbol{e}_{\perp}$ and $s_{\perp}$ be the unit tangent vector and curvilinear abcissa along an orthogonal trajectory to these lines. The component of the force (9.55) along $\boldsymbol{e}_{\perp}$ is

$$\boldsymbol{e}_{\perp} \cdot (\boldsymbol{j}_{\mathrm{P}} \times \boldsymbol{B}_{\theta}) = -\frac{\mu_0}{8\pi r^2}\, \frac{\mathrm{d}J_{\mathrm{P}}^2}{\mathrm{d}s_{\perp}}. \tag{9.56}$$

This so-called hoop stress, or pinching force, is directed to the axis in its vicinity because J_{P}^2 vanishes on the axis. From Ampere's law, J_{P} also vanishes wherever the toroidal field does. This happens at null surfaces where the poloidal magnetic field vanishes. For a dipolar type of symmetry, the equatorial plane is a null surface. In a close vicinity of a null surface, the force (9.56) points to it. The field-aligned component of (9.55) is

$$\boldsymbol{e}_{\|} \cdot (\boldsymbol{j}_{\mathrm{P}} \times \boldsymbol{B}_{\theta}) = -\frac{\mu_0}{8\pi r^2}\, \frac{\mathrm{d}J_{\mathrm{P}}^2}{\mathrm{d}s_{\|}}. \tag{9.57}$$

It is accelerating or decelerating according to whether poloidal current lines leave or enter the magnetic surface. Consider for example the wind from an accretion disk and a magnetic configuration with a dipolar type of symmetry. Assume the magnetic field lines to be rooted in the disk's equatorial plane and to reach infinity. By symmetry, B_{θ} vanishes at the equatorial plane. In its vicinity, the force (9.57) is oriented toward the equatorial plane and remains so as long as current loops enter the magnetic surface. Since current loops must eventually close, part of this current leaves the magnetic surface at larger distances. In these regions the force (9.57) causes plasma acceleration outwards. At very large distances, the poloidal current enclosed in the magnetic surface must reach some asymptotic value, perhaps a vanishing one, and the force (9.57) then becomes weak.

10 Asymptotics of perfect MHD winds and jets

10.1 General results on the asymptotics of polytropic winds

The asymptotic shape of some magnetic surface a can be characterized by the limit reached by (r/z) when the surface is followed to infinity. The radial distance r cannot approach zero as z approaches infinity because the plasma density would then diverge. When r approaches a finite limit, $r_\infty(a)$, the magnetic surface is said to be non-flaring and of an asymptotically cylindrical shape. When r approaches infinity following the magnetic surface, it is said to be flaring. If (r/z) approaches zero, it is said to be paraboloidal and if (r/z) approaches a finite limit it is said to be conical. This doesn't mean that the magnetic surface would asymptote to a cone in this case, because it may have parabolic branches. The angle θ_∞ is defined by $\tan\theta_\infty = \lim(z/r)$. A flaring magnetic surface along which (r/z) approaches infinity is said to be asymptotically flattening. We show that this cannot occur for perfect MHD polytropic winds. Consider equation (9.35). The fourth term on its left, which represents the Poynting flux, is always positive because its numerator and denominator simultaneously change sign at the Alfvén point. As r approaches infinity on a magnetic surface, the density ρ cannot exceed the Alfvén density $\mu_0\alpha^2(a)$. The gravitational energy term asymptotically vanishes. None of the remaining positive terms on the left of equation (9.35) can diverge. Then, ρr^2 must remain finite. From equations (9.22) and (9.23) this implies that v_θ approaches zero and that the Poynting energy term in equation (9.35) remains bounded. Using equations (9.23), (9.35) asymptotically reduces, on flaring magnetic surfaces, to:

$$\frac{v_{\mathrm{P}}^2}{2} + \frac{\rho r^2 \Omega^2}{\mu_0 \alpha^2} = E. \tag{10.1}$$

This analysis is valid only when B_θ does not identically vanish. It does not hold at and very near null magnetic surfaces, where, as can be seen from equation (9.11), $\alpha(a)$ diverges. We come back later on this. Otherwise, equation (10.1) applies and both terms on its left must remain smaller than E:

$$\frac{v_{\mathrm{P}}^2}{2} \leq E \qquad\qquad \frac{\rho r^2 \Omega^2}{\mu_0 \alpha^2} \leq E. \tag{10.2}$$

The inequalities (10.2) can be turned into a constraint on the shape of poloidal magnetic field lines. Using equations (9.11), (5.6) and (10.2), an upper bound on $r|\boldsymbol{\nabla} a|$ is obtained:

$$r|\boldsymbol{\nabla} a| = \frac{\rho r^2 v_{\mathrm{P}}}{\alpha} \leq \frac{\sqrt{2}\mu_0 \alpha E^{3/2}}{\Omega^2} = \lambda(a). \tag{10.3}$$

The fact that $|\boldsymbol{\nabla} a|$ is larger than $|\partial a/\partial z|$ implies that

$$|\frac{\partial a}{\partial z}| \leq \frac{\lambda(a)}{r} \qquad \Rightarrow \qquad |\frac{\partial z}{\partial a}| \geq \frac{r}{\lambda(a)} \cdot \tag{10.4}$$

Using this, we obtain, by integrating over a, a constraint on the dependance of z on a at fixed r. For a increasing from 0 at the pole to A_{eq} at the equator, this constraint can be written as:

$$z \geq r \int_a^{A_{\mathrm{eq}}} \frac{\mathrm{d}a'}{\lambda(a')} = r\Lambda(a). \tag{10.5}$$

This shows that, asymptotically, the magnetic surface a is entirely above some cone. Asymptotic flattening is thus impossible. A stronger conclusion can be reached when ρr^2 approaches zero on a set of magnetic surfaces subtending a finite flux. Equation (10.3) can be written for these magnetic surfaces as

$$r|\boldsymbol{\nabla} a| \leq \frac{\sqrt{2}\rho r^2 E^{1/2}}{\alpha} \leq \lambda(a, r). \tag{10.6}$$

The function $\lambda(a, r)$ approaches zero when r approaches infinity at a given a in the set. This provides the following constraint on $z(a)$ at fixed r:

$$z \geq r \int_a^{A_{\mathrm{eq}}} \frac{\mathrm{d}a'}{\lambda(a', r)} = r\ \Lambda(a, r). \tag{10.7}$$

Since $\Lambda(a, r)$ diverges as $r \to \infty$, all magnetic surfaces polewards of those on which ρr^2 approaches zero are asymptotically above some paraboloid.

10.2 Asymptotic transfield equation

The Bernoulli equation provides some general constraints on the asymptotic shape of polytropic winds. However, a complete understanding of their asymptotic structure requires an analysis of the transfield equation. We now derive the asymptotic form of this equation from equation (9.29). Some simplifications occur in the case of flaring field lines, when $r \to \infty$. In this limit, the second term on the left of equation (9.29) dominates over the first because $B_{\mathrm{P}}^2/\mu_0\rho v_{\mathrm{P}}^2$, which is equal to $\rho/\mu_0\alpha^2$, approaches zero. We also use the asymptotic form of equation (9.23). Unless the Lorentz force identically vanishes, we may neglect gas pressure in equation (9.29), which leads to:

$$-\mathrm{rot}\boldsymbol{v}_{\mathrm{P}} \times \boldsymbol{v}_{\mathrm{P}} = \boldsymbol{\nabla} E + \boldsymbol{\nabla} a \left(\frac{rB_\theta\Omega'}{\mu_0\alpha} - \frac{rB_\theta\Omega\alpha'}{\mu_0\alpha^2}\right) \cdot \tag{10.8}$$

It is useful to write $\boldsymbol{v}_{\mathrm{P}}$ as $v_{\mathrm{P}}\boldsymbol{t}$, where $\boldsymbol{t}$ is a unit vector tangent to poloidal field lines. The vector $\boldsymbol{n}$ normal to poloidal field lines is $\boldsymbol{n} = -\boldsymbol{\nabla} a/|\boldsymbol{\nabla} a|$.

Let the angle between $\boldsymbol{t}$ and the equatorial plane be ψ. Using the Frénet formula, we calculate

$$-\mathrm{rot}\boldsymbol{v}_{\mathrm{P}} \times \boldsymbol{v}_{\mathrm{P}} = \boldsymbol{\nabla}\left(\frac{v_{\mathrm{P}}^2}{2}\right) - \boldsymbol{t}\left(\boldsymbol{t}\cdot\boldsymbol{\nabla}\left(\frac{v_{\mathrm{P}}^2}{2}\right)\right) - v_{\mathrm{P}}^2\left(\boldsymbol{n}\frac{\mathrm{d}\psi}{\mathrm{d}s}\right)\cdot \tag{10.9}$$

Inserting equation (10.9) in equation (10.8), we obtain:

$$v_{\mathrm{P}}^2\boldsymbol{n}\frac{\mathrm{d}\psi}{\mathrm{d}s}+\boldsymbol{t}\left(\boldsymbol{t}\cdot\boldsymbol{\nabla}\left(\frac{v_{\mathrm{P}}^2}{2}\right)\right)+\boldsymbol{\nabla}\left(E-\frac{v_{\mathrm{P}}^2}{2}\right)+\boldsymbol{\nabla}a\left(\frac{rB_\theta\Omega'}{\mu_0\alpha}-\frac{rB_\theta\Omega\alpha'}{\mu_0\alpha^2}\right)=0. \tag{10.10}$$

Let us define I, which is proportional to J_{P} (Sect. 9.11), by:

$$I = -rB_\theta/\mu_0. \tag{10.11}$$

Asymptotically, v_θ and ρ vanish and $(E - v_{\mathrm{P}}^2/2)$ can be expressed as:

$$E - \frac{v_{\mathrm{P}}^2}{2} = \frac{I\Omega}{\alpha}\cdot \tag{10.12}$$

The projection of the asymptotic form of equation (10.10) on the vector $\boldsymbol{n}$ leads to a form of the transfield equation valid in the limit of very large r/r_{A}:

$$v_{\mathrm{P}}^2|\boldsymbol{\nabla}a|\frac{\mathrm{d}\psi}{\mathrm{d}s} = \frac{\Omega}{\alpha}\left(\boldsymbol{\nabla}a\cdot\boldsymbol{\nabla}I\right). \tag{10.13}$$

Again, equation (10.13) is only valid away from null surfaces. As r approaches infinity on field line a, I approaches a limit $I_\infty(a)$, which may or may not be zero, and v_{P} approaches a limit $v_\infty(a)$. Using the asymptotic form of equations (9.23), (10.13) can be restated as:

$$r\frac{\mathrm{d}\psi}{\mathrm{d}s} = \mu_0\,\frac{I_\infty(a)I'_\infty(a)}{\alpha(a)v_\infty(a)}\cdot \tag{10.14}$$

The curvature $\mathrm{d}\psi/\mathrm{d}s$ asymptotically vanishes, but does $r\mathrm{d}\psi/\mathrm{d}s$ also vanishes? If it does not, the term on the right of equation (10.14) is, at given a, a non-vanishing constant, k say. Assuming this, the shape of the field line a can be calculated by solving the differential equation $r\mathrm{d}\psi/\mathrm{d}s = k$, which can be done by eliminating $\mathrm{d}s$ with $\mathrm{d}r = \cos\psi\mathrm{d}s$. This results in the following parametric representation of r in terms of the angle ψ:

$$r = r_0\exp\left(\frac{\sin\psi}{k}\right)\cdot \tag{10.15}$$

Note that r does not approach infinity for non-zero k, which makes the solution with non-vanishing k inconsistent with the asumption that r approaches infinity [13]. On flaring magnetic surfaces, away from the polar axis and from null surfaces, the term on the right of equation (10.14) must approach zero:

$$\nabla a \cdot \nabla I_{\infty} = 0. \tag{10.16}$$

This integrates as $I_{\infty} = I_{\infty}(b)$ where b is a label for an orthogonal trajectory to poloidal magnetic field lines. We take b to be the distance R to the origin of the axial point on this curve. When R grows to infinitely large values, the asymptotic poloidal current enclosed in the magnetic surface a approaches a constant value, independent of a and R, which may or may not be zero. This conclusion has been reached from equation (10.1), which is valid only when the Poynting energy indeed dominates over the neglected terms. This cannot be so at the polar axis, nor at a null magnetic surface, where I exactly vanishes. Therefore equation (10.14) is only piecewise valid. It holds true in regions bordered by null surfaces, but looses its validity near these surfaces and near the polar axis. The regions where equation (10.16) is valid constitute the asymptotic field. Other regions, about the polar axis and the null surfaces, are asymptotic boundary layers. The dimension perpendicular to the magnetic surfaces of these latter regions must be small because the forces which balance the Lorentz forces in these regions are subdominant. They can play a role only where the Lorentz force is weak enough, *i.e.* in the asymptotic boundary layers.

10.3 Electric circuit in the asymptotic domain

A simple picture of the poloidal current circuit in the asymptotic domain then emerges. Equation (10.16) shows that the current density vanishes everywhere except in the asymptotic boundary layers, which may carry current. Figure 6 illustrates this. Regions bounded by the polar axis and the polar-most null surface, or regions bounded on both sides by a null surface, are cells in which the current exactly closes. There is pressure balance between both sides of any of these null surface boundary layers. Since, outside of the boundary layers, the toroidal magnetic pressure dominates, the azimuthal magnetic field just changes sign at their crossing. Thus $I_{\infty}(a)$ changes sign when passing from one to the neighbouring current cell. Different situations occur according to whether $|I_{\infty}|$ is zero or not [14–16]. The jet does not really extend to infinity. It is bounded at large distances by a system of expanding shocks which separates it from an unperturbed outer medium. Poloidal current closure is achieved by currents flowing on these shock fronts. The stationnary part of the flow is well inside this shock system. The stationnary state asymptotic analysis is valid at a distance from the source much less than that of the terminating shocks and much larger than that of the Alfvén surface.

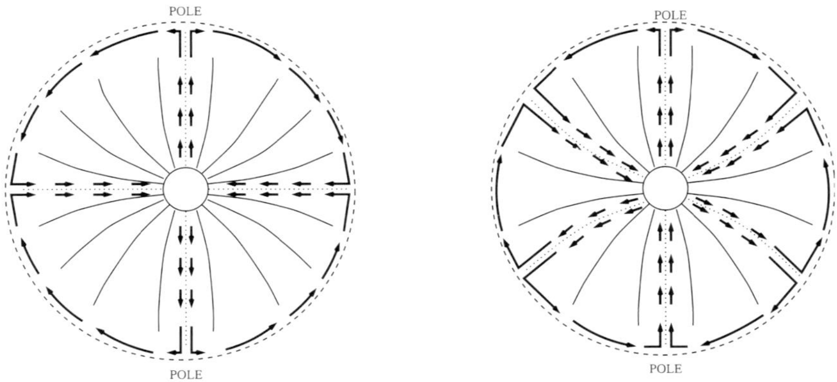

Fig. 6. Electric circuit in the asymptotic region of magnetized MHD winds. Current flows in a small vicinity of the polar axis, and returns to the source in thin sheets about null magnetic surfaces where the magnetic field vanishes. Left panel: a case when the magnetic field has a dipolar type of symmetry; the axial currents return through the equatorial plane. Right panel: a case when the magnetic field has a quadrupolar type of symmetry; the axial currents return through null surfaces. In a stationnary picture, polar and null layer currents meet at infinity. In reality they close along the expanding shock fronts which terminate the flow.

When $|I_\infty|$ is non zero, flaring magnetic surfaces are conical. A non-zero $|I_\infty|$ implies that ρr^2 approaches a finite limit on any magnetic surface of the asymptotic field. From equation (10.3), it results that $r|\boldsymbol{\nabla} a|$ approaches a finite limit $f(a)$. The shape of flaring magnetic surfaces in the asymptotic field is then determined by an Hamilton-Jacobi type of equation

$$r|\boldsymbol{\nabla} a| = f(a). \tag{10.17}$$

This equation is analoguous to the Eikonal equation for light propagation in a medium with a refraction index $N = 1/r$. Using this remark, it can be shown that the flaring magnetic surfaces are asymptotically conical when the wind subtends a finite magnetic flux [14, 15].

Equation (10.16) implies that at large distances no current flows between flaring magnetic surfaces. For non-zero total current this implies that the current that they enclose flows in some region about the polar axis, where the magnetic surfaces are cylindrical. Paraboloidal magnetic surfaces do not enclose a finite current asymptotically, unless the magnetic flux subtending the wind source is infinite [14]. A smooth matching between an asymptotically conical and an asymptotically cylindrical region is possible if, when the transition flux a_* is approached from the poleward region, $r_\infty(a)$ approaches infinity while, as it is approached from the equatorward region, $\psi(a)$ approaches $\pi/2$. This requires some special properties of the flow for a approaching a_*. In particular, $v_\infty(a)$ should approach zero [16]. Equation (10.12) shows that $\tan\psi(a)$ and $r_\infty(a)$ may both diverge as a approaches $a_*$for a non-zero I_∞ only if the latter is the absolute minimum of the function $\alpha E/\Omega$, reached at $a_* \neq 0$ [16]. When the near source dynamics causes this function to have no such minimum, I must approach zero as R approaches infinity. When $|I_\infty|$ approaches zero at infinity, none of these difficulties occur: the magnetic surfaces become all parabolic and the wind becomes kinetic energy dominated, its Poynting energy flux approaching zero.

Whatever $|I_\infty|$, any polytropic wind asymptotically suffers axial collimation. When $|I_\infty|$ is non-zero, this collimation would take the form of an asymptotically cylindrical region, possibly nested in conically flaring magnetic surfaces. When $|I_\infty|$ is zero, all magnetic surfaces become asymptotically paraboloidal. These results are valid both for classical and relativistic winds [15]. The asymptotic transfield equation generally is

$$\nabla a \cdot \nabla \left(\rho r^2 \Omega/\alpha\right) = 0. \tag{10.18}$$

For a relativistic wind with asymptotic Lorentz factor $\gamma_\infty(a)$ it reduces to

$$\frac{\mathrm{d}}{\mathrm{d}a}\left(\frac{I_\infty(a)}{\gamma_\infty(a)}\right) = 0. \tag{10.19}$$

10.4 Construction of explicit asymptotic solutions

This discussion has shown that, at very large distances, the flow consists of field regions where equations (10.16) or (10.19) are valid, separated by current-carrying boundary layers where the Lorentz force is balanced by some other force. A complete asymptotic solution can be constructed by separately solving the relevant form of the transfield equation in these different regions and then matching these pieces of solution. Equation (10.18) is easily solved when ρr^2 does not approach zero because it can be brought to the form (10.17) and solved by ray tracing [14]. In the non-relativistic case, this gives for the angle $\psi(a)$ of asymptotically conical magnetic surfaces:

$$\cos\psi(a) = \left(\cosh \int_a^A \frac{\Omega(a')\,\mathrm{d}a'}{\mu_0 I_\infty v_\infty(a')}\right)^{-1} \tag{10.20}$$

while the asymptotic radius on asymptotically cylindrical surfaces with a large terminal radius, $r_\infty(a) \gg r_\mathrm{A}(a)$, is:

$$r_\infty(a) = r_0 \exp\left(\int_0^a \frac{\Omega(a')\,\mathrm{d}a'}{\mu_0 I_\infty v_\infty(a')}\right) \cdot \tag{10.21}$$

Once the solution in the polar boundary layer is obtained, r_0 can be determined by matching. It is more difficult to solve equation (10.18) when ρr^2 approaches zero. A two-scale analysis is useful in this case. It is assumed that ρr^2 decreases to zero only very slowly. For a classical wind, this means that $I_\infty(R)$ very slowly decreases with the distance R to the origin. The shape of magnetic surfaces is given by equation (10.17), now with a term on the right, f, which weakly depends on R. Regarding f as locally constant amounts to locally approximate orthogonal trajectories to poloidal magnetic field lines by circles centered at the origin. The distribution of flux on these orthogonal trajectories slowly varies with R in a way to be determined by the matching of this field solution with the solution in the boundary layers. Whether a slow decline of I_∞ is a consistent assumption can be checked *a posteriori*. It has been in fact found to be excellent. The poloidal magnetic field line a at about a distance R from the origin makes, in the free field regions, an angle $\psi(a, R)$ with the equator which is given by equation (10.20), I_∞ being weakly dependent on R in a way that has still to be determined.

Solutions about the polar axis and about the null surfaces can also be found, owing to the boundary layer character of these regions. Which forces should then be considered when dealing with the transfield equation? An order of magnitude analysis reveals that the gas pressure forces are those which usually have the slowest decline with axial distance [14, 15]. Other possibilities can be treated similarly [14, 15]. The transfield force balance reduces in boundary layer regions to an equilibrium between the gas pressure and the Lorentz forces associated with the toroidal magnetic field. It is assumed that the first integrals α, Ω E and Q are almost constant in the polar boundary layer and equal to α_0, Ω_0 E_0 and Q_0. This is slightly restrictive [14]. The transfield equation reduces near the polar axis to

$$\frac{\mathrm{d}}{\mathrm{d}r}\left(\frac{\gamma Q_0 \rho^{\gamma-1}}{\gamma-1}\right) + \frac{\Omega_0}{\alpha_0}\frac{\mathrm{d}}{\mathrm{d}r}\left(\frac{\rho r^2 \Omega_0}{\mu_0 \alpha_0}\right) = 0. \tag{10.22}$$

In this model, the boundary layer about the polar axis has the structure of a pressure-supported plasma column. Equation (10.22) has a first integral, implying that the current parameter $|I_\infty|$ is related to the pressure on the axis by a Bennet pinch relation:

$$\frac{\gamma}{\gamma-1} Q_0 \rho_0^{\gamma-1}(R) = \frac{|I_\infty(R)|\Omega_0}{\alpha_0}. \tag{10.23}$$

On an orthogonal trajectory to poloidal field lines, that is, near $z = R$, the first integral of equations (10.22) and (10.12) can be simultaneously solved to give a representation of a and r in terms of the parameter $x = \rho/\rho_0$, where ρ_0, the axial density, depends on the distance R to the source:

$$a = K_{\rm a}\rho_0^{\gamma-1}\left(E_0 - \frac{\gamma Q_0\rho_0^{\gamma-1}}{\gamma-1}\right)^{1/2}\left(\ln\frac{1}{x} - \frac{2-\gamma}{\gamma-1}(1-x^{\gamma-1})\right) \tag{10.24}$$

$$r^2 = \frac{K_{\rm r}}{\rho_0^{2-\gamma}}\left(\frac{1}{x} - \frac{1}{x^{2-\gamma}}\right) \tag{10.25}$$

The factors $K_{\rm a}$, $K_{\rm r}$ are

$$K_{\rm a} = \frac{\gamma Q_0\mu_0\alpha_0}{\sqrt{2}(\gamma-1)\Omega_0^2} \qquad K_{\rm r} = \frac{\gamma}{\gamma-1}\,\frac{Q_0\mu_0\alpha_0^2}{\Omega_0^2}. \tag{10.26}$$

From equation (10.25), the width of the pressure-supported polar region is:

$$r_0^2 = \frac{\gamma}{\gamma-1}\,\frac{Q_0\ \mu_0\alpha_0^2}{\rho_0^{2-\gamma}(R)\ \Omega_0}. \tag{10.27}$$

The dependence of ρ_0 on R is still to be determined. The polar solution (10.24)–(10.25) also depends on $\rho_0(R)$, which is unknown unless it approaches a non vanishing constant given in terms of the supposedly known poloidal current by equation (10.23). It has been shown in Section (10.3), that a non-zero I_∞ should be the absolute minimum of $(\alpha E/\Omega)$. The polar solution must smoothly merge with the asymptotic field solution of equation (10.21), r being approximately $R\cos\psi(a,R)$. Asymptotic matching requires that these two solutions coincide in their region of common validity, which is where $x \ll 1$ for equations (10.24)–(10.25) and where $a \ll A$ for equation (10.21). For non-zero I_∞ this leads to equation (10.27). When $I_\infty(R)$ decreases to zero, a smooth matching requires equating two exponential functions multiplied by a factor [14]. The identity of the arguments of the exponential functions leads to equation (10.27) and the identity of their factors provides the variation with R of the axial density $\rho_0(R)$. The latter is the solution of

$$\frac{A_0}{R^2} = \rho_0^{2-\gamma}(R)\ \exp\left(-\frac{B_0}{\rho_0^{\gamma-1}(R)}\right) \tag{10.28}$$

where A_0 and B_0 are coefficients which depend on α_0, Ω_0 and Q_0. The solution of equation (10.28) shows that $I_\infty(R)$ and $\rho_0^{\gamma-1}(R)$ decline as the inverse of the logarithm of (R/ℓ), where ℓ is some reference length [14]:

$$I_\infty(R) \approx \rho_0^{\gamma-1}(R) \approx \left(\ln\left(\frac{R}{\ell}\right)\right)^{-1}. \tag{10.29}$$

This very slow decline justifies the adopted two-scale approach. A similar matching procedure can be used in the vicinity of null magnetic surfaces, which have the structure of pressure-supported sheet pinches. The analysis is in this case complicated by the fact that $\alpha(a)$ diverges at a null surface, as can be seen from (9.11). The analysis will not be developed in detail here [14]. Altogether the matching procedure provides a solution valid in all the asymptotic domain. A similar solution is obtained for relativistic winds [15].

At this point, the question of whether the asymptotic structure allows for non-zero asymptotic Poynting flux remains undecided, except when the function $\alpha E/\Omega$ has no absolute minimum at a non-zero a_*. A similar condition applies to relativistic winds [15]. Further analysis [16] of how $I_\infty(R)$ may approach a finite value as R approaches infinity reveals that it must do so by increasing, which is physically impossible in the context of an isolated wind source. So, formally, all rotating stationnary polytropic perfect MHD flows become kinetic energy dominated in the asymptotic limit. However, because of the very slow decline of $I_\infty(R)$, there is in practice a broad intermediate domain where there still is significant residual Poynting flux. The wind is likely to hit its terminal shocks before having got rid of all its Poynting flux.

It is worth stressing that any focused wind should host a current-carrying polar boundary layer, which is likely to be pressure-supported and a number of sheet-like current carrying boundary layers, also pressure-supported, near null surfaces. These regions are denser and they ow a source of free energy which may feed some radiative activity. What is called a "jet" may merely be the active polar boundary layer of a larger scale flow.

11 Small motions

11.1 Linearized dynamical equations for small motions

It is not enough to know an equilibrium or a stationary flow. The question of its stability should be addressed. Small motions may also reveal aspects of the structure of a stable equilibrium. The development of an initially small departure from an equilibrium or stationary state can be analyzed at the linear approximation. There is instability if diverging solutions exist. The dynamics of small, perfect MHD, perturbations can be cast in a simple standard form, which we derive below. Suppose we know an equilibrium with density $\rho_0(\boldsymbol{r})$, pressure $P_0(\boldsymbol{r})$, magnetic field $\boldsymbol{B}_0(\boldsymbol{r})$. Let us assume that perturbations evolve adiabatically, with an index γ. Any physical quantity departs from its equilibrium value by a small amount, denoted by a subscript 1, such as:

$$\rho(\boldsymbol{r},t) = \rho_0(\boldsymbol{r}) + \rho_1(\boldsymbol{r},t). \tag{11.1}$$

The velocity, $\boldsymbol{v}(\boldsymbol{r},t)$ is itself a small perturbation. The relevant equations are (3.1, 3.2) without the viscous force term, (3.6) and (6.5). Equation (3.3) reduces to the statement that P/ρ^γ is conserved following the motion. Self-gravitational effects are neglected: the gravity $\boldsymbol{g}(\boldsymbol{r})$ is constant in time. Taylor expanding the above-mentioned equations to first order, we otain, after a little algebra, the following linearized system:

$$\frac{\partial \rho_1}{\partial t} + \mathrm{div}(\rho_0 \boldsymbol{v}) = 0 \tag{11.2}$$

$$\rho_0 \frac{\partial \boldsymbol{v}}{\partial t} = -\boldsymbol{\nabla} P_1 + \boldsymbol{j}_0 \times \boldsymbol{B}_1 + \boldsymbol{j}_1 \times \boldsymbol{B}_0 + \rho_1 \boldsymbol{g} \tag{11.3}$$

$$\frac{\partial P_1}{\partial t} + (\boldsymbol{v}\cdot\boldsymbol{\nabla})P_0 = -\gamma P_0 \mathrm{div}\boldsymbol{v} \tag{11.4}$$

$$\mathrm{rot}\ \boldsymbol{B}_1 = \mu_0 \boldsymbol{j}_1 \tag{11.5}$$

$$\frac{\partial \boldsymbol{B}_1}{\partial t} = \mathrm{rot}\ (\boldsymbol{v}\times\boldsymbol{B}_0). \tag{11.6}$$

11.2 Reduction to an equation for the Lagrangean displacement

Equations (11.2)–(11.6) form a linear homogeneous system for ρ_1, P_1, $\boldsymbol{B}_1$ and $\boldsymbol{v}$. Let $\xi_{\mathrm{L}}(\boldsymbol{r}_0,t)$ be the Lagrangean displacement field from equilibrium position $\boldsymbol{r}_0$. The fluid element which, in the equilibrium, is at $\boldsymbol{r}_0$ is, at time t, at $\boldsymbol{r}_0 + \xi_{\mathrm{L}}(\boldsymbol{r}_0,t)$. The displacement at $t=0$, $\xi_0(\boldsymbol{r}_0)$, does not necessarily vanish. At time t' the fluid element is at $\boldsymbol{r}_0 + \xi_{\mathrm{L}}(\boldsymbol{r}_0,t')$ and has a Lagrangean velocity $\boldsymbol{v}_{\mathrm{L}}(t')$. In terms of the *Eulerian* velocity field, $\boldsymbol{v}(\boldsymbol{r},t)$, $\boldsymbol{v}_{\mathrm{L}}(t') = \boldsymbol{v}(\boldsymbol{r}_0 + \xi_{\mathrm{L}}(\boldsymbol{r}_0,t'),t')$ For small displacements, the velocity $\boldsymbol{v}$ is also small. The first order term in the Taylor expansion of $\boldsymbol{v}_{\mathrm{L}}$ in terms of ξ_{L} simply is $\boldsymbol{v}_{\mathrm{L}}(t') \approx \boldsymbol{v}(\boldsymbol{r}_0,t')$. Integrating $\boldsymbol{v}_{\mathrm{L}}(t')$ in time and adopting the simpler notation $\boldsymbol{\xi}(\boldsymbol{r}_0,t) \equiv \boldsymbol{\xi}_{\mathrm{L}}(\boldsymbol{r}_0,t)$, we obtain

$$\xi(\boldsymbol{r}_0,t) = \xi_0(\boldsymbol{r}_0) + \int_0^t \boldsymbol{v}(\boldsymbol{r}_0,t')\ \mathrm{d}t'. \tag{11.7}$$

All perturbed quantities can be expressed in terms of $\boldsymbol{\xi}(\boldsymbol{r},t)$. For example, equation (11.2), which gives the evolution of the density perturbation ρ_1, may be integrated in time, to give

$$\rho_1(\boldsymbol{r},t) = \rho_1(\boldsymbol{r},0) - \mathrm{div}\left(\int_0^t \rho_0\ \boldsymbol{v}(\boldsymbol{r},t')\ \mathrm{d}t'\right) \tag{11.8}$$

$$\rho_1(\boldsymbol{r},t) = \rho_1(\boldsymbol{r},0) - \mathrm{div}\left(\rho_0\ (\xi(\boldsymbol{r},t) - \xi(\boldsymbol{r},0))\right). \tag{11.9}$$

Consider a displacement with initial value $\xi(\boldsymbol{r},0)$, such that at time t all fluid elements have returned to their equilibrium positions, so that $\boldsymbol{\xi}(\boldsymbol{r},t) = 0$

and $\rho_1(\boldsymbol{r}, t) = 0$. From (11.9), $\rho_1(\boldsymbol{r}, 0) = -\mathrm{div}\,(\rho_0\ \xi(\boldsymbol{r}, 0))$. This relation between ρ_1 and $\xi(\boldsymbol{r}, 0)$ being general, equation (11.9) simplifies to:

$$\rho_1 = -\mathrm{div}\,(\rho_0 \xi(\boldsymbol{r}, t))\,. \tag{11.10}$$

Equations (11.4)–(11.6) can be similarly integrated in time, resulting in

$$P_1 = -\boldsymbol{\xi} \cdot \boldsymbol{\nabla} P_0 - \gamma P_0 \mathrm{div}\boldsymbol{\xi} \tag{11.11}$$

$$\boldsymbol{B}_1 = \mathrm{rot}\,(\boldsymbol{\xi} \times \boldsymbol{B}_0) \tag{11.12}$$

$$\boldsymbol{j}_1 = \frac{1}{\mu_0} \mathrm{rot}\ \mathrm{rot}\,(\boldsymbol{\xi} \times \boldsymbol{B}_0)\,. \tag{11.13}$$

Inserting this in equation (11.3), we get

$$\begin{aligned} \rho_0 \frac{\partial^2 \boldsymbol{\xi}}{\partial t^2} &= -\left(\mathrm{div}(\rho_0 \boldsymbol{\xi})\right) \boldsymbol{g} + \boldsymbol{\nabla}\left(\boldsymbol{\xi} \cdot \boldsymbol{\nabla} P_0\right) + \boldsymbol{\nabla}\left(\gamma P_0\ \mathrm{div}\boldsymbol{\xi}\right) \\ &+ \frac{1}{\mu_0}\left(\mathrm{rot}\ \mathrm{rot}\,(\boldsymbol{\xi} \times \boldsymbol{B}_0)\right) \times \boldsymbol{B}_0 + \frac{1}{\mu_0} \mathrm{rot}\ \boldsymbol{B}_0 \times \mathrm{rot}\,(\boldsymbol{\xi} \times \boldsymbol{B}_0)\,. \end{aligned} \tag{11.14}$$

This equation of small motions can be symbolically written as

$$\rho_0 \frac{\partial^2 \boldsymbol{\xi}}{\partial t^2} = \boldsymbol{\mathcal{F}}(\boldsymbol{\xi}) \tag{11.15}$$

where the operator of small motions $\boldsymbol{\mathcal{F}}$ can be seen on the right of equation (11.14). It involves space derivatives only. The operator of small motions plays, for the continuuous system, the role that the restauring force matrix plays for a system with a finite number of degrees of freedom. Let us seek for separable solutions of the form

$$\boldsymbol{\xi}(\boldsymbol{r}, t) = \boldsymbol{X}(\boldsymbol{r}) f(t). \tag{11.16}$$

At this point $\boldsymbol{X}$ and f are real. Inserting equations (11.16) in (11.15) we get:

$$\frac{\ddot{f}}{f} = \frac{\boldsymbol{\mathcal{F}}(\boldsymbol{X})}{\rho_0 \boldsymbol{X}}. \tag{11.17}$$

The function of t on the left and the function of $\boldsymbol{r}$ on the right of equation (11.17) must be equal to a common constant λ. Since both f and $\boldsymbol{X}$ are real, this constant is real. If it is negative, $f(t)$ oscillates. If it is positive, $f(t)$ exponentiates. $\boldsymbol{X}(\boldsymbol{r})$ is a solution of

$$\boldsymbol{\mathcal{F}}(\boldsymbol{X}(\boldsymbol{r})) = \lambda \rho_0(\boldsymbol{r}) \boldsymbol{X}(\boldsymbol{r}). \tag{11.18}$$

It is an eigenfunction of the operator $\boldsymbol{\mathcal{F}}/\rho_0$. It is often useful to go over to complex representation and to look for solutions of equation (11.15) proportional to $\exp(-i\omega t)$, with a real value, positive or negative, of ω^2. The associated eigenfunctions $\boldsymbol{X}_\omega(\boldsymbol{r})$ are complex. The general solution to the initial value problem for small perturbations is easily obtained when these eigenfunctions form a complete basis on which any initial perturbation $\boldsymbol{X}_0(\boldsymbol{r})$ may be expanded. A positive eigenvalue, $\lambda = -\omega^2$, of the eigenmode equation (11.18) reveals the existence of an instability. Conversely, if the set of eigenmodes forms a complete basis, the absence of any positive eigenvalue in the spectrum indicates stability.

11.3 Dispersion relation for MHD modes in an homogeneous medium

Consider an homogeneous equilibrium in the absence of gravity, with constant density, pressure, and magnetic field, ρ_0, P_0 and $B_0\boldsymbol{e}_z$. In this case, the equation of small motions (11.14) loses its first, second and last terms on the right. Dividing it by ρ_0 and given the constancy of ρ_0, P_0 and B_0, two important characteristic speeds, the sound speed $c_{\rm S}$ and the Alfvén speed $v_{\rm A}$, appear as factors in this equation. The sound speed is defined by $c_{\rm S}^2 = \gamma P_0/\rho_0$. The Alfvén speed $v_{\rm A}$ is defined in equation (3.21). The eigenmode equation for frequency ω is obtained from equation (11.18):

$$-\omega^2\boldsymbol{X}_\omega(\boldsymbol{r}) = c_{\rm S}^2\boldsymbol{\nabla}\left(\boldsymbol{\nabla}\cdot\boldsymbol{X}_\omega(\boldsymbol{r})\right) + v_{\rm A}^2\left(\boldsymbol{\nabla}\times\left(\boldsymbol{\nabla}\times\left(\boldsymbol{X}_\omega(\boldsymbol{r})\times\boldsymbol{e}_z\right)\right)\right)\times\boldsymbol{e}_z. \tag{11.19}$$

In the operator on the right of equation (11.19) space variables appear only in partial differential operators. The boundary condition is that $\boldsymbol{X}_\omega$ vanishes at infinity. This makes it possible and convenient to Fourier transform equation (11.19) with respect to space. Let X_x, X_y and X_z, be the components of the spatial Fourier-transform of $\boldsymbol{X}_\omega(\boldsymbol{r})$ and select a cartesian frame of reference such that $\boldsymbol{k} = k_\perp\boldsymbol{e}_x + k_\|\boldsymbol{e}_z$. Equation (11.19) gives rise to an algebraic, linear and homogeneous system of equations for X_x, X_y and X_z:

$$\begin{pmatrix} c_{\rm S}^2k_\perp^2 + v_{\rm A}^2k^2 - \omega^2 & 0 & c_{\rm S}^2k_\perp k_\| \\ 0 & v_{\rm A}^2k_\|^2 - \omega^2 & 0 \\ c_{\rm S}^2k_\perp k_\| & 0 & c_{\rm S}^2k_\|^2 - \omega^2 \end{pmatrix}\begin{pmatrix} X_x \\ X_y \\ X_z \end{pmatrix} = \begin{pmatrix} 0 \\ 0 \\ 0 \end{pmatrix}\cdot \tag{11.20}$$

For a non-trivial solution, the determinant of the system (11.20) should vanish. This condition provides the dispersion relation for modes propagating in the homogeneous medium. It can be written as

$$\left(v_{\rm A}^2k_\|^2 - \omega^2\right)\left(\omega^4 - \omega^2(v_{\rm A}^2 + c_{\rm S}^2)k^2 + v_{\rm A}^2c_{\rm S}^2k^2k_\|^2\right) = 0. \tag{11.21}$$

Many vectors $\boldsymbol{k}$ satisfy equation (11.21) for a given ω: there is considerable degeneracy. The dispersion relation has several branches, one of them being

$$\omega^2 = k_\|^2v_{\rm A}^2. \tag{11.22}$$

Its polarization vector, the displacement $\boldsymbol{X}(\omega, \boldsymbol{k})$, has a y-component only. Equation (11.10) shows that there is no associated plasma compression. This branch of the dispersion relation is the shear Alfvén mode. There are two other branches, which make the second factor in equation (11.21) to vanish. Their frequency is given as a function of the angle θ between $\boldsymbol{k}$ and $\boldsymbol{B}_0$ by:

$$\omega^2 = \frac{1}{2}k^2 \left((v_{\mathrm{A}}^2 + c_{\mathrm{s}}^2) \pm \sqrt{(v_{\mathrm{A}}^2 - c_{\mathrm{s}}^2)^2 + 4v_{\mathrm{A}}^2 c_{\mathrm{s}}^2 \sin^2\theta} \right) . \tag{11.23}$$

It can be seen that ω^2 is always positive. The solution with the positive sign is the fast (magnetosonic) mode and the solution with the negative sign is the slow mode. The phase velocities of these modes are denoted by v_{f} and v_{s} resp. For parallel propagation ($\theta = 0$),

$$v_{\mathrm{f}}^2 = \sup\left(v_{\mathrm{A}}^2, c_{\mathrm{S}}^2\right) \qquad v_{\mathrm{s}}^2 = \inf\left(v_{\mathrm{A}}^2, c_{\mathrm{S}}^2\right). \tag{11.24}$$

In perpendicular propagation ($\theta = \pi/2$), the phase speed of the slow mode vanishes, while the phase speed of the fast mode is given by:

$$v_{\mathrm{f}}^2 = v_{\mathrm{A}}^2 + c_{\mathrm{S}}^2. \tag{11.25}$$

It can be shown that $v_{\mathrm{s}}(\theta) \leq v_{\mathrm{A}}|\cos\theta| \leq v_{\mathrm{f}}(\theta)$. The polarization vector of magnetoacoustic modes is in the plane of the wave vector and the magnetic field. The ratio of its x and z components can be obtained from either the x- or the z-component of equation (11.20). It is left to the reader to analyse this polarization in different limit cases.

12 Stability

12.1 Perfect MHD potential energy

For a dissipationless perfect gas with a constant adiabatic index γ in a fixed gravitational potential Φ_{g}, the energy equation (2.19) becomes

$$\frac{\partial}{\partial t}\left(\frac{1}{2}\rho v^2 + \frac{P}{\gamma - 1} + \frac{B^2}{2\mu_0} + \rho\Phi_{\mathrm{g}}\right) + \mathrm{div}\left(\frac{1}{2}\rho v^2 \boldsymbol{v} + \frac{\gamma P}{\gamma - 1}\,\boldsymbol{v} + \rho\Phi_{\mathrm{g}}\boldsymbol{v} + \frac{\boldsymbol{E}\times\boldsymbol{B}}{\mu_0}\right) = 0. \tag{12.1}$$

The system is said to be conservative when there is no energy flux through its boundaries. Its total energy E_{t} then remains constant:

$$E_{\mathrm{t}} = \int_V \mathrm{d}^3 r\, \left(\frac{1}{2}\rho v^2 + \frac{P}{\gamma - 1} + \frac{B^2}{2\mu_0} + \rho\Phi_{\mathrm{g}}\right) . \tag{12.2}$$

The first term on the right is its kinetic energy content. The other terms can then be viewed as representing some potential energy, W.

$$W = \int_V \mathrm{d}^3\boldsymbol{r} \left(\frac{P}{\gamma - 1} + \frac{B^2}{2\mu_0} + \rho\Phi_{\mathrm{g}} \right) \tag{12.3}$$

W qualifies as a potential energy, because it reaches an extremum when the system is in equilibrium and it is a local minimum at a stable equilibrium. This is shown below. That W reaches an extremum at equilibrium can be shown by calculating its first order change $W^{(1)}$, under a small departure from this equilibrium, characterized by a displacement field $\boldsymbol{\xi}$. Using equations (11.10)–(11.12), it is found that

$$W^{(1)} = -\int_V \mathrm{d}^3\boldsymbol{r} \ \left(\boldsymbol{\xi} \cdot (-\boldsymbol{\nabla} P_0 + \rho_0 \boldsymbol{g} + \boldsymbol{j}_0 \times \boldsymbol{B}_0)\right) \tag{12.4}$$

which shows that $W^{(1)}$ vanishes when the reference state is an equilibrium.

12.2 Change of potential energy and operator of small motions

The small change of the potential energy about an equilibrium configuration is then a quadratic functional $W^{(2)}(\boldsymbol{\xi})$ of $\boldsymbol{\xi}(\boldsymbol{r}, t)$. When it is positive for all $\boldsymbol{\xi}$, the system is stable. Indeed, there can be no indefinite growth of $\boldsymbol{\xi}$ because this would imply an indefinite growth of $W^{(2)}$, which is in this case a positive-definite quadratic functional, and then of the total energy. This is however impossible. Then

$$W^{(2)}(\boldsymbol{\xi}) \geq 0 \ \forall \ \boldsymbol{\xi} \quad \Rightarrow \quad \text{Stable.} \tag{12.5}$$

This implies that when there is instability, a displacement field $\boldsymbol{\xi}$ exists, for which $W^{(2)}$ is negative. This statement is different from the reciprocal proposition to equation (12.5), which would state that

$$\exists \ \boldsymbol{\xi} \text{ such that } W^{(2)}(\boldsymbol{\xi}) < 0 \quad \Rightarrow \quad \text{Unstable.} \tag{12.6}$$

This latter statement is nevertheless true when the eigenmodes form a complete basis set, as we now show. The proof uses an expression of $W^{(2)}$ in terms of the operator of small motions. Dotting equation (11.15) by $\mathrm{d}\boldsymbol{\xi}/\mathrm{d}t$ and integrating over the volume of the system, we get:

$$\frac{\partial}{\partial t}\left(\int_V \frac{1}{2}\rho v^2 \right) - \int_V \frac{\partial \boldsymbol{\xi}}{\partial t} \cdot \boldsymbol{\mathcal{F}}(\boldsymbol{\xi}) = 0. \tag{12.7}$$

The first term is the rate of change of the kinetic energy. Since the total energy is conserved, the second term is $\mathrm{d}W^{(2)}/\mathrm{d}t$. It can be shown, at the expense of some vector calculus, that the operator $\mathcal{F}$ is self-adjoint. This means that for any two real displacement fields $\boldsymbol{\xi}$ and $\boldsymbol{\eta}$ the volume integrals of $(\boldsymbol{\xi} \cdot \boldsymbol{\mathcal{F}}(\boldsymbol{\eta}))$ and of $(\boldsymbol{\eta} \cdot \boldsymbol{\mathcal{F}}(\boldsymbol{\xi}))$ are equal. Using this in the second term of equation (12.7) and integrating over time, we obtain

$$W^{(2)} = -\frac{1}{2} \int_V \boldsymbol{\xi} \cdot \boldsymbol{\mathcal{F}}(\boldsymbol{\xi}). \tag{12.8}$$

From equation (11.14), we obtain the following expression of $W^{(2)}$:

$$\begin{aligned} W^{(2)} = &-\frac{1}{2} \int_V \left[\frac{1}{\mu_0} \left(\mathrm{rot}(\boldsymbol{\xi} \times \boldsymbol{B}_0)\right)^2 - \gamma P_0 \left(\mathrm{div}\boldsymbol{\xi}\right)^2 + (\boldsymbol{\xi} \cdot \boldsymbol{\nabla} P_0)\, \mathrm{div}\boldsymbol{\xi} \right] \\ &- \frac{1}{2} \int_V \left[\, (\boldsymbol{\xi} \cdot \boldsymbol{g})\ \mathrm{div}\,(\rho_0 \boldsymbol{\xi}) - (\boldsymbol{j}_0 \times \mathrm{rot}(\boldsymbol{\xi} \times \boldsymbol{B}_0)) \cdot \boldsymbol{\xi} \,\right]. \end{aligned} \tag{12.9}$$

12.3 Normal modes and potential energy

Expressing the self-adjointness relation for two eigenmodes $\boldsymbol{\xi}_m$ and $\boldsymbol{\xi}_n$ with eigenvalues λ_m and λ_n we find that

$$(\lambda_m - \lambda_n)\langle \boldsymbol{\xi}_m \mid \boldsymbol{\xi}_n \rangle \;=\; 0. \tag{12.10}$$

This shows that normal modes with different eigenvalues are orthogonal to the product

$$\langle \boldsymbol{\xi} \mid \boldsymbol{\eta} \rangle \;=\; \int_V \rho_0 \; (\boldsymbol{\xi} \cdot \boldsymbol{\eta}) \; \mathrm{d}^3\boldsymbol{r}. \tag{12.11}$$

When there is degeneracy, an orthogonal basis of the degenerate eigensubspace can be found. From equation (12.8) we find that if $\boldsymbol{\xi}_n$ is an eigenmode,

$$W^{(2)}(\boldsymbol{\xi}_n) = -\frac{\lambda_n}{2} \langle \boldsymbol{\xi}_n | \boldsymbol{\xi}_n \rangle. \tag{12.12}$$

An unstable eigenmode (positive eigenvalue) causes a decrease of the potential energy, while a stable one causes an increase. When normal modes form a complete basis, any displacement field $\boldsymbol{\xi}$ may be expanded on it. Using the orthogonality property (12.10), $W^{(2)}(\boldsymbol{\xi})$ can then be expressed as

$$W^{(2)}(\boldsymbol{\xi}) = -\frac{1}{2} \sum_{n,m} \int_V a_n \boldsymbol{\xi}_n \; \boldsymbol{\mathcal{F}}(a_m \boldsymbol{\xi}_m) = -\frac{1}{2} \sum_n a_n^2 \lambda_n \langle \boldsymbol{\xi}_n \mid \boldsymbol{\xi}_n \rangle. \tag{12.13}$$

Then $W^{(2)}$ cannot be negative, unless there is an unstable eigenmode. This justifies the statement (12.6). Thus, when normal modes form a complete basis, it is enough, to prove instability, to find a displacement with negative $W^{(2)}$. It can be shown that eigenmodes are extrema of the functional

$$\Lambda(\boldsymbol{\xi}) = -2 \; \frac{W^{(2)}(\boldsymbol{\xi})}{\langle \boldsymbol{\xi} \mid \boldsymbol{\xi} \rangle} \tag{12.14}$$

Λ reaches its minimum for the eigenmode with the algebraically largest eigenvalue λ. This remark provides a mean to find bounds on the maximum growth rate without actually calculating the eigenmodes.

13 Some usual instabilities

The stability of equilibria can be assessed either from an analysis of the sign of the change of potential energy about this equilibrium, as described in Section 12, or by a direct determination of the eigenvalues of normal modes. For stationnary flows, no general energy principle is available and the study of eigenmodes often is the only available tool. Usual instabilities come in a few general classes: the Rayleigh-Taylor class, the Kelvin-Helmholtz class and the Current-Driven class. Rayleigh-Taylor instabilities occur when heavy fluid sits over lighter fluid. The light fluid may however consist of such things as photons or magnetic field and inertia forces may be a substitute for gravity. Kelvin-Helmholtz instabilities develop from shear flows. Current-Driven instabilities grow from Lorentz force perturbations in a medium supporting electric currents. Instabilities are often studied in the framework of perfect MHD, but real fluids are viscous and resistive. Dissipative effects usually have a stabilizing effect. However, dissipation may allow motions forbidden by perfect MHD and thus promote instability, as in tearing modes. These modes can be viewed as the linear stage of development of reconnection motions. In the absence of resistivity they simply do not grow. We now turn to a few simple examples, chosen for illustrating different methods to study stability.

13.1 Stability of a stratified compressible gas

Consider a plane-parallel stratified, unmagnetized, fluid in a constant gravity, $\boldsymbol{g} = -g\boldsymbol{e}_z$. The pressure and density distribution at equilibrium are $P_0(z)$ and $\rho_0(z)$. Using equation (12.9) in this particular situation, we obtain

$$W^{(2)} = \frac{1}{2}\int \left(\gamma P_0 \left(\mathrm{div}(\boldsymbol{\xi})\right)^2 + (\boldsymbol{\xi}\cdot\boldsymbol{\nabla}P_0)\,\mathrm{div}\boldsymbol{\xi} + (\boldsymbol{\xi}\cdot\boldsymbol{g})\,\mathrm{div}(\rho_0\boldsymbol{\xi})\right). \quad (13.1)$$

Denoting $(\mathrm{d}\rho_0/\mathrm{d}z)$ by ρ_0' and using the equilibrium equation $P_0' = -g\rho_0$, this can be written as:

$$W^{(2)} = \frac{1}{2}\int \left(\gamma P_0 \left(\mathrm{div}(\boldsymbol{\xi})\right)^2 - 2g\rho_0\xi_z \left(\mathrm{div}(\boldsymbol{\xi})\right) - \xi_z^2 g\rho_0'\right). \quad (13.2)$$

It is enough, to prove instability, to find a displacement field with a negative $W^{(2)}$. To prove stability, it must be shown that $W^{(2)}$ cannot be negative, whatever $\boldsymbol{\xi}$. The integrand in equation (13.2) is a quadratic function of div$\boldsymbol{\xi}$. For a given $\xi_z(\boldsymbol{r})$ it is possible to chose $\xi_x(\boldsymbol{r})$ and $\xi_y(\boldsymbol{r})$ such that div $\boldsymbol{\xi} = (g\rho_0\xi_z)/(\gamma P_0)$. This value of div$\boldsymbol{\xi}$ minimizes the integrand in equation (13.2) for the given $\xi_z(\boldsymbol{r})$. The corresponding change of potential energy is

$$W^{(2)}_{\mathrm{min,div}} = -\int \left(\frac{g^2\rho_0^2}{\gamma P_0} + g\rho_0'\right)\xi_z^2. \tag{13.3}$$

If the right of equation (13.3) can be made negative by a suitable choice of $\xi_z(\boldsymbol{r})$, the system is unstable. This is so if, in some interval of z, the function in the parenthesis on the right of equation (13.3) is positive, because a $\xi_z(\boldsymbol{r})$ which vanishes everywhere except in this interval could be chosen, which would make the integral in equation (13.3) negative. Then,

$$\exists\ z \text{ such that } \frac{g\rho_0}{\gamma P_0} + \frac{\rho_0'}{\rho_0} > 0 \quad \Rightarrow \qquad \text{unstable.} \tag{13.4}$$

When ρ_0' is negative for all z the condition (13.4) can be given the form of Schwarzschild instability condition:

$$\exists\ z \text{ such that } \frac{\mathrm{d}\ln P_0}{\mathrm{d}\ln\rho_0} > \gamma \quad \Rightarrow \quad \text{unstable.} \tag{13.5}$$

When the function in the parenthesis of equation (13.3) is everywhere negative, the minimizing potential energy (13.3) is positive for all ξ_z, which indicates stability. Thus (13.4) is a necessary and sufficient condition for instability.

13.2 Interchange instability of confined plasmas

It is tempting to discover instabilities by checking the sign of $W^{(2)}(\boldsymbol{\xi})$ for very simple displacements, such as "exchange" displacements, which consist in exchanging two elementary fluid elements. Element 1 is moved so as to fill the volume initially occupied by the element 2 and conversely. If the sign of $W^{(2)}$ under this exchange is negative, the system is unstable. This test provides a sufficient condition for instability but cannot conclude to stability. For this, it should be shown that $W^{(2)}$ is positive for *any* displacement field, not for exchange displacements only.

Let us illustrate the exchange method by considering plasma trapped in supposedly closed flux tubes, such as, for example, a plasma column embedded in an azimuthal magnetic field. Consider a displacement field exchanging two neighbouring flux tubes of a very small cross section. Since the motion supposedly obeys the laws of perfect MHD, the exchanged fluid elements carry their embedded flux with them. Let V_1 and V_2 be their volumes. After the exchange, the matter and the flux which fitted in V_1

occupy the volume V_2 and conversely. The pressure in the fluid element (1), initially P_1, becomes P_1' after the exchange and the pressure in element (2) changes from P_2 to P_2'. The gas internal energy of these two elements is, before and after the exchange resp.:

$$E_{\rm g,bef} = \frac{V_1 P_1 + V_2 P_2}{\gamma - 1} \qquad E_{\rm g,aft} = \frac{V_1 P_2' + V_2 P_1'}{\gamma - 1}. \tag{13.6}$$

The magnetic field in a thin flux tube with infinitesimal flux Φ, at a point of curvilinear abcissa l along the tube axis where the tube section is $\sigma(l)$ is $B(l) = \Phi/\sigma(l)$. The volume of a piece, of length $\mathrm{d}l$, of this tube is $\sigma \mathrm{d}l$ and its magnetic energy content is $\mathrm{d}E_m = \mathrm{d}l\ (\Phi^2/\mu_0\sigma)$. The total magnetic energy of the system of both flux tubes (1) and (2) is, before and after the exchange resp.:

$$E_{\rm m,bef} = \oint_1 \frac{\Phi_1^2\,\mathrm{d}l}{\mu_0\sigma_1(l)} + \oint_2 \frac{\Phi_2^2\,\mathrm{d}l}{\mu_0\sigma_2(l)} \quad E_{\rm m,aft} = \oint_1 \frac{\Phi_2^2\,\mathrm{d}l}{\mu_0\sigma_1(l)} + \oint_2 \frac{\Phi_1^2\,\mathrm{d}l}{\mu_0\sigma_2(l)}. \tag{13.7}$$

The total magnetic energy is unchanged when both tubes have the same flux Φ, which we now assume. Their volumes are given in terms of Φ by

$$V_1 = \Phi \oint_1 \frac{\mathrm{d}l}{B_1(l)} \qquad V_2 = \Phi \oint_2 \frac{\mathrm{d}l}{B_2(l)}. \tag{13.8}$$

The ratio V/Φ, the specific volume of an infinitesimal flux tube, is

$$V_{\rm sp} = \oint_{\rm line} \frac{\mathrm{d}l}{B(l)}. \tag{13.9}$$

The integration is along the axial field line. $V_{\rm sp}$ is a property of that field line. The change of potential energy $W^{(2)}$ then reduces to the change of gas energy (13.6) which can be calculated for adiabatic transformations, so that

$$P_1 V_1^\gamma = P_1' V_2^\gamma \qquad \text{and} \qquad P_2 V_2^\gamma = P_2' V_1^\gamma. \tag{13.10}$$

This leads to the following expression of $W^{(2)}$:

$$W^{(2)} = \frac{V_1}{\gamma - 1}\left(P_2\,\frac{V_2^\gamma}{V_1^\gamma} - P_1\right) + \frac{V_2}{\gamma - 1}\left(P_1\,\frac{V_1^\gamma}{V_2^\gamma} - P_2\right). \tag{13.11}$$

The exchanged flux tubes being very close, $\delta P = (P_2 - P_1)$ is small compared to P_1, as is $\delta V = (V_2 - V_1)$ compared to V_1. To second order, we get:

$$W^{(2)} \approx P_1\,\frac{\delta V^2}{V_1}\left(\frac{V_1\ \delta p}{P_1\ \delta V} + \gamma\right). \tag{13.12}$$

The structure is unstable near the considered flux tubes if the parenthesis in equation (13.12) is negative. This happens if some direction exists along which the specific volume of field lines and the pressure on them vary such that

$$\frac{\mathrm{d}\ln P}{\mathrm{d}\ln V_{\mathrm{sp}}}+\gamma<0 \tag{13.13}$$

equation (13.13) is a sufficient condition for the plasma-containing magnetic structure to be unstable to this pressure-driven instability, refered to as the interchange instability. The convective-like unstable motions which develop bring higher pressure flux tubes to lower pressure environment where the larger specific volume of flux tubes allows for large enough gas expansion. The sufficient condition for instability (13.13) can be given a different form when the flux tubes are close to the outer boundary of the confined plasma, where the pressure is weak and the magnetic field almost potential. In in the limit of vanishing pressure, equation (13.13) reduces to

$$\frac{\mathrm{d}P}{\mathrm{d}V_{\mathrm{sp}}}<o. \tag{13.14}$$

The interchange instability develops at the edge of the confined plasma if the specific volume increases in the “outward” direction, toward which the pressure decreases. The question boils down to evaluating whether, in the local potential field, V_{sp} increases outwards or not. Consider two neighbouring field lines, (1) and (2), a distance $\delta(l_1)$ apart, near some point M_1 at the curvilinear abcissa l_1 on line 1. The magnetic field is almost current free and derives from a magnetic potential, ψ_{m}. Consider two neighbouring magnetic equipotentials, one of them passing at M_1. These two surfaces delineate a segment $\mathrm{d}l_1$ on field line (1) and a segment $\mathrm{d}l_2$ on field line (2). The magnetic potential drop between them is $\mathrm{d}\psi_{\mathrm{m}}=B_1\mathrm{d}l_1=B_2\mathrm{d}l_2$. The difference in specific volume of field lines (1) and (2) is:

$$\mathrm{d}V_{\mathrm{sp}}=\oint_2\frac{\mathrm{d}l_2}{B_2}-\oint_1\frac{\mathrm{d}l_1}{B_1}=\oint_1 B_1\mathrm{d}l_1\left(\frac{1}{B_2^2}-\frac{1}{B_1^2}\right)\cdot \tag{13.15}$$

The length $\mathrm{d}l_2$ is larger or smaller than $\mathrm{d}l_1$ according to whether the curvature of the field lines is towards the inside or the outside of the plasma. Let $\epsilon=+1$ in the former case and -1 in the latter. If $\mathcal{R}$ is the radius of curvature of field line 1, the ratio $\mathrm{d}l_2/\mathrm{d}l_1$ is given

$$\frac{\mathrm{d}l_2}{\mathrm{d}l_1}=1+\epsilon\frac{\delta}{\mathcal{R}}\cdot \tag{13.16}$$

Using this to calculate (B_2/B_1) in equation (13.15), we obtain:

$$\mathrm{d}V_{\mathrm{sp}}=\oint_1\frac{\mathrm{d}l_1}{B_1}\left(\frac{B_1^2}{B_2^2}-1\right)=2\oint_1\frac{\mathrm{d}l_1}{B_1}\frac{\epsilon(l_1)\delta(l_1)}{\mathcal{R}(l_1)}\cdot \tag{13.17}$$

The sign of the integral depends on how ϵ varies along the field line. An everywhere positive ϵ, which corresponds to the center of curvature of the field line being to the inside of the plasma, causes instability. An everywhere negative ϵ is not possible for a closed field line. For mixed concavity, stability depends on the sign of the integral on the right of equation (13.17). If it is positive, there is instability. Otherwise, pressure driven instabilities may still develop as localized perturbations in the regions of bad curvature of the plasma boundary. Such perturbations are called ballooning modes.

13.3 Kink instability

The kink instability is a current-driven instability. Its basic mechanism can be easily understood. When, for example, a plasma column with azimuthal field only is subject to a slight bending, the magnetic field intensifies on the inner side of the kink and reduces on the outer side. This causes extra magnetic pressure to appear on the inside, amplifying the kinking. When the field also has a longitudinal component, the bending is somewhat resisted by the magnetic tension of this longitudinal field. In this case, the instability develops only for large enough axial current. Let us illustrate this by deriving the instability growth rate in the simple case of a non-magnetized plasma column, of axis z and radius a, filled with incompressible plasma and covered by a skin current flow on its surface. On the outside there is a magnetized vacuum. Cylindrical coordinates r, θ, z are used. The unperturbed magnetic field in the external vacuum is

$$\boldsymbol{B} = B_0\ \boldsymbol{e}_z + B_\theta(a)\ \frac{a}{r}\ \boldsymbol{e}_\theta. \tag{13.18}$$

In the presence of a perturbation, the plasma boundary is displaced and the outer vacuum magnetic field is perturbed. In the internal, unmagnetized, region, the equation of motion reduces to

$$\rho_0\ \frac{\partial^2 \boldsymbol{\xi}}{\partial t^2} = -\boldsymbol{\nabla} P_1. \tag{13.19}$$

Using the incompressibility condition in equation (13.19), it is found that the pressure pertubation satisfies Laplace's equation:

$$\Delta P_1 = 0. \tag{13.20}$$

We seek for eigenmodes by Fourier analyzing perturbations in z, θ and t as:

$$f(r, \theta, z, t) = \mathrm{e}^{i\omega t}\ \mathrm{e}^{-(ikz+im\theta)}\ f_{\omega k m}(r). \tag{13.21}$$

From equation (13.20), $P^{(1)}_{\omega k m}(r)$, which we denote by $\varpi(r)$, satisfies the equation

$$\frac{1}{r}\frac{\partial}{\partial r} r \frac{\partial \varpi}{\partial r} - \left(k^2 + \frac{m^2}{r^2}\right) \varpi = 0. \tag{13.22}$$

The general solution of equation (13.22) is a linear combination of modified Bessel functions, $I_{\rm m}$ and $K_{\rm m}$. The solution which remains finite ar $r = 0$ is

$$\varpi(r) = \varpi(a)\ \frac{I_{\rm m}(kr)}{I_{\rm m}(ka)}. \tag{13.23}$$

The plasma displacement in the column is obtained from equation (11.14) which reduces here to:

$$-\omega^2 \rho_0 \boldsymbol{\xi}_\omega(\boldsymbol{r}, t) = -\boldsymbol{\nabla} P_1. \tag{13.24}$$

The perturbed field in the external vacuum is potential. Its magnetic potential $\psi(r, \theta, z, t)$ satisfies Laplace's equation and $\psi_{\omega k m}(r) \equiv \hat{\psi}(r)$ satisfies equation (13.22). The solution which vanishes at infinity is

$$\hat{\psi}(r) = \hat{\psi}(a) \frac{K_{\rm m}(kr)}{K_{\rm m}(ka)}. \tag{13.25}$$

These solutions for the pressure perturbation in the column and the magnetic perturbation in the outside vacuum must match at the perturbed interface. The conditions for this matching are that the total pressure be continuous and that the external field remains tangent to the modified boundary. The pressure at a point on the boundary which has suffered a displacement $\boldsymbol{\xi}$ is

$$P_{\rm L} = P_0 + P_1(\boldsymbol{r}, t) + \boldsymbol{\xi} \cdot \boldsymbol{\nabla} P_0. \tag{13.26}$$

Here, $\boldsymbol{\nabla} P_0 = 0$. The perturbed magnetic pressure is similarly given by

$$B_{\rm L}^2 = B_0^2 + 2\boldsymbol{B}_0 \cdot \boldsymbol{b} + \boldsymbol{\xi} \cdot \boldsymbol{\nabla} B_0^2. \tag{13.27}$$

The field perturbation $\boldsymbol{b}$ is obtained from the perturbation of the magnetic potential $\psi(\boldsymbol{r}, t)$, given by (13.25). The condition that the pressure is continuous can be written as:

$$\varpi(a) \left(1 + \frac{B_\theta^2(a)}{\mu_0 \rho_0} \frac{k}{a\omega^2} \frac{I'_{\rm m}(ka)}{I_{\rm m}(ka)}\right) + i \frac{\hat{\psi}(a)}{\mu_0} \left(kB_0 + \frac{m}{a} B_\theta(a)\right) = 0. \tag{13.28}$$

The field remains tangent to the boundary if $(\boldsymbol{B} \cdot \boldsymbol{n})$ remains zero at the perturbed boundary, $\boldsymbol{n}$ being the unit vector normal to the perturbed interface at the displaced point, which we may write as $\boldsymbol{n}_0 + \boldsymbol{n}_1$. Let M_0 be a point on the unperturbed boundary and M the point to which it is displaced. The position of the tangent plane to the displaced boundary at M may be defined by considering the variations of $(\boldsymbol{n}_0 \cdot \boldsymbol{\xi})$ about the point M_0. Such an analysis shows that $\boldsymbol{n}_1$, which is perpendicular to $\boldsymbol{n}_0$ because $\boldsymbol{n}$ is also a unit vector, is given by

$$\boldsymbol{n}_1 = -\boldsymbol{\nabla}(\boldsymbol{n}_0 \cdot \boldsymbol{\xi}) + \boldsymbol{n}_0(\boldsymbol{n}_0 \cdot \boldsymbol{\nabla}(\boldsymbol{n}_0 \cdot \boldsymbol{\xi})). \tag{13.29}$$

Using equations (13.24) and (13.23), this gives:

$$\boldsymbol{n}_1 = i\,\frac{k\varpi(a)}{\rho_0\omega^2}\frac{I'_{\rm m}(ka)}{I_{\rm m}(ka)}\left(\frac{m}{a}\boldsymbol{e}_\theta + k\boldsymbol{e}_z\right)\cdot \tag{13.30}$$

The first order Lagrangian perturbation of $(\boldsymbol{B}\cdot\boldsymbol{n})$ then is

$$(\boldsymbol{B}\cdot\boldsymbol{n})_{\rm L} - \boldsymbol{B}_0\cdot\boldsymbol{n}_0 = \boldsymbol{B}_0\cdot\boldsymbol{n}_1 + \boldsymbol{b}\cdot\boldsymbol{n}_0 + (\boldsymbol{\xi}\cdot\boldsymbol{\nabla})\boldsymbol{B}_0\cdot\boldsymbol{n}_0. \tag{13.31}$$

Using equations (13.23)–(13.25), the condition that this change vanishes can be written as:

$$i\,\frac{k\varpi(a)}{\rho_0\omega^2}\,\frac{I'_{\rm m}(ka)}{I_{\rm m}(ka)}\left(kB_0 + \frac{m}{a}B_\theta(a)\right) + k\hat{\psi}(a)\,\frac{K'_{\rm m}(ka)}{K_{\rm m}(ka)} = 0. \tag{13.32}$$

Equations (13.28) and (13.32) form an homogeneous linear system for $\varpi(a)$ and $\hat{\psi}(a)$. The condition that it has a non-trivial solution gives the dispersion relation, which can be written, denoting ka by x, as:

$$\omega^2 = -\frac{I'_{\rm m}(x)K_{\rm m}(x)}{I_{\rm m}(x)K'_{\rm m}(x)}\frac{1}{\mu_0\rho}\left(kB_0 + \frac{m}{a}B_\theta\right)^2 - \left(\frac{I'_{\rm m}(x)}{xI_{\rm m}(x)}\right)\frac{k^2B_\theta^2}{\mu_0\rho_0}\cdot \tag{13.33}$$

In equation (13.33), the magnetic field is taken at $r = a$. This equation simplifies in the long wavelength limit ($ka \ll 1$). For non-vanishing m it becomes:

$$\omega^2 = \frac{1}{\mu_0\rho}\left(\left(kB_0 + \frac{m}{a}B_\theta(a)\right)^2 - \frac{mB_\theta^2(a)}{a^2}\right) \tag{13.34}$$

and for $m = 0$ it becomes:

$$\omega^2 = -\frac{k^2}{2}\frac{B_\theta^2(a)}{\mu_0\rho}\cdot \tag{13.35}$$

This shows that the mode $m = 0$ is in this case always unstable to a deformation of the column which causes it to pinch at some places along the axis and to expand at other places. This is the saussage instability. Helical modes with non-zero m's are unstable too. The largest growth rate for a given m is obtained for a value of k such that

$$kB_0 + \frac{m}{a}B_\theta(a) = 0. \tag{13.36}$$

In this case the phase of the perturbation is constant along an unperturbed field line on the boundary of the plasma column. When an axial magnetic field is present in the plasma column, the system is less unstable. Consider the case when the amplitude of this internal axial field, B_0, is the same in the column and in the outer vacuum. Similar calculations show that the dispersion relation for non-zero m is, in the limit $ka \ll 1$,

$$\omega^2 = \frac{k^2 B_0^2}{\mu_0 \rho} + \frac{1}{\mu_0 \rho}\left(\left(kB_0 + \frac{m}{a}B_\theta(a)\right)^2 - \frac{mB_\theta^2(a)}{a^2}\right) \tag{13.37}$$

and for $m = 0$ it is:

$$\omega^2 = \frac{k^2 B_0^2}{\mu_0 \rho} - \frac{k^2}{2}\frac{B_\theta^2(a)}{\mu_0 \rho}. \tag{13.38}$$

For a given, non-zero, m, the minimum value of $\omega^2(k,m)$ is reached for an axial wave number $k = k_- = -mB_\theta(a)/2aB_0$. An infinitely long column can support modes with any k, in particular modes with $k < k_-$. Then,

$$\omega^2_{\min}(m) = \omega^2(k_-) = \frac{m^2 - 2m}{2a^2}\frac{B_\theta^2}{\mu_0 \rho}. \tag{13.39}$$

The mode $m = 1$ is unstable in this case, but not the modes of larger m's. A real plasma column, however, has a finite length L. In this finite column the modulus of the axial wave number k must be larger than $k_{\mathrm{L}} = 2\pi/L$. When k_- is smaller than k_{L}, the minimum value of ω^2 is reached for $m = 1$ and $k = k_{\mathrm{L}}$, and is given by

$$\omega^2_{\min} = \frac{8\pi^2 B_0^2}{\mu_0 \rho_0 L^2} - \frac{4\pi B_0 B_\theta(a)}{\mu_0 \rho_0 L a}. \tag{13.40}$$

This mode is stable when $\omega^2_{\min} > 0$, that is when

$$q = \frac{2\pi a B_0}{L B_\theta(a)} > 1. \tag{13.41}$$

This condition means that the field line at the edge must make less than one turn about the plasma column when followed over a distance L parallel to the axis. The quantity q is called the security factor and the condition (13.41) is the Kruskal-Shafranoff stability condition.

14 Some instabilities in rotating fluids

14.1 Stability of the inviscid hydrodynamical Couette flow

Rotating fluids suffer the Coriolis and the centrifugal force. Since the latter acts like a gravity, instabilities of the Rayleigh Taylor type may be expected. When rotation has a shear flow character, instabilities of the Kelvin Helmholtz type may also develop. Accretion disks are rotating astrophysical fluids of special significance. Understanding their stability is of the utmost importance because it bears on the problem of the origin of turbulence in these systems [5]. The stability of Couette flows, described in Section 8.1 is of interest in relation with this, although the supposedly cylindrical structure of these flows does not make justice of the finite vertical extent of accretion disks. The study of the stability of inviscid hydrodynamical Couette flows is a classical problem. For subsonic motions, incompressibility may be assumed and the inviscid form of the Navier Stokes equation (3.12) may be used. Only axisymmetrical motions are considered, even in the perturbed state. For a uniform density fluid, the relevant hydrodynamical equations can be written, in cylindrical coordinates r, θ, z, as:

$$\frac{1}{r}\frac{\partial(rv_{\rm r})}{\partial r}+\frac{\partial v_z}{\partial z}=0 \tag{14.1}$$

$$\left(\frac{\partial}{\partial t}+v_{\rm r}\frac{\partial}{\partial r}+v_z\frac{\partial}{\partial z}\right)v_{\rm r}=-\frac{1}{\rho}\frac{\partial P}{\partial r}+\frac{v_\theta^2}{r} \tag{14.2}$$

$$\left(\frac{\partial}{\partial t}+v_{\rm r}\frac{\partial}{\partial r}+v_z\frac{\partial}{\partial z}\right)v_\theta=-\frac{v_{\rm r}v_\theta}{r} \tag{14.3}$$

$$\left(\frac{\partial}{\partial t}+v_{\rm r}\frac{\partial}{\partial r}+v_z\frac{\partial}{\partial z}\right)v_z=-\frac{1}{\rho}\frac{\partial P}{\partial z}. \tag{14.4}$$

Any $v_\theta(r)$ provides a stationnary inviscid solution. The associated pressure distribution balances the centrifugal force, so that $\mathrm{d}P/\mathrm{d}r=\rho v_\theta^2/r$. When the viscosity vanishes, any ring of matter conserves its angular momentum. This can be shown by multiplying equation (14.3) by r. The total energy of this dissipationless system is conserved. Denoting by U the internal energy density, this energy can be written as

$$\mathcal{E}=\int \mathrm{d}^3\,\boldsymbol{r}\left(\frac{1}{2}\rho\,\left(v_{\rm r}^2+v_z^2+v_\theta^2\right)+U\right). \tag{14.5}$$

For an incompressible motion, the total internal energy is constant, because pressure forces perform no work. In the case of an instability, poloidal motions would grow. These must tap on the kinetic energy of the fluid's rotation. The total kinetic energy being conserved, this is possible only if the total rotational kinetic energy can be lowered by displacing the fluid. This can be discussed by taking an interchange approach. Suppose that two neighbouring rings of fluid of a very small cross section are exchanged. The fluid being incompressible, the matter of one ring can fit in the volume initially occupied by the other one only if the rings have the same volume and, then, the same mass m. The angular momentum of each ring is conserved. Before the exchange, ring (1) was at r_1 and ring (2) was at r_2, with angular momentum mJ_1 and mJ_2 resp. The total rotational kinetic energy was $K^\theta_{\rm bef}$. After the exchange, ring (1) is at r_2 and ring (2) is at r_1, their angular momenta having remained the same. The total rotational kinetic energy of the system has become $K^\theta_{\rm aft}$.

$$K^\theta_{\rm bef} = m\left(\frac{J_1^2}{r_1^2} + \frac{J_2^2}{r_2^2}\right) \qquad K^\theta_{\rm aft} = m\left(\frac{J_1^2}{r_2^2} + \frac{J_2^2}{r_1^2}\right). \tag{14.6}$$

The change of rotational energy is

$$K^\theta_{\rm aft} - K^\theta_{\rm bef} = m\left(\frac{1}{r_1^2} - \frac{1}{r_2^2}\right)\left(J_2^2 - J_1^2\right). \tag{14.7}$$

The conservation of total, rotational and poloidal, kinetic energy implies that there can be an instability only when the change in rotational kinetic energy is negative. The rotational kinetic energy of angular momentum-conserving motions plays here the role of a potential energy. This provides us, from equation (14.7), with a sufficient condition for 2D stability. There can be an axisymmetric instability only when $J^2(r)$ is not a monotonically increasing function of r. Thus,

$$\frac{\mathrm{d}\left(r^4\Omega^2(r)\right)}{\mathrm{d}r} > 0 \;\forall\; r \qquad \Rightarrow \qquad \text{stable.} \tag{14.8}$$

The condition does not apply to three-dimensional motions, however. Some systems in which $J(r)$ sharply increases with r nevertheless locally develop non-axisymmetric instabilities of the Kelvin-Helmholtz type. Equation (14.8) is the so-called Rayleigh stability condition. It can be shown to be also a necessary condition for stability to axisymmetric motions. The criterium (14.8) may be used to discuss inviscid perturbations on the flow described by equation (8.3). The derivative of J^2 with respect to r is in this case given by:

$$\frac{\mathrm{d}J^2(r)}{\mathrm{d}r} = \frac{4r^3\left(\Omega_2R_2^2 - \Omega_1R_1^2\right)^2}{(R_2^2 - R_1^2)^2}\left(1 + \frac{R_1^2R_2^2\left(\Omega_1 - \Omega_2\right)}{r^2(\Omega_2R_2^2 - \Omega_1R_1^2)}\right). \tag{14.9}$$

By a choice of the orientation of the z axis, Ω_1 can always be made positive. $J^2(r)$ cannot be monotonically increasing with r when $J^2(R_2) < J^2(R_1)$, or when $J(r)$ vanishes between R_1 and R_2, that is, when:

$$(\Omega_2R_2^2 - \Omega_1R_1^2) < 0. \tag{14.10}$$

When, on the other hand, this difference is positive, the sign of $\mathrm{d}J^2/\mathrm{d}r$ can be discussed from equation (14.9), with the result that it is everywhere positive in this case. The resulting stability diagram, for given R_1 and R_2 and variable Ω_1 and Ω_2 is shown in Figure 7. The linearly stable domain may however be subject to non-linear instabilities, which only grow when

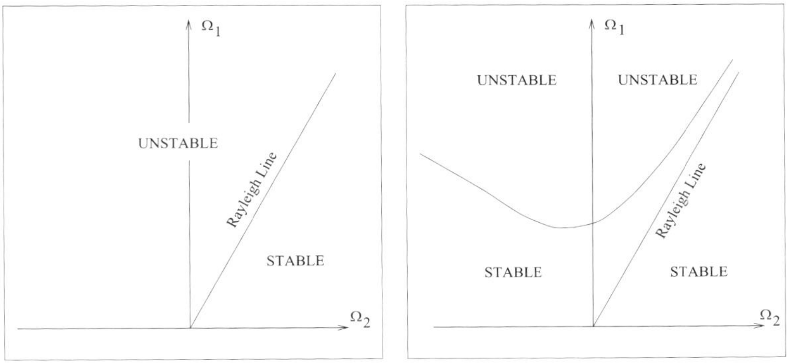

Fig. 7. The regions of stability of the flow between two cylinders (Eq. (8.3)) in the Ω_1–Ω_2 plane. Inviscid fluid on the left panel, viscous fluid on the right panel. The Rayleigh line is defined by $\Omega_1 R_1^2 = \Omega_2 R_2^2$.

the initial amplitude of the perturbation exceeds some threshhold. Taken at face value, these results show that a keplerian accretion disk is linearly stable to these kind of instabilities, because, for a Keplerian rotation profile, $\Omega^2 \sim r^{-3}$, so that $J^2 = r^4\Omega^2$ is an *increasing* function of r. Taking the effect of viscosity into account reduces the domain of linear instability [7]. This is shown in Figure 7.

14.2 Magnetorotational instability

The presence of a magnetic field drastically modifies the stability properties of rotating flows [7]. An accretion disk, unlike Couette flows, has a finite thickness. The analysis can nevertheless be kept simple by considering perturbations of a vertical wavelength λ_z much smaller than the disk thickness H and by resorting to the local WKB approximation for dealing with the vertical and radial inhomogeneity. To lowest order in λ_z/H, the local values of the unperturbed quantities are regarded as being constant and a local dispersion relation is obtained, which we establish below. Resistivity and viscosity are ignored. The plasma is a perfect gas of adiabatic index γ. The unperturbed magnetic field only has axial and azimuthal components, B_{z0} and $B_{\theta 0}$. The unperturbed motion is a rotation flow with an angular velocity profile $\Omega(r)$. We consider only axisymmetrical perturbations. The effect of the vertical component of gravity can be neglected because $\lambda_z \ll H$. With these assumptions, the unperturbed state locally reduces to a stationnary cylindrical flow, the equilibrium of which is described by:

$$-\rho_0 \frac{v_\theta^2}{r} = -\frac{\partial P_0}{\partial r} - \frac{\partial}{\partial r}\left(\frac{B_{\theta 0}^2 + B_{z0}^2}{2\mu_0}\right) - \frac{B_{\theta 0}^2}{\mu_0 r} + \rho_0 g_{\rm r} . \qquad (14.11)$$

In the presence of a perturbation, the physical quantities become:

$$\begin{aligned} \boldsymbol{v} &= r\Omega(r)\ \boldsymbol{e}_\theta + \boldsymbol{u}(r,z,t) \qquad & \boldsymbol{B} &= \boldsymbol{B}_0(r) + \boldsymbol{b}(r,z,t) \\ P &= P_0(r) + p(r,z,t) & \rho &= \rho_0(r) + \rho_1(r,z,t). \end{aligned} \tag{14.12}$$

Self-gravity is ignored and the fluid is assumed to be almost incompressible, so that the gravitational force perturbation can be neglected. The eigenmodes can be found by Fourier analyzing in z. Any perturbation $\tilde{f}$ is expanded as $\tilde{f}(r,z,t) = f(r)\exp(ikz - i\omega t)$ where ω is the (complex) eigenfrequency. The expression of MHD equations in cylindrical coordinates can be found, for example, in [7]. In writing their linearized form about this stationnary flow the radial derivatives of perturbed quantities are neglected, which is the essence of the local, WKB, approximation. A prime indicating a derivative with respect to r, we obtain the following linear homogeneous system for ρ_1, u_r, u_θ, u_z, b_r, b_θ, b_z and p:

$$-i\omega\rho_1 = -ik\rho_0 u_z \tag{14.13}$$
$$-i\omega u_\mathrm{r} = +2\Omega u_\theta + ikB_z b_\mathrm{r}/(\mu_0\rho_0) \tag{14.14}$$
$$-i\omega u_\theta = -\left(r\Omega' + 2\Omega\right)\ u_\mathrm{r} + ikB_z b_\theta/(\mu_0\rho_0) \tag{14.15}$$
$$-i\omega u_z = -ikp/\rho_0 - ikB_\theta b_\theta/(\mu_0\rho_0) \tag{14.16}$$
$$-i\omega b_\mathrm{r} = +ikB_z u_\mathrm{r} \tag{14.17}$$
$$-i\omega b_\theta = +\left(r\Omega'\right) b_\mathrm{r} + ikB_z u_\theta - ikB_\theta u_z \tag{14.18}$$
$$-i\omega b_z = 0 \tag{14.19}$$
$$p/P_0 = \gamma\left(\rho_1/\rho_0\right). \tag{14.20}$$

From equations (14.18) and (14.17), b_θ can be expressed as:

$$\omega b_\theta = kB_\theta u_z - kB_z u_\theta - i\frac{kB_z}{\omega}\left(r\Omega'\right) u_\mathrm{r}. \tag{14.21}$$

It is convenient to introduce the epicyclic frequency, κ, defined by:

$$\kappa^2 = \frac{1}{r^3}\frac{\mathrm{d}}{\mathrm{d}r}\left(r^2\Omega\right)^2 = 2\Omega\left(2\Omega + r\Omega'\right). \tag{14.22}$$

Note that the Rayleigh criterium for the satbility of inviscid hydrodynamical Couette flow is simply that the epicyclic frequency be real. The characteristic Alfvén and sound velocity are denoted as in (9.39), $v_{\mathrm{A}z}$ being associated with B_z. Using equation (14.21), the system (14.13)–(14.20) can be reduced to only three equations for the velocity components:

$$(\omega^2 - k^2 v_{\mathrm{A}z}^2)u_\mathrm{r} - 2i\omega\Omega u_\theta = 0 \tag{14.23}$$

$$i\left(\frac{\kappa^2\omega}{2\Omega} - \frac{k^2 v_{\mathrm{A}z}^2}{\omega}\left(\frac{\mathrm{d}\Omega}{\mathrm{d}\ln r}\right)\right) u_\mathrm{r} + (\omega^2 - k^2 v_{\mathrm{A}z}^2)u_\theta + k^2 v_{\mathrm{A}z} v_{\mathrm{A}\theta} u_z = 0 \tag{14.24}$$

$$\left(ik^2 v_{\mathrm{A}z} v_{\mathrm{A}\theta} \frac{1}{\omega}\frac{\mathrm{d}\Omega}{\mathrm{d}\ln r}\right) u_{\mathrm{r}} + k^2 v_{\mathrm{A}z} v_{\mathrm{A}\theta} u_\theta + (\omega^2 - k^2(v_{\mathrm{A}\theta}^2 + c_{\mathrm{S}}^2))u_z = 0. \quad (14.25)$$

The condition that the system (14.23)–(14.25) has a non-vanishing solution gives the dispersion relation:

$$(\omega^2 - k^2 v_{\mathrm{A}z}^2)\left(\omega^4 - \omega^2 k^2 (c_{\mathrm{S}}^2 + v_{\mathrm{A}}^2) + k^4 v_{\mathrm{A}z}^2 c_{\mathrm{S}}^2\right) =$$
$$\left(\kappa^2\omega^4 - \omega^2\left(\kappa^2 k^2 (c_{\mathrm{S}}^2 + v_{\mathrm{A}\theta}^2) + k^2 v_{\mathrm{A}z}^2 \frac{\mathrm{d}\Omega^2}{\mathrm{d}\ln r}\right) + k^4 v_{\mathrm{A}z}^2 c_{\mathrm{S}}^2 \frac{\mathrm{d}\Omega^2}{\mathrm{d}\ln r}\right) \cdot \quad (14.26)$$

In the absence of rotation, κ^2 and Ω^2 vanish and the dispersion relation reduces equation (11.21). In the case of keplerian rotation, $\Omega^2(r) = GM_*/r^3$, $\kappa^2(r) = \Omega^2(r)$ and $\mathrm{d}\Omega^2/\mathrm{d}\ln r = -3\Omega^2$. Equation (14.26) then becomes:

$$\omega^6 - \omega^4[k^2(c_{\mathrm{S}}^2 + v_{\mathrm{A}}^2 + v_{\mathrm{A}z}^2) - \Omega^2] + \omega^2[k^4 v_{\mathrm{A}z}^2(v_{\mathrm{A}}^2 + 2c_{\mathrm{S}}^2) + ..$$
$$.. + \Omega^2 k^2 (v_{\mathrm{A}\theta}^2 - 3v_{\mathrm{A}z}^2 + c_{\mathrm{S}}^2)] - k^4 c_{\mathrm{S}}^2 v_{\mathrm{A}z}^2 (k^2 v_{\mathrm{A}z}^2 - 3\Omega^2) = 0. \quad (14.27)$$

The Alfvén mode and the two magnetosonic modes, which are independent in the absence of rotation, now become coupled in the sense that the dispersion relation (14.27) does not factor out into independent terms. The physical reason for this is that the motions in the radial direction, which, in this geometry, is the natural polarization of the Alfvén mode, are under the influence of the centrifugal force perturbation, which is the term proportional to u_θ in equation (14.23). In the small rotation regime, the system has three, usually distinct, real positive roots for $s = \omega^2$. This is so for zero rotation and remains so for small enough Ω by continuity. For Ω larger than some threshold value, one of the real roots becomes negative. This is signaled by a change of the sign of the product of roots, $s_1 s_2 s_3 = k^4 c_{\mathrm{S}}^2 v_{\mathrm{A}z}^2 (k^2 v_{\mathrm{A}z}^2 - 3\Omega^2)$. The root which is responsible for this is the smallest of the three, *i.e.*, if we keep, by continuity, the same terminology as for non rotating fluids, it is the root which is associated with the slow mode. For weak compressibility, the slow mode esentially reduces to an Alfvén wave. Instability sets in when

$$\Omega^2 > \frac{k^2 v_{\mathrm{A}z}^2}{3} \cdot \quad (14.28)$$

We therefore reach the important conclusion that, when this k-dependent threshold rotation is reached, Alfvénic motions become unstable to the so-called magnetorotational instability [3, 7, 30]. Equation (14.28) indicates that, for a given k, a minimum rotation rate is needed for the instability to develop and, conversely, that, for a given rotation rate, the instability develops only for k smaller than a maximum value, $k_{\max}$, given by $k_{\max}^2 = 3\Omega^2/v_{\mathrm{A}z}^2$. Up to now, we have ignored the finite extent of the disk in the

z-direction. However an accretion disk has a finite thickness $H \approx c_S/\Omega$. This can be taken into account by demanding that physically acceptable solutions of equation (14.27) be such that at least half a vertical wavelength fits in the thickness of the disk. This implies that k be no less than $k_{\mathrm{disk}} \approx \pi\Omega/c_s$ For the instability to develop, k_{disk} should be less than $k_{\max}$. This requires that the Alfvén speed be small enough, namely that

$$v_A^2 < \frac{3c_S^2}{\pi^2}. \tag{14.29}$$

This shows that the magnetorotational instability is a weak field instability. The physical reason for the instability is well understood in the case of an unperturbed field parallel to the rotational axis and an incompressible fluid. Then, the Lorentz force developed in the perturbation can be written as:

$$\boldsymbol{f}_{\mathcal{L}} = -\rho_0 k^2 v_{Az}^2 \boldsymbol{\xi} \tag{14.30}$$

where $\boldsymbol{\xi}$ is the displacement of the fluid element from its corotating rest position. This force is similar to a spring force. An analysis of the motion of the fluid element under the effect of this force and the centrifugal, Coriolis and tidal forces, reveals that the displacement $\boldsymbol{\xi}$ evolves in a way entirely similar to the separation of two masses connected by a spring and orbiting in the given gravitational field. This illuminating analogy is described in [5] and [2].

References

[1] J.J. Aly, *A&A* **143** (1985) 19.
[2] S.A. Balbus and J.F. Hawley, *Rev. Mod. Phys.* **70** (1998) 1.
[3] S.A. Balbus and J.F. Hawley, *ApJ* **376** (1991) 214.
[4] V. Beskin, 2003, these proceedings.
[5] O. Blaes, 2003, these proceedings.
[6] S.I. Braginskii, *Rev. Plasma Phys.* **1** (Consultants Bureau, New York, 1965).
[7] S. Chandrasekhar, Hydrodynamic and Hydromagnetic stability (Dover Publications, New York, 1981).
[8] T.G. Cowling, Magnetohydrodynamics (Interscience, London, 1957).
[9] H.P. Fürth, J. Killeen and M. Rosenbluth, *Phys. Fluids* **6** (1963) 459.
[10] L. Landau and E.M. Lifshitz, *Fluid Mech.* (Pergamon Press, 1959).
[11] J. Heyvaerts, J.M. Lasry, M. Schatzman and P. Witomsky, *A&A* **111** (1982) 104.
[12] J. Heyvaerts, in "Late stages of Stellar evolution", edited by C. de Loore, École EADN de Ponte de Lima, *Lect. Notes Phys.* (Springer Verlag, 1991) 313.
[13] J. Heyvaerts and C.A. Norman, *ApJ* **347** (1989) 1055.
[14] J. Heyvaerts and C.A. Norman, *ApJ* (2003) in press.
[15] J. Heyvaerts and C.A. Norman, *ApJ* (2003 in press.
[16] J. Heyvaerts and C.A. Norman, *ApJ* (2003) in press.

[17] J. Heyvaerts, in Plasma Astrophysics, edited by C. Chiuderi and G. Einaudi, *Lect. Notes Phys.* (Springer Verlag, 1996) 31.

[18] J. Heyvaerts, in Transport and Energy Conversion in the heliosphere, edited by J.P. Rozelot, L. Klein and J.C. Vial, *Lect. Notes Phys.* (Springer Verlag, 2000) 1.

[19] C.W. Misner, K.S. Thorne and J.A. Wheeler Gravitation (W.H. Freeman & Co., San Francisco, 1970).

[20] F.C. Michel, *ApJ*, **158** (1969) 727.

[21] K. Moffatt, Magnetic Field Generation in Electrically Conducting Fluids (Cambridge Univ. Press, 1978).

[22] H. Petschek, in AAS-NASA Symp. on Solar Flares (1964).

[23] E.R. Priest "Solar Magnetohydrodynamics", *Geophys. Astrophys. Monogr.* (D. Reidel Publishing Comp., 1982).

[24] E.R. Priest and T. Forbes, MHD reconnection (Cambridge Univ. Press, 2000).

[25] M.R.E. Proctor and A.D. Gilbert, Lectures on Solar and Planetary Dynamos (Cambridge Univ. Press, 1994).

[26] R. Pudritz (2003), these proceedings.

[27] B.U.Ö. Sonnerup and E.R. Priest, *J. Plasma. Phys.* **14** (1975) 283.

[28] L. Spitzer, Physics of fully ionized gases (Interscience publishers, New York, 1962).

[29] M. Steenbeck, F. Krause and K.H. Rädler, *Z. Naturforsh* **21a** (1966) 369.

[30] E.P. Velikhov, *Sov. Phys. JETP* **36** (1959) 995.

[31] E.J. Weber and L. Davis, *ApJ* **148** (1967) 207.

[32] Y.B. Zel'dovich, and Y.P. Raizer, Physics of Shock Waves and High Temperature Phenomena (I) (Academic Press, New York London, 1966).

COURSE 2

2D TRANSONIC HYDRODYNAMICS IN GENERAL RELATIVITY

V. BESKIN

Lebedev Physical Institute,
Moscow, Russia

Contents

2D TRANSONIC HYDRODYNAMICS IN GENERAL RELATIVITY

V. Beskin[1]

Abstract

The goal of my lecture is to present the introduction into the hydrodynamical version of the Grad-Shafranov equation. Although not so well-known as the full MHD one, it allows us to clarify the nontrivial structure of the Grad-Shafranov approach as well as to discuss the simplest version of the $3+1$-split language – the most convenient one for the description of the ideal flows in the vicinity of a rotating black hole.

1 Introduction

Axisymmetric stationary flows in the vicinity of a central compact body have been studied for a long time in connection with many astrophysical sources. Accretion onto ordinary stars and black holes [1, 2], axially symmetrical stellar (solar) wind [3], jets from young stellar objects [4], outflow from axisymmetric magnetosphere of rotating neutron stars [5–7] – all of them are flows of the considered type.

Let me stress that the necessity of taking into account the effects of General Relativity is not so obvious for many compact sources. For instance, one cannot exclude that the black hole plays only a passive role in the jet formation process, and the effects of General Relativity in this case may be inimportant for flow description in the region of jet formation. At the same time gravitational effects make, apparently, a noticeable contribution to the determination of physical conditions in compact objects. First, this is indicated by the hard spectra and the annihilation line observed in galactic X-ray sources, which are believed to be solar mass black holes. Such

This work is supported by Grant No. 1603.2003.2 by Russian Foundation for Basic Researches.

[1]Lebedev Physical Institute, Moscow.

characteristics are never observed in the X-ray sources which are firmly established to show accretion not onto a black hole but onto a neutron star. Another indication comes from superluminal motion in quasars which may be due to the relativistic electron-positron plasma flow ejected along with the weakly relativistic jet [8]. All this testifies in favour of the existence of an additional mechanism for particle creation and acceleration, for which the effects of General Relativity may be of principal importance. So, it is undoubtedly interesting to consider the flow structure in the most general conditions, *i.e.*, in the presence of a rotating black hole.

There are several reasons why I restrict my consideration to the pure hydrodynamic flows. First of all, the hydrodynamical version of the Grad-Shafranov equation is not so popular as the full MHD one. On the other hand, it contains all the features of the full MHD version in the simplest form. In particular, within the hydrodynamic approach one can introduce the $3+1$-split language – the most convenient one for the description of the ideal flows in the vicinity of a rotating black hole.

Thus, in this lecture the basic equations describing a steady axisymmetric hydrodynamical flow in the vicinity of Kerr black hole are given. Starting with the well-known set of equations describing the nonrelativistic ideal flow [9], we will go step by step to more complicated cases up to the most general one corresponding to the axisymmetric stationary flow in the Kerr metric. Finally, several examples will be considered which demonstrate how the approach under consideration can be used to obtain the quantitative description of the real transonic flows in the vicinity of rotating black holes.

2 Ideal hydrodynamics – some fundamental results

2.1 Basic equations

To start from the very beginning, let us wright down the equations of ideal stationary ($\partial/\partial t = 0$) hydrodynamics in a flat space. They are [10]:

- continuity equation

$$\nabla \cdot (n\vec{v}) = 0, \tag{2.1}$$

- Euler equation

$$(\vec{v}\nabla)\vec{v} = -\frac{\nabla P}{m_{\rm p} n} - \nabla \varphi_{\rm g}, \tag{2.2}$$

- isentropy condition

$$\vec{v} \cdot \nabla s = 0, \tag{2.3}$$

- equation of state

$$P = P(n, s). \tag{2.4}$$

The last expression can be rewritten in the form

$$\mathrm{d}P = m_{\mathrm{p}} n \mathrm{d}w - nT\mathrm{d}s. \tag{2.5}$$

Here n (cm^{-3}) is the concentration, s is the entropy per particle (undimensional), w (cm^2/s^{-2}) is the specific enthalpy, m_{p} (g) is the mass of particle, T (erg) is the temperature in energetic unit, and c_{s} (cm s^{-1}) is the velocity of sound. For the polytropic equation of state

$$P = k(s) n^{\Gamma}, \tag{2.6}$$

used for simplicity in what follows, one can obtain for $\Gamma \neq 1$

$$c_{\mathrm{s}}^2 = \frac{1}{m_{\mathrm{p}}} \left(\frac{\partial P}{\partial n} \right)_{\mathrm{s}} = \frac{1}{m_{\mathrm{p}}} \Gamma k(s) n^{\Gamma - 1}, \tag{2.7}$$

$$w = \frac{c_{\mathrm{s}}^2}{\Gamma - 1}, \tag{2.8}$$

$$T = \frac{m_{\mathrm{p}}}{\Gamma} c_{\mathrm{s}}^2. \tag{2.9}$$

Now one can make several remarks.

- The Euler equation (2.2) together with (2.3) and (2.5) can be rewritten as the energy equation

$$\nabla \cdot \left[n\vec{v} \left(\frac{v^2}{2} + w + \varphi_{\mathrm{g}} \right) \right] = 0. \tag{2.10}$$

  Now using the continuity equation (2.1), one can obtain

$$\vec{v} \cdot \nabla E = 0, \tag{2.11}$$

  where

$$E = \frac{v^2}{2} + w + \varphi_{\mathrm{g}}. \tag{2.12}$$

  This is the well-known Bernoulli integral.

- The energy equation (2.10) together with the Euler equation (2.2) can be rewritten as a four-component energy-momentum equation

$$\nabla_{\alpha} T^{\alpha\beta} = 0, \tag{2.13}$$

where for $\varphi_{\rm g} = 0$

$$T^{\alpha\beta} = \begin{pmatrix} nm_{\rm p}v^2/2 + n\varepsilon & nm_{\rm p}\vec{v}\left(v^2/2 + w\right) \\ m_{\rm p}nv^i & P\delta^{ik} + nm_{\rm p}v^iv^k \end{pmatrix}. \tag{2.14}$$

Here and below the Greek indices α and β correspond to four-dimensional values, while the Latin i, j, and k are three-dimensional.

- As a result, hydrodynamics contains five equations over five unknown values.

Exercise.

1. *Prove expressions (2.10)–(2.14)*

2.2 An example – spherically symmetrical flow

As a most simple but very important example let us consider the spherically symmetrical flow. Since the basic (ideal) hydrodynamic equations have the form of the conservation laws, one can find

- the continuity equation

$$\Phi = 2\pi r^2 n(r) v(r) = \text{const}, \tag{2.15}$$

- the entropy equation

$$s = \text{const}, \tag{2.16}$$

- the energy equation

$$E = \frac{v^2(r)}{2} + w(r) + \varphi_{\rm g}(r) = \text{const}. \tag{2.17}$$

As a result, knowing three parameters Φ, s, and E, one can determine all the physical characteristics of a flow. Indeed, rewriting the Bernoulli equation (2.17) as

$$E = \frac{\Phi^2}{8\pi^2 n^2 r^4} + w(n, s) + \varphi_{\rm g}(r), \tag{2.18}$$

we see that this equation contains only one unknown parameter n. Hence, this algebraic equation implicitly determines the concentration n as a function of the invariants and radius r:

$$n = n(E, s, \Phi; r). \tag{2.19}$$

Together with the entropy s, it allows us to determine all the other thermodynamic functions and the flow velocity $v(r)$.

It is necessary to stress that equation (2.18) contains a singularity on the sonic surface. To show this, let us determine the derivative $\mathrm{d}n/\mathrm{d}r$:

$$\frac{\mathrm{d}n}{\mathrm{d}r}\left[\left(\frac{\partial w}{\partial n}\right)_{\mathrm{s}} - \frac{\Phi^2}{4\pi^2 n^3 r^4}\right] - \frac{\Phi^2}{2\pi^2 n^2 r^5} + \frac{GM}{r^2} = 0.$$

As a result, using the thermodynamic relation (2.5), one can obtain for the logarithmic derivative

$$\eta_1 = \frac{r}{n}\frac{\mathrm{d}n}{\mathrm{d}r} = \frac{2v^2 - GM/r}{c_{\mathrm{s}}^2 - v^2} = \frac{2 - GM/rv^2}{-1 + c_{\mathrm{s}}^2/v^2} = \frac{N_r}{D}. \tag{2.20}$$

We see that the derivative (2.20) contains the singularity when the velocity is equal to that of sound c_* ($D = 0$). It means that to pass through the sonic surface $r = r_*$ the additional critical condition is to be valid:

$$N_r(r_*) = 2 - \frac{GM}{r_* c_*^2} = 0. \tag{2.21}$$

In other words, the transonic flows are determined by two invariants only. As shown in Figure 1, the sonic surface is an X-point on the plane distance – velocity.

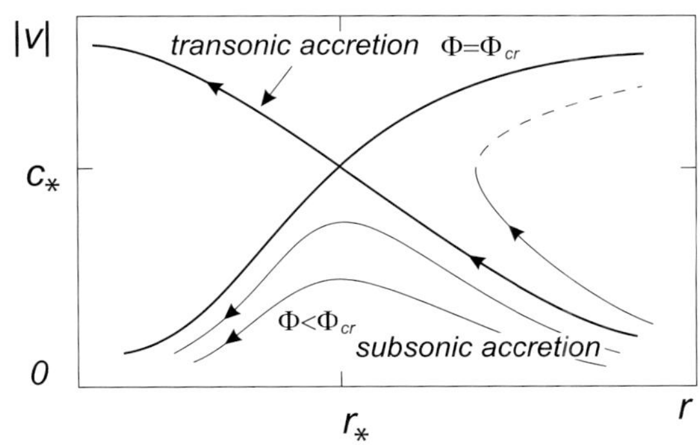

Fig. 1. Structure of the spherically symmetric accretion for a given n_∞ and c_∞, and different stream function Φ. The transonic flow corresponds to the critical accretion rate $\Phi = \Phi_{\mathrm{cr}}$ (2.34). The curves below the X-point correspond to subsonic accretion with $\Phi < \Phi_{\mathrm{cr}}$.

Exercise

1. *For spherically symmetrical transonic inflow (Bondi accretion [2]) one can determine the Bernoulli integral E through the velocity of sound at infinity*

$$E = w_{\infty} = \frac{c_{\infty}^2}{\Gamma - 1}.$$

Now using relations (2.15)–(2.17) and (2.21), obtain the well-known expressions for the velocity of sound c_* and the concentration n_* on the sonic radius r_* [2]:

$$c_*^2 = \left(\frac{2}{5 - 3\Gamma}\right) c_{\infty}^2, \tag{2.22}$$

$$n_* = \left(\frac{2}{5 - 3\Gamma}\right)^{1/(\Gamma-1)} n_{\infty}, \tag{2.23}$$

$$r_* = \left(\frac{5 - 3\Gamma}{4}\right) \frac{GM}{c_{\infty}^2}. \tag{2.24}$$

2. *Show that*

$$\eta_1(r_*) = \frac{-4 \pm \sqrt{10 - 6\Gamma}}{\Gamma + 1}, \tag{2.25}$$

the sign plus *corresponding to accretion and* minus *– to ejection.*

3. *Find that for the spherically symmetric accretion*

   - *for $r \gg r_*$ (subsonic regime) the flow is approximately incompressible:*

$$n(r) \approx const, \tag{2.26}$$

$$v(r) \propto r^{-2}. \tag{2.27}$$

   - *for $r \ll r_*$ (supersonic regime) the particle motion corresponds to a free fall:*

$$n(r) \propto r^{-3/2}, \tag{2.28}$$

$$v(r) \approx (2GM/r)^{1/2}. \tag{2.29}$$

4. *Find that for the spherically symmetric transonic outflow (Parker ejection [3]):*

   - *Physical parameters on the sound surface $r = r_*$, where again*

$$r_* = \frac{GM}{2c_*^2}, \tag{2.30}$$

have the following relations to the ones on the star surface $r = R$

$$c_*^2 = \left(\frac{2}{5-3\Gamma}\right) c_{\rm R}^2 + \left(\frac{\Gamma-1}{5-3\Gamma}\right)\left(v_{\rm R}^2 - \frac{2GM}{R}\right), \quad (2.31)$$

$$n_* = n_{\rm R}\left(\frac{c_*^2}{c_{\rm R}^2}\right)^{1/(\Gamma-1)}. \quad (2.32)$$

- *The (radial) velocity on the star surface is to be*

$$v_{\rm R} = c_*\left(\frac{c_*^2}{c_{\rm R}^2}\right)^{1/(\Gamma-1)}\left(\frac{r_*}{R}\right)^2. \quad (2.33)$$

Although fairly simple, the radial 1D flow allows us to formulate here several important properties; some of them, as we shall see, retains the common properties of the Grad-Shafranov approach.

- It is possible to pass through the sonic surface in the presence of gravity only. Indeed, the nominator N_r in (2.20) can vanish for $D = 0$ in the presence of the gravity term GM/rv^2 only.
- Solutions (2.22)–(2.24) and (2.31) have singularity for $\Gamma = 5/3$. It means that for $\Gamma = 5/3$ the increase/decrease of the velocity of sound as a result of adiabatic heating/cooling equals the change of particle velocity. As a result, in the nonrelativistic case for $\Gamma \geq 5/3$ the transonic flow is not realized.
- The transonic problem is two-parametric. It means that to determine the transonic flow it is necessary to specify two boundary conditions, say, the density $m_{\rm p}n_\infty$ and the velocity of sound c_∞ at infinity. In particular, the accretion rate is fixed:

$$2\Phi_{\rm cr} = 4\pi r_*^2 c_* n_* = 4\pi\left(\frac{2}{5-3\Gamma}\right)^{(5-3\Gamma)/2(\Gamma+1)}\frac{(GM)^2}{c_\infty^3}n_\infty. \quad (2.34)$$

  On the other hand, for a given n_∞ and c_∞ there is an infinite number of subsonic solutions with $\Phi < \Phi_{\rm cr}$ (see Fig. 1).
- For a given flow structure the number of integrals is enough to determine all the characteristics of a flow from algebraic equations.

The last property is actually the key point of the approach. Indeed, the algebraic relations (2.15)–(2.17) together with the equation of state allow us to determine all the physical parameters of the flow (the flow velocity $v(r)$, the temperature $T(r)$, etc.) through the invariants E and s and the stream function Φ. This property remains true for an arbitrary 2D flow structure. But in the general case the flow structure (*i.e.*, the stream function $\Phi(r,\theta)$) itself is not known. To determine the stream function, the extra two hydrodynamic equations are to be used.

2.3 Flat potential flow

To start, let us consider the simplest (and the well-known) case of the flat potential flow without gravity. Then one can introduce potential $\phi(x,y)$ by the relation

$$\vec{v} = \nabla\phi(x,y). \tag{2.35}$$

In addition, let us consider the case

$$E = \text{const}, \qquad s = \text{const}. \tag{2.36}$$

Then the continuity equation $\nabla\cdot(n\vec{v}) = 0$ can be rewritten as

$$\nabla^2\phi + \frac{\nabla n\cdot\nabla\phi}{n} = 0. \tag{2.37}$$

Finally, using the Euler equation to determine $\nabla n\cdot\nabla\phi$

$$\vec{v}\cdot\nabla\left(\frac{v^2}{2}\right) + c_\text{s}^2\frac{\nabla n\cdot\nabla\phi}{n} = 0,$$

we obtain

$$\phi_{xx} + \phi_{yy} + \frac{(\phi_y)^2\phi_{xx} - 2\phi_x\phi_y\phi_{xy} + (\phi_x)^2\phi_{yy}}{(\nabla\phi)^2 D} = 0. \tag{2.38}$$

Here again

$$D = -1 + \frac{c_\text{s}^2}{v^2}. \tag{2.39}$$

This well-known equation can be found in any textbook (*e.g.*, see [10]).

For us it is necessary to stress here the following properties.

- To determine c_s^2, equation (2.38) is to be supplemented with the Bernoulli equation. For the polytropic equation of state it can be solved for the velocity of sound c_s:

$$c_\text{s}^2 = (\Gamma-1)E - \frac{\Gamma-1}{2}(\nabla\phi)^2. \tag{2.40}$$

- Together with the Bernoulli equation, equation (2.38) contains the potential $\phi(x,y)$ and the invariant E only (it does not contain the entropy s at all, but s is necessary to determine n).

- For $n = \mathrm{const}$ ($c_{\mathrm{s}}^2 \to \infty$) the equation becomes linear.
- It is nonlinear in the general case, but linear for second derivatives.
- The equation is elliptical for a subsonic flow $D > 0$.
- The equation is hyperbolic for a supersonic flow $D < 0$.
- For a given flow structure (for a given ϕ, E, and s) all the physical parameters are determined by algebraic relations.
- Equation (2.38) does not contain coordinates x and y.

The latter property was very widely used in the approach of the hodograph transformation, *i.e.*, the transformation from a physical plane (x, y) to a hodograph plane (v, θ), where $v_x = v \sin\theta$, $v_y = v\cos\theta$. In this case it is possible to introduce another potential $\phi_v(v, \theta)$ so that $\vec{r} = \nabla_{\vec{v}}\phi_v$. As a result, equation (2.38) can be rewritten as

$$\frac{\partial^2 \phi_v}{\partial \theta^2} + \frac{v^2}{1 - v^2/c_{\mathrm{s}}^2}\frac{\partial^2 \phi_v}{\partial v^2} + v\frac{\partial \phi_v}{\partial v} = 0. \tag{2.41}$$

This is the linear Chaplygin equation (1902).

The hodograph transformation approach was the main direction of exploration through the XX-th century [11, 12]. Here I formulate two results obtained in this field, to be used in what follows.

- For the transonic flow it is impossible to solve the direct problem (*i.e.*, to determine the flow structure from the given shape of the boundary, say, knowing the shape of nozzle or wing).
- On the other hand, one can solve the reverse problem. This approach is based on the fundamental theorem: the transonic flow is analytical at the critical point (the point where the sonic surface is orthogonal to the flow line, see Fig. 2) [10].

Let me comment these two statements. The most visible argument clarifying the absence of the regular (not iterative) procedure for the transonic flow is as follows. As is known, the number of boundary conditions b for an arbitrary (not only for a pure hydrodynamical) flow can be determined from the following condition [13, 14]

$$b = 2 + i - \sigma. \tag{2.42}$$

Here i is the number of invariants and σ is the number of singular surfaces. In pure hydrodynamics the only singular surface is the sonic one. Hence,

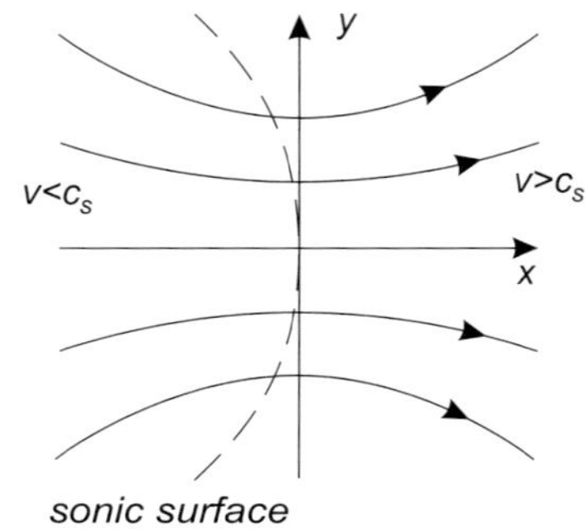

Fig. 2. Structure of the "analytical nozzle" in the vicinity of the critical point $x = y = 0$ – the only point where the sonic surface is orthogonal to the flow line. As the term $\propto y^2$ in (2.44) is absent (*i.e.*, $v_y(0, y) = 0$), the plane $x = 0$ corresponds to minimum cross-section of the stream surfaces. The sonic surface $v = c_*$ has the standard parabolic form $x = -[k(\Gamma - 1)/2c_*]y^2$.

for the transonic flow $\sigma = 1$. Furthermore, for planar geometry we have two invariants, E and s, so that $i = 2$. Thus, to determine the structure of the transonic flow it is necessary to specify three boundary conditions on a surface. These may be two thermodynamic functions and one component of the velocity. The second (the last in the planar case) component of the velocity is to be determined from the solution. But to solve equation (2.38), it is necessary to know the Bernoulli integral $E = v^2/2 + w$, *i.e.*, both components of the velocity on this surface. Hence, in the general case even the equation describing the flow structure cannot be formulated. For subsonic and supersonic flows $\sigma = 0$ (so that $b = 4$) and this difficulty is absent.

On the other hand, the structure of the transonic flow can be found directly by expansion of the solution in the vicinity of the critical point (where we put $x = y = 0$). Indeed, in addition to the invariants E (which determines the velocity of sound on the sonic surface) and s (which is necessary for the determination of the concentration n) one can specify the x component of the velocity $v_x(x, 0)$ along x-axis. For our purpose it is enough to know the first two terms in the expansion

$$v_x(x, 0) = c_* + kx + \ldots \tag{2.43}$$

Here $c_*^2 = 2E(\Gamma - 1)/(\Gamma + 1)$. As a result, as one can check directly, the first terms in the expansion of the potential $\phi(x, y)$ look like [10]

$$\phi(x, y) = c_* x + \frac{1}{2}\, kx^2 + \frac{1}{2c_*}\, k^2(\Gamma + 1)xy^2 + \frac{1}{24c_*^2}\, k^3(\Gamma + 1)^2 y^4 + \ldots \tag{2.44}$$

Knowing all the coefficients in the expansion (2.43), one can restore the potential ϕ with any precision.

Thus, in the general case equation (2.38) cannot be solved directly. Let me stress that this property is common; it takes place for the Grad-Shafranov equation as well. On the other hand, the planar approach has three extra difficulties:

- It is difficult to consider the case $E \neq$ const, $s \neq$ const.
- It is impossible to consider the nonpotential flow with $\nabla \times \vec{v} \neq 0$.
- It is impossible to include gravity (which is not planar).

3 Axisymmetric stationary flow – nonrelativistic case

3.1 Basic equations

3.1.1 Stream function

Now let us see how a similar procedure can be realized for the axisymmetric stationary flows. It means that we assume all the values to depend on two variables – r and θ. In this sense the flow remains two-dimensional. But now none of the three components of the velocity are equal to zero. Thus, axisymmetric stationary flows are more rich than planar ones.

In the axisymmetric stationary case one can introduce potential $\Phi(r,\theta)$ which is connected with the poloidal velocity $\vec{v}_{\rm p}$ as

$$n\vec{v}_{\rm p} = \frac{\nabla\Phi \times \vec{e}_{\varphi}}{2\pi r \sin\theta}. \tag{3.1}$$

Such a definition results in the following properties:

- The continuity equation is valid automatically: $\nabla \cdot (n\vec{v}) = 0$.
- On can check that $\mathrm{d}\Phi = n\vec{v}\mathrm{d}\vec{S}$, where $\vec{S}$ is the area. Thus, the potential $\Phi(r,\theta)$ is the flux through the circuit $r, \theta, 0 < \varphi < 2\pi$. In particular, the total flux through the sphere of radius r is $\Phi_{\rm tot} = \Phi(r,\pi)$.
- As $\vec{v} \cdot \nabla\Phi = 0$, the relations $\Phi(r,\theta) = $ const describe the flow surfaces.

3.1.2 Integrals of motion

As earlier, the components $\beta = t$ and $\beta = v_{\parallel}$ of the energy-momentum conservation law $\nabla_{\alpha}T^{\alpha\beta} = 0$ give

$$E = E(\Phi) = v^2/2 + w + \varphi_{\rm g}, \tag{3.2}$$

$$s = s(\Phi). \tag{3.3}$$

But as we see, it is now much easier to describe the case when integrals are different for different flow surfaces.

New information appears from the $\beta = \varphi$ component of the energy-momentum equation (or, which is the same, from the φ component of the Euler equation)

$$\nabla_\varphi \left(\frac{v^2}{2}\right) - [\vec{v} \times (\nabla \times \vec{v})]_\varphi + \frac{\nabla_\varphi P}{m_{\rm p} n} + \nabla_\varphi \varphi_{\rm g} = 0. \tag{3.4}$$

As we consider the axisymmetric case, all the gradients vanish. The last term $[\vec{v} \times (\nabla \times \vec{v})]_\varphi$ can be rewritten in the form of the conservation law

$$\vec{v} \cdot \nabla(r \sin\theta v_\varphi) = 0. \tag{3.5}$$

Hence, in the axisymmetric case the z-component of the angular momentum

$$L(\Phi) = v_\varphi r \sin\theta \tag{3.6}$$

is the third integral of motion.

3.1.3 Mathematical interlude – the covariant description

As we are going to generalize our equations to General Relativity, it is convenient to rewrite these relations right now in the covariant form. For this reason, recall that the flat 3D metric g_{ik} ($\mathrm{d}l^2 = g_{ik}\mathrm{d}x^i\mathrm{d}x^k$) for spherical coordinates $x^1 = r$, $x^2 = \theta$, and $x^3 = \varphi$ has a form

$$g_{rr} = 1, \qquad g_{\theta\theta} = r^2, \qquad g_{\varphi\varphi} = r^2 \sin^2\theta, \tag{3.7}$$

all the other components being zero. Using now expression (2.14)

$$T_i^k = P\delta_i^k + (nm_{\rm p})v^k v_i, \tag{3.8}$$

one can obtain from the φ component of the energy-momentum equation

$$\begin{aligned}\nabla_k T_\varphi^k &= \nabla_k(\delta_\varphi^k P) + \nabla_k(nm_{\rm p}v^k v_\varphi) = \frac{1}{r}\frac{\partial P}{\partial\varphi} \\ &+ \frac{\partial}{\partial x^k}(nm_{\rm p}v^k v_\varphi) + \Gamma_{ik}^k(nm_{\rm p})v^i v_\varphi - \Gamma_{\varphi i}^k(nm_{\rm p})v^i v_k = 0.\end{aligned} \tag{3.9}$$

Here Γ_{jk}^i are the Christoffel coefficients.

As one can check, the last term in (3.9) vanishes: $\Gamma_{\varphi i}^k(nm_{\rm p})v^i v_k = 0$. The first term vanishes because of axisymmetry of the problem (no φ – dependence). Using now the continuity equation

$$\nabla_k(nm_{\rm p}v^k) = \frac{1}{\sqrt{g}}\frac{\partial}{\partial x^k}(\sqrt{g}nm_{\rm p}v^k) = \frac{\partial}{\partial x^k}(nm_{\rm p}v^k) + \Gamma_{ik}^k nm_{\rm p}v^i = 0, \tag{3.10}$$

where $g = \det g_{ik} = g_{rr}g_{\theta\theta}g_{\varphi\varphi}$, we see that (3.9) can again be written down in the form of the conservation law $\nabla_k T^k_\varphi = (nm_\mathrm{p})\vec{v}\cdot\nabla v_\varphi$. Hence, the third invariant looks like

$$L(\Phi) = v_\varphi. \tag{3.11}$$

Exercise.

1. *Check equations (3.9)–(3.11).*

2. *Is there a contradiction between (3.6) and (3.11)?*

To understand the difference between (3.6) and (3.11), it is necessary to return to the main definitions of the covariant approach. Up to now we have dealt with the physical components only. Below in the relativistic expressions in Sect. 4 we will mark the physical components by hats, so that $v_{\hat{\varphi}} = v^{\hat{\varphi}}$ is the physical component of the toroidal velocity, and its dimension is $\mathrm{cm\,s^{-1}}$. But in the covariant expressions (3.9)–(3.11) we encounter other objects – contravariant components v^i and covariant components v_k. As the definition of the vector length (which is square of the physical component) has the form $\vec{v}^2 = g_{ik}v^iv^k = g^{ik}v_iv_k$, one can write down for the diagonal metric (3.7)

$$(v_{\hat{\varphi}})^2 = g_{\varphi\varphi}(v^\varphi)^2 = g^{\varphi\varphi}(v_\varphi)^2, \tag{3.12}$$

and the same for the other components. Thus,

$$v^\varphi = \frac{1}{\sqrt{g_{\varphi\varphi}}}v_{\hat{\varphi}}, \tag{3.13}$$

$$v_\varphi = \sqrt{g_{\varphi\varphi}}v_{\hat{\varphi}}. \tag{3.14}$$

In particular, it means that the dimension of the covariant and contravariant components may differ from the dimension of the physical one. Comparing now (3.14) with (3.6) and (3.11), we can understand the difference: (3.6) actually involves the physical component of the toroidal velocity while (3.11) includes the covariant one.

3.2 The stream equation

3.2.1 Grad-Shafranov equation

To formulate the stream equation (*i.e.*, the equation describing the stream function $\Phi(r,\theta)$) it is necessary to return to the poloidal component of the Euler equation. One can check that together with the definitions of the invariants $E(\Phi)$, $L(\Phi)$, and $s(\Phi)$ this vector equation can be written as the product of the scalar equation by the vector $\nabla\Phi$: $[\mathrm{Euler}]_\mathrm{p} = [\mathrm{GS}]\cdot\nabla\Phi$. For this reason, in many MHD papers the stream equation $[\mathrm{GS}] = 0$ was obtained as the projection of the poloidal equation onto the gradient $\nabla\Phi$. The equivalent hydrodynamic expression has the form

$$\frac{1}{(\nabla\Phi)^2}\nabla\Phi\cdot\left[(\vec{v}\nabla)\vec{v}+\frac{\nabla P}{nm_{\rm p}}+\nabla\varphi_{\rm g}\right]=0. \tag{3.15}$$

Using now the definitions (3.1), (3.2), and (3.6), we have

$$-\varpi^2\nabla_k\left(\frac{1}{\varpi^2}\nabla^k\Phi\right)+\frac{1}{n}\nabla_k n\cdot\nabla^k\Phi$$
$$-4\pi^2L\frac{\mathrm{d}L}{\mathrm{d}\Phi}+4\pi^2\varpi^2n^2\frac{\mathrm{d}E}{\mathrm{d}\Phi}-4\pi^2\varpi^2n^2\frac{T}{m_{\rm p}}\frac{\mathrm{d}s}{\mathrm{d}\Phi}=0. \tag{3.16}$$

Here and to the very end

$$\varpi=\sqrt{g_{\varphi\varphi}}, \tag{3.17}$$

so that for the flat metric $\varpi=r\sin\theta$.

To close the system, *i.e.*, to determine the product $(\nabla n\cdot\nabla\Phi)$, the stream equation (3.16) is to be supplemented with the Bernoulli equation (3.2). It can now be rewritten in the form (*cf.* (2.18))

$$E=\frac{(\nabla\Phi)^2}{8\pi^2\varpi^2n^2}+\frac{1}{2}\frac{L^2}{\varpi^2}+w(n,s)+\varphi_{\rm g}. \tag{3.18}$$

As previously, the Bernoulli equation (3.18) contains, besides n, the invariants E, L, and s, and the stream function Φ only. Hence, it again gives the implicit expression for n:

$$n=n(\nabla\Phi;E,L,s;r,\theta). \tag{3.19}$$

On the other hand, the implicit Bernoulli equation can be presented in the differential form

$$\nabla_k n=n\frac{N_k}{D}, \tag{3.20}$$

where now

$$D=-1+\frac{c_{\rm s}^2}{v_{\rm p}^2}, \tag{3.21}$$

and

$$N_k=-\frac{\nabla^i\Phi\cdot\nabla_i\nabla_k\Phi}{(\nabla\Phi)^2}+\frac{1}{2}\cdot\frac{\nabla_k\varpi^2}{\varpi^2}-4\pi^2\varpi^2n^2\frac{\nabla_k\varphi_{\rm g}}{(\nabla\Phi)^2}$$
$$-4\pi^2n^2L\frac{\mathrm{d}L}{\mathrm{d}\Phi}\frac{\nabla_k\Phi}{(\nabla\Phi)^2}+2\pi^2n^2L^2\frac{\nabla_k\varpi^2}{\varpi^2(\nabla\Phi)^2} \tag{3.22}$$
$$+4\pi^2\varpi^2n^2\frac{\mathrm{d}E}{\mathrm{d}\Phi}\frac{\nabla_k\Phi}{(\nabla\Phi)^2}-4\pi^2\varpi^2n^2\left[\frac{T}{m_{\rm p}}+\frac{1}{m_{\rm p}n}\left(\frac{\partial P}{\partial s}\right)_n\right]\frac{\mathrm{d}s}{\mathrm{d}\Phi}\frac{\nabla_k\Phi}{(\nabla\Phi)^2}.$$

As a result, the stream equation can be written down as [13]

$$
\begin{aligned}
& -\varpi^2\nabla_k\left(\frac{1}{\varpi^2}\nabla^k\Phi\right) - \frac{\nabla^i\Phi\cdot\nabla^k\Phi\cdot\nabla_i\nabla_k\Phi}{(\nabla\Phi)^2 D} + \frac{\nabla\varpi^2\cdot\nabla\Phi}{2D\varpi^2} \\
& -4\pi^2\varpi^2 n^2\frac{\nabla\varphi_{\mathrm{g}}\cdot\nabla\Phi}{D(\nabla\Phi)^2} - 4\pi^2 n^2\frac{D+1}{D}L\frac{\mathrm{d}L}{\mathrm{d}\Phi} \\
& +2\pi^2 n^2\frac{\nabla(\varpi^2)\cdot\nabla\Phi}{D\varpi^2(\nabla\Phi)^2}L^2 + 4\pi^2\varpi^2 n^2\frac{D+1}{D}\frac{\mathrm{d}E}{\mathrm{d}\Phi} \\
& -4\pi^2\varpi^2 n^2\left[\frac{D+1}{D}\frac{T}{m_{\mathrm{p}}} + \frac{1}{Dm_{\mathrm{p}}n}\left(\frac{\partial P}{\partial s}\right)_n\right]\frac{\mathrm{d}s}{\mathrm{d}\Phi} = 0,
\end{aligned}
\tag{3.23}
$$

or, in a compact form (*cf.* [15]), as

$$
\begin{aligned}
-\varpi^2\nabla_k\left(\frac{1}{\varpi^2 n}\nabla^k\Phi\right) - 4\pi^2 nL\frac{\mathrm{d}L}{\mathrm{d}\Phi} \\
+ 4\pi^2\varpi^2 n\frac{\mathrm{d}E}{\mathrm{d}\Phi} - 4\pi^2\varpi^2 n\frac{T}{m_{\mathrm{p}}}\frac{\mathrm{d}s}{\mathrm{d}\Phi} = 0.
\end{aligned}
\tag{3.24}
$$

At first glance, the stream equation (3.23) is much more complicated than the planar one (2.38). Nevertheless, one can easily see that these equations have many similarities. As (2.38), the stream equation (3.23) starts with the linear elliptic term and the nonlinear term with an analogous form. The third term does not, of course, exist in (2.38) – it results from non-cartesian geometry. But all the other terms are not complification. They allow us to include into consideration not only gravity, but a much wider class of flows with different invariants onto different flow surfaces.

In other respects the stream equation is quite similar to the planar equation (2.38):

- The stream equation (3.23) is to be supplemented with the Bernoulli equation.
- Together with the Bernoulli equation, equation (3.23) contains the potential $\Phi(r,\theta)$ and the invariants E, L, and s only (*i.e.*, it has the Grad-Shafranov form).
- For n = const ($c_{\mathrm{s}}^2 \to \infty$), E = const, s = const, and $L = 0$ the equation becomes linear.
- It is nonlinear in the general case, but linear for second derivatives.
- The equation is elliptical for a subsonic flow $D > 0$.

- The equation is hyperbolic for a supersonic flow $D < 0$.
- For a given flow structure (*i.e.*, for a given Φ) and for the invariants $E(\Phi)$, $L(\Phi)$, and $s(\Phi)$ all the physical parameters are determined by algebraic relations.

The following point should be stressed. The expression for the denominator $D = -1 + c_{\rm s}^2/v_{\rm p}^2$ (3.21) involves the poloidal rather than the total velocity. It means that the sonic surface exists when the poloidal, not the total velocity becomes equal to that of sound. This fact results from our basic assumption: as we consider axisymmetric flows only, the disturbances (waves) are to have the same symmetry as well. Hence, the disturbances can propagate in the poloidal direction only. For this reason, a singularity appears at the moment when the particle velocity coincides with the velocity of disturbance.

3.2.2 Linear Operator $\varpi^2\nabla_{\rm k}(\varpi^{-2}\nabla^{\rm k})$

In what follows we shall use the linear operator

$$\hat{\mathcal{L}} = \varpi^2\nabla_k\left(\frac{1}{\varpi^2}\nabla^k\right) = \frac{\partial^2}{\partial r^2} + \frac{\sin\theta}{r^2}\frac{\partial}{\partial\theta}\left(\frac{1}{\sin\theta}\frac{\partial}{\partial\theta}\right) \tag{3.25}$$

and shall therefore consider it in more detail. First, examine the angular operator

$$\hat{\mathcal{L}}_\theta = \sin\theta\frac{\partial}{\partial\theta}\left(\frac{1}{\sin\theta}\frac{\partial}{\partial\theta}\right). \tag{3.26}$$

It has the following eigenfunctions

$$Q_0 = 1 - \cos\theta, \tag{3.27}$$
$$Q_1 = \sin^2\theta, \tag{3.28}$$
$$Q_2 = \sin^2\theta\cos\theta, \tag{3.29}$$
$$\ldots$$
$$Q_m = \frac{2^m m!(m-1)!}{(2m)!}\sin^2\theta P'_m(\cos\theta), \tag{3.30}$$

and eigennumber values

$$q_m = -m(m+1). \tag{3.31}$$

Here P_m are Legendre polynomials and prime means their derivative. As a result, for the full operator $\hat{\mathcal{L}}$ we have the following set of eigenfunctions

1. $m = 1$
   - $\Phi_1^{(1)} = r^2\sin^2\theta$ – homogeneous flow,

- $\Phi_1^{(2)} = \sin^2\theta/r$ – dipole flow.

2. $m = 2$

- $\Phi_2^{(1)} = r^3\sin^2\theta\cos\theta$ – zero point,
- $\Phi_2^{(2)} = \sin^2\theta\cos\theta/r^2$ – quadrupole flow.

3. ...

At first glance, this point is absolutely clear so that it is impossible to encounter here any trouble. Nevertheless, it is not so. Indeed, let us consider the eigenfunctions corresponding to $m = 0$. The first one is clear: it is the function

$$\Phi_0^{(1)} = 1 - \cos\theta \tag{3.32}$$

which describes the spherically symmetric accretion/ejection. By the way, this harmonics alone determines the accretion/ejection rate as for all the other eigenfunctions with $m > 0$ we have $\Phi_m(r, \pi) = 0$. The uncertainty is due to the second eigenfunction

$$\Phi_0^{(2)} = r(1 - \cos\theta) \tag{3.33}$$

for which $\Phi_0(r, \pi) \neq \text{const}$. It means that this harmonics can be realized only if in the volume (not in the origin or at infinity) there are the sources or the sinks of matter. In all other cases the second eigenfunction for $m = 0$ is to be dismissed.

3.3 Examples

3.3.1 Bondi-Hoyle accretion

As a first example we consider the accretion onto a moving gravity center (Bondi-Hoyle accretion [1]). This is a classical problem of modern astrophysics. The nature of Active Galactic Nuclei and Quasars, the nature of jets, the activity of some galactic X-ray sources are believed to be associated with the accretion of a gas onto compact objects – neutron stars and black holes [4, 16–18]. Nevertheless, only a few exact solutions describing the accretion flow (even for the simplest adiabatic case) are now known, *e.g.*, the solution for the spherically symmetric flow we have already discussed.

To construct the nonspherical solution, one may assume that a small perturbation of a spherically symmetric flow cannot change strongly the structure of the accretion [19]. So, it is possible to seek the solution of the stream equation as a perturbation of the spherically symmetric solution. First of all, let me recall the main results of the qualitative theory. It is more convenient to perform the calculations in the reference frame moving with the gravity center with a velocity v_∞. Comparing the Bondi accretion rate $2\Phi = 4\pi r_*^2 n_* c_* \sim (GM)^2 n_\infty / c_\infty^3$ (2.22)–(2.24) with the flow $\Phi \sim \pi R_c^2 n_\infty v_\infty$ captured within the capture radius R_c, one can evaluate the R_c value as

$$R_c \sim \varepsilon_1^{-1/2} r_*, \tag{3.34}$$

where

$$\varepsilon_1 = \frac{v_\infty}{c_\infty}. \tag{3.35}$$

Hence, for $\varepsilon_1 \ll 1$ the capture radius is much larger than the sonic one, and we can assume that for $r \ll R_c$ the flow structure is similar to the spherically symmetric accretion. Thus, one can seek the solution of the stream equation (3.23) in the form

$$\Phi(r,\theta) = \Phi_0[1 - \cos\theta + \varepsilon_1 f(r,\theta)]. \tag{3.36}$$

For a nonmoving gravity center we return to the spherically symmetric flow.

As the stream equation (3.23) now contains $i = 3$ invariants, so that $b = 2 + 3 - 1 = 4$, it is necessary to specify four boundary conditions, say

1. $v_\infty = \text{const}$,
2. $v_\varphi = 0$ (and hence $L = 0$),
3. $s_\infty = \text{const}$,
4. $E_\infty = c_\infty^2/(\Gamma + 1)$.

In the last relation we neglect the terms $\sim\varepsilon_1^2$.

As a result, the stream equation can be linearized:

$$-\varepsilon_1 D \frac{\partial^2 f}{\partial r^2} - \frac{\varepsilon_1}{r^2}(D+1)\sin\theta \frac{\partial}{\partial\theta}\left(\frac{1}{\sin\theta}\frac{\partial f}{\partial\theta}\right) + \varepsilon_1\left(\frac{2}{r} - \frac{GM}{c_s^2 r^2}\right)\frac{\partial f}{\partial r} = 0. \tag{3.37}$$

This equation has the following properties.

- It is linear.
- The angular operator coincides with $\hat{\mathcal{L}}_\theta$ (3.26).
- As all the terms contain a small ε_1 value, the functions D, c_s, etc. can be taken from the zero solution.
- As for the spherically symmetric flow the functions D, c_s, etc. do not depend on θ, the solution of equation (3.37) can be expanded in eigenfunctions of the operator $\hat{\mathcal{L}}_\theta$.

Thus, the solution can be presented in the form

$$f(r,\theta) = \sum_{m=0}^{\infty} g_m(r) Q_m(\theta), \tag{3.38}$$

the equations for the radial functions $g_m(r)$ being

$$r^2 D \frac{\mathrm{d}^2 g_m}{\mathrm{d}r^2} + \left(2r - \frac{GM}{c_\mathrm{s}^2}\right) \frac{\mathrm{d}g_m}{\mathrm{d}r} + m(m+1)(D+1) g_m = 0. \tag{3.39}$$

As to the boundary conditions, they can be formulated as follows:

1. No singularity on the sonic surface (where $r^2 N_r = 2r - GM/c_\mathrm{s}^2 = 0$, $D = 0$)

$$g_m(r_*) = 0. \tag{3.40}$$

2. A homogeneous flow $\Phi = \pi n_\infty v_\infty r^2 \sin^2\theta$ at infinity, which gives

$$g_1 \longrightarrow \frac{1}{2} \frac{n_\infty c_\infty}{n_* c_*} \frac{r^2}{r_*^2}, \qquad g_2, g_3, \dots = 0. \tag{3.41}$$

As a result, the complete solution can be presented in the form

$$\Phi(r,\theta) = \Phi_0[1 - \cos\theta + \varepsilon_1 g_1(r) \sin^2\theta)], \tag{3.42}$$

where the radial function $g_1(r)$ is the solution of the ordinary differential equation (3.39) for $m = 1$ with the boundary conditions (3.40) and (3.41).

At the present level of personal computers it means that we succeed in constructing the analytical solution of the problem in question. It allows us to obtain all the information concerning the flow structure. In particular, the sonic surface now has non-spherical form:

$$r_*(\theta) = r_* \left(1 + \varepsilon_1 \frac{\Gamma + 1}{5 - 3\Gamma} k_1 \cos\theta\right), \tag{3.43}$$

where the numerical coefficient $k_1 = r_* g_1'(r_*)$ can be obtained from the solution (for more details see [19]). As one can see from Figure 3, the analytical solution is in full agreement with the numerical calculations [20] although the parameter $\varepsilon_1 = 0.6$ is not too small. Finally, the condition

$$g_0 = 0 \tag{3.44}$$

(to obtain which an additional consideration is necessary) shows that the accretion rate is not changed in this approximation.

In connection with this solution, it is necessary to clarify one point. As one can easily see, outside the capture radius, our main assumption –

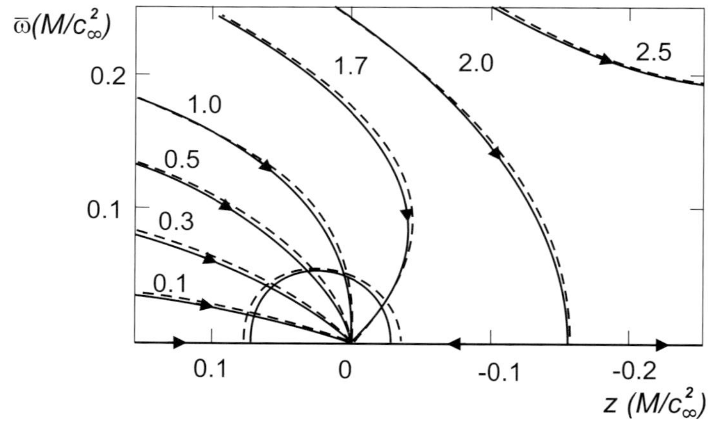

Fig. 3. Flow structure and shape of the sonic surface for $\Gamma = 4/3$, $\varepsilon_1 = 0.6$ [19]. Labels on the curves denote the values of Φ/Φ_0, and dashed curves indicate stream lines and the sonic surface obtained numerically in [20].

smallness of the disturbance of the spherically symmetric flow – is not valid. Nevertheless, the constructed solution is correct. This beautiful property is connected with the already mentioned fact that for an (approximately) constant concentration n the stream equation becomes linear. But as we know from the Bondi accretion (see (2.26)), far from the sonic surface $r \gg r_*$ the flow density is actually constant. The same is true for the homogeneous flow. As a result, for $R_c \gg r_*$, *i.e.*, for the case $\varepsilon_1 \ll 1$ under consideration, in the vicinity of and outside the capture radius (where the "disturbance" $\sim\varepsilon_1 g_1(r)$ becomes of the same order as the zero approximation ~ 1), the stream equation is linear. As a result, the sum of two solutions, homogeneous and spherically symmetric, remains a solution as well.

3.3.2 Ejection from a slowly rotating star

Another interesting nonrelativistic example is the transonic ejection from a slowly rotating star [21,22]. It is necessary to stress from the very beginning that this example is only illustrative because in reality an important role is played by the radiation pressure which cannot be included into consideration within our approach. Nevertheless, the analysis of this problem helped us to clarify some important features of the Grad-Shafranov approach [23].

As a zero approximation we consider the well-known Parker transonic outflow (2.30)–(2.32). It means that we assume all the parameters of the spherically symmetric outflow (the sonic radius r_*, the velocity of sound on the sonic surface c_*, the radial velocity $v_{\rm R}$ on the star surface $r = R$, etc.)

to be known. As before, we will seek the solution in the form

$$\Phi(r,\theta) = \Phi_0[1 - \cos\theta + \varepsilon_2^2 f(r,\theta)], \tag{3.45}$$

where the small parameter is now

$$\varepsilon_2^2 = \frac{\Omega^2 R^3}{GM}. \tag{3.46}$$

Here Ω is the angular velocity of a star.

The problem under consideration needs all the $i = 3$ invariants. Hence, $b = 2 + 3 - 1 = 4$, and it is necessary to specify four boundary conditions on the star surface $r = r_{\rm R}(\theta)$ which now differs from the sphere

$$r_{\rm R}(\theta) = R[1 + \varepsilon_2^2 \rho(\theta)]. \tag{3.47}$$

Here we introduce the dimensionless parameter $\rho(\theta) \approx 1$.

Let me stress that at first glance there is a disagreement as we add one degree of freedom (the toroidal rotation with $v_\varphi \neq 0$) while it needs two extra functions in comparison with the spherically symmetric outflow. This question will be clarify below.

It is important that for small ε_2 four boundary conditions can be determined through real physical parameters on the star surface, *e.g.*, through two thermodynamic functions (say, T and n) and two components of the velocity (say, v_r and v_φ). As a result, one can express these boundary conditions through four dimensionless functions $\tau(\theta)$, $\eta(\theta)$, $\omega(\theta)$, and $h(\theta)$:

$$T(r_{\rm R},\theta) = T_{\rm R}[1 + \varepsilon_2^2 \tau(\theta)], \tag{3.48}$$

$$n(r_{\rm R},\theta) = n_{\rm R}[1 + \varepsilon_2^2 \eta(\theta)], \tag{3.49}$$

$$v_\varphi(r_{\rm R},\theta) = \varepsilon_2 \left(\frac{GM}{R}\right)^{1/2} \omega(\theta)\sin\theta, \tag{3.50}$$

$$v_r(r_{\rm R},\theta) = v_{\rm R}[1 + \varepsilon_2^2 h(\theta)]. \tag{3.51}$$

Here $\omega(\theta)$, determined as

$$\Omega(r_{\rm R},\theta) = \Omega\omega(\theta), \tag{3.52}$$

describes the differential rotation of the star surface.

Using now the thermodynamic relation

$$\mathrm{d}s = \frac{1}{\Gamma - 1}\frac{\mathrm{d}T}{T} - \frac{\mathrm{d}n}{n}, \tag{3.53}$$

one can obtain for the first invariant $s(\theta)$

$$\delta s(\theta) = \varepsilon_2^2 \left[\frac{1}{\Gamma - 1}\tau(\theta) - \eta(\theta)\right]. \tag{3.54}$$

Accordingly, two other invariants can be determined through the boundary conditions as well

$$\delta E(\theta) = \varepsilon_2^2 v_{\rm R}^2 h(\theta) + \varepsilon_2^2 \frac{GM}{2R}\omega^2(\theta)\sin^2\theta + \frac{\varepsilon_2^2\Gamma}{\Gamma-1}\frac{T}{m_{\rm p}}\tau(\theta) + \delta\varphi_{\rm g}, \quad (3.55)$$

$$L^2(\theta) = \varepsilon_2^2 R^2 \frac{GM}{R}\omega^2(\theta)\sin^2\theta. \quad (3.56)$$

Here we use the following expression for the disturbance of the gravitational potential on the star surface

$$\delta\varphi_{\rm g}(r_{\rm R},\theta) = \varepsilon_2^2 \frac{GM}{R}\rho(\theta). \quad (3.57)$$

It is of great importance that the possibility of taking the next step, *i.e.*, writing down the Grad-Shafranov equation, is connected with the simple geometry of the zero approximation. Since in the zero approximation the stream function is $\Phi = 1 - \cos\theta$, *i.e.*, it is the function of θ only (and as all the derivatives $\mathrm{d}E/\mathrm{d}\Phi$, $\mathrm{d}L/\mathrm{d}\Phi$, and $\mathrm{d}s/\mathrm{d}\Phi$ as well as L itself vanish for a nonrotating outflow), one can use the relation $\mathrm{d}\Phi = \Phi_0 \sin\theta \mathrm{d}\theta$. It allows us to determine the derivatives $\mathrm{d}E/\mathrm{d}\Phi$, $\mathrm{d}L/\mathrm{d}\Phi$, and $\mathrm{d}s/\mathrm{d}\Phi$ in the whole space and not only on the star surface.

As a result, the stream equation can be written as

$$\begin{aligned} -\varepsilon_2^2\Phi_0 D\frac{\partial^2 f}{\partial r^2} - \frac{\varepsilon_2^2}{r^2}\Phi_0(D+1)\sin\theta\frac{\partial}{\partial\theta}\left(\frac{1}{\sin\theta}\frac{\partial f}{\partial\theta}\right) + \varepsilon_2^2\Phi_0 N_r\frac{\partial f}{\partial r} = \\ -4\pi^2 n^2 r^2\sin^2\theta(D+1)\frac{\mathrm{d}E}{\mathrm{d}\Phi} + 4\pi^2 n^2(D+1)L\frac{\mathrm{d}L}{\mathrm{d}\Phi} \quad (3.58) \\ -4\pi^2 n^2\frac{\cos\theta}{\Phi_0\sin^2\theta}L^2 + 4\pi^2 n^2 r^2\sin^2\theta\left[(D+1)\frac{T}{m_{\rm p}} + \frac{\Gamma-1}{\Gamma}c_{\rm s}^2\right]\frac{\mathrm{d}s}{\mathrm{d}\Phi}. \end{aligned}$$

Here $N_r = 2/r - 4\pi^2 n^2 r^2 GM/\Phi_0^2$.

The properties of this equation are the same as before.

- It is linear.
- The angular operator coincides with $\hat{\mathcal{L}}_\theta$ (3.26).
- As all the terms contain a small ε_2^2 value, the functions D, $c_{\rm s}$, n, etc. can be taken from the zero approximation.
- As for the spherically symmetric flow the functions D, $c_{\rm s}$, n, etc. do not depend on θ, the solution of equation (3.58) can be expanded in eigenfunctions of the operator $\hat{\mathcal{L}}_\theta$.

Hence, we can again seek the solution in the form

$$f(r,\theta) = \sum_{m=0}^{\infty} g_m(r) Q_m(\theta). \tag{3.59}$$

Introducing the dimensionless variables

$$x = \frac{r}{r_*}, \quad u = \frac{n}{n_*}, \quad a = \frac{c_{\rm s}^2}{c_*^2}, \tag{3.60}$$

one can write down the following ordinary differential equations describing the radial functions $g_m(r)$

$$\begin{aligned}
&(1 - x^4 a u^2)\frac{\mathrm{d}^2 g_m}{\mathrm{d}x^2} + 2\left(\frac{1}{x} - x^2 u^2\right)\frac{\mathrm{d}g_m}{\mathrm{d}x} + m(m+1)x^2 a u^2 g_m = \\
&\kappa_m \frac{R^2}{r_*^2} x^4 a u^4 - \lambda_m \frac{R^2}{r_*^2} u^2 - \sigma_m x^6 a u^4 \\
&+\frac{1}{\Gamma}\nu_m x^6 a^2 u^4 + \frac{\Gamma - 1}{\Gamma}\nu_m x^2 a u^2,
\end{aligned} \tag{3.61}$$

where κ_m, λ_m, σ_m, and ν_m values are defined as the expansion coefficients

$$\sin\theta \frac{\mathrm{d}E}{\mathrm{d}\theta} = \varepsilon_2^2 c_*^2 \sum_{m=0}^{\infty} \sigma_m Q_m(\theta), \tag{3.62}$$

$$\frac{\cos\theta}{\sin^2\theta} L^2 = \varepsilon_2^2 c_*^2 r_*^2 \sum_{m=0}^{\infty} \lambda_m Q_m(\theta), \tag{3.63}$$

$$\frac{L}{\sin\theta}\frac{\mathrm{d}L}{\mathrm{d}\theta} = \varepsilon_2^2 c_*^2 r_*^2 \sum_{m=0}^{\infty} \kappa_m Q_m(\theta), \tag{3.64}$$

$$\sin\theta \frac{\mathrm{d}s}{\mathrm{d}\theta} = \varepsilon_2^2 \sum_{m=0}^{\infty} \nu_m Q_m(\theta). \tag{3.65}$$

Finally, the functions $a(x)$ and $u(x)$ corresponding to the spherically symmetric flow for the polytropic equation of state (2.6) are related as $a = u^{\Gamma-1}$ and the function $u(x)$ due to (3.20)–(3.22) can be found from the ordinary differential equation

$$\frac{\mathrm{d}u}{\mathrm{d}x} = -2\frac{u}{x} \cdot \frac{1 - x^3 u^2}{1 - x^4 a u^2} \tag{3.66}$$

with the boundary conditions (*cf.* (2.25))

$$u(1) = 1, \qquad \left(\frac{\mathrm{d}u}{\mathrm{d}x}\right)_{x=1} = -\frac{4 + \sqrt{10 - 6\Gamma}}{\Gamma + 1}. \tag{3.67}$$

As to the boundary condition to the set of equations (3.61), they are quite similar to the Bondi-Hoyle accretion (for more details see [23]):

- Condition on the star surface. As

$$\mathrm{d}\Phi = 2\pi r^2 n v_r \sin\theta \mathrm{d}\theta = 2\pi R^2 n_{\mathrm{R}} v_{\mathrm{R}}[1+\varepsilon_2^2(\eta+h+2\rho)]\sin\theta\mathrm{d}\theta, \quad (3.68)$$

  we have

$$g_m(R/r_*) = \frac{(2m)!}{2^m(m+1)!m!}(\eta_m + h_m + 2\rho_m). \quad (3.69)$$

  Here η_m, h_m, and ρ_m are expansion coefficients in Legendre polynomials, *e.g.*, $\eta(\theta) = \sum_m \eta_m P_m(\cos\theta)$.

- The absence of singularity on the sonic surface $N_\theta = 0$. This condition gives

$$\varepsilon_2^2 g_m(1) = \frac{(2m)!}{2^m(m+1)!m!}\left[\frac{(\delta E)_m}{c_*^2} - (\delta s)_m - \frac{(\delta L/\sin^2\theta)_m}{2c_*^2 r_*^2}\right], \quad (3.70)$$

  where again $(\dots)_m$ means the expansion in Legendre polynomials which can be found from expressions (3.54)–(3.56).

As a result, equations (3.61) taken together with the boundary conditions (3.69) and (3.70) allow solution of the direct problem, *i.e.*, determination of the flow structure from the physical boundary condition on the star surface.

Here it is necessary to stress two important points:

1. The reason we succeeded in writing the regularity condition $N_\theta = 0$ on the sonic surface (and hence to solve the direct problem) is again a very simple geometry of the zero approximation. In particular, in the problem in question the position of the sonic surface can be taken from the spherically symmetric solution. In the general case it is not so and it is impossible to write down the condition $N_\theta = 0$ using the known functions $(\delta E)_m$, $(\delta s)_m$, etc. on the star surface. The position of the sonic surface is unknown and is to be found from the solution.

2. We can now understand the nature of the "extra" boundary condition. The point is that for $m = 0$ one can take, as was already stressed, one fundamental solution only, namely, $g_0 = \text{const}$. Another fundamental solution is unphysical. Hence, there is an additional relation

$$g_0(R) = g_0(r_*). \quad (3.71)$$

This relation determines h_0 which, as we see, is not a free parameter. In other words, we are not fully free in the boundary condition $h(\theta)$: its zero harmonics is to be found from relation (3.71). But as was demonstrated earlier, it is g_0 that determines the ejection rate. Hence, the ejection rate is a function of three parameters only, namely, zero harmonics of two thermodinamic functions η_0 and τ_0, and v_φ. For a spherically symmetric flow $v_\varphi = 0$, and we return to two functions which determine the ejection rate. As to higher harmonics with $m > 0$, they are free, and to determine them it is necessary to know four functions on the star surface. Thus, the spherically symmetric case is degenerate and it is necessary to be very careful when extending its properties on the 2D flows.

At the end of this section let me formulate some results which can be obtained under the following simplified assumptions:

- Almost the whole of the star mass is in its center, *i.e.*, $\varphi_{\rm g} = -GM/r$.
- No differential rotation, *i.e.*, $\omega(\theta) = 1$.
- Von Zeipel law: $T(R, \theta) \propto g_{\rm eff}^{1/4}$, where $\varphi_{\rm eff} = \varphi_{\rm g} + L^2/r^2$.
- No meridional convection, *i.e.*, $v_\theta(r_{\rm R}, \theta) = 0$ (it means that we specify here $v_\theta(r_{\rm R}, \theta)$ instead of $v_r(r_{\rm R}, \theta)$, *i.e.*, the coefficients h_0, h_1, etc. are to be found from the solution).

Exercise

1. *Find that the disturbances of the star radius $\rho(\theta)$ in (3.47) and the temperature $\tau(\theta)$ in (3.48) have the form*

$$\rho(\theta) = \frac{1}{2}\sin^2\theta, \qquad \tau(\theta) = -\frac{1}{2}\sin^2\theta. \tag{3.72}$$

2. *Show that the only nonzero terms in the expansion (3.59) correspond to $m = 0$ and $m = 2$, the expansion coefficients in (3.62)–(3.65) being*

$$\sigma_2 = 2\frac{r_*}{R} - \frac{5-3\Gamma}{2(\Gamma-1)} + \frac{1}{2}\frac{v_{\rm R}^2}{c_*^2} - 3\frac{v_{\rm R}^2}{c_*^2}h_2, \tag{3.73}$$

$$\lambda_2 = 2\frac{R}{r_*}, \quad \kappa_2 = 4\frac{R}{r_*}, \quad \nu_2 = -\frac{\Gamma}{\Gamma-1}, \tag{3.74}$$

and $\sigma_0, ..., \nu_0 = 0$. Remember that $h(\theta) = h_0 + h_1\cos\theta + h_2 P_2(\cos\theta) + \dots$.

Table 1.

model	$1+h_0$	h_2	q_2	b_0	b_2
$r_*/R = 1.1$, $\Gamma = 4/3$	2.9	−0.8	−0.40	2.2	−0.41
$r_*/R = 2.0$, $\Gamma = 4/3$	3.2	−3.3	−0.71	2.5	−0.47
$r_*/R = 10$, $\Gamma = 4/3$	8.0	−56.0	−2.18	5.7	−0.92
$r_*/R = 1.1$, $\Gamma = 1.1$	1.7	−0.8	−0.15	1.8	−0.37
$r_*/R = 2.0$, $\Gamma = 1.1$	2.1	−2.3	−0.18	2.2	−0.40
$r_*/R = 10$, $\Gamma = 1.1$	7.4	−26.0	−0.40	7.2	−0.58

As a result, solving the stream equation (3.61) for $m = 2$, one can find that

1. The ejection rate can be presented as

$$\Phi_{\rm tot} = 2\Phi_0 \left[1 + \frac{\Omega^2 R^3}{GM}(1+h_0)\right]. \tag{3.75}$$

Here h_0 can be obtained from relation (3.71) (see Table 1)

$$h_0 = -\frac{1}{6} + \frac{2}{3}\frac{r_*/R - R/r_*}{1 - v_{\rm R}^2/c_*^2}. \tag{3.76}$$

As we see, the rotation increases the ejection rate.

2. Far from the sonic surface $r \gg r_*$ the stream function has the form

$$\lim_{r\to\infty}\frac{\Phi(r,\theta)}{\Phi_0} = (1-\cos\theta) + \frac{\Omega^2 R^3}{GM}(1+h_0)(1-\cos\theta) + \frac{\Omega^2 R^3}{GM} q_2 \sin^2\theta\cos\theta, \tag{3.77}$$

where the coefficient q_2 is tabulated in Table 1 as well. As there is no r-dependence, the outflow becomes purely radial at large distances.

3. Accordingly, the asymptotic expression for the concentration n has the form

$$\lim_{r\to\infty}\frac{n(r,\theta)}{n_*} = \frac{c_*}{v_\infty}\frac{r_*^2}{r^2}\left[1 + \frac{\Omega^2 R^3}{GM}b_0 + \frac{1}{2}\frac{\Omega^2 R^3}{GM}b_2(3\cos^2\theta - 1)\right], \tag{3.78}$$

where $v_\infty^2 = v_{\rm R}^2 - 2GM/R$. As one can see from Table 1, $b_2 < 0$. It means that the rotation results in the appearance of a dense disk in the equatorial plane; this result is well-known [22], but previously it was obtained by numerical calculations only. Negative q_2 values in (3.77) demonstrate that for $\varepsilon_2 \geq 1$ the most mass outflow is concentrated in the vicinity of the equatorial plane as well.

4 Axisymmetric stationary flow – general relativity

4.1 Basic equations

We shall show how the Grad-Shafranov approach can be applied to the axisymmetric stationary flows in the vicinity of a rotating black hole. Remember that the main difficulty of the General Relativity is the necessity to work with four-dimensional objects. As a result, we cannot use our three-dimensional intuition in considering the relativistic processes.

But there is a convenient language – 3 + 1-split – which allows work with three-dimensional vectors even in General Relativity [24]. One can find the detailed introduction into this approach in the book "Black Holes. The Membrane Paradigm" by K. Thorne *et al.* [26]. The main idea is that for stationary metric the proper time τ and "the time at infinity" t are in one-to-one correspondence. It allows the time t to be separated from spatial coordinates x^i $(i = 1, 2, 3)$. As a result, all the equations can be written in the simple 3D form, their physical meaning remaining clear. Below I shall give the main relations of this approach (see [25] as well).

4.1.1 Kerr metric

The Kerr metric is the metric of a rotating black hole. In the Boyer-Lindquist coordinates t, r, θ, and φ it has the form

$$\mathrm{d}s^2 = -\alpha^2\mathrm{d}t^2 + g_{ik}(\mathrm{d}x^i + \beta^i\mathrm{d}t)(\mathrm{d}x^k + \beta^k\mathrm{d}t), \tag{4.1}$$

where

$$\alpha = \frac{\rho}{\Sigma}\sqrt{\Delta} \tag{4.2}$$

(α is the lapse function or the red shift), and

$$\beta^r = \beta^\theta = 0, \qquad \beta^\varphi = -\omega = -\frac{2aMr}{\Sigma^2} \tag{4.3}$$

(ω is the Lense–Thirring angular velocity; remember that β^φ is the contravariant component). Finally, it is important that the 3D metric g_{ik} in (4.1) be diagonal

$$g_{rr} = \frac{\rho^2}{\Delta}, \qquad g_{\theta\theta} = \rho^2, \qquad g_{\varphi\varphi} = \varpi^2. \tag{4.4}$$

Here M and a are respectively the black hole mass and the angular momentum per unit mass ($a = J/M$), and we introduce the standard notation

$$\Delta = r^2 + a^2 - 2Mr, \qquad \rho^2 = r^2 + a^2\cos^2\theta,$$
$$\Sigma^2 = (r^2 + a^2)^2 - a^2\Delta\sin^2\theta, \qquad \varpi = \frac{\Sigma}{\rho}\sin\theta. \tag{4.5}$$

Units where $c = G = 1$ are used below.

This metric has the following properties.

- The Kerr metric is axisymmetric and stationary. It is precisely what is needed for using the Grad-Shafranov approach.
- The Kerr metric is two-parametric, *i.e.*, it depends on two parameters: the mass M and the rotation parameter a.
- It transforms to the Schwarzschild metric for a nonrotating black hole: $g_{rr} = \alpha^{-2}$, $g_{\theta\theta} = r^2$, $g_{\varphi\varphi} = r^2\sin^2\theta$. Here $\alpha^2 = 1 - 2M/r$.
- In the limit $r \gg 2M$ Boyer-Lindquist coordinates coincide with the spherical ones: $g_{rr} = 1$, $g_{\theta\theta} = r^2$, $g_{\varphi\varphi} = r^2\sin^2\theta$.
- Lapse function α
  - The lapse function α describes the lapse (the red shift) between the proper time τ and the time at infinity t: $\mathrm{d}\tau = \alpha \mathrm{d}t$.
  - The condition $\alpha = 0$ determines the position of the horizon

$$r_{\mathrm{g}} = M + \sqrt{M^2 - a^2}. \tag{4.6}$$

  - Boyer-Lindquist coordinates do not describe the space-time inside the horizon; for $r = r_{\mathrm{g}}$ the metric has a coordinate singularity.
- Lense-Thirring angular velocity ω
  - The Lense-Thirring angular velocity ω corresponds to the proper motion of the space around a black hole.
  - By definition, $\Omega_{\mathrm{H}} = \omega(r_{\mathrm{g}})$ is the black hole angular velocity (does not depend on θ).
  - $\omega \propto a$, $\Omega_{\mathrm{H}} r_{\mathrm{g}} = a/(2M)$.
- Convenient reference frame – ZAMO
  - ZAMO (Zero Angular Momentum Observers) [26] are located at a constant radius $r = \mathrm{const}$, $\theta = \mathrm{const}$, but they rotate with the Lense–Thirring angular velocity $\mathrm{d}\varphi/\mathrm{d}t = \omega$.

- For ZAMO the four-dimensional metric $g_{\alpha\beta}$ is diagonal, the spatial 3D metric g_{ik} coinciding with (4.4).
- No gyroscope rotation in the local experiment.

To clarify the physical meaning of the values α and ω, let us consider the motion of particles in the gravitational field of a rotating black hole. Then, one can rewrite the four-dimensional equation of motion

$$\frac{\mathrm{d}^2 x^\alpha}{\mathrm{d}s^2} + \Gamma^\alpha_{\beta\gamma} \frac{\mathrm{d}x^\beta}{\mathrm{d}s} \frac{\mathrm{d}x^\gamma}{\mathrm{d}s} = 0 \tag{4.7}$$

in the simple 3D form

$$\frac{\mathrm{d}p_i}{\mathrm{d}\tau} = \frac{m_\mathrm{p}}{\sqrt{1-v^2}} g_i + H_{ik} \frac{m_\mathrm{p} v^k}{\sqrt{1-v^2}}, \tag{4.8}$$

where

$$\vec{g} = -\frac{1}{\alpha}\nabla\alpha, \tag{4.9}$$

$$H_{ik} = \frac{1}{\alpha}\nabla_i \beta_k. \tag{4.10}$$

Remember that

- The Greek indices α, β, and γ are four-dimensional, while the Latin i, j, and k are three-dimensional.
- τ is the proper time, and all three-dimensional vectors are measured by ZAMO.
- ∇_i means the covariant derivative in the three-dimensional metric (4.4).

In a weak gravitational field, *i.e.*, far from a black hole there is very nice analogy between the gravitational and electrodynamic equations. Indeed, the equation of motion (4.8) can be rewritten as

$$m_\mathrm{p} \frac{\mathrm{d}^2 \vec{r}}{\mathrm{d}\tau^2} = m_\mathrm{p} \left(\vec{g} + \frac{\mathrm{d}\vec{r}}{\mathrm{d}\tau} \times \vec{H} \right), \tag{4.11}$$

where

$$\vec{g} = -\nabla\alpha, \qquad \vec{H} = \nabla \times \vec{\beta}, \tag{4.12}$$

α and $\vec{\beta}$ playing the role of the scalar and vector potentials, respectively. Moreover, the Einstein equations in a weak gravitational field are quite similar to the Maxwell equations

$$\nabla \cdot \vec{g} = -4\pi\rho_{\rm m}, \tag{4.13}$$

$$\nabla \times \vec{g} = 0, \tag{4.14}$$

$$\nabla \cdot \vec{H} = 0, \tag{4.15}$$

$$\nabla \times \vec{H} = -16\pi\rho_{\rm m}\vec{v}. \tag{4.16}$$

In other words, the gravitational field $\vec{g}$ is analogous to the electric field while the new (so-called gravitomagnetic) field $\vec{H}$ – to the magnetic one which is proportional to the angular velocity of a black hole. The sources of the gravitoelectric field $\vec{g}$ are masses, and the sources of the gravitomagnetic field $\vec{H}$ are mass currents.

For example, for a rotating sphere with mass M and angular momentum $\vec{J}$ the fields outside the sphere are [26]

$$\vec{g} = -\frac{M}{r^2}\vec{e}_{\hat{r}}, \tag{4.17}$$

$$\vec{H} = 2\frac{\vec{J} - 3\vec{e}_{\hat{r}}(\vec{J}\vec{e}_{\hat{r}})}{r^3}, \tag{4.18}$$

i.e., the rotation induces a dipole gravitomagnetic field around the rotating body. The appearance of an additional gravitomagnetic force is the most important consequence of the black hole rotation.

To summarize, the 3 + 1 language allows the description of physical processes in a clear 3D form. If we simultaneously use ZAMO as a reference frame, no expressions will contain extra terms. In a sense, ZAMO is an inertial frame, at any case in the φ direction. As a result, within the 3 + 1-split

- All three-dimensional vectors are to be determined from the local experiment by ZAMO.

- All the calculations are to be made in the 3D diagonal metric (4.4), *e.g.*,

$$\nabla \cdot \vec{A} = \frac{1}{\sqrt{g}}\frac{\partial}{\partial x^i}\left(\sqrt{g}A^i\right), \tag{4.19}$$

$$\nabla \times \vec{A} = \frac{1}{\sqrt{g}}\begin{pmatrix} \sqrt{g_{rr}}\vec{e}_{\hat{r}} & \sqrt{g_{\theta\theta}}\vec{e}_{\hat{\theta}} & \sqrt{g_{\varphi\varphi}}\vec{e}_{\hat{\varphi}} \\ \partial/\partial r & \partial/\partial\theta & \partial/\partial\varphi \\ \sqrt{g_{rr}}A_{\hat{r}} & \sqrt{g_{\theta\theta}}A_{\hat{\theta}} & \sqrt{g_{\varphi\varphi}}A_{\hat{\varphi}} \end{pmatrix}. \tag{4.20}$$

- All the vector relations remain the same as in the flat space, *e.g.*, $\nabla \times (\nabla a) = 0$, $\nabla \cdot (\nabla \times \vec{A}) = 0$.

4.1.2 Thermodynamics

On the other hand, all thermodynamic functions within the $3+1$ approach are determined in the comoving reference frame, which is the only invariant one. For this reason, one need no think about the transformation connected with different reference frames. Actually, there is the only complification: in the relativistic case one should work with the relativistic enthalpy μ including the particle rest mass

$$\mu = \frac{\rho_{\rm m} + P}{n} \approx m_{\rm p}c^2 + m_{\rm p}w + \dots \tag{4.21}$$

Here $\rho_{\rm m}$ is the internal energy density. For the polytropic equation of state $P = k(s)n^{\Gamma}$ (2.6) we have ($c = 1$)

$$\mu = m_{\rm p} + \frac{\Gamma}{\Gamma - 1}k(s)n^{\Gamma-1}, \tag{4.22}$$

$$c_{\rm s}^2 = \frac{1}{\mu}\left(\frac{\partial P}{\partial n}\right)_{\rm s} = \frac{\Gamma}{\mu}k(s)n^{\Gamma-1}. \tag{4.23}$$

Finally, the relativistic energy-momentum tensor has a symmetrical form

$$T^{\alpha\beta} = \begin{pmatrix} \varepsilon & \vec{S} \\ \vec{S} & T^{ik} \end{pmatrix} = \begin{pmatrix} (\rho_{\rm m} + Pv^2)\gamma^2 & (\rho_{\rm m} + P)\gamma\vec{u} \\ (\rho_{\rm m} + P)\gamma\vec{u} & (\rho_{\rm m} + P)u^i u^k + Pg^{ik} \end{pmatrix}. \tag{4.24}$$

Remember that γ is the Lorentz-factor of a flow measured by ZAMO.

Now, using the relativistic version of the energy-momentum conservation law $\nabla_\alpha T^{\alpha\beta} = 0$, one can obtain [24]

$$-\frac{1}{\alpha}(\vec{\beta}\nabla)\varepsilon = -\frac{1}{\alpha^2}\nabla \cdot (\alpha^2\vec{S}) + H_{ik}T^{ik}, \tag{4.25}$$

$$\nabla_k T_i^k + \frac{1}{\alpha}S_\varphi\frac{\partial\omega}{\partial x^i} + (\varepsilon\delta_i^k + T_i^k)\frac{1}{\alpha}\frac{\partial\alpha}{\partial x^k} = 0. \tag{4.26}$$

Here the additional terms in the energy (4.25) and momentum (4.26) equations are due to the gravitomagnetic forces.

4.1.3 Stream function, etc.

As in the flat space, one can introduce the stream function $\Phi(r,\theta)$ through the poloidal component of the four-velocity of the flow $\vec{u}_{\rm p}$

$$\alpha n\vec{u}_{\rm p} = \frac{\nabla\Phi \times \vec{e}_{\hat{\varphi}}}{2\pi\varpi}. \tag{4.27}$$

It means the following relations for the physical components

$$\alpha n u_{\hat{r}} = \frac{1}{2\pi\varpi}(\nabla\Phi)_{\hat{\theta}}, \tag{4.28}$$

$$\alpha n u_{\hat{\theta}} = -\frac{1}{2\pi\varpi}(\nabla\Phi)_{\hat{r}}. \tag{4.29}$$

The definition (4.27) gives the continuity equation in the form

$$\nabla \cdot (\alpha n \vec{u}) = 0. \tag{4.30}$$

Let me clarify the extra factor α in (4.30). The point is that the 3D continuity equation (4.30) results from the 4D one

$$\nabla_\beta N^\beta = \frac{1}{\sqrt{g_{tt}g}}\frac{\partial}{\partial x^\beta}(\sqrt{g_{tt}g}N^\beta), \tag{4.31}$$

where g_{tt} is equal to α^2.

As a result, using the definitions (4.24) and (4.27), one can rewrite the energy equation (4.25) and the φ component of the momentum equation (4.26) as

$$\vec{u} \cdot \nabla(\alpha\mu\gamma) + \mu u_\varphi \vec{u} \cdot \nabla\omega = 0, \tag{4.32}$$

$$\vec{u} \cdot \nabla(\mu u_\varphi) = 0. \tag{4.33}$$

Hence, two integrals of motion can now be present as

$$E(\Phi) = \alpha\mu\gamma + \mu\omega\varpi u_{\hat{\varphi}}, \tag{4.34}$$

$$L(\Phi) = \mu\varpi u_{\hat{\varphi}}. \tag{4.35}$$

Exercise

Obtain expressions (4.32)–(4.35).

Expressions (4.34) and (4.35) are the extension of the nonrelativistic relations (3.2) and (3.6) to the case of a rotating black hole. Indeed, for $\omega = 0$ we have for Bernoulli integral

$$E = \gamma\mu\alpha \approx \left(1 + \frac{1}{2}\frac{v^2}{c^2}\right)(m_\mathrm{p}c^2 + m_\mathrm{p}w)\left(1 - \frac{GM}{c^2 r}\right)$$
$$\approx m_\mathrm{p}c^2 + m_\mathrm{p}\left(\frac{v^2}{2} + w + \varphi_\mathrm{g}\right) + \dots \tag{4.36}$$

As to the third invariant, we again have

$$s = s(\Phi). \tag{4.37}$$

4.1.4 Grad-Shafranov equation

Using now three invariants E, L, and s, one can write the poloidal component of the relativistic Euler equation [30]

$$
\begin{aligned}
& n u^b \nabla_b(\mu u_a) + \nabla_a P - \mu n (u_{\hat{\varphi}})^2 \frac{1}{\varpi} \nabla_a \varpi \\
& + \frac{1}{\alpha} \mu n \gamma (\varpi u_{\hat{\varphi}}) \nabla_a \omega + \frac{1}{\alpha} \mu n \gamma^2 \nabla_a \alpha = 0
\end{aligned}
\tag{4.38}
$$

(here the indices a and b are r and θ only) as $[\mathrm{Euler}]_{\mathrm{p}} = [\mathrm{GS}]\nabla\Phi$, where the stream equation $[\mathrm{GS}] = 0$ now looks like

$$
\begin{aligned}
& -\mathcal{M}^2 \left[\alpha \varpi^2 \nabla_k \left(\frac{1}{\alpha \varpi^2} \nabla^k \Phi \right) + \frac{\nabla^a \cdot \Phi \nabla^b \cdot \Phi \nabla_a \nabla_b \Phi}{(\nabla\Phi)^2 D} \right] \\
& + \frac{\mathcal{M}^2 \nabla'_k F \cdot \nabla^k \Phi}{2 (\nabla\Phi)^2 D} \\
& + \frac{32\pi^4}{\mathcal{M}^2} \frac{\partial}{\partial \Phi} \left[\varpi^2 (E - \omega L)^2 - \alpha^2 L^2 \right] - 16\pi^3 \alpha^2 \varpi^2 n T \frac{\mathrm{d}s}{\mathrm{d}\Phi} = 0.
\end{aligned}
\tag{4.39}
$$

Here

$$
F = \frac{64\pi^4}{\mathcal{M}^4} \left[\varpi^2 (E - \omega L)^2 - \alpha^2 L^2 - \varpi^2 \alpha^2 \mu^2 \right],
\tag{4.40}
$$

and we introduce the thermodynamic function

$$
\mathcal{M}^2 = \frac{4\pi\mu}{n}.
\tag{4.41}
$$

Next, the operator ∇'_k acts on all the variables except $\mathcal{M}^2$, and $\partial/\partial\Phi$ on the invariants $E(\Phi)$, $L(\Phi)$, and $s(\Phi)$ only. Finally, now the denominator D is

$$
D = -1 + \frac{1}{u_{\mathrm{p}}^2} \frac{c_{\mathrm{s}}^2}{1 - c_{\mathrm{s}}^2}.
\tag{4.42}
$$

Here the physical component of the poloidal four-velocity u_{p} can be found from (4.27). In a compact form the relativistic stream equation looks like

$$
\begin{aligned}
& -\alpha \varpi^2 \nabla_k \left(\frac{\mathcal{M}^2}{\alpha \varpi^2} \nabla^k \Phi \right) + \frac{32\pi^4}{\mathcal{M}^2} \frac{\partial}{\partial \Phi} \left[\varpi^2 (E - \omega L)^2 - \alpha^2 L^2 \right] \\
& - 16\pi^3 \alpha^2 \varpi^2 n T \frac{\mathrm{d}s}{\mathrm{d}\Phi} = 0.
\end{aligned}
\tag{4.43}
$$

The hydrodynamical version of the Grad-Shafranov equation in the Kerr metric was first formulated in [27] using four-dimensional notation and in [28] within the 3 + 1-split language.

As before, the Grad-Shafranov equation is to be supplemented with the Bernoulli equation ($\gamma^2 = 1 + u_{\hat{\varphi}}^2 + u_{\rm p}^2$) which can now be written as

$$(E - \omega L)^2 = \alpha^2 \mu^2 + \frac{\alpha^2}{\varpi^2} L^2 + \frac{\mathcal{M}^4}{64\pi^4 \varpi^2} (\nabla\Phi)^2. \tag{4.44}$$

Remember that in (4.39) and (4.44) the relativistic enthalpy μ is to be considered as a function of $\mathcal{M}^2$ and s: $\mu = \mu(\mathcal{M}^2, s)$. In the general case the appropriate differential connection is [28]

$$\mathrm{d}\mu = -\frac{c_{\rm s}^2}{1 - c_{\rm s}^2} \mu \frac{\mathrm{d}\mathcal{M}^2}{\mathcal{M}^2} + \frac{1}{1 - c_{\rm s}^2} \left[\frac{1}{n} \left(\frac{\partial P}{\partial s} \right)_n + T \right] \mathrm{d}s. \tag{4.45}$$

In particular, as in the nonrelativistic case, the Bernoulli equation (4.44) determines $\mathcal{M}^2$ in the implicit form through the flux Φ and three integrals of motion: $\mathcal{M}^2 = \mathcal{M}^2(\nabla\Phi; E, L, s)$.

Finally, it is convenient to use another form of the poloidal four-velocity $u_{\rm p}$:

$$u_{\rm p}^2 = \frac{E^2 - \alpha^2 L^2/\varpi^2 - \alpha^2 \mu^2}{\alpha^2 \mu^2}. \tag{4.46}$$

We see that $u_{\rm p} \to \infty$ as α^{-1} at the horizon. As was already stressed, it results from our choice of the reference frame which has a coordinate singularity for $r = r_{\rm g}$. Relation (4.46) suggests a very important conclusion. The flow in a close vicinity of a black hole is to be supersonic ($u_{\rm p} > c_{\rm s}$).

4.2 Examples

4.2.1 Exact solutions

1. Spherically symmetric accretion

   If the flow velocity at infinity is equal to zero ($\gamma_\infty = 1$) and thermodynamic conditions are homogeneous, then $E = \mu_\infty = \mathrm{const}$ and $s = s_\infty = \mathrm{const}$. So, the two thermodynamic functions at infinity determine two integrals of motion E and s. Finally, for a spherically symmetric flow one can put $L = 0$. Under such conditions the stream equation (4.39) has a trivial solution $\Phi = \Phi_0(1 - \cos\theta)$, and the accretion rate $2\Phi_0$ is to be determined from the regularity conditions on the sonic surface $r = r_*$.

   As a result, we have the following expression for the sonic radius

$$r_* = \frac{M}{2} \cdot \left(\frac{1}{c_*^2} + 3 \right), \tag{4.47}$$

so that for $c_*^2 \ll 1$ we return to the nonrelativistic expression (2.30). As to the relation between c_*^2 and c_∞^2, it can be found from the condition [23]

$$\frac{1}{[1 - c_\infty^2/(\Gamma - 1)]^2} = \frac{1 - 4c_*^2/(1 + 3c_*^2)}{[1 - c_*^2/(\Gamma - 1)]^2 \cdot (1 - c_*^2)}. \tag{4.48}$$

In the limit $c_*^2 \ll 1$ we return to the well-known relation (2.22). Next, the values $\mathcal{M}_*^2$ and μ_* on the sonic surface are

$$\mathcal{M}_*^2 = \mathcal{M}_\infty^2 \left(\frac{c_\infty^2}{c_*^2}\right)^{1/(\Gamma-1)} \left(\frac{\Gamma - 1 - c_*^2}{\Gamma - 1 - c_\infty^2}\right)^{(2-\Gamma)/(\Gamma-1)}, \tag{4.49}$$

$$\mu_* = \mu_\infty \frac{\Gamma - 1 - c_\infty^2}{\Gamma - 1 - c_*^2}. \tag{4.50}$$

Thus, the accretion rate can be written as $2m_\mathrm{p}\Phi_0$, where we now have $\Phi_0 = 8\pi^2 r_*^2 E c_*/M_*^2$.

Exercise

Show that in General Relativity a transonic flow takes place even for $\Gamma = 5/3$.

Finally, in the supersonic region $r \ll r_*$ we have

$$\frac{\mathcal{M}^2}{\mathcal{M}_*^2} \simeq 2\left(\frac{r}{r_*}\right)^{3/2}, \tag{4.51}$$

$$\frac{c_\mathrm{s}^2}{c_*^2} \simeq \frac{1}{2^{\Gamma-1}} \cdot \left(\frac{r}{r_*}\right)^{-3(\Gamma-1)/2}. \tag{4.52}$$

In particular, at the horizon

$$c_\mathrm{s}^2(r_\mathrm{g}) = \frac{1}{16^{\Gamma-1}}(c_*)^{5-3\Gamma}. \tag{4.53}$$

Hence, for $c_*^2 \approx c_\infty^2 \ll 1$ the velocity of sound remains small ($c_\mathrm{s} \ll 1$) up to the horizon.

It is necessary to stress that an accretion onto black holes differs sufficiently from the nonrelativistic case. The point is that all the subsonic trajectories taking place for an accretion onto ordinary stars (see Fig. 1) have unphysical singularity $v(r \to r_\mathrm{g}) = 0$, $n(r \to r_\mathrm{g}) = \infty$ on the horizon. In other words, to support the subsonic flow the infinite gravitational force in the vicinity of the horizon is to be balanced by the infinite pressure gradient. Hence, the only physically reasonable regime of the accretion onto a black hole is the transonic one.

2. Accretion of a dust ($P = 0$)

   For a dust, the flow lines must coincide with trajectories of particles freely moving in the gravitational field of a Kerr black hole. For the case of the motion with $L = 0$ ($u_\varphi = 0$) and zero kinetic energy at infinity $\gamma_\infty = 1$, such trajectories are "straight lines" $\theta =$ const for an arbitrary rotation parameter a [30]. Moreover, for $P = 0$ the density of the flow lines can be arbitrary as well. In other words, the arbitrary function

$$\Phi = \Phi(\theta) \tag{4.54}$$

   must be a solution to the stream equation. In particular, it means that the accretion rate is free. It is not surprising, for the flow is supersonic in the entire space.

Exercise

Using the Bernoulli equation (4.44) and the compact form of the relativistic stream equation (4.43), check that for $E = \mu =$ const, $L = 0$, and $s = 0$ the arbitrary function $\Phi(\theta)$ is a solution.

3. Accretion of a gas with $c_\mathrm{s} = 1$ [29]

   As one can see from (4.42), for $c_\mathrm{s} = 1$ we have $D^{-1} = 0$. Hence, for $E =$ const, $L = 0$, and $s =$ const the stream equation becomes linear

$$\frac{\Delta}{\rho^2}\frac{\partial^2\Phi}{\partial r^2} + \frac{1}{\rho^2}\sin\theta\frac{\partial}{\partial\theta}\left(\frac{1}{\sin\theta}\frac{\partial\Phi}{\partial\theta}\right) = 0. \tag{4.55}$$

   As a result, the solution can again be expanded in eigenfunctions of the operator $\hat{\mathcal{L}}_\theta$. For example, for a moving black hole one can obtain [29]

$$\Phi = \Phi_0(1 - \cos\theta) + \pi n_\infty v_\infty (r^2 - 2r_\mathrm{g}r)\sin^2\theta. \tag{4.56}$$

   Here the accretion rate $2\Phi_0$ is arbitrary for another reason – the flow remains subsonic up to the horizon.

4.2.2 Bondi-Hoyle accretion – relativistic version

First of all, let us consider the relativistic version of the Bondi-Hoyle accretion, *i.e.*, accretion onto a moving nonrotating black hole. The small parameter of the problem is again $\varepsilon_1 = v_\infty/c_\infty$. In the relativistic case the linearized stream equation for the stream function $\Phi = \Phi_0[1-\cos\theta+\varepsilon_1 f(r,\theta)]$ can be written as [19]

$$-\varepsilon_1\alpha^2 D\frac{\partial^2 f}{\partial r^2} - \frac{\varepsilon_1}{\rho^2}(D+1)\sin\theta\frac{\partial}{\partial\theta}\left(\frac{1}{\sin\theta}\frac{\partial f}{\partial\theta}\right) + \varepsilon_1\alpha^2 N_r\frac{\partial f}{\partial r} = 0, \tag{4.57}$$

where now

$$N_r = \frac{2}{r} - \frac{\mu^2}{E^2 - \alpha^2\mu^2}\frac{M}{r^2}. \qquad (4.58)$$

We see that this equation have the same properties as the nonrelativistic equation (3.37), namely

- Equation (4.57) is linear.
- The values of μ, N_r, and D should be determined from the unperturbed Schwarzschild metric for a spherically symmetric flow.
- As $\mu = \mu(r)$, $N_r = N_r(r)$, $D = D(r)$, $\alpha^2 = 1 - 2\mathcal{M}/r$, and $\rho = r$, one can expand the solution in eigenfunctions of the operator $\hat{\mathcal{L}}_\theta$.

But equation (4.57) has another very important property. According to (4.42) and (4.46),

$$D + 1 = \frac{\alpha^2\mu^2}{E^2 - \alpha^2\mu^2} \cdot \frac{c_s^2}{1 - c_s^2}, \qquad (4.59)$$

so the factor α^2 is contained in the all addenda of equation (4.57). Hence, equation (4.57) has no singularity at the horizon. In particular, it means that it is not necessary to specify any boundary conditions for $r = r_g$. It is not surprising because the horizon corresponds to the supersonic region which cannot affect the subsonic flow.

As a result, one can seek the solution of the stream equation in the form (for more details see [13, 19])

$$\Phi(r, \theta) = \Phi_0[1 - \cos\theta + \varepsilon_1 g_1(r)\sin^2\theta], \qquad (4.60)$$

the equation for the radial function $g_1(r)$ being

$$-D\frac{d^2 g_1}{dr^2} + N_r\frac{dg_1}{dr} + 2\frac{\mu^2}{E^2 - \alpha^2\mu^2} \cdot \frac{c_s^2}{1 - c_s^2} \cdot \frac{g_1}{r^2} = 0. \qquad (4.61)$$

We see that, as in the nonrelativistic case, the accretion rate is not changed in the first order of ε_1.

As to the boundary conditions, they are

1. The regularity condition on the sonic surface. It gives $g_1(r_*) = 0$.
2. At infinity

$$g_1(r) \to K(\Gamma)\frac{r^2}{r_*^2}, \qquad (4.62)$$

where (see Table 2)

$$K(\Gamma) = \frac{1}{2}\frac{M_*^2}{M_\infty^2}\frac{c_\infty}{c_*}. \qquad (4.63)$$

Table 2.

Γ	1.01	1.1	1.2	1.333	1.5	1.6
$K(\Gamma)$	0.49	0.09	0.07	0.044	0.016	0.003
$k_1(\Gamma)$	3.00	0.56	0.46	0.31	0.12	0.023
$K_{\rm in}(\Gamma)$	−0.74	−0.09	−0.03	0.025	0.0081	0.0002

Equation (4.61) with boundary conditions 1 and 2 determines the flow structure of the Bondi-Hoyle accretion onto a nonrotating black hole. In particular, as in the nonrelativistic case, the shape of the sonic surface has the form

$$r_*(\theta) = r_*\left[1 + \varepsilon_1 \frac{\Gamma+1}{5 - 3\Gamma + 9c_*^2} k_1(\Gamma)\cos\theta\right], \tag{4.64}$$

where again $k_1(\Gamma) = g_1'(r_*)r_*$. The $k_1(\Gamma)$ values are given in Table 2.

But in general, the flow structure remains the same as for the nonrelativistic accretion (see Fig. 3). Here one can stress only a single interesting feature. For $r \ll r_*$ the radial function $g_1(r)$ has the asymptotics

$$g_1(r) \approx K_{\rm in}(\Gamma)\left(\frac{r}{r_*}\right)^{-1/2} \tag{4.65}$$

($K_{\rm in}$ is given in Table 2 as well). Hence, for $\varepsilon_1 > (M/r_*)^{1/2}$ in the vicinity of the black hole (*i.e.*, for $r < \varepsilon_1^2 K_{\rm in}^2 r_*$) the linear approximation is violated, so here one should solve the complete nonlinear equation (4.39). Since the sign of the coefficient $K_{\rm in}(\Gamma)$ depends on the polytropic index Γ, the region of thickening of the flow lines will either appear on the front side for $\Gamma > 1.27$ or on the rear side for $\Gamma < 1.27$. However, it takes place in the supersonic region which does not affect the subsonic flow for $r > r_*$.

4.2.3 Accretion onto a slowly rotating black hole

Let us now consider the accretion of a gas with $L = 0$ (*i.e.*, $\sigma = 2$ and $b = 2 + 2 - 1 = 3$) onto a slowly rotating black hole for which the small parameter is

$$\varepsilon_3 = \frac{a}{M} \ll 1. \tag{4.66}$$

In this case the metric g_{ik} (4.4) differs from the Schwarzschild one by the values $\sim\varepsilon_3^2$. One can assume that the thermodynamic functions at infinity, s_∞ and μ_∞, remain the same as for spherically symmetric accretion. Hence, one can again seek the solution of the stream equation (4.39) in the form

$$\Phi(r,\theta) = \Phi_0\left[1 - \cos\theta + \varepsilon_3^2 f(r,\theta)\right], \tag{4.67}$$

where Φ_0 is the flux constant corresponding to a spherically symmetrical flow. Inserting this form into the stream equation (4.39), we have in the first order of ε_3^2

$$-\varepsilon_3^2\alpha^2 D\frac{\partial^2 f}{\partial r^2} - \frac{\varepsilon_3^2}{\rho^2}(D+1)\sin\theta\frac{\partial}{\partial\theta}\left(\frac{1}{\sin\theta}\frac{\partial f}{\partial\theta}\right) + \varepsilon_3^2\alpha^2 N_r\frac{\partial f}{\partial r}$$
$$= \frac{a^2}{r^4}\left(1-\frac{2M}{r}\right)\cdot\left(1-2\frac{\mu^2}{E^2-\alpha^2\mu^2}\frac{M}{r}\right)\sin^2\theta\cos\theta, \tag{4.68}$$

where N_r is given by (4.58).

The properties of equation (4.68) are quite similar to those of equation (4.57). In particular, equation (4.68) has no singularity at the horizon. As a result, one can show that the flux $\Phi(r,\theta)$ can be presented as [19]

$$\Phi(r,\theta) = \Phi_0\left[(1-\cos\theta) + \varepsilon_3^2 g_0(1-\cos\theta) + \varepsilon_3^2 g_2(r)\sin^2\theta\cos\theta\right]. \tag{4.69}$$

Hence, as for the ejection from a slowly rotating star, the flux $\Phi(r,\theta)$ contains two harmonics, $m=0$ and $m=2$, the radial function $g_2(r)$ satisfying equation

$$-D\frac{\mathrm{d}^2 g_2}{\mathrm{d}r^2} + N_r\frac{\mathrm{d}g_2}{\mathrm{d}r} + 6\frac{\mu^2}{E^2-\alpha^2\mu^2}\cdot\frac{c_\mathrm{s}^2}{1-c_\mathrm{s}^2}\cdot\frac{g_2}{r^2} = \frac{M^2}{r^4}\left(1-2\frac{\mu^2}{E^2-\alpha^2\mu^2}\frac{M}{r}\right). \tag{4.70}$$

The regularity condition on the sonic surface $N_\theta(r_*)=0$ gives

$$g_2(r_*) = -\frac{1}{2}\frac{M^2}{r_*^2}\alpha^2(r_*).$$

On the other hand, the condition at infinity (which is the third boundary condition after s_∞ and μ_∞) gives $g_2(r\to\infty)=0$. As a result, the radial function $g_2(r)$ for $r \ll r_*$ can be written in the form

$$g_2(r) = -G(r)\frac{M^2}{r_*^2}\left(\frac{r}{r_*}\right)^{(1-3\Gamma)/2}, \tag{4.71}$$

where $G(r)\sim 1$. Hence, at the horizon $g_2(r_\mathrm{g}) \sim (M/r_*)^{(5-3\Gamma)/2}$, so that the disturbance is small ($\varepsilon_3^2 g_2(r) \ll 1$) everywhere outside a black hole. On the other hand, as $g_2(r)<0$, the rotation of a black hole results in the concentration of the flow lines in the equatorial plane. Finally, an additional consideration demonstrates that [19]

$$g_0 = -2\frac{M^3}{r_*^3}. \tag{4.72}$$

Thus, the rotation of a black hole diminishes the accretion rate. In reality $c_\infty^2 \ll 1$ and hence $M/r_* \ll 1$. As a result, the effects of a black hole rotation in the vicinity of the sonic surface are actually very small.

To summarize the last two sections, one can say that the simple zero approximation, namely, a spherically symmetric transonic accretion, allows us to find analytically 2D structure for very important astrophysical flows. A similar approach was used for the construction of an analytical solution for

- accretion of a gas with small angular momentum L (the small parameter $\varepsilon_4^2 = (L/Er_\mathrm{g})^2$) onto a nonrotating black hole [27,31],
- accretion of a gas with a nonrelativistic temperature ($c_\infty \ll 1$) and without angular momentum ($L = 0$) onto an arbitrary rotating black hole [32].

At present they are the only examples where the analytical solution was found.

4.2.4 Thin transonic disk

As the last example, let us consider the internal 2D structure of a thin transonic disk. Such a flow can be realized if the intrinsic angular momentum of a gas accreting onto a black hole is large enough ($\varepsilon_4 \gg 1$) [17]. Here for simplicity we consider a nonrotating (Schwarzschild) black hole only (for more details see [33]).

According to the standard disk model [34–37], for $\varepsilon_4 \gg 1$ the matter forms a thin balanced disk and performs a nearly circular motion with Keplerian velocity $v_\mathrm{K}(r) \approx (GM/r)^{1/2}$. The disk is thin provided that the accreting gas temperature is sufficiently low and the disk thickness is determined by the pressure gradient

$$H \approx r\frac{c_\mathrm{s}}{v_\mathrm{K}}. \tag{4.73}$$

Introducing the viscosity parameter $\alpha_\mathrm{SS} \leq 1$, relating the stress tensor t_φ^r and the pressure as $t_\varphi^r = \alpha_\mathrm{SS}P$ [35], one can obtain

$$\frac{v_r}{v_\mathrm{K}} \approx \alpha_\mathrm{SS}\frac{c_\mathrm{s}^2}{v_\mathrm{K}^2}. \tag{4.74}$$

Hence, for $c_\mathrm{s} \ll v_\mathrm{K}$ the radial velocity v_r remains much smaller than both the Keplerian velocity v_K and the velocity of sound c_s.

The General Relativity effects result in two important properties:

- The absence of stable circular orbits for $r < r_0 = 3r_\mathrm{g}$.

- The transonic regime of accretion.

The first point means that the accreting matter passing a marginally stable orbit approaches the black hole horizon sufficiently fast, namely, in the dynamical time $\tau_{\rm d} \sim [v_r(r_0)/c]^{-1/3} r_{\rm g}/c$. It is important that such a flow is realized in the absence of viscosity. The second statement results from the fact that according to (4.74) up to the marginally stable orbit the flow is subsonic while at the horizon the flow is to be supersonic.

Up to now in the majority of works the procedure of vertical averaging was used, where the vertical four-velocity $u_{\hat{\theta}}$ was assumed to be zero [38]. As a result, the vertical component of the dynamic force $nu^b\nabla_b(\mu u_a)$ in (4.38) was postulated to be inimportant up to horizon. For this reason the disk thickness was determined by the pressure gradient even in the supersonic region near the black hole [39]. Here I am going to demonstrate that the assumption $u_{\hat{\theta}} = 0$ is not correct. As in the Bondi accretion, the dynamic force is to be important in the vicinity of the sonic surface.

First of all, let us consider the subsonic region in a close vicinity of the marginally stable orbit $r_0 = 3r_{\rm g}$ where the poloidal velocity $u_{\rm p}$ is much smaller than that of sound. Then equation (4.39) can be significantly simplified by neglecting the terms proportional to $D^{-1} \sim u_{\rm p}^2/c_{\rm s}^2$. As a result, we have

$$-\mathcal{M}^2\frac{1}{\alpha}\nabla_k\left(\frac{1}{\alpha\varpi^2}\nabla^k\Phi\right) + \frac{64\pi^4}{\alpha^2\varpi^2\mathcal{M}^2}\left(\varpi^2 E\frac{{\rm d}E}{{\rm d}\Phi} - \alpha^2 L\frac{{\rm d}L}{{\rm d}\Phi}\right) - 16\pi^3 nT\frac{{\rm d}s}{{\rm d}\Phi} = 0. \quad (4.75)$$

This equation describing the subsonic flow is elliptical.

To determine the structure of a two-dimensional subsonic flow ($\sigma = 0$, *i.e.*, $b = 2 + 3 - 0 = 5$) one needs to specify five quantities (three velocity components and two thermodynamic functions) on an arbitrary surface $r = r_0(\theta)$. Naturally, we choose it as the surface of the last stable orbit $r_0 = 3r_{\rm g}$ where $\alpha_0 = \alpha(r_0) = \sqrt{2/3}$, $u_{\hat{\varphi}}(r_0) = 1/\sqrt{3}$, and $\gamma_0 = \gamma(r_0) = \sqrt{4/3}$ [17]. For the sake of simplicity we consider below the case where the radial velocity is constant on the surface $r = r_0$ and the toroidal velocity is exactly equal to $u_{\hat{\varphi}}(r_0)$:

$$u_{\hat{r}}(r_0, \Theta) = -u_0, \quad (4.76)$$

$$u_{\hat{\Theta}}(r_0, \Theta) = \Theta u_0, \quad (4.77)$$

$$u_{\hat{\varphi}}(r_0, \Theta) = 1/\sqrt{3}. \quad (4.78)$$

Here $u_{\hat{\Theta}} \ll |u_{\hat{r}}|$ corresponds to the plane flow at the marginally stable orbit, and we introduced the new angular variable $\Theta = \pi/2 - \theta$ ($\Theta_{\rm disk} \sim c_0$) which is counted off from the equator in the vertical direction. Next, we suppose that the velocity of sound is also a constant on the surface $r = r_0$

$$c_{\rm s}(r_0, \Theta) = c_0. \tag{4.79}$$

For the polytropic equation of state (2.6) it means that both the temperature $T_0 = T(r_0)$ and the relativistic enthalpy $\mu_0 = \mu(r_0)$ are also constant on this surface. According to (4.74), one can find that for a nonrelativistic temperature $c_{\rm s} \ll 1$ the small parameter of this problem is

$$\varepsilon_5 = \frac{u_0}{c_0} \sim \alpha_{\rm SS} c_0 \ll 1. \tag{4.80}$$

Finally, as the last, fifth boundary condition it is convenient to specify the entropy $s(\Phi)$.

Introducing the values $E_0 = \alpha_0\gamma_0 = \sqrt{8/9}$ and $L_0 = u_{\hat{\varphi}}(r_0)r_0 = \sqrt{3}r_{\rm g}$, one can write the invariants $E(\Phi)$ and $L(\Phi)$ in the following form

$$E(\Phi) = \mu_0 E_0 = \text{const}, \tag{4.81}$$

$$L(\Phi) = \mu_0 L_0 \cos\Theta_m. \tag{4.82}$$

Here $\Theta_m = \Theta_m(\Phi)$ is the angle for which $\Phi(r_0, \Theta_m) = \Phi(r, \Theta)$. In other words, the function $\Theta_m(r, \Theta)$ has the meaning of a theta angle on the last stable orbit. This orbit is connected with a given point (r,Θ) by a line of flow $\Phi(r, \Theta) = \text{const}$. In particular, $\Theta_m(r_0, \Theta) = \Theta$.

First of all, we see that condition $E = \text{const}$ (4.81) allows us to rewrite equation (4.75) in a simpler form

$$\frac{\partial^2\Phi}{\partial r^2} + \frac{\cos\Theta}{\alpha^2 r^2}\frac{\partial}{\partial\Theta}\left(\frac{1}{\cos\Theta}\frac{\partial\Phi}{\partial\Theta}\right) = -4\pi^2 n^2 \frac{L}{\mu^2}\frac{{\rm d}L}{{\rm d}\Phi} - 4\pi^2 n^2 r^2\cos^2\Theta\frac{T}{\mu}\frac{{\rm d}s}{{\rm d}\Phi}. \tag{4.83}$$

Next, as is shown in Appendix A, for $r = r_0$, the r.h.s. of equation (4.83) describes the transverse balance of a pressure gradient and an effective potential, whereas the l.h.s. corresponds to the dynamic term $(\vec{v}\nabla)\vec{v}$. At the marginally stable orbit it is of the order of u_0^2/c_0^2 and may be dropped. It is therefore natural to choose the entropy $s(\Phi)$ from the condition of a transverse balance on the surface $r = r_0$

$$r_0^2\cos^2\Theta_m\frac{{\rm d}s}{{\rm d}\Theta_m} = -\frac{\Gamma}{c_0^2}\frac{L}{\mu_0^2}\frac{{\rm d}L}{{\rm d}\Theta_m}, \tag{4.84}$$

where $L(\Theta_m)$ is determined from the boundary condition (4.82). Thus, we have

$$s(\Theta_m) = s(0) - \frac{\Gamma}{3c_0^2}\ln(\cos\Theta_m). \tag{4.85}$$

Owing to (4.23), one can show that for s_0 = const relation (4.85) corresponds to the standard concentration profile

$$n(r_0, \Theta) \approx n_0 \exp\left(-\frac{\Gamma}{6c_0^2}\Theta^2\right). \tag{4.86}$$

Exercise

Using relations (2.5)–(2.6), show that the exact expression for $n(r_0, \Theta)$ is $n(r_0, \Theta) = n_0(\cos\Theta)^{\Gamma/3c_0^2}$.

Finally, the definition (4.27) results in the following connection between functions Φ and Θ_m

$$\mathrm{d}\Phi = 2\pi\alpha_0 r_0^2 n(r_0, \Theta_m) u_0 \cos\Theta_m \mathrm{d}\Theta_m. \tag{4.87}$$

Hence, due to (4.82), (4.86), and (4.87), the invariant $L(\Phi)$ can be directly determined from the boundary conditions as well.

Equation (4.83) together with the boundary conditions (4.82), (4.85), (4.86), and (4.87) and relationship (4.77) specifying the derivative $\partial\Phi/\partial r$ determines the structure of the inviscid subsonic flow inside the marginally stable orbit. For example, for a nonrelativistic temperature $c_\mathrm{s} \ll 1$ we obtain

$$u_\mathrm{p}^2 = u_0^2 + w^2 + \frac{1}{3}\left(\Theta_m^2 - \Theta^2\right) + \frac{2}{\Gamma - 1}(c_0^2 - c_\mathrm{s}^2) + \dots \tag{4.88}$$

Here the quantity w, where

$$w^2(r) = \frac{E_0^2 - \alpha^2 L_0^2/r^2 - \alpha^2}{\alpha^2} \approx \frac{1}{6}\left(\frac{r_0 - r}{r_0}\right)^3, \tag{4.89}$$

depending on the radius r only is a poloidal four-velocity of a free particle having zero poloidal velocity for $r = r_0$. As we see, w^2 increases very slowly when moving away from the last stable orbit. Therefore, the contribution of w^2 turns out to be negligibly small and the term may be safely dropped in most cases.

An important conclusion can be drown directly from (4.88) in which for the equatorial plane we have $\Theta_m = \Theta = 0$. Assuming $u_\mathrm{p} = c_\mathrm{s} = c_*$ and neglecting w^2, we find that the velocity of sound c_* on the sonic surface $r = r_*$, $\Theta = 0$ is of the same order of magnitude as the velocity of sound on the last stable orbit c_0

$$c_* \approx \sqrt{\frac{2}{\Gamma + 1}}\, c_0. \tag{4.90}$$

Since the entropy s remains constant along the flow lines, the gas concentration remains approximately constant ($n_* \sim n_0$) as well. In other words, in agreement with the Bondi accretion, the subsonic flow can be considered as incompressible.

On the other hand, since the density remains almost constant and for $\varepsilon_5 \ll 1$ the radial velocity increases from u_0 to $c_* \sim c_0$, *i.e.*, changes over several orders of magnitude, the disk thickness H should change in the same proportion owing to the continuity equation (see Fig. 4)

$$H(r_*) \approx \frac{u_0}{c_0} H(3r_{\rm g}). \tag{4.91}$$

As a result, a rapid decrease of the disk thickness should be accompanied by the appearance of the vertical component of velocity which also should be taken into account in the Euler equation (4.38).

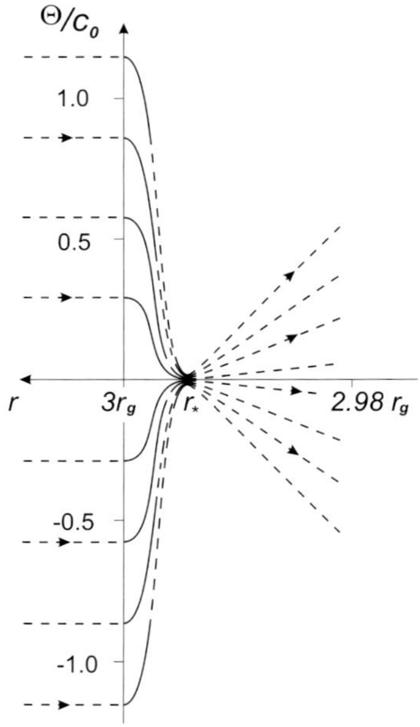

Fig. 4. The structure of the thin accretion disk (actual scale) after passing the marginally stable orbit $r = 3r_{\rm g}$ obtained by numerical solving equation (4.83) for $c_0 = 10^{-2}$, $u_0 = 10^{-5}$. The solid lines correspond to the range of parameters $u_{\rm p}^2/c_0^2 < 0.2$, where the solution should not differ greatly from the solution of the complete equation (4.39). The dashed lines indicate an extrapolation of the solution to the sonic-surface region. In the vicinity of the sonic surface the flow has a form of an ordinary nozzle.

Indeed, as one can find analyzing the asymptotics of equation (4 83) [33], in the vicinity of the sonic surface located at

$$r_* = r_0 - \Lambda u_0^{2/3} r_0, \tag{4.92}$$

where the logarithmic factor $\Lambda = (3/2)^{2/3}[\ln(c_0/u_0)]^{2/3} \approx 5{-}7$, the components of the velocity and the pressure gradient can be presented as

$$u_{\hat{\Theta}} \to -\frac{c_0}{u_0}\Theta, \tag{4.93}$$

$$u_{\hat{r}} \to -c_*, \tag{4.94}$$

$$-\frac{\nabla_{\hat{\theta}} P}{\mu} \to \frac{c_0^2}{u_0^2}\frac{\Theta}{r}. \tag{4.95}$$

On the other hand, near the sonic surface the radial scale δr determining the radial derivatives becomes as small as the transverse dimension of a disk: $\delta r \approx H(r_*) \approx u_0 r_0$, so that

$$\eta_1 = \frac{r}{n}\frac{\partial n}{\partial r} \approx u_0^{-1}. \tag{4.96}$$

As a result, both components of the dynamic force

$$\frac{u_{\hat{\Theta}}}{r}\frac{\partial u_{\hat{\Theta}}}{\partial \Theta} \to \frac{c_0^2}{u_0^2}\frac{\Theta}{r}, \tag{4.97}$$

$$u_{\hat{r}}\frac{\partial u_{\hat{\Theta}}}{\partial r} \to \frac{c_0^2}{u_0^2}\frac{\Theta}{r}, \tag{4.98}$$

become of the order of the pressure gradient (4.95).

To check our conclusions one can consider the flow structure in the vicinity of the sonic surface in more detail. Using again the expansion theorem (the smooth transonic flow is analytical at a singular point), one can write down (*cf.* (2.43)–(2.44))

$$n = n_*\left(1 + \eta_1 h + \frac{1}{2}\eta_3\Theta^2 + \ldots\right), \tag{4.99}$$

$$\Theta_m = a_0\left(\Theta + a_1 h\Theta + \frac{1}{2}a_2 h^2\Theta + \frac{1}{6}b_0\Theta^3 + \ldots\right), \tag{4.100}$$

where $h = (r - r_*)/r_*$. Here we assume that all three invariants E, L, and s are already given, *i.e.*, $i = 0$ and $b = 2 + 0 - 1 = 1$. Hence, as in the planar case, the problem needs one extra boundary condition. Now comparing the appropriate coefficients in the Bernoulli (4.44) and the full stream equation (4.43), one can obtain neglecting terms $\sim u_0^2/c_0^2$

$$a_0 = \left(\frac{2}{\Gamma+1}\right)^{(\Gamma+1)/2(\Gamma-1)} \frac{c_0}{u_0}, \tag{4.101}$$

$$a_1 = 2 + \frac{1-\alpha_*^2}{2\alpha_*^2} \approx 2.25, \tag{4.102}$$

$$a_2 = -(\Gamma+1)\eta_1^2, \tag{4.103}$$

$$b_0 = \left(\frac{\Gamma+1}{6}\right)\frac{a_0^2}{c_0^2}, \tag{4.104}$$

$$\eta_3 = -\frac{2}{3}(\Gamma+1)\eta_1^2 - \left(\frac{\Gamma-1}{3}\right)\frac{a_0^2}{c_0^2}, \tag{4.105}$$

where $\alpha_*^2 = \alpha^2(r_*) \approx 2/3$. In comparison with the planar case, all the coefficients are expressed here through the radial logarithmic derivative η_1 (4.96).

Let me stress that it is rather difficult to connect the sonic characteristics $\eta_1 = \eta_1(r_*)$ with physical boundary conditions on the marginally stable orbit $r = r_0$ (for this it is necessary to know all the expansion coefficients in (4.99) and (4.100)). In particular, it is impossible to formulate the restriction on five boundary conditions (4.76)–(4.79) and (4.85) resulting from the critical condition on the sonic surface. Nevertheless, the estimate (4.96) makes us sure that we know the parameter η_1 to a high enough accuracy. Then, according to (4.101)–(4.105), all the other coefficients can be determined exactly.

The coefficients (4.101)–(4.105) have clear physical meaning. So, a_0 gives the compression of flow lines: $a_0 = H(r_0)/H(r_*)$. In agreement with (4.91) we have $a_0 \approx c_0/u_0$. Further, a_1 corresponds to the slope of the flow lines with respect to the equatorial plane. As $a_1 > 0$, in a close vicinity of the sonic surface the compression of stream lines finishes, so inside the sonic radius $r < r_*$ the stream lines diverge. On the other hand, as $a_1 \ll u_0^{-1}$, for $r = r_*$ the divergency is still very weak. Hence, in the vicinity of the sonic surface the flow has a form of an ordinary nozzle (see Fig. 2). Finally, as $a_2 \sim \eta_3 \sim b_0 \sim u_0^{-2}$, one can conclude that the transverse scale of the transonic region $H(r_*)$ does the same as the longitudinal one. The latter point suggests a very important consequence that the transonic region is essentially two-dimensional, and so it is impossible to analyze it within the standard one-dimensional approximation.

Using now the expansions (4.99) and (4.100), one can obtain all the other physical parameters of the transonic flow. In particular, we have

$$u_{\rm p}^2 = c_*^2 \left[1 - 2\eta_1 h + \frac{1}{6}(\Gamma - 1)\,\frac{a_0^2}{c_0^2}\Theta^2 + \frac{2}{3}(\Gamma + 1)\eta_1^2\Theta^2\right],$$
$$c_{\rm s}^2 = c_*^2 \left[1 + (\Gamma - 1)\,\eta_1 h + \frac{1}{6}(\Gamma - 1)\,\frac{a_0^2}{c_0^2}\Theta^2 - \frac{1}{3}(\Gamma - 1)(\Gamma + 1)\eta_1^2\Theta^2\right].$$

As a result, the shape of the sonic surface $u_{\rm p} = c_{\rm s}$ has the standard parabolic form

$$h = \frac{\Gamma + 1}{3}\eta_1\Theta^2. \tag{4.106}$$

Thus, the analysis of the hydrodynamic stream equation (4.39) allows us to find a nontrivial structure of the thin transonic disk. As was shown, the diminishing disk thickness inevitably leads to an emergence of the vertical velocity component of the accreting matter. As a result, the dynamic term $(\vec{v}\nabla)\vec{v}$ in the vertical balance equation cannot be omitted. In this sense the situation is completely analogous to the spherically symmetric Bondi accretion for which the contribution of the dynamic term becomes significant near the sonic surface and dominant for a supersonic flow.

However, there is an important difference. For the Bondi accretion the dynamic term $(\vec{v}\nabla)\vec{v}$ has only one component $v_r\partial v_r/\partial r$ which in the vicinity of the sonic surface becomes of the same order of magnitude as both the pressure and the gravity gradients. As to the thin accretion disk, both components of the dynamic term $[(\vec{v}\nabla)\vec{v}]_\theta$, (4.97) and (4.98), become of the same order of magnitude as the pressure gradient, the role of the effective gravity gradient being unimportant: $\nabla_\theta\varphi_{\rm eff} \sim \Theta/r$, *i.e.*, it is c_0^2/u_0^2 times smaller than the leading terms. As a result, the structure of a thin transonic disk is quite similar to an ordinary planar nozzle shown in Figure 2. For this reason the critical condition on the sonic surface does not restrict the accretion rate.

5 Conclusion

Thus, in my lecture I have tried to demonstrate the possibilities and difficulties of the Grad-Shafranov approach. As we have seen, in some simple cases it does allow us to construct the analytical solutions. In particular, the Grad-Shafranov approach is very suitable in considering the (analytical) properties of the flow in the vicinity of the sonic surface and in determining the number of boundary conditions.

On the other hand, it has been shown that in the general case the regular procedure does not exist. The point is that the critical conditions are to be specified on singular surfaces which are not known from the very beginning and are themselves to be determined from the solution. Moreover, it is impossible to extend this approach to nonideal, nonstationary, and non axially symmetric flows. For this reason it is not surprising that those interested in astrophysics more than mathematics, passed from the Grad-Shafranov approach to numerical calculations analyzing absolutely another class of equations, namely, the time-dependent ones. The only thing one can wish is not to forget the basic physical results of the Grad-Shafranov approach which remain true irrespective on the method of calculation.

Appendix

A From Euler to Grad-Shafranov – the simplest way

Let me show how the nonrelativistic version of the stream equation (4.83) can be directly derived from the Euler equation. For the sake of simplicity we consider here only a case of nonrelativistic velocities and small angles $\Theta = \pi/2 - \theta$ in the vicinity of the equatorial plane.

The θ-component of the Euler equation is

$$v_r \frac{\partial v_\theta}{\partial r} + v_\theta \frac{\partial v_\theta}{r\partial\theta} + \frac{v_r v_\theta}{r} - \frac{v_\varphi^2}{r}\cot\theta = -\frac{\nabla_\theta P}{m_{\rm p} n} - \nabla_\theta \varphi_{\rm g}. \tag{A.1}$$

Adding $v_\varphi \partial v_\varphi / r\partial\theta$ to both sides and adding and subtracting $v_r \partial v_r / r\partial\theta$ in the left-hand side, we get

$$v_r \frac{\partial v_\theta}{\partial r} - v_r \frac{\partial v_r}{r\partial\theta} + \nabla_\theta \left(\frac{v^2}{2}\right) + \frac{v_r v_\theta}{r} = \frac{v_\varphi^2}{r}\cot\theta + v_\varphi \frac{\partial v_\varphi}{r\partial\theta} - \frac{\nabla_\theta P}{m_{\rm p} n} - \nabla_\theta \varphi_{\rm g}. \tag{A.2}$$

Using the definition of the Bernoulli integral $E = v^2/2 + w + \varphi_{\rm g}$ and the thermodynamic relationship ${\rm d}P = m_{\rm p} n {\rm d}w - nT{\rm d}s$, we get for $E = {\rm const}$

$$v_r \frac{\partial v_\theta}{\partial r} - v_r \frac{\partial v_r}{r\partial\theta} + \frac{v_r v_\theta}{r} = \frac{r v_\varphi \sin\theta}{\varpi^2} \frac{\partial}{r\partial\theta}(r v_\varphi \sin\theta) + \frac{T}{m_{\rm p}} \frac{\partial s}{r\partial\theta}. \tag{A.3}$$

It follows from the definition (3.1) that

$$v_r = \frac{1}{2\pi n r^2 \sin\theta} \frac{\partial\Phi}{\partial\theta}, \qquad v_\theta = -\frac{1}{2\pi n r \sin\theta} \frac{\partial\Phi}{\partial r}. \tag{A.4}$$

Now, assuming $n \approx$ const (this is the case for a subsonic flow), we obtain

$$-\frac{1}{4\pi^2 n^2}\left(\frac{\partial\Phi}{\partial\theta}\right)\left[\frac{\partial^2\Phi}{\partial r^2}+\frac{\sin\theta}{r^2}\frac{\partial}{\partial\theta}\left(\frac{1}{\sin\theta}\frac{\partial\Phi}{\partial\theta}\right)\right]= \\ rv_\varphi\sin\theta\frac{\partial}{\partial\theta}(rv_\varphi\sin\theta)+\varpi^2\frac{T}{m_\mathrm{p}}\frac{\partial s}{\partial\theta}\cdot \quad (A.5)$$

Finally, dividing both sides by $-(\partial\Phi/\partial\theta)$, we get (4.83).

Hence, whereas the first term in the l.h.s. of (4.83) does correspond to the component $v_r\partial v_\theta/\partial r$, and the last term in the r.h.s. (for c_s = const) corresponds to the pressure gradient, the role of the term $\propto L\partial L/\partial\Phi$ proves to be less trivial. It contains both the effective potential gradient and, in fact, component $v_\theta\partial v_\theta/\partial\theta$. The former is the leading one near the marginally stable orbit $r \approx 3r_\mathrm{g}$, whereas the latter becomes important only when approaching the sonic surface.

References

[1] H. Bondi and F. Hoyle, *MNRAS* **104** (1944) 272.
[2] H. Bondi, *MNRAS* **112** (1952) 195.
[3] E.N. Parker, *ApJ* **128** (1958) 664.
[4] M.C. Begelman, R.D. Blandford and M.J. Rees, *Rev. Mod. Phys.* **56** (1984) 225.
[5] F.C. Michel, Theory of Neutron Star Magnetosphere (The University of Chicago Press, Chicago, 1991).
[6] V.S. Beskin, A.V. Gurevich and Ya.N. Istomin, Physics of the Pulsar Magnetosphere (Cambridge University Press, Cambridge, 1992).
[7] L. Mestel, Stellar magnetism (Clarendon Press, Oxford, 1999).
[8] H. Sol, G. Pelletier and E. Asseo, *MNRAS* **237** (1989) 411.
[9] J. Heyvaerts, this volume.
[10] L.D. Landau and E.M. Lifshits, *Fluid Mechanics* (Pergamon Press, Oxford, 1987).
[11] R. von Mises, Mathematical Theory of Compressible Fluid Flow (Academic, New York, 1958).
[12] K.G. Guderley, Theorie Schallnaher Stromungen (Springer, Berlin, 1957).
[13] V.S. Beskin, *Physics Uspekhi* **40** (1997) 659.
[14] S.V. Bogovalov, *A&A* **323** (1997) 634.
[15] J. Heyvaerts, "Rotating MHD Winds", in Plasma Astrophysics, edited by C. Chiuderi and G. Einaudi (Springer, Berlin, 1996), 31-76.
[16] Ya.B. Zeldovich and I.D. Novikov, Stars and Relativity (The University of Chicago Press, Chicago, 1971).
[17] S.L. Shapiro and S.A. Teukolsky, Black Holes, White Dwarfs, and Neutron Stars (A Wiley–Interscience Publication, New York, 1983).
[18] V.M. Lipunov, Astrophysics of Neutron Star (Springer, Berlin, 1992).
[19] V.S. Beskin and Yu.N. Pidoprygora, *Physics JETP* **80** (1995) 575.
[20] R. Hunt, *MNRAS* **198** (1979) 83.
[21] J.-L. Tassoul, Theory of Rotating Star (Princeton University Press, Princeton, 1978).

[22] H.J.G.L.M. Lammers and J.P. Cassinelli, Introduction to stellar wind (Cambridge University Press, Cambridge, 1999).

[23] V.S. Beskin and Yu.N. Pidoprygora, *Astron. Rep.* **42** (1998) 71.

[24] K.S. Thorne and D. Macdonald, *MNRAS* **198** (1982) 339.

[25] M. Camenzind, this volume.

[26] K.S. Thorne, R.H. Price and D.A. Macdonald, Black Holes. The Membrane Paradigm (Yale University Press, New Haven, 1986).

[27] M. Anderson, *MNRAS* **239** (1989) 19.

[28] V.S. Beskin and V.I. Pariev, *Phys. Uspekhi* **36** (1993) 529.

[29] L.I. Petrich, S.L. Shapiro and S.A. Teukolsky, *Phys. Rev. Lett.* **60** (1988) 1781.

[30] V.P. Frolov and I.D. Novikov, Black Hole Physics (Kluwer Academic Publishers, Dordrecht, 1998).

[31] V.S. Beskin and L.M. Malyshkin, *Astron. Lett.* **22** (1996) 475.

[32] V.I. Pariev, *MNRAS* **283** (1986) 1264.

[33] V.S. Beskin, R.Yu. Kompaneetz and A.D. Tchekhovskoy, *Astron. Lett.* **28** (2002) 543.

[34] N.I. Shakura, *Sov. Astron.* **16** (1973) 756.

[35] N.I. Shakura and R.A. Sunyaev, *A&A* **24** (1973) 337.

[36] I.D. Novikov and K.S. Thorne, Astrophysics of black holes, in Black Holes, edited by C. DeWitt and B. DeWitt (Gordon and Breach, New York, 1973), 343-450.

[37] O. Blaes, this volume.

[38] B. Paczynski and G.S. Bisnovatyi-Kogan, *Acta Astron.* **31** (1981) 283.

[39] M.A. Abramowicz, A. Lanza and M.J. Percival, *ApJ* **479** (1997) 179.

COURSE 3

PHYSICS FUNDAMENTALS OF LUMINOUS ACCRETION DISKS AROUND BLACK HOLES

O.M. BLAES

Department of Physics,
University of California,
Santa Barbara, CA 93106, U.S.A.

Contents

PHYSICS FUNDAMENTALS OF LUMINOUS ACCRETION DISKS AROUND BLACK HOLES

O.M. Blaes

Abstract

These lectures provide an overview of the theory of accretion disks with application to bright sources containing black holes. I focus on the fundamental physics of these flows, stressing modern developments and outstanding questions wherever possible. After a review of standard Shakura-Sunyaev based models and their problems and uncertainties, I describe the basic principles that determine the overall spectral energy distribution produced by the flow. I then describe the physics of angular momentum transport in black hole accretion disks, stressing the important role of magnetic fields. Finally, I discuss the physics of radiation magnetohydrodynamics and how it might affect the overall flow structure in the innermost regions near the black hole.

1 Introduction

Accretion disk theory was developed in a flurry of activity in the early 1970's in a series of important, seminal papers [38, 72, 75, 88, 90, 95, 102, 103]. It has met with considerable success in explaining the observations we see in a number of classes of accretion-powered sources, particularly cataclysmic variables. However, it does a rather poor job of predicting and explaining the observations of black hole sources, both in the context of X-ray binaries and active galactic nuclei. This is particularly disappointing in view of the fact that much of the early development of the theory was done with this application in mind. Accretion disk theory has always suffered from a severe flaw: angular momentum transport and energy dissipation in observed flows must be due to nonlinear physics ("turbulence"), and this problem is swept under the carpet by parameterizing it away in terms of an anomalous "viscosity". Over the past thirty years, this sad state

This work is supported by NSF grant AST 9970827 and NASA grant NAG5-7075.

of affairs has persisted, although a major breakthrough was made in 1991 with the discovery that the magnetorotational instability (MRI) might be a generic source of turbulence in disks [16, 52]. Thanks to this insight and the increasing sophistication and capabilities of numerical simulations, we are now beginning to address virtually *all* aspects of accretion disk theory from first principles physics. The coming decade should see such models applied to observation, finally getting us to the point where we are able to test the basic physics, and not fudge factors. We are therefore living and working in an exciting time!

In these lectures, I will give an overview of the current state of the theory. Although I will be discussing accretion disks in some generality, I will for the most part concentrate on disks around black holes. Unless explicitly noted, I will always use Newtonian physics, even though the flows around black holes necessarily involve relativistic effects, both special and general. I do this purely for the sake of pedagogy. Relativity is not difficult – in fact, it is the *easy* part of black hole accretion disk physics. As we will see, all the really nasty (albeit interesting!) issues lie in the radiation and plasma physics, and the basic principles involved here are most quickly understood within a Newtonian framework. Almost everything I discuss has been worked out in full general relativity somewhere in the literature, and I give references to such work wherever possible.

Many good textbooks exist which discuss aspects of the basic physics of accretion disks, *e.g.* [42, 67, 104], and the reader is encouraged to consult them for a more elaborate treatment of the now-standard ideas. As I already mentioned, however, many of these ideas will soon be superseded by much more sophisticated models based on real physics. The best overview of the MRI is the review by Balbus & Hawley [19], although some advances have taken place since that article was written.

I will begin in Section 2 with a brief overview of standard models based on the Shakura-Sunyaev α-prescription for the anomalous stress that must exist in these flows, as these are still the models that people use for comparison to observational data. I will focus on recent developments in this area, including the role of advection in "slim disks", the role of torques exerted across the innermost stable circular orbit, and the vertical transfer of mechanical energy out of the disk into a corona. In Section 3, I will discuss the calculation of photon spectra from these models. I will then return to first principles in Section 4 by examining the basic origin of anomalous angular momentum transport in disks, emphasizing the role of magnetic fields. Finally, in Section 5, I will discuss the role that radiation magnetohydrodynamics may play in determining the dynamics and thermodynamics of the innermost regions of the flow, where most of the observed radiation is thought to originate.

2 Shakura-Sunyaev based models

Most, but not all, models of accretion-powered sources assume that rotation is an important source of dynamical support in the accretion flow. Gas accreting onto a compact object from an orbiting companion star is endowed with considerable angular momentum arising from the orbital motion of the binary itself. The huge dynamic range in radii between the fuel source and the radius of the central object in young stellar objects (YSO's) and active galactic nuclei (AGN) also strongly suggests that rotation is important. If material at large radii has some nonzero angular momentum with respect to the central object, and if that angular momentum is conserved on the way in, then the material is likely to become rotationally supported. This is simply because centripetal acceleration then depends on distance R from the rotation axis as R^{-3}, steeper than the R^{-2} dependence of gravitational acceleration. (This statement neglects general relativistic effects. Sufficiently close to the event horizon, the gravitational acceleration will dominate all rotational support and even material with conserved angular momentum will flow inward all the way to the singularity.) Direct observational support for the existence of rotationally supported accretion flow structures exists in many images of YSO disks and in the maser disk of the low luminosity AGN NGC 4258 [82].

Angular momentum will generally *not* be conserved in the flow, and the mechanisms of outward angular momentum transport and how they are connected to the conversion of orbital mechanical energy into other forms are central to understanding how accretion power works. Ordinary fluid viscosity is far too weak to be a significant factor, and something else must exert torques (either internal or external, or both) on the flow.

Virtually all models that actually attempt to make contact with the observations are based on a phenomenological prescription of a (vertically-averaged) anomalous internal stress introduced by Shakura & Sunyaev in 1973 [102]:

$$\tau_{\mathrm{R}\phi} = \alpha P, \tag{2.1}$$

where P is a vertically-averaged pressure. Many workers in the field continue to think about this stress as a form of "viscosity", and the stress prescription is often explicitly introduced as exactly that! (As we shall see below, such naive thinking is very dangerous and can be extremely misleading.) For example, a commonly used modification of the Shakura-Sunyaev prescription is to assume a kinematic viscosity of the form

$$\nu \simeq \alpha c_{\mathrm{s}} H \simeq \alpha c_{\mathrm{s}}^2 / \Omega_{\mathrm{K}}, \tag{2.2}$$

where $c_{\mathrm{s}}^2 \simeq P/\rho$ is the vertically averaged sound speed (and ρ is a vertically averaged density), $H = H(R)$ is the vertical half-thickness of the disk, and

$\Omega_{\rm K}$ is the angular velocity of test-particle circular orbits (the "Keplerian" angular velocity). When inserted into the standard viscous form of the stress, this gives

$$\tau_{{\rm R}\phi} = \rho\nu R\frac{{\rm d}\Omega}{{\rm d}R} = \alpha P\frac{R}{\Omega_{\rm K}}\frac{{\rm d}\Omega}{{\rm d}R}. \tag{2.3}$$

Here $\Omega = \Omega(R)$ is the actual angular velocity in the flow, which may differ from $\Omega_{\rm K}$ if the flow is not completely geometrically thin. If $\Omega \sim \Omega_{\rm K}$, then within factors of order unity, equation (2.3) gives the same stress as equation (2.1). The only difference is that the stress now depends explicitly on the shear in the flow, just as an ordinary viscous stress would. This simple change completely alters the mathematical character of the critical point problem in stationary advective flows [12], but it is not clear that any of this is real.

Yet another uncertainty that has plagued black hole accretion disk models with Shakura-Sunyaev stress prescriptions from the very beginning is what pressure (gas or radiation or...?) to stick into equations (2.1) or (2.3) when radiation pressure is comparable to or greater than gas pressure. This occurs in the central parts of standard disk models around black holes accreting at anywhere near the Eddington rate (see Eqs. (5.1)–(5.2) below), right where most of the power is generated. Moreover, the first thing that one would try (just inserting the total, gas plus radiation, pressure) leads to a disk that *cannot* be stationary because it is thermally and "viscously" unstable [72, 103].

Let us briefly trace the argument for the radiation pressure driven thermal instability, which acts on a much faster time scale $\sim 1/(\alpha\Omega)$ in geometrically thin disks than the "viscous" instability which acts on the radial flow time scale $\sim R^2/(H^2\alpha\Omega)$. Assuming that cooling of the disk proceeds through radiative diffusion, the local emergent flux at radius R is given by

$$F^- \sim \frac{acT^4}{\tau}, \tag{2.4}$$

where a is the radiation density constant, c is the speed of light, T is a measure of the disk interior temperature, and τ is half the total vertical optical depth. The inner parts of black hole accretion disks have opacities that are generally dominated by electron scattering, so $\tau \sim \kappa_{\rm T}\Sigma/2$, where Σ is the surface density of the disk, which is constant on these time scales very much less than the radial flow time scale. The Thomson opacity $\kappa_{\rm T}$ is also constant, being independent of temperature provided there is already sufficient ionization (and there is), so the optical depth is independent of temperature and the cooling rate per unit area is $F^- \propto T^4$. Note that I have made the usual assumption here that F^- is a real heat flux and does not involve other forms of energy.

The dissipation rate per unit area is

$$F^+ \sim RH\tau_{\mathrm{R}\phi}\frac{\mathrm{d}\Omega}{\mathrm{d}R}. \qquad (2.5)$$

Vertical hydrostatic equilibrium implies that the disk half thickness $H \sim 2P/(\Omega_{\mathrm{K}}^2\Sigma)$, so that equation (2.5) becomes

$$F^+ \sim \frac{2RP\tau_{\mathrm{R}\phi}}{\Omega_{\mathrm{K}}^2\Sigma}\frac{\mathrm{d}\Omega}{\mathrm{d}R}. \qquad (2.6)$$

In the radiation pressure dominated inner region, $P \simeq aT^4/3$, so that equations (2.6) and (2.1) or (2.3) imply that $F^+ \propto T^8$! Hence a perturbative increase in temperature increases both the local cooling and heating rates, but the heating rate increases much faster, leading to a thermal runaway.

Such thermal (and "viscous") instabilities arising from opacity variations in hydrogen ionization zones rather than radiation pressure have been applied with considerable success to understanding the outburst behavior of dwarf novae (*e.g.* [89]) and soft X-ray transients (*e.g.* [65]). The reality of these instabilities in radiation-pressure dominated zones has never been established, however. It has been argued [98] that the stress should be proportional to the gas pressure only, in which case $\tau_{\mathrm{R}\phi} \propto \rho T \propto \Sigma^2\Omega_{\mathrm{K}}^2 T/(4P) \propto T^{-3}$ in the radiation pressure dominated inner zone. This implies that F^+ depends only linearly on temperature, producing a thermally (and in fact "viscously" as well) stable flow. We will examine this argument in Section 5 below. In addition, we will look at how radiation magnetohydrodynamic instabilities might affect thermal instabilities in accretion flows.

Observationally motivated models of accretion flows generally start with the assumption that the flow is stationary and axisymmetric about the rotation axis, even though whatever is responsible for the anomalous angular momentum transport must almost certainly involve time-dependent, non-axisymmetric fluctuations. (Time-dependent models are often constructed to model the response of the disk to thermal and "viscous" instabilities.) Another simplifying assumption is that vertically-integrated flow equations provide a reasonably accurate description of the behavior of conserved quantities, even if the flow is not very geometrically thin. The resulting steady-state conservation laws are those for mass,

$$\dot{M} = 4\pi RH\rho v, \qquad (2.7)$$

radial momentum,

$$\rho v\frac{\mathrm{d}v}{\mathrm{d}R} = \rho(\Omega^2 - \Omega_{\mathrm{K}}^2)R - \frac{\mathrm{d}P}{\mathrm{d}R}, \qquad (2.8)$$

angular momentum,

$$\dot{M}\frac{\mathrm{d}\ell}{\mathrm{d}R} = \frac{\mathrm{d}}{\mathrm{d}R}(4\pi R^2 H\tau_{\mathrm{R}\phi}), \tag{2.9}$$

and energy,

$$-4\pi RF_{\mathrm{adv}} \equiv \dot{M}\left[\frac{\mathrm{d}U}{\mathrm{d}R} + P\frac{\mathrm{d}}{\mathrm{d}R}\left(\frac{1}{\rho}\right)\right] = 4\pi R^2 H\tau_{\mathrm{R}\phi}\frac{\mathrm{d}\Omega}{\mathrm{d}R} + 2\pi R(2F^-). \tag{2.10}$$

Here $\dot{M}$ is the constant accretion rate through the flow, R is the distance from the rotation axis, $v(R)$ is the inward radial flow speed, $\ell(R)$ is the angular momentum per unit mass, $U(R)$ is the internal energy per unit mass, and $F^-(R)$ is the energy flux leaving the flow on each of the two vertical surfaces.

Note that the anomalous stress $\tau_{\mathrm{R}\phi}$ is assumed to enter the angular momentum and energy equations in exactly the same way as an ordinary fluid viscous stress: torques are exerted in a dissipative fashion. This need not be the case. External torques exerted on the disk by a global, ordered magnetic field in a magnetohydrodynamical (MHD) wind can in principle extract energy and angular momentum in precisely the right ratio necessary to allow material to accrete with no dissipation [31]. The disk can be ice cold and still accreting! Because of possibilities like this, I have been careful to call F^- an energy flux and *not* a heat flux. Even if internal turbulence is responsible for angular momentum transport, it is still not clear that all the accretion power should be converted into heat. Instead, some of that power may be converted into bulk kinetic and magnetic energy, and F^- must include these non-radiative contributions.

Equations (2.8) and (2.10) differ from those of standard geometrically thin accretion disk theory (*e.g.* [102]) by the inclusion of advective and radial pressure support terms which have received huge theoretical attention in recent years in the guise of advection dominated accretion flows (ADAFs, [2, 86]). In the absence of these effects, one gets the standard equations of thin disk theory, *i.e.* that the angular velocity is Keplerian,

$$\Omega(R) = \Omega_{\mathrm{K}}(R), \tag{2.11}$$

and local internal dissipation is balanced entirely by local vertical cooling,

$$F^- = -RH\tau_{\mathrm{R}\phi}\frac{\mathrm{d}\Omega_{\mathrm{K}}}{\mathrm{d}R}. \tag{2.12}$$

The angular momentum equation (2.9) can be integrated from an inner radius of the flow R_{in} (usually taken to be the innermost stable circular orbit, or ISCO, radius for black hole accretion disks out to an arbitrary radius R to give

$$\dot{M}(\ell - \ell_{\rm in}) = 4\pi R^2 H \tau_{{\rm R}\phi} - 4\pi R_{\rm in}^2 H_{\rm in} \tau_{{\rm R}\phi{\rm in}}. \tag{2.13}$$

Combining this with the local energy balance condition (2.12) for geometrically thin disks allows us to write down an expression for the emergent flux which is independent of the anomalous stress throughout most of the flow,

$$F^- = \frac{3GM\dot{M}}{8\pi R^3}\left[1 - \left(\frac{R_{\rm in}}{R}\right)^{1/2} + \frac{4\pi R_{\rm in}^2 H_{\rm in}\tau_{{\rm R}\phi{\rm in}}}{\dot{M}(GMR)^{1/2}}\right]. \tag{2.14}$$

This is the most beautiful result of standard accretion disk theory: that steady state accretion disks have an emergent flux that is independent of the details of the anomalous stress. Of course, accretion flows around black holes are observed to vary on all sorts of time scales, so steady state may not be a good assumption. Moreover, there is still a dependence on the stress $\tau_{{\rm R}\phi{\rm in}}$ at the inner edge of the flow. Until recently, this stress was almost always assumed to vanish near the ISCO, due to the fact that the inward transonic flow inside this point was presumed to rapidly decouple from the disk. This assumption has been challenged recently [44,68], and the physics of angular momentum transport near the ISCO has since received considerable theoretical scrutiny. We will examine this further in Section 4 below, and for now, we will keep the inner stress in our equations.

Equation (2.14) can be integrated to give the total luminosity of the disk,

$$L = \int_{R_{\rm in}}^{\infty} 2\pi R(2F^-){\rm d}R = \frac{GM\dot{M}}{2R_{\rm in}} + (4\pi R_{\rm in}^2 H_{\rm in}\tau_{{\rm R}\phi{\rm in}})\Omega_{\rm in}. \tag{2.15}$$

This has a simple physical interpretation. The first term is the binding energy per unit mass of material at the inner radius, times the accretion rate. The second is the rate at which work is being done on the disk at the inner radius. By defining the accretion efficiency in the usual way, $\eta \equiv L/(\dot{M}c^2)$, with $\eta_{\rm SS}$ being the efficiency under the standard (Shakura-Sunyaev) assumption of a no-torque inner boundary condition and $\Delta\eta \equiv \eta - \eta_{\rm SS}$ being the additional efficiency due to the inner torque, equation (2.14) can be written in a physically appealing way (*e.g.* [7]),

$$F^- = \frac{3GM\dot{M}}{8\pi R^3}\left[1 - \left(\frac{R_{\rm in}}{R}\right)^{1/2} + \Delta\eta\left(\frac{R_{\rm in}}{r_{\rm g}}\right)\left(\frac{R_{\rm in}}{R}\right)^{1/2}\right]. \tag{2.16}$$

Here $r_{\rm g} \equiv GM/c^2$ is the gravitational radius. (The full relativistic version of (2.16) may be found in [7].) Note that the effects of the inner boundary on the flux decay with radius as $R^{-7/2}$, only slightly faster than the overall asymptotic decay of the flux at large radius, $\propto R^{-3}$.

So far we have been neglecting the advective terms in our conservation laws (2.7)–(2.10). If we include them, equation (2.14) becomes

$$F^- = \left(-\frac{\dot{M}\ell}{4\pi R}\frac{\mathrm{d}\Omega}{\mathrm{d}R}\right)\left(1 - \frac{\ell_{\mathrm{in}}}{\ell} + \frac{4\pi R_{\mathrm{in}}^2 H_{\mathrm{in}} \tau_{\mathrm{R}\phi\mathrm{in}}}{\dot{M}\ell}\right) - F_{\mathrm{adv}}. \qquad (2.17)$$

Advection, which becomes increasingly important as one moves inward in the flow, has two effects: the rotation profile is no longer Keplerian, and the orbital energy dissipated at any radius is partly advected (lost!) radially inward. The latter effect is evident from the last term in equation (2.17). The emerging energy flux F^- is less than it would be if advection was not included.

For luminous accretion disks, advection starts to play a big role when the overall luminosity starts to get above some fraction ($\sim$0.3 in standard models) of the Eddington luminosity L_{Edd}. The flow is then geometrically thicker, becoming a "slim" or "thick" accretion flow. Such flows are thermally and "viscously" stable [1,4], because F_{adv} has a temperature dependence strong enough to beat the temperature dependence of F^+. The radial and vertical structure of these flows is no longer decoupled, and the emergent flux F^- now depends on the very uncertain assumptions that are usually made to solve for the vertical structure. For a discussion of the fully relativistic equations of advective accretion flows, see *e.g.* [3,5,92], and [26]. Fully relativistic models of optically thick, "slim" accretion disks have been constructed in [27]. Models of hyper-Eddington accretion flows onto stellar mass black holes relevant for gamma-ray burst models have also been constructed [93]. By far and away most of the energy dissipated in such flows is advected into the black hole until the accretion rate exceeds $\sim 10^{-2}\ M_\odot\ \mathrm{s}^{-1}$, where neutrino losses can cool the gas on the flow time scale.

Before leaving this section, I should mention the considerable attention that has been devoted recently to optically thin, radiatively inefficient ADAF solutions around black holes. Because we are focusing on luminous accretion flows here, I will have little to say about these solutions, but it is important to recognize that they may nevertheless arise over certain ranges of radii even in otherwise luminous flows. In particular, it is noteworthy that such flow solutions exist that are thermally and "viscously" stable over much the same range of accretion rates that geometrically thin, optically thick disk solutions exist (*e.g.* [33]), and it is not clear why nature should choose one flow solution over the other. The presence or absence of an inner ADAF flow might explain the different spectral states observed in black hole X-ray binaries [39]. One should bear in mind, however, that all the models discussed in this section are based on the α stress prescription, and it is fair to say that the microphysics of optically thin ADAFs is even more poorly understood than that of optically thick flows (*e.g.* [25, 96]).

3 Spectral formation

As we saw in the last section, if we neglect the effects of advection, then the energy flux emerging from each face of a stationary accretion disk can be written

$$F^{-}(R) = \frac{3GM\dot{M}}{8\pi R^3}\mathcal{I}_{a/M}(R/r_{\rm g}), \tag{3.1}$$

where $\mathcal{I}_{a/M}(x)$ is a function which in full general relativistic treatments depends on the spin parameter a/M of the hole but still approaches unity at large radii. In standard models with a no-torque inner boundary condition, $\mathcal{I}_{a/M}$ also vanishes at the ISCO, but as we have seen, this need not be the case. Once again, let me remind the reader that although F^{-} is usually assumed to be the emerging radiation flux from the disk surface, this need not be the case. It is actually the total energy flux, and may include contributions from magnetic and bulk kinetic energy. We will return to this point shortly, but for now, let us make the standard assumption that F^{-} really does represent an emerging radiation flux.

What, then, is the emerging spectral energy distribution? An observer here on earth, viewing the disk at a distance d away at an inclination angle i to the rotation axis, will measure a spectral energy flux

$$F_\nu = \frac{\cos i}{d^2}\int_{R_{\rm in}}^{R_{\rm out}} 2\pi R I_\nu(R, i){\rm d}R, \tag{3.2}$$

where I_ν is the angle (limb darkened or brightened!) and radius-dependent local emergent specific intensity, and we have introduced a finite outer radius $R_{\rm out}$ to the disk. (Once again, we are using Newtonian physics here, and neglecting *e.g.* Doppler shifts, gravitational redshifts, the gravitational bending of light rays, and the relativistic proper emitting area, but these are all straightforward to account for using relativistic "transfer functions", *e.g.* [34, 108].)

The simplest (and therefore most often used!) way of estimating the shape of the spectrum is to just naively assume that each annulus radiates like an isotropic local blackbody, *i.e.* $I_\nu(R, i) = B_\nu[T_{\rm e}(R)]$, where B_ν is the Planck function with temperature given by the effective temperature determined from equation (3.1), $T_{\rm e}(R) = [F^{-}(R)/\sigma]^{1/4}$. If, say, the radial dependence of the emerging radiation flux can be approximated as a power law, $F^{-}(R) \propto R^{-\beta}$, then a simple change of integration variable to

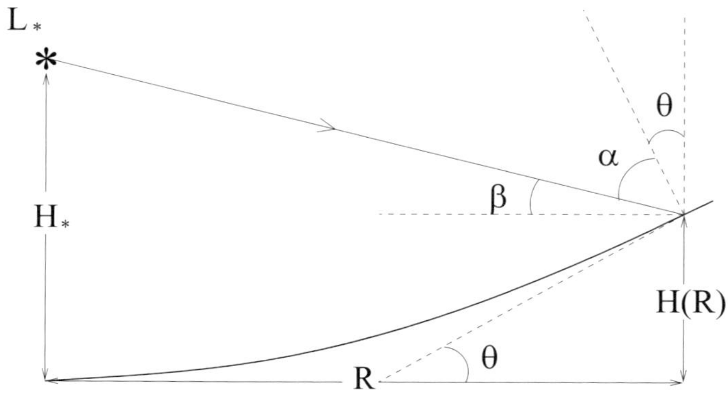

Fig. 1. Illumination geometry for a source on the symmetry axis at height $H_\star$ above the equatorial plane. The angle α between the disk normal and the incoming light ray is simply $\pi/2 - (\beta + \theta)$, where θ is the angle between the tangent to the local disk surface and the equatorial plane, and β is the angle between the illuminating light ray and the equatorial plane. For large R, β and θ are both small angles, with $\beta \simeq (H_\star - H)/R$ and $\theta \simeq \mathrm{d}H/\mathrm{d}R$.

$x \equiv h\nu/(kT)$ implies that the shape of the observed spectrum is given by

$$F_\nu \propto \nu^{3-8/\beta} \int_{h\nu/(kT_{\rm ein})}^{h\nu/(kT_{\rm eout})} \frac{x^{8/\beta-1}\mathrm{d}x}{\mathrm{e}^x - 1}. \tag{3.3}$$

As one would expect, at low frequencies satisfying $h\nu \ll kT_{\rm eout}$ we simply end up with $F_\nu \propto \nu^2$ (the Rayleigh-Jeans portion of the spectrum emitted by the outermost annulus), while at high frequencies satisfying $h\nu \gg kT_{\rm ein}$ we have $F_\nu \propto \nu^3 \exp[-h\nu/(kT_{\rm ein})]$ (the Wien portion of the spectrum emitted by the innermost annulus). In between, for $kT_{\rm eout} \ll h\nu \ll kT_{\rm ein}$, the observed spectrum is a superposition of blackbodies from many different radii, producing a power-law spectrum $F_\nu \propto \nu^{3-8/\beta}$. Now, at large radii, equation (3.1) implies that $\beta = 3$, giving the famous result that $F_\nu \propto \nu^{1/3}$ for a multi-temperature blackbody accretion disk.

There are many ways in which this standard result can be modified. One possibility is that energy liberated at small radii in the flow is reprocessed by the disk at larger radii. By way of illustration, consider a source of luminosity $L_\star$ located on the rotation axis at height $H_\star$ above the equatorial plane. Figure 1 illustrates the geometry of the problem.

Neglecting general relativistic effects, in particular light bending, the irradiating flux on the outer disk at radius R will be given by

$$F_{\rm irr} = \frac{L_\star(1-a)}{4\pi R^2}\cos\alpha, \tag{3.4}$$

where a is the albedo of the outer disk, averaged over the illuminating spectrum, and α is the angle between the disk normal and the incoming light ray. Taking the illuminating radius R to be much larger than the local disk height H, the source height $H_\star$, and also $\mathrm{d}H/\mathrm{d}\ln R$, then a little trigonometry gives

$$F_{\rm irr} = \frac{L_\star(1-a)}{4\pi R^2}\left(\frac{H}{R}\right)\left(\frac{\mathrm{d}\ln H}{\mathrm{d}\ln R} - 1 + \frac{H_\star}{H}\right). \tag{3.5}$$

For a source located near the equatorial plane, *e.g.* the surface of the inner disk itself, we have $H_\star \ll H(R)$. Equation (3.5) then implies that significant reprocessing will only occur if the outer disk geometry flares up, with H/R an increasing function of R, or is warped [94]. In this case, proportionately more power is transferred to larger radii, and this decreases the value of β and reddens the spectrum. Standard models of disks work in just this way, with $H \propto R^{21/20}$ and $R^{9/8}$ for the so-called "middle" and "outer" regions, respectively. This gives an irradiating flux approximately proportional to R^{-2}, which must therefore dominate the more rapidly decaying ($\propto R^{-3}$) local dissipation, and giving $F_\nu \propto \nu^{-1}$ (*e.g.* [42]).

For a source located far above the equatorial plane, with $H_\star \gg H(R)$, equation (3.5) gives an irradiating flux $\propto R^{-3}$, producing the same $F_\nu \propto \nu^{1/3}$ spectrum as a standard disk but with enhanced power. One way in which this might be done is by having a geometrically thick, hot corona or ADAF in the inner parts of the flow. General relativistic light bending can also produce a similar effect: an observer located on the outer disk can see the opposite side of the disk at an effectively high $H_\star$ [7]. This effect alone can produce significant additions to the locally dissipated power in the case of rapidly rotating Kerr black holes [35].

Even without reprocessing, the emergent spectrum can be affected by alterations in the radial distribution of the local radiative flux F^-. For example, if an inner torque across the ISCO produces a large increase in the accretion efficiency, equation (2.16) implies that a substantial range of inner radii could have $F^- \propto R^{-7/2}$, giving a spectral energy distribution from equation (3.3) of the form $F_\nu \propto \nu^{5/7}$, *bluer* than $\nu^{1/3}$! On the other hand, gravitational light bending from the much brighter inner disk enhances the amount of reprocessed flux at larger radii, giving $F_{\rm irr} \propto R^{-3}$ and enforcing a more usual $F_\nu \propto \nu^{1/3}$ spectrum at longer wavelengths [7].

Advection also alters the radial distribution of F^-, causing it to rise less steeply toward smaller radii because more of the accretion power is advected into the black hole rather than being radiated. This in turn implies that F^- falls less steeply when moving out in radius, implying a redder spectrum than $F_\nu \propto \nu^{1/3}$ [115]. In fact, models of slim disks can give flux distributions as flat as $F^- \propto R^{-2}$, and equation (3.3) then gives $F_\nu \propto \nu^{-1}$ [81, 121, 122].

The global effect of advection on the spectrum is that the luminosity is less than you would expect for the given accretion rate, because advection necessarily reduces the radiative efficiency.

Most models of the vertical structure of accretion disks around black holes imply that local blackbody emission is likely to be a *very* poor approximation. The most serious problem with it is the neglect of electron scattering, which is often far greater than true thermal absorption opacity. Again, a crude approach to handling this has been used from the very beginning of accretion disk modeling [102], and that is to use a so-called local modified blackbody spectrum at each radius,

$$I_\nu = \frac{2B_\nu(T)\epsilon_\nu^{1/2}}{1+\epsilon_\nu^{1/2}}, \tag{3.6}$$

where

$$\epsilon_\nu = \frac{\chi_\nu^{\rm th}}{n_{\rm e}\sigma_{\rm T}+\chi_\nu^{\rm th}} \tag{3.7}$$

is the photon destruction probability, with $\chi_\nu^{\rm th}$ being the thermal absorption coefficient, $n_{\rm e}$ the electron number density, and $\sigma_{\rm T}$ the Thomson cross-section.

The physics behind equation (3.6) lies in the idea that, once created by thermal emission processes, photons random walk out of the atmosphere. If absorption opacity is smaller than electron scattering opacity, this random walk does not destroy photons generated within an effective optical depth $\tau_{\rm eff} = (\tau_{\rm abs}\tau_{\rm T})^{1/2}$ of the surface, where $\tau_{\rm abs}$ is the true absorption optical depth and $\tau_{\rm T}$ is the Thomson depth. Assuming LTE and integrating the emission over this vertical layer immediately gives $I_\nu \sim B_\nu\epsilon_\nu^{1/2}$, in rough agreement with equation (3.6). Provided the accretion disk is *effectively* thick ($\tau_{\rm eff} > 1$ at the frequency of interest), then the spectrum may be viewed as coming from a depth in the disk corresponding to $\tau_{\rm eff} = 1$. As shown in Figure 2, the innermost regions of some black hole accretion disk models are not even effectively thick, however, and the observed spectrum is formed throughout the entire geometrical thickness of the disk. In such cases the spectrum will be directly sensitive to the assumptions that went into creating the entire vertical structure. Even when $\tau_{\rm eff} > 1$, the spectrum will depend on the ambient densities and temperatures at the $\tau_{\rm eff} = 1$ effective photosphere, and these will in turn depend on the vertical structure of the disk model.

Note that the modified blackbody spectrum in equation (3.6) has $I_\nu \leq B_\nu(T)$. This makes sense, as deviations from blackbody imply that the disk is a thermodynamically less efficient radiator. Because it must still radiate the same total flux, the ambient gas temperature must be hotter

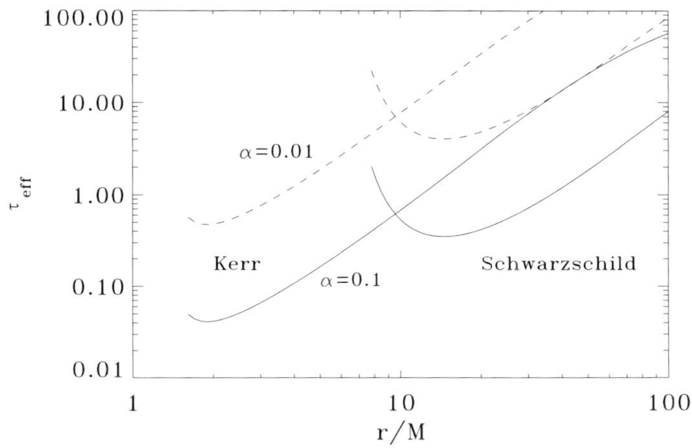

Fig. 2. The midplane effective optical depth as a function of radius in relativistic accretion disk models around a ten solar mass black hole with $L = 0.3L_{\rm Edd}$ and $\tau_{{\rm R}\phi} = \alpha P_{\rm tot}$, where $P_{\rm tot}$ is the total (gas plus radiation) pressure. Both Kerr ($a/M = 0.998$) and Schwarzschild models are shown, and a no-torque boundary condition is assumed at the ISCO, which is why the curves rise upward at small radii. The effective optical depth is actually a function of frequency, and what is depicted here is an average using the Rosseland mean free-free absorption opacity. Solid and dashed curves assume $\alpha = 0.1$ and 0.01, respectively. The lower anomalous stress of the latter implies higher surface density and higher effective optical depth in both the Kerr and Schwarzschild cases.

than the effective temperature, and the resulting spectrum therefore extends to higher photon energies. When integrated over the disk, the spectral energy distribution must therefore flatten below the canonical $F_\nu \propto \nu^{1/3}$ as photons are redistributed to higher energies. These effects are illustrated in Figure 3, which compares stellar atmosphere calculations of relativistic disk spectra with the local blackbody assumption.

In addition to the effects of Thomson scattering on the spectrum, Compton scattering can also be important in hot disks. The net Compton heating rate of the plasma per unit volume is

$$n_{\rm e}\sigma_{\rm T}c\left(\frac{h\bar{\nu}}{m_{\rm e}c^2} - \frac{4kT}{m_{\rm e}c^2}\right)\left(\frac{4\pi J}{c}\right), \tag{3.8}$$

where J is the frequency-integrated mean intensity of the radiation field,

$$J \equiv \int_0^\infty {\rm d}\nu J_\nu = \frac{1}{4\pi}\int_0^\infty {\rm d}\nu \oint {\rm d}\Omega I_\nu, \tag{3.9}$$

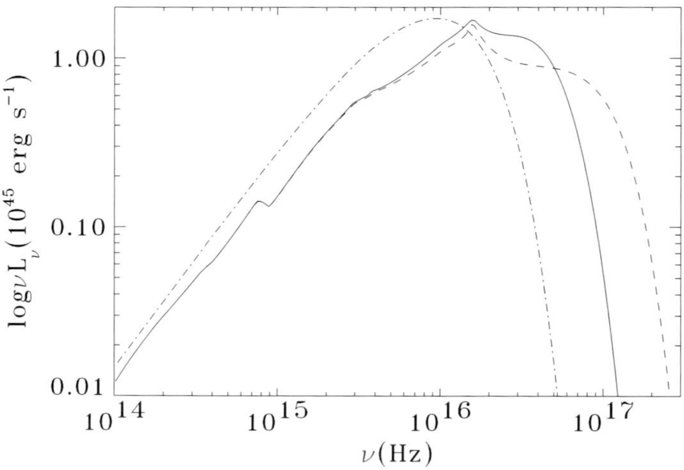

Fig. 3. Spectral energy distributions of relativistic accretion disk models around a $1.25 \times 10^8\ M_\odot$ Kerr ($a/M = 0.998$) hole, accreting at $0.25\ M_\odot\ \mathrm{yr}^{-1}$, viewed at 60° from the rotation axis [62]. The dashed curve, which extends to the highest energies, assumes that electron scattering is coherent (Thomson scattering). The solid curve allows for the effects of Compton scattering. Compton downscattering pulls the high energy tail of the spectrum down, while Compton upscattering increases the spectral luminosity at lower energies. The dot-dashed curve is the spectrum that results from assuming blackbody emission at the local effective temperature at every radius, and is clearly a very poor approximation.

and $\bar{\nu}$ is a specially defined average frequency,

$$\bar{\nu} \equiv \frac{1}{J} \int_0^\infty \mathrm{d}\nu \nu J_\nu \left(1 + \frac{J_\nu c^2}{2h\nu^3}\right). \tag{3.10}$$

(The second term in parentheses in Eq. (3.10) represents the effects of stimulated scattering.) The Compton heating rate can be compared with the net heating rate from thermal absorption and emission. Assuming LTE, this is

$$c \int_0^\infty \mathrm{d}\nu \chi_\nu^{\mathrm{th}} \left(\frac{4\pi}{c}\right) (J_\nu - B_\nu). \tag{3.11}$$

Defining the Thomson opacity as

$$\kappa_\mathrm{T} \equiv \frac{n_\mathrm{e} \sigma_\mathrm{T}}{\rho} \tag{3.12}$$

and the Planck mean opacity as

$$\kappa_\mathrm{P} \equiv \frac{1}{\rho a T^4} \int_0^\infty \mathrm{d}\nu \chi_\nu^\mathrm{th} \left(\frac{4\pi B_\nu}{c} \right), \tag{3.13}$$

we see on comparing equations (3.8) and (3.11) that Compton scattering will dominate the thermal coupling between gas and radiation if

$$\kappa_\mathrm{T} \left(\frac{4kT}{m_\mathrm{e} c^2} \right) \gtrsim \kappa_\mathrm{P}. \tag{3.14}$$

If, in addition, the effective Compton y-parameter,

$$y_\mathrm{eff} = \frac{4kT}{m_\mathrm{e} c^2} \max \left(\tau_\mathrm{T}, \tau_\mathrm{T}^2 \right), \tag{3.15}$$

exceeds unity, at the particular frequency of interest, then Compton scattering will modify the spectrum, either by downscattering if $h\nu > 4kT$, or upscattering if $h\nu < 4kT$. (In Eq. (3.15), τ_T is evaluated at the $\tau_\mathrm{eff} = 1$ surface.) Figure 3 illustrates these effects on the spectrum.

Numerous authors have attempted detailed calculations of the spectral energy distributions emerging from accretion disks around black holes using stellar atmosphere modeling at every radius of the disk (see *e.g.* [37,61–64, 71,97,101,106,107,112,113] for models in the AGN context). Many of these models fully incorporate general relativistic effects on the disk structure, and also include relativistic Doppler shifts, gravitational redshifts, and gravitational bending of light rays. These latter effects play an important role in smearing out atomic absorption and/or emission features in the spectrum, as illustrated in Figure 4. Many models also incorporate sophisticated treatments of the radiative transfer, Compton scattering, and non-LTE effects in the atomic level populations of numerous elements and ions.

As sophisticated as these models are, however, they still rest on variants of the Shakura-Sunyaev anomalous stress prescription, and are subject to all the uncertainties we discussed above in Section 2. Moreover, just as a stellar model must specify a distribution of nuclear energy generation, a model of the vertical structure of an accretion disk must specify the distribution of turbulent heating. The presence of turbulence in the disk also begs the question as to whether heat is transported radiatively, or whether bulk transport in the turbulence itself also plays a role. As we discuss below in Section 5, standard assumptions often used to model the inner, radiation pressure dominated regions lead to equilibria that are convectively unstable [28], although it is far from clear how that convection would manifest itself in a flow that already requires turbulent angular momentum transport.

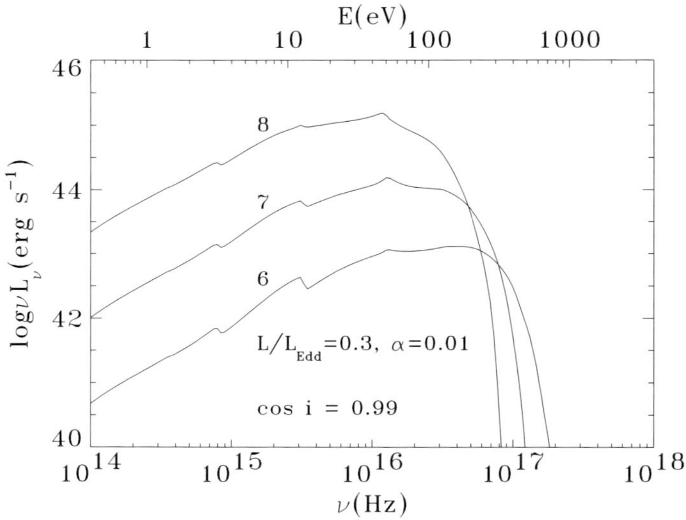

Fig. 4. Spectral energy distributions of accretion disk models around Kerr ($a/M = 0.998$) black holes at fixed Eddington ratio viewed nearly face on, for various black hole masses: 10^8 (top), 10^7 (middle), and 10^6 $M_\odot$ (bottom) [62]. Note the increased relativistic smearing of the Balmer and Lyman edges of hydrogen for higher black hole mass. This is because the temperature of the disk is lower at higher mass, so these features are formed closer to the black hole.

Perhaps the most serious question plaguing current Shakura-Sunyaev based spectral models is the partitioning of the dissipation within the optically thick and thin regions of the disk. There is widespread empirical evidence for the existence of Comptonizing hot plasma in both black hole X-ray binaries and AGN, and one possibility is that this plasma exists as a magnetized corona above the disk photosphere. After all, the turbulent convection zone of our own sun generates a hot corona, so why shouldn't a turbulent accretion disk have one as well? (This is one version of including non-radiative energy in F^-.) Models have even been considered in which most of the turbulent dissipation is assumed to occur in the corona, and not in the disk interior [49]! If a substantial fraction of the accretion power is in fact dissipated outside the disk, then there are numerous consequences [114]: the disk becomes geometrically thinner and denser, more effectively optically thick, and more gas pressure dominated.

Accretion disk physics is uncertain enough, but the physics of the hot plasma that must exist in real sources is even more insecure. Whether or not the hard X-rays observed in black hole X-ray binaries or AGN are generated in a disk corona or elsewhere in the flow (*e.g.* an inner ADAF that we discussed at the end of Sect. 2), some of these X-rays will illuminate the disk from the outside, and thereby be reprocessed.

One of the most interesting X-ray reprocessing features is fluorescent iron Kα line emission. The great excitement surrounding this line is that in at least three sources (MCG-6-30-15 [41, 116, 123], Mrk 766 [78], and NGC 3516 [84, 120]), it is clearly observed to be relativistically broadened. The resulting line profile is a convolution of the spatial emissivity profile of the line with relativistic Doppler shifts, gravitational redshifts, and gravitational light bending [70]. In other words, the line profile gives us the unprecedented opportunity to map out the equatorial test-particle orbit structure in a black hole spacetime, assuming the emitting material (the disk) is geometrically thin.

The Kα line is produced because hard X-ray photons with energies above the K-edge of iron can knock inner shell electrons out of the atom. The resulting ion is born in an excited state and can return to the ground state by having an L-shell electron drop down to the vacancy in the K-shell, emitting either a Kα line photon or one or more Auger electrons. The strength of the resulting emission line is sensitive to the ionization state of iron in the outer layers of the disk. At low ionization levels, where both the K and L shells of most iron ions have no vacancies, the material is transparent to Kα photons which then escape more or less freely from the disk once they have been created by X-ray fluorescence. Once iron is sufficiently ionized that vacancies in the L-shell start to appear (FeXVIII to FeXXIV), then a Kα photon can resonantly scatter off the iron atoms, enhancing the probability of its destruction through the Auger process. The emerging line strength is therefore reduced. This continues until iron is sufficiently ionized that only FeXXV and higher is present. The Kα line photons still resonantly scatter with such ions, but the Auger effect cannot happen because there are no L-shell electrons to be ejected. At very high ionization when iron is fully stripped and the recombination rate is small because of high temperatures, no photoionization can occur and the emission line disappears. For an excellent review of the physics of the iron Kα line, see [40].

The energy powering the iron Kα line emission comes from photoelectric absorption of X-rays above the iron K-edge. Such photoelectric absorption by iron and other heavy elements present in the disk is in fact an efficient means of converting illuminating X-ray power below $\sim$10 keV into heat in the outermost layers of the disk [48, 73]. Figure 5 shows the albedo of an illuminated slab of material as a function of incident photon energy [77]. Below $\sim$10 keV, most of the incident X-rays are absorbed due to photoionization of heavy elements, and the level of that absorption depends on the ionization state of the disk. Above $\sim$10 keV, the albedo quickly rises to

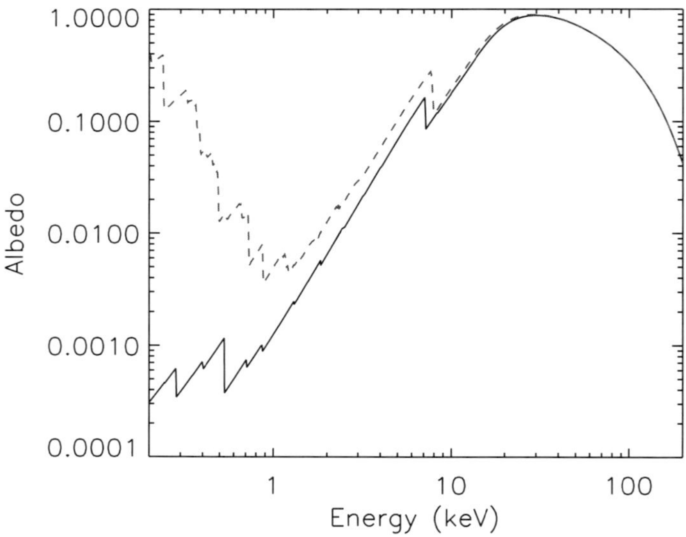

Fig. 5. Energy-dependent albedo of an X-ray illuminated, cold slab with solar abundances, viewed at twenty degrees from the slab normal. The solid curve assumes the slab is neutral, apart from hydrogen and helium which are fully ionized. The dashed curve is for an ionized slab, with ionization parameter $\xi = 100$. In both cases the illumination was assumed to be by a power-law spectrum with photon index -2 and exponential cutoff energy of 100 keV. (This figure was kindly generated by Shane Davis, courtesy of the XSPEC routines PEXRAV and PEXRIV [77].)

near unity as the incident X-rays are simply reflected back by Thomson scattering. This is widely believed to explain the characteristic upturn of the X-ray spectrum that is observed at these energies in both AGN and black hole X-ray binaries [85, 124], and is strong evidence for the presence of relatively cold material in these sources. (Note that especially in the case of AGN, however, this cold material need not be the disk, but instead could be gas much further out from the source.) At higher energies the albedo is again reduced to well below unity, independent of the ionization state of the slab, because X-rays are Compton downscattered and therefore lose energy.

Just as the thermal emission spectra from accretion disks can depend sensitively on the details of the vertical disk structure, X-ray reprocessing and reflection features also depend on the assumptions one makes in describing the physics of the illuminated slab. Models that incorporate hydrostatic equilibrium produce surface temperature profiles containing a number of sharp discontinuities due to transitions between different regions of thermally stable material [22, 87], and the resulting reflection/fluorescent spectra are altered by this fact.

In addition to the very uncertain geometry of the illumination itself, one wonders whether even this treatment of the physics is enough. The bottom line of this section is that many aspects of our models of observed accreting black hole sources are fundamentally limited by uncertainties in the basic physics of the accretion flow itself. For the rest of these lectures we will switch gears and address this physics itself, explaining how to go beyond Shakura-Sunyaev type models.

4 The physics of angular momentum transport

A system with fixed total angular momentum in thermal equilibrium must be in a state of rigid body rotation [69]. This fact lies at the very heart of the way accretion disks work, because it shows that differential rotation is a source of thermodynamic free energy in the flow. As discussed long ago by Lynden-Bell & Pringle [75], one can always reduce the total rotational kinetic energy of an accretion disk by fixing its mass distribution and redistributing the angular momentum so that it is uniformly rotating. However, such a state is generally incompatible with mechanical equilibrium in the gravitational field of the central object, which requires that the angular velocity decrease outward. In this case, pairs of orbiting fluid elements moving with different angular velocities can lower their mechanical energy by shifting angular momentum from the element with higher angular velocity to the element with lower angular velocity [75]. This transports angular momentum *outward* through the disk, causing material to lose rotational support and flow in to regions of greater binding energy. This general behavior is simply the action of the second law of thermodynamics, and is entirely analogous to the one-way flow of heat from regions of high temperature to low temperature. Ordinary microscopic viscosity acts to transport angular momentum in exactly this fashion, but this is not sufficient for astrophysical accretion disks.

Some form of non-microscopic, "anomalous" angular momentum transport mechanism or mechanisms must therefore be at work, and it is important to recognize that it need not be the same in all accretion flows in the universe. Inherent in the Shakura-Sunyaev "guess" as to the form of the anomalous stress $\tau_{r\phi}$ is the existence of some form of turbulence in the flow: ideally a form of turbulence that is generated by the differential rotation itself. Turbulence is often, though not always, generated when a laminar flow is dynamically unstable to perturbations. These perturbations may be infinitesimal, so that the slightest touch of a feather will produce instability. The initial growth of the instability may then be understood from linearized equations about the equilibrium flow, which is then said to be linearly unstable. Linear equations are *much* easier to solve than nonlinear equations, and much of the theoretical work has therefore been devoted to linear instabilities. Understanding the response of the flow to nonlinear, finite amplitude perturbations is a lot harder, but there do exist laminar flows that are nonlinearly unstable even while being linearly stable. Of course to theoretically understand how a flow really changes in response to the development of even a linear instability requires a treatment of the nonlinear equations, which in practice is usually accomplished through numerical simulation.

The simplest way to examine the linear stability of a fluid is to infinitesimally perturb a fluid element locally and examine whether the forces exerted on the element act to return it to its original position. Stability criteria derived in this way are necessary, but not sufficient because it is also possible for the global structure of the flow to produce instabilities. Unfortunately, analyzing the global stability of an accretion flow is technically much more difficult, even within linear theory. One must solve a global eigenvalue problem for the linear wave modes of the system, fully accounting for the boundary conditions on the flow. In order to derive a stability criterion, one must also ensure that the modes examined form a complete set so that any initial perturbation data can be represented in terms of them.

Global stability analyses generally recover the results of local stability analyses in the limit of very short wavelength perturbations (often called the WKB limit in the literature). If global unstable wave modes are found, it may be that changes in the assumed boundary conditions can stabilize the modes. This can never be true of local instabilities, however, which care only about local conditions in the flow. In this sense local instabilities are much more robust and generic than truly global instabilities.

In my discussion of accretion disk instabilities in this section, I will always neglect the self-gravity of the accreting plasma itself. Accretion rates in both AGN and black hole X-ray binaries are thought to be sufficiently small that the tidal field of the black hole completely overwhelms the self-gravity of the disk, at least near the hole where most of the luminosity is generated. Self-gravity is however more important further out in the disk in AGN, and may also be important in hyper-Eddington flows invoked in gamma-ray burst models.

Within ideal hydrodynamics, local linear stability of an axisymmetric rotating flow is guaranteed if the Høiland criterion is satisfied (*e.g.* [117]):

$$N^2 + \kappa^2 > 0 \tag{4.1}$$

and

$$(\boldsymbol{\nabla} P \times \hat{\mathbf{R}}) \cdot (\boldsymbol{\nabla} \ell^2 \times \boldsymbol{\nabla} S) > 0, \tag{4.2}$$

where

$$\kappa^2 = \frac{1}{R^3}\frac{\partial \ell^2}{\partial R} = \frac{1}{R^3}\hat{\mathbf{R}}\cdot\boldsymbol{\nabla}\ell^2 \tag{4.3}$$

is the square of the epicyclic frequency and

$$N^2 = \frac{1}{\rho}\boldsymbol{\nabla}P\cdot\left(\frac{1}{\rho}\boldsymbol{\nabla}\rho - \frac{1}{\gamma P}\boldsymbol{\nabla}P\right) = \frac{-1}{\rho c_{\mathrm{P}}}\boldsymbol{\nabla}P\cdot\boldsymbol{\nabla}S \tag{4.4}$$

is the square of the Brunt-Väisälä frequency. Other quantities in these equations are the pressure P, the density ρ, the entropy per unit mass S, the heat capacity per unit mass at constant pressure c_{P}, the ratio of heat capacities γ, the angular momentum per unit mass ℓ, the distance from the rotation axis R, and the corresponding cylindrical polar coordinate radial unit vector $\hat{\mathbf{R}}$. (A relativistic generalization of the Høiland criterion can be found in [100].)

The Høiland criterion is easy to understand in two limits. For non-rotating equilibria (*e.g.* a non-rotating star), the criterion reduces to the Schwarzschild criterion that the entropy must not increase inward for stability against convection. Or, perhaps in more familiar form, the temperature must not increase inward faster than the adiabatic (constant entropy!) temperature gradient. Provided this is true, local fluid elements will simply oscillate at the local Brunt-Väisälä frequency under stable buoyancy forces, and these restoring forces are responsible for supporting gravity waves (g-modes) in stars. To isolate the effects of rotation, consider an equilibrium that has constant entropy everywhere. Then the Høiland criterion reduces to the Rayleigh criterion: the specific angular momentum must not decrease outward for stability. Physically, if one perturbs a fluid element radially outward, it conserves its own specific angular momentum. If the ambient specific angular momentum decreases outward, then the fluid element will be rotating too fast to stay in its new position, and centrifugal forces will fling it further outward. Stability in this case implies that perturbed fluid elements will oscillate back and forth at the local epicyclic frequency, and organized epicyclic oscillations form the basis for Rossby waves (r-modes) in rotating stars. It is no accident that g-mode and r-mode physics appear together in the Høiland criterion, as these modes are fundamentally intertwined when both angular momentum and entropy gradients are present.

The Høiland criterion is a huge disappointment for understanding why turbulence might exist in accretion disks. The piece of it that is related to the rotation is connected to specific angular momentum gradients, and these are strongly stable for rotation profiles that do not differ too much from Keplerian. Of course, the Høiland criterion is only a local stability criterion. There do exist global, linear hydrodynamic instabilities in accretion disks. The one I am most familiar with is the Papaloizou-Pringle instability [91] but, like all global instabilities, its existence is sensitive to the assumed boundary conditions [29]. Its nonlinear development also appears to produce large scale spiral waves rather than local turbulence [50], although such waves are perfectly capable of producing "anomalous" angular momentum transport. The Høiland criterion also only addresses linear stability. Hydrodynamic, planar shear flows can be violently unstable to finite amplitude perturbations even if they are linearly stable, and this suggests that something similar might happen in accretion disks. However, extensive searches for such instabilities with numerical simulation have consistently failed to find them in linearly stable flows, except very close to the limit of marginal stability [20, 56]. It is noteworthy that this is true even though the same simulation hardware *easily* and correctly finds the instabilities in planar shear flows [20, 56]. As a point of principle, it is possible that the simulations are somehow not accessing an unknown destabilizing mechanism (see [74] for a recent argument in this direction). However, one should keep in mind that there is an important physical difference between linearly stable planar shear flows and differentially rotating flows: the latter have a strong local restoring force arising from specific angular momentum gradients [20].

There is a key piece of physics that is central to the Høiland criterion: perturbed fluid elements conserve their own entropy and specific angular momentum. This is *not* true when we include the effects of magnetic fields, as they fundamentally alter the local dynamics of perturbed fluid elements. To order of magnitude, the characteristic frequency associated with magnetic fields, within the MHD approximation anyway, is simply $kv_{\rm A}$, where k is the perturbation wavenumber and $v_{\rm A} = B/(4\pi\rho)^{1/2}$ is the Alfvén speed. Magnetic fields will alter the character of hydrodynamic r- and g-modes whenever $kv_{\rm A}$ is of order the relevant frequency ($|\kappa|$ or $|N|$) of these modes. (This is true even when the hydrodynamic modes are unstable, *e.g.* magnetic fields can strongly affect hydrodynamic convective instabilities which have imaginary N.) At first sight one would think that a weak magnetic field (low Alfvén speed) would therefore have no effect. However, no matter how weak the magnetic field is, one can always make it important by going to sufficiently high wavenumbers (short length scales), *e.g.* $k \sim |\kappa|/v_{\rm A}$ in the case of r-modes. Indeed, for $k \gg |\kappa|/v_{\rm A}$, one expects rotation to be irrelevant as the modes become magnetically dominated, no matter how weak is the magnetic field. This all assumes *ideal* MHD, in particular that magnetic field lines are frozen into the fluid. Resistive effects, which allow the field lines to "slip" with respect to the fluid, become more important on small scales. If the field is so weak that it only becomes important below resistive length scales, then we are back to hydrodynamics on all scales.

The basic physics as to how weak magnetic fields destabilize accretion disks is most transparent using a mechanical analogy developed by Balbus & Hawley [17]. Consider two identical test particles of mass m moving together (right on top of each other) in a circular orbit in an external gravitational potential $\Phi(R)$. The angular velocity Ω of the particles' motion will of course be given by

$$\Omega^2 = \frac{1}{R}\frac{\partial\Phi}{\partial R}. \tag{4.5}$$

Now suppose we perturb the particles away from their common circular orbit, but keep them in the same orbital plane. (Moving them out of the plane just introduces linearly independent vertical oscillation modes that do nothing of interest for us here.) If there were no other forces in the system, the particles would just execute epicyclic oscillations at frequency κ about their original circular orbit. To make things more interesting, let us tether the particles together with a spring of spring constant K and *zero* equilibrium length, so that the spring always exerts a tension which in and of itself tries to pull the particles back together. Viewed in a Cartesian frame rotating at the original circular orbit angular velocity, with origin at the unperturbed particle position, x-axis in the radial direction and y-axis in the azimuthal direction, the linearized perturbed equations of motion of the particles are easy to write down:

$$\frac{\mathrm{d}^2x_1}{\mathrm{d}t^2} = 2\Omega\frac{\mathrm{d}y_1}{\mathrm{d}t} - R\frac{\mathrm{d}\Omega^2}{\mathrm{d}R}x_1 + \frac{K}{m}(x_2 - x_1), \tag{4.6}$$

$$\frac{\mathrm{d}^2y_1}{\mathrm{d}t^2} = -2\Omega\frac{\mathrm{d}x_1}{\mathrm{d}t} + \frac{K}{m}(y_2 - y_1), \tag{4.7}$$

and two similar equations for the accelerations of the second particle which are identical to these except with the indices 1 and 2 flipped. The first terms on the right hand sides of these equations represent the Coriolis acceleration, the last terms represent the spring accelerations, and the middle term in equation (4.6) is the result of the combined gravitational and centrifugal acceleration.

It is easy to find the normal modes of these coupled oscillator equations. Adding corresponding equations together, we get

$$\frac{\mathrm{d}^2}{\mathrm{d}t^2}(x_1 + x_2) = 2\Omega\frac{\mathrm{d}}{\mathrm{d}t}(y_1 + y_2) - R\frac{\mathrm{d}\Omega^2}{\mathrm{d}R}(x_1 + x_2), \tag{4.8}$$

$$\frac{\mathrm{d}^2}{\mathrm{d}t^2}(y_1 + y_2) = -2\Omega\frac{\mathrm{d}}{\mathrm{d}t}(x_1 + x_2). \tag{4.9}$$

Assuming a time dependence $\propto \exp(-i\omega t)$, we immediately find that the oscillation frequency ω is given by

$$\omega^2 = R\frac{\mathrm{d}\Omega^2}{\mathrm{d}R} + 4\Omega^2 = \kappa^2, \tag{4.10}$$

which is hardly surprising. The particles are moving together as one, with an unstretched spring, executing epicyclic oscillations. (There is also a zero frequency mode corresponding to a constant azimuthal displacement of the particles.) The other normal modes are far more interesting. Subtracting corresponding equations, we get

$$\frac{\mathrm{d}^2}{\mathrm{d}t^2}(x_1 - x_2) = 2\Omega\frac{\mathrm{d}}{\mathrm{d}t}(y_1 - y_2) - R\frac{\mathrm{d}\Omega^2}{\mathrm{d}R}(x_1 - x_2) - \frac{2K}{m}(x_1 - x_2), \quad (4.11)$$

$$\frac{\mathrm{d}^2}{\mathrm{d}t^2}(y_1 - y_2) = -2\Omega\frac{\mathrm{d}}{\mathrm{d}t}(x_1 - x_2) - \frac{2K}{m}(y_1 - y_2). \quad (4.12)$$

Hence the oscillation frequencies for these normal modes satisfy

$$0 = \omega^4 - \left(\frac{4K}{m} + \kappa^2\right)\omega^2 + \frac{2K}{m}\left(\frac{2K}{m} + R\frac{\mathrm{d}\Omega^2}{\mathrm{d}R}\right). \quad (4.13)$$

This immediately implies that one of the solutions of this quadratic equation for ω^2 will be negative, implying an unstable/damped pair of frequencies, if and only if

$$\frac{2K}{m} + R\frac{\mathrm{d}\Omega^2}{\mathrm{d}R} < 0. \quad (4.14)$$

All accretion disks around black holes have angular velocities that decrease outward. Hence a sufficiently *weak* spring *always* catalyzes an instability in the particle motion. Only if the spring is strong does it snap the particles back to their equilibrium position.

In it simplest form, the MRI is exactly like this mechanical problem. Consider an equilibrium, differentially rotating fluid flow in which there is a uniform vertical component of magnetic field, as well as an azimuthal (toroidal) component which, while nonuniform, is too weak to exert significant stresses on the equilibrium flow. Like both r- and g-modes, the MRI involves essentially incompressible motions, so let us assume for now that the flow can be treated as completely incompressible ($\boldsymbol{\nabla} \cdot \mathbf{v} = 0$). Also, to keep things simple, let us neglect buoyancy forces for now by assuming that the equilibrium has uniform entropy. Now consider local (*i.e.* wavelengths much less than equilibrium length scales) perturbations with wavenumbers purely in the vertical direction. The incompressibility condition then immediately implies that $\delta v_z = 0$, *i.e.* the fluid motions are entirely horizontal. Similarly, the condition $\boldsymbol{\nabla} \cdot \mathbf{B} = 0$ implies that there is no perturbation in the vertical field component, $\delta B_z = 0$.

The perturbed flux freezing equations are

$$\frac{\partial \delta B_{\mathrm{R}}}{\partial t} = B_z \frac{\partial \delta v_{\mathrm{R}}}{\partial z} \quad (4.15)$$

and

$$\frac{\partial \delta B_\phi}{\partial t} = R\frac{\mathrm{d}\Omega}{\mathrm{d}R}\delta B_\mathrm{R} + B_z\frac{\partial \delta v_\phi}{\partial z}. \tag{4.16}$$

These equations describe the dynamics of the field lines (the spring!) themselves. Vertical gradients of radial and azimuthal velocity stretch the equilibrium vertical field out into radial and azimuthal field, respectively. In addition, the background differential rotation shears out the perturbed radial field into azimuthal field.

The perturbed fluid momentum equations are

$$\frac{\partial \delta v_\mathrm{R}}{\partial t} - 2\Omega\delta v_\phi = \frac{B_z}{4\pi\rho}\frac{\partial \delta B_\mathrm{R}}{\partial z}, \tag{4.17}$$

$$\frac{\partial \delta v_\phi}{\partial t} + \frac{\kappa^2}{2\Omega}\delta v_\mathrm{R} = \frac{B_z}{4\pi\rho}\frac{\partial \delta B_\phi}{\partial z}, \tag{4.18}$$

and

$$0 = -\frac{\partial \delta P}{\partial z} - \frac{B_\phi}{4\pi\rho}\frac{\partial \delta B_\phi}{\partial z}. \tag{4.19}$$

The radial and azimuthal momentum equations (4.17) and (4.18) are very similar to the spring equations (4.6) and (4.7), the differences being due to the fact that we are now using Eulerian rather than Lagrangian variables. In terms of the Lagrangian displacement $\boldsymbol{\xi}$, the Eulerian velocity perturbations are

$$\delta\mathbf{v} = \frac{\partial\boldsymbol{\xi}}{\partial t} + \mathbf{v}\cdot\boldsymbol{\nabla}\boldsymbol{\xi} - \boldsymbol{\xi}\cdot\boldsymbol{\nabla}\mathbf{v} = \frac{\partial \xi_\mathrm{R}}{\partial t}\hat{\mathbf{R}} + \left(\frac{\partial \xi_\phi}{\partial t} - \xi_\mathrm{R} R\frac{\mathrm{d}\Omega}{\mathrm{d}R}\right)\hat{\boldsymbol{\phi}}. \tag{4.20}$$

Note that vertical gradients in the radial and azimuthal field are producing tension forces in equations (4.17) and (4.18) very analogous to the spring forces considered earlier. The vertical momentum equation (4.19) is rather interesting. Vertical gradients in the azimuthal field create vertical magnetic pressure gradients which, for an incompressible fluid, are immediately balanced by vertical fluid pressure gradients. This equation just determines the fluid pressure, and is otherwise decoupled from the other equations. Note that in the linear system the vertical pressure balance equation (4.19) becomes trivial if the equilibrium azimuthal field B_ϕ vanishes, and in fact this is the only place where B_ϕ appears. (We will see in Sect. 5 that radiative diffusion can affect the MRI through this vertical pressure balance equation, however.)

Assuming a space-time dependence of the perturbations $\propto \exp[i(kz - \omega t)]$, the four equations (4.15)–(4.18) can be solved to give the dispersion relation

$$0 = \omega^4 - \left(2kv_\mathrm{A} + \kappa^2\right)\omega^2 + kv_\mathrm{A}\left(kv_\mathrm{A} + R\frac{\mathrm{d}\Omega^2}{\mathrm{d}R}\right), \tag{4.21}$$

where $v_{\mathrm{A}z} = B_z/(4\pi\rho)^{1/2}$ is the Alfvén speed corresponding to the vertical field component. Equation (4.21) is identical to equation (4.13), if we simply replace the natural frequency $2K/m$ of the spring with $kv_{\mathrm{A}z}$!

It is easy to show from equation (4.21) that the maximum instability growth rate is

$$|\omega|_{\max} = \frac{1}{2}R\left|\frac{\mathrm{d}\Omega}{\mathrm{d}R}\right|, \tag{4.22}$$

and occurs for wavenumbers satisfying

$$(kv_{\mathrm{A}z})_{|\omega|\max} = \Omega^2 - \frac{\kappa^4}{16\Omega^2}. \tag{4.23}$$

Note that this is all in agreement with our statement above that r-mode physics will be substantially modified when the natural frequency of our magnetic "spring" kv_{A} is of order κ.

The analysis we have done so far recovers the simplest form of the MRI. It turns out that this mode is also an exact solution of the *nonlinear* local MHD equations, although it is itself vulnerable to further instabilities [47]. It is often called the "channel solution", as it is characterized by rapid counterstreaming motions in horizontal planes [52, 53].

One can of course generalize the analysis to include general axisymmetric wavenumbers $\mathbf{k} = k_{\mathrm{R}}\hat{\mathbf{R}} + k_z\hat{\mathbf{z}}$ and arbitrary equilibrium field directions. One can also add buoyancy by introducing entropy gradients. Most relevant to geometrically thin disks are vertical entropy gradients. In this case the MRI dispersion relation can be written as [16]

$$0 = \frac{k^2}{k_z^2}\tilde{\omega}^4 - \left(\kappa^2 + \frac{k_{\mathrm{R}}^2}{k_z^2}N^2\right)\tilde{\omega}^2 - 4\Omega^2(\mathbf{k}\cdot\mathbf{v}_{\mathrm{A}})^2, \tag{4.24}$$

where $\tilde{\omega}^2 \equiv \omega^2 - (\mathbf{k}\cdot\mathbf{v}_{\mathrm{A}})^2$ and $\mathbf{v}_{\mathrm{A}} = \mathbf{B}/(4\pi\rho)^{1/2}$ is the vector Alfvén speed corresponding to the equilibrium field $\mathbf{B}$. Note that magnetic fields are therefore directly connected to both r- and g-modes. The linear analysis of the MRI can also be extended to localized, nonaxisymmetric perturbations, which are again subject to unstable growth [18]. A fully general relativistic linear analysis of the MRI has recently been done by Araya-Góchez [8].

The general local stability criterion in ideal MHD for rotating, axisymmetric flows turns out to be the same as the Høiland criterion (Eqs. (4.1) and (4.2)), but with angular momentum gradients replaced by gradients in the angular velocity Ω [13]:

$$N^2 + R\frac{\partial\Omega^2}{\partial R} > 0 \tag{4.25}$$

and

$$(\boldsymbol{\nabla}P \times \hat{\mathbf{R}})\cdot(\boldsymbol{\nabla}\Omega^2 \times \boldsymbol{\nabla}S) > 0. \tag{4.26}$$

Magnetic fields appear to facilitate the existence of instabilities which are more directly tuned into the thermodynamic sources of free energy in the flow. Indeed, Balbus [14] has shown that in situations where heat conduction exists and is restricted to flow along magnetic field lines, the entropy gradients in the MHD Høiland criterion are replaced by temperature gradients! This situation arises in cases where charged particles conduct the heat and the collisional mean free path exceeds the Larmor radius of the particles. In a non-rotating, hydrostatic system, the Schwarzschild stability criterion that the entropy increase upward is replaced by the criterion that the *temperature* increase upward! While interesting, it should perhaps be remembered that any temperature or angular velocity gradient constitutes a thermodynamic source of free energy, but magnetic fields are destabilizing for only certain directions of these gradients.

The MRI relies on weak magnetic fields. If the field is too strong at the wavelength considered, then magnetic tension will overcome the effects of magnetic torquing and the flow will stabilize. The stronger the field, the longer the wavelength required for instability. This immediately suggests that laminar accretion disk flow models will be unstable provided an unstable MRI wavelength $\sim v_{\rm A}/\Omega$ can fit inside the vertical thickness of the disk. This corresponds to initial field energy densities that are less than the thermal pressure. That this does indeed provide an upper limit to the field strength in order for the MRI to exist is confirmed by global linear analyses [36, 45].

Since its discovery, many numerical simulations have been done by a number of groups that confirm that the nonlinear development of the MRI in an initially weakly magnetized medium leads to sustained MHD turbulence in the accretion flow. Although the simplest MRI modes are axisymmetric, it is essential that these simulations be fully three dimensional. Local axisymmetric simulations with an initial net poloidal field are inevitably dominated by the channel solution, which tends to break up in three dimensional simulations. Axisymmetric simulations with no initial net poloidal field only develop a transient phase of turbulence that eventually decays away. This is a consequence of Cowling's anti-dynamo theorem, which in one version states that magnetic fields cannot be sustained by fluid motions under conditions of axisymmetry [83].

Three dimensional nonlinear simulations of MRI turbulence have been done in a number of geometries, the simplest being the "shearing box" [57, 58]. Here one zooms in on a tiny region of the disk and explores the local nonlinear behavior of the MRI, neglecting background gradients in the initial equilibrium except in the radial direction where one has to take into account the all important differential rotation. Periodic boundary conditions are adopted in the vertical and azimuthal directions, and also in the radial direction in coordinates that shear with the background differential rotation. (An excellent discussion of the shearing box can be found in [57].) These shearing box simulations show clearly that the nonlinear development of the MRI is one of fully three dimensional, anisotropic MHD turbulence that exhibits strongly correlated fluctuations in azimuthal and radial components of velocity and magnetic field. Angular momentum is transported outward by a sum of Reynolds (fluid) and Maxwell (magnetic) stresses, *i.e.*

$$\tau_{\mathrm{R}\phi} = \rho v_{\mathrm{R}} \delta v_{\phi} - \frac{B_{\mathrm{R}} B_{\phi}}{4\pi}, \tag{4.27}$$

where δv_{ϕ} is the azimuthal velocity minus the mean background orbital velocity. The Maxwell stress generally dominates the Reynolds stress by factors of 3 to 4 [57], and their values averaged over the computational volume both exhibit substantial variability on the orbital time scale.

Unfortunately, shearing box simulations cannot tell us the magnitude of the anomalous stress, even in an averaged sense, appropriate for accretion disks in nature. The neglect of vertical stratification and buoyancy prevents the gas pressure from having much dynamical significance in the turbulent state. In fact, shearing box simulations carried out with an adiabatic equation of state have monotonically growing pressure generated by heating in the turbulence. However, this increasing pressure does not affect the average value of $\tau_{\mathrm{R}\phi}$ [57], perhaps because it acts simply to enforce incompressibility in the turbulence to a greater and greater degree. Instead of the pressure, the anomalous stress is more closely related to the overall magnetic energy density in the turbulent state, but this turns out to depend on the size of the computational box, the initial magnetic field strength if there is a net mean field in any direction in the box, the resistivity (numerical or physical) and the size of the artificial viscosity used in the simulations [57, 58]. The latter two quantities determine in particular the rate at which magnetic field "reconnects" in the simulations. The initial field topology also affects the level of turbulent transport. Simulations with no initial net poloidal field result in a stress that is roughly an order of magnitude below that obtained when there is a net initial poloidal field [57, 58].

Simulations have also been done in vertically stratified shearing boxes [32,80,110]. Such simulations zoom in on a small vertical slice through the disk, fully taking into account the effects of vertical buoyancy and a finite scale height of the disk set by vertical hydrostatic equilibrium. This latter fact is important, as it connects the local pressure to the vertical size of the disk. In simulations that start out with a weak magnetic field with no net poloidal component, three dimensional MHD turbulence is again the generic result, at least within the disk interior. The time and space averaged anomalous stress generated by this turbulence turns out to be proportional to the pressure at the midplane of the disk, in excellent agreement with the

Shakura-Sunyaev prescription (2.1). The reason for this appears to be that the midplane pressure sets the vertical scale height. The thermodynamics of these simulations is relatively crude: in simulations run with an isothermal equation of state, so the gas cannot heat or cool, the turbulent stress saturates to a constant average value. In simulations run with an adiabatic equation of state, so the gas continually heats as a result of turbulent dissipation, the scale height of the disk slowly increases in response to the heating, and the average turbulent stress also increases in lock step with the increasing midplane pressure [110]. In agreement with the shearing box simulations, the stress is dominated by magnetic fields rather than velocity fields.

The actual value of the proportionality constant α, as defined by $\tau_{\mathrm{R}\phi}$ divided by the midplane pressure, is typically $\sim 10^{-2}$, although different simulations can produce somewhat larger or smaller values [32, 80, 110]. It is still not clear that this value is to be fully trusted, and it should certainly not be assumed to be a dimensionless constant of nature! In all the simulations that have been done so far, it appears that vertical buoyancy of the magnetic field does not play a dominant role in setting the saturation level of the turbulence, although this conclusion might be altered once the effects of radiative diffusion are incorporated (see Sect. 5 below). Instead, local balance between magnetic field amplification by the MRI and dissipation appears to be the most important factor in setting the overall stress level, just as in the shearing box simulations. Unfortunately, the dissipation is often largely numerical in nature, and the actual value of α produced by the simulation can depend on the grid resolution. (For example, doubling the grid resolution in one of the most recent simulations [80] increased the value of α by 1.5.)

One dramatic effect that appears to be a generic feature of all the stratified shearing box simulations is the formation of a strongly magnetized corona above the disk [32, 80, 110]. This has been explored most extensively by Miller & Stone [80] who considered an isothermal disk (resulting in a vertically Gaussian profile in pressure and density) within a computational domain that covered plus and minus five scale heights H_z around the disk midplane. Figure 6 depicts the magnetic field lines in the turbulent state in one of their simulations. Within two scale heights of the midplane, MRI turbulence dominates and the field has a highly chaotic structure. Above two scale heights, however, the field becomes much more coherent as the field energy density dominates the gas pressure in the magnetized corona. Approximately one quarter of the energy generated within two scale heights of the midplane is transferred to the corona in this particular simulation, mostly in the form of Poynting flux associated with buoyant magnetic field lines. MRI turbulence therefore appears capable of self-consistently generating an

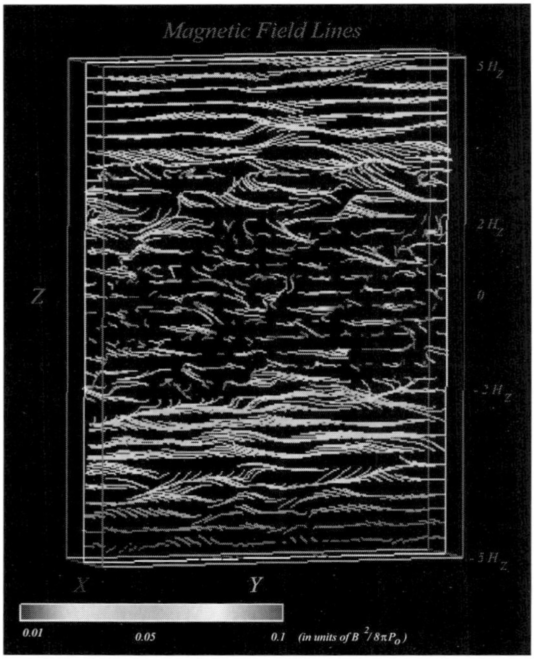

Fig. 6. Magnetic field line structure generated by the MRI in a small section of an accretion disk, from a simulation by Kristen Miller. The figure shows a slice in the Z (vertical) $-$ Y (azimuthal) plane of the disk (X is the radial coordinate). The color of the lines represents the strength of the field locally, normalized by the initial pressure at the disk midplane. MRI turbulence near the midplane of the disk ($Z \leq 2H_z$) results in tangled small-scale field structures. In the lower density ($Z > 2H_z$) regions, the field lines become smoother and show large scale coherence. (This figure was kindly generated by Kristen Miller. The field lines in this plot were generated by starting at points in a plane parallel to the $X-Z$ plane and then integrating the magnetic field line vectors. This integration eventually becomes dominated by numerical uncertainties, particularly in the inner regions of the disk where the field is so chaotic, at which point the computation of the field line stops for this particular plot. Hence the appearance of a nonzero divergence in the field is an artifact of the way the field lines were computed: $\nabla \cdot B = 0$ is satisfied in the simulations.)

energetically important corona, in agreement with the observed fact that the hard X-rays in black hole sources can carry a substantial fraction of the total accretion power.

Stratified shearing box simulations have also been done with an initial net poloidal field, but the results of the most recent such simulations produce a catastrophic change to a magnetically dominated disk structure that is inconsistent with the shearing box assumptions [80]. Large scale, ordered magnetic fields that might help generate an MHD outflow or jet also cannot be formed within the confines of a narrow shearing box. Global simulations of the entire disk are therefore of great interest, and some have now been completed. Examples can be found in [9, 51, 76, 109]. Such simulations are quite challenging, and have their own problems when it comes to making inferences about how real flows in nature work. Because they attempt to simulate the entire disk, they are necessarily limited in spatial resolution and total duration in time. They also tend to suffer from transient effects because of the presence of an outer radial boundary condition or an imposed finite radial extent of the disk itself. Nevertheless, these simulations have generated some rather interesting results.

First, time and space-averaged values of the Shakura-Sunyaev alpha stress parameter range from ~ 0.01 to ~ 0.2, consistent with results obtained from the local simulations. The reader should once again bear in mind all the caveats, however. Second, strongly magnetized coronal regions are generally formed outside the disk. Third, outflows are generated, although it is not yet clear whether a disk that starts out with a weak magnetic field and develops MRI turbulence can generate a globally ordered field in the corona that can collimate and accelerate jets. There are recent intriguing results where rotating conical outflows that are externally confined by magnetic pressure are produced near the rotation axis [54, 55], but whether such outflows can produce the apparently self-collimating jets we observe on large scales in nature is far from clear. Fourth, significant stresses are exerted on the flow near the ISCO, a point that I mentioned in Section 2 and to which I will return shortly.

Before doing that, I would like to first address the issue as to whether and how far all this MRI physics can be used to justify the standard accretion disk equations that we wrote down in Section 2. This has been carefully examined in the literature by Balbus and collaborators [15,21] in the context of geometrically thin accretion disks, and I have made an attempt here to generalize their approach to include the effects of radial pressure support and advection. Our objective here is to see how far we can sweep all the MHD physics into a simple anomalous stress prescription.

Using cylindrical polar coordinates (R, ϕ, z), the standard MHD equations can be manipulated to express local conservation laws in mass,

$$\frac{\partial \rho}{\partial t} + \boldsymbol{\nabla} \cdot (\rho \mathbf{v}) = 0, \tag{4.28}$$

radial momentum,

$$\frac{\partial}{\partial t}(\rho v_{\rm R}) + \nabla\cdot\left(\rho v_{\rm R}\mathbf{v} - \frac{B_{\rm R}}{4\pi}\mathbf{B}\right) = -\frac{\partial}{\partial R}\left(P + \frac{B^2}{8\pi}\right) + \rho\left(\frac{v_\phi^2}{R} - \frac{\partial\Phi}{\partial R}\right) - \frac{B_\phi^2}{4\pi R}, \tag{4.29}$$

angular momentum,

$$\frac{\partial}{\partial t}(R\rho v_\phi) + \nabla\cdot\left[R\rho v_\phi\mathbf{v} - \frac{RB_\phi}{4\pi}\mathbf{B}_{\rm p} + R\left(P + \frac{B_\phi^2}{8\pi}\right)\hat{\boldsymbol{\phi}}\right] = 0, \tag{4.30}$$

and energy,

$$\frac{\partial}{\partial t}\left(\frac{1}{2}\rho v^2 + \rho\Phi + \rho U + \frac{B^2}{8\pi}\right) + \nabla\cdot\left[\left(\frac{1}{2}\rho v^2 + \rho\Phi + \rho U + P\right)\mathbf{v} + \frac{1}{4\pi}\mathbf{B}\times(\mathbf{v}\times\mathbf{B}) + \mathbf{F}\right] = 0. \tag{4.31}$$

Here $\mathbf{B}_{\rm p} = B_{\rm R}\hat{R} + B_z\hat{z}$ is the poloidal piece of the magnetic field and $\mathbf{F}$ is the radiative heat flux.

There is nonaxisymmetric, time-dependent turbulence that is very complicated, but presumably there is some way of averaging over this turbulence to get equations describing macroscopic, mean flow properties. It turns out that it is possible to (nearly) recover the vertically integrated radial disk structure equations (2.7)–(2.10) by such averaging, *provided* we make the following assumptions. First, the flow velocity is dominated by rotation, on top of which there are velocity fluctuations with root mean square amplitudes that are much larger than the mean radial infall velocity:

$$\mathbf{v} = R\Omega\hat{\boldsymbol{\phi}} + \mathbf{u} \quad \text{with} \quad R\Omega \gg \langle u^2\rangle^{1/2} \gg |\langle u_{\rm R}\rangle|. \tag{4.32}$$

Also, the turbulence is MHD in character and highly subsonic, with

$$\left\langle\frac{B^2}{4\pi\rho}\right\rangle \sim \langle u^2\rangle \ll \langle c_{\rm s}^2\rangle \sim \frac{\langle P\rangle}{\langle\rho\rangle}. \tag{4.33}$$

In addition, we assume that there is negligible mean magnetic field, *i.e.* $|\langle\mathbf{B}\rangle| \ll \langle B^2\rangle^{1/2}$. Finally, we assume that negligible fluxes of mass, angular momentum, and linear momentum leave the upper or lower disk surfaces, and the only flux of energy leaving these surfaces is radiative in character, *i.e.* contained in the radiation flux $\mathbf{F}$. We already know that this last assumption is dangerous, as simulations clearly show substantial energy transfer into the disk corona. Provided it is only energy and not mass, angular momentum or linear momentum, we will be able to get away with this by calling **F** an energy flux as I did in Section 2. However, this immediately precludes the possibility of developing large scale outflows that carry away mass and angular momentum from the disk.

Now, following [21], define a vertically integrated, density weighted, spatially averaged value of a quantity X at radius R in the disk by

$$\langle X\rangle \equiv \frac{\hat{A}(\rho X)}{\Sigma} \equiv \frac{1}{2\pi\Delta R\Sigma}\int_{-H}^{H}\mathrm{d}z\int_{R-\Delta R/2}^{R+\Delta R/2}\mathrm{d}R\int_{0}^{2\pi}\mathrm{d}\phi X\rho, \tag{4.34}$$

where $\hat{A}$ is a vertical integration and averaging operator, $\Delta R \sim H \ll R$ is a small radial scale length over which we are averaging (assuming the turbulence causes variations on size scales smaller than a measure of the disk half-thickness H), and $\Sigma = \hat{A}\rho$ is the surface density. Applying the operator $\hat{A}$ to the continuity equation (4.28), and using the boundary conditions of vanishing mass flux on the disk surface and azimuthal periodicity, we obtain

$$\frac{\partial\Sigma}{\partial t} + \frac{1}{R}\frac{\partial}{\partial R}(R\Sigma\langle u_{\mathrm{R}}\rangle) = 0, \tag{4.35}$$

where the radial derivative is now defined by differencing over the range $(R-\Delta R/2, R+\Delta R/2)$. Now, by a stationary flow I mean one in which the average quantities I have constructed are independent of time. Then $R\Sigma\langle u_{\mathrm{R}}\rangle$ is a constant, which I may write as

$$\dot{M} = -2\pi R\Sigma\langle u_{\mathrm{R}}\rangle\cdot \tag{4.36}$$

Using the same procedure on the rest of the conservation equations (4.29)–(4.31), I obtain

$$0 = \bar{\rho}(\Omega^2 - \Omega_{\mathrm{K}}^2)R - \frac{\mathrm{d}\bar{P}}{\mathrm{d}R}, \tag{4.37}$$

$$\dot{M}\frac{\mathrm{d}\ell}{\mathrm{d}R} = \frac{\mathrm{d}}{\mathrm{d}R}\left[2\pi R^2\left(\frac{\Sigma}{\bar{\rho}}\right)\tau_{\mathrm{R}\phi}\right], \tag{4.38}$$

and

$$\dot{M}\left[\frac{\mathrm{d}\bar{U}}{\mathrm{d}R} + \bar{P}\frac{\mathrm{d}}{\mathrm{d}R}\left(\frac{1}{\bar{\rho}}\right)\right] = 2\pi R^2\left(\frac{\Sigma}{\bar{\rho}}\right)\tau_{\mathrm{R}\phi}\frac{\mathrm{d}\Omega}{\mathrm{d}R} + 4\pi RF^{-}(R), \tag{4.39}$$

where

$$\tau_{\mathrm{R}\phi} \equiv \bar{\rho}\left(\langle u_{\mathrm{R}}u_{\phi}\rangle - \left\langle\frac{B_{\mathrm{R}}B_{\phi}}{4\pi\rho}\right\rangle\right)\cdot \tag{4.40}$$

I have been forced to define other averages here, in particular

$$\bar{U} \equiv \frac{\langle v_{\mathrm{R}}U\rangle}{\langle u_{\mathrm{R}}\rangle}, \tag{4.41}$$

$$\frac{1}{\bar{\rho}}\frac{\partial \bar{P}}{\partial R} \equiv \frac{1}{\Sigma}\frac{\partial}{\partial R}\left(\left\langle \frac{P}{\rho}\right\rangle \Sigma\right), \tag{4.42}$$

and

$$\frac{\bar{P}}{\bar{\rho}} \equiv \frac{\langle P v_{\mathrm{R}}/\rho\rangle}{\langle u_{\mathrm{R}}\rangle}. \tag{4.43}$$

The last two equations may not be rigorously consistent, but this is the usual consequence of attempting to describe a two-dimensional average flow with one-dimensional, vertically integrated equations.

If I write $\Sigma = 2\bar{\rho}H$, then equations (4.36)–(4.39) become identical to equations (2.7)–(2.10) that we discussed in Section 2, with one important exception. Note that we recover the true meaning of the anomalous stress in equation (4.40): the correlated radial and azimuthal velocity and magnetic field fluctuations that we saw in equation (4.27)! We have (almost) succeeded in hiding the underlying MHD physics in the average flow by placing it entirely inside the anomalous stress.

The only problem is that I was unable to recover the left hand side of the radial momentum equation (2.8) in equation (4.37), *i.e.* the term $\rho v \mathrm{d}v/\mathrm{d}R$. The reason is that I chose to neglect the magnetic terms in the original exact equation (4.29). I am justified in doing this, and still retaining the radial pressure gradient term, because the turbulence is subsonic. The same reasoning, however, forces me to neglect the poloidal velocity terms in this equation. If I want to retain them, then I must also retain the magnetic terms.

In hydrodynamic models of accretion disks with radial advection, the $\rho v \mathrm{d}v/\mathrm{d}R$ term only becomes important near the sonic point, but it is precisely here that the assumptions underlying the averaging process giving rise to equations (4.36)–(4.39) fail utterly. The poloidal velocities are transonic and the character of the underlying turbulence *must* change. Magnetic fields are likely to be important here as they are stretched out and sheared by the large inflow velocities, and we cannot neglect them. It is this physics that is the source of inner torques exerted on the disk across the ISCO [44, 68]. Rather than localized turbulence over which we can average, we have torques that are generated by more coherent magnetic field lines formed by flux freezing in the rapidly infalling fluid. Averages cannot be performed here because the time and space dependence of the "turbulence" is comparable to that of the macroscopic flow itself.

Numerical simulations that probe the nature of these *non*-MRI inner torques have been done by a number of authors [10, 51, 59]. The most recent such simulation, which has the highest resolution, confirms the existence of significant stresses in the plunging region inside the ISCO, with matter having ten percent more binding energy and ten percent less angular momentum [60]. The Shakura-Sunyaev α parameter rises to huge values in the plunging region, but this is a mis-application of that prescription because it is due merely to the drop in pressure in this region of the flow. The flow is highly time-dependent and cannot be characterized by stationary models flowing through critical points of ordinary differential equations. As always, there are still caveats in these simulations and further work is needed. In particular, all the simulations that have been done so far lack a full general relativistic treatment. Performing a simulation in a Kerr spacetime in which the black hole can exert magnetic torques on the accretion disk, thereby supplementing accretion power with spin power, is likely to be a profitable research direction!

I have focussed in this section on angular momentum transport in black hole accretion flows, but the existence of turbulence also begs the question as to whether or not there could be substantial anomalous heat transport as well. Unfortunately, none of the simulations that I discussed here can address this question, as they do not include radiation transport. In fact, the thermodynamics of the fluid is generally treated very crudely by simply adopting an adiabatic ($P \propto \rho^{5/3}$) or isothermal ($P \propto \rho$) equation of state. Dissipation in the turbulence is not properly accounted for, and energy is generally not conserved. It turns out that incorporating radiation transport directly into the physics of the MHD turbulence leads to very interesting results, and this will be the topic of the next and final lecture.

5 The role of radiation magnetohydrodynamics

Shakura-Sunyaev based models of standard accretion disks around black holes are always radiation pressure dominated if the disk luminosity is anywhere near Eddington. If we take $\tau_{\mathrm{R}\phi} = \alpha P$, with P being the total (gas plus radiation) pressure, and assume that vertical heat transport proceeds through radiative diffusion, then the ratio of radiation to gas pressure is roughly

$$\frac{P_{\mathrm{rad}}}{P_{\mathrm{gas}}} \sim 10^5 \alpha^{1/4} \eta^{-2} (1-f)^{9/4} \left(\frac{M}{M_\odot}\right)^{1/4} \left(\frac{L}{L_{\mathrm{Edd}}}\right)^2 \left(\frac{R}{r_{\mathrm{g}}}\right)^{-21/8} \mathcal{I}^2, \quad (5.1)$$

where f is the fraction of local accretion power that is not dissipated in the disk interior [114]. As I discussed previously, there has always been uncertainty as to whether the anomalous stress scales more with radiation pressure or gas pressure, or indeed something else, and this has led to ambiguity about the thermal and "viscous" stability of the inner parts of black hole accretion flows. Even if we take $\tau_{\mathrm{R}\phi} = \alpha P_{\mathrm{gas}}$, which is thermally and "viscously" stable, the ratio of radiation to gas pressure is still large:

$$\frac{P_{\rm rad}}{P_{\rm gas}} \sim 10^4 \alpha^{1/5} \eta^{-8/5} (1-f)^{9/5} \left(\frac{M}{M_\odot}\right)^{1/5} \left(\frac{L}{L_{\rm Edd}}\right)^{8/5} \left(\frac{R}{r_{\rm g}}\right)^{-21/10} \mathcal{I}^{8/5}. \tag{5.2}$$

Note that radiation pressure is most important for models of active galactic nuclei and quasars that contain supermassive black holes, but it can still be important for stellar mass black hole X-ray binary models. As noted in Section 3 above, allowing much of the locally dissipated power to vertically escape from the disk interior in non-radiative fashion ($f \to 1$) reduces the importance of radiation pressure [114].

In addition to the issue of secular stability, there are also other uncertainties in this inner portion of the flow. Radiation pressure support against the tidal field of the black hole implies that

$$\frac{n_{\rm e}\sigma_{\rm T}}{c} F \simeq \Omega_{\rm K}^2 \rho z. \tag{5.3}$$

The electron number density $n_{\rm e} \propto \rho$, so that the mechanical equilibrium condition (5.3) gives $F \propto z$, or $\mathrm{d}F/\mathrm{d}z$ equals a constant. On the other hand, it has long been assumed that the vertical distribution of turbulent dissipation per unit mass is probably constant in the disk interior, *i.e.* that the power dissipated per unit volume is proportional to the local density. Radiative transport of heat then implies that

$$\boldsymbol{\nabla} \cdot \mathbf{F} \simeq \frac{\mathrm{d}F}{\mathrm{d}z} \propto \rho. \tag{5.4}$$

Hence mechanical and thermal equilibrium immediately lead to the conclusion that the density ρ is independent of height z. Such an equilibrium is violently unstable to vertical convection [28]. For example, eliminating the density gradient in equation (4.4) leads to an imaginary Brunt-Väisälä frequency. More physically, consider an upward adiabatic perturbation of a fluid element in a constant density background. Because the pressure drops with height, the fluid element expands and becomes less dense than the surrounding constant density background. It is therefore buoyant and unstable.

We are therefore lead to the conclusion that heat is transported out of the disk convectively, not radiatively. However, this whole argument is intimately wrapped up in the physics of the MRI. As noted in the last section, the MRI modes are tightly coupled to the gravity modes that are responsible for hydrodynamic convection. How does convective heat transport work in the presence of MRI turbulence? Moreover, the convective instability arises because of our assumptions about the mass distribution of dissipation, and this too is determined by the MRI turbulence, so how does this really work in a vertically stratified, radiation pressure dominated medium?

It turns out that radiation MHD profoundly affects the MRI, the Parker instability, and also introduces *new* types of dynamical instability in the flow. Radiation MHD is just like standard MHD, except that the photon "gas" is treated as a separate fluid that couples to the plasma through absorption, emission, and scattering. I summarize the basic Newtonian equations of radiation MHD that are useful in studying the physics of the radiation pressure dominated regions of high luminosity black hole accretion flows in the Appendix. The reason that we have to treat the photons in a more careful way than just writing the total pressure $P_{\rm tot}$ as the sum $P_{\rm gas}+P_{\rm rad}$ is that, in addition to being effectively thin ($\tau_{\rm eff}<1$) in some cases, the innermost regions of black hole accretion flows do not have tremendous total optical depth either. For $\tau_{{\rm R}\phi}=\alpha P_{\rm rad}$, the vertical Thomson depth of the disk midplane is

$$\tau_{\rm T}\sim\alpha^{-1}\eta(1-f)^{-2}\left(\frac{L}{L_{\rm Edd}}\right)^{-1}\left(\frac{R}{r_{\rm g}}\right)^{3/2}\mathcal{I}^{-1}. \tag{5.5}$$

On the other hand, $\tau_{{\rm R}\phi}=\alpha P_{\rm gas}$ produces a much denser, more optically thick disk,

$$\tau_{\rm T}\sim 10^4\alpha^{-4/5}\eta^{-3/5}(1-f)^{-1/5}\left(\frac{M}{M_\odot}\right)^{1/5}\left(\frac{L}{L_{\rm Edd}}\right)^{3/5}\left(\frac{R}{r_{\rm g}}\right)^{-3/5}\mathcal{I}^{3/5}. \tag{5.6}$$

Even in this case, photons, which are providing the dominant pressure support, are highly diffusive, especially on small length scales.

Even in a perfectly electrically conducting fluid, significant radiation pressure with radiative diffusion alters MHD in two important ways. First, the compressibility of the fluid is greatly enhanced on small scales beyond what would naively be expected based on a very high radiation sound speed. Even highly subsonic motions [speeds $\ll(4aT^4/9\rho)^{1/2}$] will be compressible if they occur on a scale small enough for photons to diffuse during the motion. Second, temperature fluctuations in the gas tend to be smoothed out by the radiation field. These effects have important consequences for the MRI and the Parker instability, and they also produce wholly new classes of instabilities. The nonlinear, turbulent state of the radiation pressure dominated region of an accretion disk is therefore likely to be very different from that envisaged by standard accretion disk theory, and the sorts of modeling we have been discussing up to now may be doing a very poor job of describing this, energetically most important, region of the flow.

Radiation MHD effects on the Parker instability were considered quite early in the development of accretion disk theory [98]. Consider an isolated, straight, horizontal tube of magnetic flux immersed in the vertically stratified, radiation pressure dominated environment of a black hole accretion disk. If the tube has small diameter, radiative diffusion will tend to smooth out temperature (and therefore radiation pressure) differences between the interior and exterior of the tube. The excess magnetic pressure inside the tube must therefore be balanced largely by a deficit of gas pressure caused by a lower interior density. The flux tube therefore has a tendency to be buoyant. The standard Parker instability in pure gas disks implies that flux tubes will be buoyant if $B^2/(8\pi) \gtrsim P$, and this then suggests that tubes with diameters less than the radiation diffusion length over an unstable mode growth time will be unstable if $B^2/(8\pi) \gtrsim P_{\rm gas}$. That this is in fact true is supported by numerical simulations of flux tube dynamics in radiation pressure dominated environments [99]. Specifically, if the flux tubes are strong enough to self-consistently maintain an anomalous stress $\tau_{{\rm R}\phi} = \alpha P_{\rm tot}$, they are extremely buoyant. If instead the flux tubes are only as strong to maintain the lower stress $\tau_{{\rm R}\phi} = \alpha P_{\rm gas}$, then they can be retained in the medium for much longer time scales. This is at least partly due to the fact that the ambient medium is then much denser and more optically thick, so that photon diffusion across the flux tubes is much slower. Unfortunately, the results of this study are not conclusive, as they fail to take into account the MRI (whose application to accretion disks was discovered two years later). The question remains as to whether the MRI can either generate the field sufficiently fast to counteract rapid buoyant escape, or directly modify the buoyant dynamics itself.

Radiation MHD also modifies the behavior of the MRI. Recall that the MRI is nearly incompressible. Even in the nonlinear regime, $B^2/(8\pi)$ is substantially less than the thermal pressure in the disk interior and the turbulent motions are highly subsonic. However, radiative diffusion will modify this. From equations (4.22) and (4.23), the characteristic time scale associated with the MRI is the orbital period and the characteristic length scale is $\sim v_{\rm A}/\Omega$. The turbulence will therefore remain incompressible only if the orbital frequency Ω times the diffusion time $t_{\rm diff}$ over a distance $v_{\rm A}/\Omega$ is substantially greater than unity. Now,

$$\Omega t_{\rm diff} \sim \Omega \left(\frac{\kappa_{\rm T}\rho}{c}\right)\left(\frac{v_{\rm A}^2}{\Omega^2}\right) \sim \frac{\tau_{{\rm R}\phi}\kappa_{\rm T}}{c\Omega}, \tag{5.7}$$

where the last equality comes from equation (4.40) and the fact that MRI turbulence is dominated by magnetic rather than Reynolds stresses. On the other hand, if heat is transported vertically by radiative diffusion and the disk is in thermal equilibrium, then equations (2.4) and (2.12) imply

$$\frac{\tau_{{\rm R}\phi}\kappa_{\rm T}}{c\Omega} \sim \frac{P_{\rm rad}^2}{\Sigma^2\Omega^2} \sim 1, \tag{5.8}$$

the last equality coming from vertical hydrostatic equilibrium. Hence in a radiation pressure dominated disk in which heat is transported radiatively and the turbulent stress is of order the magnetic stress, then $\Omega t_{\rm diff} \sim 1$ and the turbulence is therefore expected to be compressible [119]. Note that this argument is rather general, and in particular is completely independent of the Shakura-Sunyaev stress assumption.

Compressibility affects the MRI in interesting ways. Consider again the channel solution that we discussed in Section 4. We noted that the equilibrium azimuthal field does not affect the linear growth of the channel solution at all. In fact this is true for all the linear axisymmetric MRI modes, whose growth depends on the equilibrium field only through $\mathbf{k} \cdot \mathbf{v}_{\rm A}$ from equation (4.24). If there is some equilibrium B_ϕ (and if there is not, there soon will be!), then the channel solution rapidly develops vertical magnetic pressure gradients. Normally these do not affect anything as the fluid pressure is able to balance these through equation (4.19). However, if the radiation is diffusive on these length scales, then the radiation pressure is lost. If the azimuthal field energy density exceeds the true gas pressure, then not even the gas pressure can balance the vertical magnetic pressure gradients. As a result, vertical motions must be excited which take energy out of the growth of the MRI. This effect does *not* kill the MRI, but it can dramatically reduce its linear growth rate [30]. If the azimuthal field energy density exceeds the true gas pressure and if $B_\phi > B_z$, then the growth rate of the MRI is reduced below the orbital frequency by the ratio of the gas sound speed to the azimuthal Alfvén speed. This *suggests* that growth of magnetic stresses above the gas pressure may be sluggish, but numerical simulations are really required to answer this question.

Such simulations have now been done in *non*-stratified shearing boxes, both in two [118] and three [119] dimensions. Three dimensional simulations that have a net vertical magnetic flux through the box develop magnetic stresses that are independent of radiation diffusion effects – while linear growth of the channel solution may become sluggish, it still grows! On the other hand, it appears that radiation diffusion does diminish the turbulent magnetic stresses that develop when there is no net vertical magnetic flux through the box. Unfortunately however, for the same reasons that we discussed in Section 4, shearing box simulations cannot provide definitive estimates of the level of stress that will exist in real flows, and vertically stratified and global simulations including radiative diffusion will need to be done.

Perhaps the most interesting result that has emerged from the recent shearing box simulations is that radiative diffusion does indeed make the nonlinear turbulence highly compressible, and large density fluctuations (by more than a factor 20 [119]!) form and reform in a highly time-dependent manner. One important consequence of this fact is that some of the turbulent energy is *directly* dissipated into photon energy on scales that are completely resolved by the simulation. This occurs because photons diffuse out of compressed regions and into rarefied regions, damping compressive motions in the fluid [6]. Hence not only are we now getting a handle on the behavior of the anomalous stress with respect to angular momentum transport in the radiation pressure dominated region, we are also learning about how the mechanical energy is dissipated into heat. Ironically, this is easier to do here than when gas pressure dominates!

There are also entirely new classes of instabilities that are expected to be present due to radiation pressure and radiative diffusion when the vertical gravity is included. One of these is the "photon bubble" instability, first examined by Arons in the context of accreting X-ray pulsars [11]. Here density fluctuations in the presence of a strong magnetic field are driven buoyantly unstable by photons diffusing into and heating (out of and cooling) underdense (overdense) regions. A local MHD instability that may have similar physics has also been found to exist in the radiation pressure dominated region of black hole accretion disks [43]. Fast and slow magnetosonic waves at short wavelengths where radiative diffusion causes the loss of radiation pressure support of the wave can also be unstable to periodic driving by the equilibrium vertical forces (radiation pressure gradients, gas pressure gradients, and gravity), at least for waves propagating in certain directions [30]. The characteristic length scale for these instabilities is the gas pressure scale height (tiny in a radiation pressure dominated medium) and the growth rate is much larger than the orbital frequency, by a factor of order the square root of the radiation to gas pressure ratio. These instabilities can only exist in Thomson scattering dominated media if magnetic tension forces help support the wave, which always occurs for wave vectors that are neither parallel or perpendicular to the equilibrium magnetic field. (In some cases, the small absorption opacity can drive even purely hydrodynamic acoustic waves unstable [46].)

Simulations have yet to be done that track the nonlinear development of these instabilities in an accretion disk environment, but Begelman [23] has constructed a one-dimensional, nonlinear periodic shock train solution that may describe the outcome of the magnetosonic wave instabilities. This solution produces a medium that is highly dynamic and porous to radiation diffusion, and can in fact support fluxes that are super-Eddington by factors of 10–100 *without* driving an outflow [24] (see also [105])!

Both the density fluctuations generated in the shearing box simulations of the MRI and these compressive wave instabilities strongly suggest that radiative diffusion generates a highly inhomogeneous, very time dependent, somewhat radiatively porous inner accretion flow. This speculation needs to be verified by further work, but if true, there are numerous observational implications, including possibly super-Eddington fluxes and modification of local spectra due to increased thermalization of the photons with the denser phases of the medium. The standard reasoning behind the putative existence of thermal/"viscous" instabilities in this part of the flow will also be affected. Suppose for example that a perturbative increase in temperature increased the porosity of the medium so that photons could escape more easily. Such a process could stabilize the disk against a thermal runaway. It has even been suggested that the radiation pressure dominated region of the disk may break up into actual discrete dense clumps interpenetrated by a much hotter, more optically thin medium capable of producing hard X-rays by Comptonization [66]. Clearly, further numerical simulations are likely to produce very interesting results!

6 Conclusions

After these lectures, particularly the last one in Section 5, a student may be feeling a strong urge to run screaming from the incredibly complex physics that must be understood before we can truly build realistic models of accreting black hole sources. Let me urge the reader that this was not my intent. The physics is indeed complex (and fascinating), and this essential fact is the reason why it has proved so hard to move beyond the Shakura-Sunyaev stress prescription introduced almost three decades ago. However, this is not the time to leave the field, but to enter it! For the first time we are beginning to investigate the basic physical principles that underly these flows, and we also have (or at least will have in the very near future) the simulation hardware that can provide definitive answers to the concrete questions we are now posing. The time is now ripe for genuine and definitive theoretical breakthroughs that will lead to predictive models that can extract truly useful physical information from observations of accreting black hole sources. I encourage you to join in and take part in these exciting developments.

I would like to thank Shane Davis and Kristen Miller for preparing some of the figures in these lectures, as well as my collaborators (Eric Agol, Ivan Hubeny, Julian Krolik, and Aristotle Socrates) for their scientific insights over the years. I also acknowledge very useful exchanges with Ian George, John Hawley, Pierre-Yves Longaretti, Patrick Ogle, and Neal Turner. I am grateful to the organizers of the Les Houches summer school for the opportunity to participate in this excellent program!

A The equations of radiation magnetohydrodynamics

For completeness, I summarize here the basic (Newtonian) equations of radiation magnetohydrodynamics (*e.g.* [111]) that are relevant to black hole accretion disk applications. These are the mass continuity equation,

$$\frac{\partial \rho}{\partial t} + \boldsymbol{\nabla}\cdot(\rho \mathbf{v}) = 0, \tag{1.1}$$

the gas momentum equation,

$$\rho\left(\frac{\partial \mathbf{v}}{\partial t} + \mathbf{v}\cdot\boldsymbol{\nabla}\mathbf{v}\right) = -\boldsymbol{\nabla}p + \rho\mathbf{g} + \frac{1}{4\pi}(\boldsymbol{\nabla}\times\mathbf{B})\times\mathbf{B} + \frac{\kappa_{\mathrm{F}}\rho}{c}\mathbf{F}, \tag{1.2}$$

the total internal energy equation,

$$\frac{\partial(u+E)}{\partial t} + v_j\nabla_j(u+E) + (u+E)\nabla_j v_j = -p\nabla_j v_j - P_{ij}\nabla_j v_i - \nabla_j F_j, \tag{1.3}$$

the radiation energy equation,

$$\begin{aligned}\frac{\partial E}{\partial t} + v_j\nabla_j E + E\nabla_j v_j &= -P_{ij}\nabla_j v_i - \nabla_j F_j + \kappa_{\mathrm{P}}\rho c a T^4 - \kappa_{\mathrm{E}}\rho c E \\ &\quad + \kappa_{\mathrm{T}}\rho c\left(\frac{4kT}{m_{\mathrm{e}}c^2} - \frac{h\bar{\nu}}{m_{\mathrm{e}}c^2}\right)E,\end{aligned} \tag{1.4}$$

the radiation momentum equation,

$$\frac{1}{c^2}\left(\frac{\partial F_i}{\partial t} + v_j\nabla_j F_i + F_i\nabla_j v_j\right) = -\nabla_j P_{ij} - \frac{\kappa_{\mathrm{F}}\rho}{c}F_i, \tag{1.5}$$

and the flux-freezing equation,

$$\frac{\partial \mathbf{B}}{\partial t} = \boldsymbol{\nabla}\times(\mathbf{v}\times\mathbf{B}). \tag{1.6}$$

Equations (1.3)–(1.5) have been written in Cartesian tensor notation, with summation implied over repeated indices. Although all the equations above are Newtonian equations, v/c effects have been included in the radiation terms, and all radiation quantities are defined in the local fluid rest frame. The quantities ρ, u, p, and T are the density, internal energy per unit volume, pressure, and temperature in the gas, respectively. They are related by equations of state,

$$u = \frac{p}{\gamma - 1}, \tag{1.7}$$

where $\gamma = 5/3$ is the ratio of specific heats for an ionized gas, and

$$p = \frac{\rho k T}{\mu}, \tag{1.8}$$

where k is Boltzmann's constant and μ is the mean molecular weight of the gas particles. For simplicity, I have neglected the effects of ionization and recombination in the gas. The quantity $\mathbf{v}$ is the gas velocity and $\mathbf{g}$ is the local gravitational acceleration presumed to arise exclusively from the central object (the black hole).

The radiation energy density E, the radiation flux $\mathbf{F}$, and the radiation pressure tensor P_{ij} are defined as frequency-integrated angular moments of the specific intensity in the local fluid rest frame:

$$E = \int_0^\infty \mathrm{d}\nu E_\nu = \frac{1}{c}\int_0^\infty \mathrm{d}\nu \oint \mathrm{d}\Omega I_\nu(\mathbf{n}), \tag{1.9}$$

$$F_i = \int_0^\infty \mathrm{d}\nu F_{\nu i} = \int_0^\infty \mathrm{d}\nu \oint \mathrm{d}\Omega n_i I_\nu(\mathbf{n}), \tag{1.10}$$

and

$$P_{ij} = \frac{1}{c}\int_0^\infty \mathrm{d}\nu \oint \mathrm{d}\Omega n_i n_j I_\nu(\mathbf{n}). \tag{1.11}$$

The equations can be closed by integrating the radiative transfer equation directly to solve for the tensor Eddington factor f_{ij}, defined by

$$P_{ij} = f_{ij} E. \tag{1.12}$$

The various opacity factors (all in units of cm^2/g) are defined in terms of different frequency averages over the radiation field. The energy mean, Planck mean, and flux mean opacities are defined by

$$\kappa_\mathrm{E} \equiv \frac{1}{\rho E}\int_0^\infty \mathrm{d}\nu \chi_\nu^\mathrm{th}(\rho, T) E_\nu, \tag{1.13}$$

$$\kappa_\mathrm{P} \equiv \frac{4\pi}{\rho a c T_\mathrm{g}^4}\int_0^\infty \mathrm{d}\nu \chi_\nu^\mathrm{th}(\rho, T) B_\nu(T), \tag{1.14}$$

and

$$\kappa_\mathrm{F}\mathbf{F} \equiv \frac{1}{\rho}\int_0^\infty \mathrm{d}\nu [\chi_\nu^\mathrm{th}(\rho, T) + n_\mathrm{e}\sigma_\mathrm{T}]\mathbf{F}_\nu, \tag{1.15}$$

respectively. Here $\chi_\nu^\mathrm{th}(\rho, T)$ is the thermal absorption coefficient (in units of cm^{-1}) at frequency ν, n_e is the electron number density, and σ_T is the Thomson cross-section. The Thomson opacity is

$$\kappa_\mathrm{T} \equiv \frac{n_\mathrm{e}\sigma_\mathrm{T}}{\rho}. \tag{1.16}$$

Local thermodynamic equilibrium in the gas has been assumed, so that the thermal source function in the gas is given by the Planck function $B_\nu(T)$ at the gas temperature.

Due to the fact that it dominates the opacity in the inner portions of black hole accretion disks, I have incorporated the effects of electron scattering in the equations. This includes Compton scattering, which can dominate the thermal coupling between the gas and the radiation in some cases. I treated this in the frequency diffusion (Kompaneets) limit by approximating the radiation field as isotropic in the local rest frame (see *e.g.* [62]). The quantity $\bar{\nu}$ in the radiation energy equation (1.4) is defined as a frequency average over the local rest frame radiation field in the following way,

$$\bar{\nu} = \frac{1}{E}\int_0^\infty \mathrm{d}\nu \nu E_\nu \left(1 + \frac{E_\nu c^3}{8\pi h \nu^3}\right) . \tag{1.17}$$

(The second term inside the parentheses in this equation represents the effects of stimulated scattering, which is necessary to include if one wants zero heat exchange between the gas and radiation when they reach exact thermal equilibrium.)

For a superb introduction to radiation hydrodynamics (*sans* magnetic fields!), see the book by Mihalas & Mihalas [79].

References

[1] M.A. Abramowicz, *Nature* **294** (1981) 235.
[2] M.A. Abramowicz, X. Chen, S. Kato, J.-P. Lasota and O. Regev, *ApJ* **438** (1995) L37.
[3] M.A. Abramowicz, X.-M. Chen, M. Granath and J.-P. Lasota, *ApJ* **471** (1996) 762.
[4] M.A. Abramowicz, B. Czerny, J.-P. Lasota and E. Szuszkiewicz, *ApJ* **332** (1988) 646.
[5] M.A. Abramowicz, A. Lanza and M.J. Percival, *ApJ* **479** (1997) 179.
[6] E. Agol and J. Krolik, *ApJ* **507** (1998) 304.
[7] E. Agol and J.H. Krolik, *ApJ* **528** (2000) 161.
[8] R.A. Araya-Góchez, *MNRAS* (2002) submitted [astro-ph/0204003].
[9] R. Arlt and G. Rüdiger, *A&A* **374** (2001) 1035.
[10] P.J. Armitage, C.S. Reynolds and J. Chiang, *ApJ* **548** (2001) 868.
[11] J. Arons, *ApJ* **388** (1992) 561.
[12] I.V. Artemova, G.S. Bisnovatyi-Kogan, I.V. Igumenshchev and I.D. Novikov, *ApJ* **549** (2001) 1050.
[13] S.A. Balbus, *ApJ* **453** (1995) 380.
[14] S.A. Balbus, *ApJ* **562** (2001) 909.
[15] S.A. Balbus, C.F. Gammie and J.F. Hawley, *MNRAS* **271** (1994) 197.
[16] S.A. Balbus and J.F. Hawley, *ApJ* **376** (1991) 214.
[17] S.A. Balbus and J.F. Hawley, *ApJ* **392** (1992) 662.
[18] S.A. Balbus and J.F. Hawley, *ApJ* **400** (1992) 610.
[19] S.A. Balbus and J.F. Hawley, *Rev. Mod. Phys.* **70** (1998) 1.
[20] S.A. Balbus, J.F. Hawley and J.M. Stone, *ApJ* **467** (1996) 76.
[21] S.A. Balbus and J.C.B. Papaloizou, *ApJ* **521** (1999) 650.

[22] D.R. Ballantyne, R.R. Ross and A.C. Fabian, *MNRAS* **327** (2001) 10.
[23] M.C. Begelman, *ApJ* **551** (2001) 897.
[24] M.C. Begelman, *ApJ* **568** (2002) L97.
[25] M.C. Begelman and T. Chiueh, *ApJ* **332** (1988) 872.
[26] A.M. Beloborodov, M.A. Abramowicz and I.D. Novikov, *ApJ* **491** (1997) 267.
[27] A.M. Beloborodov, *MNRAS* **297** (1998) 739.
[28] G.S. Bisnovatyi-Kogan and S.I. Blinnikov, *A&A* **59** (1977) 111.
[29] O.M. Blaes, *MNRAS* **227** (1987) 975.
[30] O. Blaes and A. Socrates, *ApJ* **553** (2001) 987.
[31] R.D. Blandford and D.G. Payne, *MNRAS* **199** (1982) 883.
[32] A. Brandenburg, Å Nordlund, R.F. Stein and U. Torkelsson, *ApJ* **446** (1995) 741.
[33] X. Chen, M.A. Abramowicz, J.-P. Lasota, R. Narayan and I. Yi, *ApJ* **443** (1995) L61.
[34] C.T. Cunningham, *ApJ* **202** (1975) 788.
[35] C. Cunningham, *ApJ* **208** (1976) 534.
[36] C. Curry, R.E. Pudritz and P.G. Sutherland, *ApJ* **434** (1994) 206.
[37] T. Dörrer, H. Riffert, R. Staubert and H. Ruder, *A&A* **311** (1996) 69.
[38] D.M. Eardley and A.P. Lightman, *ApJ* **200** (1975) 187.
[39] A.A. Esin, J.E. McClintock and R. Narayan, *ApJ* **489** (1997) 865.
[40] A.C. Fabian, K. Iwasawa, C.S. Reynolds and A.J. Young, *PASP* **112** (2000) 1145.
[41] A.C. Fabian, S. Vaughan, K. Nandra, *et al.*, *MNRAS* **335** (2002) L1.
[42] J. Frank, A. King and D. Raine, Accretion Power in Astrophysics, 3rd Ed. (Cambridge University Press, 2002).
[43] C.F. Gammie, *MNRAS* **297** (1998) 929.
[44] C.F. Gammie, *ApJ* **522** (1999) L57.
[45] C.F. Gammie and S.A. Balbus, *MNRAS* **270** (1994) 138.
[46] W. Glatzel and S. Mehren, *MNRAS* **283** (1996) 339.
[47] J. Goodman and G. Xu, *ApJ* **432** (1994) 213.
[48] P.W. Guilbert and M.J. Rees, *MNRAS* **233** (1988) 475.
[49] F. Haardt and L. Maraschi, *ApJ* **380** (1991) L51.
[50] J.F. Hawley, *ApJ* **381** (1991) 496.
[51] J.F. Hawley, *ApJ* **528** (2000) 462.
[52] J.F. Hawley and S.A. Balbus, *ApJ* **376** (1991) 223.
[53] J.F. Hawley and S.A. Balbus, *ApJ* **400** (1992) 595.
[54] J.F. Hawley and S.A. Balbus, *ApJ* **573** (2002) 738.
[55] J.F. Hawley, S.A. Balbus and J.M. Stone, *ApJ* **554** (2001) L49.
[56] J.F. Hawley, S.A. Balbus and W.F. Winters, *ApJ* **518** (1999) 394.
[57] J.F. Hawley, C.F. Gammie and S.A. Balbus, *ApJ* **440** (1995) 742.
[58] J.F. Hawley, C.F. Gammie and S.A. Balbus, *ApJ* **464** (1996) 690.
[59] J.F. Hawley and J.H. Krolik, *ApJ* **548** (2001) 348.
[60] J.F. Hawley and J.H. Krolik, *ApJ* **566** (2002) 164.
[61] I. Hubeny, E. Agol, O. Blaes and J.H. Krolik, *ApJ*, **533** (2000) 710.
[62] I. Hubeny, O. Blaes, J.H. Krolik and E. Agol, *ApJ* **559** (2001) 680.
[63] I. Hubeny and V. Hubeny, *ApJ* **484** (1997) L37.
[64] I. Hubeny and V. Hubeny, *ApJ* **505** (1998) 558.

[65] A.R. King and H. Ritter, *MNRAS* **293** (1998) L42.

[66] J.H. Krolik, *ApJ* **498** (1998) L13.

[67] J.H. Krolik, Active Galactic Nuclei (Princeton University Press, Princeton, NJ, 1999).

[68] J.H. Krolik, *ApJ* **515** (1999) L73.

[69] L.D. Landau and E.M. Lifshitz, section 10 in Statistical Physics Part 1, 3rd Ed. (Pergamon Press, Oxford, 1980).

[70] A. Laor, *ApJ* **376** (1991) 90.

[71] A. Laor and H. Nezter, *MNRAS* **238** (1989) 897.

[72] A.P. Lightman and D.M. Eardley, *ApJ* **187** (1974) L1.

[73] A.P. Lightman and T.R. White, *ApJ* **335** (1988) 57.

[74] P.-Y. Longaretti, *ApJ* **576** (2002) 587.

[75] D. Lynden-Bell and J.E. Pringle, *MNRAS* **168** (1974) 603.

[76] M. Machida, M.R. Hayashi and R. Matsumoto, *ApJ* **532** (2000) L67.

[77] P. Magdziarz and A.A. Zdziarski, *MNRAS* **273** (1995) 837.

[78] K.O. Mason *et al.*, *ApJL* (2002) in press [`astro-ph/0209145`].

[79] D. Mihalas and B.W. Mihalas, Foundations of Radiation Hydrodynamics (Oxford University Press, 1984).

[80] K.A. Miller and J.M. Stone, *ApJ* **534** (2000) 398.

[81] S. Mineshige, T. Kawaguchi, M. Takeuchi and K. Hayashida, *PASJ* **52** (2000) 499.

[82] M. Miyoshi, J. Moran, J. Herrnstein *et al.*, *Nature* **373** (1995) 127.

[83] H.K. Moffatt, section 6.4 in Magnetic Field Generation in Electrically Conducting Fluids (Cambridge University Press, 1978).

[84] K. Nandra, I.M. George, R.F. Mushotzky, T.J. Turner and T. Yaqoob, *ApJ* **523** (1999) L17.

[85] K. Nandra and K.A. Pounds, *MNRAS* **268** (1994) 405.

[86] R. Narayan and I. Yi, *ApJ* **452** (1995) 710.

[87] S. Nayakshin, D. Kazanas and T.R. Kallman, *ApJ* **537** (2000) 833.

[88] I.D. Novikov and K.S. Thorne, in Black Holes, edited by C. DeWitt and B. DeWitt (Gordon & Breach, New York, 1973) 343.

[89] Y. Osaki, *PASP* **108** (1996) 39.

[90] D.N. Page and K.S. Thorne, *ApJ* **191** (1974) 499.

[91] J.C.B. Papaloizou and J.E. Pringle, *MNRAS* **208** (1984) 721.

[92] J. Peitz and S. Appl, *MNRAS* **286** (1997) 681.

[93] R. Popham, S.E. Woosley and C. Fryer, *ApJ* **518** (1999) 365.

[94] J.E. Pringle, *MNRAS* **281** (1996) 357.

[95] J.E. Pringle and M.J. Rees, *A&A* **21** (1972) 1.

[96] E. Quataert and A. Gruzinov, *ApJ* **520** (1999) 248.

[97] R.R. Ross, A.C. Fabian and S. Mineshige, *MNRAS* **258** (1992) 189.

[98] P.J. Sakimoto and F.V. Coroniti, *ApJ* **247** (1981) 19.

[99] P.J. Sakimoto and F.V. Coroniti, *ApJ* **342** (1989) 49.

[100] F.H. Seguin, *ApJ* **197** (1975) 745.

[101] M. Sincell and J.H. Krolik, *ApJ* **496** (1998) 737.

[102] N.I. Shakura and R.A. Sunyaev, *A&A* **24** (1973) 337.

[103] N.I. Shakura and R.A. Sunyaev, *MNRAS* **175** (1976) 613.

[104] S.L. Shapiro and S.A. Teukolsky, Black Holes, White Dwarfs, and Neutron Stars (Wiley, New York, 1983).

[105] N.J. Shaviv, *ApJ* **494** (1998) L193.

[106] T. Shimura and F. Takahara, *ApJ* **419** (1993) 78.

[107] T. Shimura and F. Takahara, *ApJ* **440** (1995) 610.

[108] R. Speith, R. Riffert and H. Ruder, *Comp. Phys. Comm.* **88** (1995) 109.

[109] A. Steinacker and T. Henning, *ApJ* **554** (2001) 514.

[110] J.M. Stone, J.F. Hawley, C.F. Gammie and S.A. Balbus, *ApJ* **463** (1996) 656.

[111] J.M. Stone, D. Mihalas and M.L. Norman, *ApJS* **80** (1992) 819.

[112] H. Störzer, P.H. Houschildt and F. Allard, *ApJ* **437** (1994) L91.

[113] W.-H. Sun and M.A. Malkan, *ApJ* **346** (1989) 68.

[114] R. Svensson and A.A. Zdziarski, *ApJ* **436** (1994) 599.

[115] E. Szuszkiewicz, M.A. Malkan and M.A. Abramowicz, *ApJ* **458** (1996) 474.

[116] Y. Tanaka *et al.*, *Nature* **375** (1995) 659.

[117] J.-L. Tassoul, Theory of Rotating Stars (Princeton University Press, 1978).

[118] N.J. Turner, J.M. Stone and T. Sano, *ApJ* **566** (2002) 148.

[119] N.J. Turner, J.M. Stone, J.H. Krolik and T. Sano, *ApJ* (2002), in preparation.

[120] T.J. Turner, R.F. Mushotzsky, T. Yaqoob *et al.*, *ApJ* **574** (2002) L123.

[121] J.-M. Wang, E. Szuszkiewicz, F.-J. Lu and Y.-Y. Zhou, *ApJ* **522** (1999) 839.

[122] K. Watarai and J. Fukue, *PASJ* **51** (1999) 725.

[123] J. Wilms, C.S. Reynolds, M.C. Begelman *et al.*, *MNRAS* **328** (2001) L27.

[124] A.A. Zdziarski, P. Lubiński and D.A. Smith, *MNRAS* **303** (1999) L11.

COURSE 4

ACCRETION-EJECTION MODELS OF ASTROPHYSICAL JETS

R.E. PUDRITZ

Department of Physics and Astronomy, McMaster University, Hamilton, ON L8S 4M1, Canada

Contents

ACCRETION-EJECTION MODELS OF ASTROPHYSICAL JETS

R.E. Pudritz

Abstract

Astrophysical jets are almost always associated with accreting objects, be they young stellar objects or massive black holes. Their role in the evolution of accretion disks is tied to the efficiency with which hydromagnetic flows can transport angular momentum from their sources. These lectures constitute a review of the latest developments in the observation, theory, and numerical simulation of astrophysical jets. Among the most basic results in this field is the observed relation between the jet mass loss rate and the disk accretion rate $\dot{M}_{\rm w}/\dot{M}_{\rm a} \simeq 0.1$ which is observed in many different systems and arises in numerical simulations of all kinds. This connection is a consequence of angular momentum extraction from the disk by magneto-centrifugal winds. These systems can be regarded as integrated accretion-ejection structures – a notion that is explored in some detail. This paper also presents new results on numerical simulations of 2D and 3D jets for a wide variety of initial magnetic configurations. Astrophysical jets are shown to have both highly collimated as well as wide-angle outflow characteristics, and they are robustly stable in 3D.

1 Introduction

Astrophysical jets are among the most fascinating and ubiquitous phenomena in astronomy. These high speed, well-collimated, episodic outflows are associated with young stellar objects (YSOs) of all masses [1, 2]; accreting stellar mass black holes or microquasars [3]; massive black holes in active galactic nuclei (AGNs) [4]; planetary nebulae [5]; and compact isolated objects such as the Crab pulsar [6]. One invariably finds that jets are associated with accretion disks around their central objects. We shall see that

This work is supported by a grant from the National Science & Engineering Council of Canada.

the evidence strongly suggests that accretion disks are the drivers for most nonrelativistic or mildly relativistic outflows.

Jets are not merely exotic oddities in these diverse systems. Their astrophysical importance derives from the high efficiency with which they can extract and transport gravitational potential energy and angular momentum from the central engine. The evidence suggests that jets in most of these sytems are powered by accretion – the gravitational binding energy that is released by the accretion of gas onto the central object. Thus jet properties such as their terminal velocities naturally scale with the depth of the gravitational potential well of their central objects. Astrophysical jets are also highly robust structures in that they are able to propagate through inhomogenous gaseous environments (*e.g.* clumpy gas in molecular clouds, AGNs) and out to very large physical scales.

In these lectures, I shall explore the link between accretion disks and jets in these seemingly disparate physical systems. Theoretical and numerical work strongly suggests that the simultaneous accretion of gas through a magnetized accretion disk and the ejection of a portion of it in a magnetized wind lie at the heart of the jet phenomenon. I start by discussing the key observational results on the relation between accretion disks and jets from YSOs to quasars. I then review the general theory of hydromagnetic disk winds. Finally, I review the rapidly growing body of numerical work that is contributing new insights into these processes. I also present some new simulations of jets from a wide variety of initial magnetized disks, including Blandford & Payne (1982), as well as Pelletier & Pudritz (1992). Many new results about the structure and dynamics of jets have arisen from studies of YSOs and this is reflected in the contents of this review (see also [7,8]).

2 Observations and basic physics

2.1 Molecular outflows and jets in YSOs

Bipolar molecular outflows are the most obvious and ubiquitous sign-posts of star formation. There are now more than 200 molecular outflows known [9] and over 40 of these are associated with well collimated jets. Molecular CO outflows trace the interaction between an underlying jet and the ambient molecular cloud. The gas mass in the molecular flows ranges from $10^{-2}\ M_\odot$ to 200 $M_\odot$ (Mon R2) while their sizes range from $<$0.1 pc to 5 pc. Outflows are associated with both low-mass T-Tauri stars as well as with high mass star formation. It appears that 45% of all massive star formation regions are associated with high velocity bipolar outflows whose underlying stars correspond to mid to early B type stars [10].

Some of the best evidence for the hydromagnetic nature of the engine that drives these outflows is the well documented correlation between the thrust ($F_{\rm CO} \simeq \dot{M}_{\rm w} v_{\rm w}$) of the material in the swept-up CO flow and the radiative thrust associated with bolometric luminosity of the central source ($F_{\rm rad} \simeq L_{\rm bol}/c$) [11]. Here $\dot{M}_{\rm w}$ is the mass loss rate observed in the CO flow and $v_{\rm w}$ is the typical outflow speed.

The data show a nearly linear relation between these two quantities and that the thrust is observed to be two of orders of magnitude larger than what could be provided by a radiative drive from the central star:

$$F_{\rm CO}/F_{\rm rad} = 250(L_{\rm bol}/10^3 L_\odot)^{-0.3}. \tag{2.1}$$

This empirical result excludes the driving of YSO outflows by thermal or radiative pressure alone. The mechanical luminosity of outflows, $L_{\rm CO} \simeq (1/2)\dot{M}_{\rm w} v_{\rm w}^2$, is observed to be small in comparison with the luminosity of the central stars

$$L_{\rm CO}/L_{\rm bol} = 0.01(L_{\rm bol}/10^3 L_\odot)^{-0.2}. \tag{2.2}$$

It is remarkable that these relations are valid over nearly six orders of magnitude in $L_{\rm bol}$ which strongly suggests that the outflows from both low and high mass YSOs share a similar underlying mechanism.

The velocity and morphological structure of CO outflows provide useful constraints on the physics of the underlying jet. Jets can often be traced by their shocked molecular hydrogen emission so that by comparing H_2 and CO maps of outflows one may infer how a jet interacts with the ambient medium. Two basic models for jet-driven CO outflows have had some observational success: i) jet-driven bowshocks [12, 13]; or ii) wind-driven shells that are pushed by an underlying wide-angle wind [14]. These models are compared in [15] and [2]. The former mechanism predicts that CO outflows delineate the working surfaces and bowshocks of underlying well collimated jets with the ambient medium, while the latter mechanism proposes that a wide-angle wind provides a largely undetected source of momentum that pushes on the walls of outflow cavities.

Tests of these two models have been performed by high spatial resolution OVRO observations of several molecular outflows [16] Symmetric 2-sided H_2 jets are observed in at least one case, HH 212. They are 0.6 pc long and consist of matched knots and bow shocks along the jet which is perpendicular to an extended $C^{18}O$ disk. The CO is well aligned with the shocked jet – the tips of the CO lobes connect to the bow shocks of the jet as well as to the inner shocks. This source provides a good fit to the jet-driven bow-shock model. However, many of the other outflows in the Lee *et al.* [16] sample are best fit by the wide-angle wind-driven shells.

Protostellar outflows are initiated in the earliest stages of star formation and persist for as long as accretion disks are present. A quantitative measure of the prevalence of a protostellar disk in any system is obtained by measuring the system's infrared spectral energy distribution (SED) – classified as Class 0 to Class 3. The infrared emission that is in excess of that arising from the underlying central YSO is produced by the accretion disk [17].

The most embedded sources (Class 0's) are so well hidden that even IRAS point sources are hidden at wavelengths $\lambda < 10\ \mu$ m [18]. Class 0 sources have infalling envelopes whose masses are comparable with those of their central objects. These are probably bona-fida protostars – objects undergoing gravitational collapse – and persist for roughly 10^4 years. Class 0's have fast CO outflows and bright submillimetre emission. The SED of a Class 0 source is essentially that of a simple black body at the temperature of the dense, obscuring envelope. Class I sources consist of an evolved protostar that is surrounded with a thick disk and a jet. Their SEDs are dominated by disk emission. This phase lasts for 10^5 year. For Class 2 objects – the so-called classical T-Tauri stars (CTTSs) – the SED is dominated by emission from the star though infrared emission from the disk is still present and one still observes jets. The characteristic lifetime for this phase is 10^6–10^7 yr. Finally, the last stage – Class 3 – is called a weak lined T-Taur star (WTTS). It has no obvious associated disk or jet and lasts on the order of 10^7 yr.

There is good evidence to show that accretion disks are present in even the early Class 0 stage. Observations of one of the youngest Class 0 sources, VLA 1623 in the ρ Oph cloud, show that the source is a binary system whose disk masses are around $0.04\ M_\odot$ [19, 20]. Disks in other very young protostellar systems have also been found [21].

For class 0 protostars the bolometric luminosity of the central YSO arises primarily from accretion energy [18]. Class 0 sources are distinguished in having the most highly collimated flows CO outflows with speeds $v_{\rm CO} \geq 50\ {\rm km\,s^{-1}}$ [9]. They are also unusual in having mechanical luminosities comparable to the bolometric luminosity $L_{\rm CO} \simeq (1/2)L_*$. Given that the protostellar luminosity at this early stage is dominated by accretion, one sees that the driving mechanism of the outflow must be as efficient as it could be in extracting gravitational potential energy and converting it into the mechanical energy of the outflow (the remaining 1/2 of the gravitational binding energy is released in heating the region nearest the inner edge of the accretion disk).

The underlying jet in these systems becomes amenable to direct optical observations after most of the initial core has collapsed into the disk. As an example, the well studied source DG Tau has been observed in both H_α – a good tracer of shocked gas – as well as in the forbidden line [OI] [22]. The former indicates that several bow shock structures are present and that the jet ploughs into previously ejected material. The latter diagnostic reveals the the jet moves at 200 km s^{-1} and undergoes distinct side-to-side wiggles which may be a sign that a kink mode ($m = 1$) is operative in the jet (see Sect. 4.2).

Figure 1 shows high resolution images of H_α and [OI] emission in the DG Tau jet taken in various velocity ranges from 50–450 km s^{-1} in 125 km s^{-1} wide velocity bins [23]. The top image shows two distinct bow shocks in the mid velocity range (MV). Its density increases towards the star and axis. One also sees that there is a continuous variation in the velocity of the jet, the highest (VHV) being nearest the jet axis, and the lowest farther away. The very high speed gas (VHV) gas is the most collimated and is denser than the lower speed gas. The lower speed flow is farther from the axis and appears to broaden out and become less collimated with increasing radius in the jet. The jet can be traced to within 15 AU (0.1″) of the central source. These remarkable images correspond well to the theory and simulations of disk winds (see Sects. 3.1 and 4.1 respectively). While earlier papers suggested that jets may have two distinct velocity components (fast and high collimated, *versus* slower and wide-angled components; [24]), these data suggest a continuously varying velocity structure.

There is also good evidence that this flow is rotating clockwise when viewed from the flow towards the source [25]. The rotation speed is 6–15 km s^{-1}, depending on the position in the flow within 110 AU from the source. These authors find agreement between these measured rotation speeds with those expected in a centrifugally driven hydromagnetic wind. Using the scalings for disk wind speeds and angular momentum transport rates outlined in Sects. 3.1 and 3.2, this outflow is estimated to be carrying up to 60% of the necessary angular momentum loss from the disk at its footpoint.

HST observations provide direct information about the structure and dynamics of the underlying (optical) jets from YSOs – HH30 being one of the most outstanding examples. This jet is associated with a flared, optically thick disk with a diameter of about 450 AU. The jet is resolved with a FWHM at the base of 25 AU. Jet structure is resolved down to 0.1″ scales, proper motions are 100–300 km s^{-1}, and knots in the jet are generated at the rate of 0.4 new knots per year [26]. Jet collimation occurs within a cone of 3°, within 30 AU of the source. Moreover, the transverse velocities of the knots is consistent with zero.

The mechanism of jet collimation can be constrained by observing the opening angle of a jet as a function of its distance from the source. As an example, the jet in HH30 has a 3° opening angle between 0.5″ and 5″ from the source and is only 1.9^0 between 2.5″ to 75″. Given that any inertial confinement would be expected to diminish with distance from the source, this argues that the jet may be collimated by magnetic forces.

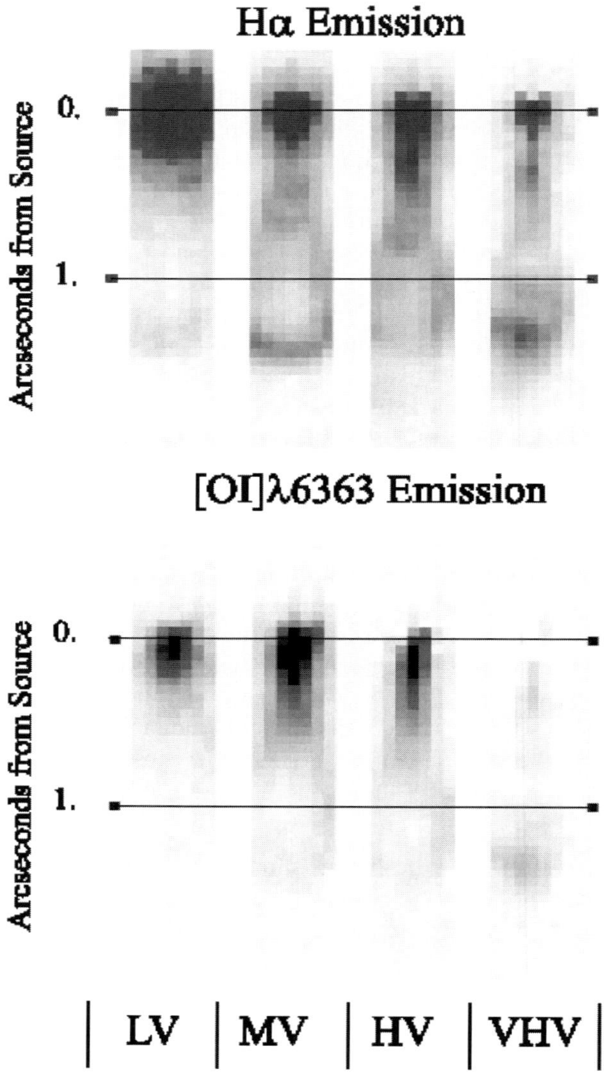

Fig. 1. Synthetic images of the DG Tau outflow in H_α and [OI] derived from 7 overlapping long-slit positions offset in the transverse direction to the flow. Courtesy of Bacciotti *et al.* [23].

Finally, jets are highly time-dependent phenomena. This has implications for how we attempt to model them. In addition to the appearance of knots on short time scales (on the order of months), jets are dominated by a series of large bow-shocks. The dynamical time scales associated with these bow shocks indicates that jets are episodic on time scales of 10^3 years. This time scale is similar to the time scale for the recurrance of FU Ori outbursts in disks around TTSs as discussed below [27].

2.2 YSO accretion disks: Links to jets

Millimetre [28, 29] as well as near IR surveys [30] showed that 50% of YSOs harbour accretion disks and that the median disk mass (modulo uncertainties about dust emission properties) is several times $10^{-2}\ M_\odot$. These disk masses are comparable to the minimum mass solar nebular models necessary to form the solar system. The HST images of silhouette disks in the Orion nebula provide striking proof that accretion disks are an essential part of star formation.

Surveys for optical and near IR jets in the Taurus region reveal that jet activity wanes with the gradual disappearance of disks – 60% of Class I objects have jets as compared with 10% in Class IIs, and none in Class III objects [31]. Observational connections between the mass of circumstellar disks and molecular outflows [32] shows that sources with well developed outflows had 20 times the (1.3) millimeter flux of sources that did not have flows. This also suggests that accretion rates ($\dot{M}_\mathrm{a}$) through disks may be higher in the earlier phases of disk evolution than later.

There is a strong correlation between the observed mass accretion rate through protostellar disks and the mass loss rates carried by their associated jets. Optical jets associated with TTS have typical mass loss rates that are consistently about a tenth of the underlying disk accretion rate. The jet in HH30, as an example, has a mass outflow rate as determined from emission line diagnostics of $5 \times 10^{-10}\ M_\odot\ \mathrm{yr}^{-1}$. The total jet mass loss rate is a factor of ten times greater than this. This in turn is 10% of the material that is accreting through the disk; $\dot{M}_\mathrm{a} \simeq 5 \times 10^{-8}\ M_\odot\ \mathrm{yr}^{-1}$. Hartmann & Calvet [33] found, for a sample of 31 TTSs, jet mass loss rates calculated from the forbidden lines in the range $3 \times 10^{-7} < \dot{M}_\mathrm{j} < 10^{-10}\ M_\odot\ \mathrm{yr}^{-1}$ with a median value of $10^{-9}\ M_\odot\ \mathrm{yr}^{-1}$ (*e.g.* review [34], and Calvet these Proceedings). These mass loss rates are a tenth of the underlying disk accretion rate in the system;

$$\dot{M}_\mathrm{j}/\dot{M}_\mathrm{a} \simeq 0.1. \qquad (2.3)$$

A class of YSOs with an especially high accretion rates are known as FU Ori systems ([27]). These systems are suspected to be accretion disks around TTSs that undergo disk instability leading to high accretion rates and disk temperatures. Jets associated with FU Ori bursts are particularly easy to study and have provided some excellent tests for theoretical models of jets. Fu Ori luminosities range from 200–500 $L_\odot$ while that of a typical TTS is only $\simeq$1 $L_\odot$. The emission in these systems is completely dominated by the disk and temperature reach $T \simeq 7000$ K in the inner regions. The wind mass-loss rates in FU Ori jets are nearly 4 orders of magnitude greater than in the TTS phase, at $\dot{M}_\mathrm{w} \simeq 10^{-5}\ M_\odot\ \mathrm{yr}^{-1}$. Even in this phase, however, the ratio of wind mass loss rate to disk accretion rate is again $\dot{M}_\mathrm{w}/\dot{M}_\mathrm{a} \simeq 0.1$ [33]. The next chapter provides a theoretical explanation of this robust result.

One can probe regions near the base of the jet, through a well-chosen set of absorption lines, to determine whether or not accretion disks accelerate these outflows (see Calvet – these Proceedings). Calvet *et al.* ([35]) searched for this connection by choosing two types of metallic absorption lines; weak lines which probe the region deep in the disk atmosphere close to its surface (where one expects to see double-peaked lines due to the Kepler rotation), and stronger absorption lines which form in the upper disk atmosphere where the outflow speed is expected to approach the local escape speed. Their observations clearly show that double peaked weak lines which dominate in the high column depth region near the disk surface give way to single peaked strong lines farther from the disk. This strongly implicates accretion disks as a central component of the driving engine.

2.3 Magnetic fields: From molecular clouds to disks around YSOs

There is abundant evidence that magnetic fields are important on a variety of spatial scales in molecular clouds. Zeeman measurements on the VLA and GBT of a sample of 27 clouds [36] find that magnetic fields dominate the thermal energy of the gas at densities $\geq 10^3\ \mathrm{cm}^{-3}$ ($\beta = P_\mathrm{gas}/P_\mathrm{mag} \simeq 0.04$). The magnetic field energy density is comparable to gravitational energy densities as predicted by the virial theorem; $M \simeq (1/2) W_\mathrm{grav}$.

The structure of the magnetic field in dense gas is best inferred by sub mm polarimetry. At these wavelengths, dust emission predominates over scattering. Spinning dust grains are aligned by the magnetic field with their long axes perpendicular to the magnetic field line so that sub mm polarimetry traces the structure of cloud's magnetic field. The observations are surprising – relatively well ordered fields are seen down to scales 0.04 pc [37–39]. The data is well fit by a theoretical model for filamentary molecular clouds in which helical field lines wrap around the filaments [40, 41].

There is some evidence for the existence of strong, ordered magnetic fields that are associated with jets. A radio map of T-Tau at 6 cm [42] shows that two lobes of opposite helicity, circularly polarized emission straddle the central star. The emission is likely to be gyrosynchotron from mildly relativistic electrons ($\gamma \simeq 2$–3). The field strength deduced is $B_\mathrm{j} \geq$ several Gauss at a distance of 10–20 AU from the source. The associated magnetic pressure $B_j^2/8\pi = 3.6 \times 10^{-1}$ $(B/3G)^2$ dyne is large compared to the ram pressure of the jet; $\rho_j v_j^2 \simeq 0.67 \times 10^{-3}$ $(n_j/10^4\ \mathrm{cm}^{-3})$ $(v_j/200\ \mathrm{km\,s}^{-1})^2$ dyne. The structure of the field is not determined by these observations. One anticipates however that since the toroidal field falls probably falls off less steeply than a poloidal field in the jet ($B_\phi \propto r^{-1}$), one may be probing a strong toroidal field at these scales. Jet collimation will arise from the associated hoop stress (see Sect. 3.3).

It is very difficult to obtain direct evidence for magnetic fields that are associated with the accretion disks at this point in time. Nevertheless, Zeeman observations of an H30α maser recombination lines at 1.3 mm have found a field strength of 22 mG at 40 AU [43]. This field strength is below equipartition strength with the gas pressure of a standard disk at these radii, but is still significant. Finally, the meteoritic evidence shows that a magnetic field strength approaching 1 Gauss (at 3 AU) must have been present in the solar nebula [44].

A prerequisite for any hydromagnetic flow phenomenon is that the fluid be a good conductor. For the dense regions that can occur in protostellar accretion disks, however, this is not generally true. Molecular clouds are heated by the flux of cosmic rays through them. More significantly, the bulk of the lower density molecular gas in clouds is probably ionized by X-rays from YSOs of all classes (with the possible exception of Class 0's). YSOs are copious X-ray producers with X-ray energies 2–5 keV (*e.g.* [45,46], Montmerle these Proceedings). At 5 kev, X-rays penetrate material down to a visual extinction of $A_\mathrm{v} \simeq 100$, which corresponds to a column density of up to 30 g cm^{-2}. This provides reasonable coupling of magnetic fields within the bulk of a molecular cloud's volume. However, penetration depth of X-rays is far less than the column density of the minimum mass solar nebula, which has typically $\Sigma_{\mathrm{min}}(1\ \mathrm{AU}) \simeq 1500$ g cm^{-2}. The interiors of these disks can only be ionized by cosmic rays (see Sect. 2.6).

2.4 The role of jets and accretion disks in star formation

Star formation involves the collapse of an initially slightly rotating molecular cloud core into an embryonic protostellar core and a surrounding accretion disk. Most of the material that forms the star probably flows through the accretion disk first so the question of how angular momentum is removed from disks is pivotal to the theory of star formation. The energy that is liberated as gas accretes through a disk depends on the depth of the gravitational potential well produced by the central object, as well as the disk accretion rate $\dot{M}_\mathrm{a}$. Any ordered magnetic field that threads through a spinning object – be it a disk or a star – exerts a torque upon it. This torque can extract the object's angular momentum and transport it away from the disk in a wind. The gravitational binding energy that is released during accretion appears as the mechanical energy of the outflow. Other competing torques that drive angular momentum transport in disks without driving a wind are due to spiral waves and MHD turbulence. Although it has been suggested that outflows and jets might help to determine the mass of a star by ejecting the material in the infalling envelope, it is difficult to see how the well collimated jets in the Class 0 phase have the necessary cross-sectional area to do the job.

Numerical simulations of the collapse of finite mass, non-singular cores, find that the infall rate varies with time with high initial rates giving way to slower infall rates as the remaining remnants of an initial core envelope are accreted onto the central object (*e.g.* [47, 48]). Gravitational collapse of the molecular cloud core in which a star forms feeds the disk accretion and ultimately, jet activity. One would therefore expect this evolution of the infall rate to be mirrored in the changing strength of the outflows. This appears to happen – outflow activity is diminished in going from Class 0 to Class I sources. The former have thrusts that are a factor of 100 larger than the latter. Measurements of the diminishing thrusts for 45 young YSOs suggest that the accretion rate in the disk decreases with time [49]. This can be seen by writing the observed thrust carried by the CO outflow as $F_{\rm CO} = f_{\rm ent}(\dot{M}_{\rm w}/\dot{M}_{\rm a})\dot{M}_{\rm a}v_{\rm w}$ where $\dot{M}_{\rm a}$ is the accretion rate through the disk. Theoretical arguments, which will be discussed in the next section, suggest that the accretion ratio, $\dot{M}_{\rm w}/\dot{M}_{\rm a} \simeq 0.1$ may remain nearly constant through various phases of disk evolution ([50, 51]), so that outflow evolution reflects the diminuation of accretion rates from $\dot{M}_{\rm a} \simeq 10^{-5}$ to 10^{-7} $M_\odot$ yr^{-1}.

The disk lifetimes for low mass TTS – measured by fitting isochrones to central TTS – is $\simeq 10^6$ yr. This implies that roughly 10^{-2} $M_\odot$ accretes onto the YSO in these later phases of TTS evolution. If TTSs typically undergo the transition to FU Ori behaviour, and these outbursts have a duration of order 2000 years (as is suggested by time scale associated with jet outbursts) then most of a star's mass may be accreted in 5–10 of these outbursts within disks [52]. Thus, the accretion rates that are mediated by jet outflows may be the predominant mode of building a central star.

2.5 Relativistic systems: Microquasars, quasars, and X-ray pulsars

The link between jets and disks is observed in many relativistic systems as well (see Sunyaev, these Proceedings). There are 200 galactic X-ray binaries systems known. X-ray emission arises from the inner parts of accretion disk around a compact primary – either a neutron star or black hole [53]. Of this population only 10% are radio loud, and 10 have relativistic jets with $v > 0.1c$ [3]. The proof that these are relativistic jets comes from the superluminal motion of discrete knots that one sees in several of these sources – most famously GRS 1915+105 [54]. The presence of two IRAS sources along the jet axis indicates its interaction with the surrounding environment. The jets associated with stellar mass black holes are like those seen in quasars – with the considerable advantage that all of the time scales in the former are shorter by the ratio of the orbital times at their respective Schwarzschild radii $\propto (M_{\rm stellar}/M_{\rm quasar}) \simeq 10^{-7}$.

Jets and accretion disks also appear to be well correlated in the context of quasars such as NGC 4258. Here, VLBI observations reveal a set of spatially resolved water masers that extend from 0.13–0.25 pc from the central object. The radial velocities of the disk-like spatial distribution of these masers fall on a Keplerian rotation curve, implying a central black hole mass of $\simeq 3.5 \times 10^7\ M_\odot$ [55]. Perpendicular to this flared disk, one can discern an extended structure which may be the base of the relativistic jet that is observed on kpc scales. This may represent the shocks that are seen in relativistic flows [56].

Why is the jet frequency apparently so low in microquasars and quasars (perhaps $\simeq 10\%$) and yet so high YSO disk/jet systems? Accretion disks around black holes are expected to be geometrically thick and to have a very different underlying dynamics (*e.g.* review [57]). In the so-called ADAF solutions, most of the gravitational energy release goes into heating the tenuous gas of the disk which, it is proposed, is then swallowed by the black hole without radiating very much [58]. On the other hand, there are good arguments to suggest that such disks ought to have winds that carry off this hot, tenuous gas before it has the chance of being so devoured – the so-called ADIOS solutions [59]. In this latter case, relativistic jets may be as common as their YSO counterparts, but they are simply too tenuous to be readily detected.

A beautiful example of a jet from a compact object that does not have an associated accretion disk is the Crab pulsar. The theory of magnetized winds was originally developed to explain the braking of magnetized stars (see next section). Sufficiently rapidly rotating isolated objects such as pulsars therefore, ought to show outflows that are unrelated to accretion disks, but that are nevertheless collimated. The X-ray emission from the Crab was recently mapped using the Chandra observatory [6]. The X-ray emission is observed to be more compact than that from the optical nebula (about 40% as large). The data show both an X-ray torus as well as a knotted jet and counterjet that is perpendicular to the semi-major axis of the torus ($\simeq 0.37$ pc – with a semi-minor axis of 0.17 pc). The X-ray jet emission indicates continuous emssion from highly relativistic particles that are flowing parallel to the pulsar magnetic axis. Axisymmetric solutions for outflows from rapidly rotating magnetized stars show that such outflows can become highly collimated along their rotation axes [60]. In such situations, it is the spin energy of the rapidly rotating neutron star that ultimately powers the jet.

2.6 Magnetized accretion disks: General physical processes

What general characteristics are expected of a magnetized disk and its jet? In particular, how are magnetized disks heated – and why do they drive off jets? Is a disk corona required to initiate such a wind (as in the solar wind)? Given that the only astrophysical MHD laboratory that we can directly observe is the Sun, what general physical processes seen on the Sun will be important for understanding disk-jet systems?

The activity of the solar corona provides some important insights. Three major mechanisms have been proposed for heating the solar corona: (i) heating by current dissipation during magnetic reconnection of the loops, (ii) damping of Alfvén waves, and (iii) MHD turbulence. Observations from two X-ray imaging satellites, Yohko and Soho, as well as the imaging ultraviolet satellite TRACE resolve the physics of the solar corona on physical scales ranging from 60 $R_\odot$ down to 5×10^{-4} $R_\odot$. The magnetic field on the Sun consists of a heirarchy of magnetic loops in which magnetic reconnection and flares occur. Flares have energies ranging over 7 decades in magnitude down to so-called nanoflares which release $\geq 10^{24}$–10^{27} ergs. The contribution of flare energy can be ascertained by measuring the flare frequency distribution. By combining the hard and soft X-ray data, as well as the EUV emission from 281 nanoflares, Aschwanden *et al.* [61] find the power-law relation $N(E) \simeq 10^{-46}(E/10^{24}\ \mathrm{erg})^{-1.8}\ \mathrm{s}^{-1}\ \mathrm{cm}^{-2}\ \mathrm{erg}^{-1}$. This result shows that there is insufficient power of nanoflares to heat the solar corona. Non-luminous energy carriers such as waves may provide the dominant mode of heating the solar corona.

These solar results may be relevant to accretion disks. Accretion disk models typically assume that turbulent heating occurs in disks, although there is not much known about energy deposition in a disk corona. Thin accretion disks are highly sheared flows. Magnetic loops cannot have footpoints that are widely separated in disk radius since the dominant Kepler shear would stretch these out leading to reconnection. Magnetic buoyancy would ensure that a forest of loops would protrude from the disk surface and that shear would orient these in largely toroidal direction. MHD turbulence in the body of the disk is the expected end-state of the operation of the magneto-rotational (MRI) instability ([62, 63], Blaes – these proceedings). This instability results in the exponential growth of magnetic field on the Oort dynamical time-scale, and saturation occurs when the Alfvén velocity $v_{\mathrm{A}} \leq c_{\mathrm{s}}$ in the disk.

Accretion disks and their coronae could be heated by a combination of flares, an inverse cascade of turbulent (MHD) energy, or by dissipation of Alfvén waves. It is still unknown as to how much of the gravitational potential energy that is released locally from accretion would actually be dissipated within the body of the disk as compared to dumping it in a corona. Global simulations of the MRI instability in thin disks show that considerable wave energy flows into a disk corona and where it is damped in shocks [64] (see Sect. 5).

The production and collimation jets may require that accretion disks be threaded by an ordered magnetic field. What is the source of this field? The two possibilities are that the field pervades the larger-scale medium (*e.g.* protostellar cloud core) and is advected along with the material that forms the accretion disk – or – that the field it is generated *in situ* by local dynamo action or MRI instability.

Consider the first scenario. The dynamics of a mean field in a disk with a turbulent viscosity ν_{T} and magnetic diffusivity η_{T} is controlled by the magnetic Prandtl number defined as their ratio. We follow the convention of Lubow *et al.* [65] and write it as $Pr_{\mathrm{m}} \equiv \eta_{\mathrm{T}}/\nu_{\mathrm{T}}$. In the limit of small Prandtl number ($Pr_{\mathrm{m}} << 1$) viscosity dominates field slippage and the field is dragged inwards by the radial flow ($v_{\mathrm{r}} \propto \nu_{\mathrm{T}}$). In the opposite limit, $Pr_{\mathrm{m}} >> 1$, the large magnetic diffusivity that arises from poor conductivity of the gas leads to outward slippage of the field which dominates over its inward advection by the accretion flow. Stationary solutions in idealized thin disk models show that the basic governing parameter for field evolution is $D \equiv (H/r)Pr_{\mathrm{m}}$ where $H(r)$ is the disk pressure scale height. The radial inflow speed is $v_{\mathrm{r}} = -3\nu_{\mathrm{T}}/2r$ and the ratio of the radial to vertical components of the disk field is given by $B_{\mathrm{r}}/B_z = \tan^{-1}\theta \simeq D^{-1}$ where θ is the opening angle of the field with respect to the vertical axis (Lubow *et al.* 1994). The opening angle of the field is very important for the dynamics of hydromagnetic winds, so this result shows that disk physics and outflow dynamics are intimately coupled with one another. Centrifugally driven outflows are favoured when $D \simeq 1$.

In the second scenario, magnetic fields are generated *in situ*. The MRI instabiltiy may feed local dynamo action ([66,67]) wherein the energy in the sheared Kepler flow is converted into magnetic energy. While traditional dynamo actions required helical turbulence in order to produce a poloidal field from a toroidal one as an example, Balbus and Hawley emphasize that the MRI instability does not require helical turbulence. It is also interesting to note that the MRI instability can also be tapped by radial global modes in a disk which grow at somewhat slower but still significant rates [68]. Global simulations of disk confirm that these global modes are generated [69].

All of the discussion on accretion-ejection structures has assumed that accretion disks are good conductors and that therefore the magnetic field is well coupled to the gas. While the disks in FU Ori systems, quasars and microquasars are hot enough to guarantee this, the dense disks around TTSs may not be well coupled throughout their volume. Turbulent transport of angular momentum requires that the MRI instability be operative. In a resistive disk, there will be some scale λ whose growth rate through MRI instability ($\simeq v_A/\lambda$) is just balanced by the dissipation rate (η/λ^2) prescribed by the microphysical magnetic diffusivity η. Thus, magnetic viscosity will be completely damped out on physical scales whose magnetic Reynold's numbers are $Re_m \equiv v_A \lambda/\eta \leq 1$ [70]. The magnetic diffusivity of the gas depends on its degree of ionization, such that $\eta = 1.5\times 10^{-12} T_{500}^{1/4} \zeta_{-17}^{1/2} n_{H,13}^{-1/2}$ cm^{-2} s^{-1} where the temperature is in units of 500 K, the cosmic ray ionization rate is in units of particles per 10^{17} s, and the density is in units of 10^{13} cm^{-3}. The resulting "dead-zone" (where material is decoupled from the field) for a protostellar accretion disk may extend out to several AU and leave only a relatively thin well-coupled surface layer along which accretion flow can still occur. This zone also has significant consequences for planet formation [71].

Having discussed the observational and physical underpinnings of jets from accretion disks, we now review some of the essentials of the large body of theoretical work on the physics of jet production.

3 Theoretical models

3.1 Axisymmetric hydromagnetic winds: Conservation laws

The theory of hydromagnetic winds originated with the early work on winds from rotating magnetized stars [72, 73]. These papers developed 1-D, axisymmetric models of hydromagnetic flows from rotating stars. They showed that a centrifugally driven, stellar wind could efficiently transport stellar angular momentum, leading to the spin down of the central star. The application of this idea to self-similar accretion disks was carried out in the seminal paper by Blandford & Payne (BP) [74].

Consider the simplest possible description of a magnetized gas, rotating gas threaded by a large-scale field. The equations of stationary, ideal MHD are the conservation of mass (continuity equation); the equation of motion with conducting gas of density ρ undergoing gas pressure (p), gravitational (from the central object whose gravitational potential is ϕ), as well as the Lorentz force (from the field $\boldsymbol{B}$); the induction equation for the evolution of the magnetic field in the moving fluid; as well as the conservation of magnetic flux. These equations were written down by Chandrasekhar, Mestel, and many others;

$$\nabla.(\rho \boldsymbol{v}) = 0 \tag{3.1}$$

$$\rho \boldsymbol{v}.\nabla \boldsymbol{v} = -\nabla p - \rho \nabla \phi + \frac{1}{4\pi}(\nabla \times \boldsymbol{B}) \times \boldsymbol{B} \tag{3.2}$$

$$\nabla \times (\boldsymbol{v} \times \boldsymbol{B}) = 0 \tag{3.3}$$

$$\nabla . \boldsymbol{B} = 0. \tag{3.4}$$

It is convenient to decompose all magnetic fields and velocity fields into poloidal and toroidal field components $\boldsymbol{B} = \boldsymbol{B_p} + B_\phi \hat{\boldsymbol{e}}_\phi$ and $\boldsymbol{v} = \boldsymbol{v_p} + v_\phi \hat{\boldsymbol{e}_\phi}$.

The solution of the induction equation is particularly simple;

$$\boldsymbol{v_p} \times \boldsymbol{B_p} = \nabla \psi \tag{3.5}$$

where ψ is akin to an "electrostatic" potential. This solution has two important ramifications. The first arises from the toroidal component of the equation which states that $\partial \psi / \partial \phi = 0$; a simple consequence of axisymmetry. Therefore, the poloidal velocity in axisymmetric and stationary flows must be oriented along magnetic field lines, ie, parallel to the poloidal field component $\boldsymbol{v_p} || \boldsymbol{B_p}$. Another obvious consequence of the induction equation follows by taking the dot product of it with $\boldsymbol{B}$, which gives $\boldsymbol{B_p} . \nabla \psi = 0$. Thus, ψ is a constant along field lines and is nothing other than the (poloidal) magnetic flux.

Let us turn now to the equations (3.1) and (3.4) that express the conservation of mass and magnetic flux. By comparing the form of these two equations, we see that the poloidal field and the poloidal mass flux along the field line must be proportional to one another along a field line. Defining a function k that is a constant along a magnetic field line and which we will call the "mass load" of the wind, then;

$$\rho \boldsymbol{v_p} = k \boldsymbol{B_p}. \tag{3.6}$$

This mass load function can be revealingly cast by noting that the wind mass loss rate passing through an annular section of the flow of area $\mathrm{d}A$ through the flow is $\mathrm{d}\dot{M}_\mathrm{w} = \rho v_\mathrm{p} \mathrm{d}A$, while the amount of poloidal magnetic flux through this same annulus is $\mathrm{d}\Phi = B_\mathrm{p} \mathrm{d}A$. Thus, the mass load per unit time, per unit magnetic flux of the wind that is preserved along each streamline along the flow emanating from the rotor (a disk in this case) is

$$k = \frac{\rho v_\mathrm{p}}{B_\mathrm{p}} = \frac{\mathrm{d}\dot{M}_\mathrm{w}}{\mathrm{d}\Phi}. \tag{3.7}$$

The mass load is determined by the physics of the underlying rotor, which is its source. As we whall see, this source boundary condition plays an important role in the physics of the resulting magnetized wind.

We can extract additional information from the induction equation by examining its poloidal component; $\boldsymbol{v_p} \times \boldsymbol{B_\phi} + \boldsymbol{v_\phi} \times \boldsymbol{B_p} = \nabla \psi |_{\boldsymbol{p}}$. This may be simplified by using the relation between v_p and B_p, to find that $\psi = \Omega - (k B_\phi / \rho)$ where $\Omega(r)$ is the local angular velocity of the gas at any point in the flow. In order to evaluate ψ, note that at the mid-plane of the disk, the toroidal field *must* vanish by symmetry. Thus $\psi = \Omega_{\rm o}$, the angular velocity of the disk at mid-plane. Thus, one has found the relation between the toroidal field and the angular velocity in a rotating flow;

$$B_\phi = \frac{\rho r}{k}(\Omega - \Omega_{\rm o}). \tag{3.8}$$

This result has an interesting consequence for jet collimation – the larger the value of k along a field line, the smaller is the value of the toroidal field. This is an inertial effect; it implies that jets that have low density with respect to the magnetic flux that threads them will have stronger toroidal fields and will hence be more easily collimated by the resulting hoop stress.

Consider the **angular momentum equation** for axisymmetric flows. This is described by the ϕ component of equation (3.2), which yields;

$$\rho \boldsymbol{v_p}.\nabla(r v_\phi) = \frac{\boldsymbol{B_p}}{4\pi}.\nabla(r B_\phi). \tag{3.9}$$

Let us apply this equation to two different situations – (i) the transport of angular momentum along a field line and (ii) the transport of angular momentum out of a disk, by an outflow (which is analyzed in the following subsection).

By applying equations (3.6) to (3.9) we find that angular momentum flow along a field line reduces to the statement that

$$\boldsymbol{B_p}.\nabla\left(r v_\phi - \frac{r B_\phi}{4\pi k}\right) = 0. \tag{3.10}$$

This equation says that there is a quantity,

$$l = r v_\phi - \frac{r B_\phi}{4\pi k}, \tag{3.11}$$

which is a constant along a field-line in the flow. This is just the total angular momentum per unit mass. The form for l reveals that the total angular momentum is carried by both the rotating gas (first term) as well by the twisted field (second term).

The solution for the toroidal components of the outflow velocity may now be found by eliminating the toroidal field between equations (3.8) and (3.11);

$$r v_\phi = \frac{l m^2 - r^2\Omega_{\rm o}}{m^2 - 1} \tag{3.12}$$

where the so-called Alfvén Mach number of the flow is $m^2 = v_{\rm p}^2/v_{\rm A}^2$ and $v_{\rm A} = B_{\rm p}/\sqrt{(4\pi\rho)}$ is the Alfvén speed of the flow. The Alfvén surface is

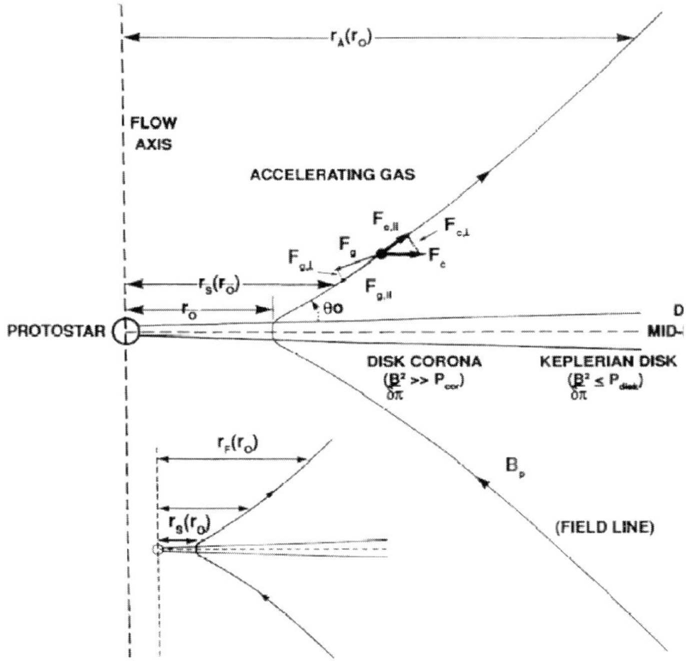

Fig. 2. Geometry and dynamics of a disk wind from a thin accretion disk. The three critical points corresponding to the slow magnetosonic r_S, Alfvénic, r_A, and fast magnetosonic r_F conditions in the jet are shown. From Pelletier & Pudritz [51].

that point $r = r_\mathrm{A}$ at each field line in the wind where $m = 1$. The flow may be regarded as kept in co-rotation with the rotor until this point is reached.

We see that in order that no spurious singular behaviour occur at the Alfvén surface ($m = 1$), the value of the specific angular momentum is then fixed;

$$l(a) = \Omega_\mathrm{o} r_\mathrm{A}^2 . \tag{3.13}$$

For a field line starting at a point r_o on the rotor (disk in our case), the Alfvén radius is $r_\mathrm{A}(r_\mathrm{o})$ and constitutes a lever arm for the flow. The true critical points of the outflow are where speed v_p equals the speed of the slow and fast magnetosonic modes in the flow (at the so-call SM and FM surfaces). The Alfvén point is not a true critical point in the flow; flow passing steadily through SM and FM surfaces must go through the Alfvén surface. These 3 critical points are illustrated in Figure 2.

Finally, a generalized version of Bernoulli's theorem may be derived by taking the dot product of the equation of motion with $\boldsymbol{B_p}$. Thus, there is a specific energy $E(a)$ that is a constant along field lines, given by (*e.g.* Pelletier & Pudritz (PP92) [51]);

$$E(a) = \frac{1}{2}(v_{\rm p}^2 + \Omega^2 r^2) + \phi + h + \Omega_{\rm o}(\Omega_{\rm o} r_{\rm A}^2 - \Omega r^2). \tag{3.14}$$

The terminal speed $v_{\rm p} = v_\infty$ is such that the gravitational potential and rotational energy of the flow are negligible. For cold flows, the pressure may also be ignored. It follows that

$$v_\infty \simeq 2^{1/2}\Omega_{\rm o} r_{\rm A} \tag{3.15}$$

which is a result first obtained by Michel [75] for 1-D flows. The important point regarding outflow speeds from disks is that $v_\infty/\Omega_{\rm o} r_{\rm o} = 2^{1/2} r_{\rm A}/r_{\rm o}$; the poloidally directed flow is larger than the rotor speed by a factor $r_{\rm A}(r_{\rm o})/r_{\rm o}$ which is just the lever arm of the magnetic wind torque on the disk.

3.2 MHD wind torques on disks: Jets as the mechanical manifestation of accretion power

Let us now carry out our second application of the angular momentum equation (3.9), this time to find the torque that is exerted upon a thin accretion disk by an external field $\boldsymbol{B}$. Any vertical flow in the thin disk is negligible so only the $v_{\rm r}$ contribution matters, and the rotation speed v_ϕ is the Kepler speed if disks are thin. On the right hand side, both the radial and vertical magnetic contributions come into play; thus

$$\rho v_{\rm r} \frac{\partial(r_{\rm o} v_\phi)}{\partial r_{\rm o}} = \frac{B_{\rm r}}{r_{\rm o}} \frac{\partial(r_{\rm o} B_\phi)}{\partial r_{\rm o}} + B_z \frac{\partial B_\phi}{\partial z}. \tag{3.16}$$

One sees that specific angular momentum that is stripped out of the radial inward accretion flow is due to the action of two types of magnetic torque. The first term on the RHS represents radial outward transport of angular momentum due to the radial shear of the toroidal field, while the second term is vertical tranport due to the vertical shear of the toroidal field as it threads the disk. In a thin disk, and for a typical field opening angle, the second term will dominate. Note that the second term vanishes at the disk mid-plane because symmetry considerations force $B_{\rm r} = 0$ on the mid-plane. We then vertically integrate the resulting equation, noting that the disk accretion rate is $\dot{M}_{\rm a} = -2\pi\Sigma v_{\rm r} r_{\rm o}$. The result is the angular momentum equation for the accretion disk under the action of an external magnetic torque;

$$\dot{M}_{\rm a} \frac{{\rm d}(r_{\rm o} v_{\rm o})}{{\rm d}r_{\rm o}} = -r_{\rm o}^2 B_\phi B_z|_{r_{\rm o},{\rm H}}. \tag{3.17}$$

This result shows that angular momentum is extracted out of disks threaded by magnetic fields. It is ultimately attached to either toroisional Alfvén waves, or, with magnetic field lines inclined by more than 30° from the vertical, to a wind. By solving for $rB_\phi = k(rv_\phi - l)$ and relations (3.7 and 3.13) for k and l the disk angular momentum equation (3.17) can be cast into its most fundamental form;

$$\dot{M}_\mathrm{a}\frac{\mathrm{d}(\Omega_\mathrm{o}r_\mathrm{o}^2)}{\mathrm{d}r_\mathrm{o}} = \frac{\mathrm{d}\dot{M}_\mathrm{w}}{\mathrm{d}r_\mathrm{o}}\Omega_\mathrm{o}r_\mathrm{A}^2.(1-(r_\mathrm{o}/r_\mathrm{A})^2). \tag{3.18}$$

This equation shows that there is a crucial link between the mass outflow in the wind, and the mass accretion rate through the disk;

$$\dot{M}_\mathrm{a} = (r_\mathrm{A}/r_\mathrm{o})^2\dot{M}_\mathrm{w}. \tag{3.19}$$

This result simply states that the rate at which the disk loses angular momentum, $\dot{J}_\mathrm{d} = \dot{M}_\mathrm{a}\Omega_\mathrm{o}r_\mathrm{o}^2$, is exactly the rate that is carried away by the wind, $\dot{J}_\mathrm{w} = \dot{M}_\mathrm{w}\Omega_\mathrm{o}r_\mathrm{A}^2$. The crucial point about the magnetic fields in any MHD wind is that they force gas to co-rotate with the rotor out to a distance of the order r_A.

The value of the Alfvén lever arm $r_\mathrm{A}/r_\mathrm{o} \simeq 3$ for most theoretical and numerical models that we are aware of, so that one finds $\dot{M}_\mathrm{w}/\dot{M}_\mathrm{a} \simeq 0.1$. Thus, the observed link between accretion and outflow has a natural explanation in the context of disk wind theory. This can be traced to the fact that magnetic wind torques have lever arms that are *external* to the disk, and therefore potentially very large. This is quite distinct from the scaling that arises from viscous torques which have lever arms that are no more than αH where α is the S-S parameter for the torque (the value is about 10^{-2} in the MHD simulations of Balbus-Hawley turbulence). The ratio of an MHD disk wind torque to a viscous disk torque may be written as $(B_z^2/4\pi P).(r_\mathrm{A}/\alpha H)$ [51] which shows that even if the magnetic energy density within a disk is far less than the gas pressure P, its lever arm is still sufficiently large to beat viscous stresses in the competition to transport disk angular momentum. The only natural competitor for magnetic wind torques upon disks may be spiral waves which also have long lever arms.

We may now connect this theoretical analysis with the observations of momentum and energy transport in the molecular outflows reviewed in Sect. 2.1. First we may use equation (3.18) to calculate the total mechanical energy that is carried by the jet;

$$L_\mathrm{w} = \frac{1}{2}\int_{r_\mathrm{i}}^{r_\mathrm{j}} \mathrm{d}\dot{M}_\mathrm{w}v_\infty^2 \simeq \frac{1}{2}\frac{GM_*\dot{M}_\mathrm{a}}{r_\mathrm{i}}[1-(r_\mathrm{i}/r_\mathrm{j})^2] \simeq \frac{1}{2}L_\mathrm{acc}. \tag{3.20}$$

This results states that the wind luminosity taps the gravitational energy release through accretion in the gravitational potential of the central object.

The last equality – which pertains to the case in which all of a protostar's energy release is still due to accretion – matches the observations of jet luminosity of the class 0 objects (see Sect. 2.1).

We may also calculate the total momentum flux (thrust) carried by a jet launched from an accretion disk. For simplicity, assume that the Alvén radius of flow along a field line whose footpoint is located at r_o obeys a power law scaling $r_\mathrm{A}(r_\mathrm{o})/r_\mathrm{o} \propto r_\mathrm{o}^{-\beta}$. The thrust is then

$$\frac{F_\mathrm{w}}{L_\mathrm{bol}/c} = \frac{c}{L_\mathrm{bol}} \cdot \int_{r_\mathrm{i}}^{r_\mathrm{j}} \mathrm{d}\dot{M}_\mathrm{w} v_\infty = f(\beta) \frac{r_\mathrm{i}}{r_\mathrm{A,i}} \frac{c}{v_\mathrm{esc,i}} \frac{L_\mathrm{w}}{L_\mathrm{bol}} \tag{3.21}$$

where $f(\beta) \equiv (4/[1+2\beta])$ and the index "i" denotes the inner disk radius, and $v_\mathrm{esc,i}$ is the escape speed from r_i. This result accounts for the fact that the observed thrust of the CO is larger than L_bol/c since $c/v_\mathrm{esc,i} \simeq 10^3$ while $L_\mathrm{w}/L_\mathrm{bol} \simeq 10^{-2}$. It also shows that the thrust scales with $L_\mathrm{w}/L_\mathrm{bol}$ and will therefore have the largest value for the largest values of the luminosity ratio – which occurs during the Class 0 stage.

3.3 Force-balance in jets; the mechanism of jet collimation

The most stiking observational feature of jets is that they remain so well collimated over many decades in physical scales as they work their way through inhomogenous media far from their source. How is this accomplished?

In the standard picture of hydromagnetic winds, collimation of an outflow occurs because of the increasing toroidal magnetic field in the flow as one moves through its various critical points. From equation (3.8) one deduces that at the Alfvén surface $B_\phi \simeq B_\mathrm{p}$, while in the far field (assuming that $r_\mathrm{j} >> r_\mathrm{A}$) this ratio is of the order $B_\phi/B_\mathrm{p} \simeq r/r_\mathrm{A}$. Therefore, collimation can in principle be achieved by the tension force associated with the toroidal field which leads to a radially inwards directed component of the Lorentz force (or "z-pinch"); $F_\mathrm{Lorentz,z} \simeq J_\mathrm{z} B_\phi$.

The detailed radial structure of the outflow is found from the condition of force balance perpendicular to the field lines in the flow, which is known as the Grad-Shafranov equation. This is one of the most complicated nonlinear equations in theoretical physics (see contributions by Heyvaerts and Beskin – these Proceedings). No general solutions are known [76]. Because of the mathematical difficulties, analytic studies have been dominated by simplified approaches which focus mainly on finding stationary, self-similar solutions of the flow ([74, 77–80]), or on analyzing the asymptotic limits of the general equations ([51, 81]).

Heyvaerts & Norman [81] show that two types of solution are present depending upon the asymptotic behaviour of the total current intensity $I = 2\pi \int_0^r J_\mathrm{z}(r', z')\mathrm{d}r' = (c/2) r B_\phi \propto r^2 \rho \Omega_\mathrm{o}$ where, $\boldsymbol{J} = (c/4\pi)\nabla \times B$ is the current density. In the limit that $I \rightarrow 0$ as $r \rightarrow \infty$, then the field lines are paraboloids which fill space. On the other hand, if the current is finite in this limit, then the flow is collimated to cylinders. The character of the flow therefore depends upon the boundary conditions on the disk.

Self-similar solutions have provided considerable insight into the physics of disk winds. The model of Blandford & Payne [74] treats the disk accretion rate as a constant which implies that the mass of the disk carried away by the wind is unimportant. The only velocity in the self-similar problem is the Kepler speed of the disk. Therefore, one anticipates that the various velocities in the problem scale as $v_{\rm A} \propto c_{\rm s} \propto v_{\rm r} \propto v_\infty \propto v_{\rm K}$ where one has the Alfvén, sound, radial inflow, terminal wind speed, and Kepler speed all being proportional to one another. Now, the hydrostatic balance condition for thin disk gives $H(r)/r = c_{\rm s}/v_{\rm K}$. Therefore, the self-similar scaling $c_{\rm s} \propto v_{\rm K}$ implies that $H \propto r$; the disk is wedge-like in its spatial structure. This also implies that the disk temperature follows the virial scaling $T \propto r^{-1}$. The density scaling for the disk follows from the definition of the disk accretion rate, $\dot{M}_{\rm a} = 2\pi(2H\rho)v_{\rm r}.r$, which given the scaling of the radial inflow velocity $v_{\rm r} \propto v_{\rm K}$ and the disk scale height relation implies that $\rho \propto r^{-3/2}$. The wind mass loss rate in the wind follows since the mass loss per unit area along the surface of the disk is $\mathrm{d}\dot{M}_{\rm w}/\mathrm{d}A = \rho_{\rm o} v(w,o) \propto r_{\rm o}^{-2}$, which imples that $M_{\rm w}(r_{\rm o}) \propto \ln r_{\rm o}$.

Next, note that the scaling of the disk Alfvén speed $v_{\rm A} \propto v_{\rm K}$, together with the density implies a particular scaling of the disk poloidal field, $B_{\rm p} \propto r^{-5/4}$. In the BP model, the toroidal field on the disk should therefore scale as $B_\phi/B_{\rm p} = $ const, so that $B_\phi \propto r^{-5/4}$. This presents a bit of a dilema however because the lowest energy toroidal field would be expected to be a force-free condition which has a different scaling $B_\phi \propto r^{-1}$. The Alfvén surface in this model is then $v_\infty/v_{\rm K} \propto r_{\rm A}/r_{\rm o} = $ const; *i.e.*, a cone. Finally, given the scaling for the disk magnetic field, we can now calculate the mass load of the wind; $k_{\rm o} \propto r_{\rm o}^{-3/4}$.

What properties does a self-similar, collimated flow have? In a self-similar outflow with cylindrical asymptotics, and for a wind with a (spherical) density profile (where R is the spherical radius) $\rho \propto r^{-q}$, then the Alfvén radius scales as (in cylindrical radius) $r_{\rm A}/r_{\rm o} = [(4-2q)/(3-2q)]^{-1/2}$ [78]. For flow that is collimated to radial cylinders at infinity, the axial current scales as $r^{(1-q)/2}$ ([80]). It turns out that the hoop stress can collimate such a flow only if $q < 1$. Specific solutions can be constructed which are constrainted to collimate to a cylindrical radius r_1 subject to the power law density profile characterized by the value of q. As an example, for $q = 0.75$ and $r_1/r_{\rm o} = 2.86$, the Alfvén surface is found to occur at the radius $ir_{\rm A}/r_{\rm o} = 2.41$. The asymptotic velocities for these solutions are too small however, to account for observed jets. With the parameters above,

$v_\infty/v_{\rm K,o} \simeq$ a few 0.1. The Alfvén speed in the toroidal field is much larger than in the poloidal component however, with $v_{\rm A,\phi}/v_{\rm A,p} >> 1$. These results define an outflow that is driven primarily by a magnetic pressure gradient primarily in the toroidal component of the field. This can be thought of as a coiled, compressed spring that pops vertically upwards, pushing matter up the outflow axis (as in the simulations of Uchida & Shibata [82]).

Jets need not be self-similar structures however, and it is perhaps more natural to think of them as minimum energy configurations instead. We have seen above that for the magnetic configuration in which the toroidal field scales as $B_\phi \propto r_{\rm o}^{-1}$, the current intensity is finite everywhere. The BP self-similar solution leads to diverging currents which corresponds to a higher energy state for the magnetic configuration. The disk-wind solutions elucidated by Pelletier & Pudritz [51] were designed to characterize non self-similar, but minimum energy-state jets. To derive this solution, we first write down the general Grad-Shafranov (GS) equation for stationary, axisymmetric flow, for the surfaces of magnetic flux $a(r,z)$ in the flow. Magnetic field lines are restricted to surfaces of constant magnetic flux, $a(r,z) =$ const, where the the poloidal magnetic field follows from the relation $\boldsymbol{B_p} = (1/r)\nabla(a)\times\hat{\phi}$. The GS equation can be written in a highly interesting form:

$$\frac{r^4}{m^2}\left(\nabla.\frac{m^2-1}{r^2}\nabla a\right) = \lambda(a)\bar{\eta}(a)a \tag{3.22}$$

where the function $\bar{\eta}(a) \simeq$ const and the function

$$\lambda(a) \equiv \frac{\Omega_{\rm o}^2 r_{\rm A}^2}{v_{\rm A}^2} c_{\rm A} = \frac{4\pi\rho_{\rm A}(a)\Omega_{\rm o}^2 r_{\rm A}^6}{a^2}\left(\frac{{\rm d}\ln r_{\rm A}}{{\rm d}\ln a}\right)^2 \tag{3.23}$$

is a constant on a surface of constant magnetic flux and depends upon the density at the Alfvén point, $\rho_{\rm A}$. Solving this last relation for the Alfvén radius, we see that $r_{\rm A} \propto \lambda^{1/6}$ which is such a weak dependence that solutions with $\lambda =$ const were sought ([51]).

The disk in this picture imposes an important set of boundary conditions on the GS equation that include the value of the magnetic flux function across its surface; $a(r,z) = a_{\rm o}(r_{\rm o}, z_{\rm o})$. Without loss of generality one may assume that the flux takes on a power-law form on the disk; $a_{\rm o} \propto r_{\rm o}^{3/2\alpha}$. Using the Ansatz that $\lambda =$ const, the preceeding equations show that the Alfvén surface takes the form $r_{\rm A}/r_{\rm o} \propto a^{2-\alpha} \propto r_{\rm o}^{3/\alpha-3/2}$ while the current takes the form $I(r,z) \propto a^{(\alpha/3)-1}$. The minimum energy state for the current arises from the solutions with $\alpha = 3$, for which $I =$ const, $B_{\rm p} \propto r_{\rm o}^{-3/2}$, and the Alfvén surface behaves as $r_{\rm A}/r_{\rm o} \propto r_{\rm o}^{-1/2}$. The terminal speed in this solution scales as $v_\infty \propto r_{\rm o}^{-1}$. Finally, the mass loss rate in the wind for the

PP92 solution is $\dot{M}_{\rm w} \propto [(r_{\rm o}/r_{\rm i}) - 1]$, where $r_{\rm i}$ is the inner edge of the disk and the mass loading of the wind scales as $k \propto r_{\rm o}^{-1/2}$. This mass loading is therefore a less steeply raked function of disk radius than in the BP self-similar case (wherein $k \propto r_{\rm o}^{-3/4}$), which in turn suggests that the BP outflow may be more easily collimated than the PP92 outflow. (The BP solution is obtained for the choice $\alpha = 2$ for which $I \propto r_{\rm o}^{-1/4}$ and $r_{\rm A}/r_{\rm o} = $ const.) We shall compare 2D time-dependent simulations of the BP with the PP92 boundary conditions in Sect. 4.1.

3.4 Accretion-ejection structures

The initiation of a disk wind may be different than occurs for winds from magnetized stars. For the latter, a star has a corona that is sufficiently well pressurized so as to help drive the magnetized wind. For disk winds however, a corona is not required as long as the field lines that emerge from the disk make an angle $\theta_{\rm o} \leq 60°$ ([74]). In this eventuality, it is the upper levels of the disk rather than the corona that is the mass source for the wind. The launch condition is found by applying Bernoulli theorem (Eq. (3.14)) at the footpoint of a field line on the disk surface and comparing the change in the effective gravitational potential with the change in kinetic energy of a particle as it is moved an infinitesimal amount outward along the field line; $\tan(\theta_{\rm o}) = \delta z/\delta r \leq \sqrt{3}$ ([74]).

How is material launched from the disk in this cold, centrifugally driven wind picture? Without a corona, material from some height above the disk midplane must move upwards and eventually be accelerated outwards in a jet. Therefore, there must be a direct link between the physics of magnetized disks, and the origin of outflows – hence the idea of an "accretion-ejection" structure.

In order to see why this happens, let us follow a field line from the midplane of the disk out through its surface and then to large distances in a collimated outflow (see Figs. 2 or 3). At the disk mid-plane, the poloidal field must be perpendicular to it ($B_{\rm r} = 0$). If this were not so, there would be a singular current sheet on the mid-plane that would be unstable and would rapidly heat the disk until the current is dissipated. As one follows the field line from the mid-plane, it curves away from the rotation axis (convex) which is a consequence of the dragging of the field due to the inward accretion flow (see Sect. 2.7) which, as we have seen, is determined by the value of the magnetic Prandtl number. For a collimated flow to exist, this field line must begin to curve back towards the rotation axis (concave) of the system somewhere near the surface of the disk. Thus, the field changes its curvature from convex to concave as it traces through the disk. The magnetic tension associated with this curved field line exerts a pinch force down on the disk,

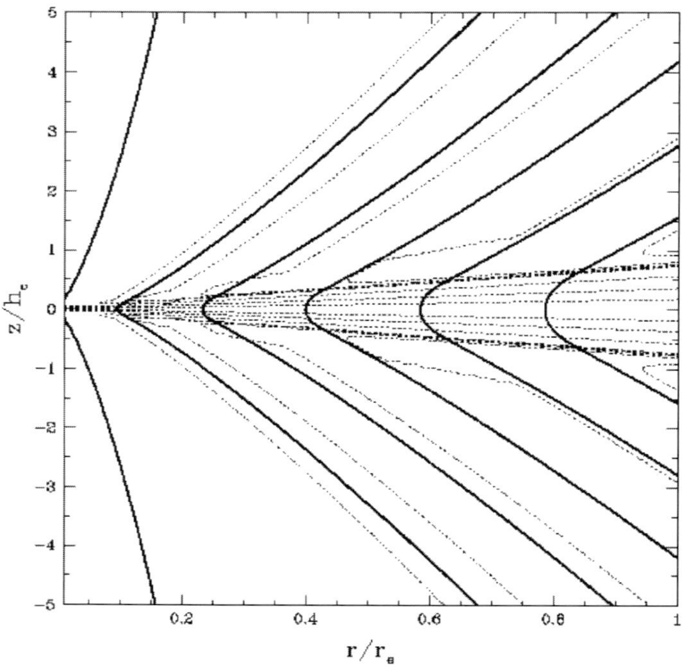

Fig. 3. Field lines (bold) and streamlines (dotted) of a self-similar accretion-ejection structure. Model parameters are $\epsilon = 0.1$, $\alpha_{\rm m} = 1$, and ejection efficiency $\zeta = 0.004$ from Ferriera [84]. Courtesy of Jonathan Ferriera.

compressing it at all heights (denoted by $z_{\rm curv}$) below which the field-line curvature is convex. Above $z_{\rm curv}$ however, the (concave) curvature of the field exerts a tension force upwards, pushing the material slightly away from the disk mid-plane. It is this type of pressure imbalance which feeds gas in the upper layers of a disk into a cold disk wind [83, 84].

The formal expression of this physical picture is obtained from the condition of vertical hydrostatic balance ([83], [84]);

$$\frac{\partial P}{\partial z} = -\rho \Omega_{\rm K}^2 z + F_{\rm z} \tag{3.24}$$

where the vertical component of the Lorentz force is $F_{\rm z} = J_{\rm r} B_{\phi} - J_{\phi} B_{\rm r}$. A deviation from this equilibrium solution will lead to matter either falling onto the disk mid-plane, or being lifted off the disk by a slight excees in pressure. The vertical profile of the radial current $J_{\rm r}$ thus plays a central role since a decrease of this current with height leads to a reversal of the direction of this force component.

The initiation of the outflow in the upper part of the disk above $z_{\rm curv}$ can be gleaned from the detailed angular momentum balance equation. The toroidal component of the Lorentz force on material rotating in the disk is $F_\phi = J_{\rm z}B_{\rm r} - J_{\rm r}B_{\rm z}$. One can see that as the radial current falls with height (z) in the disk, this component of the Lorentz force component also changes sign. Below the height $z_{\rm curv}$, this force opposes the gas motion slowing it down slightly below Keplerian value and thereby causing it to fall to a slightly smaller disk radius; *i.e.* accretion is affected by the magnetic torque. At heights in the disk $z > z_{\rm curv}$ however, the force accelerates the gas slightly. It is this magnetic acceleration and upward direction of the force that gets the upper layers of the disk moving, eventually to become a cold, centrifugally driven disk wind. These types of models employ six disk parameters only four of which are independent ([84]): the disk profile $\epsilon \equiv h(r)/r$; the magnetic Reynold's number $Re_{\rm m} = rv_{\rm r}/\eta_{\rm T}$; the strength of the magnetic turbulence $\alpha_{\rm m} = \eta_{\rm T}/v_{\rm A}h$, which taken together, define the magnetic Prandtl number discussed in Sect. 2.7; the disk magnetic field strength $\mu = B^2/4\pi P_{\rm gas}$; the "ejection index" $\zeta = {\rm d}\ln \dot{M}_{\rm a}/{\rm d}\ln r$ which describes the radial accretion flow and measures how much mass is ejected from each radius of the disk (akin to the mass loading); and finally, the vertical current profile $\Gamma \equiv (r[\partial\Omega/\partial r]B_{\rm r}/J_{\rm r}[\partial\eta_{{\rm T},\phi}/\partial z])|_{z=0}$ that controls the vertical force balance and the details of mass acceleration from the surface layers of the disk. Self-consistent flow solutions that evolve through the 3 critical points in the flow have been found which give a good correspondence with observed jets. The parameter ranges giving rise to these solutions are: $Re_{\rm m} \simeq \epsilon^{-1}$, with $0.005 \le \zeta \le 0.08$; $0.0005 < \epsilon \le 0.1$; and $0.1 < \alpha_{\rm m} \le 1$. Figure 3 gives a side view of a solution which shows both the magnetic field lines and particle trajectories in the disk and wind, for $\epsilon = 0.1$, $\alpha_{\rm m} = 1$, and an accretion efficiency of $\zeta = 0.004$. While the previous solutions pertain to disks with purely magnetic torques, self-similar solutions that include viscous disk torques have also been found ([85]).

The assumption that disk winds get a cold start may not be entirely general since there are mechanisms for pumping energy into a disk corona. Extrapolating from what is observed in solar physics (Sect. 2.7), the expectation is that loops of magnetic flux as well as a flux of Alfvén waves propagate into the upper reaches of the disk atmosphere and deposit the energy released by accretion, there. Global simulations (in vertical scale) [64] of the MRI instability in accretion disks show that waves propagate into the upper reaches of disk, steepen, and then undergo strong shocks wherein their energy is converted into heat. These shocks occur at Mach numbers >2. The simulations are highly time-dependent but in some cases they settle down into a disk-coronal structure in which the disk is weakly magnetized below two pressure scale heights (with $\beta = P_{\rm gas}/(B^2/8\pi) \simeq 50$) and the corona is strongly magnetized above ($\beta \simeq 0.1$).

A resonant coupling picture for pumping Alfvén wave power into the disk corona has also been proposed [86, 87]. The resonant coupling of the MRI instability with Rossby vortices that are located at the co-rotation radius of the disk leads to the extraction of considerable gravitational binding energy interior to the co-rotation radius and its transmission into the corona by a flux of torsional Alfvén waves. Both of the above mechanisms suggest that there may be a much wider class of solutions for "accretion-ejection" structures in which accretion disk coronae, rather than the upper reaches of the disk, are the launch pads for disk winds.

Finally, a class of inflow-outflow solutions has been found that may combine gravitational collapse with accretion-ejection [88]. In these solutions, the region around a protostar is divided into three regions; a jet, an infalling envelope, and a circulation region in which matter rains down towards the central region and is reflected outwards again from a pressure gradient near the central region. This circulation region is interpreted as being the molecular outflow which is predicted not to arise not from the entrainment of surrounding molecular material by a fast moving jet, but by a separate hydrodynamic circulation flow in its own right. These solutions have a quadropolar symmetry with jets moving up the axes, and a ratio of wind outflow to accretion of the correct order; $\dot{M}_{\rm a}/\dot{M}_{\rm w} \simeq 0.1$.

3.5 Disk – magnetosphere coupling

For typical conditions in TTSs disks, accretion rates are low enough that a stellar magnetosphere rooted to the central star truncates the disk at an inner radius. The magnetopause radius of a dipole field interacting with an accretion disk has the well known scaling $R_{\rm mag} \propto \mu^{4/7}\dot{M}_{\rm a}^{-2/7}$, which is several stellar radii for typical TTS magnetic fields (here μ is the magnetic dipole moment of the star) and accretion rates. The interaction of this magnetosphere with the disk has been extensively analyzed by Shu and his collaborators (*e.g.* [14]), and reviewed in this volume (see Calvet's, and Hartmann's contributions). The stress of an inner disk that is not exactly spinning at the rate of the central star should pump energy into the dipole field and open it up into an open structure [89]. This would drive a disk wind from a single annulus on the disk; its inner edge. This wind has most of its density collimated along the rotation axis of the outflow, and exerts a wide-angle outflow component as well. As we shall see in Sect. 4.1, disk winds can also have this feature.

It is interesting to apply this analysis to FU Ori systems where the accretion rate is four orders of magnitude greater than in the TTS stage. The scaling above then suggests that the magnetosphere will most likely be totally crushed to the stellar surface under these conditions and that the accretion disk will butt against the stellar surface [52]. These systems are still observed to have strong jets and outflows however. This is a clear indication that disk winds must be active in the FU Ori stage.

4 Numerical simulations

The necessity of using excellent MHD codes to explore the physics of magnetized disks and jets is self-evident when one has such analytically intractable equations as the Grad-Shafranov equation to solve. The fact that jets are also highly episodic is yet another reason why time-dependent simulations of astrophysical MHD have rapidly become an essential research tool in the field.

The simulations have given important insights into how jets are launched and collimated, as well as into the coupling between jets and physical processes in the underlying disk such as the MRI instability. The published simulations mirror these two types of questions: i) *MHD disks as boundary conditions*, the underlying accretion disk remains constant and provides fixed boundary conditions for the outflow simulation, whereas; ii) *global simulations*, include both the dynamics of the disk and its jet. While the latter approach attempts to solve the entire accretion-ejection problem in one huge simulation, the rich variety of physical processes involved has given great importance to strategy (i) as well. We examine these in turn.

4.1 2D simulations: Jets from stationary Keplerian disks

Several groups have exploited this approach including [90–97]. The published simulations differ in their assumed initial conditions, such as the magnetic field distribution on the disk, the plasma β ($\equiv P_{\rm gas}/P_{\rm mag}$) above the disk surfaces, the state of the initial disk corona, and the handling of the gravity of the central star. Broadly speaking however, all of the existing calculations show that winds from accretion disks can indeed be launched and accelerated, much along the lines suggested by the theory presented above. This is reassuring and emphasizes the robustness of the theoretical model discussed in the previous section.

Many groups (including ours) employ the ZEUS 2-D code to study the origin and collimation of disk winds. This code is arguably the best documented and utilized MHD code in the literature (Stone & Norman [98]). It is an explicit, finite difference code that runs on a staggered grid. The evolution of the magnetic field is followed by the method of constrained transport [99]. In this approach, if $\nabla.\boldsymbol{B} = 0$ holds for the initial magnetic configuration, then it remains so for all later times to machine accuracy. The obvious way of securing this is to use an intial vector potential $\boldsymbol{A(r, z, t = 0)}$ that describes the desired initial magnetic field.

I will for illustrative purposes, first focus the discussion on the approach we have taken, as in [91–94] and documented in Ouyed & Pudritz [92]. The accretion disk in these papers is taken to be initially surrounded by a polytropic corona ($\gamma = 5/3$) that is in hydrostatic balance with the central object. The accretion disk at the base of the corona is given a density profile which it maintains at all times since the disk boundary conditions are applied to the "ghost zones" and are not part of the computational domain. It is chosen so that the disk surface is in pressure balance with the corona above it (*i.e.* $\rho \propto r_{\rm o}^{-3/2}$ for a $\gamma = 5/3$ model – as in BP). This hydrostatic state has a simple analytic solution which was used as the initial state for all of our simulations. We tested our numerical methods by ensuring that this eqilibrium state could be maintained for arbitrary amounts of machine time. We always use an unsoftened gravitational potential from the central object. The corona is ultimately swept away by the disk wind in all of these simulations so that the mass source for our jets becomes dominated by the material injected from the fixed disk below.

The initial magnetic configurations in our studies are chosen so that no Lorentz force is exerted on the initial (non-rotating) hydrostatic corona described above; more specifically, we used initial current-free configurations $\boldsymbol{J} = 0$ in the computational domain. The initial vector potential is subject to the boundary condition on the disk that the poloidal field on the disk surface $(r_{\rm o}, 0)$ has a power-law form $B_{\rm z}(r_{\rm o}, 0) = br_{\rm o}^{\mu-1}$. The solution for $\mu = 0$ is analytic and is the so-called "potential" configuration which we used in many of our papers. This initial field configuration is shown in the upper left hand panel of Figure 4.

Recent work by Pudritz *et al.* (2003, PRO [100]) is designed to simulate 2D disk winds from a greater variety of magnetized disks. One may accomodate other initial magnetic configurations by using a Hankel transform technique to solve for the initial vector potential that is subject to these constraints. In particular, we have now done simulations for an initial BP model ($\mu = -1/4$), PP92 ($\mu = -1/2$), and a yet more steeply declining magnetic field such as $\mu = -3/4$. These intial configurations are plotted out in the left panels of Figure 4. We also have simulated jets in the case of an initially uniform vertical field because of its unfavourable opening angle for launching a wind at all radii. The disk has an inner edge at radius $r_{\rm i}$, inside of which no material is rotating, and a uniform, vertical field that is everywhere parallel to the disk rotation axis.

These simulations of thin disks require that we specify five physical quantities at all points of the disk surface at all times. There are the disk density $\rho(r_{\rm o})$; components of the vertical and toroidal magnetic field, $B_{\rm z}(r_{\rm o})$ and

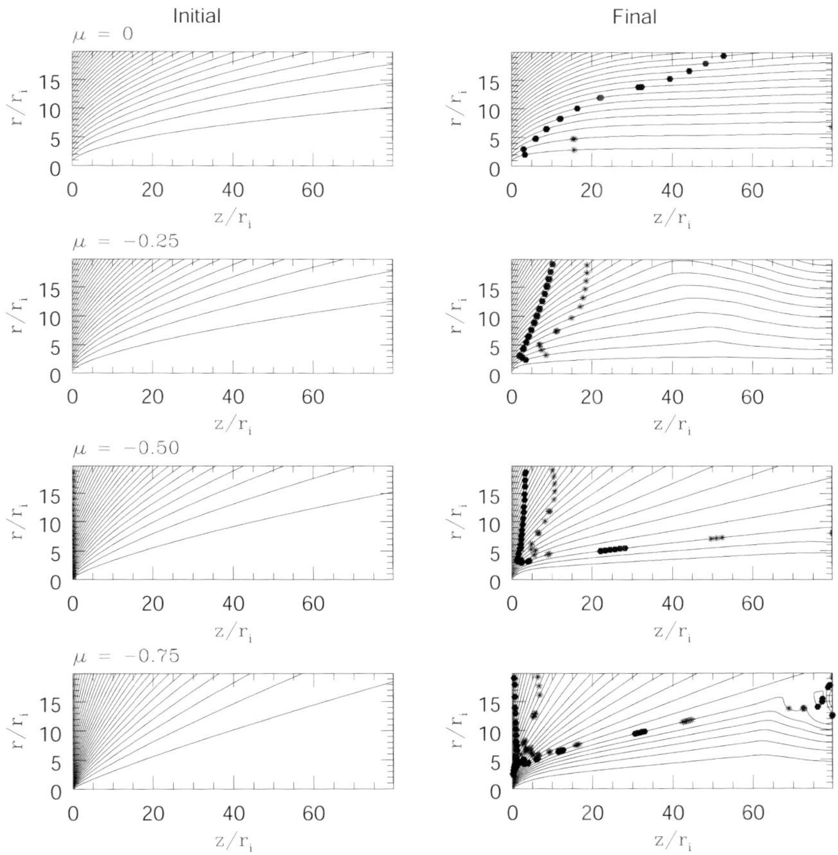

Fig. 4. Left panels: initial magnetic field configurations for winds with initial potential ($\mu = 0$), Blandford & Payne ($\mu = -0.25$), Pelletier & Pudritz ($\mu = -0.5$), and steeper ($\mu = -0.75$) configurations shown. Right panels: final magnetic field configurations (at $t = 400$) for each case, with Alfvén points (filled circles) and FM points (stars) marked. From PRO [100]. Note the more open magnetic – and stream line structures as the initial magnetic configuration (value of μ) changes.

$B_\phi(r_o)$; and velocity components in the disk, $v_z(r_o)$ and $v_\phi(r_o)$. The remaining field component $B_r(r_o)$ is determined by the solenoidal condition while the radial inflow speed through the disk $v_r(r_o) \simeq 0$ for all practical considerations since it is far smaller than the sound speed in a real disk. The model is described by five parameters, whose values describe conditions at r_i. Three of these parameters describe the initial corona: β_i which is typically 1.0 (and which therefore falls with radius); the density jump across the corona/disk boundary η_i which is typically 100; and ratio of the Keplerian to thermal energy density δ_i. Two additional parameters describe the disk physics. The important one is the injection speed v_{inj} of the material from the disk into the base of the corona. We scale the injection speed with the local Keplerian velocity at each radius of the disk where typically $v_{inj} = 10^{-3}\ v_K$ at any point on the disk. This scaling and that of the magnetic field fixes the all-important mass loading at the footpoint of each field line on the disk;

$$k(r_o) = \rho_o v_{p,o}/B_{p,o} \propto r_o^{-3/2} r_o^{-1/2}/r_o^{-1+\mu} \propto r_o^{-1-\mu}. \tag{4.1}$$

The second parameter, μ_i, which scales the toroidal field in the disk $B_\phi = \mu_i/r_o$ is unimportant. Even if it is set to zero there is little change in the results of the simulations since the disk will quickly produce a toroidal field in the computational domain just above the disk when its rotation is initiated.

We ran high resolution simulations in which (500×200) spatial zones were used to resolve a region $20r_i$ in radius and $80r_i$ in the z direction. Our simulations ran up to $400t_i$ (where t_i is the Kepler time for an orbit at the inner edge of the disk, r_i). The standard parameter settings were

$$(\eta_i, \mu_i, v_{inj}, \delta_i, \beta_i) = (100.0, 1.0, 0.001, 100.0, 1.0). \tag{4.2}$$

The details of the outflow in the "potential" configuration ($\mu = 0$) are found in [91, 92]. The disk immediately begins to generate a centrifugally driven outflow that drives a bow shock through the corona. The highest velocity and densest material is found near to the rotation axis in agreement with the prediction of equation (3.15) and the observations shown in Figure 1, Sect. 2.1. In the final time frame at $t = 400$ (the bowshock clears the grid earlier), the flow is highly collimated. The magnetic configuration at this time is shown in the top right panel of Figure 4 where one sees that the field lines have been bent from an initial parabola like configuration into a state in which the field lines are almost parallel to the outflow axis in this computational domain. This panel of Figure 4 also shows the position of the Alfvén surface and a few FM points as well. The position of the Alfvén surface matches the theoretical prediction very well [92]. The toroidal field generated by the jet is responsible for collimating the outflow.

The outflow begins to collimate just beyond the Alfvén surface and is largely accomplished by the time that the FM surface is reached. The ratio of the toroidal to the poloidal field strength at the FM surface is between 2–3 for many field lines. We also find that the opening angles of field lines as they emerge from the disk and along which an outflow is driven, obey $\theta_o \leq 60^0$ again as predicted by steady state theory.

Outflows also occur in an initially uniform, vertical magnetic field configuration. This is possible because the pressure gradient $\partial(B_\phi^2/8\pi)/\partial r$ generated in the jet pushes the fields lines radially outwards above the disk surface thus allowing the wind to be accelerated and ejected. The gradient arises from the fact that the strongest toroidal magnetic field is generated in the inner region of the flow where the Kepler speed is the highest.

What systematic differences can be expected from jet models as the magnetic fields, or other physical boundary conditions, are varied. To what extent is the structure of the jet controlled by the physics of the accretion disk? This question is pursued by PRO [100] in a series of simulations in which the initial magnetic configurations are varied but which otherwise adopt the same initial conditions as OPS.

Figure 4 (from PRO) plots the magnetic field lines of both the initial states of each model (left panels) as well as the final states of simulations. In addition to the potential case, simulations of the Blandford-Payne, Pelletier-Pudritz, as well as an even more steeply raked initial magnetic configuration are shown. It is evident that the field lines in these right hand panels are progressively less collimated towards the axis suggestive of a wide angle flow at larger radii. Also marked on these field lines are the positions of the Alfvén surfaces for each configurations. The BP final states shows that the Alfvén surface is indeed nearly a cone as expected while the PP simulation qualitatively agrees with the expected scaling $r_\mathrm{A} \propto r_\mathrm{o}^{1/2}$ (see Sect. 3.3).

Figure 5 shows snapshots of the evolution of the density of the jet for the BP (left panels) and PP (right panels) models at 4 different times ($t = 60, 120, 200, 400$). In each case we see the jet driving a bow shock through the quiescent gas in the first frame. In the second frame the shock has just moved out of the grid. The densest gas in the jet is always well collimated and moving nearly parallel to the axis. Comparison with Figure 4 though shows that there is material that is moving outwards in a less collimated fashion, following the less well collimated field lines.

The poloidal jet velocities for these two simulations of BP and PP initial conditions are shown in Figures 6 and 7 respectively. The four frames shown in each of these panels correspond to the 4 times shown in Figure 5. In both Figures 6 and 7 we see that the highest velocity gas is the best collimated and occurs closest to the rotation axis. In both cases, the velocity falls off with radial distance, reflecting the fact that the footpoints of these field lines rotate more and more slowly as expected for Keplerian rotation. The high-speed well collimated gas grades continuosly into the lower-speed, less collimated flow at larger radial distance from the outflow axis. Finally, we see that the jet recollimates in each case, as was pointed out by BP and PP. These two figures bear strong relations to the observations shown in

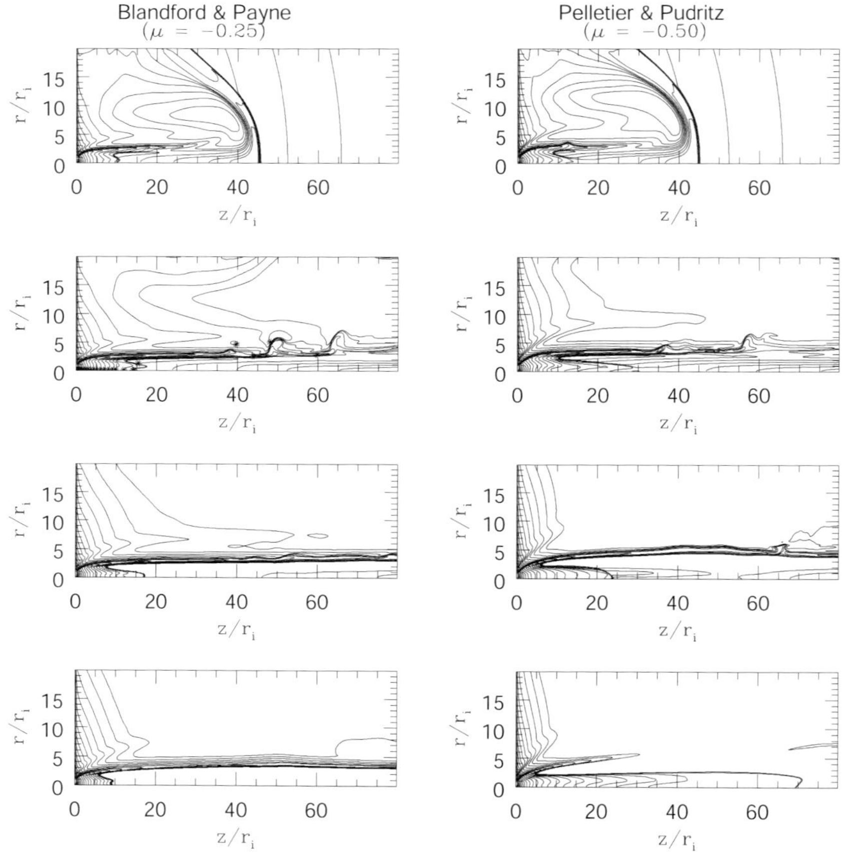

Fig. 5. Density structure of jets from Blandford-Payne (left panels), or Pelletier-Pudritz (right panels) initial conditions. Contours mark surfaces of constant density in the jet at four different times (t = 60, 120, 200, 400). Note the densest material moves along the jet axis. From PRO.

Figure 1. It is apparent that a wide-angle aspect to this outflow exists for disk winds – not only X-wind solutions [14].

These simulations of a variety of disk winds have all the attributes of jets suggested by the observations (see Sect. 2.1). There is no need to postulate that observed outflows have separate high speed cores and wide-angle disk winds that are produced by different mechanisms. Rather, it is a reflection of the physics of the jet launched from the disk, and on the distribution of magnetic flux across the surface of the disk. The degree of collimation of

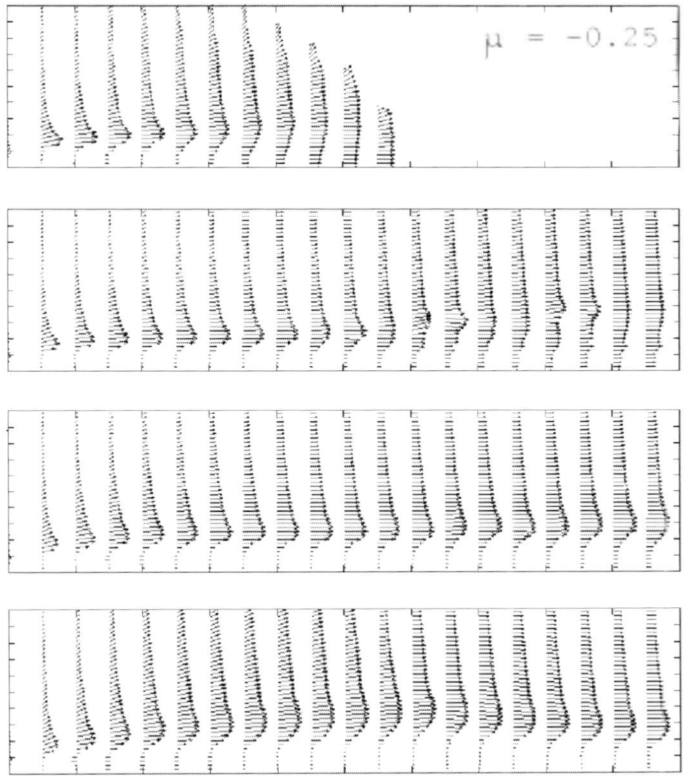

Fig. 6. Poloidal velocity field of a Blandford-Payne jet, shown at 4 different times as in Figure 5. Note that the highest velocities are nearest the outflow axis, and that the flow gently recollimates. From PRO.

the field lines in these flows is probably controlled by their mass loading. In going from BP to PP models as an example, k increases from $k \propto r_o^{-3/4}$ to $k \propto r_o^{-1/2}$ which implies that it is harder for the toroidal field to collimate the increasingly mass-loaded outflow (Eq. (3.8) and discussion).

The robustness of these results follows from the fact that simulations with very different initial conditions produce qualitatively the same picture. Thus Krasnopolsky *et al.* [97] set up fiducial models with a constant density disk; a poloidal magnetic flux distribution on the disk idential to PP ($\mu = -1/2$); and a mass flux density $\rho v_z \propto r_o^{3/2}$ which leads to a mass loading $k =$ const across the disk. An important difference in their boundary conditions is their injection of a thin, axial jet along the rotation axis chosen to mimic

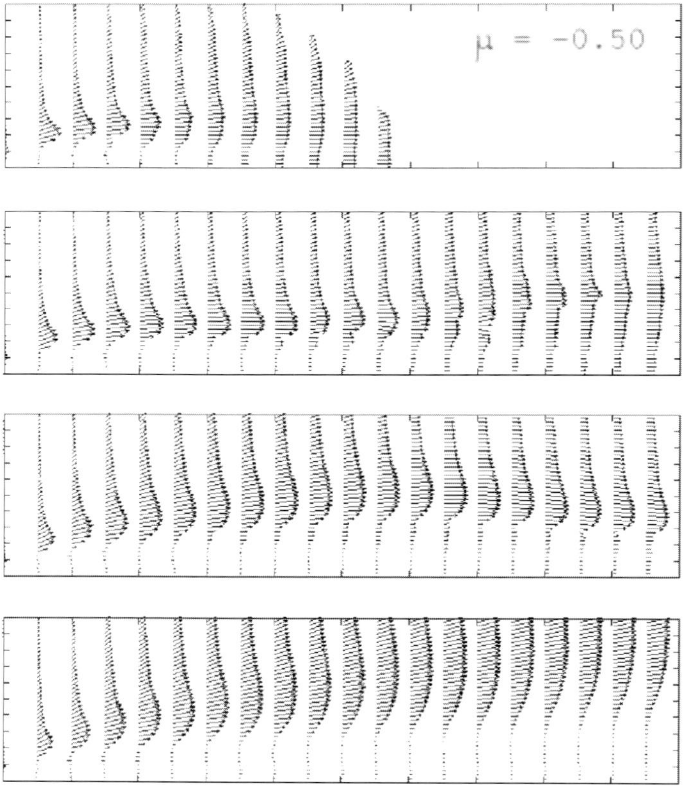

Fig. 7. Same as Figure 6 except showing the Pelletier-Pudritz jet. The flow is less well collimated while the highest speed material (which is also the densest) is closest to the outflow axis. This shows a wide-angle behaviour to the flow. From PRO.

the possible production of a relativistic jet by a central spinning object. In [92] the coronal material that is initially in equilibrium along the rotation axis is squeezed outward by the jet when it is formed, which leaves a "hole" at the centre of their disk-wind. Once again however, collimation of the flow is observed. Also, the outflow is insensitive to the density or injection speed at the base of the wind as long as the mass flux is left unaffected.

Jets launched in the most extreme initial magnetic field geometry on the disk – a so-called split monopole configuration – have also been simulated [95]. The field lines in this configuration constitute a very steeply decreasing function of radius. In these simulations too, the densest gas has the highest speeds and is closest to the rotation axis. The field lines however do not collimate and some of the material does flow outwards radially.

It appears therefore that a common theme is established with all the disk-wind simulations. In all cases one sees the densest material moving at the highest speeds parallel to the axis, with lower density gas moving off in a less collimated way. This behaviour is also observed in the X-wind model – discussed in Sect. 3.5. It appears therefore that the X-wind is not a unique explanation of wide-angled outflows. Disk winds with appropriately falling field strengths can also do the job (see PRO for details).

Finally, how does the apparently universal, episodic behaviour of jets arise? There are indications that disk winds naturally do this. We have also discovered episodic jets in our simulations – found by sufficiently "starving" the acceleration zone of material from the disk [93,94]. Episodic behaviour of jets can be initiated by sufficiently reducing the injection speed of material from the disk into its corona and thereby reducing the mass loading of field lines. These episodic states appear to be independent of the magnetic configuration that is employed (we find them for both uniform field and potential configurations), and they *never* become stationary. The region of knot generation is located at a distance of $z \simeq 5{-}7r_{\rm i}$ from the central source, beyond the Alfvén surface.

Knots in 2D simulations are produced on a time scale that is expected for sausage instability, the so-called $m = 0$ mode. Our simulations [93] show that if $r_{\rm jet}$ is the radial size of the jet region, then the time scale for knot production at the generator region is, roughly $\tau_{\rm knot} \simeq r_{\rm jet}/v_{{\rm A},\phi}$. The knots are coherent structures that propagate down the length of the jet. The transition from episodic to stationary jets occurs for sufficiently well mass loaded field lines in any given magnetic configuration. The transition to episodic behaviour can be quantified by the value of the quantity $N \equiv B_\phi^2/8\pi\rho v_{\rm p}^2$. In general, stationary flows involve low values for N (high mass load), while episodic jets typically occur for higher values for N (low mass load) [94].

4.2 3D simulations: Self-regulatory stability of jets

One of the central challenges to the theory of jets is in accounting for their 3D stability. Why don't outflows that are dominated by toroidal field undergo catastrophic kink mode instabilities that would tear them apart? While 3D computational studies of jet stability have been performed in the context of extragalactic radio jets [101] most of these simulations are done for jets that propagate from orifices. What about astrophysical jets that are produced by accretion disks?

The Kelvin-Helmholtz instability is one of the primary threats to the propagation of jets in 3D. There are several ways in which jets can survive this however. Linear instability analysis of magnetized jets showed that long wavelength kink modes ($m = 1$) are stable for all wavelengths greater than the jet radius R (*i.e.*, for $kR \leq 1$) as long as the flows are sub-Alfvén (*i.e.* $m_{\rm a} \leq 1$) [102]. This was confirmed in 3D simulations of "equilibrium jets" – models with uniform density [101]. Analytic work also suggests that jets are increasingly stabilized at progressive larger scales as the fast magnetosonic Mach number becomes large. Finally, the cores of jets from accretion disks will have "backbones" of poloidal magnetic field that can resist kinks [103]. These arguments suggest that jets should be particularly vulnerable to instability in the regime of moderate super Alfvénic Mach numbers but with lower FM Mach numbers – that is in the region beyond the Alfvén surface but before the FM Mach numbers have gotten too large.

3D simulations of jets with boundary conditions identical to those in the 2D simulations of [93] (constant initial vertical field), were performed by Ouyed *et al.* (2003, [104]). These simulations – although performed on a Cartesian rather than axisymmetric grid, run at lower resolution, and subject to symmetry breaking by the effective wobbling of the disk – were otherwise identical to [93]. During the early rather stable evolution of the jet until times $t \simeq 100$, the jet behaves much like its 2D counterpart undergoing sausage ($m = 0$) modes but otherwise fairly stable. After this time, these modes merge together and the ($m = 4$) mode symptomatic of the symmetry breaking appears. At the time $t = 160$, the $m = 1$ "kink" mode is strong and the jet appears to be severely stressed. However, the jet survives this and by $t = 200$ to the end of the simulation at $t = 240$, the jet settles down and runs in a stable way with a low amplitude $m = 1$ perturbation. The body of the jet is displaced from the axis and winds around it in a stable "corksrew" morphology. This stable end-state is shown in Figure 8. The upper frame shows is a map of the jet column density while the lower frame maps the quantity $\nabla.\boldsymbol{v}$ which measures the compression of the jet. These jets survive growing instability and adopt a final state that is often seen in YSO jets. Why?

The nonlinear saturation of the jet's Kelvin-Helmholtz (KH) instability arises as a response to the buckling of the jet by the instability. The KH instability sets in when the jet becomes fairly super-Alfvénic as previous papers have suggested. This drives the $m = 1$ mode which becomes predominant at $t = 160$. This mode stretches and twists the poloidal magnetic field, thereby strengthening it through the tension force. Thus, energy drained from the kinetic energy by the instability bolsters the poloidal field, and thereby reduces the Alfveén Mach number m. The instability shuts itself off when the jet Alfvén Mach number is reduced to below unity, which kills the KH instability. Thus saturation occurs not because of a large reduction of the flow speed, but rather through the increase of the Alfvén velocity

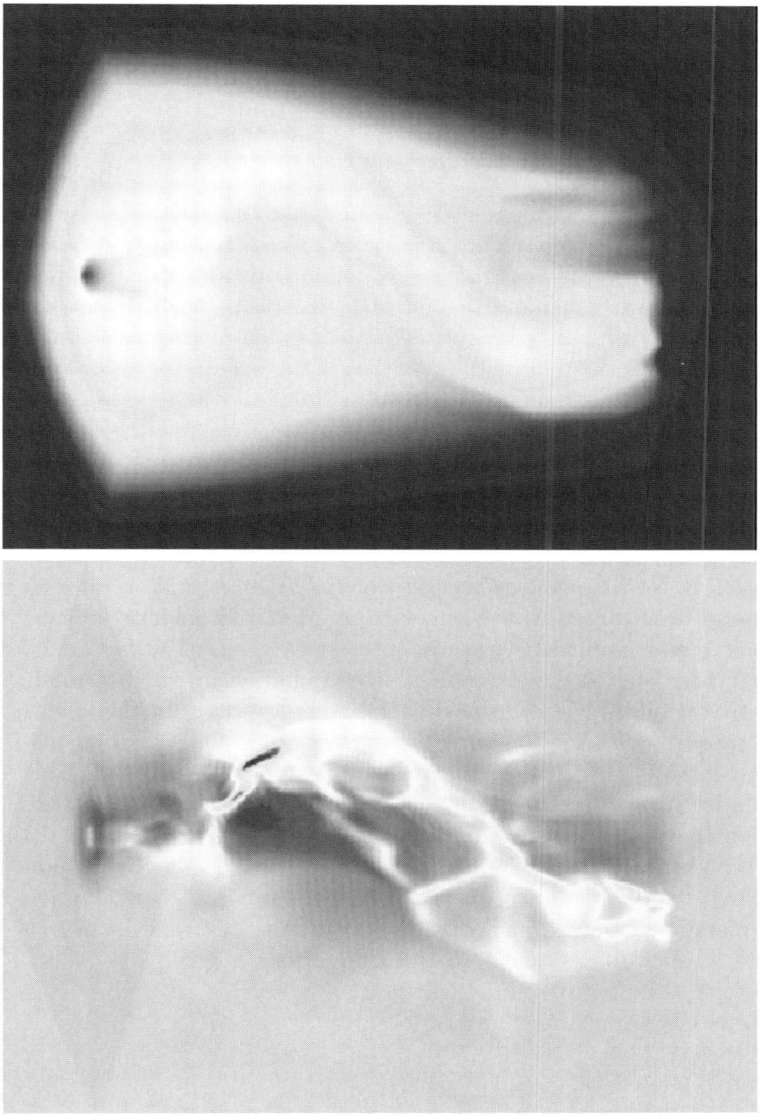

Fig. 8. A corkscrew jet from an accretion disk (shown on the left-hand side of each panel, in projection). Top frame is the column density through the jet, and the bottom frame is a measure of its compression. This jet survives because of the nonlinear saturation of KH instabilities, which prevent jet velocities from exceeding the Alfvén speed. From Ouyed *et al.* [104].

due the the increased poloidal field strength. These results are robust in that stabilization occurs for a variety of jet boundary conditions and initial perturbuations.

4.3 Global simulations of accretion-ejection structures

The first disk and wind numerical calculations were published by Uchida & Shibata [82]. These pioneering simulations were carried out for magnetized disks with sub-Keplerian rotation, designed to mimic the collapse of an initially more extended and slowly rotating object. Their models developed a rapid radial collapse in which the initially poloidal field threading the disk winds up due to the differential rotation of the collapsing disk. The vertical Alfvén speed being smaller than the free-fall collapse speed implies that a strong, vertical toroidal field pressure gradient $\partial B_{\mathrm{z}}^2/\partial z$ must rapidly build up. The resulting vertical pressure gradient results in the transient ejection of coronal material above and below the disk as the spring uncoils (see also [105]). These outflows, while very suggestive, represent more transient responses to the initial state.

Stone & Norman [105] ran a series of ZEUS 2-D simulations for a uniform magnetic field threading a wedge-shaped disk of constant opening angle. Among their simulations was the case of a Kepler disk with an initially weak field. Due to the extremely strong channel instability arising from the MRI instability in 2D, the simulation discovered that the disk collapses very rapidly in the radial direction without a strong toroidal field being formed (also [106]). These highly unstable channel flows arising from MRI instability in 2D become turbulent in 3D and the rate of outward angular momentum transport is considerably reduced (*e.g.* [107]). This makes 3D simulations indispensible for this problem. More recent global simulations of the MRI instability in a torus threaded by a weak uniform, vertical initial field show that accretion flow, MRI instability, as well as mass ejection rapidly ensue [108]. Although the simulations are run only until $t = 5$, the relationship between mass accretion and outflow rates discussed earlier manages to establish itself. Here too, $\dot{M}_{\mathrm{w}}/\dot{M}_{\mathrm{a}} \simeq 0.1$!

We close with a return to the large picture of star formation in the Class 0 stage wherein collapse, disk accretion, and outflow are all simultaneously in play. Tomisaka [109] has attacked this problem with a subtle use of nested grids in order to gain the required large dynamical range. His simulations show that the collapse of an intially magnetized filament produces a magnetized disk that drives an outflow with many of the characteristics that we have cited here. Much more remains to be done in this fertile global numerical domain, particularly with the advent of adaptive mesh refinement codes.

5 Conclusion

These lectures have shown that the accretion-ejection mechanism for jets has an elegant simplicity. Astrophysical jets are inevitable because magnetic fields are inevitable in astrophysical plasmas. Whether these fields are dragged in from elsewhere, or generated *in situ* by the MRI instability, disks cannot help but becoming substantially magnetized. The MHD conservation laws then assure us that the torques that are exerted by such ordered fields upon their disks are extremely efficient in transporting angular momentum and gravitational binding energy in an ordered and well-collimated manner. The fact that jets are seen in black hole systems that have no central dipole magnetic fields, as well as in YSOs which do, suggests that it is their accretion disks rather than the central objects themselves that may be playing the central role in non-relativistic or mildly relativistic outflows.

Gravity is the prime mover in this model of jets. One of the most promising aspects of the disk wind model for jets is their prediction that it is the gravitational binding energy that is liberated due to accretion that ultimately powers jets in the vast majority of disk-jet systems. This provides a powerful unifying idea for all jets – their central source is largely irrelevant and only the depth of the gravitational potential that they imprint really matters. Many aspects of MHD jets including their stability and the presence of both collimated and wide-angle outflow sectors have also been confirmed in numerical experiments.

The richness of any scientific field is measured by the number of new and exciting ideas and questions that it generates. Jets are an efficient means of transporting angular momentum in a wide range of astrophysical systems. Nothing could be more fundamental to their evolution. By this measure, the study of astrophysical jets should enjoy a long and excellent future.

I am indebted to the organizers of this wonderful school for the invitation to participate and for their patience during the preparation of my written lectures. I enjoyed many conversations with the students of the school, as well as Vasily Beskin, Max Camenzind, Nuria Calvet, Lee Hartmann, Jean Heyvaerts, François Menard, Guy Pelletier and Rachid Sunyaev. I sincerely thank my collaborators, postdocs, and students over the years – David Clarke, Ana Gomez de Castro, Arieh Königl, Sean Matt, Colin Norman, Guy Pelletier, Conrad Rogers, Jim Stone, and especially Rachid Ouyed – for their insights, hard work, and friendship. Special thanks to Sean Matt for useful comments on this manuscript. My research is supported through a grant from the Natural Sciences and Engineering Research Council of Canada.

References

[1] B. Reipurth and J. Bally, *ARA&A* **39** (2001) 403.

[2] J.S. Richer, D.S. Shepherd, S. Cabrit, R. Bachiller and E. Churchwell, "Molecular outflows from young stellar objects", in Protostars and Planets IV, edited by V. Mannings, A.P. Boss and S.S. Russell (Tucson: Univ. Arizona Press, 2000) 867.

[3] I.F. Mirabel and L.F. Rodriguez, *Nature* **392** (1998) 673.

[4] M.C. Begelman, R.D. Blandford and M. Rees, *Rev. Mod. Phys.* **56** (1984) 255.

[5] B. Balick and A. Frank, *ARA&A* **40** (2002) 439.

[6] M.C. Weisskopf *et al.*, *ApJ* **536** (2000) L81.

[7] A. Königl and R.E. Pudritz, "Disk Winds and the Accretion-Outflow Connection", in Protostars & Planet IV, V. Mannings, edited by A.P. Boss and S.S. Russell (Tucson: Univ. of Arizona Press, 2000) 759.

[8] H.C. Spruit, in NATO ASI Ser. C. 477, Evolutionary processes in binary stars, R.A.M.J. Wijers, M.B. Davies and C.A. Tout (Dordrecht: Kluwer, 1996) 249.

[9] R. Bachillor, *ARA&A* **34** (1996) 111.

[10] D.S. Shepherd and E. Churchwell, *ApJ* **472** (1996) 225.

[11] S. Cabrit and C. Bertout, *A&A* **261** (1992) 274.

[12] A. Raga and S. Cabrit, *A&A* **278** (1993) 267.

[13] C.R. Masson and L.M. Chernin, *ApJ* **414** (1993) 230.

[14] F.H. Shu, J. Najita, H. Shang and Z.-Y. Li, "X-Winds: Theory and Observations", in Protostars and Planets IV, edited by V. Mannings, A.P. Boss and S.S. Russell (Tucson: Univ. Arizona Press, 2000) 789.

[15] S. Cabrit, A.C. Raga and F. Gueth, in Herbig-Haro Flows and the Birth of Low Mass Stars, edited by B. Reipurth and C. Bertout (Dordrecht: Kluwer, 1997) 163.

[16] C.-F. Lee, L.G. Mundy, B. Reipurth, E.C. Ostriker and J.M. Stone, *ApJ* **542** (2000) 925.

[17] F.C. Adams, C.J. Lada and F.H. Shu, *ApJ* **321** (1987) 788.

[18] P. André, D. Ward-Thompson and M. Barsony, *ApJ* **406** (1993) 122.

[19] R.E. Pudritz, C.D. Wilson, J.E. Carlstrom, O.P. Lay, R.E. Hills and D. Ward-Thompson, *ApJ* **470** (1996) L123.

[20] L.W. Looney, L.G. Mundy and W.J. Welch, *ApJ* **529** (2000) 477.

[21] C.J. Chandler and J.S. Richer, "Observations of circumstellar discs and jets", in Astrophysical Discs, ASP Conf. Ser. 160, edited by J.A. Sellwood and J. Goodman (ASP, San Francisco, 1999) 91.

[22] C. Dougados, S. Cabrit, C. Lavalley and F. Ménard, *A&A* **357** (2000) L61.

[23] F. Bacciotti, T.P. Ray, R. Mundt, J. Eilöffel, J. Solf and M. Camenzind, *ApJ* **537** (2000) L49.

[24] J. Kwan and E. Tademaru, *ApJ* **332** (1988) L41.

[25] F. Bacciotti, T.P. Ray, R. Mundt, J. Eilöffel and J. Solf, *ApJ* **576** (2002) 222.

[26] C.J. Burrows *et al.*, *ApJ* **473** (1996) 437.

[27] L. Hartmann and S.J. Kenyon, *ARA&A* **34** (1996) 207.

[28] S.V.W. Beckwith, A.I. Sargent, R.S. Chini and R. Güsten, *AJ* **99** (1990) 924.

[29] P. André and T. Montmarle, *ApJ* **420** (1994) 837.

[30] C.J. Lada, E.T. Young and T.P. Greene, *ApJ* **408** (1993) 471.

[31] M. Gomez, B.A. Whitney and S.J. Kenyon, *AJ* **114** (1997) 1138.

[32] S. Cabrit and P. André, *ApJ* **379** (1991) L25.

[33] L. Hartmann and N. Calvet, *AJ* **109** (1995) 1846.

[34] N. Calvet, "Properties of the Winds of T-Tauri Stars", in Herbig-Haro Flows and the Birth of Low Mass Stars, IAU Symp. 182, edited by B. Reipurth and C. Bertout (Dordrecht: Kluwer, 1997) 417.

[35] N. Calvet, L. Hartmann and S.J. Kenyon, *ApJ* **402** (1993) 623.

[36] R.M. Crutcher, *ApJ* **520** (1999) 706.

[37] J.S. Greaves, A.G. Murray and W.S. Holland, *MNRAS* **284** (1994) L19.

[38] D.A. Schleuning, *ApJ* **493** (1998) 811.

[39] B. Matthews and C.D. Wilson, *ApJ* **531** (2000) 868.

[40] J. Fiege and R.E. Pudritz, *MNRAS* **311** (2000a) 85.

[41] J. Fiege and R.E. Pudritz, *ApJ* **544** (2000b) 830.

[42] T. Ray, T.W.B. Muxlow, D.J. Axon *et al.*, *Nature* **385** (1997) 415.

[43] C. Thum and D. Morris, *A&A* **344** (1999) 923.

[44] E.H. Levy and C.P. Sonnett, "Meteorite magnetism and early solar system magnetic fields", in Protostars & Planets, edited by T. Gehrels (Tucson: Univ. of Arizona Press, 1978) 516.

[45] E.D. Feigelson and T. Montmerle, *ARA&A* **37** (1999) 363.

[46] A.E. Glassgold, E.D. Feigelson and T. Montmerle, "Effects of Energetic Radiation in Young Stellar Objects", in Protostars and Planets IV, edited by V. Mannings, A.P. Boss and S.S. Russell (Tucson: Univ. Arizona Press, 2000) 429.

[47] P.N. Foster and R.A. Chevalier, *ApJ* **416** (1993) 303.

[48] M.A. Reid, R.E. Pudritz and J. Wadsley, *ApJ* **570** (2002) 231.

[49] S. Bontemps, P. André, S. Terebey and S. Cabrit, *A&A* **311** (1996) 858.

[50] R.E. Pudritz and C.A. Norman, *ApJ* **301** (1986) 571.

[51] G. Pelletier and R.E. Pudritz, *ApJ* **394** (1992) 117.

[52] L. Hartmann, in Herbig-Haro Flows and the Birth of Low Mass Stars, edited by B. Reipurth and C. Bertout (Dordrecht: Kluwer, 1997) 391.

[53] J. van Paradijs and N. White, *ApJ* **447** (1995) L33.

[54] I.F. Mirabel and L.F. Rodriquez, *Nature* **371** (1994) 46.

[55] M. Miyoshi *et al.*, *Nature* **373** (1995) 127.

[56] J.R. Herrnstein *et al.*, *ApJ* **475** (1997) L17.

[57] R.D. Blandford, "Relativistic Accretion", in Astrophysical Discs, ASP Conf. Ser. 160, edited by J.A. Sellwood and J. Goodman (ASP, San Francisco, 1999) 265.

[58] R. Narayan and I. Yi, *ApJ* **428** (1994) L13.

[59] R.D. Blandford and M.C. Begelman, *MNRAS* **303** (1999) L1.

[60] T. Sakurai, *A&A* **152** (1985) 121.

[61] M.J. Aschwanden *et al.*, *ApJ* **535** (2000) 1047.

[62] S. Chandrasekhar, Hydrodynamic and hydromagnetic stability (Clarendon: Oxford, 1961).

[63] S.A. Balbus and J.F. Hawley, *ApJ* **376** (1991) 214.

[64] K.A. Miller and J.M. Stone, *ApJ* **534** (2000) 398.

[65] S.H. Lubow, J.C.B. Papaloizou and J. Pringle, *MNRAS* **267** (1994) 235.

[66] C.A. Tout and J.E. Pringle, *MNRAS* **281** (1996) 219.

[67] A. Brandenburg, A. Nordlund, R.F. Stein and U. Torkelsson, *ApJ* **446** (1995) 741.

[68] C. Curry, R.E. Pudritz and P.G. Sutherland, *ApJ* **434** (1994) 206.

[69] P.J. Armitage, *ApJ* **501** (1998) L189.

[70] C.F. Gammie, *ApJ* **457** (1996) 355.

[71] S. Matsumura and R.E. Pudritz 2003, *ApJ*, submitted.

[72] L. Mestel, *MNRAS* **138** (1968) 359.

[73] E.J. Weber and L. Davis, *ApJ* **148** (1967) 217.

[74] R.D. Blandford and D.G. Payne, *MNRAS* **199** (1982) 883.

[75] F.C. Michel, *ApJ* **158** (1969) 727.

[76] M. Heinemann and S. Olbert, *J. Geophys. Res.* **83** (1978) 2457.
[77] K. Tsinganos and E. Trussoni, *A&A* **231** (1990) 270.
[78] Z.-Y. Li, *ApJ* **444** (1995) 848.
[79] A. Königl, *ApJ* **342** (1989) 208.
[80] E. Ostriker, *ApJ* **486** (1998) 291.
[81] J. Heyvaerts and C.A. Norman, *ApJ* **347** (1989) 1055.
[82] Y. Uchida and K. Shibata, *PASJ* **37** (1985) 31.
[83] J. Ferreira and G. Pelletier, *A&A* **295** (1995) 807.
[84] J. Ferreira, *A&A* **319** (1997) 340.
[85] F. Casse and J. Ferreira, *A&A* **353** (2000) 1115.
[86] M. Tagger and R. Pellat, *A&A* **349** (1999) 1003.
[87] P. Varniere and M. Tagger, *A&A* **394** (2002) 329.
[88] T. Lery, R.N. Henriksen, J.D. Fiege *et al.*, *A&A* **387** (2002) 187.
[89] A.P. Goodson, R.M. Winglee and K-.H. Böhm, *ApJ* **489** (1997) 199.
[90] G.V. Ustyugova *et al.*, *ApJ* **500** (1998) 703.
[91] R. Ouyed, R.E. Pudritz and J.M. Stone, *Nature* **385** (1997) 409.
[92] R. Ouyed and R.E. Pudritz, *ApJ* **482** (1997a) 712.
[93] R. Ouyed and R.E. Pudritz, *ApJ* **484** (1997b) 794.
[94] R. Ouyed and R.E. Pudritz, *MNRAS* **309** (1999) 233.
[95] M.M. Romanova, G.V. Ustyugova, A.V. Koldoba, V.M. Chechetkin and R.V. Lovelace, *ApJ* **482** (1997) 708.
[96] D.L. Meier, S. Eddington, P. Godon, D.G. Payne and K.R. Lind, *Nature* **388** (1997) 350.
[97] R. Krasnopolsky, Z.-Y. Li and R.D. Blandford, *ApJ* **526** (1999) 631.
[98] J.M. Stone and M.L. Norman, *ApJS* **80** (1992) 753.
[99] C.R. Evans and J.F. Hawley, *ApJ* **332** (1988) 659.
[100] R.E. Pudritz, R. Rogers and R. Ouyed 2003, in preparation (PRO).
[101] P.E. Hardee and A. Rosen, *ApJ* **524** (1999) 650.
[102] T.P. Ray, *MNRAS* **196** (1981) 195.
[103] S. Appl and M. Camenzind, *A&A* **256** (1992) 354.
[104] R. Ouyed, D.A. Clarke and R.E. Pudritz, *ApJ* **582** (2003) 292.
[105] J.M. Stone and M.L. Norman, *ApJ* **433** (1994) 746.
[106] A.R. Bell and S.G. Lucek, *MNRAS* **277** (1995) 1327.
[107] J. Goodman and Xu, *ApJ* **432** (1994) 213.
[108] T. Kudoh, R. Matsumoto and K. Shibata, *PASJ* **54** (2002) 121.
[109] K. Tomisaka, *ApJ* **502** (1998) L163.

Section II: High energy astrophysics

COURSE 5

ULTRA HIGH ENERGY COSMIC RAYS

V. BEREZINSKY

INFN,
Laboratori Nazionali del Gran Sasso,
67010 Assergi (AQ),
Italy

Contents

ULTRA HIGH ENERGY COSMIC RAYS

V. Berezinsky

Abstract

The origin of Ultra High Energy Cosmic Rays (UHECR) is reviewed. The protons and nuclei with energy higher than (3–10) $\times 10^{19}$ eV are not confined by Galactic magnetic field and are expected to have large disc anisotropy in contrast to observations. Extragalactic protons and nuclei from the remote sources must have cutoff in energy spectra due to interaction with microwave photons (GZK cutoff [1]). The absence of such cutoff in the observed spectrum presents the UHECR problem. The recent HiRes spectrum [2] perfectly agrees with the GZK cutoff, while the AGASA data [3] show the significant excess of highest energy events. If the HiRes spectrum is correct, the conventional astrophysical models (*e.g.* UHECR from AGN) explain all observed spectrum. In case the AGASA excess is real, and the observed spectrum does not exhibit the GZK cutoff, the new physics or astrophysics is needed. In each case there are strong indications that up to 1×10^{20} the observed flux is composed by UHE protons from AGN.

1 Introduction

The problem with UHECR is known longer than 30 years. It consists in observation of primary particles with energies up to 2–3 $\times 10^{20}$ eV (for the recent review of observations see [4]. If these particles are extragalactic protons and their sources are distributed uniformly in the Universe, their spectrum must expose steepening due to interaction with microwave photons, known as the Greisen-Zatsepin-Kuzmin (GZK) cutoff [1].

The spectrum of UHECR according to AGASA observations [3] is shown in Figure 1 together with the spectrum calculated for uniform distribution of the sources in the Universe under the assumption that the generation spectrum is proportional to $E^{-2.3}$ with an assumed galactic contribution at $E < 1 \times 10^{19}$ eV. The excess of the observed events above the GZK cutoff is clearly seen at the highest energies.

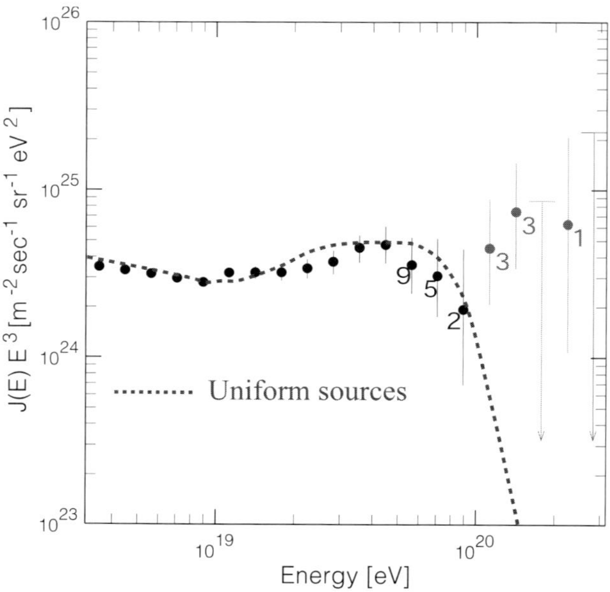

Fig. 1. AGASA spectrum [3] compared with "astrophysical" spectrum calculated under assumptions that the sources distributed uniformly in the Universe and have generation spectrum $\sim E^{-2.3}$.

If UHE primaries are nuclei, their spectrum also must exhibit steepening due to interaction with microwave photons approximately at the same energies as protons. The UHE photons have too small absorption length in the universe, and neutrinos with the standard interaction have too small cross-section for interaction in the atmosphere.

The observational data can be shortly summarized as follows.

- At $E \geq 10^{19}$ eV the spectrum is flatter than at lower energies and it extends up to $2\text{–}3 \times 10^{20}$ eV;
- At the highest energies the primaries are favored as protons, though UHE photons are not excluded;
- Data are consistent with isotropy, but close angular pairs (doublets) and triplets compose about 20% of all events at $E \geq 4 \times 10^{19}$ eV (22 events in doublets and triplets from 92 total [5];

- There are correlations between the arrival directions of the highest energy particles $E \geq 4 \times 10^{19}$ eV and directions to AGN (BL Lacs) [6];
- At energies $E \geq 1 \times 10^{18}$ eV according to preliminary HiRes data the mass composition is dominated by protons [7].

It is natural to think that the UHECR have extragalactic origin, though, in principle, very large halo with regular magnetic field can confine particles of these energies, especially if they are heavy nuclei. As was already mentioned above, the signature of extragalactic UHECR is the GZK cutoff. This phenomenon is caused by energy losses of UHE protons due to pion production in collisions with microwave photons. The energy losses start sharply increasing at $E_{\rm GZK} \sim 3 \times 10^{19}$ eV. This energy is connected with energy of the spectrum steepening ("cutoff") in the model-dependent way. In case the sources are distributed uniformly in the Universe (standard assumption), the steepening starts at $E_{\rm GZK} \approx 3 \times 10^{19}$ eV. The flux at $E > E_{\rm GZK}$ is produced by nearby sources. If there is a local enhancement of the sources, $E_{\rm GZK}$ increases [8]; in case the sources are located at large distances, $E_{\rm GZK}$ decreases and steepening becomes exponential. It is more convenient to characterise steepening by energy $E_{1/2}$, where the flux becomes half of the power-law extrapolation of unmodified flux. In case of uniform distribution of the sources $E_{1/2} \approx 5.3 \times 10^{19}$ eV (see below) for a wide range of exponents $\gamma_{\rm g}$ of generation spectrum.

2 Extragalactic astrophysical origin of UHECR

In this section we shall consider the extragalactic origin of UHECR: acceleration, propagation of the particles with energy losses taken into account, and some possible astrophysical model.

Astrophysical extragalactic models for the observed UHECR must satisfy some constraints. They include the acceleration to ultra high energies in the sources, the energy-release requirement, and the fit to the observed spectrum.

2.1 Extragalactic acceleration sources

Shock acceleration (including ultra-relativistic shocks) and unipolar induction are the "standard" acceleration mechanisms to UHE, considered in the literature. These mechanisms can operate in the various astrophysical objects, such as Active Galactic Nuclei (AGN), large scale structures (*e.g.* the shocks in AGN jets or shocks in the clusters of galaxies), in gamma-ray bursters (ultra-relativistic shocks), in the accretion discs around massive black holes (due to large electric potentials produced by unipolar induction) etc. A comprehensive list of possible sources was thoroughly studied in Ref. ([9]) with a conclusion, that maximum energy of acceleration does not exceed 10^{19}–10^{20} eV (see also Ref. ([10]) with a similar conclusion). The most promising source from this list is a hot spot in radiogalaxy produced by a jet [11–13].

A powerful jet ejected from the AGN supplies energy to a gigantic radiolobe. A hot spot observed at the termination of the jet is interpreted as a location of a standing shock. This is an ideal place for acceleration of protons to very high energies: magnetic field is strong and the energy density of radiation, responsible for proton energy losses, is relatively small. The maximum energy can be estimated as [11]

$$E_{\max} = (1\text{--}3) \times 10^{20} H_{-4} (R/1 \text{ kpc}) v_{\rm j} \text{ eV}, \tag{2.1}$$

where H_{-4} is the magnetic field in units of 10^{-4} G, R is radius of the shock and $v_{\rm j}$ is velocity of the jet in units of sound speed. However, the powerful radio sources are at large distances from our Galaxy, and the maximum energy is strongly attenuated. The discussed sources can provide acceleration up to energies 6–7×10^{19} eV.

Unipolar induction produces very large potential drop in the accretion discs around massive black holes (see Ref. ([14, 15] and also [8] for a discussion). The electrical potential for a rotating disc, at the distance r from a black hole is

$$\phi(r) = \frac{1}{\sqrt{6}} H_{\rm c} r_{\rm c} \ln R/r, \tag{2.2}$$

where $r_{\rm c} = 9 \times 10^5 (M_{\rm h}/M_\odot)$ is the radius of the last stable orbit for a black hole with mass $M_{\rm h}$, $H_{\rm c}$ is magnetic field at the last stable orbit, and R is a radius of accretion disc. The maximum potential given by equation (2.2) is $\phi_{\rm c} \sim 3 \times 10^{21}$ V for $M_{\rm h} \sim 1 \times 10^9\, M_\odot$ and $H_{\rm c} \sim 1 \times 10^4$ G. This mechanism is attractive because it can operate not only in AGN, but also for the old black holes, which lost their activity ($\dot{M}$ in the disc is small). Such sources can be located nearby, *e.g.* in the Local Supercluster, from where UHE protons can reach us without appreciable energy losses.

Relativistic shocks can in principle provide very high maximum energy. A particle reflecting from a relativistically moving mirror increases its energy by a factor proportional to Γ^2, where Γ is a Lorentz factor of the mirror. This acceleration phenomenon is known from the time of the pioneering work by E. Fermi [16]. The less known phenomenon is capturing of accelerated particles behind relativistically moving shock front. One can easily reconstruct this principle considering head-on collision of a particle with a transverse relativistic shock (magnetic field is perpendicular to the shock normal). Let us assume that magnetic field behind the shock is homogeneous on the scale of the particle Larmor radius. Then at the moment a particle finishes semi-circle, the shock front run away to a distance $cT_{\rm L}/2$

from a particle, where T_L is the Larmor period. The confinement described above can be illustrated by numerical simulation, *e.g.* in Ref. ([17]).

The capturing mechanism described above, does not exclude completely the Γ^2-regime of acceleration at relativistic shocks; it restricts the incident angles at which particles escape and thus the flux of accelerated particles. The most interesting objects where Γ^2-mechanism might operate are gamma-ray bursters [18, 19]. The Lorentz factor of the shock here can reach 10^2–10^3. However, the capturing properly taken into account might dramatically decrease the output of accelerated particles. As to the explanation of the observed UHECR, the protons from cosmologically remote gamma-ray bursts strongly degrade in energy on the way to our Galaxy.

2.2 Energy losses and propagation

We shall start with energy losses of *protons* due to pair production, $\mathrm{p} + \gamma_{\mathrm{CMBR}} \to \mathrm{p} + \mathrm{e}^+ + \mathrm{e}^-$, and pion production $\mathrm{p} + \gamma_{\mathrm{CMBR}} \to N +$ pions, where γ_{CMBR} is a microwave photon.

The energy losses of UHE proton per unit time due to its interaction with CMBR with temperature T is given by

$$-\frac{1}{E}\frac{\mathrm{d}E}{\mathrm{d}t} = \frac{cT}{2\pi^2\Gamma^2}\int_{\epsilon_{\mathrm{th}}}^{\infty} d\epsilon_{\mathrm{r}}\sigma(\epsilon_{\mathrm{r}})f(\epsilon_{\mathrm{r}})\epsilon_{\mathrm{r}}\left\{-\ln\left[1-\exp\left(-\frac{\epsilon_{\mathrm{r}}}{2\Gamma T}\right)\right]\right\}\cdot \tag{2.3}$$

where Γ is the Lorentz factor of the proton, ϵ_{r} is the energy of background photon in the system where the proton is at rest, ϵ_{th} is the threshold of the considered reaction in the rest system of the proton, $\sigma(\epsilon_{\mathrm{r}})$ is the cross-section, $f(\epsilon_{\mathrm{r}})$ is the mean fraction of energy lost by the proton in one $\mathrm{p}\gamma$ collision in the laboratory system.

From equation (2.3) one can see that the mean fraction of energy lost by the proton in lab system in one collision, $f(\epsilon_{\mathrm{r}}) = \langle 1-x\rangle = (E_{\mathrm{p}} - E'_{\mathrm{p}})/E_{\mathrm{p}}$, is the basic quantity needed for calculations of energy losses. The threshold values of these quantities are well known:

$$f_{\mathrm{pair}} \approx \frac{2m_{\mathrm{e}}}{m_{\mathrm{p}}}, \qquad f_{\mathrm{pion}} \approx \frac{\epsilon_{\mathrm{r}}}{m_{\mathrm{p}}}\frac{1+\mu^2/2\epsilon_{\mathrm{r}}m_{\mathrm{p}}}{1+2\epsilon_{\mathrm{r}}/m_{\mathrm{p}}}, \tag{2.4}$$

where f_{pair} and f_{pion} are the threshold fractions for $\mathrm{p}+\gamma \to \mathrm{p} + \mathrm{e}^+ + \mathrm{e}^-$ and $\mathrm{p} + \gamma \to N + \pi$, respectively, and μ is the pion mass.

For the accurate calculations of energy losses the fraction f properly averaged over differential cross-section is needed.

Pair production loss has been calculated in many papers. In Figure 2 we present these energy losses according to our recent calculations [20] compared with the recent calculations of Stanev *et al.* [21]. The results of our calculations are presented in Figure 2 in terms of relative energy losses per

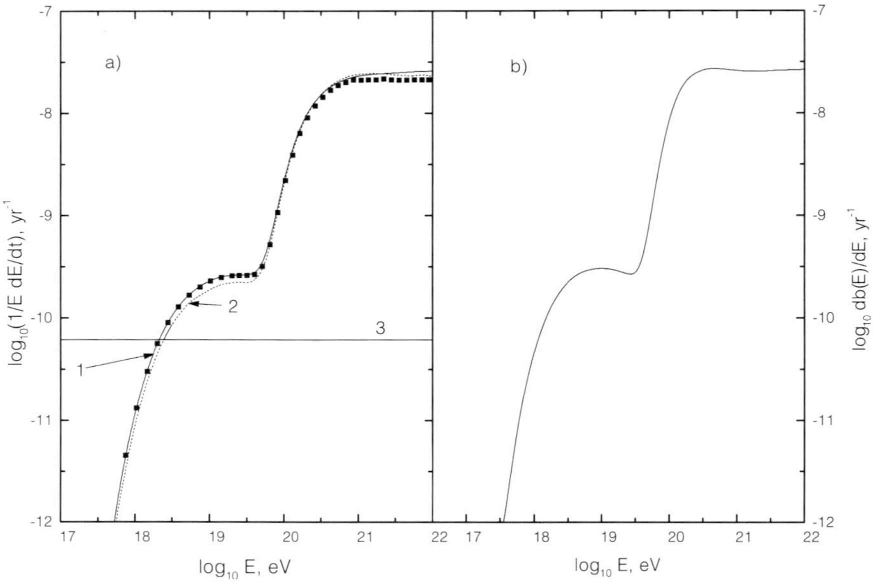

Fig. 2. a) UHE proton energy losses $E^{-1}\mathrm{d}E/\mathrm{d}t$ (present work: curve 1; Berezinsky & Grigorieva (1988) [22]: curve 2; Stanev *et al.* 2000 [21]: black squares). The line 3 gives energy losses due to redshift ($H_0 = 65$ km s^{-1} Mpc). **b)** The derivative $\mathrm{d}b_0(E)/\mathrm{d}E$, where $b_0(E) = \mathrm{d}E/\mathrm{d}t$ at present epoch $z = 0$.

unit time $E^{-1}\mathrm{d}E/\mathrm{d}t$ as function of energy (curve 1). Also plotted is the derivative $\mathrm{d}b_0(E)/\mathrm{d}E$, where $b_0(E) = \mathrm{d}E/\mathrm{d}t$ (Fig. 2b). This quantity is needed for calculation of differential energy spectrum (see Sect. III). In Figure 2 we plot for comparison the energy losses as calculated by Berezinsky & Grigorieva (1988) [22] (dashed curve 2). The difference in energy losses due to pion production is very small, not exceeding 5% in the energy region relevant for comparison with experimental data($E \leq 10^{21}$ eV). The difference with energy losses due to pair production is larger and reaches maximal value 15%. The results of calculations by Stanev *et al.* [21] are shown by black squares. These authors have performed the detailed calculations for both aforementioned processes, though their approach is somewhat different from ours, especially for photopion process.

Apart from GZK cutoff, there may be two more signatures of extragalactic cosmic rays: a bump and a dip in differential spectrum which precede the cutoff [22]. The bump is a consequence of a number conservation of protons in the spectrum: protons loose energy and are accumulated before the cutoff. The dip is formed due to pair-production (e^+e^-) energy losses of UHE

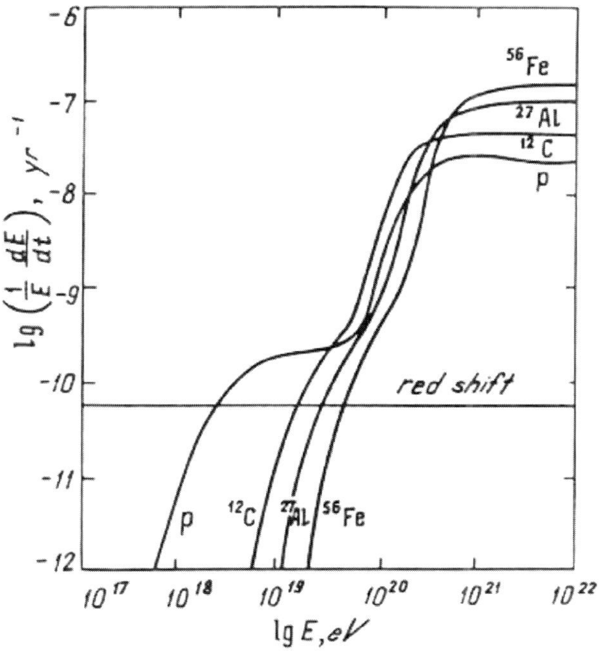

Fig. 3. Energy losses of UHE protons and nuclei [8].

proton. The both features show up most clearly in the differential spectrum of a single distant source in the case of a flat generation spectrum. In diffuse spectra (from many sources) the dip usually is seen, while the bump is absent for steep enough spectra.

UHE nuclei spectra exhibit steepening ("cutoff") approximately at the same energy as protons, though due to different physical processes (see [23] and [8] for a review). The relevant energy losses are caused by e^+e^- pair production and photo-disintegration of nuclei at collisions with microwave photons, and the steepening energy is determined by energy, when photo-disintegration energy-losses start to dominate over adiabatic ones (Fig. 3). For recent calculations see [24, 25]. *UHE photons* with $E_\gamma \sim 10^{19}$–10^{22} eV have an absorption length less than 10 Mpc, mainly due to interaction with radio background [26, 27].

2.3 Uniform distribution of UHECR sources and GZK cutoff

The GZK cutoff is a model–dependent feature of the spectrum, *e.g.* the GZK cutoff for a single source depends on the distance to the source. At

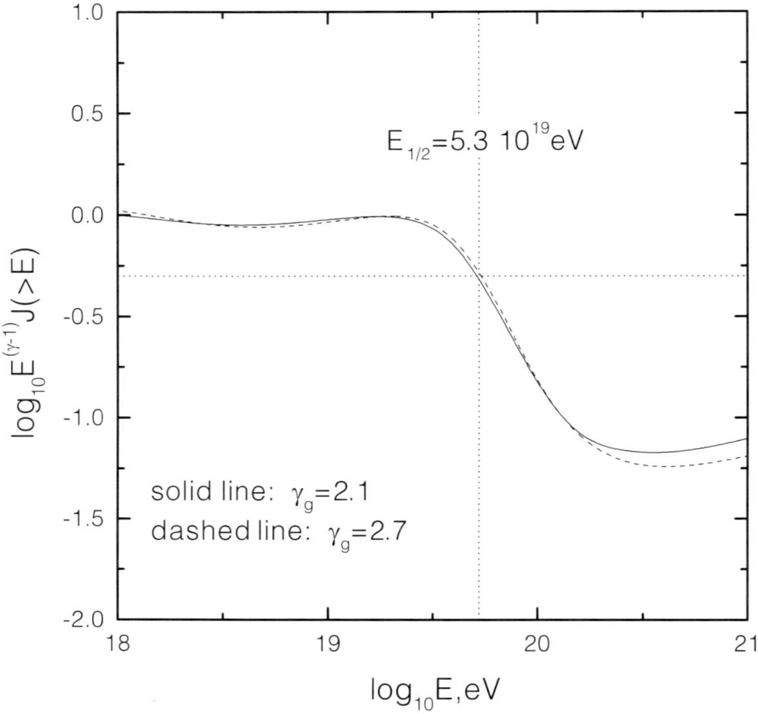

Fig. 4. Integral UHECR spectra with indices of generation spectra $\gamma_g = 2.1$ and $\gamma_g = 2.7$ (solid and dashed line, respectively). The vertical dotted line shows the energy $E_{1/2}$, where the calculated flux becomes two times lower than power-law extrapolation.

large distance the cutoff is exponential. A common convention is that the GZK cutoff is defined for diffuse flux from the sources uniformly distributed over the universe. In this case one can give two definitions of the GZK cutoff. In the first one the cutoff is determined as the energy, $E_{\rm GZK} \approx 3 \times 10^{19}$ eV, where the steep increase in the energy losses starts (see Fig. 2). The GZK cutoff starts at this energy. The corresponding pathlength of a proton is $R_{\rm GZK} \approx (E^{-1}{\rm d}E/{\rm d}t)^{-1} \approx 1.3 \times 10^3$ Mpc. The advantage of this definition of the cutoff energy is independence on spectrum index, but this energy is too low to judge about presence or absence of the cutoff in the measured spectrum.

More practical definition is $E_{1/2}$, where the flux with cutoff becomes lower by factor 2 than power-law extrapolation. This definition is convenient to use for the integral spectrum, which is better approximated by power-law function, than the differential one. In Figure 4 the function $E^{(\gamma-1)}J(>E)$, where $J(>E)$ is calculated integral diffuse spectrum, is plotted as function of energy. Note, that here and in Figure 4 $\gamma >$ (it is larger than $\gamma_{\rm g}$) is an effective index of power-law approximation of the spectrum modified by energy losses. For wide range of generation indices $2.1 \leq \gamma_{\rm g} \leq 2.7$ the cutoff energy is the same, $E_{1/2} \approx 5.3 \times 10^{19}$ eV. The corresponding proton pathlength is $R_{1/2} \approx 800$ Mpc. It means that in case of rectilinear propagation the beginning of the cutoff is formed at very large distance from an observer.

Using energy losses given in Section 2, one can calculate the diffuse spectra for the model when sources are distributed uniformly in the universe, using formalism of Ref. [22].

$$J_{\rm p}(E) = (\gamma_{\rm g} - 2)\frac{c}{4\pi}\frac{\mathcal{L}_0}{H_0}E^{-\gamma_{\rm g}}\int_0^{z_{\max}} {\rm d}z_{\rm g}(1+z_{\rm g})^{m-5/2}\lambda^{-\gamma_{\rm g}}(E,z_{\rm g})\frac{{\rm d}E_{\rm g}(z_{\rm g})}{{\rm d}E}, \tag{2.5}$$

where $z_{\rm g}$ is a redshift at generation and $E_{\rm g}(z_{\rm g})$ is energy of a proton at generation, if at present ($z = 0$) its energy is E: $E_{\rm g}(z_{\rm g}) = \lambda(E, z_{\rm g})E$ and $\lambda(E, z_{\rm g})$ is calculated numerically using energy losses ${\rm d}E/{\rm d}t$ accounted for their time evolution; $\mathcal{L}_0 = n_0 L_{\rm p}$ is CR emissivity at $z = 0$ (n_0 and $L_{\rm p}$ are space density of the sources and their CR luminosity, respectively). As the general case we assume cosmological evolution of the sources given by $\mathcal{L}(z) = \mathcal{L}_0(1+z)^m$, where the absence of evolution corresponds to $m = 0$. All energies in equation (2.5) are given in GeV and luminosities in GeV/s. Dilation of energy interval is calculated in Ref. [22].

$$\frac{{\rm d}E_{\rm g}(z_{\rm g})}{{\rm d}E} = (1+z_{\rm g})\exp\left[\int_0^{z_{\rm g}}\frac{{\rm d}z}{H_0}(1+z)^{1/2}\left(\frac{{\rm d}b_0(E')}{{\rm d}E'}\right)_{E'=(1+z)E_{\rm g}(z_{\rm g})}\right], \tag{2.6}$$

where $b_0(E) = {\rm d}E/{\rm d}t$ is energy loss due to interaction with CMBR photons at $z = 0$ (adiabatic energy loss due to redshift must not be included!). Derivative ${\rm d}b_0(E)/{\rm d}E$ at $z = 0$ is given in Figure 2b.

2.4 The astrophysical AGN model for UHECR

We shall discuss here a natural UHECR model with extragalactic protons. The sources are assumed to be AGN. The acceleration for them has been considered in Section 2.1. We assume that the generation spectrum is the standard one, *i.e.* $\propto E^{-2}$ at $E \leq E_{\rm c}$ and becomes steeper at the higher energies $\propto E^{-\gamma_{\rm g}}$ at $E \geq E_{\rm c}$. The model is restricted to explanation of events with energies up to 1×10^{20} eV, while eleven events observed at by AGASA at higher energies are assumed to have different origin.

More specifically the generation spectrum is given by

$$Q_{\rm g}(E_{\rm g},z)=\frac{L_{\rm p}(z)}{\ln\frac{E_{\rm c}}{E_{\rm min}}+\frac{1}{\gamma_{\rm g}-2}}q_{\rm gen}(E_{\rm g}), \tag{2.7}$$

where $L_{\rm p}(z)$ is a luminosity of a source and $q_{\rm gen}$ is given by

$$q_{\rm gen}(E_{\rm g})=\begin{cases}1/E_{\rm g}^2 & \text{at } E_{\rm g}\le E_{\rm c}\\ E_{\rm c}^{-2}(E_{\rm g}/E_{\rm c})^{-\gamma_{\rm g}} & \text{at } E_{\rm g}\ge E_{\rm c}.\end{cases} \tag{2.8}$$

The diffuse spectrum can be readily calculated at as

$$J_{\rm p}(E)=\frac{cH_0^{-1}}{4\pi}\,\mathcal{L}_0\,\frac{1}{\ln\frac{E_{\rm c}}{E_{\rm min}}+\frac{1}{\gamma_{\rm g}-2}}\int_0^{z_{\rm max}}{\rm d}z_{\rm g}(1+z_{\rm g})^{m-5/2}q_{\rm gen}(E_{\rm g})\frac{{\rm d}E_{\rm g}}{{\rm d}E}. \tag{2.9}$$

The fluxes given by equation (2.9) are displayed in Figure 5 and compared with AGASA, HiRes, Fly's Eye and Yakutsk data. The solid curve shows the spectrum for non-evolutionary model ($m=0$) with $E_{\rm c}=1\times10^{18}$ eV and $\gamma_{\rm g}=2.7$, while the broken line refers to evolutionary model with $m=3$, $\gamma_{\rm g}=2.5$ and $E_{\rm c}=3\times10^{17}$ eV. The required CR emissivity for both models is $\mathcal{L}_0=(2.5\text{–}3.5)\times10^4$ 6 erg/Mpc^3 yr for the AGASA data. The required CR emissivity meets well the local emissivity of AGN, *e.g.* that of Seyfert galaxies is of order $\mathcal{L}_{\rm Sy}\sim n_{\rm Sy}L_{\rm Sy}\sim10^{48}$ erg/Mpc^3 yr. In fact, in all four panels of Figure 5 the same spectrum is displayed, but to fit the data of HiRes, Fly's Eye and Yakutsk the the emissivity $\mathcal{L}_0$ has been scaled by factors 0.63, 0.80, and 1.7, respectively.

The AGASA excess at $E\ge1\times10^{20}$ eV needs for explanation another component of cosmic rays. It can be produced by decays of superheavy dark matter particles (see Sect. 3). The observed AGASA spectrum at $E\ge1\times10^{20}$ eV is described well by the calculated decay spectrum. It is shown by dashed curve labeled SHDM in Figure 5. As one can see the data of all detectors confirm the dip, which is produced due to e^+e^- production in interaction of extragalactic UHE protons with CMB.

Three kinds of the data support our model.

(i) The recent preliminary HiRes mass composition [7] favor the proton primaries at $E\ge1\times10^{18}$ eV.

(ii) The correlation of arrival direction of primaries at energies $(4\text{–}8)\times10^{19}$ eV with BL Lacs [6].

(iii) The explanation of small-angle clustering by statistically occasional arrival of two or three particles from a compact source [28]. For this explanation the rectilinear propagation of primaries is required.

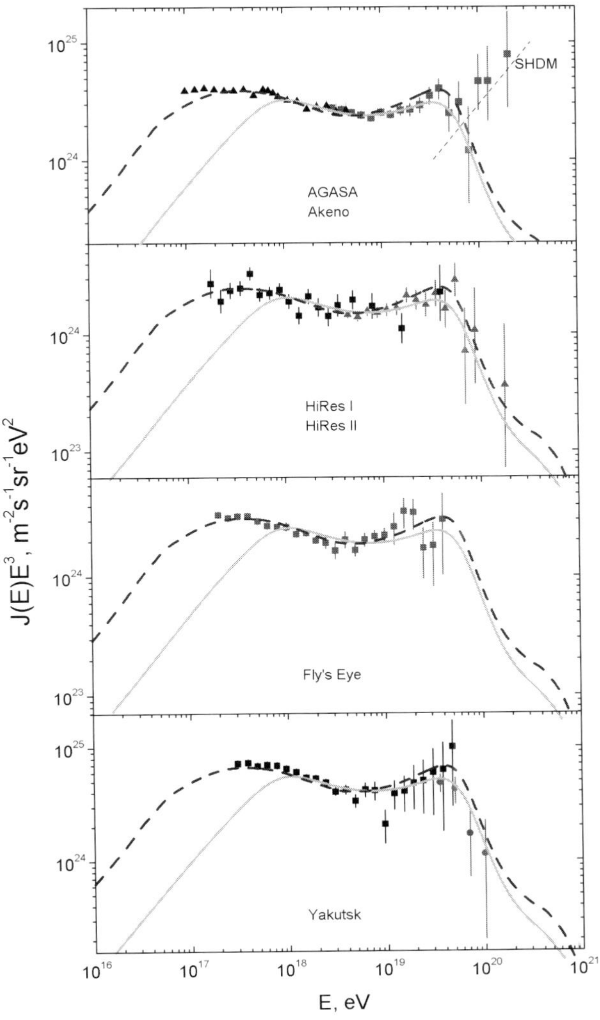

Fig. 5. The calculated spectra for non-evolutionary model (full lines), and evolutionary model (dotted lines) with parameters indicated in the text. Both curves were first normalized to the AGASA data at $E = 1 \times 10^{18}$ eV adjusting the emissivity $\mathcal{L}_0$. To fit the data of HiRes, Fly's Eye and Yakutsk the emissivity $\mathcal{L}_0$ has been scaled by factors 0.63, 0.80, and 1.7, respectively.

Since the proton origin of UHECR is almost proved, the correlation with BL Lacs directly implies the rectilinear propagation of UHE protons. Thus, *these correlations become supersensitive tools to measure extragalactic magnetic fields.* Below we shall discuss what is the scale of this field and whether this field can be already excluded.

Magnetic field must not produce the angular deflection larger than angular resolution of sources in the detectors, which is typically $\theta_{\rm res} \approx 2.5^\circ$. The correlation is found in the energy range $(4-8)\times 10^{19}$ eV, for which the largest attenuation length is $l_{\rm att} \sim 1000$ Mpc. The required upper limit for the magnetic field, which is homogeneous on this scale, is $B_l \leq 2\times 10^{-12} l_{1000}^{-1}$ G, where l_{1000} is attenuation length for 4×10^{19} eV protons in units of 1000 Mpc. For a magnetic field with small homogeneity length $l_{\rm hom}$ the required upper limit is

$$B \leq \frac{E\theta_{\rm res}}{e\sqrt{l_{\rm att} l_{\rm hom}}} \sim 6\times 10^{-10}\ {\rm G}, \tag{2.10}$$

where the numerical value is given for $l_{\rm att} \sim 1000$ Mpc and $l_{\rm hom} \sim 10$ kpc.

We argue that these fields are not excluded.

The observed Faraday rotations give only the upper limits on large scale extragalactic magnetic field [29]. All known mechanisms of generation of the large scale cosmological magnetic field results in extremely weak magnetic field $\sim 10^{-17}$ G or less (for a review see [30]). The strong magnetic field can be generated in compact sources, presumably by dynamo mechanism, and spread away by the flow of the gas. These objects thus are surrounded by magnetic halos, where magnetic field can be estimated or measured. The strong magnetic fields of order of 1 μG are indeed observed in galaxies and their halos, in clusters of galaxies and in radiolobes of radiogalaxies.

As an example one can consider our local surroundings. Milky Way belongs to the Local Group (LG) entering the Local Supercluster (LS). LG with a size $\sim$1 Mpc contains 40 dwarf galaxies, two giant spirals (M 31 and Milky Way) and two intermediate size galaxies. The galactic winds cannot provide the appreciable magnetic field inside this structure. LS with a size of 10–30 Mpc is a young system where dynamo mechanism cannot strengthen the primordial magnetic field. In fact LS is filled by galactic clouds submerged in the voids. The vast majority of the luminous galaxies reside in a small number of clouds: 98% of all galaxies in 11 clouds [31].

Thus, accepting the hypothesis of generation of magnetic fields in compact sources, one arrives at the perforated picture of the universe, with strong magnetic fields in the compact objects and their halos (magnetic bubbles produced by galactic winds) and with extremely weak magnetic fields outside. However, even in this picture there is a scattering of UHE protons off the magnetic bubbles and the scattering length is $l_{\rm sc} \sim 1/\pi R^2 n$, where R is the radius ofa magnetic bubble and n is their space density. Among different structures, the largest contribution is given by galaxy clusters which can provide $l_{\rm sc} \sim (1–2)\times 10^3$ Mpc.

3 Elementary particle solutions to UHECR problem

Many ideas were put forward to explain the absence of the GZK cutoff, using new particles and phenomena. They include:

- Superheavy Dark Matter (SHDM) particles born at inflationary stage and decaying to UHE particles at present epochs [32–34]. As any other cold DM particles, they are accumulated in the halo of our galaxy with very large overdensity, and thus UHECR from their decay do not exhibit the GZK cutoff. Note, that the decay spectrum of superheavy particles describe well the UHECR spectrum of AGASA excess (see Fig. 5);
- Topological defects produce superheavy particles [35, 36]. The spectra of UHECR from their decay have weak GZK cutoff because the production spectra of UHE particles are flat [37];
- There can be the signal carriers which are not absorbed on the microwave radiation, such as glueballino [38,39], neutrino with a strong interaction with nucleons (*e.g.* in the theories with extra dimensions etc.);
- Z-bursts produced by very high energy neutrinos on DM neutrinos: $\nu + \nu_{\rm DM} \rightarrow Z^0 \rightarrow$hadrons. The GZK cutoff is weak due to steep spectrum of produced hadrons [40];
- Breaking of Lorentz invariance can shift the GZK cutoff to much larger energy [41].

We shall not consider here these ideas in any detail.

4 Conclusions

Spectrum of UHECR extends up to (2–3) $\times 10^{20}$ eV. According to AGASA the GZK cutoff is absent, while recent data of HiRes agree with presence of the GZK cutoff. The primary particles at $E \geq 1 \times 10^{18}$ eV are protons according to HiRes data. The AGASA data are consistent with isotropy in arrival of the particles, but about 20% of particles at $E \geq 4 \times 10^{19}$ eV arrive as doublets and triplets within $\sim 2^\circ$.

The galactic origin of UHECR due to conventional (acceleration) sources is disfavoured: the maximal observed energies are higher than that calculated for the galactic sources, and the strong Galactic disc anisotropy is predicted even for the extreme magnetic fields in the disc and halo.

The signatures of extragalactic UHE protons are dip, due to e^+e^- production on CMBR, and GZK cutoff due to photopion production. The dip is clearly observed in all four detectors (AGASA, HiRes, Fly's Eye and Yakutsk), the GZK cutoff is observed in HiRes but absent in AGASA. The protons as primaries are confirmed by HiRes observations. The correlations of arrival directions of primaries with AGN (BL Lacs) imply rectilinear propagation of primary protons, which can occur in case of weak extragalactic magnetic fields. The particles with energy higher than 1×10^{20} eV must have another origin. This is obvious in case of AGASA data, because they show the absence of GZK cutoff, but this is also true for the Fly's Eye and HiRes events at $E > 1 \times 10^{20}$ eV, because the attenuation length for these particles is shorter than 100 Mpc (30 Mpc for $E = 2 \times 10^{20}$ eV) and no AGN are seen in these directions.

The AGASA excess can be explained by new physics: superheavy dark matter, topological defects, new particles – signal carriers, Z-bursts and Lorentz-invariance violation. The energy spectrum of AGASA excess is well observed by UHE particles from superheavy DM decays.

I am grateful to the organisers of the school, and especially to Profs. G. Henri and G. Pelletier, fot hospitality and excellent organisation of the school. This work is partially supported by INTAS through grant 99-1065.

References

[1] K. Greisen, *Phys. Rev. Lett.* **16** (1966) 748.
G.T. Zatsepin and V.A. Kuzmin, *JETP Lett.* **4** (1966) 78.

[2] T. Abu-Zayyad *et al.* (HiRes collaboration), [astro-ph/0208243], [astro-ph/0208301], [astro-ph/0208024].

[3] M. Takeda *et al.*, *Phys. Rev. Lett.* **81** (1998) 1163 [astro-ph/0209422].

[4] M. Nagano and A.A. Watson, *Rev. Mod. Phys.* **72** (2000) 689.

[5] Y. Uchihori *et al.* [astro-ph/9908193].

[6] P.G. Tinyakov and I.I. Tkachev, *JETP Lett.* **74** (2001) 445.

[7] P. Sokolsky, Proceedings of SPIE Conference on Instrumentation for Particle Astrophysics (Waikoloa, Hawaii, 2002).

[8] V.S. Berezinsky, S.V. Bulanov, V.A. Dogiel, V.L. Ginzburg and V.S. Ptuskin, Astrophysics of Cosmic Rays, chapter 4 (North-Holland, 1990).

[9] C.A. Norman, D.B. Melrose and A. Achtenberg, *ApJ* **454** (1995) 60.

[10] R.D. Blandford [astro-ph/9906026].

[11] P.L. Biermann and P.A. Strittmatter, *ApJ* **322** (1987) 643.

[12] W.H. Ip and W.I. Axford, Astrophysical Aspects of the Most Energetic Cosmic Rays, edited by M. Nagano (World Scientific, 273, 1991).

[13] J.P. Rachen and P.L. Biermann, *A&A* **272** (1993) 161.

[14] R.D. Blandford, *MNRAS* **176** (1976) 465.

[15] R.V.E. Lovelace, *Nature* **262** (1976) 649.

[16] E. Fermi, *Phys. Rev.* **75** (1949) 1169.

[17] M.C. Begelman and J.G. Kirk, *ApJ* **353** (1990) 66.

[18] M. Vietri, *ApJ* **453** (1995) 883.

[19] E. Waxman, *Phys. Rev. Lett.* **75** (1995) 386.

[20] V. Berezinsky, A.Z. Gazizov and S.I. Grigorieva [hep-ph/0107306].

[21] T. Stanev, R. Engel, A. Muecke, R.J. Protheroe and J.P. Rachen, *Phys. Rev. D* **62** (2000) 093005.

[22] V.S. Berezinsky and S.I. Grigorieva, *A&A* **199** (1988) 1.

[23] V.S. Berezinsky, S.I. Grigorieva and G.T. Zatsepin, Proc. 14th Int. Cosm. Ray Conf. (Munich) 2, 91975, 711.

[24] F.W. Stecker, *Phys. Rev. Lett.* **80** (1998) 1816.

[25] L.N. Epele, S. Mollerach and E. Roulet, *JHEP* **03** (1999) 017.

[26] V.S. Berezinsky, *Sov. J. Nucl. Phys.* **11** (1970) 222.

[27] R.J. Protheroe and P.L. Biermann, *Astrophys. Phys.* **6** (1996) 45.

[28] S.L. Dubovsky, P.G. Tinyakov and I.I. Tkachev, *Phys. Rev. Lett.* **85** (2000) 1154; Z. Fodor and S.D. Katz, *Phys. Rev. D* **63** (2000) 023002.

[29] P.P. Kronberg, *Rep. Prog. Phys.* **57** (1994) 325; J.P. Vallee, *Fund. Cosmic Phys.* **19** (1997) 1.

[30] D. Grasso, H. Rubinstein, *Phys. Rep.* **348** (2001) 163.

[31] R.B. Tully, *ApJ* **257** (1982) 389.

[32] V. Berezinsky, M. Kachelriess and A. Vilenkin, *Phys. Rev. Lett.* **79** (1997) 4302.

[33] V.A. Kuzmin and V.A. Rubakov, *Yadern. Fiz.* **61** (1998) 1122.

[34] M. Birkel and S. Sarkar, *Astrop. Phys.* **9** (1998) 297.

[35] C.T. Hill, D.N. Schramm and T.P. Walker, *Phys. Rev. D* **36** (1987) 1007;

[36] P. Bhattacharjee, C.T. Hill and D.N. Schramm, *Phys. Rev. Lett.* **69** (1992) 567; G. Sigl, D.N. Schramm and P. Bhattacharjee, *Astropart. Phys.* **2** (1994) 401.

[37] V. Berezinsky, P. Blasi and A. Vilenkin, *Phys. Rev. D* **58** (1998) 103515.

[38] D.J.H. Chung, G.R. Farrar and E.W. Kolb, *Phys. Rev. D* **57** (1998) 3579.

[39] V. Berezinsky, M. Kachelriess and S. Ostapchenko, *Phys. Rev. D* **65** (2002) 083004.

[40] D. Fargion, B. Mele and A. Salis, *ApJ* **517** (1999) 725; T. Weiler *Astrophys. Phys.* **11** (1999) 303.

[41] D.A. Kirzhnitz and V.A. Chechin, *Pisma ZHETP* **14** (1972) 261; S.L. Glashow, *Nucl. Phys. B (Proc. Suppl)* **70** (1999) 180.

COURSE 6

GAMMA-RAY BURSTS

F. DAIGNE

Institut d'Astrophysique de Paris,
98 bis boulevard Arago,
75014 Paris, France

Contents

GAMMA-RAY BURSTS

F. Daigne[1]

Abstract

This lecture is an introduction to the study of gamma-ray bursts. Observations of the burst and its afterglow are first reviewed, with the identification of a few key observationnal facts that lead to a three-step theoretical scenario. The physical processes at work in each step are then presented, focussing on the internal/external shock model. The last part briefly mention the possible non-photonic emission of gamma-ray bursts.

1 Introduction

Gamma-ray bursts (hereafter GRBs) have been discovered by the *VELA* satellites at the end of the 60's Klebesadel *et al.* (1973). A GRB is a short (from a few milliseconds to a few hundreds seconds) but very intense (peak flux up to hundreds of photons per square centimeters and per second) burst of gamma-rays (typically between 100 keV and 1 MeV). After their discovery, GRBs have been observed by many satellites with gamma-ray capabilities. I will not review here the results obtained in the 70's and the 80's. At the end of this period, the total number of detected GRBs was about five hundreds but the GRB distance scale remained undetermined, due to the poor localization capabilities of gamma-ray instruments which did not allow the identification of any counterpart. In most models, GRBs were related to Galactic neutron stars and the apparent isotropy of the GRB distribution was interpreted as a lack of sensibility, limiting the detection to local sources only. It was then believed that the next generation of detectors, thanks to a better sensibility, would show an accumulation of GRBs in the Galactic disk.

The *Burst And Transient Source Experiment* was launched on board the *Compton Gamma-Ray Observatory* in 1991 and observed GRBs until 2000 with a much better sensibility than all previous detectors. *BATSE* detected about one GRB per day during nine years and provided us with a catalog

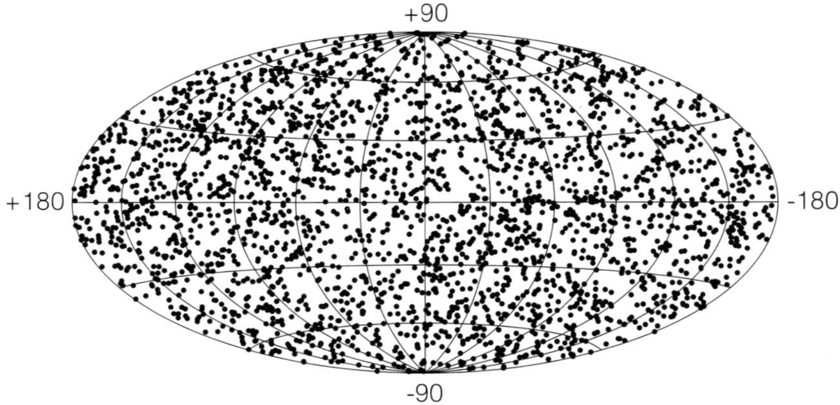

Fig. 1. Skymap of the 2704 BATSE GRBs in galactic coordinates (from Paciesas *et al.* 1999).

of about three thousands bursts. The most important *BATSE* result is probably that despite the detection of much weaker GRBs, the distribution over the sky remained almost perfectly isotropic (see Fig. 1). This was the first strong evidence in favor of an extragalactic origin of GRBs.

The next major step was achieved in 1997 thanks to the *Beppo-SAX* satellite which discovered the X-ray afterglow of GRBs, *i.e.* a X-ray fading counterpart detected a few hours after the burst. This X-ray afterglow could be localized by *Beppo-SAX* with an accuracy of about one arcminute (instead of a few degrees in gamma-rays) and the corresponding position was distributed in ground-based observatories within a delay of a few hours. This led to the discovery of fading counterparts at other wavelengths: the optical and radio afterglows. The identification of highly redshifted absorption and emission lines in the spectrum of the optical afterglow and/or its host galaxy is the direct proof that GRBs are produced at cosmological distances (observed redshifts $z \sim 0.3$–4.5).

This direct determination of the distance scale has allowed rapid progress in the physical understanding of GRBs. Today most of the about 130 models of GRBs which were listed in 1994 Nemiroff (1994) have been eliminated and only a few models remain, which sometimes differ only by one aspect of the theoretical scenario (for instance the central source).

In this lecture, I will first review the main observed properties of GRBs (Sect. 2), starting from 1991 and the results of *BATSE*. Then I will describe

[1]Institut d'Astrophysique de Paris, France.

the three step theoretical scenario which is adopted by most models (Sect. 3). The three next sections will introduce the basic concepts for each of these three steps in the context of the most-discussed model, often called the fireball model. Section 7 will briefly discuss the very interesting possibility that GRBs may be the source of non-photonic emission (cosmic rays, neutrinos and gravitational waves). The conclusions in the last section will mainly list the most important theoretical issues that are unsolved and describe the promising observational prospects in the field.

2 Observations

2.1 *Prompt emission*

Apart from the gamma-rays, the prompt emission (*i.e.* the burst itself) has been observed in X-rays in a few cases and in no other spectral band, with the notable exception of GRB 990123[1] were an optical flash was also detected. In this subsection I summarize the main properties of this prompt emission, focussing for the gamma-rays on the results of *BATSE*, which, from 1991 to 2000, has observed 2704 GRBs (not including un-triggered events), corresponding to a rate of about one event per day.

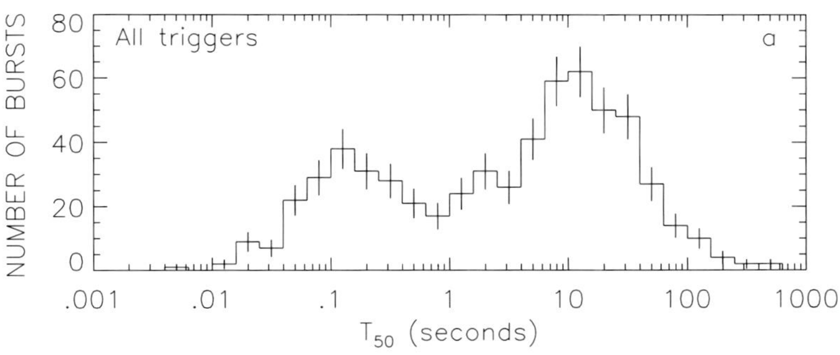

Fig. 2. Number of GRBs as a function of their duration (from Paciesas *et al.* 1999).

2.1.1 Duration

Figure 2 shows the distribution of the GRB duration T_{90} in the *BATSE* catalog. The duration T_{90} is defined as $T_{90} = t_{95} - t_5$, where t_{95} (resp. t_5)

[1] GRBs are usually named after the day they were detected, here the 23th January 1999, followed by a, b, c, d, ... in case of several detections in the same day.

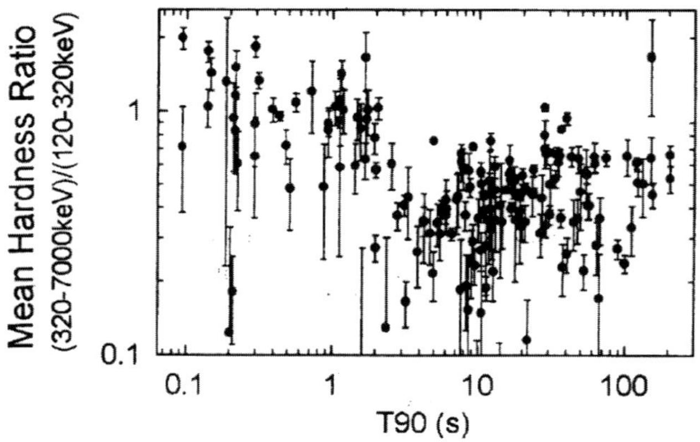

Fig. 3. Evolution of the hardness ratio with the burst duration (from Dezalay *et al.* 1996).

is the time at which 95% (resp. 5%) of the total fluence in the 2+3 *BATSE* band (50–300 keV) has been received. This duration is about 1–5 ms for the shortest bursts and can reach 1000 s. It is clear in Figure 2 that there are at least two groups of GRBs: a group of short GRBs (about one third of the total, typical duration $T_{90} \sim 100$ ms) and a group of long GRBs (about two thirds, typical duration $T_{90} \sim 10$ s). Figure 3 shows the burst hardness ratio as a function of the duration T_{90}. The hardness ratio is the ratio of the count number in a high energy band (here 320–7000 keV) over the count number in a low energy band (here 120–320 keV). This figure corresponds to GRBs observed with the PHEBUS experiment Dezalay *et al.* (1996) but *BATSE* shows very similar results Kouveliotou *et al.* (1993): there is a clear evolution of the hardness from the group of short bursts to the group of long bursts, the short bursts being harder.

2.1.2 Time profile

One remarkable property of GRBs is the extreme diversity of their time profiles. A few examples are given in Figure 4. Some time profiles are well structured, with a few well separated pulses (*e.g.* BATSE trigger 7469), whereas some others have extremely complex light curves (*e.g.* BATSE trigger 2533). However, it has been suggested that most of these profiles could be analysed as the sum of a few individual pulses Norris *et al.* (1996). An example of such a decomposition is given in Figure 5. These individual pulses usually show an asymmetric profile with a fast rise and a slower decay. The pulse shape also changes with the duration, becoming more

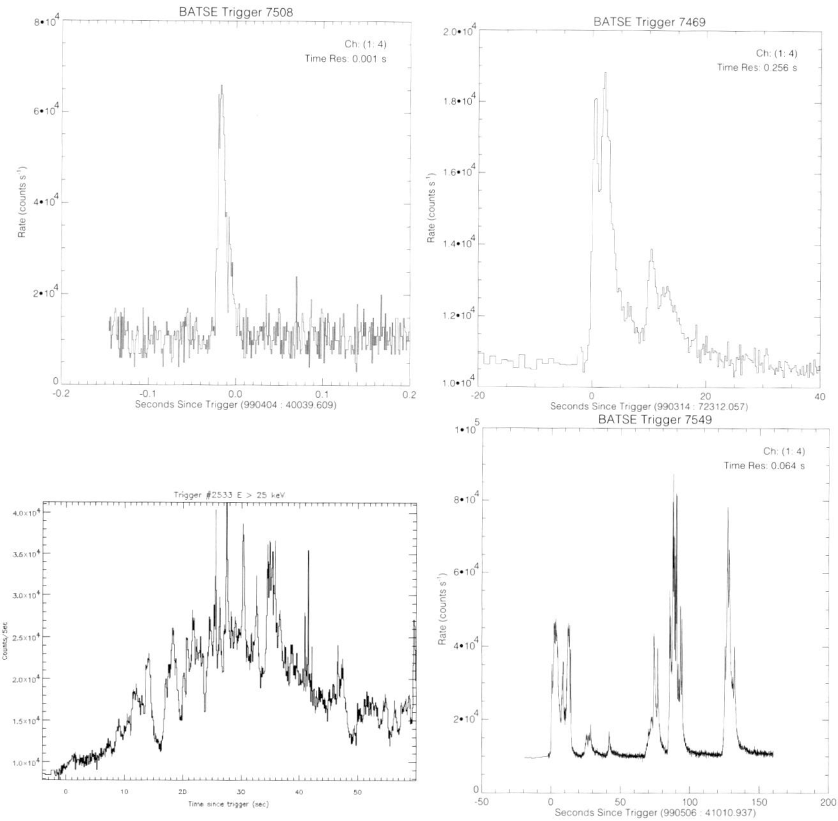

Fig. 4. A few GRB time profiles (from the BATSE catalog, Paciesas *et al.* 1999).

symmetric for small durations, and with the energy band: at high energy, the pulse peaks earlier and is narrower. This "pulse paradigm" can be quantified Norris *et al.* (1996): for instance the width $W(E)$ of the pulse at energy E is found to follow a power-law: $W(E) \propto E^{-0.4}$. In the X-ray range, the light curve is similar to what is observed in the gamma-ray range, the width of the pulses being even larger, as shown for example in Figure 6. The possible X-ray precursors will be discussed later. A last important aspect to note is that the variability in the lightcurve appears to be present on several timescales and that the shortest timescales can be as short as 1 ms. This provides an important constraint for the models.

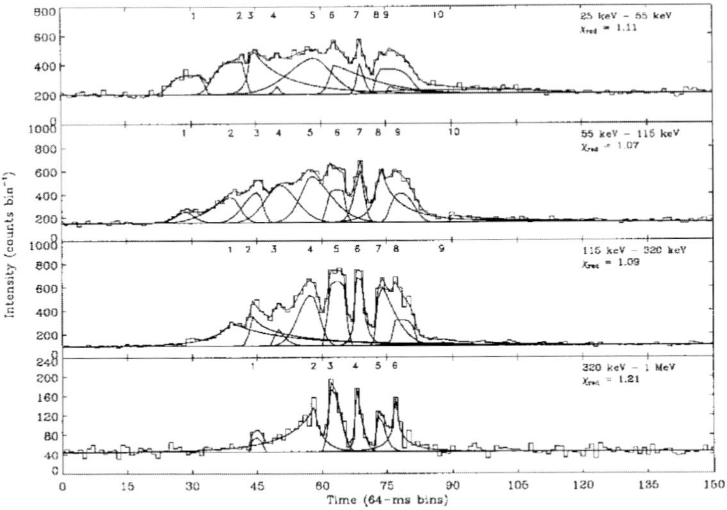

Fig. 5. Analysis of a complex GRB lightcurves as the sum of several pulses (BATSE data, from Norris *et al.* 1996).

2.1.3 Spectrum

The time-integrated GRB spectra do not show the same diversity as time profiles. They are non-thermal and are usally well fitted by the so-called GRB function Band *et al.* (1993):

$$n(E) = \frac{E_\gamma}{E_\mathrm{p}^2} \times \mathcal{B}\left(\frac{E}{E_\mathrm{p}}\right), \tag{2.1}$$

where E_γ is the total energy radiated in gamma-rays, E_p is the peak energy of the burst, *i.e.* the energy at which $E^2 n(E)$ (or νF_ν) is maximum, and the function $\mathcal{B}(x)$ is made of two power-laws with a smooth transition:

$$\mathcal{B}(x) = \begin{cases} x^\alpha \exp\left(-(2+\alpha)x\right) & x \le x_\mathrm{b} \\ x^\beta x_\mathrm{b}^{\alpha-\beta} \exp\left(-(2+\alpha)x_\mathrm{b}\right) & x \ge x_\mathrm{b}. \end{cases} \tag{2.2}$$

The transition occurs at $x_\mathrm{b} = (\alpha - \beta)/(2+\alpha)$. The distribution of the three parameters α, β and E_p (or the break energy $E_\mathrm{b} = x_\mathrm{b} E_\mathrm{p}$) has been studied for a sample of long bright GRBs Preece *et al.* (2000). The results are plotted in Figure 7. It is found that the averaged value of the low-energy and high-energy slopes are $\alpha \sim -1.0$ and $\beta = -2.25$ and that the break energy has a very narrow distribution, between 100 keV and 1 MeV. The

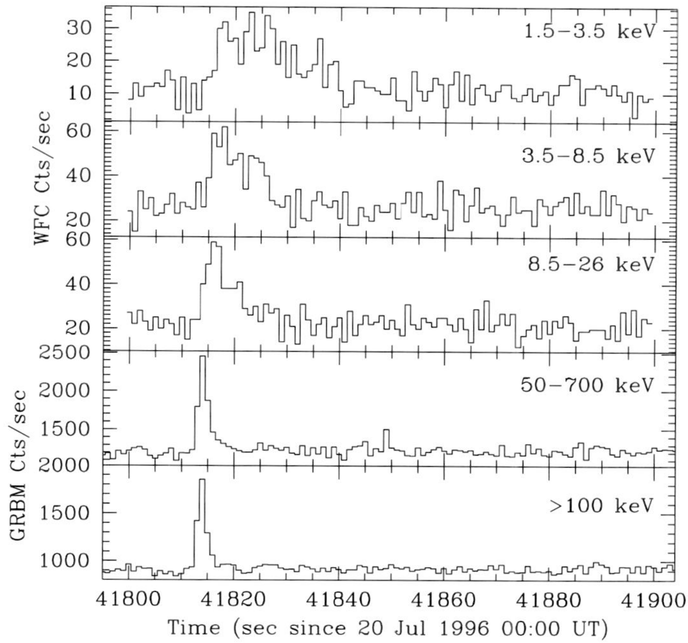

Fig. 6. Gamma- and X-ray profiles of GRB 960720 (Beppo-SAX data, from Piro *et al.* 1998).

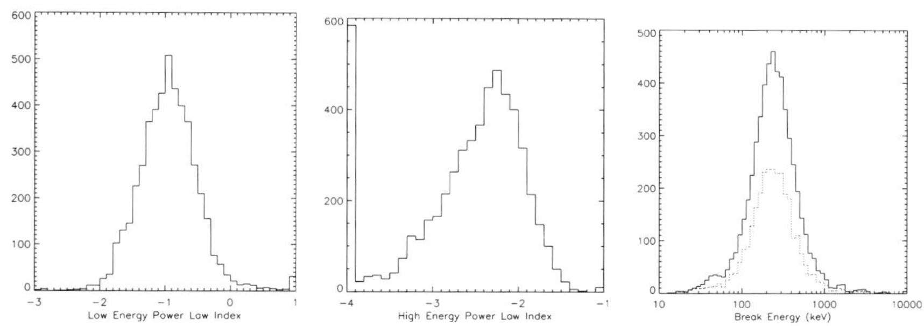

Fig. 7. Distribution of the low- and high-energy slopes and the break energy of the Band function (from Preece *et al.* 2000).

spectrum of GRB 990123 is shown as an example in Figure 8. Notice that this bright, long, hard burst has been detected by the four instruments of *CGRO*, up to about 10–20 MeV. The GRB function seems to fit correctly the spectrum over this whole spectral range, with no indication of high energy

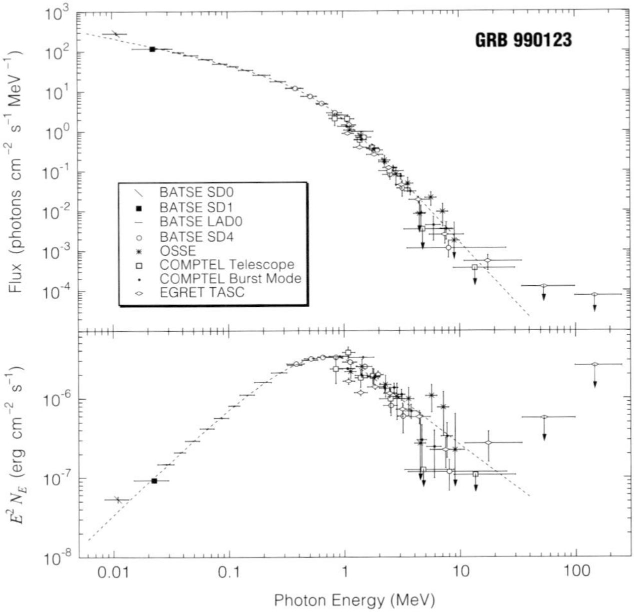

Fig. 8. The time integrated spectrum of GRB 990123 observed with the four instruments of CGRO: $\alpha \sim -0.6$, $\beta \sim -3.1$ and $E_{\rm p} \sim 720$ keV (from Briggs *et al.* 1999).

breaks that could be associated with the photon-photon annihilation. The lack of such a break will give another interesting constraint for the models. Notice also that no convicing line has ever been detected in the prompt emission spectrum.

2.1.4 Spectral evolution

The spectrum is strongly evolving during the burst. A global hard-to-soft evolution is observed in many GRBs (*e.g.* BATSE trigger 543 in Fig. 9), as well as a hardness-intensity (HIC) Kargatis *et al.* (1995) and a hardness-fluence (HFC) Liang & Kargatis (1996) correlation in the decay phase of pulses:

$$E_{\rm p}(t) \propto N(t)^{\delta} \tag{2.3}$$

$$E_{\rm p}(t) \propto \exp\left(-\Phi(t)/\Phi_0\right), \tag{2.4}$$

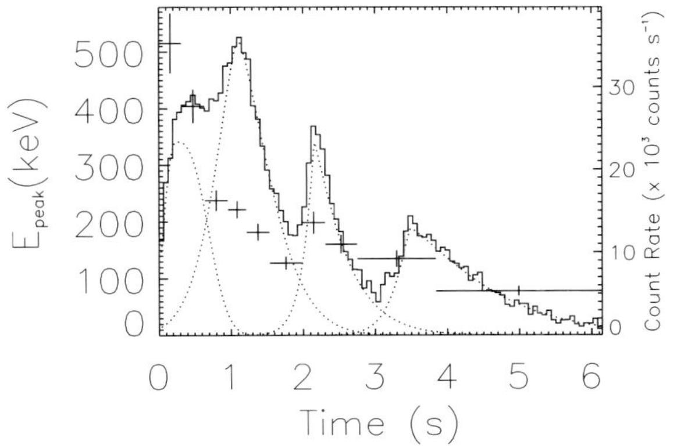

Fig. 9. The spectral evolution of BATSE trigger 543. A global hard-to-soft evolution as well as a hardness-intensity correlation in the decay phase of pulses are clearly identified (from Crider *et al.* 1999).

where $N(t)$ is the intensity (photon count number), $E_{\rm p}(t)$ is the peak energy at time t and

$$\Phi(t) = \int_0^t N(t')\mathrm{d}t'$$

is the photon fluence. The index δ is usually smaller than unity. When these two correlations are verified, it can be shown that the intensity and the peak energy follow Ryde & Svensson (2000):

$$N(t) = \frac{N(0)}{1+t/\tau}, \tag{2.5}$$

$$E_{\rm p}(t) = \frac{E_{\rm p}(0)}{(1+t/\tau)^{\delta}}, \tag{2.6}$$

where $\tau = \delta\Phi_0/N(0)$. This behaviour has been indeed observed in a sample of 25 long and bright GRBs verifying both the HIC and the HFC (see Fig. 10) Ryde & Svensson (2002). Such a behaviour is a severe constraint for all models of the GRB prompt emission.

2.1.5 Optical flash

There is only one detection of the prompt emission of a GRB outside the X- and γ-ray bands. GRB 990123 was a long ($\sim$150 s) and very bright GRB and was observed by a robotic telescope (ROTSE) which succeeded in measuring the optical magnitude of the burst at three times before the end of the gamma-ray emission. The resultat is amazing: GRB 990123 was

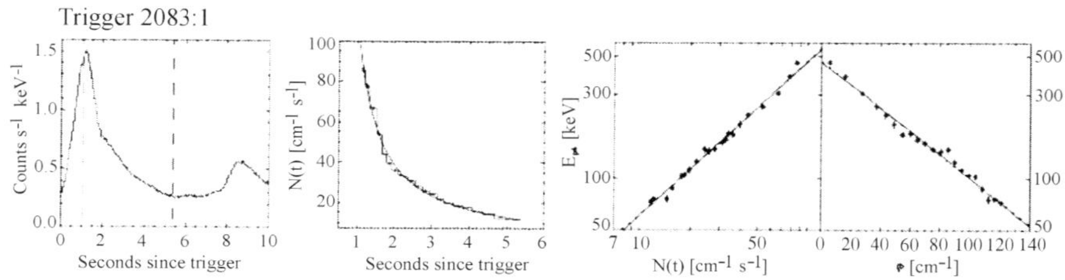

Fig. 10. The HFC and the HIC in the two pulses of BATSE trigger 2082 Ryde & Svensson (2002).

also very bright in optical, peaking at magnitude 8.9 Akerlof *et al.* (1999). As the measured redshift was $z = 1.6$ (see later), the same source located at 1 kpc would been as bright as the sun! Figure 11 shows the three ROTSE images as well as the time profile of GRB 990123.

2.2 Afterglow

2.2.1 Observational strategy

The typical BATSE GRB error box has a radius of a few degrees which makes very difficult the search for any counterpart at other wavelengths. The Beppo-SAX satellite was able to reduce this radius to a few arcminutes when a GRB is detected by the Wide Field Cameras (WFC) in the 2–28 keV band. Once per orbit, Beppo-SAX data were transmitted to the ground so that in case of detection, the GRB position can be distributed to ground-based observatories within a few hours. This observational strategy has allowed in 1997 the first detection of an optical counterpart to a GRB (GRB 970228). This first discovery has been followed by many others[2], in several spectral bands (X-ray, optical and radio bands). The Beppo-SAX mission has ended in April 2002 and GRB alerts are now provided by the HETE-2 satellite, which has been joined by INTEGRAL since October 2002. The SWIFT mission (to be lauched at the end of 2003) will increase the rate of GRB localizations up to about 150 per year.

These couterparts are observed after the end of the prompt gamma-ray emission (typically a few hours after the GRB in the case of the Beppo-SAX detections) and their luminosity is decreasing down to the limit of detection (see Fig. 12). They are named "afterglows" to distinguish them from the prompt burst emission. An important issue that should be solved in

[2]Jochen Greiner is maintaining a web page with a very useful list of all well located GRBs and their detected afterglows: `http://www.mpe.mpg.de/~jcg/grbgen.html`

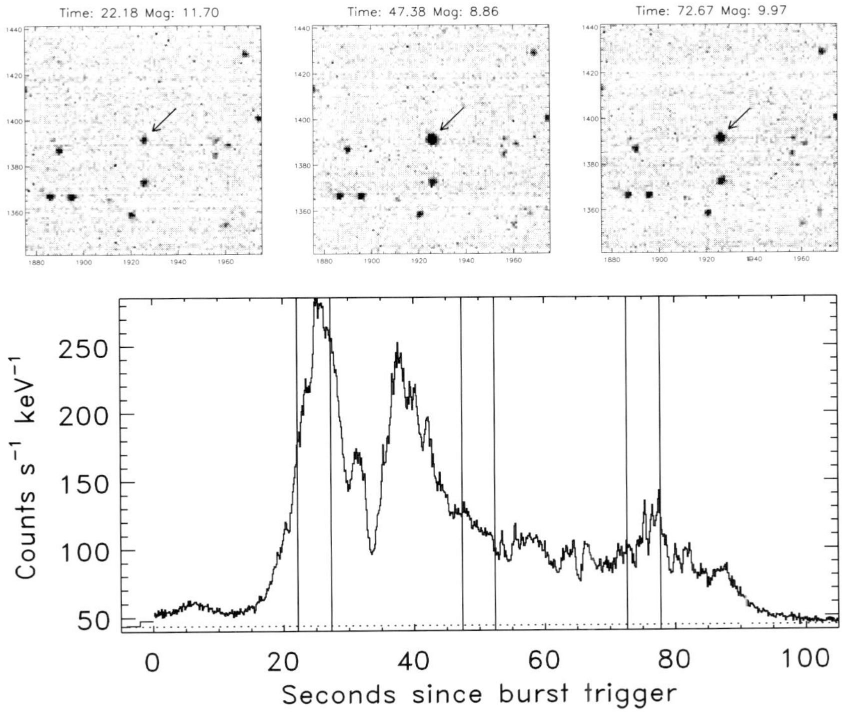

Fig. 11. GRB 990123: *upper panel:* the three ROTSE prompt optical images Akerlof *et al.* (1999). *Lower panel:* BATSE time profile. The vertical lines correspond to the three optical observations.

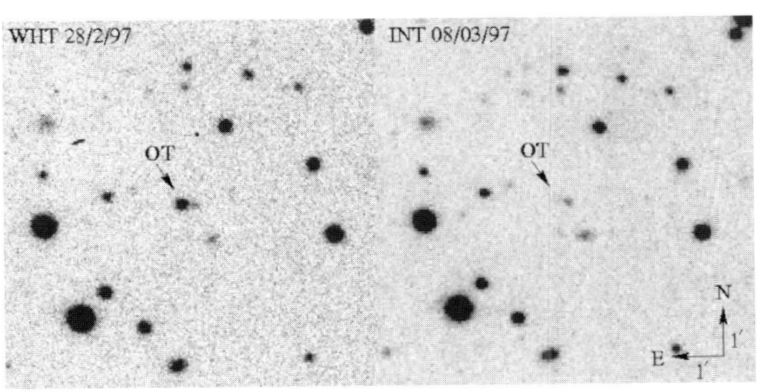

Fig. 12. GRB 970228 afterglow: the fading optical transient (OT) (from Groot *et al.* 1997).

the coming years thanks to HETE-2, INTEGRAL and SWIFT is the understanding of the transition period between the burst and its afterglow.

2.2.2 Lightcurve and spectrum

In a first approximation, the flux of the afterglow at a given frequency is decreasing as a power-law. For instance, Figure 13 shows the lightcurve of the afterglow of GRB 970228 in the R band. In most cases, the lightcurve cannot be represented by a unique power-law but is better discribed as a succession of power-laws with different slopes, separated by one or several breaks (see Fig. 14). The spectrum is also made of several power-laws with evolving break frequencies (see for instance the spectrum of the afterglow of GRB 970507 after 12 days in Fig. 15). Then, on a large time interval and over a broad energy range, the flux of a GRB afterglow is

$$F_\nu(t) \propto \nu^{-\alpha} \times t^{-\beta}, \tag{2.7}$$

α and β depending on the spectral range and the period of observation. Typically, the observed β is in the interval 0.6–2.2. We will see in Section 4 that this behaviour is very well understood as the synchrotron emission of relativistic electrons accelerated in a strong ultra-relativistic shock, due to the deceleration of a relativistic shell by its environment.

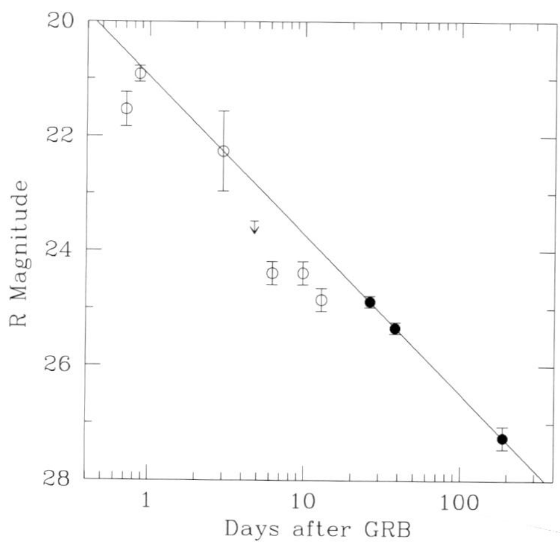

Fig. 13. GRB 970228 afterglow: evolution of the R band magnitude of the OT during 200 days Fruchter *et al.* (1999).

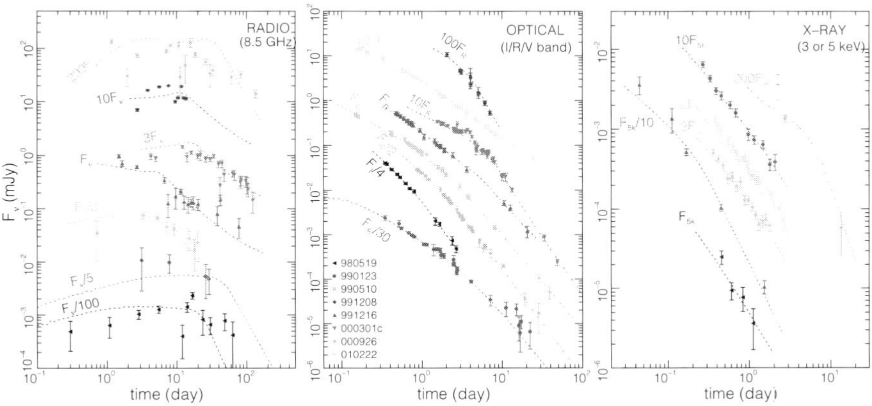

Fig. 14. Radio, optical and X-ray lightcurves of a few GRBs (from Panaitescu & Kumar 2001a).

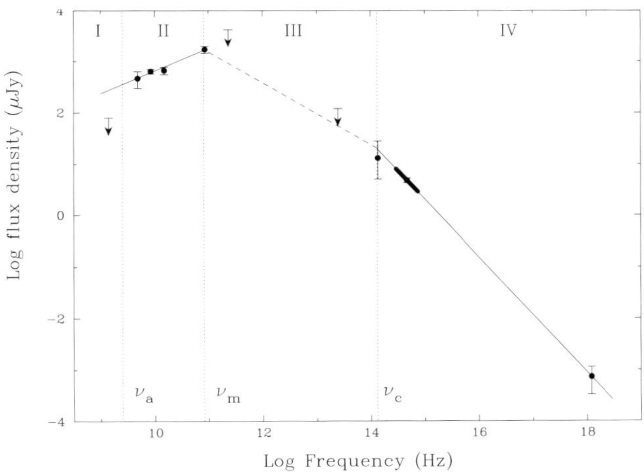

Fig. 15. Spectrum of the afterglow of GRB 970507 at 12 days (from Galama *et al.* 1998).

2.2.3 Redshift

The first important consequence of the detection of the optical afterglow is the possibility to take a good-quality spectrum of the object and find lines allowing the determination of the redshift of the source. The first redshift determination occured in the case of the afterglow of GRB 970508 thanks to the presence of several absorption lines in the spectrum of the optical

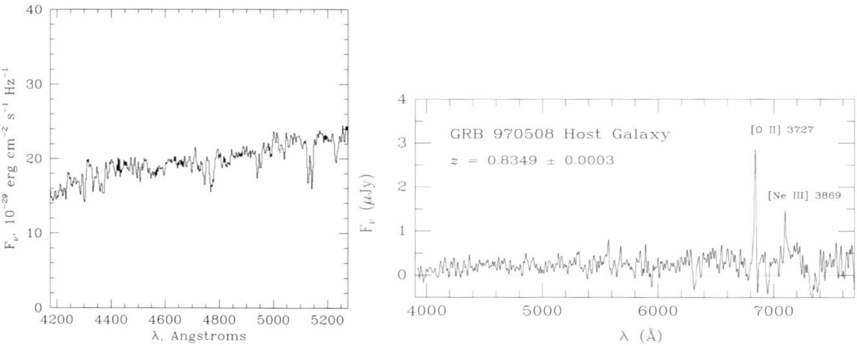

Fig. 16. Redshift of GRB 970508; *left*: spectrum of the optical transient, showing FeII and MgII absorption lines at $z = 0.835$ Metzger *et al.* (1997); *right*: spectrum of the host galaxy, showing [OII] and [NeIII] emission lines at $z = 0.835$ Bloom *et al.* (1998).

transient (see Fig. 16, left). It was shown that the most distant absorbing medium was located at $z = 0.835$ Metzger *et al.* (1997). Later, emission lines have been identified at the same redshift in the spectrum of the host galaxy (see Fig. 16, right), making this redshift determination very reliable Bloom *et al.* (1998). This is of course a very important result, providing the only direct proof of the cosmological origin of (at least long duration) GRBs! About 25 redshifts have now been obtained directly from the OT and/or from the host galaxy spectrum, with redshifts going from $z \sim 0.36$ (GRB 990712) to $z \sim 4.5$ (GRB 000131): see Table 1. An important consequence of this very distant origin is that the energy released during the GRB is huge: the equivalent isotropic energy computed from the distance (assuming "standard" cosmological parameters) and the observed gamma-ray fluence is

$$E_{\gamma,4\pi} = 4\pi D_{\mathrm{L}}^2(z)\ \frac{\mathcal{F}_\gamma}{1+z} \simeq 10^{51} \rightarrow 10^{54}\ \mathrm{erg}. \tag{2.8}$$

We will see in Section 4 that some of the breaks observed in the afterglow lightcurves are good evidences in favor of a beamed emitting ejecta. Then the values of equation (2.8) have to be multiplied by a factor $\Omega/4\pi$, where Ω is the opening angle of the emitting ejecta. The method for the determination of $\Omega/4\pi$ from the observed break time will be explained in Section 4. The result will be to decrease the total gamma-ray energy to about 10^{51} erg, with a large uncertainty. This is still equivalent to the total kinetic energy of a supernova radiated only in the 20 keV–20 MeV energy range!

Table 1. The redshift of GRB afterglows detected from 1997 to 2002 (measured from the OT and/or host galaxy spectrum).

1997	GRB	970228	970508	970828	971214		
	z	0.695	0.835	0.958	3.42		
1998	GRB	980613	980703				
	z	1.096	0.966				
1999	GRB	990123	990510	990705	990712	991208	991216
	z	1.60	1.619	0.86	0.434	0.706	1.02
2000	GRB	000131	000301C	000418	000911	000926	
	z	4.5	2.03	1.118	1.058	2.066	
2001	GRB	010222	010921	011121	011211		
	z	1.477	0.45	0.36	2.14		
2002	GRB	020405	020813	021004	021211		
	z	0.69	1.25	2.3	1.01		

2.2.4 Host galaxy

The study of the host galaxies of GRB afterglows is a very interesting emerging topic, as it is a very peculiar sample of distant galaxies, selected with totally unusual methods. Their main properties have been summarized by Djorgoski *et al.* in a recent review Djorgovski *et al.* (2001): they are faint ($\langle R\rangle \sim 25$) galaxies at redshift $\langle z\rangle \sim 1$, have a large range of luminosities and morphologies, are broadly typical for the normal, evolving, actively star-forming galaxies at comparable z and R magnitude. There are some evidence of a somewhat elevated SFR per unit luminosity and some spectroscopic hints of massive star formation. The typical observed SFR in these galaxies is about a few $M_\odot$/yr but a high fraction of the total star formation may be obscured by dust.

An important question regarding the identification of the GRB progenitors is the location of the GRB inside its host galaxy. We will see in Section 6 that the two main families of models are (i) the merger of two compact objects and (ii) the collapse of very massive stars (collapsars). The second case leads to GRBs located in the star-forming region of the host galaxy whereas the first case should produce GRBs distributed at larger distances from the central region of the galaxy, due to a long delay between the binary system formation and the merger. As can be seen in Figure 17, the study of the location of the optical transient in its host galaxy seems to favor an association with central star-forming region Bloom *et al.* (2002) and then to favor the collapsar model. Of course, this concerns only the long-duration

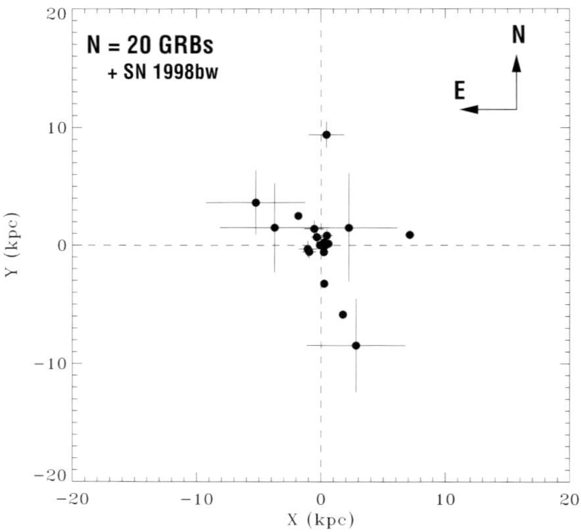

Fig. 17. Position of the optical transient inside its host galaxy (from Bloom *et al.* 2002).

bursts, as no afterglow has ever been detected in association with a short GRB.

2.2.5 Other aspects

We list here a few observational facts that have not been developped elsewhere in this section.

The supernova connection. In several classes of models, GRBs are associated with a supernova, exploding either before or simultaneously to the GRB. There are a few observationnal evidences in favor of such an association: (i) a type Ic supernova (SN 1998bw) has been found in the error box of GRB 980425 Galama *et al.* (1998). The analysis of the lightcurve and spectrum of this supernova shows that it is a very energetic one, with probably an ejected enveloppe which was initially midly relativistic. The main question regarding the SN1998bw–GRB 980425 association is that it leads to a very low redshift as SN1998bw is located at $z = 0.0085$. As GRB 980425 has gamma-ray properties which are absolutely normal (based on the properties of the more than 2500 BATSE GRBs), especially concerning the peak energy and the fluence, it implies that GRB 980425 is a very peculiar underluminous gammma-ray burst, with an equivalent isotropic energy many orders of magnitude below the usual values (10^{51}–10^{54} erg);

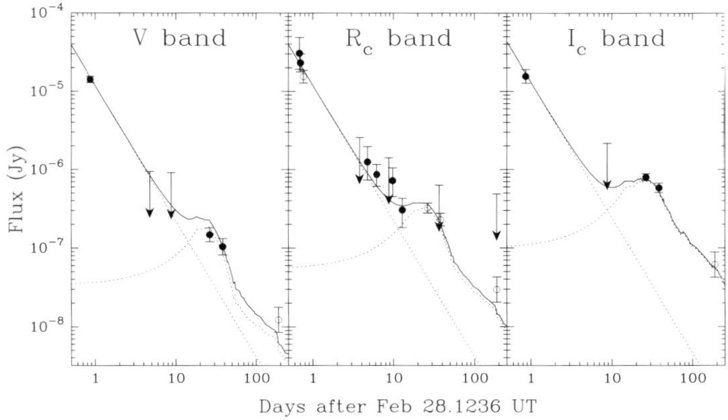

Fig. 18. Lightcurve of the optical afterglow of GRB 970228 (from Galama *et al.* 2000).

(ii) a SN lightcurve could be hiden behind the afterglow lightcurve and appear at late time as a small "bump". Such bumps have been observed in several cases (see Fig. 18) when the afterglow is very faint. However, it seems that there are other interpretations which are not related to a supernova but for instance to inhomogeneities in the external medium or to a late heating from decaying neutrons Beloborodov (2002).

Lines in the X-ray afterglow? Evidence for spectral lines have been found in a few X-ray afterglows. See for instance the Chandra spectrum of the X-ray afterglow of GRB 991216 in Figure 19 and the XMM-Newton spectrum of the X-ray afterglow of GRB 011211 in Figure 20. Should these lines be confirmed, they would provide strong constraints on GRB models, especially on the initial event responsible for the burst. They would indeed show that the close environment of the source is very metal rich, as for instance in a supernova remnant, enforcing the possible supernova-gamma ray burst connection.

The afterglow polarization. Polarization has been measured in a few optical afterglows, typically at one day after the burst, showing a linear polarization of a few percents Covino *et al.* (1999); Wijers *et al.* (1999). The interpretation of this polarization, which is evolving in time is not very simple but seems to be in agreement with the general picture (described in Sect. 4) of an afterglow produced by the synchrotron radiation of a beamed relativistic ejecta (see *e.g.* Sari & Piran 1999).

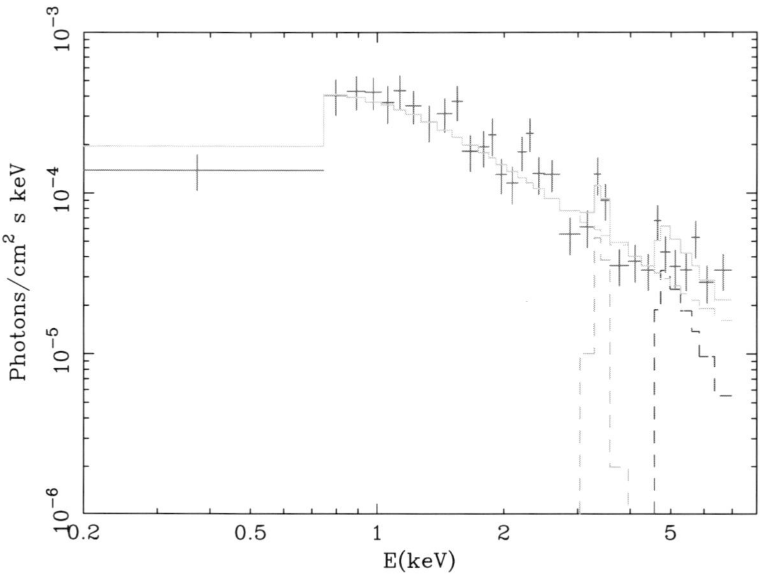

Fig. 19. Chandra observation of GRB 991216 (from Piro *et al.* 2000).

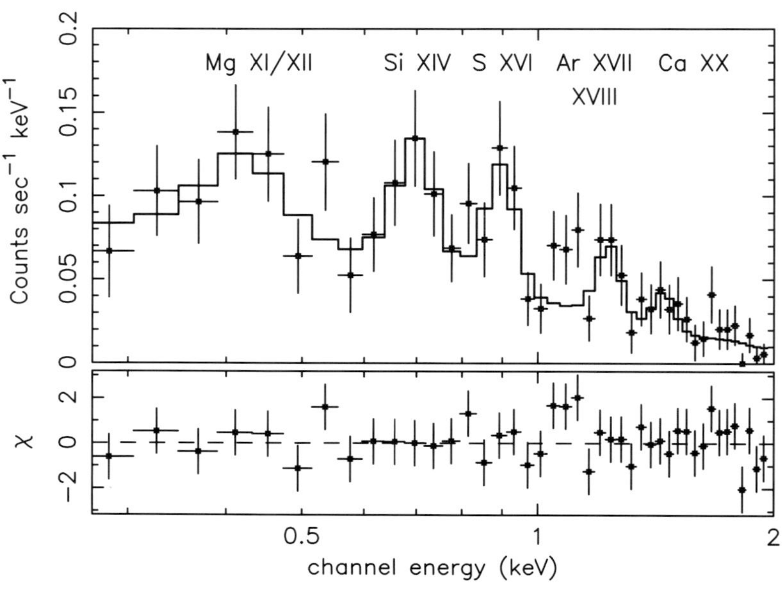

Fig. 20. XMM-Newton observation of GRB 011211 (from Reeves *et al.* 2002).

3 Theoretical scenario

In this section, we describe the key arguments leading to the most discussed theoretical scenario for cosmological GRBs.

3.1 Distance scale

In this lecture, we will only consider the case of GRBs produced at cosmological distances. We know from the redshift measurements in about 25 afterglows (see Sect. 2) that this is certainly the case for a large fraction of the long duration GRBs ($T_{90} \gtrsim 2$ s). The isotropic distribution over the sky of the GRBs detected by BATSE (Fig. 1) as well as the observed value of the $\langle V/V_{\max} \rangle$ indicator,

$$\left\langle \frac{V}{V_{\max}} \right\rangle \simeq 0.328 \pm 0.012 \text{ (first 601 BATSE GRBs)}, \tag{3.1}$$

clearly smaller than the 0.5 value corresponding to a homogeneous distribution in a Euclidian space, are clearly strong evidences that most GRBs if not all (short and long duration bursts) are indeed produced at large redshifts. Then all acceptable models of GRBs must consider extragalactic sources. An important issue is of course the distribution of GRBs with redshift, as it can put some constraints on their progenitors. It would in particular be very important to know if the rate of GRBs (or at least of a sub-group of GRBs, especially the long ones) is following the star formation rate (SFR). The number of redshift measurements ($\sim$25) is still too low to be able to construct directly this rate. The SWIFT mission will increase the number of afterglow detection and should greatly improve the present situation, where the information has to be extracted from other data. The most useful tool is the $\log N$–$\log P$ diagram, where the number N of GRBs with a peak flux larger that P is plotted as a function of P (see Fig. 21). The method consists in fitting this curve by assuming a GRB rate, a GRB luminosity function and the "standard" values for the cosmological parameters. The main problem is that it is impossible to discriminate between models as different as a GRB rate constant in time or proportional to the SFR. In Figure 21, we show an example of a fit obtained in the last case (the GRB rate follows the SFR), which is the situation expected in the collapsar model where the GRB progenitors are massive stars. The rate with is obtained is about one GRB for 10^6 type II supernovae Porciani & Madau (2001). This estimate of course assumes an isotropic emission. As the emission is more probably beamed with an opening angle Ω, this rate has to be multiplied by a factor $(\Omega/4\pi)^{-1}$ which could be of the order of 500–1000 according to recent estimates Frail *et al.* (2001).

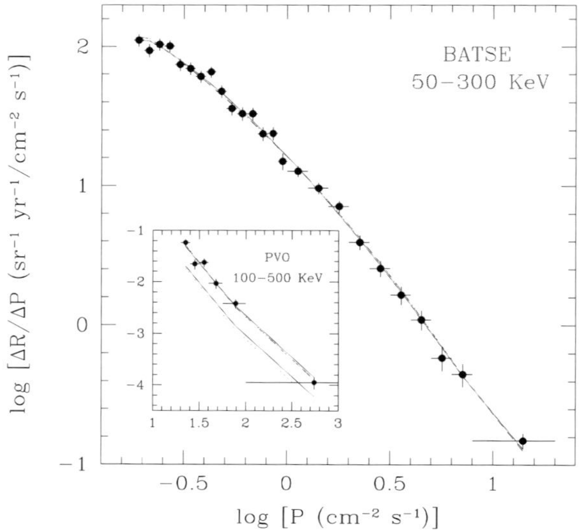

Fig. 21. The $\log N$–$\log P$ diagram of BATSE GRBs (some untriggered bursts being included Kommers *et al.* 2000) and the fit obtained assuming that the GRB rate follows the star formation rate Porciani & Madau (2001).

3.2 Compact source – relativistic motion

After the distance scale, the second key ingredient in every realistic model is to assume that the source of radiation is compact but relativistically moving. The basic argument is very simple: as we have seen in Section 2, the energy radiated in gamma-rays by a GRB is

$$E_\gamma \geq f_\Omega \; 10^{51} \text{ erg}, \tag{3.2}$$

where $f_\Omega = \Omega/4\pi$ and Ω is the opening angle of the emitting source. On the other hand, the rapid variability observed in the GRB time profiles down to timescales

$$t_{\rm var} \leq 10 \text{ ms} \tag{3.3}$$

implies that the source of radiation is compact with a typical size

$$R \leq c t_{\rm var} \leq 3000 \text{ km}. \tag{3.4}$$

It is then easy to estimate the corresponding opacity $\tau_{\gamma\gamma}$ to pair production ($\gamma\gamma \to e^+e^-$):

$$\tau_{\gamma\gamma} \simeq \frac{f_{\gamma\gamma} E_\gamma \sigma_{\rm T}}{4\pi R^2 m_{\rm e} c^2} \geq 7 \times 10^{14} \; f_\Omega \; f_{\gamma\gamma}, \tag{3.5}$$

where $f_{\gamma\gamma}$ is the fraction of the gamma-ray photons above the pair production threshold (see *e.g.* Piran 1999). As most GRB spectra have a large fraction of high-energy gamma-ray photons and as current estimates of f_Ω are of the order of 10^{-2}–10^{-3}, it is impossible to avoid very large values of the optical depth $\tau_{\gamma\gamma}$. Then the number of pairs produced in the source should increase very rapidly, resulting in a very large optical depth for all photons. However, the non-thermal nature of GRB spectra indicate that the sources are certainly optically thin. This so-called "compactness problem" can be easily solved if the emitting material is assumed to be moving relativistically. If the emitting matter moves with a large Lorentz factor $\Gamma = 1/\sqrt{1-v^2/c^2}$, then two effects will reduce dramatically the optically depth $\tau_{\gamma\gamma}$: (i) the photon energies in the comoving frame of the source are reduced by a factor $\sim\Gamma$, resulting in a much softer spectrum (typically peaking in the X-ray domain for $\Gamma \sim 100$) where the fraction $f_{\gamma\gamma}$ of high-energy photons above the pair production threshold is much lower. If the observed high energy spectral index is β ($\beta \sim -2.25$), $f_{\gamma\gamma}$ is divided by $\sim\Gamma^{-2\beta-2}$. (ii) The size of the emitting region allowed to produce variability on a time scale $t_{\rm var}$ is multiplied by $\sim \Gamma^2$ (see Sect. 5), leading to

$$R \lesssim \Gamma^2 c t_{\rm var} = \left(\frac{\Gamma}{100}\right)^2\ 3\times 10^7\ \mathrm{km}\ . \tag{3.6}$$

Then the new optical depth for pair production in the relativistically moving case is

$$\tau_{\gamma\gamma} \simeq \frac{\Gamma^{2\beta+2}\ f_{\gamma\gamma} E_\gamma \sigma_{\rm T}}{4\pi\Gamma^4 R^2 m_{\rm e} c^2} \geq 7\times 10^{14}\ f_\Omega\ f_{\gamma\gamma}\ \Gamma^{-(2-2\beta)}\ . \tag{3.7}$$

It is clear that for

$$\Gamma \gtrsim \left(7\times 10^{14}\ f_\Omega\ f_{\gamma\gamma}\right)^{\frac{1}{2-2\beta}} \simeq 200\ \left(f_\Omega f_{\gamma\gamma}\right)^{0.15}\ \text{for } \beta = -2.25, \tag{3.8}$$

the optical depth $\tau_{\gamma\gamma}$ is smaller than unity. Recently, precise determinations of the minimum Lorentz factor required to garanty an optically thin source have been obtained for a few GRBs for which the redshift z was known (allowing a precise estimate of E_γ) and for which a good-quality high-energy spectrum was available (allowing a precise estimate of $f_{\gamma\gamma}$). The authors have considered not only the pair production process but also the scattering of photons by the pair-created electrons and positrons and the scattering of photons by the ambient electrons (accompanying the baryons). They find Lorentz factors in the interval 24–420, most minimum values being above 100 Lithwick & Sari (2001). This indicates that GRBs are associated to the most relativistic flows known in the Universe (micro-quasars or blazars having Lorentz factors of a few tens only).

3.3 Fireballs

The last argument leads to the concept of fireball, which is used in the most discussed models of GRBs, even if this is not exactly with the same definition that was first introduced by Paczynski (1986) and Goodman (1986). Assume that a large amount of energy E is released in a small spherical volume of radius R_0. Whatever the dominant form of this energy (photons, electrons/positrons, magnetic energy, ...) is, the medium is initially completely optically thick (see previous subsection) and will first follow a purely adiabatic spherical expansion in the ambient medium, during which part of the internal energy will be converted into kinetic energy. The special relativistic equations of hydrodynamics in one dimension and spherical symmetry are (with $c = 1$)

$$\frac{\partial D}{\partial t} + \frac{1}{R^2}\frac{\partial}{\partial R}\left(R^2 D v\right) = 0 \text{ (mass conservation)}, \tag{3.9}$$

$$\frac{\partial S}{\partial t} + \frac{1}{R^2}\frac{\partial}{\partial R}\left(R^2 S v\right) = -\frac{\partial P}{\partial R} \text{ (momentum conservation)}, \tag{3.10}$$

$$\frac{\partial \tau}{\partial t} + \frac{1}{R^2}\left(R^2(\tau + P)v\right) = 0 \text{ (energy conservation)}, \tag{3.11}$$

where t and R are the time and the radius in a fixed frame linked to the central source, D, S and τ are the mass, momentum and energy density (the rest-mass energy being substracted) in the frame frame. These quantities are related to the velocity v, the Lorentz factor $\Gamma = 1/\sqrt{1-v^2}$ and comoving quantities by

$$D = \rho\Gamma, \tag{3.12}$$

$$S = \rho h \Gamma^2 v, \tag{3.13}$$

$$\tau = \rho h \Gamma^2 - P - D, \tag{3.14}$$

where ρ is the mass density, P is the pressure, $h = 1 + \epsilon + P/\rho$ the specific enthalpy and ϵ the specific energy in the comoving frame. If the medium behaves as a perfect fluid of adiabatic index γ, we have

$$\epsilon = \frac{1}{\gamma - 1}\frac{P}{\rho}\cdot \tag{3.15}$$

In this case, the last equation (energy conservation) can be rewritten as

$$\frac{\partial}{\partial t}\left(P^{\frac{1}{\gamma}}\Gamma\right) + \frac{1}{R^2}\left(R^2 P^{\frac{1}{\gamma}}\Gamma v\right) = 0\cdot \tag{3.16}$$

In our context, it is very useful to use a new set of coordinates, where t is replaced by the "arrival time" $t_{\mathrm{a}} = t - R/c$, which is the time at which photons emitted at time t and radius R are detected by an observer located at distance D from the source, fixing the origin of time at the arrival time of photons emitted by the source ($R=0$) at $t=0$. The equations become

$$\frac{1}{R^2}\frac{\partial}{\partial R}\left(R^2\rho\Gamma v\right) = -\frac{\partial}{\partial t_{\rm a}}\left(\rho\Gamma(1-v)\right), \quad (3.17)$$

$$\frac{1}{R^2}\frac{\partial}{\partial R}\left(R^2\left(\rho h\Gamma^2 v^2\right)\right) = -\frac{\partial}{\partial t_{\rm a}}\left(\rho h\Gamma^2 v(1-v)\right) + \frac{\partial P}{\partial t_{\rm a}} - \frac{\partial P}{\partial R}, \quad (3.18)$$

$$\frac{1}{R^2}\frac{\partial}{\partial R}\left(R^2 P^{\frac{1}{\gamma}}\Gamma v\right) = -\frac{\partial}{\partial t_{\rm a}}\left(P^{\frac{1}{\gamma}}\Gamma(1-v)\right)\cdot \quad (3.19)$$

If $\Gamma \gg 1$, the velocity becomes very close to the speed of light, with

$$1 - v \simeq \frac{1}{2\Gamma^2}\cdot \quad (3.20)$$

It is then easy to show that all terms in the right hand side of the three equations above are negligible for small radii compared to the left hand side terms. This leads to three conservation equations:

$$R^2\rho\Gamma \simeq \text{constant}, \quad (3.21)$$

$$R^2\rho h\Gamma^2 \simeq \text{constant}, \quad (3.22)$$

$$R^2 P^{\frac{1}{\gamma}}\Gamma \simeq \text{constant}. \quad (3.23)$$

These equations are valid for each shells of the relativistic ejecta, that evolve independently in this regime, as these equations are now only R dependent. This approximation breaks at the spreading radius that is defined in equation (3.35) below. A general solution to this system of three equations is (see *e.g.* Piran 1999):

$$\left(\frac{R}{R_0}; \frac{\rho}{\rho_0}; \frac{P}{P_0}\right) = f\left[\gamma; \frac{\Gamma_\infty}{\Gamma_0}; \quad \frac{\Gamma}{\Gamma_0}\right], \quad (3.24)$$

with

$$\frac{R}{R_0} = \left(\frac{\Gamma_\infty}{\Gamma_0} - 1\right)^{\frac{1}{2(\gamma-1)}} \left(\frac{\Gamma_\infty}{\Gamma} - 1\right)^{-\frac{1}{2(\gamma-1)}} \left(\frac{\Gamma}{\Gamma_0}\right)^{-1/2}, \quad (3.25)$$

$$\frac{\rho}{\rho_0} = \left(\frac{\Gamma_\infty}{\Gamma_0} - 1\right)^{-\frac{1}{\gamma-1}} \left(\frac{\Gamma_\infty}{\Gamma} - 1\right)^{\frac{1}{\gamma-1}}, \quad (3.26)$$

$$\frac{P}{P_0} = \left(\frac{\Gamma_\infty}{\Gamma_0} - 1\right)^{-\frac{\gamma}{\gamma-1}} \left(\frac{\Gamma_\infty}{\Gamma} - 1\right)^{\frac{\gamma}{\gamma-1}}, \quad (3.27)$$

where R_0, ρ_0, P_0 and Γ_0 are the initial values of R, ρ, P and Γ when the relativistic expansion starts and Γ_∞ is given by

$$\Gamma_\infty = h_0\Gamma_0 \text{ with } h_0 = 1 + \frac{\gamma}{\gamma-1}\frac{P_0}{\rho_0}\cdot \quad (3.28)$$

The ejecta experiences successively two phases:

- **The radiation dominated phase:** the internal energy term dominates in the specific enthalpy so that (the adiabatic index γ is of course extremely close to 4/3 is this phase):

$$\frac{\Gamma}{\Gamma_0} \simeq \frac{R}{R_0}, \tag{3.29}$$

$$\frac{\rho}{\rho_0} \simeq \left(\frac{R}{R_0}\right)^{-3}, \tag{3.30}$$

$$\frac{P}{P_0} \simeq \left(\frac{R}{R_0}\right)^{-4}. \tag{3.31}$$

As $\Gamma \propto R$ and $P/\rho \propto R^{-1}$, this is clearly an acceleration phase where the internal energy is converted into kinetic energy.

- **The matter dominated phase:** the rest-mass energy term now dominates in the specific enthalpy so that

$$\Gamma \simeq \Gamma_\infty, \tag{3.32}$$

$$\frac{\rho}{\rho_0} \simeq \left(\frac{R}{R_0}\right)^{-2}, \tag{3.33}$$

$$\frac{P}{P_0} \simeq \left(\frac{R}{R_0}\right)^{-2\gamma}. \tag{3.34}$$

In this phase, the Lorentz factor has reached its terminal value Γ_∞ and is constant. It is possible to show that the width Δ of the ejecta remains constant up to the spreading radius

$$R_{\mathrm{spreading}} \simeq \frac{\Gamma_\infty^2}{\Gamma_0}\Delta \simeq 4.8\times 10^{16}\ \Gamma_0 \left(\frac{\Gamma_\infty/\Gamma_0}{400}\right)^2 \left(\frac{\Delta/c}{10\ \mathrm{s}}\right)\ \mathrm{cm}. \tag{3.35}$$

At the spreading radius, the approximation corresponding to equations (3.21)–(3.23) is not valid anymore and the complete hydrodynamics equations have to be considered.

The transition between the two phases occurs at the saturation radius R_{sat} where Γ reaches the terminal value Γ_∞, *i.e.*

$$R_{\mathrm{sat}} \simeq R_0 \frac{\Gamma_\infty}{\Gamma_0} \simeq 1.2\times 10^9 \left(\frac{\Gamma_\infty/\Gamma_0}{400}\right)\ \mathrm{cm}. \tag{3.36}$$

The complete solution for $\gamma = 4/3$ and $\Gamma_\infty/\Gamma_0 = 400$ is plotted in Figure 22.

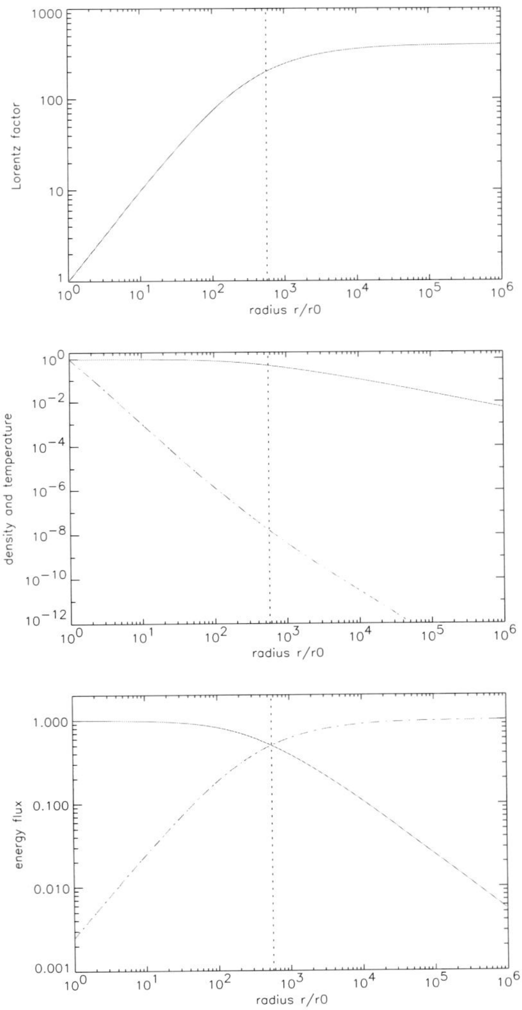

Fig. 22. The fireball adiabatic evolution for $\Gamma_\infty/\Gamma_0 = 400$. The dotted vertical line indicates the saturation radius $R_{\rm sat}$. First panel: evolution of the Lorentz factor. Second panel: evolution of the comoving density (dashed line) and the temperature in the observer frame (solid line). Third panel: evolution of the kinetic (dashed line) and internal (solid line) energy fluxes (the sum being constant).

The energy fluxes are given by

$$\dot{E} = \dot{E}_{\rm kin} + \dot{E}_{\rm int}, \tag{3.37}$$

$$\dot{E} = 4\pi R^2 \rho h \Gamma^2 c = \Gamma_\infty \dot{M} c^2 = \text{constant (total energy flux)}, \tag{3.38}$$

$$\dot{M} = 4\pi R^2 \rho \Gamma c = \text{constant (mass flux)}. \tag{3.39}$$

The kinetic and internal energy fluxes $\dot{E}_{\rm kin}$ and $\dot{E}_{\rm int}$ plotted in Figure 22 are given by

$$\dot{E}_{\rm kin} = \Gamma \dot{M} c^2 = \frac{\Gamma}{\Gamma_\infty} \dot{E}, \tag{3.40}$$

$$\dot{E}_{\rm int} = (\Gamma_\infty - \Gamma)\, \dot{M} c^2 = \left(1 - \frac{\Gamma}{\Gamma_\infty}\right) \dot{E}. \tag{3.41}$$

The temperature in the observer frame is $T = \Gamma T'$ where the comoving temperature is $T' = (3P/a)^{1/4}$. Then

$$\frac{T}{T_0} = \left(\frac{\Gamma_\infty}{\Gamma_0} - 1\right)^{-1} \left(\frac{\Gamma_\infty}{\Gamma} - 1\right) \left(\frac{\Gamma}{\Gamma_0}\right), \tag{3.42}$$

$$\simeq 1 \text{ (radiation dominated phase)}, \tag{3.43}$$

$$\simeq \left(\frac{R}{R_0}\right)^{-1} \text{ (matter dominated phase)}. \tag{3.44}$$

The initial temperature of the fireball is related to the released energy by

$$kT_0 \simeq 1.1\ \Gamma_0^{1/2} \left(1 - \frac{\Gamma_0}{\Gamma_\infty}\right)^{1/4} \left(\frac{\dot{E}}{10^{51}\ \text{erg/s}}\right)^{1/4} \left(\frac{R_0}{30\ \text{km}}\right)^{-1/2} \text{MeV}. \tag{3.45}$$

It can be shown that the optical depth for pair creation falls below unity when the comoving temperature T' drops below $T'_{\rm pairs} \sim 20$ keV. This uusally occurs in the radiation dominated phase, at radius (see *e.g.* Piran 1999)

$$R_{\rm pairs} \simeq 1.7 \times 10^8\ \Gamma_0^{-1/2} \left(\frac{\dot{E}}{10^{51}\ \text{erg/s}}\right)^{1/4} \left(\frac{R_0}{30\ \text{km}}\right)^{1/2} \left(\frac{T'_{\rm pairs}}{20\ \text{keV}}\right)^{-1} \text{cm}. \tag{3.46}$$

In the initial version of the fireball model, the ejecta was assumed to be purely leptonic and all the energy was radiated at this stage to produce the GRB. The predicted spectrum was of course thermal, which is in contradiction with the observations. For this reason, it is now assumed that there is a small baryonic pollution in the initial stage of energy release. Then the fireball remains optically thick up to the beginning of the matter-dominated phase, mainly due to the scattering of photons by the electrons accompanying the baryons. These bayons of course dominate the mass flux $\dot{M}$. As the terminal Lorentz factor is given by $\Gamma_\infty/\Gamma_0 = \dot{E}/\dot{M}c^2$, the baryonic load must remain at a very low level ($\dot{M}c^2/\dot{E} < 10^{-2}$) to reach the Lorentz factors that are required ($\Gamma_\infty > 100$, see previous subsection). This is not easy to obtain and puts a severe constraint on models for the GRB central engine (see Sect. 6).

In the matter-dominated phase ($\Gamma \simeq \Gamma_\infty$ and $\rho \propto R^{-2}$), photons emitted at radius R inside the relativistic ejecta will escape at radius $R_{\rm esc}(R) \simeq R + 2\Gamma^2\Delta$. Then the Thomson optical depth at radius R is given by (see Daigne & Mochkovitch 2002a for a complete derivation):

$$\tau(R) \simeq \int_R^{R_{\rm esc}(R)} \frac{\sigma_{\rm T}\rho}{m_{\rm p}}\frac{{\rm d}r}{2\Gamma} \simeq \frac{\sigma_{\rm T}\rho_0 R_0 \Gamma_0}{2\Gamma_\infty^2 m_{\rm p}}\frac{R_0}{R}\frac{2\Gamma_\infty\Delta}{R+2\Gamma_\infty\Delta}. \quad (3.47)$$

For usual GRB parameters $R \ll 2\Gamma_\infty^2\Delta$ when the fireball becomes optically thin, and then the photospheric radius $R_{\rm ph}$ is given by

$$R_{\rm ph} \simeq 9.2\times 10^9\ \Gamma_0^{-3}\left(\frac{\dot{E}}{10^{51}\ {\rm erg/s}}\right)\left(\frac{\Gamma_\infty/\Gamma_0}{400}\right)^{-3}\ {\rm cm}. \quad (3.48)$$

As the acceleration is essentially completed well before this radius, the remaining internal energy that can be radiated at the photosphere is small. It can however produce a thermal precursor to the GRB which will be discussed in Section 5.

In conclusion, the baryonic loaded fireball provides a natural way to convert the initial energy realeased into a highly relativistic ejecta that becomes optically thin at large distance from the source, where new processes can operate to convert a fraction of the kinetic energy into radiation (with an optically thin spectrum) and produce the GRB and its afterglow.

3.4 A three-step scenario

All these considerations lead to a three-step theoretical scenario for GRBs:

1. Central engine: an initial event is responsible for a huge energy release. This energy is first injected into a highly relativistic ejecta. This step is probably the less understood today and will be discussed in Section 6.
2. Production of the prompt emission: at large distance from the source, the relativistic ejecta becomes transparent and can radiate part of its kinetic energy to produce the GRB. Several mechanisms have been proposed. We will discuss in Section 5 the most popular one: the "internal shock" model.
3. Production of the afterglow: at even larger radii, the deceleration of the relativistic ejecta by the surrounding medium becomes important. If the remaining kinetic energy after the second step is still high, a strong shock appears (the so-called "external shock") and propagates within the

Table 2. The three-step theoretical scenario for cosmological gamma-ray bursts.

Central engine (see Sect. 6)	R_0	10^6–10^7 cm	Initial energy release → baryonic loaded ejecta → becomes relativistic
	R_{sat}	10^8–10^{10} cm	End of the acceleration → $\Gamma \sim \Gamma_\infty \geq 100$
Prompt emission (see Sect. 5)	R_{ph}	10^{10}–10^{12} cm	Transparency → thermal precursor?
	R_{GRB}		Kinetic energy → radiation → prompt GRB
Afterglow (see Sect. 4)	R_{dec}	10^{13}–10^{18} cm	Deceleration starts → "external shock" → afterglow

external medium. The electrons accelerated behind this strong shock radiate and produce the observed afterglow. This will be described in Section 4.

This three-step scenario apply to most GRB models, which usually have in common the last step (which is the most constrained by the observations) and differ in the second or in the first step. Notice that the deceleration radius at which the third step starts can be easily estimated (see Sect. 4). For a uniform medium of density n it is given by

$$R_{\mathrm{dec}} \simeq 4.6 \times 10^{16} \left(\frac{E}{10^{53}\ \mathrm{erg}}\right)^{1/3} \left(\frac{n}{1\ \mathrm{cm}^{-3}}\right)^{-1/3} \left(\frac{\Gamma_\infty}{400}\right)^{-2/3}\ \mathrm{cm}. \quad (3.49)$$

Table 2 presents a summary of this scenario with all important radii.

4 Afterglow

In this section, we describe the production of the afterglow. We assume that a source has produced a highly relativistic ejecta and that after the gamma-ray burst has been produced, the mean Lorentz factor is Γ_0 and the equivalent isotropic kinetic energy is E_0.

4.1 The external shock

At large radii, when the swept-up mass becomes large, the relativistic ejecta is decelerated by the external medium. Two shocks are formed: (i) a forward shock (the so-called "external shock"), which propagates within the external medium and (ii) a reverse shock which propagates backwards in the relativistic ejecta. The model where the forward shock is responsible for the afterglow works very well Meszaros & Rees (1997) and will be detailed below. The emission associated with the reverse shock is not so well understood and is still largely debated. It will be discussed in Section 5.

A key ingredient of afterglow models is the assumption about the density profile of the external medium. In particular, mergers of compact object should occur in low density media as they are located far from the central regions of the galaxy, whereas collapsars should have a high density environment (they occur in star-forming regions), possibly dominated by the dense wind of the massive star in its late stages. Therefore the density profile is assumed to be

$$\rho(r) = \frac{A}{r^s}, \tag{4.1}$$

with $s = 0$ (uniform medium) and $A = nm_\mathrm{p}$, where the density n is of order 10^{-3}–$10\ \mathrm{cm}^{-3}$ for mergers and 1–$10^3\ \mathrm{cm}^{-3}$ for collapsars. In the last case, if a dense wind is present, parameters become $s = 2$ and $A = \dot{m}/4\pi v_\infty$ where typical values for a Wolf-Rayet star are a mass loss rate $\dot{m} \simeq 3\times 10^{-5}\ M_\odot/\mathrm{yr}$ and a terminal velocity $v_\infty \simeq 2000\ \mathrm{kms}^{-1}$, so that $A \simeq 7.6 \times 10^{11}\ \mathrm{g\,cm}^{-1}$. With such a density profile, the swept-up mass when the relativistic ejecta has reached radius R is

$$M_\mathrm{ext}(R) = \int_0^R 4\pi r^2 \rho(r)\mathrm{d}r = \frac{4\pi}{3-s} A R^{3-s}. \tag{4.2}$$

The dynamics of the external shock is described by the self-similar solution of Blandford & McKee (1976), which is the relativistic counterpart of the well-known self-similar Sedov solution that applies for supernova remnants. We describe here the main properties of the solution. Assume that the relativistic ejecta initially has a Lorentz factor Γ_0 and a mass $M_0 = E_0/\Gamma_0 c^2$. The deceleration starts when the swept-up mass becomes of the order of M_0/Γ_0. This yields a simple estimate of the deceleration radius:

$$R_\mathrm{dec} \simeq \left(\frac{3-s}{4\pi}\frac{M_0}{A\Gamma_0}\right)^{\frac{1}{3-s}}, \tag{4.3}$$

corresponding for a uniform medium ($s = 0$) of density n to:

$$R_\mathrm{dec} \simeq 4.6\times 10^{16}\ \left(\frac{E}{10^{53}\ \mathrm{erg}}\right)^{1/3} \left(\frac{n}{1\ \mathrm{cm}^{-3}}\right)^{-1/3} \left(\frac{\Gamma}{400}\right)^{-2/3}\ \mathrm{cm} \tag{4.4}$$

and for a dense stellar wind ($s = 2$) to:

$$R_{\rm dec} \simeq 7.3 \times 10^{13} \left(\frac{E}{10^{53}\ {\rm erg}}\right) \left(\frac{A}{7.6 \times 10^{11}\ {\rm g\ cm^{-1}}}\right)^{-1} \left(\frac{\Gamma}{400}\right)^{-2}\ {\rm cm}. \tag{4.5}$$

In the ultra-relativistic limit of a strong shock, the comoving density ρ_*, the comoving specific internal energy density ϵ_* in the shocked external medium and the Lorentz factor $\Gamma_{\rm shock}$ of the forward shock are given as a function of the external density $\rho = \rho(R)$ and the Lorentz factor $\Gamma = \Gamma(R)$ of the ejecta at radius $R \geq R_{\rm dec}$ by:

$$\rho_* \simeq (4\Gamma + 3)\,\rho \simeq 4\Gamma\rho, \tag{4.6}$$

$$\epsilon_* \simeq (\Gamma - 1)\,c^2 \simeq \Gamma c^2, \tag{4.7}$$

$$\Gamma_{\rm shock} \simeq \sqrt{\frac{(\Gamma+1)(4\Gamma-1)^2}{(8\Gamma+10)}} \simeq \sqrt{2}\Gamma. \tag{4.8}$$

Equation (4.7) shows that initially (when $\Gamma \gtrsim 100$), the internal energy in the shocked material can be as high as ~100 GeV per proton! Energy conservation combined with these shock conditions leads to:

$$\Gamma_0 M_0 c^2 + M_{\rm ext}(R)c^2 \simeq \Gamma(R) M_0 c^2 + \Gamma^2(R) M_{\rm ext}(R) c^2. \tag{4.9}$$

With the expression of $M_{\rm ext}(R)$ given in equation (4.2), we get the following approximation of the Blandford-McKee solution (see *e.g.* Panaitescu & Kumar 2000):

$$\Gamma(x) \simeq \Gamma_0 \frac{\sqrt{1 + 4x^{3-s} + \left(2x^{3-s}/\Gamma_0\right)^2} - 1}{2x^{3-s}}\ \text{with}\ x = R/R_{\rm dec}. \tag{4.10}$$

We can distinguish three regimes: (i) $R \ll R_{\rm dec}$ ($x \ll 1$) and $\Gamma(R) \simeq \Gamma_0$ (deceleration has not started); (ii) $R_{\rm dec} \ll R \ll R_{\rm Sedov}$ ($1 \ll x \ll x_{\rm Sedov}$) and $\Gamma(R) \propto \Gamma_0 \left(R/R_{\rm dec}\right)^{-(3-s)/2}$ (deceleration phase); (iii) $R \gg R_{\rm Sedov}$ ($x \gg x_{\rm Sedov}$) and $\Gamma(R) \simeq 1$ and $v(R) \propto R^{-(3-s)/2}$ (non-relativistic Sedov phase). The transition to the Sedov phase occur for

$$\frac{R_{\rm Sedov}}{R_{\rm dec}} = x_{\rm Sedov} \simeq \left(\frac{\Gamma_0^2}{3}\right)^{\frac{1}{3-s}}. \tag{4.11}$$

This solution has been plotted in Figure 23 for $s = 0$ and $n = 1$ cm^{-3} (left) and $s = 2$, $\dot{m} = 3\ 10^{-5}\ M_\odot$/yr and $v_\infty = 2000$ kms^{-1} (right). The relativistic ejecta parameters are $\Gamma_0 = 400$, $E_0 = 10^{53}$ erg and $M_0 = E_0/\Gamma_0 c^2 = 1.4 \times 10^{-4}\ M_\odot$ in both cases. It is clear that deceleration starts

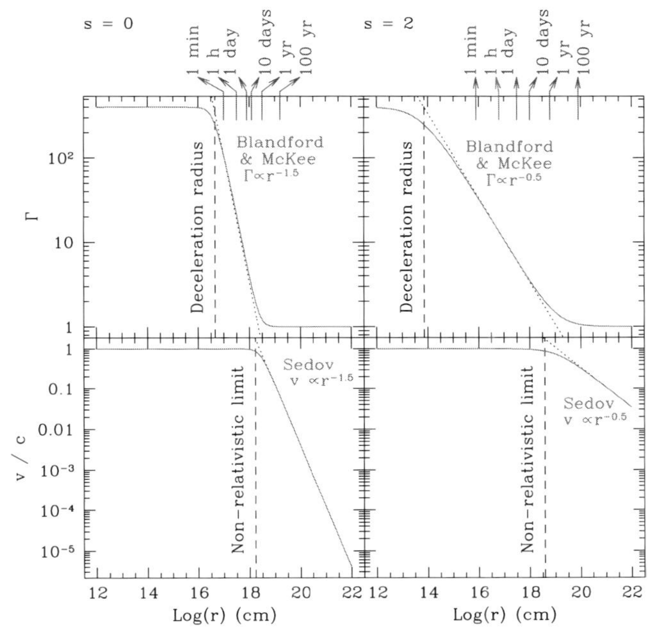

Fig. 23. Lorentz factor and velocity of the external shock as a function of radius for a uniform medium (left) and a non uniform medium with $s = 2$ (right).

much earlier in the dense wind case. But due to the $\rho \propto r^{-2}$ density profile, the efficiency of the deceleration decreases rapidly and the transition to the non relativistic phase ($\Gamma \sim 1$) occurs approximatively at the same radius in both cases. This shows that only early observations of the afterglow can be used to distinguish between different possible environments. The arrival time t_{a} of afterglow photons emitted when the shock is at radius R has been indicated on the same figure. It is given by

$$t_{\mathrm{a}} = t - \frac{R}{c} = \frac{1}{c}\left(\int_0^R \frac{\mathrm{d}r}{v/c} - R\right)\cdot \qquad (4.12)$$

4.2 *Radiative processes*

Physical conditions in the shocked external medium: once the dynamics of the forward shock has been specified, one needs to estimate the distribution of the electrons as well as the magnetic field in the shocked external medium. We assume that a fraction α_{B} of the internal energy density

$\rho_* \epsilon_*$ created behind the shock wave is converted into magnetic energy:

$$\frac{B^2}{8\pi} \simeq \alpha_{\rm B} \, \frac{4\Gamma A}{R^s} \, \Gamma c^2 \tag{4.13}$$

(from Eqs. (4.1), (4.6) and (4.7)), so that the local field is amplified up to:

$$B \simeq 4\Gamma c \left(2\pi\alpha_{\rm B}\frac{A}{R^s}\right)^{1/2} \simeq 8\left(\frac{\alpha_{\rm B}}{0.01}\right)^{1/2}\left(\frac{\Gamma}{100}\right)\left(\frac{n}{1\ {\rm cm}^{-3}}\right)^{1/2}\ {\rm G}. \tag{4.14}$$

We also assume that a fraction $\alpha_{\rm e}$ of the internal energy is injected in the electrons:

$$n_{\rm e}\bar{\Gamma}_{\rm e} m_{\rm e} c^2 \simeq \alpha_{\rm e} \, \frac{4\Gamma A}{R^s} \, \Gamma c^2, \tag{4.15}$$

where the total electron density $n_{\rm e}$ is given by $n_{\rm e} = \rho_*/m_{\rm p} \simeq 4\Gamma A/(m_{\rm p} R^s)$. Therefore the electrons get a mean Lorentz factor:

$$\bar{\Gamma}_{\rm e} \simeq \alpha_{\rm e}\frac{m_{\rm p}}{m_{\rm e}}\Gamma \simeq 1.8\times 10^4 \, \left(\frac{\alpha_{\rm e}}{0.1}\right)\left(\frac{\Gamma}{100}\right). \tag{4.16}$$

We assume that the electron distribution just behind the shock follows a power-law starting at Lorentz factor $\Gamma_{\rm i}$:

$$n(\Gamma_{\rm e}) \propto \Gamma_{\rm e}^{-p} \ \text{for}\ \Gamma_{\rm e} \geq \Gamma_{\rm i}, \tag{4.17}$$

$$\text{with}\ \Gamma_{\rm i} = \frac{p-2}{p-1}\bar{\Gamma}_{\rm e}. \tag{4.18}$$

Of course these estimates should be confirmed by a more detailed description of the physical conditions in the shocked medium, including the microphysics leading to the magnetic field amplification and the electrons acceleration.

Synchrotron radiation: the electrons moving in the magnetic field radiate by the synchrotron process. The synchrotron power of an electron with Lorentz factor $\Gamma_{\rm e}$ is given by Rybicki & Lightman (1979):

$$P_{\rm syn}\left(\Gamma_{\rm e}\right) = \frac{4}{3}\sigma_{\rm T} c \, \frac{B^2}{8\pi} \, \Gamma_{\rm e}^2 \tag{4.19}$$

and the typical energy of the emitted photons is (in the observer frame)

$$h\nu_{\rm syn}\left(\Gamma_{\rm e}\right) = \Gamma \, \frac{3he}{2\pi m_{\rm e} c} B\Gamma_{\rm e}^2 \tag{4.20}$$

$$\simeq 9.0\left(\frac{\alpha_{\rm B}}{0.01}\right)^{1/2}\left(\frac{\alpha_{\rm e}}{0.1}\right)^2\left(\frac{\Gamma}{100}\right)^4\left(\frac{n}{1\ {\rm cm}^{-3}}\right)^{1/2}\ {\rm keV}. \tag{4.21}$$

Then the radiative timescale, defined as the time necessary for an electron to lose all its energy by radiation, is (in the comoving frame):

$$t'_{\mathrm{syn}}\left(\Gamma_{\mathrm{e}}\right) = \frac{\Gamma_{\mathrm{e}} m_{\mathrm{e}} c^2}{P_{\mathrm{syn}}\left(\Gamma_{\mathrm{e}}\right)} \simeq 770 \left(\frac{B}{10\ \mathrm{G}}\right)^{-2} \left(\frac{\Gamma_{\mathrm{e}}}{10^4}\right)^{-1}\ \mathrm{s}. \tag{4.22}$$

This timescale has to be compared with the adiabatic cooling time, defined as the time necessary for the shocked medium to cool adiabatically due to spherical expansion:

$$t'_{\mathrm{ex}} \simeq \frac{R}{\Gamma c} \simeq 3.3\times 10^4 \left(\frac{R}{10^{17}\ \mathrm{cm}}\right) \left(\frac{\Gamma}{100}\right)^{-1}\ \mathrm{s}. \tag{4.23}$$

As pointed by Sari *et al.* (1998), there are two possible regimes: (i) initially, the radiative timescale is very short and all electrons radiate efficiently ("fast cooling" regime); (ii) once the magnetic field and the electron energy have decreased, the radiative timescale of most of the electrons becomes larger than the adiabatic cooling time and only the highest energy electrons (which have the shortest radiative timescale) can still radiate efficiently ("slow cooling" regime). A given electron will radiate efficiently as long as $t'_{\mathrm{syn}} \ll t'_{\mathrm{ex}}$, which gives the condition:

$$\Gamma_{\mathrm{e}} \gg \Gamma_{\mathrm{c}} = \frac{6\pi m_{\mathrm{e}} c}{\sigma_{\mathrm{T}} B^2 t'_{\mathrm{ex}}}. \tag{4.24}$$

The "fast cooling" regime corresponds to $\Gamma_{\mathrm{i}} \gg \Gamma_{\mathrm{c}}$ – all electrons can radiate their total energy in less than the adiabatic cooling time – and the "slow cooling" regime corresponds to $\Gamma_{\mathrm{i}} \ll \Gamma_{\mathrm{c}}$ – only high-energy electrons with $\Gamma_{\mathrm{e}} > \Gamma_{\mathrm{c}}$ can radiate effiently; most of the energy, which is contained in low-energy electrons, is not radiated. These two regimes lead to two different spectral shapes, described in Figure 24. The resulting observed flux at frequency ν follows

$$f_\nu \propto \nu^\alpha t^\beta, \tag{4.25}$$

which is in good agreement with the observations (see Sect. 2). The slopes α and β are given in Figure 24.

4.3 Additional effects

Geometric beaming: the basic description which has been made in the last two subsections needs a few improvements to become more realistic. The most important effect is due to the probable non-spherical symmetry of the relativistic ejecta, as first pointed out by Rhoads (1997). Figure 25 illustrates this effect: due to relativistic beaming, the observer can only

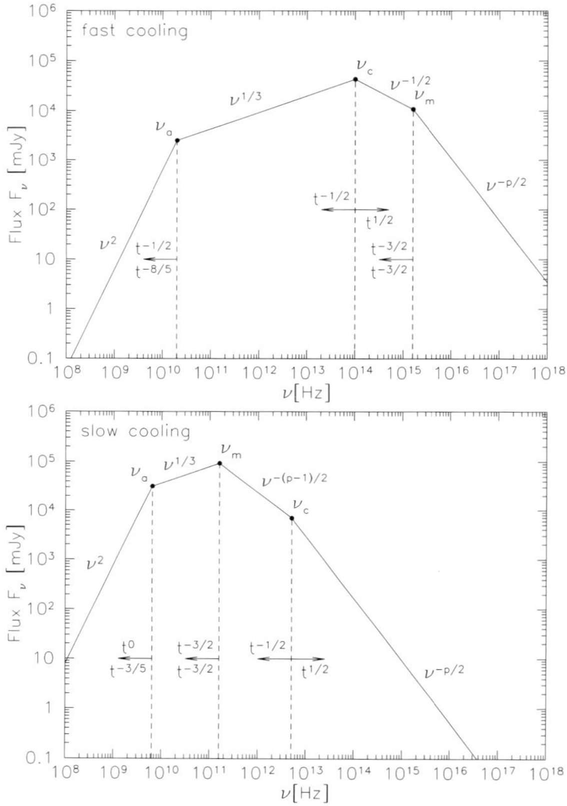

Fig. 24. The synchrotron spectrum. *Upper panel:* "fast cooling" regime; *lower panel:* "slow cooling" regime. The spectral breaks ν_m and ν_c corresponds respectively to the synchrotron frequency of electrons with Lorentz factor Γ_i and Γ_c (see text). The frequency ν_a corresponds to the break due to synchrotron self-absorption. The time dependance of these frequencies is given for a uniform medium $s = 0$ (above the arrows) and for a stellar wind $s = 2$ if different (bellow the arrows) (from Sari *et al.* 1998).

detect photons coming from a small fraction of the emitting surface, this region being centered on the line of sight and having an opening angle $\sim 1/\Gamma$. Initially, except for a highly beamed ejecta, this opening angle is small compared to the geometrical opening angle of the ejecta θ_0. Therefore, the observer cannot see any difference in this period between a spherical and a beamed emitting surface (the prompt GRB is entirely produced during this period). However, the deceleration due to the external medium makes the

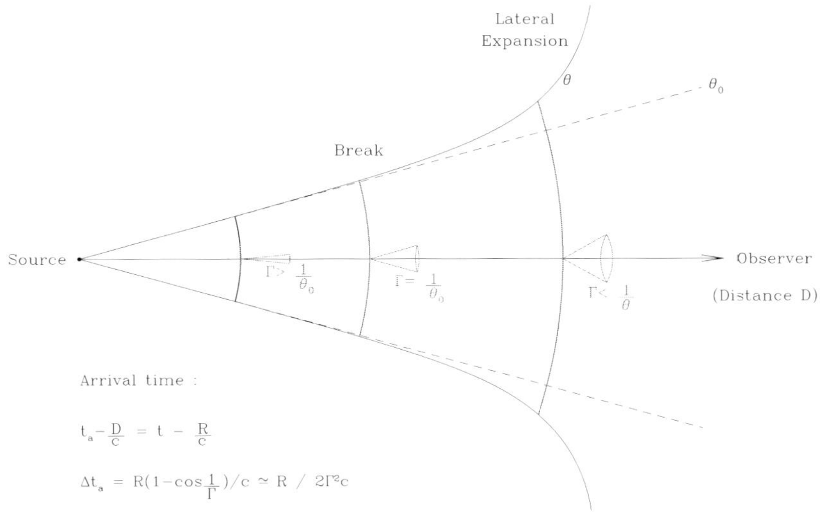

Fig. 25. Geometrical beaming *vs.* relativistic beaming.

Lorentz factor Γ decrease and then the angle $1/\Gamma$ increase. The observer can see the whole emitting surface for the first time in the history of the GRB at the time when $1/\Gamma \simeq \theta_0$. This should appear as a break in the lightcurve since a fraction of the photons emitted on the edge of the surface are now not suffiently beamed to be detected. An additional physical effect has to be taken into account and makes this picture a little more complex: the lateral expansion of the ejecta. The geometrical opening angle of the relativistic ejecta is not kept constant to θ_0 during the whole evolution. However, it can be shown that this lateral expansion becomes important more or less at the same period when when $1/\Gamma \sim \theta_0$.

The observations indeed show breaks in the lightcurve that can be interpreted as resulting from this effect (one important signature which confirms that a given break is due to this geometrical effect is that it is expected to be achromatic, *i.e.* observed at the same time at all frequencies). If the break time is precisely identified, one can then try to estimate the corresponding Lorentz factor and the initial beaming angle of the relativistic ejecta $\theta_0 \simeq 1/\Gamma$, which in the case of a uniform medium ($s = 0$) is given by

$$\theta_0 \simeq 2^\circ \left(\frac{t_{\mathrm{a}}(\mathrm{break})}{0.8\,(1+z)\ \mathrm{days}}\right)^{3/8} \left(\frac{E}{10^{53}\ \mathrm{erg}}\right)^{-1/8} \left(\frac{n}{1\ \mathrm{cm}^{-3}}\right)^{-1/8} . \quad (4.26)$$

This allows to estimate the beaming fraction f_Ω defined in Section 2 and then the true energy radiated during the burst. This has been done for a few

Table 3. Sample of 17 afterglows studied by Frail *et al.* (2001): redshift, isotropic equivalent energy radiated in gamma-rays, break time, opening angle and true energy radiated in gamma-rays.

GRB	Redshift	$E_{\gamma,4\pi}$ (10^{51} erg)	$t_{\rm a}$(break) (days)	θ_0 (°)	E_γ (10^{51} erg)
970228	0.695	22.4	...	...	...
970508	0.835	5.46	25.	16.8	0.234
970828	0.958	220.	2.2	4.13	0.575
971214	3.418	211.	>2.5	>3.21	>0.333
980613	1.096	5.67	>3.1	>7.28	>0.045
980703	**0.966**	**60.1**	**7.5**	**7.73**	**0.544**
990123	1.600	1440	2.04	2.86	1.80
990506	1.30	854	...	...	...
990510	**1.619**	**176.**	**1.20**	**3.04**	**0.248**
990705	0.84	270.	~ 1	3.09	0.389
990712	0.433	5.27	>47.7	>23.5	>0.445
991208	0.706	147	<2.1	<4.53	<0.455
991216	1.02	535	1.2	2.92	0.695
000131	4.500	1160	<3.5	<2.69	<1.30
000301C	**2.034**	**46.4**	**5.5**	**6.02**	**0.256**
000418	**1.119**	**82.0**	**25.**	**11.3**	**1.60**
000926	2.037	297	1.45	2.92	0.379

GRBs listed in Table 3 Frail *et al.* (2001). The result seems to indicate that the large dispersion of the isotropic equivalent energy radiated in gamma-rays comes mainly from a large dispersion in the opening angle, the true energy radiated in gamma-rays being remarkably concentrated around 5×10^{50}–10^{51} erg. This study suffers of course many limitations, the main one being that the break is clearly identified in several spectral bands in only four cases (marked in bold face in Table 3). In all other cases, either the break is detected in only one band, which is not unambiguous, or even worse, only a limit on the break time could be derived, by explaining the observed slope of the lightcurve as a pre- or post-break slope.

Other effects: a realistic afterglow model has to take into account many other processes: effect of radiative losses on the shock dynamics, changes in the photon spectrum due to inverse compton scatterings, modification of the high-energy spectrum due to the maximum Lorentz factor of the electrons, etc. An exemple of a detailed calculation taking into account all these effects can be found in Panaitescu & Kumar (2001b). The result obtained

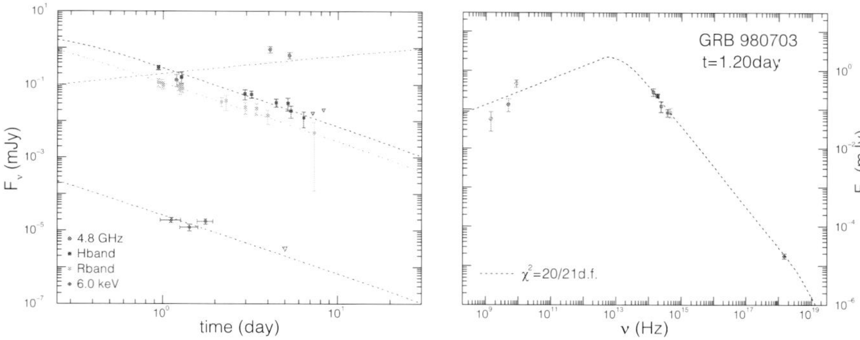

Fig. 26. GRB 980713: result of a detailed multiwavelength fit of the afterglow Panaitescu & Kumar (2001b). The best-fit parameters are $E = 2.9 \times 10^{54}$ erg, $\theta_0 \geq 2.7°$, $s = 0$, $n = 7.8 \times 10^{-4}$ cm^{-3}, $\alpha_{\mathrm{B}} = 4.6 \times 10^{-4}$, $\alpha_{\mathrm{e}} = 0.075$ and $p = 3.08$. The authors mention that a good fit with $s = 2$ is also possible.

in the case of GRB 980703 is plotted in Figure 26. Such multiwavelength fits could in principle be used to estimate all important parameters, such as the density of the external medium, the initial energy and Lorentz factor of the relativistic ejecta, etc. However, it seems that the available data can only poorly constrain most of these parameters. In particular, it is often difficult to distinguish between a $s = 0$ or a $s = 2$ external medium. This is mainly due to the lack of early multiwavelength observations and should be improved in the coming years.

Long-term evolution: hydrodynamical 2D-simulations of the long-term evolution of the GRB remnant in the non-relativistic phase have been carried by Ayal & Piran (2001). Their main result is that an initially beamed relativistic ejecta becomes spherical at a time

$$t_{\mathrm{a}} \simeq 3000 \left(\frac{E}{10^{51}\ \mathrm{erg}}\right)^{1/3} \left(\frac{n}{1\ \mathrm{cm}^{-3}}\right)^{-1/3} \mathrm{yr}. \tag{4.27}$$

The radius of the sphere is then

$$R \simeq 12 \left(\frac{E}{10^{51}\ \mathrm{erg}}\right)^{1/3} \left(\frac{n}{1\ \mathrm{cm}^{-3}}\right)^{-1/3} \mathrm{pc}. \tag{4.28}$$

At this stage, it is very difficult to distinguish the GRB remnant from a supernova remnant.

5 Prompt emission

If most of the current models of GRBs agree about the physical origin of the afterglow – deceleration of an ultra-relativistic ejecta by its environment as described in Section 4 – the situation is not so clear concerning the origin of the GRB itself. However, we will not describe all proposed models for the prompt emission in this lecture but rather concentrate on a particular one, the internal shock model, which is probably the model which has been studied most extensively and which seems to correctly reproduce most of the observed properties of GRBs. We will then discuss the possible contribution of the reverse shock, which is formed together with the external shock as mentioned in Section 4, to the prompt emission and we will end this section by an estimate of the thermal precursor emitted when the relativistic ejecta becomes transparent at the end of the acceleration phase.

5.1 Internal shocks

5.1.1 A simple model

The internal shock model has been proposed in 1994 by Rees & Meszaros (1994). The main assumption is that the relativistic ejecta emitted by the central source is produced with a highly variable distribution of the Lorentz factor, varying on timescales possibly as short as the dynamical timescale of the central engine. As the central object is most probably a stellar-mass black hole of mass M_{BH} surrounded by an accretion disk (see Sect. 6), this timescale is of the order of

$$t_{\mathrm{var}} \simeq \frac{6GM_{\mathrm{BH}}}{c^3} \simeq 0.3 \left(\frac{M_{\mathrm{BH}}}{10\ M_{\odot}}\right)\ \mathrm{ms}. \tag{5.1}$$

The fastest shells in the relativistic ejecta (highest Γ) catch up with the slowest ones (lowest Γ) which leads to the formation of shock waves propagating within the relativistic ejecta, the so-called "internal shocks". The hot matter behind the shocks radiate and produce the prompt GRB emission. This phase naturally stops when the ejecta is ordered, the Lorentz factor decreasing from the front to the back of the ejecta. This is illustrated in Figure 27.

Description of the relativistic outflow. In a first simple version of the model Daigne & Mochkovitch (1998), we consider that the relativistic ejecta is made of shells emitted regularly by the central source on a timescale t_{var}. Each shell has a Lorentz factor Γ_{i}, a mass M_{i} and an energy $E_{\mathrm{i}} = \Gamma_{\mathrm{i}} M_{\mathrm{i}} c^2$. Here all quantities have their isotropic equivalent values. We also assume that the central engine is active for a total duration t_{w} so that the number

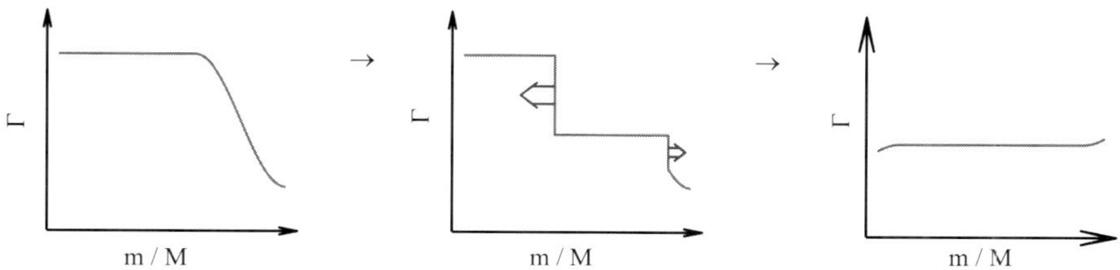

Fig. 27. The internal shock model: *left:* in this simple example, the relativistic ejecta is initially produced with a "slow" part ahead of a "rapid" part. *Center:* the rapid part catches up with the slow one and two shocks appear: a forward and a reverse shock. These two "internal shocks" propagate within the ejecta and the shocked matter radiate to produce the GRB. *Right:* once the two waves have crossed the ejecta, the Lorentz factor is ordored with rapid layers at the front: the internal shock phase stops. The burst produced in this example is made of a single pulse burst. The complex observed profiles are due to relativistic ejecta with much more variable Lorentz factors.

of shells is $N = t_{\mathrm{w}}/t_{\mathrm{var}}$ and that the total width of the relativistic ejecta is initially $\Delta \sim ct_{\mathrm{w}}$. The total mass and energy are $M = \sum_{\mathrm{i}} M_{\mathrm{i}}$ and $E = \sum_{\mathrm{i}} E_{\mathrm{i}}$ and the averaged value of the Lorentz factor $\bar{\Gamma} = E/Mc^2$. The Lorentz factor Γ_{i} is highly variable with the condition that $\bar{\Gamma} \gtrsim 100$, which was derived from the spectral observations in Section 2. These relativistic shells are "solid": they only interact by direct collisions so that all pressure waves are neglected. This very crude assumption has been validated by more realistic hydrodynamical simulations Daigne & Mochkovitch (2000). The succession of these direct collisions between shells represent the propagation of shock waves within the relativistic ejecta, *i.e.* internal shocks.

Typical radius and observed timescales. If a shell 2 (Lorentz factor Γ_2) is emitted a time t_{var} after a shell 1 (Lorentz factor $\Gamma_1 < \Gamma_2$), a collision occurs a the radius

$$R_{12} \simeq 2\frac{\Gamma_1^2\Gamma_2^2}{\Gamma_2^2 - \Gamma_1^2}\ ct_{\mathrm{var}}. \tag{5.2}$$

This allows to derive the typical radius of the internal shocks:

$$R_{\mathrm{IS}} \simeq 2f^2\bar{\Gamma}^2 ct_{\mathrm{var}} \simeq 6\times 10^{14}\ f^2 \left(\frac{\bar{\Gamma}}{100}\right)^2 \left(\frac{t_{\mathrm{var}}}{1\ \mathrm{s}}\right)\ \mathrm{cm}, \tag{5.3}$$

where f is a coefficient depending on the shape of the initial Lorentz factor distribution. This coefficient is smaller than unity and of the order of

$\Gamma_{\rm min}/\bar{\Gamma}$, where $\Gamma_{\rm min}$ is the minimum value of the initial Lorentz factor. Photons emitted along the line of sight by an internal shock at radius R and time t will be observed at the "arrival time" $t_{\rm a} = t - R/c$ already defined in the previous sections (see Fig. 25). Photons emitted at the same R and t on the edge of the "visible" fraction of the emitting surface will be detected at time $t_{\rm a} + \Delta t_{\rm a}$ with $\Delta t_{\rm a} \simeq R/2c\Gamma^2$, where Γ is the Lorentz factor of the emitting material. This means that even an impulsive emission is observed with a non zero duration $\Delta t_{\rm a}$ which is then the shortest timescale that one can observe in the GRB profiles. From the estimate of the typical internal shock radius, it is easy to see that $t_{\rm a}$ as well as $\Delta t_{\rm a}$ are of the order of

$$t_{\rm a} \sim \Delta t_{\rm a} \sim f^2 t_{\rm var}. \tag{5.4}$$

The $\bar{\Gamma}^2$ factor has disappeared, which is not surprising as this property was already used in Section 3 to solve the compactness problem. Equation (5.4) has two fundamental consequences Daigne & Mochkovitch (1998); Kobayashi *et al.* (1997): (i) **the ability of the internal shock model to reproduce the observed variability.** If the shortest timescale of variation in the initial distribution of the Lorentz factor is of the order of the dynamical timescale of the central source as estimated above, then variability in the GRB time profiles can be observed down to this time scale (corrected with the f coefficient and the redshift time dilation). (ii) **The origin of the observed duration of the GRB:** if the shortest variations in the GRB time profiles are related to the shortest timescales of variation in the initial distribution of the Lorentz factor, conversely the total burst duration is related to the largest timescale, *i.e.* to the total duration of the central source activity $t_{\rm w}$. It means that in the internal shock model, a short burst of 0.1 s has been produced by a source active for about $0.1/(1+z)$ s and a long burst of 10 s by a source active for about $10/(1+z)$ s. This is of course a constraint for the models of central engine: (a) the bimodal distribution of the observed duration should correspond to two class of sources with different characteristic timescales; (b) if they are also produced by internal shocks, the longest bursts (more than 100 s) correspond to central engines active for very long durations (compared to their dynamical timescale), which is a severe constraint. Notice that equation (5.3) indicates that for $t_{\rm var}$ varying from $\sim$1 ms to $\sim$10 s, the internal shock phase occurs on a large interval of radii, from $\sim 10^{11}$ to $\sim 10^{16}$ cm, depending on the average value of the Lorentz factor.

Description of a two shell collision. The physics of the internal shocks is close to the physics of the external shock described in Section 4, with an important difference: these shocks are only midly relativistic as there are produced by the interaction between shells with relative velocities of about

0.5–0.8c. In particular, the shock conditions written in equations (4.6–4.8) are not valid any more. We will see below that the specific internal energy in the shocked material can be estimated to be much less that in the case of the external shock (see Eqs. (4.7) and (5.7) below).

In the simple description considered here, when a collision between two shells of mass and Lorentz factor Γ_1, M_1 and Γ_2, M_2 occur, the two shells merge to produce a new shell of mass $M_1 + M_2$ and Lorentz factor Γ_r. This Lorentz factor can be computed from the conservation of the energy-momentum quadri-vector (inelastic collision) and is given by

$$\Gamma_\mathrm{r} = \sqrt{\Gamma_1\Gamma_2}\sqrt{\frac{M_1\Gamma_1 + M_2\Gamma_2}{M_1\Gamma_2 + M_2\Gamma_1}}, \tag{5.5}$$

which becomes $\Gamma_\mathrm{r} = \sqrt{\Gamma_1\Gamma_2}$ when $M_1 = M_2$. The dissipated energy during the collision can then be obtained:

$$e = \left(\Gamma_1 M_1 c^2 + \Gamma_2 M_2 c^2\right) - \Gamma_\mathrm{r}\left(M_1 + M_2\right) c^2. \tag{5.6}$$

The corresponding efficiency $e/(\Gamma_1 M_1 c^2 + \Gamma_2 M_2 c^2)$ is typically a few 10%, depending on the contrast between the two Lorentz factors. This dissipated energy is initially present in the shocked material as internal energy. One can estimate the corresponding specific internal energy in the comoving frame, which for $M_1 = M_2$ is:

$$\epsilon_* \simeq \left[\frac{1}{2}\left(\sqrt{\frac{\Gamma_1}{\Gamma_2}} + \sqrt{\frac{\Gamma_2}{\Gamma_1}}\right) - 1\right] c^2 \simeq 240\ \mathrm{MeV/proton\ for}\ \Gamma_2/\Gamma_1 = 4. \tag{5.7}$$

This is much less than in the early afterglow as expected because of the midly relativistic nature of the internal shocks. As in Section 4, we assume that a fraction α_e is injected into relativistic electrons, with the difference that only a fraction ζ of the electrons is accelerated in a non-thermal power-law distribution. There are some theoretical arguments derived from the physics of non relativistic MHD shocks to expect such a behaviour Bykov & Meszaros (1996). This leads to an averaged value of the electron Lorentz factor equal to

$$\bar{\Gamma}_\mathrm{e} \simeq \frac{\alpha_\mathrm{e}}{\zeta}\frac{m_\mathrm{p}}{m_\mathrm{e}}\frac{\epsilon_*}{c^2} \simeq 40\ \left(\frac{\alpha_\mathrm{e}/\zeta}{0.1}\right)\left(\frac{m_\mathrm{p}\epsilon_*}{200\ \mathrm{MeV}}\right)\cdot \tag{5.8}$$

This indicates that the acceleration is not very efficient if the fraction ζ is high ($\bar{\Gamma}_\mathrm{e}$ in the range 10-100). On the other hand, for small values of ζ, the accelerated electrons can of course be much more energetic (Eq. (5.8) gives $\bar{\Gamma}_\mathrm{e} \sim 4 \times 10^4$ for $\alpha_\mathrm{e}/\zeta = 100$). The magnetic field strength can also be estimated by assuming that the magnetic energy corresponds to a fraction $\alpha_{\rm B}$ of the internal energy:

$$B \simeq \sqrt{8\pi\alpha_{\rm B}\rho_*\epsilon_*}. \tag{5.9}$$

The density ρ_* of the shocked material can be obtained in this simple version of the model by

$$\rho_* \simeq \frac{E}{\bar{\Gamma}^2 4\pi R_{\rm IS}^2 c^3 t_{\rm w}}, \tag{5.10}$$

leading to typical magnetic field of about $100\sqrt{\alpha_{\rm B}}$ to $10^4\sqrt{\alpha_{\rm B}}$ G. It is important to notice that the magnetic field in several scenarios can be very different from this estimate if a large scale magnetic field anchored in the central source is also present Spruit *et al.* (2001).

Radiative processes. As the observed spectra are non-thermal, this is a difficult task to understand the details of the radiative processes operating during the internal shock phase. This question is still under debate and we will only make here a few estimates and then point out in the results what are the main problems that a more realistic description of these processes should solve.

– *The "inverse Compton" case:* as the synchrotron process seems to be dominant in the afterglow phase, it is natural to imagine that it plays also a role in the internal shock phase. If all electrons are accelerated ($\zeta \sim 1$), equation (5.8) gives $\bar{\Gamma}_{\rm e} \sim 100$ and from equation (4.20), the typical energy of synchrotron photons is given by (in the observer frame):

$$E_{\rm syn} \simeq 17 \left(\frac{\Gamma_{\rm r}}{100}\right)\left(\frac{B}{1000\ {\rm G}}\right)\left(\frac{\bar{\Gamma}_{\rm e}}{100}\right)^2 \ {\rm eV}, \tag{5.11}$$

which typically corresponds to UV photons, well below the spectral range of the observed GRBs. The gamma-ray photons can then be produced by inverse Compton scattering of the UV synchrotron photons by the relativistic electron. At these energies we are in the Thomson limit and the resulting typical energy of the scattered photons is

$$E_{\rm ic} \simeq 170 \left(\frac{\Gamma_{\rm r}}{100}\right)\left(\frac{B}{1000\ {\rm G}}\right)\left(\frac{\bar{\Gamma}_{\rm e}}{100}\right)^4 \ {\rm keV}, \tag{5.12}$$

which is now in the expected spectral range corresponding to the observed peak energy of GRBs. It is of course important to check that the inverse Compton photons carry most of the emitted energy. This is done by computing the optical depth of the emitting material for this process. The result is that the fraction $f_{\rm ic}$ of the emitted energy which is radiated by the inverse Compton process is given by the following equation (for a detailed derivation, see Daigne & Mochkovitch 1998):

$$f_{\rm ic}^2 - \left(2 + \frac{3}{8}\frac{\alpha_{\rm B}}{\alpha_{\rm e}}\right) f_{\rm ic} + 1 \simeq 0. \tag{5.13}$$

Equipartition ($\alpha_{\rm e} \sim \alpha_{\rm B}$) then gives a moderate efficiency $f_{\rm ic} \sim 0.55$. A weak magnetic field is required to increase $f_{\rm ic}$: $\alpha_{\rm B}/\alpha_{\rm e} \sim 10^{-3}$ leads to $f_{\rm ic} \sim 0.98$.

– *The "synchrotron" case:* if we now assume that the fraction ζ is low, the typical Lorentz factor of the electron can be much higher than in the previous case and equation (4.20) gives

$$E_{\rm syn} \simeq 170 \left(\frac{\Gamma_{\rm r}}{100}\right) \left(\frac{B}{1000\ {\rm G}}\right) \left(\frac{\bar{\Gamma}_{\rm e}}{10^4}\right)^2 \ {\rm keV}. \tag{5.14}$$

In this case, the observed prompt gamma-ray emission can be directly produced by the synchrotron process. The inverse Compton scattered photons are now produced in the Klein-Nishina regime and have a typical energy

$$E_{\rm ic} \simeq 500 \left(\frac{\Gamma_{\rm r}}{100}\right) \left(\frac{\bar{\Gamma}_{\rm e}}{10^4}\right) \ {\rm GeV}. \tag{5.15}$$

We have of course to check that the fraction $f_{\rm syn}$ of the energy which is radiated by the synchrotron process is high. Due to the Klein-Nishina regime, the equation defining $f_{\rm syn}$ is not so simple than equation (5.13):

$$\frac{1 + \ln(2w)}{w^2} \frac{\alpha_{\rm e}}{\alpha_{\rm B}} f_{\rm syn}^2 + f_{\rm syn} - 1 = 0, \tag{5.16}$$

where $w \simeq \bar{\Gamma}_{\rm e} E_{\rm syn} / \left(\Gamma_{\rm r} m_{\rm e} c^2\right) \gg 1$. Calculations show that during most of the burst $f_{\rm syn}$ is very close to unity due to the decrease of the scattering cross section in the Klein-Nishina regime Daigne & Mochkovitch (1998).

Radiative efficiency: the fraction $f_{\rm rad}$ of the energy $\alpha_{\rm e} \times e$ injected in the non-thermal electrons that is radiated can be easily obtained from the comparision between the adiabatic cooling time $t'_{\rm ex}$ defined in Section 4 and the radiative timescale, which is $f_{\rm syn} t'_{\rm syn}$ in the two cases presented just above. Simulations show that electrons are in the "fast cooling" regime during most of the burst so that $f_{\rm rad}$ is very close to unity Daigne & Mochkovitch (1998).

Synthetic gamma-ray bursts: for each two shell collision, the dissipated energy e and the arrival timeof photons $t_{\rm a}$ can be computed. If we then assume an elementary spectral shape (*i.e.* for the synchrotron case a broken power-law spectrum, corresponding to the "fast cooling" case in Fig. 24), the elementary contribution of each single collisions to the burst is now entirely fixed. A synthetic GRB can be produced by summing all these elementary contributions. This allows to compute its lightcurve and spectrum and to compare them to the observations. The total dynamical efficiency of the internal shock phase can be also computed:

$$f_{\mathrm{d}} \simeq \frac{\sum_{\mathrm{collisions}} e}{E}, \tag{5.17}$$

and is typically of a few 10%. To get the total efficiency f_{tot} of the internal shocks, f_{d} has to be multiplied by $f_{\mathrm{rad}} \times \alpha_{\mathrm{e}} \times f_{\mathrm{BATSE}}$, where f_{BATSE} is the fraction of the radiated energy which is emitted in the BATSE spectral range. The result is a low efficiency, with f_{tot} of a few percents only...

5.1.2 Hydrodynamical model

The simple model presented above offers the advantage to allow the simulation of many synthetic bursts in a very short computing time. However the very simplifying assumptions which have been made have to be validated. Hydrodynamical relativistic simulations – where the complete evolution of an ejecta with a highly variable Lorentz factor is followed from the saturation radius R_{sat} (where the acceleration is complete) to the deceleration radius R_{dec} (where the deceleration by the external medium starts) – allow to follow properly the internal shock phase and to compute in a consistent way the density and the internal energy density in the shocked material. The results are in a very good agreement with the results of the simple model Daigne & Mochkovitch (2000), probably because the dynamics is completely dominated by the kinetic energy of the flow so that pressure waves do not play a major role.

5.1.3 Results

The simulation of synthetic GRBs and the comparison of their properties with the observations lead to the following results: (i) the temporal properties of GRBs are very well reproduced, including the variability of the time profiles and the asymmetry of the elementary pulses. In this model, the light curve can be seen as reflecting the variable activity of the central engine; (ii) the spectral properties are satisfactory, the peak energy and the global shape of the spectrum being in agreement with the observations, but a major problem remains: the predicted low energy slope is very close to $\alpha \sim -1.5$ and the distribution of α shown in Figure 7 is not reproduced. It is hoped that this problem can be solved with a more detailed description of

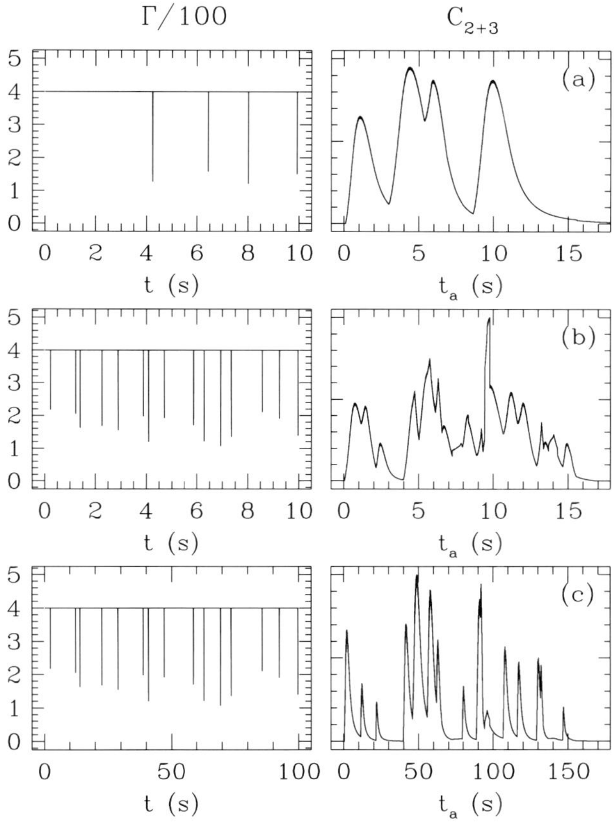

Fig. 28. Three synthetic GRB time profiles. *Left column:* the initial distribution of the Lorentz factor in the relativistic ejecta. *Right column:* the corresponding time profile in the 2+3 band of BATSE. The adopted redshift is $z = 0.5$. In the three cases, the relativistic outflow has a more or less constant Lorentz factor $\Gamma = 400$, with a few slower layers with Lorentz factors between 100 and 400. The number of slow layers is higher in the last two cases and the corresponding profile more complex. Notice that case **c)** is similar to case **b)** with all timescales shifted by a factor of ten (from Daigne & Mochkovitch 1998).

the radiative processes; (iii) most observed relations between the temporal and the spectral properties are reproduced, including the hardness ratio – duration correlation, the evolution of the pulse shape with the energy band, the photon fluence – peak energy correlation, etc. Daigne & Mochkovitch (1998, 2000). Another difficulty of the model is the low total efficiency of

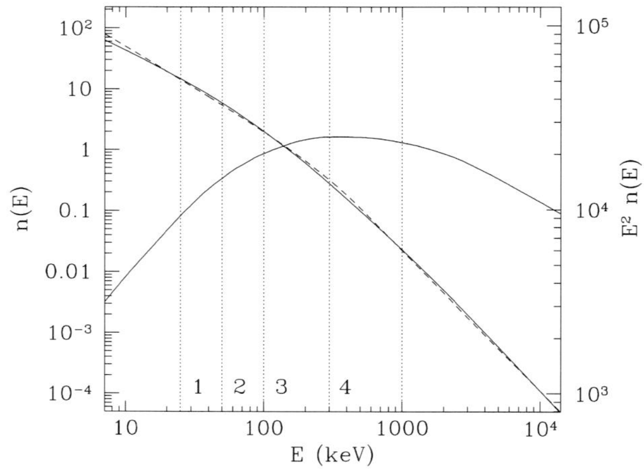

Fig. 29. Synthetic spectrum corresponding to case **a)** in Figure 29 (from Daigne & Mochkovitch 1998).

the mechanism, which implies very powerful and efficient central engines. Three examples of synthetic GRB lightcurves are presented in Figure 28. The global spectrum corresponding to case (a) in Figure 28 is shown in Figure 29. The peak energy is 365 keV and the dotted line shows the best-fit by the Band spectral model. Figure 30 shows an example of a single pulse burst which illustrates how the pulses properties are reproduced by the model. Figure 31 shows the hardness ratio H_{32} – duration T_{90} relation, obtained with sequences of models where only the duration $t_{\rm w}$ is varied.

Despite these encouraging results, many aspects of the model have still to be better understood, mainly the radiative processes. A difficulty is that because of the lack of prompt observations at other wavelength, it is very difficult to distinguish between the "inverse Compton" and the "synchrotron" cases presented above. The synchrotron case may be prefered due to a somewhat better total efficiency. Of course, the detection of a high-energy second peak in the spectrum, for instance by GLAST, would very rapidly solve this question. Another approach is to study the details of the pulses properties, especially the HIC and the HFC presented in Section 2, to put some constraints on the physical processes at work in the shocked medium. Recently, this approach has allowed to propose a small modification of the standard $\alpha_{\rm e}$, $\alpha_{\rm B}$, ζ prescription that leads to an excellent agreement between the synthetic pulse evolution and the observations Daigne & Mochkovitch (2002b).

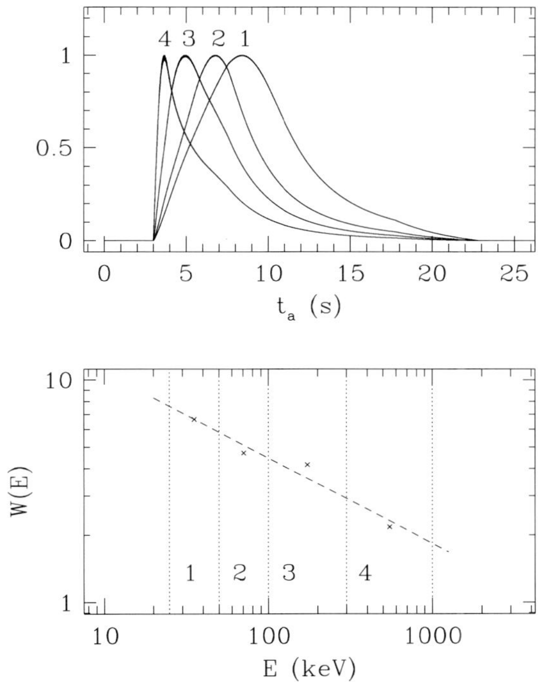

Fig. 30. A synthetic single pulse burst in the four BATSE energy bands. The pulse is peaking earlier at high energy (band 4) and its width $W(E)$ is decreasing with the energy band (lower panel), in agreement with the observations (from Daigne & Mochkovitch 1998).

5.2 Reverse shock

5.2.1 Physics of the reverse shock

Dynamics. As described in Section 4, the reverse shock appears together with the external shock, at the deceleration radius $R_{\rm dec}$. However, as it propagates within the relativistic ejecta, its physics is very close to that of internal shocks (in particular the reverse shock is usually only midly relativistic). The Lorentz factor of the reverse shock is given by Sari & Piran (1995):

$$\Gamma_{\rm RS} \simeq \left(\frac{\bar{\Gamma}}{2}\right)^{1/2} \left(\frac{\bar{\rho}}{\rho_{\rm ext}}\right)^{1/4}, \tag{5.18}$$

where $\bar{\rho}$ is the averaged density in the relativistic ejecta. Therefore, the reverse shock has entirely crossed the relativistic ejecta (whose width is

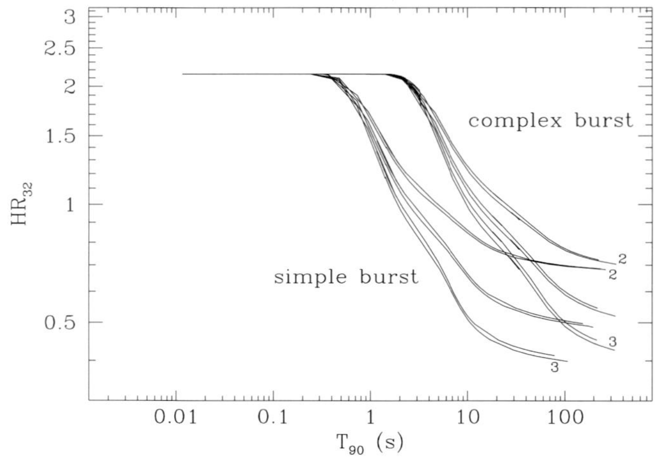

Fig. 31. Hardness ratio H_{32} as a function of the burst duration T_{90}. Each line corresponds to a sequence of models where only the duration $t_{\rm w}$ of activity of the central source is varied. Two different initial distribution of the Lorentz factor have been considered: a simple one leading to a single pulse burst ("simple burst") and a more complex one leading to a five pulses burst ("complex burst"). In each case, the sequence has been computed for three high-energy index β (-2, -2.5 and -3) and two different redshifts (the two corresponding curves are very close) (from Daigne & Mochkovitch 1998).

$\Delta \sim ct_{\rm w}$) at radius

$$R_{\rm cross} \simeq \Gamma_{\rm RS}^2 \Delta \simeq \sqrt{R_{\rm dec} R_{\rm IS}}, \tag{5.19}$$

where $R_{\rm dec}$ is given by equations (4.4) ($s = 0$) and (4.5) ($s = 2$) and $R_{\rm IS}$ is the radius of the last internal shocks:

$$R_{\rm IS} \simeq 2 f^2 \bar{\Gamma}^2 c t_{\rm w} \simeq 6 \times 10^{15}\ f^2 \left(\frac{\bar{\Gamma}}{100}\right)^2 \left(\frac{t_{\rm w}}{10\ {\rm s}}\right)\ {\rm cm}. \tag{5.20}$$

From the expression of $R_{\rm cross}$, it is clear that there is a unique condition to have distinct reverse shock and internal shock phases: $R_{\rm dec} > R_{\rm IS}$. This condition can be written as Daigne & Mochkovitch (1999)

$$Q = \left(\frac{R_{\rm dec}}{R_{\rm IS}}\right)^{3-s} > 1, \tag{5.21}$$

which leads to the following condition for a uniform medium ($s = 0$):

$$Q \simeq 7.2 \times 10^3\ f^{-6} \left(\frac{E}{10^{53}\ \mathrm{erg}}\right)\left(\frac{n}{1\ \mathrm{cm}^{-3}}\right)\left(\frac{\bar{\Gamma}}{100}\right)^{-8}\left(\frac{t_{\mathrm{w}}}{10\ \mathrm{s}}\right)^{-3} > 1, \tag{5.22}$$

and for a dense stellar wind ($s = 2$):

$$Q \simeq 0.19\ f^{-2} \left(\frac{E}{10^{53}\ \mathrm{erg}}\right)\left(\frac{A}{7.6\times 10^{11}\ \mathrm{g\,cm}^{-1}}\right)^{-1} \times \left(\frac{\bar{\Gamma}}{100}\right)^{-4}\left(\frac{t_{\mathrm{w}}}{10\ \mathrm{s}}\right)^{-1} > 1. \tag{5.23}$$

It is clear that for a uniform medium ($s = 0$), except for low external densities, or for low injected total energy E, the deceleration starts at large radii and the reverse shock propagates within a ordered relativistic ejecta where the internal shocks have already disappeared. On the other hand, in the dense stellar wind case ($s = 2$), the estimate of Q seems to indicate that in most cases the reverse shock will interact with the internal shocks.

Radiative processes. The same kind of parametrization with α_{B}, α_{e} and ζ already made for the external and the internal shocks can be used for the reverse shock. Then it is possible to estimate in the same way the magnetic field, the typical electron Lorentz factor and the synchrotron and/or inverse Compton radiation to compute the contribution of the reverse shock to the observed emission. The main differences between different models are in the assumptions concerning the α_{B} and α_{e} parameters: (i) Sari & Piran (1999) adopt the same values than for the external shock. Therefore, as the reverse shock is just midly relativistic, so that the internal energy created in the shocked medium is not so high as in the case of the external shock, the reverse shock radiates at low energy and is observed as an "optical flash", that could explain the ROTSE observations of GRB 990123. (ii) Daigne & Mochkovitch (1999) argue that because of the midly relativistic nature of the reverse shock, this shock is much closer to the internal shocks and they adopt the same parameters than in the internal shock phase. Therefore, when the reverse shock is present in the relativistic ejecta at the same moment than the internal shocks ($Q < 1$), its contribution to the emission is also in the gamma-ray range. When the reverse shock appears later ($Q > 1$), as the density and the magnetic field decrease with the radius, the spectral range goes towards the low energy bands (typically soft X-rays for $Q \gtrsim 1$).

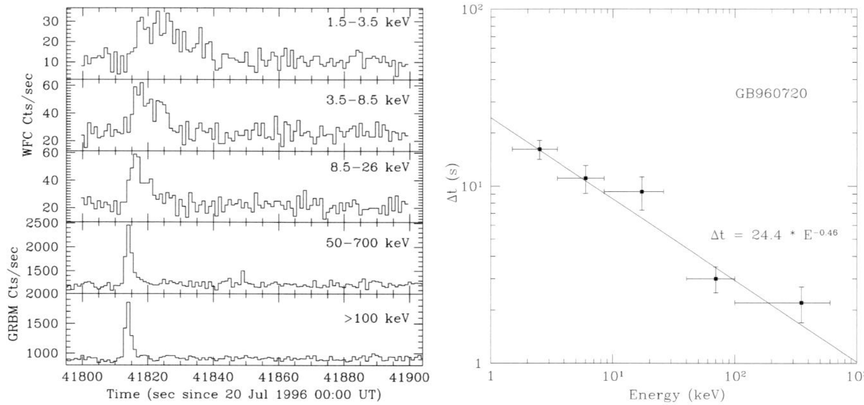

Fig. 32. Beppo-SAX observation of GRB 960720. *Left:* time profile (X- and γ-rays). *Right:* evolution of the pulse width with the energy band (from Piro *et al.* 1998).

5.3 Results

Uniform external medium: we illustrate the effect of the reverse shock for GRB 960720 in the case of a small density uniform external medium. This single pulse burst has been observed both in X-rays and gamma-rays by Beppo-SAX (see Fig. 32). It follows the power-law dependence of the pulse width with the energy band Norris *et al.* (1996) not only in the gamma-ray range but also with the same slope down to the soft X-ray bands. The gamma-ray properties of this burst can be easily reproduced by the internal shock model. Figure 33 show the result obtained for $\bar{\Gamma} = 290$, $t_{\mathrm{w}} = 4$ s and $E = 5 \times 10^{52}$ erg. It is clear that the pulse width-energy relation is reproduced only in the gamma-ray band, as the X-ray emission stops too early whereas the observations show an X-ray emission perduring after the end of the gamma-ray emission. We then include the contribution of the reverse shock due to an external medium with a uniform density $n = 10\ \mathrm{cm}^{-3}$. This contribution is only detected in X-rays and increase the pulse width in this energy band so that the observed relation is now reproduced (with the correct slope) over the whole spectral range Daigne & Mochkovitch (1999).

Dense stellar wind: the case of a dense external medium, for instance due to the presence of a stellar wind, is much more problematic. As Q is now very low, the reverse shock appears very early (possibly even before the internal shocks) and dramatically affect the internal shock phase so

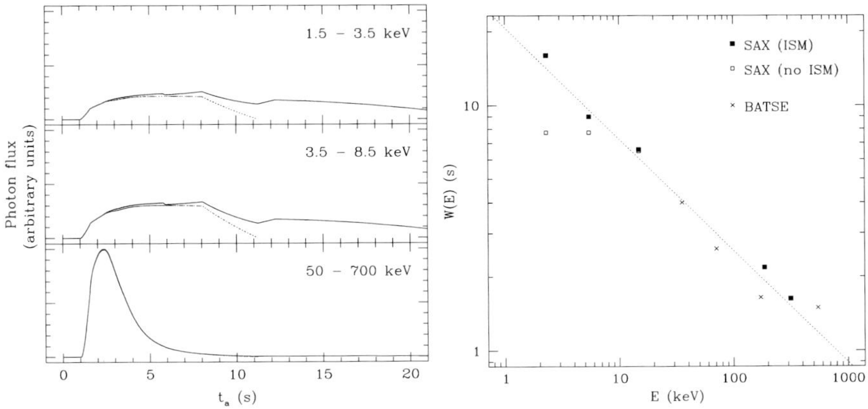

Fig. 33. Simulation of GRB 960720. Both the internal shocks and the reverse shock have been included (solid line). For comparison, the dotted line corresponds to a calculation including the internal shocks only. The effect of the reverse shock is only detected in X-rays and leads to an X-ray emission perduring after the end of the pulse observed in gamma-rays. The correct pulse width–energy relation is then reproduced from the gamma- down to the X-rays (from Daigne & Mochkovitch 1999).

that the prompt gamma-ray profile is strongly affected. The temporal and spectral properties of GRBs are therefore no longer reproduced Daigne & Mochkovitch (2001). To avoid this major problem, it is necessary to recover large values of Q. From equation (5.23), this leads to new constraints on the model, (i) either on the environment density profile: if the stellar wind is not so dense, or even better, if it disappears (because of a low metallicity of the progenitor for instance), the deceleration starts later and we come back to the previous case; (ii) or on the relativistic ejecta: if for instance the energy E increases, the ejecta can more easily penetrate within the external medium and the deceleration starts later Daigne & Mochkovitch (2001). Another possibility would be to decrease $\bar{\Gamma}$ but this is limited by the constraint $\bar{\Gamma} \gtrsim 100$.

5.4 Photosphere

We close this section on the prompt emission by a brief discussion about the emission produced when the relativistic ejecta becomes transparent at the end of the acceleration phase. According to the estimate of the photospheric radius obtained in Section 3, this emission is produced well before the internal shock phase and the deceleration phase (reverse and external shocks).

At this radius, the internal energy which is still present can be radiated. The resulting spectrum is close to a blackbody with a temperature that can be derived from the "matter-dominated" regime equations obtained in Section 3:

$$kT_{\rm ph} \simeq kT_0 \left(\frac{R_{\rm ph}}{R_{\rm sat}}\right)^{-2/3}, \tag{5.24}$$

and a luminosity

$$L_{\rm ph} \simeq \dot{E} \left(\frac{R_{\rm ph}}{R_{\rm sat}}\right)^{-2/3}. \tag{5.25}$$

The initial temperature T_0 has been computed in Section 3 for an initial radius R_0 assumed to be $6GM_{\rm BH}/c^2$. It is then possible to estimate the corresponding observed count rate in a given energy band $[E_1, E_2]$:

$$C_{12}^{\rm ph} = \frac{L_{\rm ph}}{4\pi D_{\rm L}^2} \frac{1+z}{kT_{\rm ph}} \frac{1}{\mathcal{I}_{\rm Planck}} \int_{(1+z)E_1/kT_{\rm ph}}^{(1+z)E_2/kT_{\rm ph}} \frac{x^2}{\exp x - 1} {\rm d}x, \tag{5.26}$$

where $D_{\rm L}$ is the luminosity distance of the source and $\mathcal{I}_{\rm Planck} = \pi^4/15$. This has to be compared to the count rate in the same energy band due to the non-thermal emission by internal shocks:

$$C_{12}^{\rm IS} = \frac{L_{\rm IS}}{4\pi D_{\rm L}^2} \frac{1+z}{E_{\rm p}} \frac{1}{\mathcal{I}_{\rm Band}} \int_{(1+z)E_1/E_{\rm p}}^{(1+z)E_2/E_{\rm p}} \mathcal{B}(x) {\rm d}x, \tag{5.27}$$

where $L_{\rm IS}$ is the luminosity of the internal shocks, $\mathcal{B}(x)$ is the Band function defined in Section (2) and $\mathcal{I}_{\rm Band}$ the constant $\mathcal{I}_{\rm Band} = \int_0^{+\infty} x\mathcal{B}(x){\rm d}x$. Assuming that $L_{\rm IS} = f_\gamma \dot{E}$ (f_γ is the efficiency of the internal shocks), we can compute the ratio $R_{12} = C_{12}^{\rm ph}/C_{12}^{\rm IS}$ of the two count rates:

$$R_{12} \simeq 1.6 \left(\frac{f_\gamma}{0.1}\right)^{-1} \left(\frac{\dot{E}}{10^{52}\ {\rm erg\ s^{-1}}}\right)^{-1/4} \left(\frac{M_{\rm BH}}{10\ M_\odot}\right)^{1/2} \left(\frac{E_{\rm p}}{200\ {\rm keV}}\right) \times \frac{\mathcal{I}_{\rm Band}}{\mathcal{I}_{\rm Planck}} \frac{\int_{(1+z)E_1/kT_{\rm ph}}^{(1+z)E_2/kT_{\rm ph}} \frac{x^2}{\exp x - 1} {\rm d}x}{\int_{(1+z)E_1/E_{\rm p}}^{(1+z)E_2/E_{\rm p}} \mathcal{B}(x) {\rm d}x}. \tag{5.28}$$

This ratio is plotted for two energy bands (X- and gamma-rays) and different parameters in Figure 34. It is clear that for reasonnable parameters, the thermal photospheric emission dominates or is at least very easily detectable both in X- and gamma-rays. This is of course in contradiction with the observed spectrum of GRBs. The conclusion of this study is that the "standard" fireballs are too hot and luminous. One elegant way that can be proposed to solve the problem is to assume that most of the energy

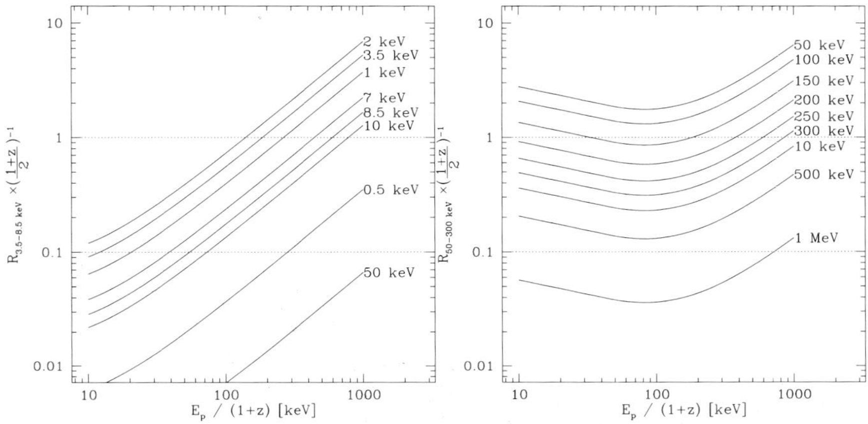

Fig. 34. The ratio of the count rate due to the photospheric emission over the count rate due to the internal shocks as a function of the peak energy of the non-thermal (internal shocks) emission for different values of the temperature (labeled after each curve). *Left:* X-ray band 3.5–8.5 keV; *right:* gamma-ray band 50–300 keV. The following parameters have been used: $\dot{E} = 10^{52}$ erg s^{-1}, $f_\gamma = 0.1$, $M_{\rm BH} = 10\ M_\odot$ (from Daigne & Mochkovitch 2002c).

is initially released under magnetic rather than thermal form. In a cold MHD outflow, the reservoir of internal energy when the matter becomes transparent can be much lower than in the previous calculation. Simple estimates show that if only a few percents of the initial energy released is in thermal energy form, the thermal-over-non-thermal radiation ratio R_{12} becomes small enough to be in agreement with the observations Daigne & Mochkovitch (2002c).

6 Central engine

The physics of the central engine in GRBs is probably the less understood among the three steps of the scenario described in Section 3. We recall here the main requirements for the sources: (i) they must be located at cosmological distance with $\langle z \rangle \gtrsim 1$; (ii) they must be compatible with a GRB rate $\mathcal{R}_{\rm GRB}$ of about (see Sect. 3)

$$\mathcal{R}_{\rm GRB} \simeq 5 \times 10^{-4} \left(\frac{\left\langle \left(\frac{\Omega}{4\pi}\right)^{-1} \right\rangle}{500} \right) \mathcal{R}_{\rm SN}, \tag{6.1}$$

where current estimates of the beaming factor $\left\langle \left(\frac{\Omega}{4\pi}\right)^{-1} \right\rangle$ are about 500 with a large uncertainty Frail *et al.* (2001); (iii) they must be able to release a huge energy of about

$$E \simeq 10^{52} \left(\frac{f_\gamma}{0.1}\right) \left(\frac{\langle \Omega/4\pi E_{\gamma,4\pi}\rangle}{10^{51}\ \mathrm{erg}}\right)\ \mathrm{erg}, \tag{6.2}$$

where estimates of the efficiency for the conversion to gamma-rays differ among the models but is probably less than 10% for internal shocks. The estimate of the true energy released in gamma-rays, $\Omega/4\pi E_{\gamma,4\pi}$, also suffers large uncertainties associated to the measure of the beaming factor; (iv) this energy must be injected in an almost baryon-free flow so that it can accelerate up to high Lorentz factors with $\bar{\Gamma} \gtrsim 100$. An additionnal requirement in the context of the internal shock model is that the source must vary on short timescales (1 ms) and the process of relativistic ejection must last for at least the typical observed duration of GRBs.

It is clear that such requirements make the question extremely complex. Only a few answers have been proposed and most of them have not yet been studied in detailed simulations. We summarize below the two main proposals: (i) merger of two compact objects; (iii) gravitational collapse of very massive stars into black holes.

6.1 Mergers

The coalescence of a binary system of two neutron stars Eichler *et al.* (1989); Paczynski (1991) and the merger of a neutron star – black hole system Mochkovitch *et al.* (1993); Narayan *et al.* (1992) have been proposed as a possible source for GRBs. The product of the merger is a black hole ($M_{\rm BH} \sim 2.5\ M_\odot$) surrounded by a debris torus ($m \lesssim 0.1\ M_\odot$) which appears to satisfy some of the requirements: (i) the variability time scale is correct, as the dynamical timescale at the inner edge of the torus is $\lesssim$1 ms; (ii) the region along the rotation axis is almost baryon-free which offers a nice possibility to have ejections with small baryonic load; (iii) the environment is very clean with a probably low density which favors the acceleration up to high Lorentz factors and corresponds to a late deceleration so that the reverse shock does not suppress the internal shock phase. However this proposal also suffers from severe problems: (i) the known rate of mergers is probably too low, except if $\langle(\Omega/4\pi)^{-1}\rangle$ has been overestimated; (ii) the energy reservoir (rotation of the black hole + accretion from the torus to the hole) is probably too low for the most luminous GRBs; (iii) mergers are supposed to occur at large distance from the central regions of the host galaxy as there is a long delay between the binary system formation and the coalescence, whereas observations of afterglows show evidences in favor of an association with the star-forming central regions. However all these elements prove that mergers still remain possible model for short duration GRBs, for which no optical afterglow has never be found.

6.2 Collapsars

The collapse of a very massive star into a black hole, sometimes called a "failed supernova" Woosley (1993) or an "hypernova" Paczynski (1998), is another possibility which solves some of the problems of the previous model and is obviously favored for the long-duration GRBs from the observational evidences of an association with star-forming regions. In the center of the collapsing object, a black hole + thick disk system is also formed, with the difference that the black hole can be more massive (10 $M_\odot$ or even more) and the torus is feeded with the still falling material coming from the external layers of the progenitor. The main advantages are that the rate of massive star collapses, for reasonable estimates of the critical mass above which a black hole is formed, is much higher than the GRB rate, even for large $\left\langle(\Omega/4\pi)^{-1}\right\rangle$ factors and that the energy reservoir is much larger than in mergers. Two problems are that it is certainly more difficult to accelerate matter inside an infalling star than in empty space and that the environment, except for low metallicity progenitors, should be dominated by the dense wind of the massive star, which is not supported by the observations as explained in Sections 4 and 5.

6.3 MHD winds?

The neutrino-antineutrino annihilation along the rotation axis of the system, *i.e.* in a region widely depleted from baryons due to the centrifugal forces, may be the process at the origin of the energy injection in GRBs Meszaros & Rees (1992); Mochkovitch *et al.* (1995, 1993) but this possibility has probably to be considered for short GRBs only ($\lesssim$1 s) as this process seems to be too unefficient to work for long bursts Ruffert *et al.* (1997). It is then often proposed that the solution could be a magnetic outflow Meszaros & Rees (1997); Spruit *et al.* (2001); Thompson (1994) where the energy of the disk and/or the black hole (*via* the Blandford-Znajek effect Blandford & Znajek 1977) is extracted by a strong magnetic field ($B \sim 10^{15}$ G). No numerical simulations of such a mechanism in the GRB context are available. Several important questions are still to be asked: (i) how can the magnetic energy be converted into kinetic energy (reconnection at large distance?) and is the process efficient enough? Daigne & Drenkhahn (2002); (ii) is it possible with such a mechanism to have a low baryonic pollution? Some recent estimates show that it is probably very difficult in the case of an outflow coming from the disk Daigne & Mochkovitch (2002a). It is probably much easier with the Blandford-Znajek effect, as long as the flow is not polluted by material coming from the disk; (iii) is the propagation of the outflow in the collapsing star possible so that it reaches terminal Lorentz factors larger than 100: recent simulations indicate that it is difficult but probably possible Aloy *et al.* (2000); MacFadyen *et al.* (2001).

7 More than photons?

The non-photonic emission of GRBs is very briefly discussed in this section.

7.1 Ultra-high energy cosmic rays

During the internal shock phase, the presence of several shocks propagating within the relativistic ejecta favors the acceleration of particles. It is possible to make basic estimates of the maximum energy $E_{\rm max}$ that one given type of particle can get (this maximum energy is reached when the acceleration time becomes longer than the radiative time and/or the escape time). For protons, with reasonable parameters regarding the size of the shells and the magnetic field, $E_{\rm max}$ can be as high as 10^{19}–10^{20} eV (see *e.g.* Waxman 2001). Therefore GRBs appear as possible candidates for the origin of UHECRs (ultra-high energy cosmic rays). However, even if GRBs have the ability to accelerate protons up to very high energies (which has still to be investigated in details), another difficulty is that the origin of UHECRs above the GZK cutoff is necessarily local, within a distance of about 100 Mpc. The local rate of GRBs is very poorly known but certainly low. Then it has to be tested that the observations, especially the coming detections by the AUGER observatory, will be compatible with only one or two local sources.

7.2 High energy neutrinos

During the internal shock phase, if protons are accelerated, they can interact with gamma-ray photons to produce pions, decaying in muons and neutrinos (see *e.g.* Waxman 2001). This is independent from the production of UHECRs as protons of only $\sim 10^{16}$ eV (!) are required to produce $\sim 10^{14}$ eV neutrinos by photo-meson interaction with MeV photons. Such neutrinos can be detected by experiments like ANTARES, IceCube, AMANDA, etc. However, current estimates of the expected number of neutrinos from GRBs are not very optimistic, typically a few per year...

7.3 Gravitational waves

The main hope for the detection of gravitational waves in association with GRBs is the possibility that short GRBs ($\lesssim$1 s) are related to compact object mergers. This justifies an important effort in the coming years to better study this subclass of bursts and identify their progenitors. Concerning the long bursts, it is very unlikely that a signal of gravitational waves can be detected with the current experiments, as the expected gravitational wave production during an (even very) asymmetric collapse of a massive star is much weaker than for a coalescence.

8 Conclusions

To conclude this lecture which is a short introduction to GRBs, I briefly list a few questions that will have to be investigated in the coming years. A first series of questions are related to the prompt emission, and especially to the spectral properties:

• What is the high-energy emission? This will be studied with several instruments (AGILE, GLAST, HESS, ...). If a cutoff in the high-energy power-law can be identified, it will provide a direct measure of the Lorentz factor of the emitting material (see Sect. 3). If the main peak observed in the gamma-ray range is due to synchrotron radiation, the identification of a second peak at high energy, due to inverse Compton scattering, would lead to the measure of important parameters such as the magnetic field and the Lorentz factor of the electrons. If the radiative process at work in GRBs is not the SSC process, the extension of the spectrum possibly up to 300 GeV would greatly help in its indentification.
• The detection of high-energy neutrinos from a (probably bright and close) GRB would be of great interest. This would provide completely new information on the processes operating in GRBs, especially on the acceleration of particles by relativistic shock waves.
• What is the optical spectrum? Only one detection of the optical prompt emission has ever been made, in GRB 990123 by ROTSE. It is a great challenge for the next generation of robotic telescopes as ARAGO Boër (2001) or for future satellites as ECLAIRs Paul *et al.* (2002) to increase this number to better constraint the theoretical interpretation of the GRB optical flashes.
• Will the narrow distribution of the peak energy E_{p} observed by BATSE be confirmed by the current and future missions (HETE-2, INTEGRAL, SWIFT). Are the X-ray flashes observed by Beppo-SAX Heise *et al.* (2001) and HETE-2 Barraud *et al.* (2002) a peculiar class of intrinsically softer GRBs?

• Do GRBs have precursors in the X-ray or optical energy ranges? Such precursors will be searched by the future satellite ECLAIRs Paul *et al.* (2002).

A second series of questions are related to the afterglow:

• Is it possible to use the almost real-time GRB alerts that are now available to perform a rapid multi-wavelength follow-up of the early afterglow. During this phase, the afterglow is probably very bright even at large redshift, which makes GRBs promising tools to explore the very distant Universe. A related question is the capability to do rapid infrared observations to detect GRB afterglows at very high redshift ($z \gtrsim 6$).
• When does the afterglow start and in which energy band does it initially peak? Early afterglow observations would give very strong constraints on the energetics of the burst and on the density profile of the environment. It is also important to understand the transition between the burst and the afterglow.
• What is the distribution of the long GRB redshifts? This is of course the main result expected from SWIFT. This observation in association with a better identification of the GRB progenitors would allow to trace the star-formation rate with GRBs.
• Is it possible to better constrain the beaming factor in GRBs? This is a fundamental parameter to know the true energy budget of the GRB phenomenon and the true rate of events. A very promising possibility is the search for orphan afterglows in large surveys. If one is able to detect afterglows off-axis, when the initial GRB was not observed, a comparison of the event rates would give a completely independant determination of the beaming factor.
• Can the GRB-supernova association be confirmed by new observations? This would have a strong impact on the models of GRB progenitors.
• Do short GRBs ($\lesssim$1 s) have afterglows and what are their progenitors? The answer could come from the location of the afterglow, either in star-forming regions or at the periphery of their host galaxies?

All these questions show that the GRB field has a very promising future, especially if it is confirmed that GRBs can occur at very high redshift. They would allow to search for the reionization epoch, and for population III stars, to trace the star-formation rate up to high redshifts, to study the intergalactic medium, etc. GRBs may then become a unique tool to probe the very distant Universe.

It is a pleasure to thank Robert Mochkovitch for his careful reading of this manuscript.

References

C. Akerlof, R. Balsano, S. Barthelemy *et al.*, *Nature* **398** (1999) 400.
M.A. Aloy, E. Müller, J.M. Ibáñez, J.M. Martí and A. MacFadyen, *ApJ* **531** (2000) L119.
S. Ayal and T. Piran, *ApJ* **555** (2001) 23.
D. Band, J. Matteson, L. Ford *et al.*, *ApJ* **413** (1993) 281.
C. Barraud, J. Olive, J.P. Lestrade (2002), *A&A* to be published [astro-ph/0206380].
A.M. Beloborodov (2002), ApJL to be published [astro-ph/0209228].
R.D. Blandford and C.F. McKee, *Phys. Fluids* **19** (1976) 1130.
R.D. Blandford and R.L. Znajek, *MNRAS* **179** (1977) 433.
J.S. Bloom, S.G. Djorgovski, S.R. Kulkarni and D.A. Frail, *ApJ* **507** (1998) L25.
J.S. Bloom, S.R. Kulkarni and S.G. Djorgovski, *AJ* **123** (2002) 1111.
M. Boër, *Astron. Nachr.* **322** (2001) 343.
M.S. Briggs, D.L. Band, R.M. Kippen *et al.*, *ApJ* **524** (1999) 82.
A.M. Bykov and P. Meszaros, *ApJ* **461** (1996) L37.
S. Covino, D. Lazzati, G. Ghisellini *et al.*, *A&A* **348** (1999) L1.
A. Crider, E.P. Liang, R.D. Preece *et al.*, *A&AS* **138** (1999) 401.
F. Daigne and G. Drenkhahn, *A&A* **381** (2002) 1066.
F. Daigne and R. Mochkovitch, *MNRAS* **296** (1998) 275.
F. Daigne and R. Mochkovitch, *A&AS* **138** (1999) 523.
F. Daigne and R. Mochkovitch, *A&A* **358** (2000) 1157.
F. Daigne and R. Mochkovitch 2001, in Gamma-ray Bursts in the Afterglow Era, Proceedins of the 2nd Workshop held in Rome, October 2000, edited by E. Costa, F. Frontera and J. Hjorth, 324.
F. Daigne and R. Mochkovitch, *A&A* **388** (2002a) 189.
F. Daigne and R. Mochkovitch (2002b), *MNRAS*, submitted.
F. Daigne and R. Mochkovitch, *MNRAS* **336** (2002c) 1271.
J.-P. Dezalay, J.P. Lestrade, C. Barat *et al.*, in Gamma-ray bursts, Proceedings of the 3rd Huntsville Symposium, edited by C. Kouveliotou, M.F. Briggs and G.J. Fishman, *AIP Conf. Proc. Ser.* **384** (1996) 37.
S.G. Djorgovski, S.R. Kulkarni, J.S. Bloom *et al.*, in Gamma-ray Bursts in the Afterglow Era, Proceedins of the 2nd Workshop held in Rome, October 2000, edited by E. Costa, F. Frontera and J. Hjorth (2001) 218.
D. Eichler, M. Livio, T. Piran and D.N. Schramm, *Nature* **340** (1989) 126.
D.A. Frail, S.R. Kulkarni, R. Sari *et al.*, *ApJ* **562** (2001) L55.
A.S. Fruchter, E. Pian, S.E. Thorsett *et al.*, *ApJ* **516** (1999) 683.
T.J. Galama, N. Tanvir, P.M. Vreeswijk *et al.*, *ApJ* **536** (2000) 185.
T.J. Galama, P.M. Vreeswijk, J. van Paradijs *et al.*, *Nature* **395** (1998) 670.
J. Goodman, *ApJ* **308** (1986) L47.
P.J. Groot, T.J. Galama, J. van Paradijs *et al.*, IAU Circ. 6584 (1997).
J. Heise, J. in't Zand, R.M. Kippen and P.M. Woods, in Gamma-ray Bursts in the Afterglow Era, Proceedins of the 2nd Workshop held in Rome, October 2000, edited by E. Costa, F. Frontera and J. Hjorth (2001) 16.
V.E. Kargatis, E.P. Liang and BATSE Team, *Ap&SS* **231** (1995) 177.
R.W. Klebesadel, I.B. Strong and R.A. Olson, *ApJ* **182** (1973) L85.
S. Kobayashi, T. Piran and R. Sari, *ApJ* **490** (1997) 92.
J.M. Kommers, W.H.G. Lewin, C. Kouveliotou *et al.*, *ApJ* **533** (2000) 696.
C. Kouveliotou, C.A. Meegan, G.J. Fishman *et al.*, *ApJ* **413** (1993) L101.

E. Liang and V. Kargatis, *Nature* **381** (1996) 49.
Y. Lithwick and R. Sari, *ApJ* **555** (2001) 540.
A.I. MacFadyen, S.E. Woosley and A. Heger, *ApJ* **550** (2001) 410.
P. Meszaros and M.J. Rees, *MNRAS* **257** (1992) 29P.
P. Meszaros and M.J. Rees, *ApJ* **482** (1997) L29.
M.R. Metzger, S.G. Djorgovski, S.R. Kulkarni *et al.*, *Nature* **387** (1997) 878.
R. Mochkovitch, M. Hernanz, J. Isern and S. Loiseau, *A&A* **293** (1995) 803.
R. Mochkovitch, M. Hernanz, J. Isern and X. Martin, *Nature* **361** (1993) 236.
R. Narayan, B. Paczynski and T. Piran, *ApJ* **395** (1992) L83.
R.J. Nemiroff, in AIP Conf. Proc. 307: Gamma-Ray Bursts (1994) 730.
J.P. Norris, R.J. Nemiroff, J.T. Bonnell *et al.*, *ApJ* **459** (1996) 393.
W.S. Paciesas, C.A. Meegan, G.N. Pendleton *et al.*, *ApJS* **122** (1999) 465.
B. Paczynski, *ApJ* **308** (1986) L43.
B. Paczynski, *Acta Astron.* **41** (1991) 257.
B. Paczynski, *ApJ* **494** (1998) L45.
A. Panaitescu and P. Kumar, *ApJ* **543** (2000) 66.
A. Panaitescu and P. Kumar, *ApJ* **560** (2001a) L49.
A. Panaitescu and P. Kumar, *ApJ* **554** (2001b) 667.
J. Paul, D. Barret and The Eclairs Collaboration, in SF2A-2002: Semaine de l'Astrophysique Française (2002).
T. Piran, *Phys. Rep.* **314** (1999) 575.
L. Piro, G. Garmire, M. Garcia *et al.*, *Science* **290** (2000) 955.
L. Piro, J. Heise, R. Jager *et al.*, *A&A* **329** (1998) 906.
C. Porciani and P. Madau, *ApJ* **548** (2001) 522.
R.D. Preece, M.S. Briggs, R.S. Mallozzi *et al.*, *ApJS* **126** (2000) 19.
M.J. Rees and P. Meszaros, *ApJ* **430** (1994) L93.
J.N. Reeves, D. Watson, J.P. Osborne *et al.*, *Nature* **416** (2002) 512.
J.E. Rhoads, *ApJ* **487** (1997) L1.
M. Ruffert, H.-T. Janka, K. Takahashi and G. Schaefer, *A&A* **319** (1997) 122.
G.B. Rybicki and A.P. Lightman, Radiative processes in astrophysics (New York, Wiley-Interscience, 1979) 393.
F. Ryde and R. Svensson, *ApJ* **529** (2000) L13.
F. Ryde and R. Svensson, *ApJ* **566** (2002) 210.
R. Sari and T. Piran, *ApJ* **455** (1995) L143.
R. Sari and T. Piran, *ApJ* **517** (1999) L109.
R. Sari, T. Piran and R. Narayan, *ApJ* **497** (1998) L17.
H.C. Spruit, F. Daigne and G. Drenkhahn, *A&A* **369** (2001) 694.
C. Thompson, *MNRAS* **270** (1994) 480.
E. Waxman 2001, in *Lect. Notes Phys.* **576**: Physics and Astrophysics of UHECRs, 122.
R.A.M.J. Wijers, P.M. Vreeswijk, T.J. Galama *et al.*, *ApJ* **523** (1999) L33.
S.E. Woosley, *ApJ* **405** (1993) 273.

COURSE 7

COSMIC RAYS AND PARTICLE ACCELERATION AT ASTROPHYSICAL SHOCKS

A. ACHTERBERG

Sterrenkundig Instituut, Utrecht University, & Center for High Energy Astrophysics, Amsterdam, The Netherlands

Guy Pelletier presenting the course prepared by Abraham Achterberg

Contents

COSMIC RAYS AND PARTICLE ACCELERATION AT ASTROPHYSICAL SHOCKS

A. Achterberg*

Abstract

The acceleration of particles in the vicinity of astrophysical shocks has become the main paradigm in astrophysics for the production of cosmic rays particles. In this review, I consider the basic theory of shock acceleration, the observational constraints, and the open questions.

1 The birth of cosmic ray physics

Historically, a good case can be made that the birth of experimental high-energy particle physics, and of high-energy astrophysics are one and the same event: the discovery of cosmic rays. At the beginning of the twentieth century, experimental physics was fascinated by all forms of "radiation". The discovery of radioactivity by Becquerel in 1896 led to many experimental investigations into naturally occurring radiation. It soon became obvious that part of natural radioactivity does not originate on Earth, but has an extra-terrestrial origin. This was convincingly demonstrated by the Austrian physicist Victor Hess in 1912 during balloon experiments. He showed that the intensity of naturally occurring radiation *inreases* again with sufficient height above the Earth's surface, after the initital decrease that one would expect if the sources of radioactivity are confined to the radioactive elements in the Earth's crust. This increase can only be explained by the decreased shielding of an *extraterrestrial* source of radiation by the remaining column of air above the balloon.

Cosmic rays, a term coined by Robert Millikan around 1930, immediately started to play an important role in elementary particle physics. The reason is simple: cosmic rays were the most energetic particles known at

* In the absence of Dr. Achterberg, this course was presented by Dr. Guy Pelletier, Laboratoire d'Astrophysique de Grenoble, France.

that time. They are a natural source of particles which can be used to study particle interactions at high energy. Particles like the positron, muon and pion were discovered in cosmic ray experiments. Cosmic rays clearly showed the importance of high-energy collisions for experimental particle physics, and thus paved the way for the construction of particle accelerators which now have become the main experimental tool for particle physics.

Even today, cosmic radiation contains the most energetic particles known to man: modern measurements using "lightbuckets" infer particles with an energy[1] up to 3×10^{20} eV (*e.g.* Bird *et al.* 1993). These light-buckets detect the Cherenkov radiation and fluoresence from atoms induced by *airshower particles* in the upper atmosphere, essentially using the Earth's atmosphere as a particle detector. These airshowers are the result of the collisions between energetic cosmic rays and atoms at the top of the Earth's atmosphere. They consist of an avalanche of elementary particles such as pions, muons and Kaons, as well as the gamma rays which result from the decay of such particles and from pair production. Airshowers where first identified some 50 years ago by Pierre Auger, who noticed that two Geiger Counters high in the French Alps where triggered simultaneously even though they where some distance apart. Experimental techniques are reviewed in Gaiser (1990) and Longair (1992a). The (astro)physics of cosmic rays is discussed in detail in Berezinskii *et al.* (1990), Longair (1992b) and Schlickeiser (2002).

Cosmic rays are the only highly energetic particles from astrophysical, and possibly cosmological, sources whose presence can be measured *directly*. As we will see, cosmic rays up to an energy of 10^{17} eV are believed to originate in our own galaxy. The flux is nearly isotropic and does not mirror the distribution of matter in the Galaxy. This simple fact is an important indication that cosmic ray particles are scattered strongly so that any clue as to the direction of their sources is erased. The typical cosmic ray flux, $\mathcal{F}(\mathcal{E}) = \mathcal{E} \mathrm{d}N_{\rm cr}/\mathrm{d}t\mathrm{d}\mathcal{E}\mathrm{d}\Omega$, decreases rapidly with particle energy (see Table 1). Typical fluxes at the top of the atmosphere are shown in the table below.

The Galactic cosmic rays (nucleons) have a simple "broken power-law" energy distribution above ~1 GeV/nucleon where the screening effect of the magnetized Solar Wind, the so-called Solar Modulation, of the cosmic ray flux becomes negligible:

$$N(\mathcal{E})\,\mathrm{d}\mathcal{E} = \kappa\,\mathcal{E}^{-s}\,\mathrm{d}\mathcal{E}, \text{ with } s \simeq \begin{cases} 2.7 & \text{for } \mathcal{E} < 10^{15.5} \text{ eV;} \\ 3.1 & \text{for } 10^{15.5} < \mathcal{E} < 10^{18.5} \text{ eV.} \end{cases} \quad (1.1)$$

[1]As a comparison: the largest man-made particle accelerators are now producing particles up to 10 TeV = 10^{13} eV.

Table 1. Cosmic ray flux at the top of the atmosphere.

Energy $\mathcal{E}$	Particle flux $\mathcal{F}$	Origin
10^{11} eV	1 particle m^{-2} s^{-1}	Galactic (Supernova Remnants)
$10^{15.5}$ eV	1 particle m^{-2} yr^{-1}	Galactic (Supernova Remnants)
$10^{18.5}$ eV	1 particle km^{-2} yr^{-1}	Extragalactic (??)
$10^{20.5}$ eV	1 particle km^{-2} $(350\ \mathrm{yr})^{-1}$	Extragalactic (??)

This spectrum varies only slightly with the nucleon species, but at the higher energies the chemical composition shows an enrichment with the heavier nuclei from the Iron group, exactly as one expects if these particles are produced "electromagnetically" (see below) so that the energy is proportional to particle charge. The break at $10^{15.5}$ eV (usually called the "knee" in the Cosmic Ray community, with the break at $10^{18.5}$ eV referred to as the "ankle") may be the result of changing propagation characteristics within the galaxy, or may be the result of a different production site (*e.g.* Biermann *et al.* 1995; Biermann *et al.* 2001).

Above $10^{18.5}$ eV, where one usually calls the particles *Ultra-High Energy Cosmic Rays* (UHECRs), the spectrum flattens to a slope $s \approx 2.7$ (with large uncertainties at energies above 10^{20} eV due to small-number statistics) and the composition changes to predominantly protons or possibly light nuclei. Particles at these energies cannot be confined magnetically in the galaxy (unless they are heavy nuclei like Iron), so the majority opinion in the community is that these particles are extragalactic (see for instance Norman *et al.* 1995 for some of the arguments). In particular, the fact that the arrival directions are nearly isotropic on the sky suggests that the sources of these particles cannot be Galactic. A brief readable review of UHECRs and their implications is Sigl (2001).

2 Radio astronomy, magnetic fields and synchrotron radiation

With the advent of radio astronomy in the years immediately following the Second World War it was found that the radio sky is not simply a mirror image of the distribution of souces in the visible sky. One of the first sources to be identified was the *Crab Nebula*, the remnant of the supernova of 1054AD. In 1953 it became clear that the optical continuum radiation of the Crab Nebula is polarised, and is therefore non-thermal in origin. Measurements by Oort, Baade and Walraven confirmed that the emission is *synchrotron radiation* emitted by relativistic electrons spiraling in a weak magnetic field. The Russian astronomer Shklovsky had already proposed synchrotron radiation as the source of the radio emission from the Crab, which was discovered in 1949. In 1968 Hewish, Bell and co-workers discovered that the Crab Nebula contains a *pulsar*: a rapidly rotating, magnetised neutron star. This pulsar apparently powers the emission from the Crab Nebula.

Many strong radio sources are very faint optically. In 1954 Baade and Minkowski were able to identify the strong radio source Cygnus A with an optical galaxy, confirming for the first time the extragalactic nature of some of the strong radio sources.

We now know that almost all non-thermal radio emission in galactic sources, such as supernova remnants and micro-quasars (*e.g.* Mirabel & Rodríguez 1999; Mirabel 2001), and from extra-galactic sources such as active galaxies and quasars, is synchrotron radiation from relativistic electrons. The typical synchrotron radiation frequency equals, in practical units,

$$\nu_{\rm s} = 16.1\, B_{\mu{\rm G}}\, \mathcal{E}_{\rm GeV}^2 \text{ MHz}. \tag{2.1}$$

Here $B_{\mu\rm G}$ is the magnetic field strength in micro-Gauss and $\mathcal{E}_{\rm GeV}$ the electron energy in GeV. In some of these sources, synchrotron emission has now been found extending all the way up to the optical wavelength range and even up to X-rays!

From the radiation spectrum, usually a power-law with the flux scaling with frequency ν as $S_\nu \propto \nu^{-\alpha}$ in the MHz-GHz range, one can infer the energy distribution of the radiating electrons, a power law in energy, $N(\mathcal{E})\,{\rm d}\mathcal{E} \propto \mathcal{E}^{-s}\,{\rm d}\mathcal{E}$, with slope $s = 2\alpha + 1 \sim 2.5$. The fact that this distribution of GeV electrons so closely resembles the energy distribution of the Galactic cosmic ray nuclei sparked a discussion about a "universal" acceleration mechanism that should operate in a wide variety of conditions and produce a very similar power-law distribution of accelerated particles.

The predominance of synchrotron radiation also focussed the attention on the importance of magnetic fields in the physics of energetic particles in astrophysics. This is particularly true for the Galactic cosmic rays, which are believed to be confined in the galaxy for $\sim 10^7$ years. This follows from the chemical composition of the *secondary* cosmic rays produced by spallation in collisions between the primary cosmic rays and nuclei in the interstellar gas, and in particular from the relative abundance of radioactive and stable isotopes of Berillium and Aluminium which can be used as radioactive clocks. This confinement must be the result of the scattering of cosmic rays by irregularities in the $\sim 1-10\ \mu$G galactic magnetic field, which reduces the mean-free-path along the mean field to $\lambda_{\rm cr} \sim 1$ pc at $\mathcal{E} \sim 1$ GeV.

3 Cosmic ray origin theories

Almost eighty years ago, Baade & Zwicky (1934) proposed that supernovae are the sources of galactic cosmic rays. Their suggestion was based on a simple energy argument which, in a modernized form, still survives. A supernova is the manifestation of the death of a massive star ($M > 10\ M_{\odot}$) which has exhausted its nuclear fuel and collapses under its own weight. Without the energy supply of nuclear fusion, gas pressure is no longer capable of supporting the core. The amount of energy liberated when the core collapses into a neutron star is essentially the gravitational binding energy of the core at the moment it "bounces". This bounce is due to the change in the equation of state of the material in the collapsing core when it is compressed to nuclear densities ($\rho \simeq \rho_{\rm nuc} = 10^{14}$ g cm^{-3}). The material becomes "stiff", preventing a further collapse. With a core mass of $M_{\rm c} \sim 1.5\ M_{\odot}$ and a core radius of $R_{\rm c} \sim 10$ km the binding energy is $E_{\rm b} \sim GM_{\rm c}^2/R_{\rm c} \approx 10^{53}$ erg. About 99% of this energy is radiated away in the form of neutrino's, which carry away the core's lepton number as the core material neutronizes (and de-leptonizes) in the reaction $\rm p + e^- \longrightarrow n + \nu_e$. The remaining ∼1% of the energy is used to drive a shock wave through the stellar envelope. The pressure forces associated with the passage of this shock reverse the gravitational collapse of the stellar mantle, leading to an expanding shell of stellar material. This shell is preceded in the surrounding environment by a strong blastwave. When this shell starts to decelerate after sweeping up roughly its own mass in surrounding material, either from the interstellar medium, a progenitor wind or both, a strong *reverse shock* propagates into the remnant (*e.g.* Draine & McKee 1993), reheating the stellar ejecta, leading to the formation of a supernova remnant. The amount of mechanical energy put into the Galaxy in this manner corresponds to a power

$$P_{\rm snr} = \mathcal{R}_{\rm snr}\, E_{\rm snr} \approx 3 \times 10^{41}\ {\rm erg/s}. \tag{3.1}$$

Here $\mathcal{R}_{\rm snr} \sim 0.01-0.03$ yr^{-1} is the supernova rate, and $E_{\rm snr} \sim 10^{51}$ erg is the mechanical energy associated with the supernova explosion. The power needed to maintain the Galactic cosmic ray population at its current level against losses due to cosmic ray escape from the Galaxy is

$$P_{\rm cr} = U_{\rm cr} \mathcal{V}_{\rm cr}/t_{\rm cr} \approx 10^{40}\ {\rm erg/s}. \tag{3.2}$$

Here $U_{\rm cr} \approx 0.5$ eV/cm^3 is the cosmic ray energy density, a quantity that can be determined from direct measurements at Earth, $\mathcal{V}_{\rm cr} \sim 400$ kpc^3 is the volume of the cosmic ray halo around the galactic disk, estimated to have a thickness of about 500 pc, and $t_{\rm cr} \sim 5 \times 10^6$ yr is the typical residence time of a cosmic ray in the galaxy. Comparing these two rates one sees that supernovae alone could maintain the cosmic ray population provided that about 3% of their mechanical energy is somehow converted into cosmic rays.

It is interesting to note that a very similar line of reasoning has been applied more recently to establish a possible connection between UHECRs above $10^{18.5}$ eV and the most powerful explosions known to man (after the Big Bang): *Gamma Ray Bursts* (GRBs). Waxman (1995) noted that the typical average energy flux in gamma rays from GRBs and the energy flux of UHECRs are of similar magnitude. He argued for a common origin, associated with the relativistic shocks which occur in the *relativistic fireballs* which occur in almost all models for Gamma Ray Bursts (for reviews of these models see Piran 1999 and Mészáros 2001, 2002), with a similar luminosity in gamma rays and UHECRs. Waxman's suggestion is not uncontroversial (*e.g.* Stecker 2000; Scully & Stecker 2002), and I will return to this point later.

4 Astrophysical particle acceleration

One of the first two things to realize about the acceleration of particles in the astrophysical context is the fact that all acceleration mechanisms are [1] *electromagnetic* in nature and [2] must involve material *motions* in the form of waves or shocks in a *magnetized* and ionized gas, such as the hot-phase interstellar medium. The first is a simple consequence of the fact that cosmic rays are collisionless in the strict sense of the word, for instance: the Galactic disk is completely transparent to cosmic rays for Coulomb collisions with nuclei in the interstellar medium. Although collisions play a role as an energy-*loss* mechanism (*e.g.* ionization losses or pion-production losses at high energy) they are totally unimportant for the acceleration of particles. The second is a consequence of the fact that astrophysical plasma's are highly conducting, so that any large-scale electric field is given by the well-known condition of ideal magnetohydrodynamics (MHD):

$$\boldsymbol{E} = -\frac{\boldsymbol{V}}{c} \times \boldsymbol{B} = -\Gamma\boldsymbol{\beta} \times \tilde{\boldsymbol{B}}. \tag{4.1}$$

Here $\boldsymbol{V} = \boldsymbol{\beta}c$ is the bulk velocity of the plasma, $\Gamma = 1/\sqrt{1-\beta^2}$ the Lorentz-factor associated with the bulk motion, $\boldsymbol{B}$ is the magnetic field in the laboratory frame, and $\tilde{\boldsymbol{B}}$ is the magnetic field in the rest frame of the material. This large-scale electric field is essential for acceleration as the magnetic component of the Lorentz force on a particle does no work: the equation of motion for a charged particle gives the change of kinetic energy $K = (\gamma - 1)mc^2$ as

$$\frac{\mathrm{d}K}{\mathrm{d}t} = q(\boldsymbol{E} \cdot \boldsymbol{v}). \tag{4.2}$$

Here q is the particle charge and $\boldsymbol{v}$ the particle velocity. As we will see below, most models still rely on a form of scattering, not by particles but by the magnetic fields associated with low-frequency waves.

The importance of bulk motion was first realized by Fermi (1949), who described two simple acceleration models which have served as "templates" for all subsequent work: *stochastic* (Fermi-II) acceleration by scattering off randomly moving magnetized clouds in a turbulent medium, and regular (Fermi-I) acceleration by reflection off shocks. In the first model particles diffuse in energy in such a way that the *mean* energy per particle increases, a simple consequence of the fact that head-on collisions which result in an energy gain are slighly more frequent than overtaking collisions in which a particle loses energy. In the second model all particles systematically gain energy. The connection between acceleration and the presence of hydromagnetic waves was already suggested for the Galactic cosmic rays by Parker (1955), long before the existence of the weak (micro-Gauss) interstellar magnetic field was established experimentally. Parker considered a simple physical realization of the Fermi-II model in which the moving clouds are replaced by MHD waves. His suggestion was followed by many stochastic acceleration models where particles diffuse in energy due to the random momentum changes induced by the electric field of low-frequency MHD waves (*e.g.* Hall & Sturrock 1967a). Such models have now gone largely out-of-fashion, and I will not consider them here.

4.1 Shocks as accelerators

Shocks were first considered in some detail as a possible source of particle acceleration by Colgate & Johnson (1960) in the context of cosmic ray production by supernova remnants. Their idea was a simple one: a shock propagating down a steep density gradient, as exists in the outer layers of an exploding supernova, accelerates to near-relativistic velocities, so that the energy per particle is large. Early calculations of the interaction between charged particles and a thin shock in a magnetised plasma were done by Shatzman (1967) Some of the first theories considered shock acceleration as a one-shot process. If the shock speed $V_{\rm s}$ is much less than the particle speed v, and if the shock can be considered infinitesimally thin, the net momentum gain after interaction with the shock follows from the conservation of the magnetic moment in the encounter (*e.g.* Hudson 1965 and Alekseyev & Kroptokin 1970),

$$\mathcal{M} = \frac{p_\perp^2}{2|\boldsymbol{B}|} = \text{constant}. \tag{4.3}$$

Here $p_\perp$ is the component of particle momentum perpendicular to the magnetic field. The energy gain results from the field compression in the shock: the downstream field $|\boldsymbol{B}_{\rm d}|$ exceeds the upstream field $|\boldsymbol{B}_{\rm u}|$. This conservation law for $\mathcal{M}$ remains valid as long as the crossing event involves many gyration loops. It can be seen as an form of adiabatic invariance (Whipple *et al.* 1986). This process is known as *Shock Drift Acceleration* (SDA): the particle drifts along the shock face in the direction of the flow-induced electric field, $\boldsymbol{E} = -(V_{\rm s}\hat{\boldsymbol{n}} \times \boldsymbol{B})/c$ with $\hat{\boldsymbol{n}}$ the shock normal. The drift exhibited by shock-intersecting orbits results from the smaller radius of curvature of a downstream section of a gyro-orbit compared to its upstream value. Shock Drift Acceleration does not have an equivalent in (ultra-)relativistic shocks: since $V_{\rm s} \sim c$ a crossing always involves a single (small) section of a gyro-orbit, and the adiabatic limit never applies. Shock drift acceleration is essentially a single-particle collisionless mechanism which occurs if particle orbits near the shock are undisturbed by the scattering effect of the fluctuating electromagnetic fields of hydromagnetic waves.

In 1977/78, a number of researchers independently realized that particles can be confined near a shock through the scattering action of hydromagnetic waves (Axford *et al.* 1977; Krimsky 1977; Bell 1978; Blandford & Ostriker 1978). This process, known as *Diffusive Shock Acceleration* (DSA), assumes that particles have a scattering mean free path $\ell_{\rm mfp}$ much larger than the shock thickness. As a result particles can cross the shock *repeatedly*.

Diffusive shock acceleration operates also for *parallel shocks*, where $\hat{\boldsymbol{n}} \parallel \boldsymbol{B}_{\rm u}$. In these shocks the magnetic field is not amplified, and the acceleration mechanism is a realization of *regular* (Fermi type II) acceleration where a particle gains energy through repeated scattering by converging scattering centers, much like the momentum gain experienced by table-tennis bal, trapped in a bouncing motion between the table and a bat while the bat moves towards the table. The momentum gain, together with the escape probability at each crossing, determines the slope of the resulting spectrum (see below).

In the general case of oblique shocks, the momentum gain can be thought of as resulting from a combination of the Fermi-II process and the drift process of SDA. It can be shown however that the *mean* momentum gain is the same as in the case of a purely parallel shock (Drury 1983) as long as frequent scattering by MHD waves keeps the particle distribution almost isotropic near the shock.

In DSA, particle gain momentum in each cycle where they cross the shock into the downstream medium, and are scattered back upstream. At the same time there is a finite escape chance per cycle: not all particles return to the shock once they enter the downstream region. The competition between the relative momentum gain $\Delta p/p$ per cycle and the escape probability $\mathcal{P}_{\rm esc}$ per cycle determines the spectrum of the accelerated particles: particles can reach a high energy by many crossings, but only a small fraction succeeds in doing so. If both are independent of momentum (or have the same momentum dependence) one finds a power-law distribution for the accelerated particles (Fermi 1949):

$$N(p)\,{\rm d}p \propto p^{-s}\,{\rm d}p, \;\text{with}\;\; s = 1 + \frac{\ln(1/\mathcal{P}_{\rm ret})}{\ln\langle p_{\rm f}/p_{\rm i}\rangle}\cdot \tag{4.4}$$

Here $\mathcal{P}_{\rm ret} \equiv 1 - \mathcal{P}_{\rm esc}$ is the return probability per crossing cycle and $\langle p_{\rm f}/p_{\rm i}\rangle$ is the average ratio of the final and initial momentum in a cycle.

One of the main attractions of the DSA is that (in absence of losses) the spectral slope of the momentum distribution of accelerated particles does not depend on the details of the flow, such as the magnetic field orientation near the shock, or on the precise mechanism of particle diffusion. It also naturally gives a power-law distribution, with a slope s which is determined solely by the shock compression ratio $r \equiv \rho_2/\rho_1$:

$$s = \frac{r+2}{r-1}\cdot \tag{4.5}$$

Here ρ_1 (ρ_2) is the density ahead (behind) the shock. This makes Diffusive Shock Acceleration -in principle- a very robust mechanism which -at first sight- fits the requirement of the aforementioned "universal" acceleration mechanism. However, as we will see, the above result is only valid as long as the back-reaction of the accelerated particles on the shock and pre-shock flow can be neglected.

I will consider these mechanisms, and the generalization to relativistic shocks, in some detail. Figure 1 below gives a cartoon of the different types of acceleration I will consider in this review.

5 Basic principles of diffusive shock acceleration

In shock acceleration, particles gain energy by repeatedly crossing a shock going through a cycle

$$\text{upstream} \Longrightarrow \text{downstream} \Longrightarrow \text{upstream}$$

This crossing cycle is made possible by efficient scattering in the fluid on both sides of the shock, presumably by hydromagnetic waves. For the moment I will assume that this scattering is elastic in the rest frames of the up- and downstream fluids. For simplicity I consider a normal shock, where the shock surface is in the $x-y$ plane, with the shock propagating in the

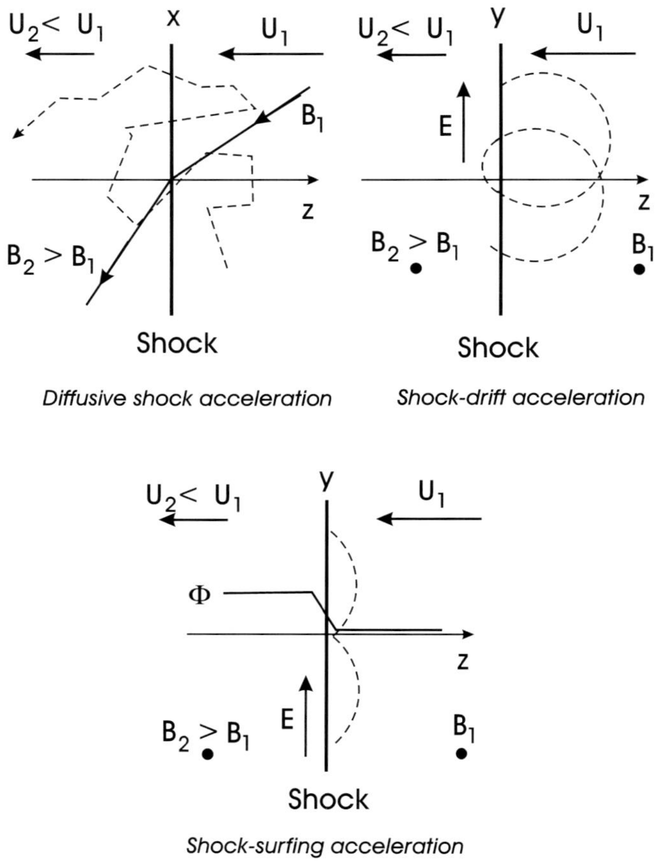

Fig. 1. Shock acceleration mechanisms considered in this review: diffusive shock acceleration (DSA), which relies on the velocity jump created by the shock together with frequent scattering of particles near the shock, collisionless shock-drift acceleration (SDA) that relies on the jump in the magnetic field in the shock, and shock-surfing acceleration which relies on reflection by a potential jump in the shock. Typical particle orbits are shown as dashed lines/curves, as well as the velocities U_1 (U_2) ahead (behind) the shock, the pre- and post-shock magnetic fields B_1 and B_2, the convective electric field $E_1 = E_2 \equiv E = VB/c$ and, in the case of shock-surfing acceleration, the electrostatic potential Φ which reflects ions at the shock. The pictures are in the *normal incidence frame* where the shock is at rest, and the plasma flows along the shock normal, which is aligned with the z-axis. The shock has been simplified as an infinitely thin surface in the $x - y$ plane, the magnetic field is in the $x - z$ plane and the convective electric field along the y-axis. This convention is adheared to throughout this review.

positive z-direction. The energy gain in each cycle can be calculated by using Lorentz-transformations between the rest frames of the upstream- and downstream fluid, where the scattering centers are at rest and where only the direction of motion changes in a scattering event, but not the particle energy. The up- and downstram frames will be denoted by K and $\overline{K}$ respectively. In a frame where the shock is stationary at $z = 0$, the frames K and $\overline{K}$ move with the velocity $-U_1$ and $-U_2$ respectively along the $z-$axis. Note that for a normal shock U_1 equals the shock velocity $V_{\rm s}$ if the upstream material is at rest. The relative velocity between the scattering centers follows from the relativistic velocity addition law:

$$U_{\rm rel} = \frac{U_1 - U_2}{1 - \dfrac{U_1 U_2}{c^2}} . \tag{5.1}$$

This quantity corresponds with the velocity with which the downstream scattering centers approach an upstream observer, in the positive z-direction. Consider a particle with energy $\mathcal{E}_{\rm i}$ which crosses the shock into the downstream region at an angle $\theta_{\rightarrow {\rm d}}$ with respect to the shock normal so that $v_z = v_{\rm i} \cos\theta_{\rightarrow {\rm d}}$. After a number of scatterings in the downstream flow, the particle again crosses the shock, re-entering the upstream medium with the velocity making an angle $\theta_{\rightarrow {\rm u}}$ with the shock normal, and with an energy $\mathcal{E}_{\rm f}$. All these quantities are measured in the frame K of an observer moving with the upstream flow. Seen from the point of view of an *down*stream observer, scattering in the frame $\overline{K}$ has changed the direction of motion of the particle, but *not* its energy. In terms of the energy $\overline{\mathcal{E}}$ measured in the downstream frame this implies:

$$\overline{\mathcal{E}}_{\rm i} = \Gamma_{\rm rel} \, \left[\mathcal{E}_{\rm i} - U_{\rm rel} \, p_{z{\rm i}}\right] = \overline{\mathcal{E}}_{\rm f} = \Gamma_{\rm rel} \, \left[\mathcal{E}_{\rm f} - U_{\rm rel} \, p_{z{\rm f}}\right] , \tag{5.2}$$

where $\Gamma_{\rm rel} = \left(1 - U_{\rm rel}^2/c^2\right)^{-1/2}$, p_z is the component of the particle momentum along the z-axis, and the subscripts "i" and "f" are used to denote the state of the particle as it enters and leaves the downstream flow. Using the relations

$$p_{z{\rm i}} = \frac{\mathcal{E}_{\rm i} v_{\rm i}}{c^2} \cos\theta_{\rightarrow {\rm d}} , \quad p_{z{\rm f}} = \frac{\mathcal{E}_{\rm f} v_{\rm f}}{c^2} \cos\theta_{\rightarrow {\rm u}} \tag{5.3}$$

one can immediately write down an expression for the ratio of the final and initial energy:

$$\boxed{\frac{\mathcal{E}_{\rm f}}{\mathcal{E}_{\rm i}} = \frac{1 - \dfrac{U_{\rm rel} v_{\rm i} \mu_{\rightarrow {\rm d}}}{c^2}}{1 - \dfrac{U_{\rm rel} v_{\rm f} \mu_{\rightarrow {\rm u}}}{c^2}} .} \tag{5.4}$$

Here I have defined $\mu_{\rightarrow\mathrm{d}} = \cos\theta_{\rightarrow\mathrm{d}}$ and $\mu_{\rightarrow\mathrm{u}} = \cos\theta_{\rightarrow\mathrm{u}}$. As $-1 \leq \mu_{\rightarrow\mathrm{d}} < U_1/v_\mathrm{i}$ and $U_1/v_\mathrm{f} < \mu_{\rightarrow\mathrm{u}} \leq 1$ one always has $\mathcal{E}_\mathrm{f}/\mathcal{E}_\mathrm{i} \geq 1$. A similar calculation from the point-of-view of a downstream observer for a cycle where a particle moves from downstream to upstream and back gives a similar result.

Equation (5.4) is an implicit equation since the left-hand-side still contains the velocities v_i and v_f. It can be solved relatively simply in two important cases.
If the shock is slow, so that $U_\mathrm{rel}v/c^2 \ll 1$, the energy- and velocity change incurred in each crossing cycle will be small. One can then expand equation (5.4) to first order in $U_\mathrm{rel}v/c^2 \ll 1$, using $v_\mathrm{f} = v_\mathrm{i} + \mathcal{O}(U_\mathrm{rel}/v)$. The energy change in one cycle is small, corresponding to:

$$\frac{\Delta\mathcal{E}}{\mathcal{E}} = \frac{\mathcal{E}_\mathrm{f} - \mathcal{E}_\mathrm{i}}{\mathcal{E}_\mathrm{i}} \approx \left(\frac{vU_\mathrm{rel}}{c^2}\right) \left[\mu_{\rightarrow\mathrm{u}} - \mu_{\rightarrow\mathrm{d}}\right]. \tag{5.5}$$

The second important case is that of ultra-relativistic particles with Lorentz-factor $\gamma \gg 1$. In that case one can put $v_\mathrm{i} = v_\mathrm{f} = c$, and one immediately finds, with $\beta_\mathrm{rel} \equiv U_\mathrm{rel}/c$:

$$\frac{\mathcal{E}_\mathrm{f}}{\mathcal{E}_\mathrm{i}} = \frac{1 - \beta_\mathrm{rel}\mu_{\rightarrow\mathrm{d}}}{1 - \beta_\mathrm{rel}\mu_{\rightarrow\mathrm{u}}}. \tag{5.6}$$

In order to calculate the *mean* energy change, one has to average over different values of $\mu_{\rightarrow\mathrm{d}}$ and $\mu_{\rightarrow\mathrm{u}}$. This must be an average over the *flux* of particles crossing the shock, a quantity most easily defined in the shock rest frame. In the shock rest-frame $\tilde{K}$, the number of particles in a momentum interval $\mathrm{d}^3\tilde{\boldsymbol{p}}$, crossing the shock per unit area and per unit time with momentum $\tilde{\boldsymbol{p}}$, equals

$$\mathrm{d}\tilde{\mathcal{F}} = \tilde{f}(\tilde{\boldsymbol{x}},\, \tilde{\boldsymbol{p}},\, \tilde{t})\, (\tilde{\boldsymbol{v}} \cdot \tilde{\boldsymbol{n}})\, \mathrm{d}^3\tilde{\boldsymbol{p}}. \tag{5.7}$$

Here a $\tilde{}$ is used to denote quantities in the shock rest frame, with $\tilde{\boldsymbol{v}}$ the particle velocity and $\tilde{\boldsymbol{n}}$ the shock normal, which points into the upstream medium. Particles crossing from downstream to upstream have $\tilde{\boldsymbol{v}} \cdot \tilde{\boldsymbol{n}} = \tilde{v}_z > 0$, and those crossing the shock in the opposite direction have $\tilde{\boldsymbol{v}} \cdot \tilde{\boldsymbol{n}} < 0$. The quantity $\tilde{f}(\tilde{\boldsymbol{x}},\, \tilde{\boldsymbol{p}},\, \tilde{t}) = \mathrm{d}\tilde{N}/(\mathrm{d}^3\tilde{\boldsymbol{x}}\, \mathrm{d}^3\tilde{\boldsymbol{p}})$ is the phase-space density of the particles. By defining a space-like four-vector[2] n^μ, which reduces to $\tilde{n}^\mu \equiv (0,\, \tilde{\boldsymbol{n}}) = (0,\, 0,\, 0,\, 1)$ in the shock rest frame, one can write this flux as

$$\mathrm{d}\tilde{\mathcal{F}} = -\tilde{f}(\tilde{\boldsymbol{x}},\, \tilde{\boldsymbol{p}},\, \tilde{t})\, \tilde{u}^\mu \tilde{n}_\mu\, \frac{\mathrm{d}^3\tilde{\boldsymbol{p}}}{\tilde{\gamma}}. \tag{5.8}$$

[2] I will denote components of four-vectors with greek indices, and use a Minkowski metric $\eta^{\mu\nu} = \mathrm{diag}(1,\, -1,\, -1,\, -1)$ so that the scalar product of two four-vectors A^μ and B^μ equals $A \cdot B = A^0 B^0 - \boldsymbol{A} \cdot \boldsymbol{B}$.

Here $\tilde{u}^{\mu} = \tilde{\gamma}(c, \tilde{\boldsymbol{v}})$ is the four-velocity of the particle and $\tilde{\gamma}$ its Lorentz factor in the shock rest frame. A remarkable property of this expression is that it is Lorentz-invariant! The phase-space density $\tilde{f}(\tilde{\boldsymbol{x}}, \tilde{\boldsymbol{p}})$ and the quantity $\mathrm{d}^3\tilde{\boldsymbol{p}}/\tilde{\gamma}$ are related to the corresponding quantities in another frame by

$$\tilde{f}(\tilde{\boldsymbol{x}}, \tilde{\boldsymbol{p}}, \tilde{t}) = f(\boldsymbol{x}, \boldsymbol{p}, t), \quad \frac{\mathrm{d}^3\tilde{\boldsymbol{p}}}{\tilde{\gamma}} = \frac{\mathrm{d}^3\boldsymbol{p}}{\gamma}, \tag{5.9}$$

provided the variables $\boldsymbol{x}$, $\boldsymbol{p}$ and t are connected through Lorentz-transformations to those in the shock rest frame (*e.g.* Rybicki & Lightman 1979). The scalar product $u^{\mu}n_{\mu}$ is an invariant by construction if n^{μ} and $\tilde{n}^{\mu}$ are related by Lorentz-transformations. In the upstream frame where the shock moves with velocity $U_1 = V_{\mathrm{s}}$ along the x-axis, the four-vector n^{μ} is

$$n^{\mu} = \Gamma_1\,(\beta_1,\, 0,\, 0,\, 1), \tag{5.10}$$

with $\beta_1 = U_1/c$ and $\Gamma_1 = 1/\sqrt{1-\beta_1^2}$. This means that the number of particles at a given momentum crossing the shock per unit area and unit time in the shock rest frame can be expressed in upstream frame variables as

$$\begin{aligned} \mathrm{d}\tilde{\mathcal{F}} &= -f(\boldsymbol{x}, \boldsymbol{p}, t)\, u^{\mu}n_{\mu}\,\frac{\mathrm{d}^3\boldsymbol{p}}{\gamma} \\ &= \Gamma_1\, f(\boldsymbol{x}, \boldsymbol{p}, t)\,\left(v\,\mu - U_1\right)\,\mathrm{d}^3\boldsymbol{p}. \end{aligned} \tag{5.11}$$

Here $\mu = \cos\theta$ is cosine of the angle between the particle velocity and the $z-$axis.

Analytical results are easily obtained in the case of non-relativistic shocks and flows, where $U_1 \ll c$ and $U_{\mathrm{rel}} \approx U_1 - U_2 \ll c$. In that case the upstream distribution can be considered as nearly isotropic in the rest frame of the medium. One can write the upstream-frame distribution just ahead of the shock as:

$$f(\boldsymbol{x} = \boldsymbol{x}_{\mathrm{s}}, \boldsymbol{p}, t)\,\mathrm{d}^3\boldsymbol{p} = f_1(p)\,p^2\mathrm{d}\phi\,\mathrm{d}\mu. \tag{5.12}$$

Here I have defined the polar angle $0 \leq \phi < 2\pi$ around the z-axis, and the phase-space density $f_1(p)$ immediately ahead of the shock. The mean relative energy gain follows from equation (5.5) as

$$\langle\langle\, \frac{\Delta\mathcal{E}}{\mathcal{E}}\,\rangle\rangle = \frac{v\,(U_1 - U_2)}{c^2}\,\left(\langle\langle\, \mu_{\rightarrow\mathrm{u}}\,\rangle\rangle - \langle\langle\, \mu_{\rightarrow\mathrm{d}}\,\rangle\rangle\right). \tag{5.13}$$

Here brackets $\langle\langle\cdots\rangle\rangle$ denote a flux average. From equations (5.11) and (5.12) it follows that the flux averages of $\mu_{\to\mathrm{d}}$ and $\mu_{\to\mathrm{u}}$ are

$$\langle\langle\, \mu_{\to\mathrm{d}}\,\rangle\rangle = \frac{\displaystyle\int_{-1}^{U_1/v} \mathrm{d}\mu\;\mu\;[\mu-(U_1/v)]}{\displaystyle\int_{-1}^{U_1/v} \mathrm{d}\mu\;[\mu-(U_1/v)]} = -\frac{2}{3}\left[1-\frac{U_1}{2v}\right], \tag{5.14}$$

and

$$\langle\langle\, \mu_{\to\mathrm{u}}\,\rangle\rangle = \frac{\displaystyle\int_{U_1/v}^{1} \mathrm{d}\mu\;\mu\;[\mu-(U_1/v)]}{\displaystyle\int_{U_1/v}^{1} \mathrm{d}\mu\;[\mu-(U_1/v)]} = \frac{2}{3}\left[1+\frac{U_1}{2v}\right]. \tag{5.15}$$

This immediately gives

$$\langle\langle\, \mu_{\to\mathrm{u}} - \mu_{\to\mathrm{d}}\,\rangle\rangle = \frac{4}{3}, \tag{5.16}$$

independent of U_1/v. Of course, these equations are only valid if $v > U_1$ so that particles can outrun the shock (until the next scattering) in the upstream flow. This yields:

$$\boxed{\left\langle\left\langle \frac{\Delta\mathcal{E}}{\mathcal{E}} \right\rangle\right\rangle = \frac{4}{3}\,\frac{v\;(U_2-U_1)}{c^2}.} \tag{5.17}$$

Because of the scattering in the upstream flow, particles diffuse in the upstream medium, and are ultimately always overtaken by the shock. However, in the downstream medium they have a finite chance that they never see the shock again if –by chance– they are advected too far away from the shock before they are scattered. The return probability $\mathcal{P}_{\mathrm{ret}}$ and the corresponding escape probability $\mathcal{P}_{\mathrm{esc}} = 1-\mathcal{P}_{\mathrm{ret}}$ can be determined by calculating the shock-frame flux $\tilde{\mathcal{F}}_+$ across the shock into the upstream medium, and the flux $\tilde{\mathcal{F}}_-$ back into the downstream medium under the assumption that frequent scattering in the downstream flow keeps the post-shock distribution nearly isotropic. Using the Lorentz-invariance properties once again, the flux due to particles with momentum $\overline{p} = |\overline{\boldsymbol{p}}|$ in the range $\overline{p}$, $\overline{p}+\mathrm{d}\overline{p}$, and with direction cosine in the range $\overline{\mu}$, $\overline{\mu}+\mathrm{d}\overline{\mu}$ equals:

$$\mathrm{d}\tilde{\mathcal{F}} = 2\pi\Gamma_2\,\overline{p}^2\,\overline{f}_2(\overline{p})\;(\overline{v}\,\overline{\mu}-U_2)\;\mathrm{d}\overline{\mu}\mathrm{d}\overline{p}, \tag{5.18}$$

with $\overline{f}_2(\overline{p})$ the downstream phase-space density immediately behind the shock, and $\Gamma_2 = 1/\sqrt{1-U_2^2/c^2}$.

Here I have integrated over the polar angle $\overline{\phi}$. The flux across the shock into the upstream medium of particles in momentum-interval $\mathrm{d}\overline{p}$ is

$$\begin{aligned}\tilde{\mathcal{F}}_{\to\mathrm{u}}(\overline{p}) &= 2\pi\Gamma_2\,\overline{p}^2\,\Gamma_2\,\overline{f}_2(\overline{p})\,\mathrm{d}\overline{p}\int_{U_2/\overline{v}}^{1}\mathrm{d}\overline{\mu}\;(\overline{v}\,\overline{\mu}-U_2)\\ &= \pi\Gamma_2\,\overline{p}^2\,\overline{v}\,\overline{f}_2(\overline{p})\,\mathrm{d}\overline{p}\left[1-\frac{U_2}{\overline{v}}\right]^2.\end{aligned} \tag{5.19}$$

A similar calculation gives the flux across the shock back into the downstream medium:

$$\begin{aligned}\tilde{\mathcal{F}}_{\to\mathrm{d}}(\overline{p}) &= 2\pi\Gamma_2\,\overline{p}^2\,\overline{f}_2(\overline{p})\,\mathrm{d}\overline{p}\int_{-1}^{U_2/\overline{v}}\mathrm{d}\overline{\mu}\;(\overline{v}\,\overline{\mu}-U_2)\\ &= -\pi\Gamma_2\,\overline{p}^2\,\overline{v}\,\overline{f}_2(\overline{p})\,\mathrm{d}\overline{p}\left[1+\frac{U_2}{\overline{v}}\right]^2.\end{aligned} \tag{5.20}$$

The minus sign reflects that this flux is the direction of negative z.

The ratio of the absolute value of these two fluxes must equal the probability that a particle, once downstream, returns to the shock to begin a new cycle:

$$\mathcal{P}_{\mathrm{ret}} = \frac{\tilde{\mathcal{F}}_{\to\mathrm{u}}}{|\tilde{\mathcal{F}}_{\to\mathrm{d}}|} = \left(\frac{\overline{v}-U_2}{\overline{v}+U_2}\right)^2 = 1-\mathcal{P}_{\mathrm{esc}}. \tag{5.21}$$

This expression was first derived by Peacock (1981) for the case $v\sim\overline{v}\approx c$. For non-relativistic shocks with $U_2\ll\overline{v}\leq c$ one has

$$\mathcal{P}_{\mathrm{ret}}\approx 1-\frac{4U_2}{\overline{v}},\quad \mathcal{P}_{\mathrm{esc}}\approx\frac{4U_2}{\overline{v}}\ll 1, \tag{5.22}$$

so that particles can complete many crossing cycles before escaping into the downstream flow.

The spectrum that results from diffusive shock acceleration now follows from a simple balance argument. Particles starting the upstream $\Longrightarrow$ downstream $\Longrightarrow$ upstream cycle with initial value of the momentum eqial to $p_{\mathrm{i}}\equiv p$ will return on average with a larger momentum $\langle\langle\, p_{\mathrm{f}}\,\rangle\rangle = p+\Delta p$. Once upstream, they will always start a new cycle at the new momentum. Given a return probability $\mathcal{P}_{\mathrm{ret}}$, this means that in a steady state the flux $\tilde{\mathcal{F}}_{\to\mathrm{d}}(p+\Delta p)$ at the next cycle, now expressed in terms of upstream variables, must be related to the flux $\tilde{\mathcal{F}}_{-}(p)$ by

$$\tilde{\mathcal{F}}_{\to\mathrm{d}}(p+\Delta p) = \mathcal{P}_{\mathrm{ret}}\,\tilde{\mathcal{F}}_{\to\mathrm{d}}(p). \tag{5.23}$$

A simple calculation similar to the one leading to equation (5.20) gives

$$\tilde{\mathcal{F}}_{\to \mathrm{d}}(p) = -\pi\Gamma_1\, p^2 v\, f_1(p)\, \mathrm{d}p\, \left[1 + \frac{U_1}{v}\right]^2 , \tag{5.24}$$

assuming once again that the upstream distribution $f_1(p)$ at the shock can be considered to be isotropic to lowest order. I will assume that $v \gg U_{1,2}$ in what follows, in fact a necessary condition in view of the isotropy assumption. This restricts this discussion to non-relativistic shocks. In that case condition (5.23) becomes, to lowest order in U_1/v:

$$\left(p^2 v f_1(p)\, \mathrm{d}p\right)_{p+\Delta p} = \left[1 - \frac{4U_2}{v}\right]\, p^2 v f_1(p)\, \mathrm{d}p, \tag{5.25}$$

with

$$\frac{\Delta p}{p} = \frac{4}{3}\, \frac{U_1 - U_2}{v} \ll 1. \tag{5.26}$$

This last relation follows directly from equation (5.17), together with $\Delta\mathcal{E} = v\Delta p$. We have assumed here that no *new* particles are added to the population by injection at the shock. Expanding to first order in Δp, using the relation

$$\mathrm{d}(p + \Delta p) = \mathrm{d}p\, \left(1 + \frac{\mathrm{d}\Delta p}{\mathrm{d}p}\right) \tag{5.27}$$

which describes the "stretching" of the momentum interval $\mathrm{d}p$ in which particles will be found, one finds a simple differential equation for the phase-space density $f_1(p)$ at the shock:

$$\frac{\mathrm{d}}{\mathrm{d}p}\, \left(p^2 v\, \Delta p\, f_1(p)\right) = -\left(\frac{4U_2}{v}\right)\, p^2 v\, f_1(p). \tag{5.28}$$

Using expression (5.26) for Δp one immediately finds:

$$\frac{\mathrm{d}}{\mathrm{d}p}\left(p^3 f_1(p)\right) = -\left(\frac{3U_2}{U_1 - U_2}\right)\, p^2 f_1(p). \tag{5.29}$$

The solution is a simple power-law in momentum, where the phase-space density at the shock satisfies

$$f_1(p) = \kappa\, p^{-\tilde{s}} \ \text{ with } \ \tilde{s} = \frac{3U_1}{U_1 - U_2}\cdot \tag{5.30}$$

The differential momentum distribution is

$$dN = 4\pi p^2 \, f_1(p) \, dp \propto p^{-s} \, dp, \tag{5.31}$$

where the slope $s = \tilde{s} - 2$ equals

$$s = \frac{U_1 + 2U_2}{U_1 - U_2} = \frac{r+2}{r-1} \equiv s_r. \tag{5.32}$$

Here I define the quantity

$$r \equiv U_1/U_2 = \rho_2/\rho_1, \tag{5.33}$$

which is the compression ratio of the shock. The last equality follows from the fact that the mass-flux across the shock, $\mathcal{J} \equiv \rho U$ with ρ the mass density in the flow, must be conserved.

The slope s depends only on the compression ratio r, and result (5.31) is a simple realization of Fermi's formula (4.4), as is easily seen by noting that for $U_{1,2} \ll v$ one has:

$$\ln(1/\mathcal{P}_{\rm ret}) \approx \frac{4U_2}{v}, \quad \ln\langle p_{\rm f}/p_{\rm i}\rangle \approx \frac{4}{3}\frac{U_1 - U_2}{v}. \tag{5.34}$$

6 Diffusion and acceleration time

The results derived so far have used a single assumption: the near-isotropy of the particle distribution near the shock. It is commonly assumed that this isotropy is the result of frequent scattering by low-frequency hydromagnetic waves, the same proces responsible for the confinement of cosmic rays in the galaxy for some 10^7 years. Whenever cosmic rays try to stream with respect to a magnetized plasma with magnetic field B and density ρ, with a net transport velocity exceeding the Alfvén velocity $V_{\rm A} = B/\sqrt{4\pi\rho}$, the cosmic rays will generate of Alfvén waves through a *gyroresonant instability*. The wavelength λ of these waves is comparable with the particle gyration radius: $\lambda \sim r_{\rm g} \sim pc/qB$, with p the particle momentum, q the particle charge and B the strength of the ambient magnetic field (Kennel & Petchek 1966; Hall & Sturrock 1967b; Wentzel 1974). The self-generated waves provide the magnetic fields which deflect the particles and limit the mean free path of a charged particle along the field to a value

$$\ell_{\rm mfp} \sim r_{\rm g}/\mathcal{I}_{\rm magn}(k \sim r_{\rm g}^{-1}), \tag{6.1}$$

where $\mathcal{I}_{\rm magn}(k)$ is the relative intensity of the magnetic fluctuations around a wavenumber $k = 2\pi/\lambda$, defined in such a way the rms amplitude $B_{\rm rms}$ of the fluctuating magnetic fields is given by

$$B_{\rm rms}^2 = B^2 \int {\rm d}\ln k \, \mathcal{I}_{\rm magn}(k). \tag{6.2}$$

This leads to the *self-confinement* of cosmic rays where the cosmic rays are coupled to the fluid motions by the self-generated waves. A modern review of the theory of cosmic ray transport in a magnetically turbulent interstellar medium can be found in Dolginov & Katz (1994).

These ideas were the basis of the work of Skilling (1975a,b) who gave a comprehensive theory for the interaction between cosmic rays and waves in a streaming plasma, allowing for the effects of diffusion, advection of the cosmic rays by the fluid, and the energy gains due to adiabatic compression. Skilling's transport equation, based on earlier work by Parker (1965) and Jokipi & Parker (1967), reads in its simplest form

$$\frac{\partial F(p,\, \boldsymbol{x},\, t)}{\partial t} + \boldsymbol{\nabla} \cdot \boldsymbol{S}(p,\, \boldsymbol{x},\, t) + \frac{\partial S_p(p,\, \boldsymbol{x},\, t)}{\partial p} = 0. \tag{6.3}$$

Here $F(p,\, \boldsymbol{x},\, t)$ is the momentum distribution, which depends only on the *magnitude* $p = |\boldsymbol{p}|$ of the particle momentum, defined in such a way that the number density of cosmic rays with momentum in the interval p, $p + {\rm d}p$ is ${\rm d}N(\boldsymbol{x},\, \boldsymbol{p},\, t) = F(p,\, \boldsymbol{x},\, t)\, {\rm d}p$, or equivalently in terms of the phase-space density: $F(p,\, \boldsymbol{x},\, t) = 4\pi p^2 f(p,\, \boldsymbol{x}\,,\, t)$. The cosmic ray flux density $\boldsymbol{S}$ in this expression consists of an advective and a diffusive term,

$$\boldsymbol{S}(p,\, \boldsymbol{x},\, t) = \boldsymbol{V} F - (\boldsymbol{\kappa} \cdot \boldsymbol{\nabla}) F, \tag{6.4}$$

with $\boldsymbol{V}(\boldsymbol{x},\, t)$ is the cosmic-ray advection velocity, usually close to the flow speed of the plasma, and $\boldsymbol{\kappa}(p,\, \boldsymbol{x},\, t)$ is a diffusion tensor which characterises the diffusion of cosmic rays with respect to the flow. Finally, the flux in momentum space is

$$S_p(p,\, \boldsymbol{x},\, t) = \left\langle \frac{{\rm d}p}{{\rm d}t} \right\rangle F, \tag{6.5}$$

where, if there are no radiation losses, the mean momentum gain equals

$$\left\langle \frac{{\rm d}p}{{\rm d}t} \right\rangle \equiv -\tfrac{1}{3}(\boldsymbol{\nabla} \cdot \boldsymbol{V})\, p. \tag{6.6}$$

This gives the momentum change resulting from adiabatic (de-)compression. The term momentum term $\partial S_p/\partial p$ in equation (6.3) is responsible for the acceleration at shocks, even though $\boldsymbol{\nabla} \cdot \boldsymbol{V}$ becomes singular at a sudden jump in streaming velocity. The momentum p in this equation is the momentum of a cosmic ray particle in the rest-frame of the flow, which explains the appearance of the momentum change term $\propto \boldsymbol{\nabla} \cdot \boldsymbol{V}$ in the transport equation (6.3). An insightfull derivation of the momentum change (6.6) has been given by Webb & Gleeson (1979).

The result (5.31) and (5.32) for the distribution of shock-accelerated particles can be derived straightforwardly from combining equations (6.3), (6.4) and (6.6) in the simple case on a one-dimensional flow along the z-axis, with an infinitely thin shock stationary located at $z = 0$. If the flow is uniform and steady on both sides of the shock, with

$$V = \begin{cases} -U_1 & \text{upstream} \quad (z > 0); \\ \\ -U_2 & \text{downstream} \quad (z < 0). \end{cases} \tag{6.7}$$

Equation (6.3) reduces in the steady state ($\partial F/\partial t = 0$) to:

$$\frac{\partial}{\partial z} \left\{ VF - \kappa \, \frac{\partial F}{\partial z} \right\} = - \frac{U_1 - U_2}{3} \, \delta(z) \left(\frac{\partial (pF)}{\partial p} \right) , \tag{6.8}$$

where the term with the Dirac delta-function $\delta(z)$ gives the effect of the velocity jump in the shock at $z = 0$. For $z \neq 0$ this leads to $\partial S/\partial z = 0$, the requirement of a spatially uniform (but *different*) flux on both sides of the shock:

$$S(p,\, z) = \begin{cases} -U_1 F - \kappa_1 \dfrac{\partial F}{\partial z} = S_1(p) & \text{upstream} \quad (z > 0); \\ \\ -U_2 F - \kappa_2 \dfrac{\partial F}{\partial z} = S_2(p) & \text{downstream} \quad (z < 0). \end{cases} \tag{6.9}$$

Here κ_1 (κ_2) is the upstream (downstream) value of the diffusion coefficient. An integration of equation (6.8) across the shock from $z = -\epsilon$ to $z = +\epsilon$ with $\epsilon \downarrow 0$ connects the up- and downstream state, and yields an equation for the distribution $F(p,\, z = 0)$ at the shock in terms of the up- and downstream flux at given momentum:

$$\frac{U_1 - U_2}{3} \, \frac{\partial [pF(p,\, 0)]}{\partial p} = S_2(p) - S_1(p) . \tag{6.10}$$

Note that this assumes that $F(p,\, z)$ is *continuous* across the shock, which requires that the shock thickness is small compared with the scattering mean free path ℓ_{mfp}.

The flux of particles into the shock results from advection towards the shock with the fluid velocity $\boldsymbol{V}$, and diffusion away from the shock due to the density gradient set up as a result of particle acceleration at the shock. If particles are produced exclusively at the shock, the *net* cosmic ray flux (Eq. (6.4)) must vanish in a steady state. In a one-dimensional, uniform upstream flow, with the z-coordinate along the shock normal $\hat{\boldsymbol{n}}$, this implies $S_1(p) = 0$, *i.e.* the advective and diffusive flux balance exactly. This requirement leads, for a plane shock in a uniform upstream flow, to an exponential *precursor* of accelerated particles ahead of the shock,

$$F(p,\, z \geq 0) = F(p,\, 0)\, \mathrm{e}^{-z/L_1}, \tag{6.11}$$

with z the distance from the shock and L_1 the upstream *diffusion length*,

$$L_1(p) \equiv \kappa_1(p)/U_1. \tag{6.12}$$

This precursor of accelerated particles plays an important role in the theory of cosmic ray modified shocks (see below).

Downstream, the only admissable (*i.e.* finite) solution[3] with S = constant is a spatially uniform distribution,

$$F(p,\, z < 0) = \text{constant} = F(p,\, 0), \tag{6.13}$$

corresponding to a downstream flux $S_2(p) = -U_2\, F(p,\, 0)$. In this case the above equation for $F(p,\, 0)$ reduces to

$$\frac{\partial}{\partial p}\left[p\, F(p,\, 0)\right] = -\frac{3U_2}{U_1 - U_2}\, F(p,\, 0). \tag{6.14}$$

Using $U_1 = r\, U_2$ it is easily seen that the solution takes the form

$$F(p,\, 0) \propto p^{-(r+2)/(r-1)}, \tag{6.15}$$

which is our earlier result (5.32). The proportionality constant must be calculated from the details of the injection process (*i.e.* $S_1(p)$) at low momenta. Even though the details of cosmic ray transport do not enter this result, they **do** determine the typical length and time scales of the shock-acceleration process, and the maximum energy a particle can reach in this process.

Scattering also determines the cycle time. Consider the case of a parallel shock with $\hat{\boldsymbol{n}} \parallel \boldsymbol{B}$. In that case, the normal diffusion coefficient corresponds to the field-aligned diffusion coefficient (denoted by $\kappa_{\parallel}$):

$$\kappa_{\mathrm{n}} = \kappa_{\parallel} = v\ell_{\mathrm{mfp}}/3. \tag{6.16}$$

[3] Assuming the flow extends to $z = -\infty$.

Upstream, once particles start to diffuse at about one mean free path ahead of the shock, they are quickly overtaken. As a result they spend a time $t_{\rm u} \sim \ell_1/U_1 \sim L_1/v$ upstream, where ℓ_1 is the upstream value of the mean free path, and where I have used the relation between the mean free path and the diffusion coefficient and the definition (6.12) of the diffusion length.

Similarly, the time spent downstream is equal to $t_{\rm d} \sim \ell_2/U_2$. A more precise calculation (*e.g.* Drury 1983, 1991) gives the cycle time, $t_{\rm cy} = t_{\rm u}+t_{\rm d}$, as

$$t_{\rm cy} = \frac{4(L_1 + L_2)}{v}, \tag{6.17}$$

where $L_1 = \kappa_1/U_1$ ($L_2 = \kappa_2/U_2$) is the upstream (downstream) diffusion length.

The momentum change (5.26) per cycle for non-relativistic shocks,

$$\frac{\Delta p}{p} = \frac{4(U_1 - U_2)}{3v}, \tag{6.18}$$

determines the associated *acceleration time*. One has

$$t_{\rm acc} \equiv \frac{t_{\rm cy}}{\Delta p/p} = \frac{3(L_1 + L_2)}{U_1 - U_2}. \tag{6.19}$$

This immediately defines the *acceleration rate* $\mathcal{R}_{\rm acc} \equiv t_{\rm acc}^{-1}$ for a particle while it interacts with the shock: a particle gains momentum at a rate

$$\left(\frac{{\rm d}p}{{\rm d}t}\right)_{\rm DSA} \approx \mathcal{R}_{\rm acc}\, p, \tag{6.20}$$

with

$$\mathcal{R}_{\rm acc} = \frac{U_1 - U_2}{3(L_1 + L_2)} = \frac{U_1 - U_2}{3(\kappa_1/U_1 + \kappa_2/U_2)}. \tag{6.21}$$

The mean free path due to scattering on hydromagnetic waves usually depends on energy, and consequently the cycle time, the acceleration time and the acceleration rate are all functions of the particle momentum, *e.g.* $\mathcal{R}_{\rm acc} = \mathcal{R}_{\rm acc}(p)$.

The acceleration time also determines the typical time scale T needed to establish a steady state distribution up to some maximum momentum $p_{\rm max}$ (*e.g.* Forman & Morfill 1979; Axford 1981; Toptygin 1985; Drury 1991):

$$T(p_{\rm max}) \sim \int_{p_{\rm inj}}^{p_{\rm max}} \frac{{\rm d}p}{({\rm d}p/{\rm d}t)_{\rm DSA}} = \int_{p_{\rm inj}}^{p_{\rm max}} t_{\rm acc}(p)\, \frac{{\rm d}p}{p}, \tag{6.22}$$

with $p_{\rm inj}$ the typical injection momentum.

Although these relations have been derived for a one-dimensional flow with the magnetic field and fluid velocity along the shock normal, these results are valid in a more general geometry, *i.e.* for oblique shocks with the magnetic field and flow speed at an arbitrary orientation, provided one defines the diffusion coefficient as

$$\kappa = \kappa_{\rm n} \equiv \hat{\boldsymbol{n}} \cdot \boldsymbol{\kappa} \cdot \hat{\boldsymbol{n}}, \tag{6.23}$$

the component of the diffusion tensor along the shock normal $\hat{\boldsymbol{n}}$, and the relevant velocities as

$$U_{1,2} = |\hat{\boldsymbol{n}} \cdot \boldsymbol{U}_{1,2}|, \tag{6.24}$$

the components of the pre- and post shock velocity along the shock normal. This simple theory does break down when the shock becomes *superluminal.* This occurs when the intersection point of the shock surface and a given magnetic field line moves with a velocity larger than the velocity of light:

$$V_{\rm i} = \frac{V_{\rm s}}{|\cos(\theta_{Bn})|} > c. \tag{6.25}$$

Here $\cos(\theta_{Bn}) = (\boldsymbol{B} \cdot \hat{\boldsymbol{n}})/|\boldsymbol{B}|$ is the cosine of the angle between the magnetic field and the shock normal. For non-relativistic shocks with $V_{\rm s} \ll c$ this only occurs if the upstream magnetic field is almost perpendicular to the shock normal.

7 The Box Model for shock acceleration

Many of the above results can be derived using a simple "Box Model", where the acceleration takes place in a "box" at a rate

$$\frac{{\rm d}p}{{\rm d}t} = \mathcal{R}_{\rm acc}(p)\, p, \tag{7.1}$$

and particles escape from the box at a rate

$$\left(\frac{{\rm d}N}{{\rm d}t}\right)_{\rm esc} = -\mathcal{R}_{\rm esc}(p)\, N. \tag{7.2}$$

Here $N(p,\, t)\, {\rm d}p$ is the number of particles in the box at time t, with momentum in the range $p,\ p + {\rm d}p$. Both $\mathcal{R}_{\rm acc}$ and $\mathcal{R}_{\rm esc}$ may be functions of momentum and/or time. The number of particles in the box evolves according to

$$\frac{\partial N}{\partial t} + \frac{\partial}{\partial p}\left(\mathcal{R}_{\rm acc}\, p\, N\right) = Q_{\rm inj}(p,\, t) - \mathcal{R}_{\rm esc}\, N. \tag{7.3}$$

Here $Q_{\rm inj}(p,\,t)$ is a source term which describes the introduction of fresh particles in the box. This equation is equivalent with the two characteristic equations

$$\begin{aligned}\frac{\mathrm{d}N}{\mathrm{d}t} &\equiv \frac{\partial N}{\partial t} + \left(\frac{\mathrm{d}p}{\mathrm{d}t}\right)\frac{\partial N}{\partial p} = Q(p,\,t) - N\left(\mathcal{R}_{\rm esc} + \mathcal{R}_{\rm acc} + p\,\frac{\partial \mathcal{R}_{\rm acc}}{\partial p}\right); \\ \frac{\mathrm{d}p}{\mathrm{d}t} &= \mathcal{R}_{\rm acc}(p)\,p.\end{aligned} \tag{7.4}$$

A formal solution of this equation (Drury *et al.* 1999) is a weighted integral over the injection rate:

$$N(p,\,t) = \int^{t} \mathrm{d}t_{\rm i}\; Q(p_{\rm i},\,t_{\rm i})\,\exp\left\{-\int_{t_{\rm i}}^{t}\mathrm{d}t'\,\left(\mathcal{R}_{\rm esc} + \mathcal{R}_{\rm acc} + p\,\frac{\partial \mathcal{R}_{\rm acc}}{\partial p}\right)_{t'}\right\}, \tag{7.5}$$

where, if $\mathcal{R}_{\rm acc}$ is time-independent, the injection time $t_{\rm i}$, injection momentum $p_{\rm i}$ and momentum p at time t are related by

$$t - t_{\rm i} = \int_{p_{\rm i}}^{p} \frac{\mathrm{d}p}{\mathcal{R}_{\rm acc}(p)\,p}. \tag{7.6}$$

In a steady state ($\partial N/\partial t = 0$), and above the momentum range where particles are injected, it is easily checked that the distribution inside the box satisfies

$$\frac{\mathrm{d}\ln N}{\mathrm{d}\ln p} = -\left(1 + \frac{\mathcal{R}_{\rm esc}}{\mathcal{R}_{\rm acc}} + \frac{\partial \ln\,\mathcal{R}_{\rm acc}}{\partial\,\ln p}\right). \tag{7.7}$$

An *outside* observer sees the escaping particles with flux $\propto \mathcal{R}_{\rm esc}\,N$, and a slope of the spectrum

$$\frac{\mathrm{d}\ln(\mathcal{R}_{\rm esc}\,N)}{\mathrm{d}\ln p} = -\left(1 + \frac{\mathcal{R}_{\rm esc}}{\mathcal{R}_{\rm acc}} - \frac{\partial\,\ln(\mathcal{R}_{\rm esc}/\mathcal{R}_{\rm acc})}{\partial\,\ln p}\right). \tag{7.8}$$

This is equivalent with Fermi's original result (4.4) if the ratio $\mathcal{R}_{\rm esc}/\mathcal{R}_{\rm acc}$ is momentum-independent.

As shown by Drury *et al.* (1999), diffusive shock acceleration can be represented by a Box Model if one defines the number of particles in the box by

$$N(p,\,t)\,\mathrm{d}p = L(p)\,F(p,\,0,\,t)\,\mathrm{d}p, \tag{7.9}$$

where the size of the box is

$$L(p) = L_1(p) + L_2(p) = \frac{\kappa_1}{U_1} + \frac{\kappa_2}{U_2}, \tag{7.10}$$

the sum of the up- and downstream diffusion lengths. This seems the "natural" choice in this case. The acceleration rate is given by equation (6.21), and the escape rate is

$$\mathcal{R}_{\rm esc} = \frac{U_2}{L} = \frac{U_2}{\kappa_1/U_1 + \kappa_2/U_2}, \tag{7.11}$$

which decribes "advection" out of the box due to particle escape into the downstream flow. The equation (7.3) for shock acceleration then reads, in absence of losses:

$$\frac{\partial}{\partial t}(L\,F) + \frac{\partial}{\partial p}\left(\frac{U_1 - U_2}{3}\,pF\right) = Q_{\rm inj} - U_2\,F. \tag{7.12}$$

The attraction of this formulation is that extra effects and processes can be added relatively simply.

If one adds energy losses, some care has to be taken because of the momentum dependence of the box size $L(p)$. Particles can now leave the box not only by advection, but also by crossing the "downstream boundary" $L_2(p)$ due to the change in momentum. In most models of shock acceleration the diffusion length increases with momentum so that ${\rm d}L_2/{\rm d}p > 0$. By loosing momentum a particle my find itself outside the box. This happens only downstream since particles, once upstream, always encounter the shock again and therefore remain in the box[4]. The escape rate due to this process is

$$\mathcal{R}_p = -\left(\frac{{\rm d}p}{{\rm d}t}\right)_{\rm loss}\frac{1}{L}\,\frac{{\rm d}L_2}{{\rm d}p}, \tag{7.13}$$

where the loss rate is defined in such a way that the mean momentum gain in the box is

$$\frac{{\rm d}p}{{\rm d}t} = \frac{U_1 - U_2}{3L}\,p - \left(\frac{{\rm d}p}{{\rm d}t}\right)_{\rm loss}. \tag{7.14}$$

As a result, the box equation in the presence of losses becomes

$$\frac{\partial}{\partial t}(L\,F) + \frac{\partial}{\partial p}\left(\frac{{\rm d}p}{{\rm d}t}\,L\,F\right) = Q_{\rm inj} - U_2\,F - \left(\frac{{\rm d}p}{{\rm d}t}\right)_{\rm loss}\frac{{\rm d}L_2}{{\rm d}p}\,F. \tag{7.15}$$

If we now define a *net* acceleration rate and escape rate by

$$\begin{aligned}
\mathcal{R}_{\rm acc} &\equiv \frac{U_1 - U_2}{3L} - \frac{1}{p}\left(\frac{{\rm d}p}{{\rm d}t}\right)_{\rm loss} \equiv \frac{U_1 - U_2}{3L} - \mathcal{R}_{\rm loss}, \\
\mathcal{R}_{\rm esc} &\equiv \frac{U_2}{L} + \left(\frac{{\rm d}p}{{\rm d}t}\right)_{\rm loss}\frac{1}{L}\,\frac{{\rm d}L_2}{{\rm d}p} = \frac{U_2}{L} + \mathcal{R}_{\rm loss}\left(\frac{1}{L}\,\frac{{\rm d}L_2}{{\rm d}\ln p}\right),
\end{aligned} \tag{7.16}$$

[4] Assuming of course that the shock is infinitely large so it can not be "missed" by particles once upstream!

equation (7.15) takes the standard form (7.3) for $N = L\,F$, and we can use the results derived above. From equation (7.7) we find the momentum distribution at the shock, with a local slope

$$s(p) \equiv -\frac{\mathrm{d}\ln F}{\mathrm{d}\ln p} = 1 + \frac{\mathcal{R}_{\rm esc}}{\mathcal{R}_{\rm acc}} + \frac{\partial \ln\,\mathcal{R}_{\rm acc}}{\partial\,\ln p} + \frac{\partial\,\ln L}{\partial\,\ln p}. \tag{7.17}$$

Using (7.16) one finds after some algebra, assuming that $\mathrm{d}U_{1,2}/\mathrm{d}p = 0$:

$$s(p) = \frac{U_1 + 2U_2 - 3\mathcal{R}_{\rm loss}L\left(1 + \dfrac{\partial\,\ln\mathcal{R}_{\rm loss}}{\partial\,\ln p} + \dfrac{1}{L}\dfrac{\partial L_1}{\partial\,\ln p}\right)}{U_1 - U_2 - 3\mathcal{R}_{\rm loss}L}. \tag{7.18}$$

The assumption that U_1 and U_2 are independent of momentum only applies if [1] the drift between the scattering centers and the fluid can be neglected, and [2] if the test-particle approximation applies. In the momentum range where energy losses can be neglected ($3\mathcal{R}_{\rm loss}L \ll U_1 - U_2$) we find the standard result for diffusive shock acceleration: $s = (r+2)/(r-1)$ with $r = U_1/U_2$.

7.1 Maximum energy

The Box Model can be used to derive simple relations for the maximum energy attainable in diffusive shock acceleration. Acceleration will only produce particles up to a momentum where the net momentum gain per cycle equals the momentum losses in a cycle time, *i.e.* acceleration will proceed up to a momentum $p_{\rm max}$ where the equality

$$U_1 - U_2 = 3\left(\mathcal{R}_{\rm loss}L\right)_{p=p_{\rm max}} \tag{7.19}$$

is satisfied. The Box Model also predicts the behaviour of the spectrum near $p_{\rm max}$. As the momentum approches $p_{\rm max}$ the denominator in expression (7.17) goes to zero. The sign of the numerator determines the behaviour of the spectrum. Using (7.19) to evaluate the numerator at $p_{\rm max}$ there are two possibilities: if

$$U_1 + 2U_2 - (U_1 - U_2)\left(1 + \frac{\partial\,\ln\mathcal{R}_{\rm loss}}{\partial\,\ln p} + \frac{1}{L}\frac{\partial L_1}{\partial\,\ln p}\right)_{p_{\rm max}} > 0, \tag{7.20}$$

the spectrum will have an exponential cut-off at $p_{\rm max}$, as $s(p) \to \infty$. If, on the other hand, the numerator has the opposite sign so that $s(p) \to -\infty$, a pile-up of particles will occur at $p_{\rm max}$. Such a pile-up will occur provided

$$s_r = \frac{r+2}{r-1} < \left(1 + \frac{\partial\,\ln\mathcal{R}_{\rm loss}}{\partial\,\ln p} + \frac{1}{L}\frac{\partial L_1}{\partial\,\ln p}\right)_{p_{\rm max}}. \tag{7.21}$$

If losses are unimportant the finite age t_s of the shock, or its finite size R_s, will limit the maximum energy according to

$$\int_{p_i}^{p_{\max}} \frac{dp}{p} \frac{3L}{U_1 - U_2} \sim t_s, \quad L(p_{\max}) \sim R_s. \tag{7.22}$$

The simple power-law predicted in the loss-free case will be terminated around $p_{\max}$.

7.2 Bohm diffusion

An often-made assumption is that the mean free path is roughly equal to the particle gyration radius, $\ell_{\rm mfp} \sim r_{\rm g}(p) \sim pc/|q|B$, and that the diffusion coefficient equals to the so-called *Bohm diffusion coefficient* $\kappa_{\rm B}$:

$$\kappa_{\rm n} \sim \kappa_{\rm B} = \frac{v r_{\rm g}(p)}{3} = \frac{pcv}{3\,|q|B}. \tag{7.23}$$

In this case the diffusion length increases with momentum as $L_{\rm diff} \sim \kappa_{\rm B}/U_{1,2} \propto pv$. This corresponds to a situation where the relative intensity of the magnetic turbulence satisfies $\mathcal{I}_{\rm magn}(k) \approx 1$ in the relevant wavenumber range, so that wave turbulence is strong in the sense that $B_{\rm rms} \sim B$. If Bohm diffusion applies to both the up- and downstream flow, so that

$$L_{1,2} = \frac{pcv}{3|q|(BU)_{1,2}}, \tag{7.24}$$

the mean momentum gain for a particle with charge $q = Ze$ satisfies according to (6.21)

$$\mathcal{R}_{\rm acc}^{\rm B} = \xi_r \, \frac{|Z|eB_1U_1^2}{pcv}, \tag{7.25}$$

where

$$\xi_r \equiv \frac{r-1}{r + r^2/r_{\rm B}}. \tag{7.26}$$

Here $r_{\rm B} \equiv B_2/B_1$ is the laminar magnetic field amplification in the shock, which satisfies

$$1 \leq r_{\rm B} = \sqrt{\cos^2\theta_{Bn} + r^2\,\sin^2\theta_{Bn}} \leq r. \tag{7.27}$$

The quantity ξ_r varies from $\xi_r = 3/20$ for a strong parallel shock ($\boldsymbol{B} \parallel \hat{\boldsymbol{n}}$) with $\theta_{Bn} = 0$, $r_{\rm B} = 1$ and $r = 4$ to $\xi_r = 3/5$ for a strong perpendicular shock shock ($\boldsymbol{B} \perp \hat{\boldsymbol{n}}$) with $\theta_{Bn} = \pi/2$ and $r = r_{\rm B} = 4$.

In Bohm diffusion, particles of the same charge gain energy at the same rate:

$$\left(\frac{d\mathcal{E}}{dt}\right)_{\rm B} = \mathcal{R}_{\rm acc}^{\rm B}\, pv = \xi_r \, \frac{|Z|eB_1U_1^2}{c}. \tag{7.28}$$

I use rate (7.28) as a "optimum situation" to calculate the maximum energy for a number of simple cases from $\mathcal{R}^{\rm B}_{\rm acc}(p) \approx \mathcal{R}_{\rm loss}(p)$:

- **Adiabatic losses.** In expanding and almost spherical shock waves, such as those found in supernova remnants, the wave-mediated coupling between the accelerated particles and the expanding downstream gas leads to expansion losses, with a loss rate

$$\mathcal{R}^{\rm exp}_{\rm loss} = \tfrac{1}{3}\,\boldsymbol{\nabla}\cdot\boldsymbol{V}. \tag{7.29}$$

  In supernova remnants older than $\sim$1000 years the expansion follows the well-known Sedov-Taylor law, $R_{\rm s}(t) \propto t^{2/5}$ for a spherical remnant expanding into a uniform interstellar medium (*e.g.* Cioffi 1990). The Sedov solution for the flow velocity $V(r)$ as a function of radius r inside the remnant gives directly behind the blast wave, for $R_{\rm s} - r \ll R_{\rm s}$,

$$\frac{V(r)}{V_{\rm s}} \approx \frac{3}{4}\left(\frac{2r}{R_{\rm s}} - 1\right) \quad\longleftrightarrow\quad \boldsymbol{\nabla}\cdot\boldsymbol{V} = \frac{1}{r^2}\frac{\partial}{\partial r}\left(r^2\,V(r)\right) \approx \frac{3V_{\rm s}}{R_{\rm s}}. \tag{7.30}$$

  Here $R_{\rm s}$ is the shock radius and $V_{\rm s}$ the shock speed, and the factor 3/4 in (7.30) is $2/(\tilde{\gamma}_{\rm g}+1)$ for $\tilde{\gamma}_{\rm g} = 5/3$. The typical loss rate is $\mathcal{R}^{\rm exp}_{\rm loss} \sim V_{\rm s}/R_{\rm s}$, and criterion (7.19) together with (7.28) for $U_1 = V_{\rm s}$ yields:

$$\mathcal{E}_{\rm max} \sim |Z|eB\,\xi_r\,\beta_{\rm s}\,R_{\rm s}, \tag{7.31}$$

  with $\beta_{\rm s} \equiv V_{\rm s}/c$. At this energy the diffusion length equals $L_{\rm diff}(p_{\rm max}) \sim R_{\rm s}$, so this is also the energy at which particles start to notice the fact that the shock front is curved. A similar value for $\mathcal{E}_{\rm max}$ is obtained from (7.22) if one assumes that the finite age of the source, $t_{\rm s} \sim 2R_{\rm s}/5V_{\rm s}$ determines the maximum attainable energy.

  These estimates are valid for non-relativistic shocks, but can be extended up to the region $\beta_{\rm s} \sim 1$, $\Gamma_{\rm s} \sim 1$. The case of acceleration at ultra-relativistic shocks is treated below.

- **Synchrotron/inverse Compton losses.** For relativistic electrons, and or for ultra-relativistic protons, emission of synchrotron radiation and inverse Compton radiation are an important loss mechanism. In the Thomson limit[5], and for relativistic particles ($v \sim c$, $\mathcal{E} \sim pc$), these losses increase with energy:

$$\mathcal{R}^{\rm sc}_{\rm loss}(\mathcal{E}) = \frac{1}{t_{\rm sc}}\left(\frac{\mathcal{E}}{mc^2}\right), \tag{7.32}$$

[5] Where the relevant cross-section for electrons is the Thomson cross section $\sigma_{\rm T} = 6.65\times10^{-25}$ cm^2.

where the time $t_{\rm sc}$ is given by

$$t_{\rm sc} = \begin{cases} \dfrac{6\pi m_{\rm e}c}{\sigma_{\rm T}B_{\rm e}^2} \equiv t_{\rm sc}^{\rm e} & \text{electrons;} \\ \\ \left(\dfrac{m_{\rm p}}{m_{\rm e}}\right)^3 \left(\dfrac{A^3}{Z^4}\right) t_{\rm sc}^{\rm e} & \text{nuclei with mass } m = A\, m_{\rm p}. \end{cases} \tag{7.33}$$

In this expression $B_{\rm e} \equiv \sqrt{B^2 + 8\pi U_{\rm rad}}$ is an effective magnetic field, which includes the effect of the ambient radiation density $U_{\rm rad}$.

If synchrotron/inverse-Compton losses are the factor determining the maximum energy one has

$$\mathcal{E}_{\max} \sim \sqrt{mc^2}\, \beta_{\rm s} \left(|Z|eB\, \xi_r\, \ell_{\rm sc}\right)^{1/2}, \tag{7.34}$$

with $\ell_{\rm sc} \equiv ct_{\rm sc}$. Many non-thermal radio sources (SNRs, jets and radio lobes associated with AGNs) are powered by the synchrotron emission of relativistic ($\mathcal{E} \sim 1$ GeV) electrons in a magnetic field ($B \sim 1$–$100\ \mu$G). The scaling of synchrotron losses, $\mathcal{R}_{\rm loss} \propto \mathcal{E}$, can cause a pile up or "bump" in the spectrum of shock-accelerated electrons near the maximum energy (Schlickeiser 1984; Webb *et al.* 1984; Melrose & Crouch 1997; Protheroe & Stanev 1999; Drury *et al.* 1999). According to criterion (7.21), with $L_{1,2} \propto \mathcal{E} \sim pc$, this occurs whenever the shock compression $r = U_1/U_2$ satisfies

$$\frac{r+2}{r-1} < 2 + \frac{L_1}{L_1 + L_2}. \tag{7.35}$$

For Bohm diffusion, where $\kappa \propto B^{-1}$, one has

$$L_1/L_2 = U_2B_2/U_1B_1 = r_{\rm B}/r. \tag{7.36}$$

- **Pion production losses.** Highly-relativistic protons and nuclei moving through a non-relativistic ambient medium, or through an intense radiation field, suffer collisions with nuclei (N) or photons (γ). Above a certain threshold energy, these collisions can lead to pion production in reactions like $p + \gamma \longrightarrow p + \pi$ and $p + N \longrightarrow p + N + \pi$. These processes can become an important loss mechanism, in particular in Active Galactic Nuclei where the radiation density is very high (*e.g.* Zdziarski 1986; Kazanas & Ellison 1986; Sikora *et al.*, 1987 and Begelman *et al.* 1990). It also is the most important mechanism for limiting the energy in possible production sites of the *Ultra High*

Energy Cosmic Rays (UHECRs) (Norman *et al.* 1995) if such particles are indeed the result of acceleration from lower energies. In the simplest case of a constant loss length ℓ_π, so that the losses are

$$\mathcal{R}^{\pi}_{\rm loss} = c/\ell_\pi , \tag{7.37}$$

the maximum energy attainable in non-relativistic shocks equals

$$\mathcal{E}_{\rm max} \sim |Z|eB\, \xi_r\, \beta_{\rm s}^2\, \ell_\pi . \tag{7.38}$$

8 Outstanding issues: Theory

Even though the general principles behind the process of diffusive shock acceleration are well-established, and the complications of the non-linear nature of the process are qualitatively understood, there are a number of open questions. I list the most important ones below:

8.1 The question of injection

Models of DSA need "seed particles" which can be picked up by the acceleration process. One either assumes that there is some pre-existing population advected into the shock from far upstream, or that these particles are produced at the shock itself. In the latter case, the question of injection is intimately connected with the microscopic structure of (collisionless) shocks. Some models circumvent the details of the shock structure by assuming that some form of leakage of shock-heated downstream particles (ions), satisfying $\boldsymbol{v} \cdot \hat{\boldsymbol{n}} > V_{\rm s}/r$ in the shock frame, provides the injection at the shock (*e.g.* Eichler 1979; Edminston *et al.* 1982).

This particular hypothesis has been re-investigated by Malkov & Völk (1995). They essentially develop a formalism for describing wave-particle interactions between shock-heated particles and self-generated waves at high anisotropies (in contrast to standard DSA theory), and solve the transport equation in the up- and downstream flow. This is in essence a low-energy extension of standard DSA, where the injection consists of supra-thermal particles crossing the shock repeatedly. They find that for parallel shocks the spectrum of injected particles is significantly steeper than the standard result up to a particle energy several times the downstream thermal energy. For strong shocks ($r \approx 4$) they find that a significant fraction of the downstream ions can be injected into the acceleration process.

The other approach to investigate the injection problem is that of direct (hybrid) simulation of collisionless shocks. These simulations are always concerned with *ions*. This approach has been reviewed recently by Scholer *et al.* (1998). These authors come to the conclusion that thermal leakage is not important, and that seed particles are produced as a result of interaction between the particles gyrating near the shock front, and electric fields along the shock front produced by shock-generated plasma waves or waves advected into the shock from upstream. Wave production by quasi-parallel shocks is well-established both theoretically and observationally, *e.g.* in the case of the Earth's bow shock.

Most of the above considerations apply to protons and nuclei. Diffusive shock-acceleration of *electrons* requires that they are pre-accelerated to mildly relativistic energies. The reason is that the mean-free-path of electrons must be large compared with the thickness of the shock. Simulations show that high-Mach number, quasi-parallel collisionless shocks ($\theta_{\rm Bn} < \pi/4$) have a thickness $\ell_{\rm s} \sim c/\omega_{\rm pi}$, where $\omega_{\rm pi} = \sqrt{4\pi e^2 n/m_{\rm i}}$ is the ion-plasma frequency. I will usually assume an almost pure hydrogen plasma with $m_{\rm i} = m_{\rm p}$. If the electron scattering mean free path is roughly the gyro-radius $r_{\rm g} \sim pc/eB$, the condition $r_{\rm g} > c/\omega_{\rm pi}$ corresponds with

$$\gamma_{\rm e}\beta_{\rm e} > \frac{m_{\rm p}}{m_{\rm e}}\,\frac{V_{\rm A}}{c} \sim 1.3\left(\frac{B}{10^{-4}\ {\rm G}}\right) n_{\rm ism}^{-1/2} . \tag{8.1}$$

Here

$$V_{\rm A} = \frac{B}{\sqrt{4\pi n_{\rm ism} m_{\rm i}}} \tag{8.2}$$

is the Alfvén velocity in the general interstellar medium with number density $n_{\rm ism}$, and $\gamma_{\rm e} = 1/\sqrt{1-\beta_{\rm e}^2}$ is the Lorentz-factor of the electrons.

A second problem with electrons concerns the hydromagnetic waves needed for scattering. Upstream from the shock, ions produce Alfvén waves due to the *cyclotron streaming instability* (*e.g.* Bell 1978). The wavelength of waves produced by ions with momentum p is of the order of the gyro-radius $r_{\rm g} \sim pc/|Z|eB$. However, the background of *thermal* ions damps these waves if their frequency, $\omega = kV_{\rm A}$, approaches the thermal gyro-frequency $\omega_{\rm ci} = eB/m_{\rm i}c$ of the thermal ions at a wavenumer $k_{\rm max} \sim \omega_{\rm ci}/V_{\rm A}$. So, ion-generated waves are only available for electron scattering with $k < k_{\rm max}$, and the since the interaction ocurs at $\lambda = 2\pi/k \sim r_{\rm g} = pc/eB$ only electons with $r_{\rm g} > 2\pi/k_{\rm max}$ can be scattered by ion-generated waves. It is easily checked that this gives the same requirement as condition (8.1) since $c/\omega_{\rm pi} = V_{\rm A}/\omega_{\rm ci}$. So, for non-relativistic electrons, the waves generated by shock-accelerated ions have too long a wavelength, and can not scatter the electrons. One can appeal to other electromagnetic plasma waves, such as *Whistlers*, which occur (in a cold plasma) in the frequency range $\omega_{\rm ci} < \omega < \omega_{\rm ce} = eB/m_{\rm e}c$, and which can be generated by the electrons themselves if the shock velocity is sufficiently high (Levinson 1995).

The pre-acceleration of electrons is in all probability closely related to the detailed physics of astrophysical collisionless shocks: the way in which collective effects (wave-particle interactions) convert the kinetic energy of the incoming upstream ions and electrons into random ion motions (*i.e.* a high-temperature state) downstream. In particular for quasi-parallel shocks ($\theta_{\rm Bn} < \pi/4$), where hot and fast downstream particles can easily stream backwards along the magnetic field into the upstream flow, one would expect that the shock thickness would be large. Simulations however show that the interaction between particles and large-amplitude hydromagnetic waves limit the leakage of hot downstream ions into the upstream medium. A detailed consideration of particle motion in such waves (Malkov 1998) has shown that only a small fraction (~5%) of downstream ions manage to return upstream. These ions can then serve as the "seed particles" for the shock acceleration process. Malkov's prescription for the injection of seed particles have been incorporated in recent numerical simulations of cosmic-ray modified shocks (Kang *et al.* 2002). For quasi-perpendicular shocks ($\theta_{\rm Bn} > \pi/4$) the gyration of particles around the ambient magnetic field limits the shock thickness to a few ion gyro-radii based on the downstream kinetic temperature, $\ell_{\rm s} \sim V_{\rm s}/\omega_{\rm ci}$, and the magnetic field acts as a natural "barrier", limiting the escape of particles back upstream. Both simulations and observations of quasi-perpendicular shocks in the Solar Wind have shown that such shocks have a turbulent precursor ("foot") in which ions reflected off the jump in the magnetic field at the shock create turbulence which scatters and accelerates incoming particles. Simulations (Shimada & Hoshino 2000; Schmitz *et al.* 2002) seem to show that in these situations electrons can be pre-accelerated by shock-associated turbulence, but it also seems that this works best if the ratio of the kinetic to magnetic pressure, $\tilde{\beta}_{\rm g} = 8\pi P_{\rm g}/B^2$, is small.

8.2 Stability: Precursor and shock stability

The wave-mediated coupling between cosmic rays and fluid in the precursor may lead to instabilities. The simplest version occurs in a quasi-par allel shock, as pointed out by Drury & Falle (1986). Compressive perturbations (short-wavelength sound waves) will be unstable if the precursor length scale satisfies

$$L \equiv \left|\frac{{\rm d}\ln P_{\rm cr}}{{\rm d}z}\right|^{-1} \sim \frac{\overline{\kappa}}{V_{\rm s}} < \frac{\overline{\kappa}\,|1 + (\partial \ln \overline{\kappa}/\partial \ln \rho)|}{\tilde{\gamma}_{\rm cr} c_{\rm s}}. \tag{8.3}$$

Here $c_{\rm s} = \sqrt{\tilde{\gamma}_{\rm g} P_{\rm g}/\rho}$ is the sound speed of the gas. This instability condition is easily satisfied in high Machnumber flows where $\mathcal{M}_{\rm s} = V_{\rm s}/c_{\rm s} \gg 1$. This particular instability, known as the *squeezing instability*, arises from the modulation of the cosmic ray diffusion expected as a result of the density perturbations. This modulation occurs naturally when the scattering waves (Alfvén waves) are compressed. This work was extended to more general situations by Zank *et al.* (1990), and by Begelman & Zweibel (1994). The latter authors include the effect of drift between cosmic rays and fluid, and of heating, and show that a drift velocity $V_{\rm d} > c_{\rm s}/(\tilde{\gamma}_{\rm g} - 1)$ will lead to an instability. The conclusion seems to be that the presence of cosmic rays can lead to an instability in the precursor. It has been suggested by Begelman & Zweibel that these instabilities trigger the formation of a series of weak shocks in the precursor due to wave steepening, analogous to shock formation for sound waves propagating down a density gradient. How this influences cosmic ray acceleration and the emergent spectrum has not been investigated in any detail yet.

Mond & Drury (1998) have considered a different kind of stability: the stability of cosmic-ray-modified shocks against the corrugational instability and the emission of sound waves. They confirm an earlier conjecture, which states that one of the solutions in the regime where the two-fluid approximation yields three possible downstream states for a given upstream state must be unstable.

9 Inclination angle dependence and quasi-perpendicular shocks

Most of the discussion so far has been confined to quasi-parallel shocks, defined operationally by the relation

$$\theta_{\rm Bn} \equiv \cos^{-1}(\hat{\boldsymbol{n}} \cdot \hat{\boldsymbol{b}}) \lesssim \pi/4, \tag{9.1}$$

with $\hat{\boldsymbol{b}} \equiv \boldsymbol{B}/|\boldsymbol{B}|$ the unit vector along the magnetic field. In quasi-perpendicular shocks, with $\theta_{\rm Bn} > \pi/4$, there are a number of effects which can modify the simple picture.

First of all, the intersection point between a planar shock and a given fieldline moves with a velocity

$$V_{\rm i} = V_{\rm s}/\cos(\theta_{\rm Bn}) = V_{\rm s}/(\hat{\boldsymbol{b}} \cdot \hat{\boldsymbol{n}}). \tag{9.2}$$

If particle motion on the microscopic level is confined to sliding along field lines, this means that only particles with velocity $v > V_{\rm i}$ can re-cross the shock into the upstream medium. In particular in thermal leakage models, where the shock-heated downstream particles from the tail of the Maxwellian distribution serve as a injection source, this means that injection efficiency will decrease with increasing $\theta_{\rm Bn}$. Monte-Carlo simulations (Baring *et al.* 1993–1995) bear this out: for high Mach numbers injection by thermal leakage drops strongly above $\theta_{\rm Bn} > 30^{\circ}$.

Secondly, cosmic ray transport changes qualitatively when the magnetic field is not aligned with the shock normal. Diffusion *across* the magnetic field becomes important, and diffusion along the shock normal is governed by an effective diffusion coefficient $\kappa_{\rm n}$ given by

$$\kappa_{\rm n} = \kappa_{\parallel} \; \cos^2(\theta_{\rm Bn}) + \kappa_{\perp} \; \sin^2(\theta_{\rm Bn}). \tag{9.3}$$

Here $\kappa_{\parallel}$ is the field-aligned diffusion coefficient, and $\kappa_{\perp}$ the diffusion coefficient in the plane perpendicular to the magnetic field. In simple collisional models, where the diffusion process is characterised by some collision time τ and the associated field-aligned mean-free-path $\ell_{\rm mfp} = v\tau$, one has

$$\kappa_{\parallel} \sim \kappa_{\rm B} \left(\frac{\ell_{\rm mfp}}{r_{\rm g}} \right) , \quad \kappa_{\perp} \sim \frac{\kappa_{\parallel}}{1 + (\ell_{\rm mfp}/r_{\rm g})^2} , \tag{9.4}$$

where $r_{\rm g} \sim pc/ZeB$ is the typical gyration radius and $\kappa_{\rm B} = vr_{\rm g}/3$ is the Bohm diffusion coefficient. This model is valid for simple elastic "hard sphere" collisions, but this relation between $\kappa_{\parallel}$, $\kappa_{\perp}$ and $\ell_{\rm mfp}/r_{\rm g}$ does *not* hold generally for wave-particle interactions.

Using this simple model for illustrative purposes, it is obvious that for weak collisionality ($\ell_{\rm mfp} \gg r_{\rm g}$) the value of $\kappa_{\rm n}$ will be much *less* than the Bohm value for nearly perpendicular shocks, a point made most forcefully by Jokipii (1982, 1987). Since the acceleration time scales as $t_{\rm acc} \sim \kappa_{\rm n}/V_{\rm s}^2$ this means that the acceleration proceeds more rapidly. However, in contrast to what is sometimes stated (*e.g.* Jones & Ellison 1991), this does not necessarily change the maximum energy $\mathcal{E}_{\rm max}$ that can be achieved. As already mentioned, the energy gain at very oblique shocks can be thought of as the result of a particle drift along the shock face. The typical electrical field strength which follows from the condition of ideal MHD, $\boldsymbol{E} \simeq -\Gamma(\boldsymbol{V} \times \boldsymbol{B})/c$, with $\boldsymbol{B}$ the magnetic field in the rest-frame of the plasma. It equals $|\boldsymbol{E}| \sim \Gamma_{\rm s}\beta_{\rm s}\,|\boldsymbol{B}|$ for nearly perpendicular shocks, with $\Gamma_{\rm s} = 1/\sqrt{1-\beta_{\rm s}^2}$ the shock Lorentz factor. If the particle drifts a small distance $\mathrm{d}s$ along the shock face in the direction of $\boldsymbol{E}$, the gain in (kinetic) energy of a particle with charge $q = Ze$ is (Webb *et al.* 1983; Jokipii 1987)

$$\mathrm{d}\mathcal{E} = Ze|\boldsymbol{E}|\,\mathrm{d}s \sim Ze|\boldsymbol{B}|\,\Gamma_{\rm s}\beta_{\rm s}\,\mathrm{d}s. \tag{9.5}$$

If the size of the shock is $R_{\rm s}$, the maximum possible energy gain in absence of losses is $\mathcal{E}_{\rm max} \sim ZeB\,\Gamma_{\rm s}\beta_{\rm s}\,R_{\rm s}$, for non-relativistic shocks essentially the same value already derived for quasi-parallel shocks under the assumption of Bohm diffusion. The maximum energy at perpendicular shocks will only be significantly larger than in the case of a quasi-parallel shock in the case where energy losses are particularly intense so that the loss time rather than shock geometry/shock age determines $\mathcal{E}_{\rm max}$ (see Eq. (7.19)).

9.1 Superluminal shocks

In oblique shocks, one must also distinguish between *subluminal* and *superluminal* shocks. Consider a shock propagating with velocity $V_{\rm s}$ into a magnetic field $\boldsymbol{B}$, defined in the upstream rest frame. For a given value of the inclination angle $\theta_{\rm Bn}$ between this field and the shock normal, the intersection point with the shock surface moves along an upstream fieldline with a speed $V_{\rm i} = V_{\rm s}/\cos\theta_{\rm Bn}$. Subluminal and superluminal shocks are defined according to the criterion

$$\text{when } V_{\rm i} = \frac{V_{\rm s}}{\cos(\theta_{\rm Bn})} \text{ is } \begin{cases} < c: & \text{subluminal shock} \\ > c: & \text{superluminal shock.} \end{cases} \tag{9.6}$$

In subluminal shocks, it is always possible to transform to a frame where the shock is stationary, and the magnetic field and the flow speed are aligned on both sides of the shock so that the electric field vanishes everywhere: (deHoffmann & Teller 1950):

$$\boldsymbol{E}_{\rm dHT} = 0. \tag{9.7}$$

The transformation from the lab frame, where the shock moves with velocity $V_{\rm s}$, to the dHT frame can be achieved in a two-step approach (*e.g.* Kirk & Heavens 1989): first one performs a Lorentz-boost along the shock normal $\hat{\boldsymbol{n}}$ with velocity $V_{\rm s}$, followed by a boost in the plane of the shock, directed along the perpendicular component $\boldsymbol{B}_{\perp}$ of the magnetic field, with a boost velocity equal to the *deHofmann-Teller velocity* $V_{\rm HT}$:

$$V_{\rm HT} = \Gamma_{\rm s} V_{\rm s}\, \tan(\theta_{\rm Bn}). \tag{9.8}$$

It is easily checked that $V_{\rm HT} < c$ as long as the shock is subluminal.

In contrast, in superluminal shocks one can transform to a frame where the shock is stationary, the flow velocity is inclined with respect to the shock normal and the magnetic field is purely perpendicular, *i.e.* $B_{\parallel} = \boldsymbol{B} \cdot \hat{\boldsymbol{n}} = 0$ in that frame. The transformation velocity along the shock face now equals

$$\widetilde{V}_{\rm HT} = \frac{c^2}{V_{\rm HT}} = \frac{c}{\Gamma_{\rm s}\beta_{\rm s}\tan(\theta_{\rm Bn})}. \tag{9.9}$$

In this frame, a planar shock crosses the whole fieldline simultaneously, and particles can not re-cross the shock once they are downstream as long as their gyrocenter is tied to a given field line for a sufficient time. In this case, cross-field diffusion, or some form of *anomalous transport process* such as *field line wandering* (Ball & Kirk 1992; Achterberg & Ball 1994) is *essential* for DSA. In absence of such processes, only Shock Drift Acceleration can occur.

Diffusive acceleration at superluminal shocks can only occur if two conditions are satisfied simultaneously (Achterberg & Ball 1994):

1. The cross-field diffusion must be fast enough so that downstream particles can "catch up" with the with the shock diffusively. If one characterises the diffusion process by a stepsize $\Delta X_{\perp}$ across the field and a characteristic time τ between steps so that $\kappa_{\perp} \sim \Delta X_{\perp}^2/2\tau$, this condition translates into

$$\Delta X_{\perp} > V_{\rm nd}\,\tau , \tag{9.10}$$

   where $V_{\rm nd} = |\boldsymbol{V}_{\rm d} \cdot \hat{\boldsymbol{n}}| = |\boldsymbol{V}_{\rm u} \cdot \hat{\boldsymbol{n}}|/r_{\rm s}$ is the magnitude of the downstream flow velocity along the shock normal and $r_{\rm s}$ is the shock compression ratio.

2. Once upstream, the direction of the particle momentum must be randomised *before* the the shock overtakes the particle once again. If this is not the case one gets SDA rather than DSA. If the upstream scattering is characterised by some scattering frequency $\nu_{\rm s} = 1/\tau_{\rm s}$, this means that the upstream residence time associated with DSA, $t_{\rm u} \sim 4\kappa_{\perp}/vV_{\rm n}$, must exceed the scattering time $\tau_{\rm s}$, leading to the condition

$$\kappa_{\perp} \gg \tfrac{1}{4}\,V_{\rm n} v \tau_{\rm s} \sim \left(\frac{V_{\rm n}}{v}\right)\,\kappa_{\|} . \tag{9.11}$$

   Here $\kappa_{\|} = \frac{1}{3}\,v^2\tau_{\rm s}$ is the field-aligned diffusion coefficient.

If collisions are weak in the sense that $\ell_{\rm mfp} \sim v\tau_{\rm s} \gg r_{\rm g}$ one has $\Delta X_{\perp} \sim r_{\rm g}$, $\tau \sim \tau_{\rm s}$ and the second condition turns out to be the most stringent of the two:

$$\frac{r_{\rm g}}{\ell_{\rm mfp}} \gg \left(\frac{V_{\rm n}}{v}\right)^{1/2} . \tag{9.12}$$

In the limit of very strong collisions, $\ell_{\rm mfp} \ll r_{\rm g}$ so that a particle is scattered many times in one gyro-period, the magnetic field is no longer very important and diffusion becomes isotropic with $\Delta X_{\perp} \sim \ell_{\rm mfp} = v\tau_{\rm s}$. Not suprisingly, in that case particles with $v \gg V_{\rm n}$ will undergo DSA. In *relativistic shocks* with $V_{\rm s} \sim c$ it becomes difficult to satisfy both these conditions, and the traditional assumption that the particle distribution is close to isotropic in the vicinity of the shock has to be abandoned.

10 Relativistic shocks

It is believed that the same principles that govern particle acceleration near slow shocks, with $V_{\rm s} \ll c$, must operate near relativistic shocks with $V_{\rm s} \sim c$.

Such shocks are present in the collimated outflows (jets) associated with powerful radio galaxies (AGNs and quasars, in particular in Blazars and in the Fanarof-Riley Class II radio galaxies), in the micro-quasars associated with some stellar-mass compact objects in our own galaxy, and in the relativistic fireballs believed to be associated with Gamma Ray Bursts.

The theory of particle acceleration at relativistic shocks is complicated by the fact that there is, at first sight, no natural approximation scheme available to describe the process: the anisotropy of particle distributions near these shocks will be large since the particle velocity and shock velocity are always of similar magnitude. Consequently, the diffusion approximation, which describes spatial particle transport in terms of some diffusion coefficient κ, does not apply. Also, at relativistic shocks the energy gain $\Delta\mathcal{E}$ per shock crossing cycle is no longer small: in terms of the particle energy $\mathcal{E} = \sqrt{p^2c^2 + m^2c^4}$ one has $\Delta\mathcal{E} \sim \mathcal{E}$.

This is best illustrated by the *exact* expression for the ratio of the inital and final energy (denoted by $\mathcal{E}_{\rm i}$ and $\mathcal{E}_{\rm f}$ respectively) in the upstream rest frame of a relativistic ($\mathcal{E} \sim pc$) particle interacting with the flow near a shock with upstream (downstream) flow velocity $V_1 \equiv \beta_1 c$ ($V_2 \equiv \beta_2 c$), both measured in the shock rest frame (Eq. (5.6)):

$$\frac{\mathcal{E}_{\rm f}}{\mathcal{E}_{\rm i}} \sim \frac{p_{\rm f}}{p_{\rm i}} = \frac{1 - \beta_{\rm rel}\mu_{\to{\rm d}}}{1 - \beta_{\rm rel}\mu_{\to{\rm u}}}, \quad \beta_{\rm rel} = \frac{\beta_1 - \beta_2}{1 - \beta_1\beta_2}. \tag{10.1}$$

Here $\mu_{\to{\rm d}}$ is the cosine of the inclination angle of the particle momentum with respect to the shock normal at the moment it crosses the shock into the downstream medium, and $\mu_{\to{\rm u}} > \mu_{\to{\rm d}}$ the same quantity at the moment it re-crosses the shock into the upstream medium after an elastic scattering event in the downstream rest frame. Both cosines are measured in the upstream rest frame. This result is easily obtained from the up-down Lorentz transformation, followed by the down-up transformation after scattering. If $\beta_{\rm rel} \approx 1$ this ratio will be of order unity (see below). In addition, the downstream escape probability becomes large, which is intuitively obvious from the fact that particles must catch up with a shock moving at a velocity comparable with the particle velocity. This can be illustrated most readily for an *isotropic* distribution in the downstream rest frame where the *return* probability equals (Eq. (5.21); Peacock 1981):

$$\mathcal{P}_{\rm ret} = 1 - \mathcal{P}_{\rm esc} = \left(\frac{1 - \beta_2}{1 + \beta_2}\right)^2. \tag{10.2}$$

For ultra-relativistic shocks one has $\beta_2 \sim 1/3$ (Blandford & McKee 1976) and the above expression yields $\mathcal{P}_{\rm ret} = 1/4$, $\mathcal{P}_{\rm esc} = 3/4$. The Table 2 below lists the most important differences between non-relativistic and relativistic shock acceleration.

Table 2. Comparison non-relativistic and relativistic DSA.

Quantity	**Non-relativistic DSA** ($\beta_{\rm s} = V_{\rm s}/c \ll 1$)	**Relativistic DSA** ($\beta_{\rm s} \sim c$, $\Gamma_{\rm s} \geq 1$)
Escape probability	$\mathcal{P}_{\rm esc} \sim \beta_{\rm s} \ll 1$	$\mathcal{P}_{\rm esc} \sim 0.5\text{–}0.6$
Energy gain	$\mathcal{E}_{\rm f}/\mathcal{E}_{\rm i} \sim 1 + \beta_{\rm s}$	$\mathcal{E}_{\rm f}/\mathcal{E}_{\rm i} \sim 2$
slope distribution	$s = \dfrac{r+2}{r-1} \sim 2$	$s \sim 2.2\text{–}2.3$
upstream anisotropy	**small:** $\sim \beta_{\rm s} \ll 1$	**strong**: particles are confined to cone with small opening angle $\theta \sim 2/\Gamma_{\rm s}$ around shock normal $\hat{\boldsymbol{n}}$.
downstream anisotropy	**none** (isotropic)	**strong**: few particles with $\hat{\boldsymbol{n}} \cdot \hat{\boldsymbol{p}} \lesssim -0.5$
spatial transport	diffusive	non-diffusive

For these reasons, one has resorted largely to two methods to calculate particle acceleration in the relativistic shock regime: *(Monte-Carlo) simulations* or a semi-analytical description in terms of *eigenfunctions* of some scattering operator. The latter approach (*e.g.* Kirk & Schneider 1987a,b; Heavens & Drury 1988; Kirk & Schneider 1988; Schneider & Kirk 1989; Kirk & Heavens 1989) solves the equation for the Lorentz-invariant particle distribution in phase space,

$$f(p,\, \mu,\, \boldsymbol{x}) \equiv \frac{{\rm d}N}{2\pi p^2\, {\rm d}p\, {\rm d}\mu\, {\rm d}^3\boldsymbol{x}}, \tag{10.3}$$

including the effect of scattering. In the one-dimensional test-particle case, where the flow is stationary in the shock rest frame and uniform on both sides of the shock, this equation reads for relativistic particles (in units with $c = 1$)

$$\left\{ \Gamma\ (\mu - \beta)\ \frac{\partial}{\partial z} - \mathcal{C} \right\} f(p,\, \mu,\, \boldsymbol{x}) = 0. \tag{10.4}$$

Here $\Gamma = 1/\sqrt{1-\beta}$ is the Lorentz-factor associated with the flow velocity β, assumed to be directed towards negative z, and $|z|$ is the distance from the shock located at $z = 0$, so that $\beta = \beta_1 = \beta_s$ for $z > 0$ and $\beta = \beta_2$ for $z < 0$, μ is the direction cosine the angle between particle momentum and the shock normal so that $v_z = v\mu$.

The operator $\mathcal{C}(f)$ describes the scattering of the particles. Particle momentum, velocity and μ are all defined in the local fluid rest frame. An often-used form for the scattering operator, valid for isotropic differential (small-angle) elastic scattering in the fluid frame, is

$$\mathcal{C}(f) = \frac{\partial}{\partial\mu}\left[\mathcal{D}_\mu \frac{\partial f}{\partial\mu}\right]. \tag{10.5}$$

Here $\mathcal{D}_\mu$ is the angle diffusion coefficient, related to the scattering frequency ν_s by

$$\mathcal{D}_\mu = \frac{1-\mu^2}{2}\,\nu_s. \tag{10.6}$$

One seeks a solution in terms of a discrete set of eigen-functions $f_n(\mu)$ (with corresponding eigenvalues λ_n) which satisfy

$$\frac{\partial}{\partial\mu}\left[\mathcal{D}_\mu \frac{\partial f_n}{\partial\mu}\right] - \lambda_n\,(\mu-\beta)\,f_n = 0. \tag{10.7}$$

Solutions are assumed to take the form

$$f(p,\,\mu,\,z) = p^{-s}\sum_{n=-\infty}^{n=+\infty} f_n(\mu)\,\mathrm{e}^{\lambda_n z/\Gamma} \tag{10.8}$$

upstream, and a similar form in the downstream rest frame. The power-law assumption for the momentum dependence is justified as the problem (without losses) has no intrinsic momentum scale. For sensible boundary conditions only negative eigenvalues are allowed upstream ($z > 0$), and only positive eigenvalues downstream ($z < 0$).

The functions $f_n(p)$ are then determined by demanding that f is continuous across the shock at $z = 0$ (Liouville's theorem once again). Together with the Lorentz invariance of f this implies

$$f(p,\,\mu,\,0) = \overline{f}(\overline{p},\,\overline{\mu},\,0), \tag{10.9}$$

where the downstream rest frame quantities (denoted by $\overline{\cdots}$) are related to the upstream rest frame quantities by

$$\overline{p} = \Gamma p\,(1-\beta_{\rm rel}\mu), \quad \overline{\mu} = \frac{\mu-\beta_{\rm rel}}{1-\beta_{\rm rel}\mu}. \tag{10.10}$$

In practical applications one has to truncate the infinite series of which f and $\overline{f}$ are comprised after a finite number of terms. Using this method for a relativistic shock,with a jump condition (valid for a strong shock with a relativistic equation of state with $\tilde{\gamma} = 4/3$ in the downstream flow)

$$\beta_1 \, \beta_2 = \tfrac{1}{3} , \tag{10.11}$$

formally corresponding to $r \approx 3$ when $\beta_1 \to 1$, Kirk & Schneider (1987) determine the slope s of the momentum power law from the boundary conditions:

$$F(p) \propto p^2 f(p) \propto p^{-2.3} \tag{10.12}$$

for $\beta_{\rm s} \sim 0.98$. A naive application if the standard (diffusive) DSA result (5.32) for $r = 3$ would give $s = 2.5$.

Kirk & Heavens (1989) have extended this discussion to oblique shocks, assuming gyro-phase independence of the distribution function f. The equations to be solved are similar, save for the replacement of $\Gamma \Longrightarrow \Gamma \, \cos\theta_{\rm Bn}$ **and** the replacement of the matching condition (10.9) by the conservation of magnetic moment *in the shock rest frame* and the associated relations for reflected and transmitted particles, *cf.* the discussion in Section 10.4 of this Review. For very oblique shocks (*i.e.* shocks where $\cos\theta_{\rm Bn} \ll 1$ and $\beta_{\rm i} \equiv \beta_{\rm s}/(\cos\theta_{\rm Bn}) \sim 1$) they find a significant hardening of the spectrum, leading to a spectrum $F(p) \propto p^{-1}$ for strong shocks. Numerical simulations of the same problem (Ellison *et al.* 1990; Ostrowski 1991) confirm these conclusions. These simulations also show that the acceleration process, and consequently the spectral slope, becomes rather sensitive to the details of the scattering process. Differential and large-angle scattering yield different results for slope and acceleration time scale at quasi-perpendicular shocks (*e.g.* Bednarz & Ostrowski 1996).

A somewhat different situation was considered by Ballard & Heavens (1992). They considered the acceleration of particles near relativistic shocks in the case where the magetic field is *disordered.* They integrate the exact particle orbits in a random up- and (shock-compressed) downstream field. They find a power-law distributions with a spectral index s in the range $s \approx 2{-}2.4$ for mildly relativistic shocks ($\beta_{\rm s} \sim 0.2{-}0.7$), steepening to a value $s \sim 3$ for $\beta_{\rm s} \approx 0.9$.

11 Ultra-relativistic shocks

Most approximations employed for mildly relativistic shocks break down in the regime of ultra-relativistic shocks with a Lorentz factor $\Gamma_{\rm s} \gg 1$, for which the shock speed satisfies[6]

$$\beta_{\rm s} \approx 1 - \frac{1}{2\Gamma_{\rm s}^2} , \tag{11.1}$$

[6]In this section I use units with $c = 1$.

and where the relative velocity between the up- and downstream flow corresponds to

$$\beta_{\rm rel} \approx 1 - \frac{1}{\Gamma_{\rm s}^2}. \tag{11.2}$$

Consider such a shock propagating with velocity $\beta_{\rm s}$ in the $z-$direction so that the shock normal $\hat{\boldsymbol{n}} = \hat{\boldsymbol{e}}_z$. A relativistic particle ($v \approx 1$) will only be able to outrun this shock if

$$v_z \approx \cos\theta > \beta_{\rm s}, \tag{11.3}$$

with θ the angle between the shock normal and the particle momentum. This condition can be written as

$$\sin\theta < 1/\Gamma_{\rm s}, \tag{11.4}$$

which defines a *loss cone* with opening angle $\theta_{\rm c} = \sin^{-1}(1/\Gamma_{\rm s})$ around the shock normal. Particles residing within the loss cone outrun the shock, particles outside the loss cone are rapidly swept up by the shock. For $\Gamma_{\rm s} \gg 1$ we can use the small angle approximation, $\sin\theta \approx \theta$, to see that only particles with

$$\theta < \theta_{\rm c} \approx 1/\Gamma_{\rm s} \tag{11.5}$$

will outrun the shock. Particles that have just crossed the shock fro downstream to upstream always reside within the loss cone. For ultra-relativistic case the shock velocity in the downstream rest frame is $\overline{\beta_{\rm s}} = \beta_2 \approx \frac{1}{3}$, (Blandford & McKee 1976), and a Lorentz transformation to the downstream frame, which moves with velocity $\beta_{\rm rel} \approx 1 - 1/2\Gamma_{\rm s}^2$, which corresponds to $\Gamma_{\rm rel} = \Gamma_{\rm s}/\sqrt{2}$, shows that the condition $\theta > \theta_{\rm c}$ corresponds to $\cos\overline{\theta} > \beta_2$.

It can be shown that any reasonable scattering or deflection mechanism in the upstream medium is only capable of deflecting a particle out of the loss cone by an amount $\Delta\theta \lesssim 1/\Gamma_{\rm s}$ before it is recaptured by the shock (Gallant & Achterberg 1999). This implies that the upstream particle distribution is very anisotropic, with an angular extent $\Delta\theta \sim 2/\Gamma_{\rm s} \ll 1$, essentially the result of *relativistic beaming*.

11.1 The case of upstream deflection

The simplest illustration is the case of deflection by a uniform upstream magnetic field $\boldsymbol{B} = (B_\perp, 0, B_\parallel)$. Here $\perp$ and $\parallel$ refer to the direction perpendicular and along the shock normal respectively. As before, I will assume that the shock normal $\hat{\boldsymbol{n}}$ is along the z-axis pointing towards positive z. The equation of motion,

$$\frac{{\rm d}\boldsymbol{\beta}}{{\rm d}t} = \Omega_{\rm g}\left(\boldsymbol{\beta} \times \hat{\boldsymbol{b}}\right) \tag{11.6}$$

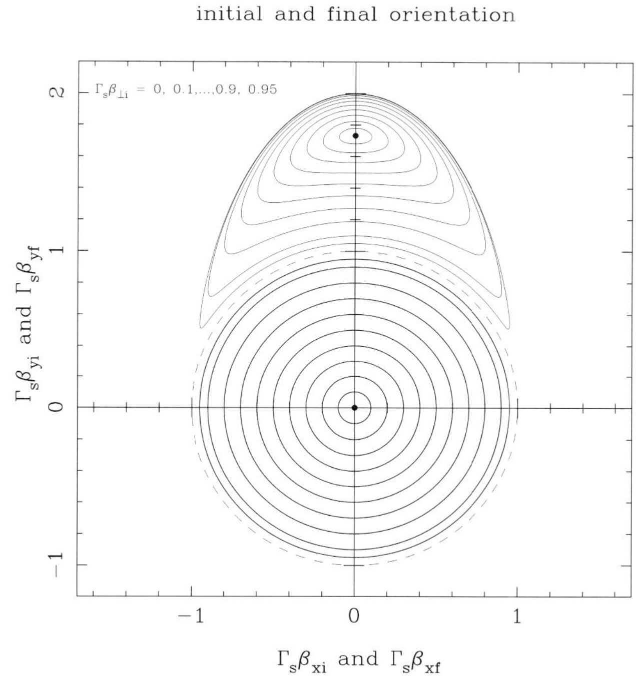

Fig. 2. The location of a particle in the $\Gamma_s\beta_x - \Gamma_s\beta_y$ plane when it is overtaken by the shock as a function of its location when it enters the upstream region. Shown is the location of particles entering on concentric circles within the loss cone centered on the shock normal. The dashed circle with radius $\Gamma_s\beta_\perp = 1$ corresponds to the edge of the loss cone. The concentric circles have radii corresponding to $\Gamma_s\beta_{\perp i}$ equal to 0, 0.1, 0.2, $\cdots$, 0.9 and 0.95 (the dot and the thick solid circles). These circles map onto the kidney-shaped curves along lines of constant β_x. The origin $\beta_{xi} = \beta_{yi} = 0$, corresponding to a particle entering the upstream flow along the shock normal, maps to $\beta_{xf} = 0$, $\Gamma_s\beta_{yf} = \sqrt{3}$. This is indicated by the thick dots. The larger changes in the angle $\theta \approx |\boldsymbol{\beta}_\perp|$ occurs for particles with $\beta_{yi} < 0$ and $\beta_{xi} = 0$, with the largest change ocurring for $\Gamma_s\beta_{yi} = -1$. These particles must turn through most of the loss cone before they leave it and are overtaken by the shock. In contrast, particles with $\beta_{yi} > 0$ change their orientation by a relatively small amount, leaving the loss cone almost immediately. This figure is for $ZeB_\perp > 0$. If the perpendicular field has the opposite sign, this figure should be mirrored with respect to the β_x-axis.

with $\Omega_g = ZeBc/\mathcal{E}$ and $\hat{\boldsymbol{b}} = \boldsymbol{B}/B$, can be solved approximately using the fact that upstream the velocity is almost along the shock normal:

$$\beta_\| = 1 + \mathcal{O}(\Gamma_s^2), \quad \beta_\perp \approx \theta = \mathcal{O}(\Gamma_s^{-1}) \ll 1. \tag{11.7}$$

Here $\beta_{\|} = \beta_z$ and $\beta_{\perp} = \sqrt{\beta_x^2 + \beta_y^2}$. Assuming that a particle enters the upstream region with velocity $\boldsymbol{\beta}_{\rm i} = (\beta_{x\rm i},\, \beta_{y\rm i},\, \beta_{z\rm i})$, and is re-captured by the shock when it has a velocity $\boldsymbol{\beta}_{\rm f}$, one finds that initial and final velocity are related by

$$\beta_{x\rm f} = \beta_{x\rm i}, \quad \beta_{y\rm f} = -\frac{\beta_{y\rm i}}{2} + \sqrt{\frac{3}{\Gamma_{\rm s}^2} - 3\beta_{x\rm i}^2 - \frac{3}{4}\,\beta_{y\rm i}^2}. \tag{11.8}$$

The parallel velocity component follows from $\beta_z = \sqrt{1 - \beta_{\perp}^2}$. The final angle at which the particle is recaptured is $\theta_{\rm f} = \sqrt{\beta_{x\rm f}^2 + \beta_{y\rm f}^2}$.

Particles entering the upstream region with $\beta_{x\rm i} = 0$ and $\beta_{y\rm i} < 0$ incur the largest deflections[7]. These particles turn through most of the loss cone before being captured by the shock. The largest deflection occurs for $\Gamma_{\rm s}\beta_{x\rm i} = 0$, $\Gamma_{\rm s}\beta_{y\rm i} = -1$, where $\Gamma_{\rm s}\beta_{y\rm f} = \Gamma_{\rm s}\theta_{\rm f} = 2$. This is illustrated in the Figure 3 below.

It turns out that, as long as $B_{\|} \ll \Gamma_{\rm s} B_{\perp}$, the time $t_{\rm u}$ of re-capture (upstream residence time) is determined only by the perpendicular field component:

$$\Omega_{\perp} t_{\rm u} = \sqrt{\frac{3}{\Gamma_{\rm s}^2} - 3\beta_{x\rm i}^2 - \frac{3}{4}\,\beta_{y\rm i}^2} - \frac{3\beta_{y\rm i}}{2}, \tag{11.9}$$

with $\Omega_{\perp} = ZeB_{\perp}c/\mathcal{E}$. It is easily checked that the upstream residence time (measured in the upstream rest frame) is of order

$$t_{\rm u} \sim 1/\Gamma_{\rm s}\Omega_{\perp}, \tag{11.10}$$

since particles need only turn through an angle $\Delta\theta \sim \Gamma_{\rm s}^{-1}$ to be overtaken by the shock.

The fact that the upstream turning angle remains small limits the energy gain that can be achieved in a crossing cycle. Assuming the particle goes through a crossing cycle, where it enters the downstream medium with an angle $\theta_{\to\rm d} \lesssim 2/\Gamma_{\rm s}$ and re-crosses the shock into the upstream medium with an angle $\theta_{\to\rm u} \le 1/\Gamma_{\rm s}$, one has:

$$\frac{\mathcal{E}_{\rm f}}{\mathcal{E}_{\rm i}} = \frac{1 - \beta_{\rm rel}\mu_{\to\rm d}}{1 - \beta_{\rm rel}\mu_{\to\rm u}} \approx \frac{2 + \Gamma_{\rm s}^2\theta_{\to\rm d}^2}{2 + \Gamma_{\rm s}^2\theta_{\to\rm u}^2}. \tag{11.11}$$

In the second expression I used the small-angle expansion for the cosines valid in the upstream region, and the relation $\beta_{\rm rel} \approx 1 - 1/\Gamma_{\rm s}^2$ valid for an ultra-relativistic shock where (asymptotically) $\beta_2 \approx 1/3$ *regardless* the

[7] Assuming without loss of generality that $ZeB_{\perp} > 0$.

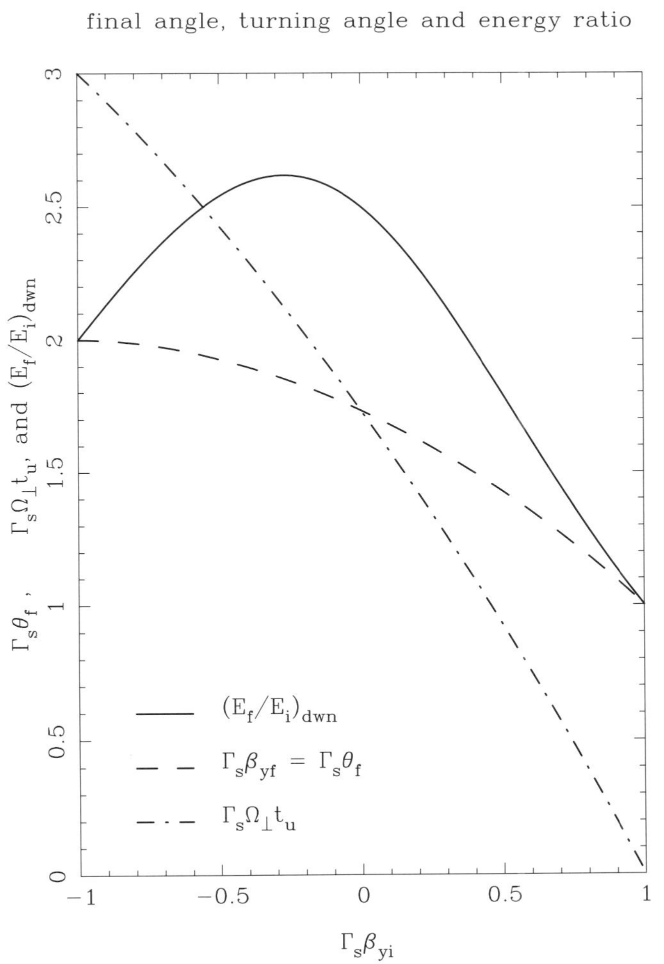

Fig. 3. The final angle $\theta_{\rightarrow d} \approx \beta_{yf}$, the upstream residence time, plotted as $\Gamma_s\Omega_\perp t_u$, and the downstream energy gain $\overline{\mathcal{E}_f}/\overline{\mathcal{E}_i} \equiv (\mathcal{E}_f/\mathcal{E}_i)_{dwn}$ in the downstream-upstream-downstream cycle starting with $\theta_{\rightarrow u} = |\beta_{yi}|$ and ending with $\theta_{\rightarrow d} = \beta_{yf}$, all as a function of β_{yi} with respect to the shock normal at the moment the particle enters the upstream flow. Curves are shown for those particles moving in the y-z plane, *i.e.* $\beta_{xi} = \beta_{xf} = 0$.

upstream conditions. For regular deflection by an uniform field upstream, one has

$$0 \leq \Gamma_s\theta_{\rightarrow u} \leq 1, \quad 1 \leq \Gamma_s\theta_{\rightarrow d} \leq 2. \tag{11.12}$$

From the above relation between $|\boldsymbol{\beta}_{\perp \mathrm{i}}| \approx \theta_{\to \mathrm{u}}$ and $|\boldsymbol{\beta}_{\perp \mathrm{f}}| \approx \theta_{\to \mathrm{d}}$ one finds that

$$1 \leq \mathcal{E}_{\mathrm{f}}/\mathcal{E}_{\mathrm{i}} \lesssim 2.62, \tag{11.13}$$

where the maximum energy gain occurs for $\Gamma_{\mathrm{s}}\beta_{y\mathrm{i}} \approx -0.27$ (see Fig. 4).

The large anisotropy upstream implies that the distribution downstream is also very anisotropic: the fact that particles upstream only occur at $0 \leq \Gamma_{\mathrm{s}}\theta_{\mathrm{u}} \leq 2$ corresponds to a downstream distribution with $-1/3 \leq \overline{\mu} \leq 1$, with particles crossing the shock from upstream confined to the range $-1/3 \leq \overline{\mu}_{\to \mathrm{d}} \leq 1/3$. It is perhaps good to point out that this regime of regular upstream deflection specifically applies to ultra-relativistic shocks, and has no equivalent for non-relativistic shocks.

11.2 The case of upstream scattering

A similar result is obtained if one replaces regular deflection upstream by isotropic diffusion of the particle direction of flight. This leads to diffusion of the angle $\theta = \cos^{-1}(\hat{\boldsymbol{p}} \cdot \hat{\boldsymbol{n}})$ between the upstream particle momentum $\boldsymbol{p} \approx (\mathcal{E}/c)\,\hat{\boldsymbol{p}}$ and the shock normal $\hat{\boldsymbol{n}}$. Such angle diffusion can be caused by random magnetic fields, provided the field coherence length (size of a magnetic cell) is sufficiently small (Achterberg *et al.* 2001):

$$\ell_{\mathrm{coh}} \ll ct_{\mathrm{u}} \sim r_{\mathrm{g}}(\mathcal{E})/\Gamma_{\mathrm{s}}. \tag{11.14}$$

In that case the direction angle θ will start to diffuse before the particle is overtaken by the shock. Using diffusion theory, one can show that the average recrossing angle is independent of the angle diffusion rate:

$$\langle\langle\, \theta^2_{\to \mathrm{d}} \,\rangle\rangle \simeq \frac{2}{\Gamma_{\mathrm{s}}^2} - \theta^2_{\to \mathrm{u}}. \tag{11.15}$$

If one substitutes this into relation (11.11) one finds that the typical upstream energy gain per cycle is

$$\frac{\mathcal{E}_{\mathrm{f}}}{\mathcal{E}_{\mathrm{i}}} \approx \frac{4 - \Gamma_{\mathrm{s}}^2\theta^2_{\to \mathrm{u}}}{2 + \Gamma_{\mathrm{s}}^2\theta^2_{\to \mathrm{u}}}. \tag{11.16}$$

This gain falls steadily from $\mathcal{E}_{\mathrm{f}}/\mathcal{E}_{\mathrm{i}} = 2$ at $\theta_{\to \mathrm{u}} = 0$ to $\mathcal{E}_{\mathrm{f}}/\mathcal{E}_{\mathrm{i}} = 1$ when $\theta_{\to \mathrm{u}} = \theta_{\mathrm{c}} = 1/\Gamma_{\mathrm{s}}$, at the edge of the loss cone.

Simulations (Bednarz & Ostrowski 1998; Achterberg *et al.* 2001) show that for an ultra-relativistic shock with $\beta_1\beta_2 = 1/3$ the resulting spectrum is nearly universal,

$$\mathrm{d}N = F(p)\,\mathrm{d}p \propto p^{-s}\,\mathrm{d}p, \quad s \approx 2.2 - 2.3, \tag{11.17}$$

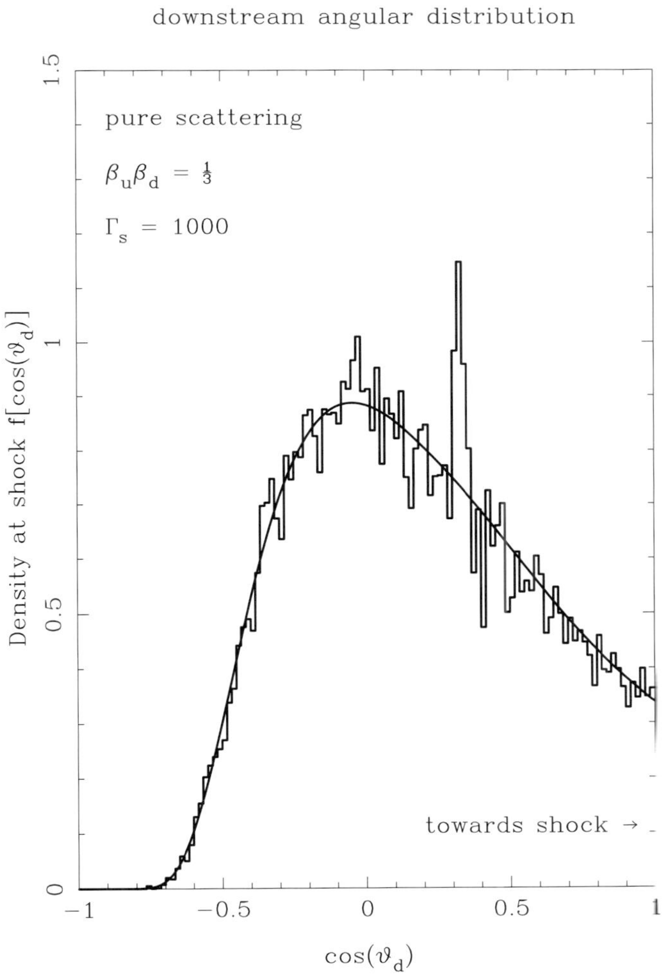

Fig. 4. The angular distribution of particles at the shock in the downstream rest frame. The angle θ_d is the angle between the particle momentum and the shock normal, with $\theta_d = 0$ corresponding to a particle crossing the shock along this normal into the upstream medium. The distribution is normalized to a unit integral. The histogram gives the results from our simulations, and the smooth curve shows the semi-analytical result from the eigenfunction approach. The downstream loss cone corresponds to $\cos\theta_d > \overline{\beta}_s \approx \frac{1}{3}$. This curve is for particles with an energy $\mathcal{E} > 30 \times \mathcal{E}_{inj}$, with $\mathcal{E}_{inj}$ the energy of a particle when it starts true (Fermi-type) shock acceleration. This ensures that the distribution has relaxed so that the initial conditions no longer influence the shape of the distribution.

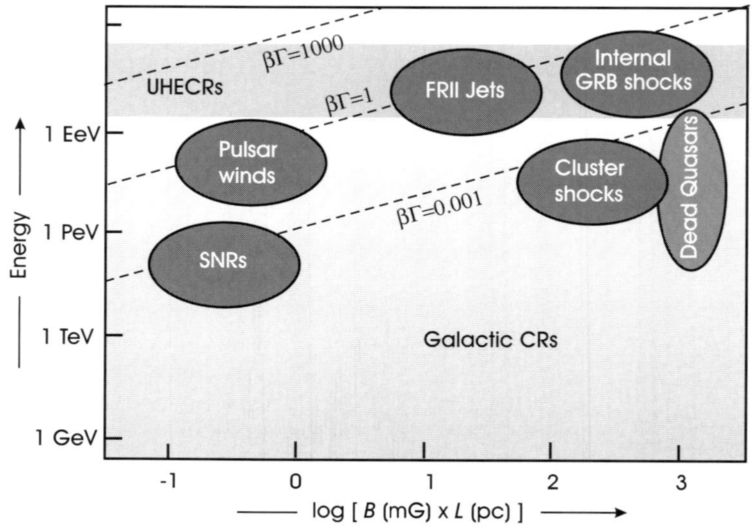

Fig. 5. A diagram showing possible cosmic ray sources. The maximum energy produced is shown as a function of $B \times L$, the product of magnetic field strength and source size. The dashed lines give the loss-free energy, $\mathcal{E}_{\max} \sim eB\, R_{\rm s}\Gamma_{\rm s}\beta_{\rm s}$, that protons can achieve in absence of losses. Losses are taken into account in the construction of this variant of the well-known *Hillas-diagram*.

with the spectral index s almost independent of the deflection/scattering mechanism in the upstream flow. The value for s is compatible with the values that have been estimated from the temporal decay of Gamma-Ray-Burst afterglows. The typical energy gain observed in these simulations corresponds to $\langle\langle\, \mathcal{E}_{\rm f}/\mathcal{E}_{\rm i}\,\rangle\rangle \sim 2$ at each crossing, in agreement with the analytical estimate above.

The eigenfunction method, decribed in the previous Section for mildly relativistic shocks, has been adapted to the case of ultra-relativistic shocks by Kirk *et al.* (2000). This method gives similar results as the simulations. It clearly shows the anisotropic character of the distribution functions at the shock, and determines the power-law index of the distribution of the accelerated particles asymptotically as $\Gamma_{\rm s} \longrightarrow \infty$ as

$$s = 2.23 \pm 0.01. \tag{11.18}$$

The Table 3 below summarizes both the simulation results and the analytical results, and Figure 5 shows the angular distribution of particles in the downstream frame directly behind the shock, comparing the results from simulations with those of the eigen-function method.

Table 3. Properties DSA at UR shocks.

Quantity	Value	reference/method
spectral index s	2.2	Bednarz & Ostrowski (1998) (simulations, pure scattering)
	2.21 ± 0.03	Achterberg *et al.* (2001) (simulations, pure scattering)
	2.3 ± 0.01	Achterberg *et al.* (2001) (simulations, deflection/scattering)
	2.23	Gallant *et al.* (2001) (simulations, pure scattering)
	2.23 ± 0.01	Kirk *et al.* (2001) (eigenfunction method, pure scattering)
energy gain/crossing $\langle \mathcal{E}_{\rm f}/\mathcal{E}_{\rm i} \rangle_{\rm u}$	1.95 ± 0.02	Achterberg *et al.* (2001) (simulations, pure scattering)
	1.65 ± 0.02	Achterberg *et al.* (2001) (simulations, deflection/scattering)
return probability $\mathcal{P}_{\rm ret}$	0.44 ± 0.01	Achterberg *et al.* (2001) (simulations, pure scattering)
	0.53 ± 0.01	Achterberg *et al.* (2001) (simulations, deflection/scattering)

It is interesting to compare the value obtained for the spectral index from a fit of the momentum distribution with the theoretical result obtained from the mean energy gain per crossing and the return probability which are calculated independently from the simulation results. Using the values of Achterberg *et al.* (2001) one has

$$s = 1 + \frac{\ln(1/\mathcal{P}_{\rm ret})}{\ln \langle \mathcal{E}_{\rm f}/\mathcal{E}_{\rm i} \rangle} = \begin{cases} 2.23 \pm 0.02 & \text{pure scattering} \\ \\ 2.31 \pm 0.02 & \text{deflection/scattering.} \end{cases} \tag{11.19}$$

All these values are consistent within the (quoted) errors.

11.3 Initial boost

So far, the discussion has been confined to particles crossing the shock repeatedly in the ultra-relativistic equivalent of DSA. In the case of shocks with $\Gamma_s \gg 1$ the *first* interaction between a particle and a shock proceeds in a qualitatively different manner, unlike the case of non-relativistic shocks. Particles encountering the shock for the first time can have any angle between particle momentum and shock normal. Let us consider such a particle, crossing the shock for the first time with energy $\mathcal{E}_i$ and at angle corresponding to $\mu_{\to d} = (\hat{\boldsymbol{p}} \cdot \hat{\boldsymbol{n}})_{\to d}$. If the particle is deflected or scattered elastically in the downstream flow, in such a manner that it re-crosses the shock into the upstream medium with angle in the downstream rest frame corresponding to $\overline{\mu}_{\to u} = (\bar{\hat{\boldsymbol{p}}} \cdot \hat{\boldsymbol{n}})_{\to u}$, the energy gain seen by an upstream observer equals

$$\frac{\mathcal{E}_f}{\mathcal{E}_i} = \Gamma_{rel}^2 \left(1 - \beta_{rel}\mu_{\to d}\right)\left(1 + \beta_{rel}\overline{\mu}_{\to u}\right). \tag{11.20}$$

The kinematics already described above require $\beta_2 \approx \frac{1}{3} < \overline{\mu}_{\to u} \leq 1$ for particles returning upstream, so the factor $(1 + \beta_{rel}\overline{\mu}_{\to u})$ is of order unity. If $\mu_{\to d}$ is more-or-less isotropically distributed, the factor $(1 - \beta_{rel}\mu_{\to d})$ will (on average) be of order unity also, and the energy gain at the first encounter is of order $\mathcal{E}_f/\mathcal{E}_i \sim \Gamma_{rel}^2 \sim \Gamma_s^2$.

A similar estimate is the basis for the claim by Vietri (1995) that shock acceleration at ultra-relativistic shocks is particularly efficient, with an energy gain corresponding to $\mathcal{E}_f/\mathcal{E}_i \sim \Gamma_s^2$ at *each* encounter. The discussion above shows that such a large energy gain is only possible at the first encounter of the particle with the shock.

The energy gain by a factor Γ_s^2 seen by an upstream observer in the initial encounter is in keeping with the general properties of ultra-relativistic shocks (Blandford & McKee 1976). In the limit $\Gamma_s \gg 1$ a shock isn a very "democratic" device: *all* energies are boosted by the same amount. The downstream energy density $\overline{e}$ (including rest energy), pressure $\overline{P}$ and the proper number density $\overline{n}$ downstream are related to the corresponding upstream quantities by

$$\overline{e} \sim 3\overline{P} = 2\Gamma_s^2 \left(e + P\right), \quad \overline{n} = 2\sqrt{2}\,\Gamma_s n. \tag{11.21}$$

This means that the typical thermal energy per particle in the downstream fluid satisfies

$$\overline{\mathcal{E}} = \frac{\overline{e}}{\overline{n}} = \tfrac{1}{2}\sqrt{2}\,\Gamma_s \left(\frac{e + P}{n}\right). \tag{11.22}$$

In the case of a cold upstream fluid ($P \ll e \sim nmc^2, \mathcal{E} \approx mc^2$) one finds $\overline{\mathcal{E}} \approx 0.7\Gamma_s\,\mathcal{E}$, and in the case of a relativistically hot upstream medium ($P \sim e/3, \mathcal{E} \approx e/n$) one finds $\overline{\mathcal{E}} \approx 0.94\Gamma_s\,\mathcal{E}$. A Lorentz-transformation back to the upstream frame adds another factor $\sim\Gamma_s$, so these downstream particles are assigned an energy $\sim\Gamma_s^2\,\mathcal{E}$ by an upstream observer, with $\mathcal{E}$ the energy per particle upstream. This result is not suprising: in order to have a shock at all some mechanism must be present to randomize the incoming flow and dissipate the kinetic energy of that flow, converting it into the random energy of the thermal motions downstream. Therefore the kinematics of all particles (thermal or otherwise) is the same at the first encounter with the shock. Simulations (Achterberg *et al.* 2001) show that after the first encounter the particles of energy $\mathcal{E}$ are boosted to an energy $\approx 0.9\,\Gamma_{\rm s}^2\,\mathcal{E}$, very close to the above estimate based on the shock jump conditions. About 10% of the particles manage to return upstream after the first shock enounter. These particles then serve as the "seed particles" for diffusive shock acceleration.

Particles undergoing diffusive shock acceleration in subsequent shock crossings are subject to the relativistic beaming described above, and the limitations that this beaming places on the energy gain per crossing cycle.

An important consequence of the initial boost is that particles subject to the ultra-relativistic variant of diffusive shock acceleration are *always* ultra-relativistic themselves, with lorentz-factor $\gamma \gtrsim \Gamma_{\rm s}^2$ in the upstream frame, which usually coincides with the observer's frame.

11.4 Cycle- and residence times and the maximum energy

The maximum energy that can be achieved in the ultra-relativistic variant of DSA can be calculated in similar fashion as in the non-relativistic case by finding the energy that can be reached in the life-time of the shock, or by equating the energy gain per unit time to the losses. Since the typical energy gain per crossing cycle is of order unity, $\Delta\mathcal{E} \sim \mathcal{E}$, the acceleration time $t_{\rm acc}$ and the cycle time, $t_{\rm cy} = t_{\rm u} + t_{\rm d}$, are similar. One has to be careful to about the reference frame used in the calculation. In particular, if the time spent in the downstream flow is $\bar{t}_{\rm d}$ in the downstream rest frame, the corresponding time in the upstream rest frame (which usually coincides with the observers frame) equals $t_{\rm d} \sim \Gamma_{\rm s}\bar{t}_{\rm d}$. Using the fact that the Lorentz-transformation between up- and downstream frame typically results in a downstream energy $\overline{\mathcal{E}} \sim \mathcal{E}/\Gamma_{\rm s}$ as long as $\mu \sim 1$ this implies:

$$t_{\rm cy} \sim t_{\rm u}(\mathcal{E}) + \Gamma_{\rm s}\,\bar{t}_{\rm d}(\mathcal{E}/\Gamma_{\rm s}), \tag{11.23}$$

up to factors of order unity.

11.4.1 Upstream residence time

For simple magnetic deflection upstream, where particles must turn through an angle $\Delta\theta \sim 1/\Gamma_{\rm s}$, one has

$$t_{\rm u}(\mathcal{E}) \approx \frac{\mathcal{E}}{ZeB_\perp\,\Gamma_{\rm s}c}. \tag{11.24}$$

If the upstream momentum direction changes due to scattering on short-wavelength waves rather than deflection, with an (energy dependent) angular diffusion coefficient upstream equal to

$$\mathcal{D}_\theta(\mathcal{E}) \equiv \frac{\langle \theta - \langle\theta\rangle\rangle^2}{2t} , \tag{11.25}$$

where θ is the angle between the shock normal and the momentum direction, the upstream residence time can be estimated as (*e.g.* Achterberg *et al.* 2001)

$$t_{\rm u} \sim \frac{1}{2\Gamma_{\rm s}^2\, \mathcal{D}_\theta(\mathcal{E})} \left(1 - \Gamma_{\rm s}^2\theta_{\rm i}^2\right) . \tag{11.26}$$

Here $\theta_{\rm i}$ is the angle with which the particle re-enters the upstream medium. The usual *gyroresonant scattering* by small-amplitude Alfvén waves, often invoked in non-relativistic DSA as the scattering agent, will not work near ultra-relativistic shocks except in extreme circumstances. The reason is simply that even the non-resonant "sloshing motion" of particles in the magnetic field of the waves is enough to deflect a particle out of the loss cone if the wave amplitude $\delta B_\perp$ in the direction perpendicular to the shock normal satisfies

$$\delta B_\perp > B/\Gamma_{\rm s}. \tag{11.27}$$

In that case the wave magnetic field acts as a "regular" field. Also, particles typically complete only a fraction ($\sim 1/\Gamma_{\rm s}$) of a gyration around the ambient upstream magnetic field before being overtaken by the shock, so there is simply no time for a gyro-resonance to be set up.

A case where the scattering approximation for the change of upstream momentum could apply occurs when the upstream magnetic field is totally disordered. Let us assume that the field consists of magnetic cells with a field amplitude B and a coherence length ($\sim\frac{1}{2}$ cell size) $\ell_{\rm coh}$, with the field direction oriented randomly in each cell. If the coherence length satisfies

$$\ell_{\rm coh} < \frac{9 r_{\rm g}}{4\Gamma_{\rm s}} = \frac{9}{4}\, \frac{\mathcal{E}}{ZeB\,\Gamma_{\rm s}} , \tag{11.28}$$

the momentum direction of upstream particles will start to diffuse before the shock overtakes them. Here $r_{\rm g}(\mathcal{E}) = \mathcal{E}/ZeB$ is the typical gyration radius of a particle with energy $\mathcal{E}$ and charge $q = Ze$ in a field with a magnetic field strength B. The random walk in angle has an amplitude $\theta_{\rm rms}$ satisfying $\theta_{\rm rms}^2 = 4\ell_{\rm coh}^2/3r_{\rm g}^2$ and a time $t_{\rm c} \sim 2\ell_{\rm coh}/c$ between successive steps. In that case one the angle diffusion coefficient is[8]

$$\mathcal{D}_\theta(\mathcal{E}) = \frac{\theta_{\rm rms}^2}{4t_{\rm c}} = \frac{c\ell_{\rm coh}}{3r_{\rm g}^2}. \tag{11.29}$$

Substituting this into equation (11.26), one finds in this case

$$t_{\rm u} \sim \frac{3\mathcal{E}}{2ZeB\,\Gamma_{\rm s}c}\left(\frac{\Gamma_{\rm s}\ell_{\rm coh}}{r_{\rm g}}\right)^{-1}. \tag{11.30}$$

Note that in this case B is the rms amplitude of the magnetic field fluctuations. If condition (11.28) is not satisfied, particles are overtaken by the shock while remaining in a single magnetic cell, and are simply deflected so that the upstream residence time is once again given by expression (11.24). Combining these two results, we find:

$$t_{\rm u} \sim \frac{\mathcal{E}}{ZeB\,\Gamma_{\rm s}c} \times \max\left[1,\ \left(\frac{\Gamma_{\rm s}\ell_{\rm coh}}{r_{\rm g}}\right)^{-1}\right]. \tag{11.31}$$

11.4.2 Downstream residence time

For diffusive shock acceleration to operate, one needs strong scattering in the downstream medium to turn the particle diffusively through an angle $\Delta\overline{\theta} \sim 1$, allowing them re-cross the shock. The downstream residence time is of order

$$\bar{t}_{\rm d} \sim \frac{1}{2\overline{\mathcal{D}_\theta}(\overline{\mathcal{E}})}. \tag{11.32}$$

If the downstream field consists simply of the compressed (large-scale) upstream field resulting from the shock jump conditions, the downstream magnetic field will be almost completely aligned with the shock surface. The perpendicular field component is amplified according to $\overline{B}_\perp \approx 2\sqrt{2}\,\Gamma_{\rm s}B_\perp$, while the parallel field component remains the same, $\overline{B}_\parallel = B_\parallel$, where all fields refer to values in the rest-frame of the up- and downstream plasma. In that case shock acceleration can only result if there is efficient cross-field diffusion of particles (Jokipii 1987; Achterberg & Ball 1994) so that particles can catch up with the shock, which moves with a speed $\overline{\beta}_{\rm s} \sim c/3$ with respect to the downstream medium. This diffusion is presumably due to (shock-induced) wave turbulence.

As long as the turbulence is not so strong that the downstream field geometry becomes completely chaotic, such a situation can only be realised

---

[8]Due to dynamical friction, the relation between the stepsize $\theta_{\rm rms}$, time $t_{\rm c}$ between steps and diffusion coefficient reads $\mathcal{D}_\theta = \theta_{\rm rms}^2/4t_{\rm c}$ rather than the standard result $\mathcal{D}_\theta = \theta_{\rm rms}^2/2t_{\rm c}$.

if the cross-field diffusion coefficient $\kappa_\perp$ is close to the maximum possible value: *Bohm diffusion* with diffusion coefficient $\overline{\kappa}_{\rm B} = c\overline{r_{\rm g}}/3$, where particles randomly move across the field with a step size of order the particle gyration radius every gyration period. This follows from the requirement that the diffusive stepsize $\Delta\overline{z}$ along the shock normal, which downstream is almost perpendicular to the downstream field, must satisfy (Achterberg & Ball 1994)

$$\Delta\overline{z} > c\overline{\beta}_{\rm s}\overline{\tau} \approx c\overline{\tau}/3, \tag{11.33}$$

with $\overline{\tau}$ the time interval between diffusive steps. For cross-field diffusion in the weak turbulence limit one has $\Delta\overline{z} \sim \overline{r_{\rm g}}/\sqrt{2}$, with $\overline{r_{\rm g}}$ the downstream gyroradius. The above condition then corresponds to $\overline{\tau} < 3\overline{r_{\rm g}}/c$, and leads to a diffusion coefficient which must satisfy

$$\overline{\kappa}_\perp \sim \frac{\overline{r_{\rm g}}^2}{2\overline{\tau}} > \frac{c\overline{r_{\rm g}}}{3\sqrt{2}} \equiv \tfrac{1}{2}\sqrt{2}\,\overline{\kappa}_{\rm B}. \tag{11.34}$$

Since the downstream flow is only mildly relativistic, we can use the results from the non-relativistic case to estimate the downstream residence time. Assuming Bohm diffusion ($\overline{\kappa} = \overline{\kappa}_{\rm B}$) one has

$$\overline{t}_{\rm d}(\overline{\mathcal{E}}) \approx \frac{4\overline{\kappa}_{\rm B}}{c^2/3} = \frac{4\overline{\mathcal{E}}}{Ze\overline{B}_\perp c}. \tag{11.35}$$

The contribution of the downstream residence time to the cycle time in this case equals (compare Eq. (11.23)):

$$t_{\rm d}(\mathcal{E}) \equiv \Gamma_{\rm s}\,\overline{t}_{\rm d}(\mathcal{E}/\Gamma_{\rm s}) \approx \frac{4\,\mathcal{E}}{Ze\xi_B\,B_\perp\,\Gamma_{\rm s}c}. \tag{11.36}$$

Here $\xi_B \equiv \overline{B}_\perp/(2\sqrt{2}\,B_\perp) \geq 1$ is a factor that takes account of the possible amplification of the post-shock magnetic field above the value following from simple compression. On the basis of estimate (11.36) I conclude that $t_{\rm d}(\mathcal{E}) \sim t_{\rm u}(\mathcal{E})/\xi_B$ in this case.

I now briefly treat the case where the upstream magnetic field consists of randomly oriented cells of size $\sim 2\ell_{\rm coh}$ and field amplitude B. The compression in the shock "squashes" any pre-existing upstream magnetic cells in the direction along the shock normal, so that they become very anisotropic, with size $\sim\overline{\ell_{\rm coh}} \sim \ell_{\rm coh}/\Gamma_{\rm s}$ in the direction normal to the shock. Also, the magnetic fields are amplified by a factor $\sim\Gamma_{\rm s}$ while the downstream particle energy is $\overline{\mathcal{E}} \sim \mathcal{E}/\Gamma_{\rm s}$. This means that the downstream gyro-radius of a particle is related to its upstream value by $\overline{r}_{\rm g} = r_{\rm g}/\Gamma_{\rm s}^2$. If the upstream field satisfies the requirement $\ell_{\rm coh} \lesssim r_{\rm g}(\mathcal{E})/\Gamma_{\rm s}$ so that upstream the direction of flight of a particle diffuses, downstream the same particle has $\overline{\ell}_{\rm c} \lesssim \overline{r}_{\rm g}(\overline{\mathcal{E}})$, and the particle momentum will diffuse once again. The corresponding downstream angular diffusion coefficient of the same particle becomes of order $\overline{\mathcal{D}}_\theta(\mathcal{E}/\Gamma_{\rm s}) \approx \Gamma_{\rm s}^3\, \mathcal{D}_\theta(\mathcal{E})$. These estimates neglect the possible generation of additional post-shock magnetic fields as a by-product of shock physics, which would shorten the downstream residence time by a factor ξ_B^{-2}. This implies that the downstream residence time contributes to the cycle time at the shock (see Eq. (11.23)) by an amount

$$t_{\rm d} = \Gamma_{\rm s}\bar{t}_{\rm d}(\mathcal{E}/\Gamma_{\rm s}) \sim \frac{\mathcal{E}}{ZeB\, \xi_B^2\, \Gamma_{\rm s}^2 c} \left(\frac{\Gamma_{\rm s}\ell_{\rm coh}}{r_{\rm g}}\right)^{-1} . \tag{11.37}$$

Note that I have expressed everything in the corresponding upstream variables. In this case one has $t_{\rm d} \sim t_{\rm u}/(\Gamma_{\rm s}\xi_{\rm B}^2)$. For a wide range of circumstances, and assuming $\Gamma_{\rm s} \gg 1$, it follows that in this case most of the cycle (as seen by an upstream observer) is spent upstream, so that $t_{\rm cy} \approx t_{\rm u}$.

11.4.3 Maximum energy

Assuming $t_{\rm cy} \sim t_{\rm u}$, and using the fact that $\Delta\mathcal{E}/\mathcal{E} \sim 1$, the typical acceleration rate at an ultra-relativistic shock equals

$$\mathcal{R}_{\rm acc} = \frac{1}{\mathcal{E}}\,\frac{{\rm d}\mathcal{E}}{{\rm d}t} \approx \frac{1}{t_{\rm u}} \sim \frac{ZeB\,\Gamma_{\rm s}c}{\mathcal{E}}\, \min\left[1,\ \frac{\Gamma_{\rm s}\ell_{\rm coh}}{r_{\rm g}(\mathcal{E})}\right] . \tag{11.38}$$

If losses are unimportant, the finite age of an expanding shock, $t_{\rm s} \sim R_{\rm s}/c$, will limit the energy to

$$\mathcal{E}_{\rm max} \approx ZeB\,\Gamma_{\rm s}\ \times\ \min\left[R_{\rm s},\ \sqrt{R_{\rm s}\ell_{\rm coh}}\,\right] . \tag{11.39}$$

Here $\ell_{\rm coh}$ is the upstream field correlation length. The case of a regular upstream field is included in this estimate as $\ell_{\rm coh}$ goes to infinity in that case. This shows that shocks where regular deflection upstream confines the particles near the shock are in principle capable of producing more energetic particles if $R_{\rm s} > \ell_{\rm coh}$.

Adiabatic (expansion) losses, which occur behind an expanding shock front, give a similar value of the maximum energy if $t_{\rm d} \sim t_{\rm u}$. In a relativistic flow, particles which are coupled collisionally to the flow by elastic scattering on hydromagnetic waves loose energy in the lab frame at a rate (*e.g.* Webb 1985):

$$\left(\frac{1}{\mathcal{E}}\,\frac{{\rm d}\mathcal{E}}{{\rm d}t}\right)_{\rm exp} = -\mathcal{R}_{\rm loss} = -\frac{1}{3\Gamma}\left[\frac{\partial\Gamma}{\partial t} + \boldsymbol{\nabla}\cdot(\Gamma\boldsymbol{V})\right] , \tag{11.40}$$

where $\Gamma = 1/\sqrt{1 - |\boldsymbol{V}|^2/c^2}$ is the lorentz-factor of the flow and $\boldsymbol{V}$ is the flow velocity. This assumes that the particle distribution is nearly isotropic in the fluid rest-frame, which should be true some distance behind the shock. These losses involve the same time-scale $t_{\rm s} \sim V_{\rm s}/R_{\rm s}$ as in the estimate (11.39), and therefore expansion losses give a similar value of $\mathcal{E}_{\rm max}$.

If **synchrotron/inverse Compton losses** (see Eqs. (7.32) and (7.33)) limit the energy gain, the typical maximum energy equals:

$$\mathcal{E}_{\rm max} \approx \min \left[\sqrt{mc^2 \, E_{\rm sc}}, \; \sqrt[3]{mc^2 \, E_* E_{\rm sc}} \right] . \tag{11.41}$$

Here I have defined two typical energy scales,

$$E_* = ZeB \, \Gamma_{\rm s} \, \ell_{\rm coh}, \quad E_{\rm sc} = ZeB \, \Gamma_{\rm s} \, ct_{\rm sc}. \tag{11.42}$$

For electrons, using expression (7.33) for $t_{\rm sc}$ and assuming $E_{\rm sc} \ll E_*$, one finds:

$$\mathcal{E}_{\rm max} \sim m_{\rm e}c^2 \left(\frac{4\pi e \, \Gamma_{\rm s}}{\sigma_{\rm T} B} \right)^{1/2} . \tag{11.43}$$

One can show that the losses in the upstream field, and in the compressed downstream field give a similar value for $\mathcal{E}_{\rm max}$.

Finally, **pion-production losses** with loss length ℓ_π give

$$\mathcal{E}_{\rm max} \approx \min \left[\mathcal{E}_\pi, \; \sqrt{\mathcal{E}_* \mathcal{E}_\pi} \right] , \tag{11.44}$$

where

$$\mathcal{E}_\pi \equiv ZeB \, \Gamma_{\rm s} \, \ell_\pi . \tag{11.45}$$

Here I have assumed for simplicity that ℓ_π does not depend on energy.

12 Maximum energy: A cosmic conspiracy?

In Table 4 I give the typical values of the energy $\mathcal{E}_{\rm max}$ attainable in astrophysical shocks for protons and electrons, together with the relevant process limiting the attainable energy in the best possible case: shock acceleration in the Bohm limit. For the calculation of $\mathcal{E}_{\rm max}$ one can express the various energies involved in terms of a number of fundamental physical parameters:

- The fine-structure constant,
  $\alpha = e^2/\hbar c \sim 1/137$;
- The magnetic field in units of the critical field,
  $B_{\rm cr} = m_{\rm e}^2 c^3/\hbar e = 4.4 \times 10^{13}$ G;

Table 4. Maximum attainable observer's frame energy in shock acceleration with diffusion at the Bohm rate.

Object	R_{s}	β Γ	B (G)	**limiting mechanism**	$\mathcal{E}_{\max}$ (in eV)
Solar Flare Type II shock	10^{10} cm	$10^{-2.5}$ 1	10	*protons:* size/age *electrons:* synchr. losses	10^{11} 10^{9}
Interplanetary Shocks	10^{13} cm	10^{-3} 1	10^{-5}	*protons:* size/age *electrons:* size/age	10^{8} 10^{8}
Supernova Remnants	10 pc	$10^{-2.5}$ 1	10^{-4}	*protons:* size/age *electrons:* synchr. losses	10^{15} 10^{12}
Superbubbles	5 kpc	10^{-4} 1	10^{-4}	*protons:* size/age *electrons:* synchr. losses	$10^{15.5}$ 10^{11}
Gamma Ray Burst, external shock:	0.01 pc	1 $10^{2}-10^{3}$	10^{-6}	*protons:* size/age *electrons:* size/age	10^{15} 10^{15}
Gamma Ray Burst, internal shocks:	0.01 pc	$\sim$1 $\sim$1	10^{2}	*protons:* size/age *electrons:* synchr. losses	$10^{20.5}$ $10^{15.5}$
AGN	10^{13} cm	0.1 1	10^{4} (?)	*protons:* pion prod. losses *electrons:* SC losses	10^{16} 10^{11}
Hot Spots radiogalaxy	10 kpc	$\sim$1 1–10	10^{-4}	*protons:* shock size *electrons:* synchr. losses	10^{20} 10^{14}
Dead Quasar magnetosphere	10^{13} cm	n.a. n.a.	10^{4} (?)	*protons:* synchr. losses *electrons:* synchr. losses	10^{18} 10^{11}
Cosmic Large- Scale Structure	10 Mpc	$10^{-2.5}$ 1	10^{-6}	*protons:* pion prod. losses *electrons:* SC losses	10^{19} 10^{14}

Note: in the case of particles produced by internal shocks in the fireballs of Gamma Ray Bursts, the energies given are the energy measured by an outside observer. Because of the relativistic bulk motion in these objects with Lorentz factor Γ_{s}, this energy is $\sim\Gamma_{\mathrm{s}}\times$ the energy in the rest-frame of the material. Taking $\Gamma_{\mathrm{s}} \sim 10^3$, this means that the internal shocks can produce protons up to $\sim 10^{17.5}$ eV (and electrons up to $\sim 10^{12.5}$ eV) in the rest frame of the fireball, provided the magnetic energy density is in equipartition with the kinetic energy density so that $B \sim 100$ G (see below). The Lorentz-factor Γ, "velocity" β and magnetic field strength B on the other hand all refer to the values measured in the frame comoving with the fireball material. The "Dead Quasar" case does *not* involve shock acceleration, but direct acceleration in magnetospheric fields, and is shown for comparison.

- The electron Compton length,
  $\lambda_{\mathrm{e}} = \hbar/m_{\mathrm{e}}c = 3.86 \times 10^{-11}$ cm.

In particular, the relation

$$eB_{\rm cr}\,\lambda_{\rm e} = m_{\rm e}c^2 \tag{12.1}$$

allows one to express the loss-free maximum energy attainable in (diffusive) shock acceleration as

$$\mathcal{E}_{\max} \approx ZeB\,\Gamma_{\rm s}\beta_{\rm s}R_{\rm s} = Z\,m_{\rm e}c^2\left(\frac{B}{B_{\rm cr}}\right)\left(\frac{\Gamma_{\rm s}\beta_{\rm s}R_{\rm s}}{\lambda_{\rm e}}\right). \tag{12.2}$$

This expression is valid for both non-relativistic and relativistic shocks provided upstream deflection is due to a "regular" field with $\ell_{\rm coh} > R_{\rm s}$ in the relativistic case. In a similar way, the electron synchrotron loss length can be expressed as $\ell(\mathcal{E}) = \ell_{\rm sc}\,(\mathcal{E}/m_{\rm e}c^2)^{-1}$, with $\ell_{\rm sc}$ equal to

$$\ell_{\rm sc} = \frac{9}{4\alpha}\left(\frac{B}{B_{\rm cr}}\right)^{-2}\lambda_{\rm e}, \tag{12.3}$$

with an associated energy scale

$$\mathcal{E}_{\rm sc} = eB\,\Gamma_{\rm s}\,\ell_{\rm sc} = m_{\rm e}c^2\,\frac{9\Gamma_{\rm s}}{4\alpha}\left(\frac{B}{B_{\rm cr}}\right)^{-1}. \tag{12.4}$$

The maximum energy attainable for electrons when synchrotron losses dominate is

$$\mathcal{E}_{\max} = \beta_{\rm s}\,\sqrt{m_{\rm e}c^2\,\mathcal{E}_{\rm sc}} = m_{\rm e}c^2\,\frac{3\beta_{\rm s}\sqrt{\Gamma_{\rm s}}}{2\sqrt{\alpha}}\left(\frac{B}{B_{\rm cr}}\right)^{-1/2}. \tag{12.5}$$

The synchrotron emission from electrons peaks at a frequency $\nu_{\max} = 0.29\nu_{\rm c}$, with $\nu_{\rm c} = (3/4\pi)(eB/m_{\rm e}c)(\mathcal{E}/m_{\rm e}c^2)^2$ (*e.g.* Rybicki & Lightman 1979). Using $\mathcal{E}_{\max}$ from the above expression, one finds that the maximum energy of synchrotron photons is *independent* of the fieldstrength (Phinney 19xx, see also Guilbert *et al.* 1983):

$$h\nu_{\max} = 21.3\,\beta_{\rm s}^2\Gamma_{\rm s}\,m_{\rm e}c^2 \sim 11\,\beta_{\rm s}^2\Gamma_{\rm s}\ {\rm MeV}. \tag{12.6}$$

The constant in the first equality is $0.29\times(27/16\pi\alpha)$. This is the synchrotron photon energy in the rest-frame of the material.

The synchrotron length $\ell_{\rm sc}$ and the associated energy scale $\mathcal{E}_{\rm sc}$ of nuclei of mass m and charge number Z is larger by a factor $Z^4\,(m/m_{\rm e})^3$ than the same quantity for electrons, and the maximum energy is larger by a factor $(Zm/m_{\rm e})^2$.

Finally, pion production losses with loss length ℓ_π can limit the maximum energy to

$$\mathcal{E}_{\max} \sim ZeB\,\Gamma_{\rm s}\beta_{\rm s}^2\,\ell_\pi = m_{\rm e}c^2\,\Gamma_{\rm s}\beta_{\rm s}^2\left(\frac{B}{B_{\rm cr}}\right)\left(\frac{\ell_\pi}{\lambda_{\rm e}}\right). \tag{12.7}$$

The loss length equals

$$\ell_\pi = \frac{1}{n_{\rm t}\sigma_\pi K_\pi}, \tag{12.8}$$

with σ_π the relevant cross section, $n_{\rm t}$ the density of target particles (nucleons or photons) and $K_\pi = |\Delta\mathcal{E}|/\mathcal{E}$ the inelasticity of the interaction. Typically $K_\pi \sim 0.5$ for high-energy nucleons far above production threshold energy. The typical cross section equals $\sigma_\pi \sim 10^{-28}$ cm^2 for photo-pion production, and $\sigma_\pi \sim 5 \times 10^{-26}$ cm^2 for pion production in collisions between nucleons and protons.

If one reviews the values for $\mathcal{E}_{\rm max}$ in Table 4, there seems to be a "Cosmic Conspiracy" in operation: with the exception of [1] the extreme case of the shocks associated with the "hot spots" in powerful radio galaxies (Fanaroff-Riley Class II galaxies), which occur at the point where the trans-relativistic jets collide with the intergalactic gas and form the extended radio lobes, [2] the extremely large ($\sim$10 Mpc) shocks one believes to be present in regions of ongoing Large-Scale Structure formation, and [3] possibly the *internal* shocks in Gamma-Ray-Burst fireballs, no objects seems to be capable of accelerating protons to energies above 10^{15-16} eV! This is a real problem for models which try to explain the observed Ultra-High Energy Cosmic Rays as the product of a conventional acceleration mechanism. In most cases, the size and/or finite age of the shock is the limiting factor: the product $B\beta_{\rm s}\Gamma_{\rm s}R_{\rm s}$ is too small to produce particles at energies exceeding 10^{16} eV. The figure below summarizes the main results graphically.

Particle acceleration associated with Gamma Ray Bursts will be discussed in some detail below. Here, I briefly consider the case of Active Galactic Nuclei which deserves some discussion. One might expect on the basis of the enormous amount of energy produced in these objects, typically $L \sim 10^{46}$ erg/s for the more powerful radio galaxies and quasars, generated in a region several AU across, that it would be the perfect environment to accelerate particles to high energies. In addition, assuming that the energy release is the product of accretion of $\sim$1 $M_\odot$/yr on a Black Hole with a mass of $\sim 10^6 - 10^8$ $M_\odot$, one could expect shocks in a trans-relativistic accretion flow with $V_{\rm s} \sim c$. If the large-scale magnetic field is close to equipartition with the radiation field ($B^2/8\pi \sim L/4\pi R_{\rm s}^2 c$) it could be very large,

$$B \sim \left(\frac{2L}{cR_{\rm s}^2}\right)^{1/2} \sim 6 \times 10^3 \left(\frac{L}{10^{46}\ {\rm erg/s}}\right)^{1/2} \left(\frac{R_{\rm s}}{10\ {\rm AU}}\right)^{-1}\ {\rm G}. \tag{12.9}$$

If one uses this estimate for B in the standard expression for $\mathcal{E}_{\rm max}$ one finds:

$$\begin{aligned} \mathcal{E}_{\rm max} &\sim Ze\,\Gamma_{\rm s}\beta_{\rm s}\left(\frac{2L}{c}\right)^{1/2} \\ &\sim 2.5\times 10^{20}\,Z\beta_{\rm s}\Gamma_{\rm s}\left(\frac{L}{10^{46}\ {\rm erg/s}}\right)^{1/2}, \end{aligned} \tag{12.10}$$

this is tantalizingly close to the typical energy of the UHECRs. Unfortunately, this scenario is sabotaged by the effect of photo-pion production in the intense radiation field, which limits the attainable energy (Norman *et al.* 1995). The photon density is very high. The typical AGN continuum emission scales with photon energy $\epsilon = h\nu$ as $L(\epsilon) \propto \epsilon^{-1}$, so

$$n_{\rm ph}(>\epsilon) \sim 10^{17}\left(\frac{L}{10^{46}\ {\rm erg/s}}\right)\left(\frac{R_{\rm s}}{10\ {\rm AU}}\right)^{-2}\left(\frac{\epsilon}{1\ {\rm eV}}\right)^{-1}\ {\rm cm}^{-3}. \tag{12.11}$$

The threshold photon energy for pion production by protons of energy $\mathcal{E}$,

$$\epsilon_{\rm th} \approx m_{\rm p}m_{\pi}c^4/\mathcal{E} \sim \left(\frac{\mathcal{E}}{10^{15}\ {\rm eV}}\right)^{-1}\ {\rm eV}, \tag{12.12}$$

together with the typical cross section for the process, $\sigma_{\pi\gamma} \sim 10^{-28}\ {\rm cm}^2$, leads to a loss length scaling as

$$\begin{aligned} \ell_{\pi}(\mathcal{E}) &= \frac{1}{n_{\rm ph}(\epsilon > \epsilon_{\rm th})\,\sigma_{\pi\gamma}} \\ &\approx 10^{11}\left(\frac{L}{10^{46}\ {\rm erg/s}}\right)^{-1}\left(\frac{R_{\rm s}}{10\ {\rm AU}}\right)^{2}\left(\frac{\mathcal{E}}{10^{15}\ {\rm eV}}\right)^{-1}\ {\rm cm}. \end{aligned} \tag{12.13}$$

If one balances the pion production losses with the acceleration rate in the Bohm limit, ${\rm d}\mathcal{E}/{\rm d}t \sim ZeBc\,\beta_{\rm s}^2\Gamma_{\rm s}$, and uses the free-fall velocity (or Keplerian rotation speed) near a black hole as an estimate for the shock speed, $\beta_{\rm s} \sim \sqrt{2GM_{\rm bh}/c^2R_{\rm s}}$, one finds:

$$\mathcal{E}_{\rm max} \sim 10^{16}\left(\frac{M_{\rm bh}}{10^8\ M_{\odot}}\right)^{1/2}\left(\frac{L}{10^{46}\ {\rm erg/s}}\right)^{-1/4}\ {\rm eV}. \tag{12.14}$$

One is forced to conclude that the strong losses prevent the production of particles with an energy more than 10^{16} eV.

Even if one is somehow able to circumvent this limit on $\mathcal{E}_{\rm max}$, one can show that high-energy particles cannot escape the AGN environment: the optical depth for photo-pion production is large:

$$\tau_\pi = \frac{\sigma_\pi L}{4\pi Rc \ <\epsilon>} \sim 10^2 \, \ell_{\rm T} \left(\frac{<\epsilon>}{1 \text{ eV}} \right)^{-1} , \tag{12.15}$$

where $<\epsilon>$ is the mean photon energy in the continuum and $\ell_{\rm T}$ is the *compactness parameter* for two-photon pair production,

$$\ell_{\rm T} \equiv \frac{\sigma_{\rm T} L}{4\pi m_{\rm e} c^3 R} . \tag{12.16}$$

The compactness parameter is estimated to be larger than unity on the basis of the observed variability of AGN emission, assuming this is intrinsic to the source and not the result of scintillation due to foreground ionized clouds in our Galaxy. What happens if a Quasar or AGN is inactive ("dead") is explored below.

13 Gamma ray bursts and ultra-high-energy cosmic rays

As an application of the theoretical concepts outlined above I will discuss two hot topics in present-day high-energy astrophysics: *Ultra-High-Energy Cosmic Rays* (UHECRs), *Gamma Ray Bursts* (GRBs), and their possible connection.

In 1993 it was discovered that the cosmic-ray spectrum extends beyond the *Greisen-Zatsepin Kuzmin* cut-off energy:

$$\mathcal{E}_{\rm GZK} \approx 10^{19.7} \text{ eV}. \tag{13.1}$$

This is the energy where one expects the flux from cosmologically distant sources ($D > 10$ Mpc) to be strongly attenuated due to photo-pion production on the photons of the *Cosmic Microwave Background* (CMWB), which pervades the Universe with a photon density of about $n_{\rm ph} \approx 400 \text{ cm}^{-3}$. Figure 6 shows the loss length $\ell(E)$ of energetic protons in the intergalactic medium as a function of energy. The mean loss length is the result of three processes: [1] the universal expansion which "redshifts" the energy of an ultra-relativistic particle in exactly the same manner as the energy of a photon, with loss length equal to the Hubble-length, $\ell_{\rm H} \sim c/H_0$, with H_0 the Hubble constant; [2] pair production on CMWB photons, with loss length $\ell_{\rm p}$ and [3] photo-pion production with loss length ℓ_π. One sees that at $\mathcal{E} \sim \mathcal{E}_{\rm GZK}$ the loss length decreases dramatically to about $\ell(\mathcal{E}) = \ell_\pi \sim 10$ Mpc. Below the GZK cut-off energy, the loss length is larger than 1 Gpc.

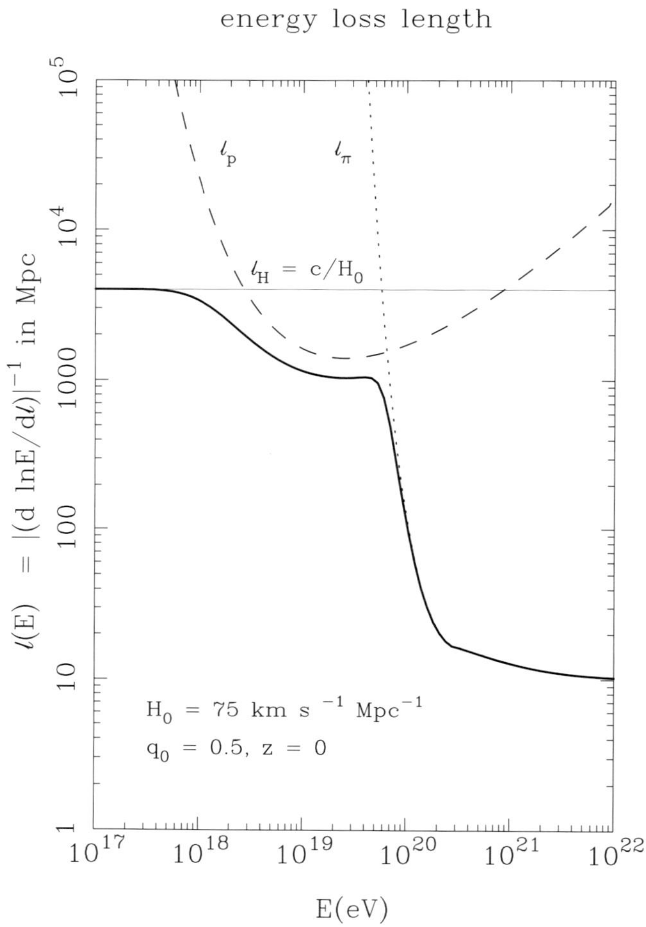

Fig. 6. The loss length $\ell(\mathcal{E})$ (thick solid curve) as a function of energy. It is the result of losses due to pion production (ℓ_π, dotted curve), pair losses ($\ell_{\rm p}$, dashed curve) and expansion losses in the Hubble flow ($\ell_{\rm H} = c/H_0$, horizontal solid line) in a flat Universe ($q_0 = \frac{1}{2}$).

It was generally expected that no cosmic rays with $\mathcal{E} > \mathcal{E}_{\rm GZK}$ would be observed, based on three arguments:

1. The gyration radius of particles at this energy is so large,

$$r_{\rm g}(\mathcal{E}) = 0.1 \left(\frac{\mathcal{E}_{20}}{Z B_{\mu {\rm G}}} \right) \text{ Mpc} \tag{13.2}$$

with $\mathcal{E}_{20} = \mathcal{E}/(10^{20}\ \mathrm{eV})$, that they cannot be confined by the $\sim\mu$G magnetic fields of the Galaxy, unless they have a large charge like particles in the Iron group ($Z \sim 20$);

2. Estimates of the maximum energy that could be produced in Galactic objects (*e.g.* Hillas 1984) seemed to rule out a local production site in our own Galaxy;

3. The most likely extragalactic production sites, such as the radio lobes associated with powerful active galaxies, are too far removed.

It therefore came as somewhat of a suprise when observations showed that there is a non-negligible flux above the GZK-cutoff (Bird *et al.* 1994; Yoshida *et al.* 1994; Takeda *et al.* 1998).

The observations suggest that UHECRs are of extra-galactic origin:

1. The energy spectrum hardens, from $J(\mathcal{E}) \propto \mathcal{E}^{-3.2\pm0.1}$ below $10^{18.5}$ eV to $J(\mathcal{E}) \propto \mathcal{E}^{-2.8\pm0.3}$ above $10^{18.5}$ eV;

2. Arrival directions do not seem to be clustered significantly along the Galactic plane, although there is an indication that they may be clustered along the supergalactic plane for $\mathcal{E} > 10^{20}$ eV (Hayashida *et al.* 1996; Stanev *et al.* 1995). A study of 92 events above 4×10^{19} eV show a significant number of double and triple coincidences on the supergalactic plane (Uchihori *et al.* 2000). However, the statistics are not as yet sufficiently good to draw firm conclusions.

3. There is a suggestion, based on the analysis of the observed cosmic-ray induced airshowers, that the composition changes from one dominated by heavy nuclei (iron group) below $10^{18.5}$ eV, as expected for a population of Galactic origin, to one dominated by light nuclei or even protons above $10^{18.5}$ eV. However, given the large uncertainty in the modelling of nuclear collisions with a center-of-mass energy $\geq 10^{19}$ eV and the resulting atmospheric cascades, this conclusion is not uncontroversial (*e.g.* Anchordoqui *et al.* 2000), and some have proposed that UHECRs are extra-galactic heavy nuclei (*e.g.* Stecker 1998; Szabelski *et al.* 2002).

Some theorists however speculate that UHECRs are produced as Iron nuclei in pulsar magnetospheres/pulsar winds associated with rapidly spinning (young) Neutron Stars (*e.g.* Blasi *et al.* 2000).

The experimental situation of UHECRs has been reviewed by Watson (2000) and Nagano & Watson (2000). A recent review of the various issues involved in the production and propagation of UHECRs can be found in reviews by Biermann (1997), Bhattcharjee & Sigl (2000) and Olinto (2000).

In an apparently unrelated development, the identification of the optical counterparts of Gamma Ray Bursts in 1997 (Van Paradijs *et al.* 1997; Djorgovski *et al.* 1997), settled a question that had been around since the serendipitous discovery of GRBs almost thirty years ago. Gamma Ray Bursts are flashes of Gamma Rays, typically 0.1–100 s long which occur on average once a day. Pre-1997 models for this phenomenon could be divided in two main classes: [1] models with local production in the Galaxy, presumably associated with neutron stars, and [2] the cosmological models where the GRBs are of extragalactic origin. The discovery of afterglows in X-rays with the BEPPO-SAX satellite, at optical wavelengths and at radio wavelengths settled this origin question: GRBs are cosmological. Afterglow observations allowed a precise position determination, the identification of some host galaxies, and a distance determination based on the redshift of the hosts. To date some 50 afterglows have been observed in X-rays, and some 30 optical afterglows. The median redshift of the host galaxies is $< z > \sim 1$, corresponding to a typical source distance of $D \sim 3$ Gpc.

The distance determination sets the energy scale of the phenomenon, based on the observed number and energy of the Gamma-Ray photons per unit area at Earth (the so-called *fluence* $\mathcal{F}_\gamma$):

$$E_{\rm GRB} \sim 10^{51}\,(1+z)^2 \left(\frac{\mathcal{F}_\gamma}{10^{-6}\ {\rm erg/cm^2}}\right) \left(\frac{D}{3\ {\rm Gpc}}\right)^2 \theta^2\ {\rm erg}. \qquad (13.3)$$

Here θ is the solid angle over which the energy is emitted, for instance: $\theta = 4\pi$ for a spherical explosion, and the factor $(1+z)^2$ corrects for the cosmological redshift of the photon energy and the arrival rate of photons, which both contribute a factor $(1+z)$.

The canonical model for the GRB phenomenon is that of a *relativistic fireball*, where most of the explosion energy is in radiation, with a small fraction of the energy residing in the (rest) energy of baryons. The observations require that the baryonic mass $M_{\rm b}$ is such that

$$\eta \equiv \frac{E_{\rm GRB}}{M_{\rm b} c^2} \sim 100\text{–}1000. \qquad (13.4)$$

The reason for this requirement is simple: both the short duration of the burst phenomenon itself, as well as the rapid fluctuations observed in some bursts, which occur at $\sim ms$ timescales, would infer a very small source region if the radiation comes from a stationary source. This, together with the observed fluence, would imply that the source is optically thick for two-photon pair production. In that case, most radiation would be converted into pairs. The source would be optically thick, and the emergent spectrum would be a Black-Body spectrum. The observations however show a decidedly non-thermal spectrum (broken power-law). The relativistic fireball model (Rees & Mészáros 1992) circumvents this compactness problem by assuming relativistic motion in the source. If the source material approaches the observer along the line-of-sight with velocity βc and Lorentz-factor $\Gamma = 1/\sqrt{1-\beta^2}$, the *observed* time-interval between events is compressed by light travel-time effects to

$$t_{\rm obs} = (1-\beta)t \approx t/2\Gamma^2 , \tag{13.5}$$

where the second equality is valid if $\Gamma \gg 1$. In addition, the photon energy is blueshifted to $\epsilon_{\rm obs} \approx \Gamma\epsilon$. As a result, the optical depth for two-photon pair production (the *compactness parameter* defined in Sect. 16) is reduced by a factor Γ^2.

The evolution of a relativistic fireball (whether spherical, or collimated in a jet) is relatively simple: initially, the radiation energy is converted into motion by pressure forces, until most of the energy resides in the kinetic energy few baryons which then move with a velocity corresponding to $\Gamma \approx \eta$. The fireball starts to decelerate when it has swept up and compressed so much mass that the energy in the compressed material becomes appreciable. We have already seen in Section 14 that relativistic shocks are extremely "democratic" machines: from the point-of-view of an upstream observer the energy per swept-up particle is always boosted by a factor $\sim\Gamma^2$. This means that deceleration must become important when an amount of mass

$$M \sim \frac{E_{\rm GRB}}{\Gamma^2 c^2} \approx \frac{M_{\rm b}}{\eta} \tag{13.6}$$

has been swept up. Here I have used $\Gamma \approx \eta$ and definition (13.4). If the GRB is a conical jet with opening angle θ, and expands into the general interstellar medium so that the swept-up mass equals $M \approx \rho_{\rm ism} R^3(\pi\theta^2/3)$, this occurs at the *deceleration radius*

$$R_{\rm d} \simeq \left(\frac{E_{\rm GRB}/\theta^2}{\eta^2 n_{\rm ism} m_{\rm p} c^2}\right)^{1/3} \simeq 10^{16}\, E_{52}^{1/3} \eta_3^{-2/3} n_0^{-1/3} \ {\rm cm}. \tag{13.7}$$

Here $E_{52} \equiv (E_{\rm GRB}/\theta^2)/(10^{52}\ {\rm erg\ sr^{-1}})$, $\eta_3 \equiv \eta/10^3$, and $n_0 \equiv n_{\rm ism}/1\ {\rm cm^{-3}}$. The deceleration radius immediately defines the typical duration of a GRB. Including the cosmological time-dilatation for a source at redshift z one has:

$$t_{\rm obs} \approx (1+z)\, \frac{R_{\rm d}}{\eta^2 c} = 0.3\,(1+z)\, E_{52}^{1/3} \eta_3^{-8/3} n_0^{-1/3} \ {\rm s}, \tag{13.8}$$

which has the correct order-of-magnitude. It is now believed on the basis of the observed variability that the Gamma Ray emission is due to internal shocks within the GRB fireball, but that most of the afterglow emission is associated with the blast wave preceding the fireball material in the interstellar medium. A review of GRB observations and physical models can be found in Piran (2000) and in Mészáros (2002).

13.1 Is there a connection between GRBs and UHECRs?

Two years before it was established that GRBs indeed occur at cosmological distances, Waxman (1995) suggested that there is a connection between GRBs and UHECRs. His idea is based on a line-of-reasoning not unlike the reasoning followed by Baade & Zwicky when they suggested the connection between Galactic cosmic rays and supernovae: an energy argument.

Waxman notes that the integrated flux per solid angle of UHECRs above 10^{20} eV,

$$\frac{\mathrm{d}F}{\mathrm{d}\Omega} \approx 5 \times 10^{-12} \text{ erg cm}^{-2} \text{ sr}^{-1} \text{ s}^{-1}, \tag{13.9}$$

is roughly of the same order of magnitude as the *mean* energy flux one expects from cosmological GRBs, provided the burst sources are distributed homogeneously in a volume with a radius corresponding with a redshift $z \approx 1$. This corresponds with a mean GRB rate equal to

$$Q_{\rm GRB} \simeq 3 \times 10^{-8}\ \theta^{-2} \left(\frac{H_0}{75 \text{ km s}^{-1} \text{ Mpc}^{-1}}\right)^{-3} \text{ Mpc}^{-3}\,\text{yr}^{-1}. \tag{13.10}$$

Note that the GRB rate scales as $1/\theta^2$, with θ the opening angle, due to the fact that only those bursts are seen where the line-of-sight to the source lies within a cone of opening angle θ. The energy production per unit volume is *independent* of the opening angle since $E_{\rm GRB} \propto \theta^2$.

If one takes account of the fact that the UHECR flux is strongly attenuated if the sources are more distant than $D_{\rm GZK} \approx 50$ Mpc, the typical *Greisen-Zatsepin-Kuzmin cut-off distance*, so that most observed UHECRs must have been generated in the "local" Universe with a volume of $\sim 5 \times 10^5$ Mpc^3, and if one assumes that a similar amount of energy is put into UHECRs and into gamma rays, one can estimate a typical mean all-sky flux $F_{\rm UHECR} = 4\pi\ \mathrm{d}F/\mathrm{d}\Omega$ due to UHECRs generated in local Gamma Ray Bursts:

$$\begin{aligned} F_{\rm GRB} &\simeq Q_{\rm GRB} E_{\rm GRB} D_{\rm GZK} \\ &\simeq 10^{-11}\, Q_{-8}\, E_{52} \left(\frac{D_{\rm GZK}}{50 \text{ Mpc}}\right) \text{ erg cm}^{-2} \text{ s}^{-1}, \end{aligned} \tag{13.11}$$

which is the right order-of-magnitude. Here

$$Q_{-8} \equiv \left(\frac{Q_{\rm GRB}\,\theta^2}{10^{-8} \text{ Mpc}^{-3}\,\text{yr}^{-1}}\right), \quad E_{52} \equiv \left(\frac{(E_{\rm CR}^{\rm GRB})/\theta^2}{10^{52} \text{ erg sr}^{-1}}\right). \tag{13.12}$$

with $E_{\rm CR}^{\rm GRB}/\theta^2$ the energy put into cosmic rays per steradian in a GRB event. If these cosmic rays are generated at the relativistic outer blast wave associated with a GRB, they are subject to the same relativistic beaming as photons, *unless* they are deflected considerably in transit to Earth (see below). This estimate for the cosmic ray flux does *not* depend on the opening angle θ of the cone in which the material is emitted, as the θ factors cancel.

There are a four main reasons why Waxman's suggestion seems rather less likely now as it once did:

1. The fact that the *local* GRB burst rate is probably at least an order of magnitude smaller than the estimate (13.10);

2. The fact that the energy yield of GRBs has been downsized as it became obvious that most GRBs are beamed with a small opening angle;

3. The fact that the *external* GRB blast waves do not seem to be capable of accelerating particles to UHECR energies, at least not in the preferred model of Gamma Ray Bursts: *hypernovae*; *internal* shocks or waves may be able to produce UHECRs *provided* the magnetic field reaches equipartition with the internal energy in the fireball;

4. The fact that the spectrum due a collection of sources with most sources at larger distances has a stronger cut-off around $\mathcal{E}_{\rm GZK}$ than the observations seem to show.

The observed distribution of GRB redshifts shows that most bursts are distant, with a median redshift $< z > \sim 1$. This means that the *local* GRB rate is small. Various estimates have been discussed in the context of UHECR production by Stecker (2000). Schmidt (1999) estimates the local ($z \approx 0$) burst rate as

$$Q_{\rm GRB} \approx 10^{-9}\,\theta^{-2}\,E_{52}^{-1}\left(\frac{H_0}{75\ {\rm km\,s^{-1}\,Mpc^{-1}}}\right)^{-3}\ {\rm Mpc^{-3}\,yr^{-1}}, \qquad (13.13)$$

about 30 times smaller than the "average" rate if $E_{52} \sim 1$. This is what one expects in a scenario where the progenitors of GRBs are massive stars, and where GRBs occur when the stellar core collapses directly into a black hole: the *hypernova* or *collapsar* scenario (Woosley 1993; Paczyński 1998; MacFadyen & Woosley 1999). Such stars are short-lived, which implies that the GRB rate should closely follow the star-formation rate (Wijers *et al.* 1998). The star formation rate rises rapidly with increasing redshift, just like the observed GRB distribution.

The light-curves of a number of GRB afterglows show a clear break, leading to a more rapid decay of the observed brightness at later times. This behaviour is expected if the "fireball" is indeed collimated in a jet. By the time that the Lorentz-factor of the emitting material becomes smaller than $\Gamma \sim 1/\theta$ one starts to notice that the emitting surface covers only a small solid angle. When $\Gamma < 1/\theta$, relativistic beaming of the emission within a cone of opening angle $\theta_{\rm em} \sim 1/\Gamma$ hides this fact from a distant observer. An analysis of the observations (Frail *et al.* 2001) provides strong evidence that most GRBs are indeed "beamed", with a typical opening angle $\theta \sim 5^\circ$, and that the energy yield in Gamma Rays is strongly clustered around $E_{\rm GRB} \sim 0.5 \times 10^{51}$ erg and $E_{\rm GRB}/\theta^2 \sim 10^{53}$ erg/sr. Using these numbers, the mean cosmic ray flux expected from GRBs is of order

$$F_{\rm UHECR} \sim 5 \times 10^{-13}\,{\rm erg\,cm^{-2}\,s^{-1}}, \tag{13.14}$$

assuming a similar amount of energy is deposited in gamma rays and UHECRs. This flux is more than an order of magnitude too small. Therefore, unless much more energy is deposited in UHECRs above 10^{20} eV than in Gamma Rays, the GRB-origin hypothesis is not viable.

Perhaps even more damaging for the GRB origin hypothesis is the limit on the energy $\mathcal{E}_{\rm max}$ that can be achieved in conventional acceleration processes like shock acceleration. The limit on the maximum energy at a relativistic shock, $\mathcal{E}_{\rm max} = ZeB\,\Gamma_{\rm s}R_{\rm s}$, yields, when applied to the *external* shock with $\Gamma_{\rm s} \sim \eta = E_{\rm GRB}/M_{\rm b}c^2$ and $R_{\rm s} \sim R_{\rm d}$ (Gallant & Achterberg 1999),

$$\mathcal{E}^{\rm ext}_{\rm max} \approx 5 \times 10^{15}\,ZB_{\mu{\rm G}}\,E_{52}^{1/3}\,\eta_3^{1/3}n_0^{-1/3}\,{\rm eV}. \tag{13.15}$$

This is well below the energy range of UHECRs for any realistic combination of parameter values. This implies that the relativistic incarnation of shock acceleration is incapable of producing UHECRs at the *external* blast wave in the general interstellar medium, in a manner comparable to the production of Galactic cosmic rays below 10^{15} eV at the external blast wave around a supernova remnant. Waxman (2000) (see also: Waxman & Bahcall 2000) has countered this argument, essentially proposing that the production of ultra-relativistic protons could proceed to UHECR energies at the *internal* shocks, which occur within the fireball, and which are believed to be responsible for the prompt Gamma-Ray emission.

Let us consider this possibility in some detail. Internal shocks are thought to be mildly relativistic, with a Lorentz factor $\Gamma_{\rm i}$ such that $\Gamma_{\rm i} - 1 \approx 1$, or equivalently $\Gamma_{\rm i}\beta_{\rm i} \approx 1$ with $\beta_{\rm i} = V_{\rm i}/c$ the shock velocity in the comoving frame in units of light speed. Shock physics also tells us that the external magnetic field is compressed by the *external* shock by a factor $\sim\Gamma_{\rm s}$ so that $B_{\rm i} \sim \Gamma_{\rm s}B$. The calculation of the maximum energy must be done carefully because of the different Lorentz-frames involved.

The mildly-relativistic internal shocks propogate through a medium which itself is moving with a Lorentz-factor $\Gamma \sim \Gamma_{\rm s}$ in the observers frame, with $\Gamma_{\rm s}$ the Lorentz-factor of the external shock. In the comoving frame, the shock-acceleration rate is maximally of order

$$\left(\frac{{\rm d}\mathcal{E}}{{\rm d}t}\right)_{\rm co} \sim ZeB_{\rm i}c\,\Gamma_{\rm i}\beta_{\rm i}^2, \tag{13.16}$$

an expression which covers both non-relativistic and relativistic diffusive shock acceleration at the Bohm rate, and relativistic shock-surfing acceleration. If the shock has a (observer's frame) radius $R_{\rm i}$, its age is $t_{\rm i} \sim R_{\rm i}/c$, where I assume $\beta_{\rm i} \approx 1$. This corresponds to an age $t_{\rm co} = R_{\rm i}/\Gamma_{\rm s}c$ in the comoving frame. Therefore, the maximum particle energy attainable without losses is in the comoving frame

$$\left(\mathcal{E}^{\rm i}_{\rm max}\right)_{\rm co} = \left(\frac{{\rm d}\mathcal{E}}{{\rm d}t}\right)_{\rm co} t_{\rm co} \approx \frac{ZeB_{\rm i}\Gamma_{\rm i}\beta_{\rm i}\,R_{\rm i}}{\Gamma_{\rm s}}. \tag{13.17}$$

Particles produced by the shock are Lorentz-boosted by the motion of the fireball to an observer's frame energy $\mathcal{E} \sim \Gamma_{\rm s}\,\mathcal{E}_{\rm co}$, which cancels the factor $1/\Gamma_{\rm s}$ in expression (13.17). Assuming that the internal shocks penetrate into the compressed swept-up material, and therefore see this compressed field with strength $B_{\rm i} \sim \Gamma_{\rm s}B$, this would give the maximum attainable observer's frame energy of particles produced at internal shocks as:

$$\mathcal{E}^{\rm i}_{\rm max} \approx ZeB_{\rm i}\,\Gamma_{\rm i}\beta_{\rm i}R_{\rm i} \approx ZeB_{\rm i}\,R_{\rm i} \sim ZeB\,\Gamma_{\rm s}\,R_{\rm i}, \tag{13.18}$$

which is always *less* than the limit (13.15) as obviously $R_{\rm i} < R_{\rm s}$.

The only way out of this argument is to allow for a significant amount of field-amplification, either through instabilities ahead of the external shock much like the Lucek & Bell mechanism, which amplify the external magnetic field, or by instabilities in (or behind) the external shock, which could be attributed to instabilities like the *Weibel instability* ocurring in the shock-transition (Medvedev & Loeb 1999). In order to reach UHECR energies, the field-amplification factor should be at least $B/\Gamma_{\rm s}B_{\rm ISM} \sim 10^5$, leading to typical field strengths of about 10^2 Gauss in the fireball (see below).

In fact, most afterglow models have to assume that the internal field within the fireball is much larger than the compressed interstellar field in order to achieve a larger radiation efficiciency, a least in those models where the afterglow is assumed to be synchrotron radiation. If one assumes that the magnetic field reaches equipartition with the downstream internal energy $e_{\rm i}$, so that[9]

$$e_{\rm i} \sim \frac{B_{\rm i}^2}{8\pi}, \tag{13.19}$$

the physics of relativistic shocks (*e.g.* Blandford & McKee 1976) tells us that $e \sim \Gamma_{\rm s}^2\, \rho$, with $\rho \approx n_{\rm ism} m_{\rm p} c^2$ the rest-mass density of the interstellar gas. In that case the required field-amplification is

$$\xi_{\rm B} = \frac{B}{\Gamma_{\rm s}\, B_{\rm ism}} \approx \frac{\sqrt{8\pi n_{\rm ism} m_{\rm p} c^2}}{B_{\rm ism}} \approx 3 \times 10^4\, \frac{n_0^{1/2}}{B_{\mu {\rm G}}}. \tag{13.20}$$

This amplification factor has the correct order of magnitude. If the field is required to reach equipartition with the electrons only, so that $B^2/8\pi \sim e_{\rm e} \approx \Gamma_{\rm s}^2 n_{\rm ism} m_{\rm e} c^2$, the field-amplification is a factor $\sqrt{m_{\rm p}/m_{\rm e}} \approx 43$ smaller, $\xi_{\rm B} \sim 10^3$.

The equipartition field can be expressed in terms of GRB parameters as

$$B_{\rm eq} \approx \sqrt{8\pi e_{\rm i}} = \begin{cases} 2 \times 10^2\, n_0^{1/2}\, \Gamma_3 \ {\rm G} & \text{equipartition with protons;} \\ \\ 5\, n_0^{1/2}\, \Gamma_3\ {\rm G} & \text{equipartition with electrons.} \end{cases} \tag{13.21}$$

Particle acceleration at the internal shocks with $\Gamma_{\rm i}\beta_{\rm i} \approx 1$ could then possibly produce protons of energy

$$\mathcal{E}^{\rm i}_{\rm max} \approx ZeB\, R_{\rm i} \approx \begin{cases} 5 \times 10^{20}\, n_0^{1/2}\, \Gamma_3\, R_{16}\ {\rm eV} & \text{equipartition with protons;} \\ \\ 10^{19}\, n_0^{1/2}\, \Gamma_3\, R_{16}\ {\rm eV} & \text{equipartition with electrons.} \end{cases} \tag{13.22}$$

Here $\Gamma_3 = \Gamma_{\rm s}/10^3$ is the normalized Lorentz-factor of the *external* shock, and R_{16} is $R_{\rm i}/10^{16}$ cm. If the internal magnetic field indeed reaches equipartition with the total post-shock energy density of protons and electrons, acceleration to UHECR-energies could occur. If, on the other hand, the field only reaches equipartition with the electrons, it becomes unlikely that highest-energy UHECRs can be produced at the internal shocks for typical GRB parameters.

There are, however, a number of reasons why the maximum energy (13.22) will not be reached in practical situations. First of all, the internal magnetic field will be highly turbulent. If the coherence length of the field in the comoving frame satisfies

$$\ell_{\rm coh} < r_{\rm g}/\Gamma_{\rm i} \approx r_{\rm g} \tag{13.23}$$

[9]The internal energy density $e_{\rm i}$ and field strength $B_{\rm i}$ of the fireball are evaluated in the local rest frame of the material.

with $r_{\rm g}$ the particle gyration radius, the acceleration rate is reduced by a factor $\sim\ell_{\rm coh}/r_{\rm g}$ (see Sect. 14.4). One can expect

$$\ell_{\rm coh} < r/\Gamma_{\rm s} \tag{13.24}$$

in the comoving frame, as regions separated by a distance $\sim r/\Gamma_{\rm s}$ are causally disconnected in the fireball. This affects particles with an energy

$$\mathcal{E}_{\rm co} \sim ZeB_{\rm i}\,\ell_{\rm coh} = \mathcal{E}^{\rm i}_{\rm max} \times \left(\frac{\Gamma_{\rm s}\ell_{\rm coh}}{R_{\rm i}}\right) . \tag{13.25}$$

If $\ell_{\rm coh} \sim R_{\rm i}/\Gamma_{\rm s}$ this energy is close to the comoving value of $\mathcal{E}_{\rm max}$, and the above estimates for the maximum energy approximately hold. If, on the other hand, plasma instabilities like the Weibel instability are responsible for the magnetic field amplification, the field correlation length will be smaller, and could be as small as the collisionless plasma skin depth (Achterberg & Wiersma 2002, in preparation):

$$\ell_{\rm i} \sim c/\omega_{\rm pe} . \tag{13.26}$$

in the comoving frame, with $\omega_{\rm pe} = \sqrt{4\pi e^2 n_{\rm co}/m_{\rm eff}}$ the plasma frequency, and $m_{\rm eff} = (e + P)/n_{\rm co}$. This would lead to

$$\Gamma_{\rm s}\ell_{\rm coh}/R_{\rm i} \sim \Gamma_{\rm s}c/\omega_{\rm pe}R_{\rm i} = 5 \times 10^{-7}\,\Gamma_3\,R_{16}^{-1}\,n_0^{-1/2} . \tag{13.27}$$

In that case, the maximum energy would be reduced by a factor

$$\sqrt{\frac{c}{\omega_{\rm pe}R_{\rm i}}} \sim 10^{-5}\,R_{16}^{-1/2}\,n_0^{-1/4} , \tag{13.28}$$

which again would put the maximum energy in the observer's frame in the range $\mathcal{E}_{\rm max} \sim 10^{16}$ eV! In these expressions n_0 is the typical density in the interstellar medium. One can show that the compression and heating of the material in the external shock produces an effective post-shock plasma frequency in the swept-up material roughly equal to the plasma frequency in the interstellar medium.

Also, in such strong fields and such high energies synchrotron-losses could become important even for protons. The maximum proton energy in the presence of synchrotron losses for $\Gamma_{\rm i}\beta_{\rm i} \approx 1$ is (see Sect. 16)

$$\mathcal{E}^{\rm co}_{\rm max} \approx \frac{3m_{\rm p}c^2}{2\sqrt{\alpha}}\left(\frac{m_{\rm p}}{m_{\rm e}}\right)^{3/2}\left(\frac{B_{\rm i}}{B_{\rm cr}}\right)^{-1/2} , \tag{13.29}$$

which is

$$\mathcal{E}^{\rm co}_{\rm max} = 9 \times 10^{20}\left(\frac{B_{\rm i}}{100\ {\rm G}}\right)^{-1/2}\ {\rm eV}. \tag{13.30}$$

This is well below the maximum energy due to the finite age of the internal shock in the comoving frame, $\mathcal{E}^{\rm co}_{\rm max} \lesssim 10^{16}$ eV, so proton synchrotron losses may be neglected for typical parameters. This same of course does not hold for electrons, whose synchrotron-limited energy is a factor $\sim(m_{\rm e}/m_{\rm p})^2 \approx 3\times10^{-7}$ smaller, putting it in the range of $\sim10^{12}$–10^{13} eV in the comoving frame.

13.2 Acceleration by trans-relativistic waves/weak shocks

It is quite possible, given our relative uncertainty about the precise conditions in GRB fireballs, that acceleration to UHECR energies is possible without resorting to shocks at all! If there is a significant amplification of magnetic fields in the post-shock flow, particles could be accelerated using relativistic hydromagnetic waves, such as Alfvén waves (Pelletier 1999; Pelletier & Kersalé 2000). Essentially, this is a relativistic variant of Fermi-II type acceleration: stochastic acceleration by waves. Particles interacting with these waves could be accelerated efficiently if there are waves propagating in *both* directions along the magnetic field. The reason for this is the same as in the non-relativistic case (Skilling 1975): if waves propagate only in one direction, scattering by the waves will make the distribution in the "wave frame" (which moves with the wave phase speed $\omega/k = V_{\rm A}$, with $V_{\rm A}$ the (relativistic) Alfvén speed, along the field) almost isotropic. Once isotropy in the wave frame is achieved, no net energy gain results in the laboratory frame[10] when averaged over all particles. If the two wave frames (with velocity $\pm\omega/k = \pm V_{\rm A}$) are "competing", there is always a residual anisotropy in both frames, and a net energy gain results in the laboratory frame.

I will estimate the acceleration rate by a simple model: specular reflection of particles in the rest-frame of a magnetic wave (or soliton), moving with velocity $\boldsymbol{V} = \pm V_{\rm A}\,\hat{\boldsymbol{b}}$ along an ambient magnetic field $\boldsymbol{B} = B\,\hat{\boldsymbol{b}}$. Here I have taken the velocity to be of order the Alfvén speed $V_{\rm A}$. Assuming relativistic particles with $\mathcal{E} \sim pc$, the energy change is

$$\frac{\Delta\mathcal{E}}{\mathcal{E}} = 2\Gamma_{\rm A}^2 \left(\frac{V_{\rm A}}{c}\right) \left(\frac{V_{\rm A}}{c} - \mu\right) \tag{13.31}$$

for a front moving with velocity $V_{\rm A}\hat{\boldsymbol{b}}$ in the lab frame. Here $\mu = (\boldsymbol{p}\cdot\hat{\boldsymbol{b}})/p$ is the cosine of the angle between the particle momentum and the magnetic field in the lab frame, and $\Gamma_{\rm A} = 1/\sqrt{1 - V_{\rm A}^2/c^2}$. If we assume that the distribution is isotropic in the laboratory frame one must calculate a *flux*

[10] In this section, the "lab frame" corresponds to the rest-frame of the fluid, the comoving frame of the GRB material.

average of the relative energy gain $\Delta\mathcal{E}/\mathcal{E}$, just as we did in the case of shock acceleration (Sect. 6). Only particles with $-1 \leq \mu \leq V_{\rm A}/c$ collide with the front, and the average gain is

$$\left\langle \frac{\Delta\mathcal{E}}{\mathcal{E}} \right\rangle = \tfrac{4}{3}\,\Gamma_{\rm A}^2 \left(\frac{V_{\rm A}}{c}\right)\left(1 + \frac{V_{\rm A}}{c}\right). \tag{13.32}$$

A similar calculation for reflection off a wave front moving in the opposite direction ($\boldsymbol{V} = -V_{\rm A}\hat{\boldsymbol{b}}$), which sweeps up particles with $-V_{\rm A}/c \leq \mu \leq 1$, gives exactly the same answer, as it should for symmetry reasons. The relativistic Alfvén speed satisfies

$$\frac{V_{\rm A}}{c} = \frac{B}{\sqrt{4\pi\,(e + P + B^2/4\pi)}}. \tag{13.33}$$

where e and P are the energy density (including rest-energy) and kinetic pressure of the plasma. This implies

$$\Gamma_{\rm A}^2 = 1 + \frac{B^2}{4\pi\,(e + P)}, \tag{13.34}$$

so $\Gamma_{\rm A}^2\;(V_{\rm A}/c) = \mathcal{O}(1)$ for GRB fireball models where the field reaches equipartition. This means that the particle energy typically doubles at each reflection. Of course, replacing weakly non-linear Alfvén waves by (weak) shocks increases the propagation speed, and the energy gain of a reflected particle. On the other hand, the assumption of specular reflection (which reverses the component of momentum along $\hat{\boldsymbol{b}}$ in the rest frame of the wave front) and the subsequent averaging procedure assumes implicitly that most particles incident on the front have a large change of momentum during the interaction with the front. We have already seen that the effects of relativistic beaming for $\Gamma_{\rm s} \gg 1$ limit the energy gain, and one can expect something similar to happen if one pushes $\Gamma_{\rm A}$ to large values. Having counter-propagating fronts (Pelletier 1999), or a "gas" of magnetised wave fronts with a random velocity distribution (essentially the situation proposed by Fermi 60 years ago for non-relativistic propagation speeds) would alleviate this constraint.

One could use this process to further accelerate particles that have been accelerated (or whose energy has been boosted) by the external shock. To reach UHECR energies in the observer's frame, one should produce particles with a typical energy

$$(\mathcal{E})_{\rm co} = 10^{17}\mathcal{E}_{20}\,\Gamma_3^{-1}\ {\rm eV} \tag{13.35}$$

in the comoving frame. Here $\mathcal{E}_{20} = \mathcal{E}/(10^{20}\ {\rm eV})$ is the observer's frame energy in units of 100 EeV, and $\Gamma_3 = \Gamma_{\rm s}/10^3$. Limit (13.15) for the maximum energy attainable from relativistic shock acceleration at the external shock corresponds to a comoving energy

$$\mathcal{E}_{\max}^{\rm co} \approx 5 \times 10^{12}\, ZB_{\mu \rm G}\, E_{52}^{1/3}\, \eta_3^{-2/3} n_0^{-1/3}\ \rm eV. \tag{13.36}$$

The boosted energy of upstream protons, $\mathcal{E}_{\rm co} = \Gamma_{\rm s} m_{\rm p} c^2 \approx 10^{12}\, \eta_3$ GeV, is only marginally smaller. In order to produce UHECRs by relativistic Fermi-II acceleration would therefore require a boost in energy of order $\sim 10^4 - 10^5$ in the comoving frame. If one takes equipartition inside the relativistic fireball, so that

$$e = 3P = \frac{B^2}{8\pi} \quad \Longleftrightarrow \quad \Gamma_{\rm A}^2 \left(\frac{V_{\rm A}}{c}\right) \approx 0.82, \tag{13.37}$$

one finds an energy change $\Delta\mathcal{E}/\mathcal{E} \approx 1.94$, so the particle energy roughly triples each interaction. In order to achieve UHECR energies one typically needs $N_{\rm int} \sim 5/\log 3 \approx 10$ interactions with a wavefront per particle. To achieve this in the available time, $t_{\rm co} \sim R_{\rm s}/\Gamma_{\rm s} c \sim 300\ R_{16}\eta_3^{-1}$ s, a particle needs to collide with wave fronts with a collision frequency exceeding $\nu_* \approx 0.03$ Hz in the comoving frame. The mean-free path of a ultra-relativistic ion would then be of order

$$\lambda_* \sim c/\nu_* \sim R_{\rm s}/10\Gamma_{\rm s} = 10^{12}\ R_{16}\, \eta_3^{-1}\ \rm cm, \tag{13.38}$$

where I put $\Gamma_{\rm s} \sim \eta$. Note that $R_{\rm s}/\Gamma_{\rm s}$ is the typical radial extent of the fireball in the comoving frame This means that particles accelerated in this manner diffuse in the fireball, with a spatial diffusion coefficient $\kappa \sim c\lambda_*/3$. They are collisionally coupled to the expanding flow, and would consequently suffer expansion losses in this flow. In the "coasting phase" of a GRB fireball, where $\Gamma_{\rm s} \approx \eta =$ constant, the proper number density in the fireball scales as (*e.g.* Piran *et al.* 1993)

$$n_{\rm i} \propto R_{\rm s}^{-2}, \tag{13.39}$$

as the fireball coasts with almost constant width in the comoving frame, so the density decrease is due to the lateral expansion of the fireball. The expansion losses of relativistic particles in a relativistic flow are described by (see Sect. 14.4.3)

$$\frac{1}{\mathcal{E}}\frac{\mathrm{d}\mathcal{E}}{\mathrm{d}t} = -\frac{1}{3\Gamma}\left(\frac{\partial \Gamma}{\partial t} + \boldsymbol{\nabla}\cdot(\Gamma \boldsymbol{V})\right). \tag{13.40}$$

The equation of particle conservation,

$$\frac{\partial(\Gamma n)}{\partial t} + \boldsymbol{\nabla}\cdot(\Gamma n \boldsymbol{V}) = 0 \tag{13.41}$$

allows us to write (13.40) as

$$\frac{1}{\mathcal{E}}\,\frac{\mathrm{d}\mathcal{E}}{\mathrm{d}t}=\frac{1}{3n}\,\frac{\mathrm{d}n}{\mathrm{d}t}. \tag{13.42}$$

Relation (13.39) then implies in the comoving frame

$$\left(\frac{1}{\mathcal{E}}\,\frac{\mathrm{d}\mathcal{E}}{\mathrm{d}t}\right)_{\mathrm{co}}=-\frac{2\Gamma_{\mathrm{s}}}{3}\,\left(\frac{1}{R_{\mathrm{s}}}\,\frac{\mathrm{d}R_{\mathrm{s}}}{\mathrm{d}t}\right), \tag{13.43}$$

where I have used that $\mathrm{d}t_{\mathrm{co}}=\mathrm{d}t/\Gamma_{\mathrm{s}}$. The acceleration rate due to interaction with wave fronts or weak (Alfvénic) shocks is:

$$\left(\frac{\mathrm{d}\mathcal{E}}{\mathrm{d}t}\right)_{\mathrm{co}}\approx 2\nu_{*}\mathcal{E}_{\mathrm{co}}, \tag{13.44}$$

with ν_{*} the encounter rate introduced in equation (13.38). The maximum energy in the comoving frame follows from balancing the acceleration rate (13.44) with the loss rate (13.43) due to expansion losses, which yields

$$\mathcal{E}^{\mathrm{co}}_{\mathrm{max}}\sim\frac{3}{\Gamma_{\mathrm{s}}\beta_{\mathrm{s}}}\,\left(\frac{\nu_{*}}{\Omega_{\mathrm{co}}}\right)\,eB_{\mathrm{i}}\,R_{\mathrm{s}}. \tag{13.45}$$

Here $\Omega_{\mathrm{co}}=eB_{\mathrm{i}}c/\mathcal{E}_{\mathrm{co}}$ is the particle gyration frequency in the comoving frame. This energy corresponds with

$$\mathcal{E}_{\mathrm{max}}\sim\Gamma_{\mathrm{s}}\,\mathcal{E}^{\mathrm{co}}_{\mathrm{max}}\sim 3\,eB_{\mathrm{i}}\,R_{\mathrm{s}}=10^{21}\,\left(\frac{B_{\mathrm{i}}}{100\ \mathrm{G}}\right)\,R_{16}\ \mathrm{eV}, \tag{13.46}$$

in the observer's frame, provided the encounter rate is close to the maximum possible, once per gyroperiod so that $\nu_{*}\sim\Omega_{\mathrm{co}}$, and where I have put $\beta_{\mathrm{s}}\approx 1$. Again, UHECR energies could be reached. In this situation particles will diffuse in the fireball at the Bohm rate. With the comoving radial extent of the fireball of order $\Delta r_{\mathrm{co}}\approx R_{\mathrm{s}}/\Gamma_{\mathrm{s}}$, it takes a particle diffusing at the Bohm rate at time

$$\Delta t_{\mathrm{co}}=\frac{\Delta r^{2}_{\mathrm{co}}}{\kappa_{\mathrm{B}}}\sim\frac{3}{2}\,\frac{eB_{\mathrm{i}}\,R^{2}_{\mathrm{s}}}{c\Gamma^{2}_{\mathrm{s}}\,\mathcal{E}_{\mathrm{co}}} \tag{13.47}$$

to diffuse to the edge of the fireball and escape. In the observer's frame this corresponds with an escape time $\Delta t=\Gamma_{\mathrm{s}}\,\Delta t_{\mathrm{co}}$, which is

$$\Delta t\approx\frac{R_{\mathrm{s}}}{2c}\,\left(\frac{\mathcal{E}}{\mathcal{E}_{\mathrm{max}}}\right)^{-1}, \tag{13.48}$$

where I have used equation (13.46). Because of the time compression due to the relativistic motion of the source material, $\Delta t_{\mathrm{obs}}\sim\Delta t/2\Gamma^{2}_{\mathrm{s}}$, this means that the UHECRs are released to an outside observer in a burst with typical duration

$$\Delta t_{\rm obs} \approx \frac{R_{\rm s}}{4\Gamma_{\rm s}^2 c} \left(\frac{\mathcal{E}}{\mathcal{E}_{\rm max}} \right)^{-1} \approx 0.1\, R_{16}\, \Gamma_3^{-2} \left(\frac{\mathcal{E}}{\mathcal{E}_{\rm max}} \right)^{-1} \ {\rm s}. \qquad (13.49)$$

As we will see below, the pulse of UHECRs will be smeared out by small deflections due to a weak intergalactic magnetic field ($B_{\rm IGM} \sim 10^{-9}$ G), so that a distant observer receives these protons over a time that is much larger than etimate (13.49) would suggest.

The conclusion of the last two sections seems to be that GRB fireballs could possibly produce GRBs if [1] acceleration, either at internal shocks or at magnetic waves, proceeds at the maximum possible rate (Bohm rate), **and** [2] if the magnetic field is amplified to equipartition with the post-shock protons, leading to an internal field strength of $\sim 10^2$ G. If equipartition is only reached with the electrons, the internal magnetic field is a factor $\sqrt{m_{\rm p}/m_{\rm e}} \sim 43$ smaller, and internal shocks or waves fall short of producing the most energetic UHECRs by about one order of magnitude in energy.

13.3 Acceleration of pulsar-wind material by the external shock

The only alternative for the *direct* production of UHECRs by the *external* shock is that the initial energy gain of particles swept up by the shock, which boosts the energy by a factor $\sim \Gamma_{\rm s}^2$, produces UHECRs. Given the fact that $\Gamma_{\rm s} \le \eta \sim 100 - 1000$, one would need very energetic seed particles, with energies in the range $\mathcal{E} \sim 10^{14} - 10^{16}$ eV. This is tantalizingly close to the typical energy of the most energetic Galactic cosmic rays, so one might think that these are boosted during their encounter with a GRB shock. Unfortunately, this creates a huge efficiency problem: only a fraction of $\sim 10^{-9}$ of all the energy in the ISM, including the rest-energy of the hydrogen and helium atoms (~ 1 GeV/nucleon), resides in cosmic rays. This means that, by the democratic nature of the boost process, only about 10^{-9} of all energy dissipated in the shock will be put into boosted cosmic rays, and even less in the small fraction of the cosmic rays with the required (high) energies (Gallant & Achterberg 1999).

The only way one could use the boost process as a UHECR production mechanism is if *most* of the energy ahead of the blast wave resides in relativistic massive particles such as protons, with a Lorentz-factor of order $\gamma \sim 10^4 - 10^6$. There *is* an environment in which a situation like this could occur: in the *relativistic pulsar winds* associated with young, rapidly rotating pulsars. Pulsar winds have been observed around single pulsars as the hot non-thermal emission regions in the centers of "filled" (plerionic) supernova remnants. The Crab Nebula is the archetypical example of such a pulsar wind nebula.

Although many of the details of the wind mechanism are not clear, it is commonly assumed that a pulsar wind consists of a relativistic pair plasma with ions mixed in as a minority species. Because of their large mass, a small admixture ($\sim$0.1%) of ions would rapidly make them the dominant carrier of mechanical energy. Although the Lorentz-factor $\Gamma_{\rm w}$ of the wind is not very well-known, estimates range from $\Gamma_{\rm w} \sim 10^4$–10^6. Such winds are terminated at a *termination shock* where the energy of the wind is dissipated. The largest uncertainty of the models is the manner in which dissipation in the wind and termination shock proceeds, since most of wind energy is carried in the form of an electromagnetic (Pointing) flux (Kennel & Coroniti 1984; Melatos & Melrose 1996; Bogalov 1999; Lyubarski & Kirk 2001; Kuijpers 2001). The best posible scenario for UHECRs would occur if a significant fraction of the wind energy ended up in the form of relativistic ions behind the termination shock. These ions are then boosted in energy when they interact with a passing relativistic blast wave.

In the alternative explanation of the GRB phenomenon, the coalescence of binary neutron stars, or other compact objects (Paczyński 1986; Goodman 1986; Eichler *et al.* 1989 and Narayan *et al.* 1992), the GRB blastwave could overrun the relativistically hot ions in the pulsar wind nebula left by the progenitors, and so create UHECRs (Gallant & Achterberg 1999). If the ions end up with a significant fraction of the wind energy density, the typical ion energy before the boost is (*e.g.* Michel 1991):

$$\mathcal{E}_{\rm i} \sim \Gamma_{\rm w} m_{\rm i} c^2 \sim e \left(\frac{L_{\rm sd}}{c}\right)^{1/2} , \tag{13.50}$$

where $L_{\rm sd}$ is the spin-down luminosity of the pulsar. Typically, this is

$$\mathcal{E}_{\rm i} \sim 5 \times 10^{13} \left(\frac{L_{\rm sd}}{10^{33}\ {\rm erg/s}}\right)^{1/2} \ {\rm eV}. \tag{13.51}$$

The boosted particles would have an energy of order 10^{20} eV if the shock has $\Gamma_{\rm s} \gtrsim 10^3$.

The binary-pulsar scenario has gone out of fashion for a number of reasons: the fact that calculations seem to indicate that the energy release in this case, coming from processes in the short-lived debris disk around the black hole that forms after coalescence, is not very efficient in the sense that too much of the mechanical energy ends up in baryons, resulting in too small a value of $\eta = E/M_{\rm b}c^2$, and that most of the *total* energy available ($E_{\rm m} \sim 5 \times 10^{53}$ erg) escapes in the form of neutrino's, not unlike what happens in an ordinary (type II) supernova. Also, the characteristic time it takes for a binary pulsar orbits to shrink due to the emission of gravitational radiation is long, $\tau \sim 10^8$ yr (Taylor *et al.* 1983). This would imply that some of the pulsar binaries could travel a significant distance from the star-forming region in which they were born. Observations on the other hand seem to indicate that GRBs occur in dense regions (molecular clouds?) (Galema *et al.* 2001), and that GRBs occur close to their host galaxies, again arguing against a long delay between the birth of the progenitor system and the GRB (Bloom *et al.* 2002).

One might argue however (and in fact it is quite likely) that pulsar/compact object-binary related merger events *do* occur. They could show up as a different phenomenon, such as short-duration GRBs and/or the short-duration X-Ray tranients, the so-called *X-Ray flashes* which are also observed about once a day, and last a few minutes. The estimated merger rate is $\sim 10^{-6}$/yr per galaxy (Narayan *et al.* 1991; Phinney 1991). The mean space-density of galaxies is ~ 0.02 Mpc^{-3}, which would lead to a mean merger rate equal to

$$Q_{\rm m} \sim 2 \times 10^{-8} \ {\rm Mpc}^{-3} \ {\rm yr}^{-1}, \tag{13.52}$$

similar to the unbeamed GRB rate. The amount of energy generated in this manner is

$$Q_{\rm m} \, E_{\rm m} \sim 10^{46} \ {\rm erg \ Mpc}^{-3} \ {\rm yr}^{-1}, \tag{13.53}$$

If a fraction $f_{\rm cr}$ of the total energy ends up in UHECRs, the mean flux due to merging events within a distance $D_{\rm GZK}$ is

$$\begin{aligned} F_{\rm UHECR} &\approx Q_{\rm m} \, E_{\rm m} D_{\rm GZK} \\ &= 2 \times 10^{-9} \, f_{\rm cr} \, (Q_{\rm m} \, E_{\rm m})_{46} \, D_{50} \ {\rm erg \ cm}^{-2} \ {\rm s}^{-1}, \end{aligned} \tag{13.54}$$

where

$$(Q_{\rm m} \, E_{\rm m})_{46} \equiv \frac{Q_{\rm m} \, E_{\rm m}}{10^{46} \ {\rm erg \ Mpc}^{-3} \ {\rm yr}^{-1}}, \quad D_{50} = \frac{D_{\rm GZK}}{50 \ {\rm Mpc}}. \tag{13.55}$$

The observations then require $f_{\rm cr} \gtrsim 0.03$. This is rather large, given the fact that most of the energy escapes in the form of neutrino's.

An interesting consequence of the boost model for UHECR production is that it predicts a specific spectrum for the particles once the relativistic blast wave starts to decelerate. In the deceleration phase for $R > R_{\rm d}$, the behaviour of the shock Lorentz-factor $\Gamma_{\rm s}$ of the external blast wave follows from

$$E_{\rm GRB} \approx \Gamma_{\rm s}^2 E_{\rm sw} = {\rm constant}, \tag{13.56}$$

where $E_{\rm sw}$ is the total energy (including rest energy) of the swept-up material. If the material is cold it simply equals $E_{\rm sw} = Mc^2$ with M the swept-up mass. If the material is relativistically hot, such as in a pulsar wind nebula behind the termination shock, it equals

$$E_{\rm sw} = N\,\mathcal{E}_0, \tag{13.57}$$

with N the number of particles that have crossed the shock and $\mathcal{E}_0$ the mean energy/particle. In the latter case, the typical energy per particle in the post-shock medium as seen from the observer's frame equals

$$\mathcal{E} \approx \Gamma_{\rm s}^2\,\mathcal{E}_0 = \left(\frac{E_{\rm GRB}}{E_{\rm sw}}\right)\mathcal{E}_0. \tag{13.58}$$

If we combine equations ((13.56)–(13.58)) it is easily seen that the diffential spectrum is

$$\frac{{\rm d}N}{{\rm d}\mathcal{E}} = \frac{{\rm d}N/{\rm d}E_{\rm sw}}{{\rm d}\mathcal{E}/{\rm d}E_{\rm sw}} = -\frac{E_{\rm sw}^2}{E_{\rm GRB}\mathcal{E}_0^2} \propto \mathcal{E}^{-2}. \tag{13.59}$$

Here I have assumed $\mathcal{E}_0$ to be constant. So the boost mechanism naturally produces a power-law distribution of particles with slope $s = 2$.

13.4 UHECR spectrum

I now briefly discuss the problem of the observed UHECR spectrum. Figure 7 illustrates the effect of photo-pion production losses on a cosmic rays which are created in a mono-energetic distribution with an energy equal to 3×10^{21} eV. It shows that for distances $\le D_{\rm GZK} \sim 50$ Mpc, particles still remain with energies above $E_{\rm GZK}$. This is due to the fact that the interaction between UHECRs and CMWB photons is subject to Poisson statistics, and there are always some particles that –by chance– have not interacted with many photons, and lost a lot of energy. This "Poisson tail" in the energy-distribution becomes less-and-less prominent with increasing source distance.

This is illustrated in the Figure 8 below, which shows the *mean* flux from bursting sources as a function of arrival energy, as obtained in a simulation. It clearly shows that particles with $\mathcal{E} > \mathcal{E}_{\rm GZK}$ *are* present in the spectrum, but that there is still a significant reduction of the flux at the cut-off energy. The spectrum shown is for an integration time $t_{\rm int} = 100$ yr, but the spectra for shorter integration time are similar, but with more Poisson noise.

To illustrate how large the variation in spectra can be, Figure 9 below shows two spectra, one slected for having a large flux above $\mathcal{E}_{\rm GZK}$, and one with almost no flux. Both are the results of a simulation with the same UHECR propagation code, but with a (randomly chosen) different distribution of bursting sources in the local Universe ($D \le D_{\rm GZK}$). Nevertheless, there is almost always a significant reduction in the flux above $\mathcal{E}_{\rm GZK}$.

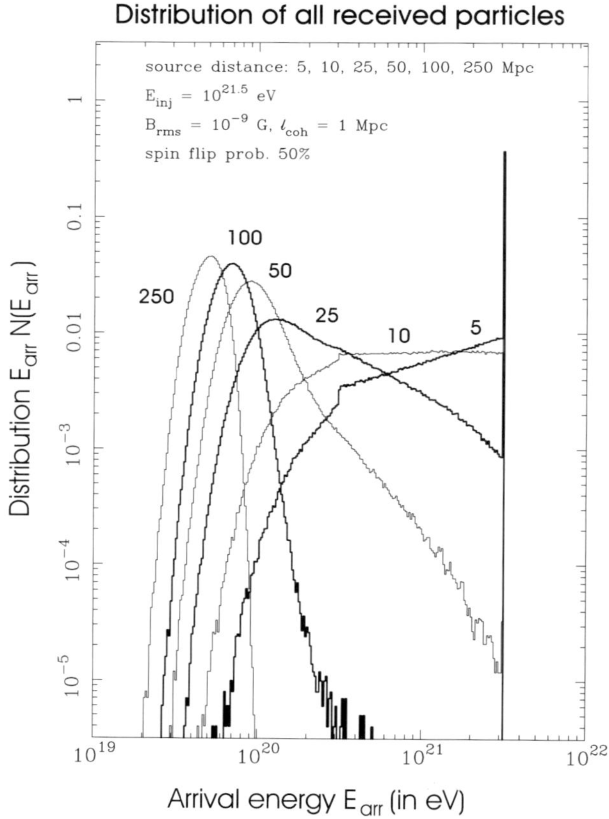

Fig. 7. Distribution of arrival energies at Earth, for mono-energetic injection at an energy of $10^{21.5}$ eV at the source. Shown are six source distances between 5 and 250 Mpc. The spectra have been multiplied by an arbitrary factor $\propto D^2$, and the curves have been labelled with the source distance. At a source distance equal to $D = 10$ Mpc there is still a significant fraction of particles at the injection energy, giving rise to a distinct spike. This spike has almost completely vanished for $D = 50$ Mpc. Once the particle energy falls below 10^{20} eV, where the loss length $\ell(E)$ increases by almost two orders of magnitude, particles drift slowly to lower energies, and collect in a narrow distribution.

The observations on the other hand do not seem to show such a reduction (Fig. 10), even though the small number of events above 10^{20} eV makes the precise spectral shape of the UHECR-flux rather uncertain. Nevertheless, the conclusion seems to be that "typical" realizations of the GRB scenario

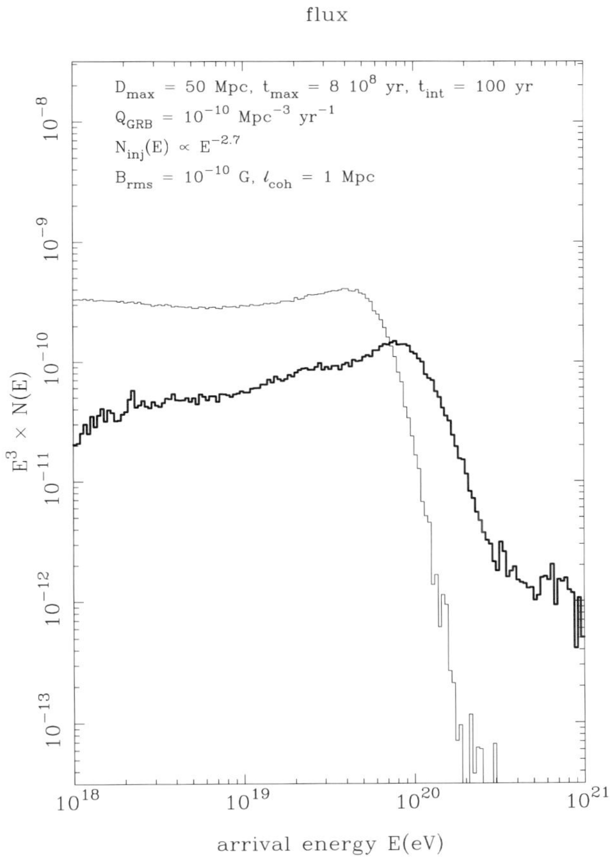

Fig. 8. The mean flux from bursting sources for a burst rate equal to $Q_{\rm GRB} = 10^{-10}$ Mpc^{-3} yr^{-1}. Shown is the number of particles received over a period (integration time $t_{\rm int}$) of 100 years. Two contributions are shown: the thick histogram is due to sources within a distance of 50 Mpc, the GZK cut-off distance. The source distances and source ages has been generated by a Monte-Carlo method extending up to 8×10^8 years in the past. The thin histogram, which almost exclusively contains UHECRs below the cut-off energy $E_{\rm GZK} \sim 10^{19.7}$ eV, is the (steady) background of sources at larger distances. It was assumed the particles were injected at the source with an energy distribution $N_{\rm inj}(\mathcal{E})\,{\rm d}\mathcal{E} \propto \mathcal{E}^{-2.7}\,{\rm d}\mathcal{E}$.

will produce a spectrum with a more prominent decrease in the flux above $10^{19.7}$ eV than the observations seem to show.

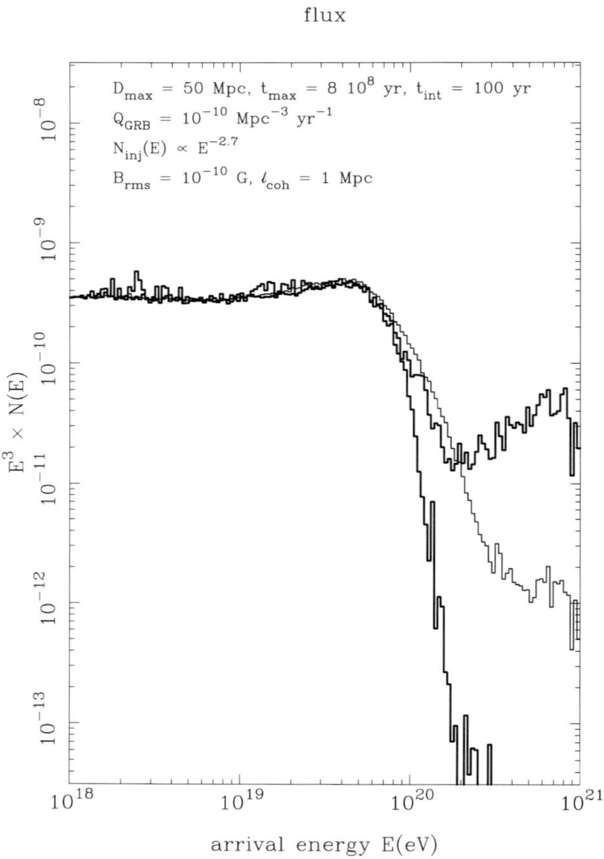

Fig. 9. The flux received as a function of energy for two different realizations of the distribution of sources with a distance of 50 Mpc (thick histograms). The mean flux (thin histogram) is shown for comparison. In one case, there are ~2 relatively close-by sources, which give an appreciable flux above the GZH cut-off energy $\mathcal{E}_{\rm GZK}$. In the other case, there are –by chance– no closeby sources which went off in the relatively recent past, and almost no flux is received above $\mathcal{E}_{\rm GZK}$.

13.5 Conclusion

The different models, and the limits on the acheivable energy, propagation properties etc. show that the origin of UHECRs in bottom-up scenario's is far from clear. A number of production mechanisms (internal shocks in Gamma Ray Bursts, near-by radio galaxies, pulsar winds, the Galactic Center) come tantalizingly close energy-wize, but in most cases one has to

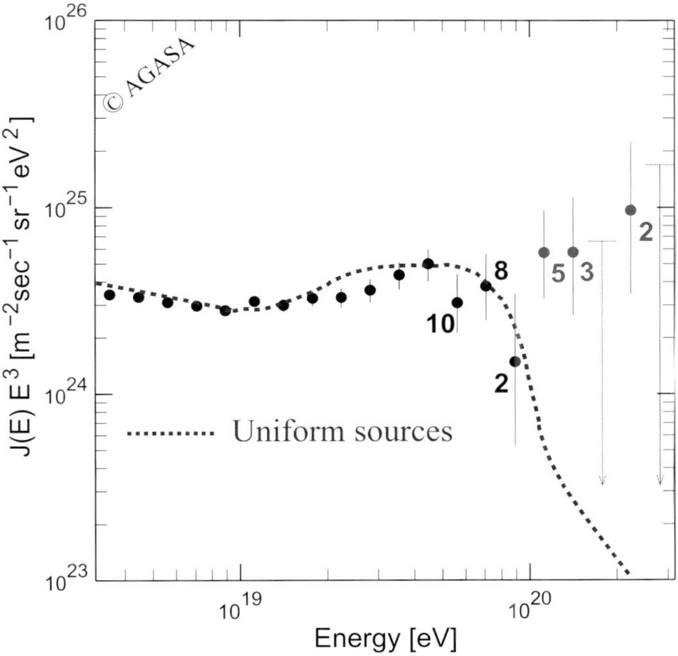

Fig. 10. The UHECR spectrum as observed by the *Ageno Giant Air Shower Array* (AGASA). The dotted curve gives the predicted spectrum for a homogeneous distribution of non-bursting sources in the local Universe.

push the physical parameters, like field strength and size, to the limit of the allowable values. It is not suprising that particles of such extreme energy can only be produced in extreme conditions. I believe that at this time no clear favorite mechanism can be selected on observational or theoretical grounds. Optimistically speaking, the advent of new observing facilities like the Auger Observatory, with their order-of-magnitude larger data rate, will lead to the resolution of many of the uncertain details about UHECRs: arrival direction statistics, "bumps" and "dips" in the spectrum near the GZK cut-off, composition etc., and the origin of UHECRs may be clarified. On the other hand, almost eighty years of research of Galctic cosmic rays has not unequivocably resolved their origin and production mechanisms. So a pessimistic view could be that a similar situation will arise for the far less numerous UHECRs. Only time will tell.

References

[1] A. Achterberg, *A&A* **76** (1979) 276.
[2] A. Achterberg, *A&A* **98** (1981) 195.
[3] A. Achterberg, *A&A* **231** (1990) 251.
[4] A. Achterberg, in: Galactic High-Energy Astrophysics, High-Accuracy Timing and Positional Astronomy, edited by J. van Paradijs and H.M. Maitzen (Springer Verlag, Berlin, 1993) 37.
[5] A. Achterberg, R.D. Blandford and V. Periwal, *A&A* **132** (1984) 97.
[6] A. Achterberg, R.D. Blandford and S.P. Reynolds, *A&A* **281** (1994) 220.
[7] A. Achterberg and L. Ball, *A&A* **285** (1994) 687.
[8] I.I. Alekseyev and A.P. Kroptokin, *Geomagn. Aeron.* **10** (1970) 755.
[9] A. Alibés, J. Labay and R. Canal, *ApJ* **571** (2002) 326.
[10] L.A. Anchordoqui *et al.*, *Phys. Lett. B* **492** (2000) 237.
[11] T.P. Armstrong, M.E. Pesses and R.B. Decker, in Collisionless Shocks in the Heliosphere: Reviews of Current Research, edited by B.T. Tsurutani & R.G. Stone, American Geophysical Union, *Geophys. Monogr.* **35** (1985).
[12] B. Aschenbach and D.A. Leahy, *A&A* **341** (1999) 602.
[13] W.I. Axford, *Ann. NY. Acad. Sci.* **375** (1981) 297.
[14] W.I. Axford, E. Leer and G. Skadron, Proc. 15 Int. Cosmic Ray Conf. (Plovdiv), **11** (1977) 132.
[15] W.I. Axford, E. Leer and J.F. McKenzie, *A&A* **111** (1982) 317.
[16] W. Baade and F. Zwicky, *Proc. Nat. Acad. Sci.* **20** (1934) 259.
[17] K.R. Ballard and A.F. Heavens, *MNRAS* **259** (1992) 89.
[18] R. Balescu, H.-D. Wang and J.H. Misguich, *Phys. Plasmas* **1** (1994) 1070.
[19] L. Ball and J.G. Kirk, *ApJ* **396** (1992) L42.
[20] L. Ball and D.B. Melrose, *Publ. Astron. Soc. Aust.* **18** (2001) 361.
[21] M.G. Baring, D.C. Ellison and F.C. Jones, *ApJ* **409** (1993) 327.
[22] M.G. Baring, D.C. Ellison and F.C. Jones, *ApJS* **90** (1994) 547.
[23] M.G. Baring, D.C. Ellison and F.C. Jones, *ApJ* **453** (1995) 873.
[24] M.G. Baring, D.C. Ellison, S.P. Reynolds, I.A. Grenier and P. Goret, *ApJ* **513** (1999) 311.
[25] J. Bednarz and M. Ostrowski, *MNRAS* **283** (1996) 447.
[26] J. Bednarz and M. Ostrowski, *Phys. Rev. Lett.* **80** (1998) 3911.
[27] M. Begelman and D.F. Cioffi, *ApJ* **345** (1989) L21.
[28] M. Begelman and J.G. Kirk, *ApJ* **353** (1990) 66.
[29] M.C. Begelman, D. Rudak and M. Sikora, *ApJ* **363** (1990) 38.
[30] M.C. Begelman and E.G. Zweibel, *ApJ* **431** (1994) 689.
[31] A.R. Bell, *MNRAS* **182** (1978) 147.
[32] A.R. Bell, *MNRAS* **257** (1992) 493.
[33] A.R. Bell and S.G. Lucek, *MNRAS* **283** (1996) 1083.
[34] E.G. Berezhko, V.K. Yelshin and L.T. Ksenofontov, *Astrop. Phys.* **2** (1994) 215.
[35] E.G. Berezhko and H.J. Völk, *Astrop. Phys.* **7** (1997) 183.
[36] E.G. Berezhko, *Astrop. Phys.* **5** (1996) 367.
[37] V.S. Berezinskii, S.V. Bulanov, V.A. Dogiel, V.L. Ginzburg and V.S. Ptuskin, Astrophysics of Cosmic Rays (North Holland Publ. Co., 1990).

[38] P. Bhattacharjee and G. Sigl, *Phys. Rep.* **327** (2000) 109.
[39] D.J. Bird *et al.*, *ApJ* **441** (1995) 144.
[40] P.L. Biermann, J. Phys. G **23** (1997) 1.
[41] P.L. Biermann, T.K. Gaisser and T. Stanev, *Phys. Rev. D* **51** (1995) 3450.
[42] P.L. Biermann, N. Langer, E.-S. Deo and T. Stanev, *A&A* **369** (2001) 269.
[43] R.G. Blandford and C.F. McKee, *Phys. Fl.* **19** (1976) 1130.
[44] R.D. Blandford and D. Eichler, *Phys. Rep.* **154** (1987) 1.
[45] R.D. Blandford and J.P. Ostriker, *ApJ* **221** (1978) L29.
[46] P. Blasi, *Astrop. Phys.* **16** (2002) 429.
[47] P. Blasi and A.V. Olinto, *Phys. Rev. D* **59** (1998) 023001-1.
[48] P. Blasi, R.I. Epstein and A.V. Olinto, *ApJ* **533** (2000) L123.
[49] J.S. Bloom, S.R. Kulkarni and S.G. Djorgovski, *AJ* **123** (2002) 1111.
[50] E. Boldt and P. Ghosh, *MNRAS* **307** (1999) 491.
[51] S.V. Bogovalov, *A&A* **349** (1999) 1017.
[52] B.J. Burn, *A&A* **45** (1975) 435.
[53] R.A. Chevalier, in: Supernovae, edited by D.N. Schramm (D. Reidel Publ. Co., Dordrecht, Holland, 1977) 53.
[54] D.F. Cioffi, in: Physical Processes in Hot Cosmic Plasmas, edited by W. Brinkmann, A.C. Fabian & F. Giovanelli, NATO ASI Series Vol. 305 (Kluwer Acad. Publ., 1990).
[55] T.E. Clarke, P.P. Kronberg and H. Böringer, *ApJ* **547** (2001) 111.
[56] S.A. Colgate and M.H. Johnson, *Phys Rev. Lett.* **5** (1960) 235.
[57] T. deHoffmann and E. Teller, *Phys. Rev.* **80** (1950) 692.
[58] S.G. Djorgovski *et al.*, *Nature* **387** (1997) 876.
[59] A.Z. Dolginov and M.E. Katz, *Phys. Rep.* **239** (1994) 285.
[60] B.T. Draine and C.F. McKee, *ARA&A* **31** (1993) 373.
[61] L.O'C. Drury, *Rep. Progr. Phys.* **46** (1983) 963.
[62] L.O'C. Drury, *MNRAS* **251** (1991) 340.
[63] L.O'C. Drury and S.A.E.G. Falle, *MNRAS* **223** (1986) 353.
[64] L.O'C. Drury and H.J. Völk, *ApJ* **248** (1981) 344.
[65] L.O'C. Drury, F.A. Aharonian and H.J. Völk, *A&A* **287** (1994) 959.
[66] L.O'C. Drury, H.J. Völk and E.G. Berezhko, *A&A* **299** (1995) 222.
[67] L.O'C. Drury, P. Duffy, D. Eichler and A. Mastichiadis, *A&A* **347** (1999) 370.
[68] P. Duffy, L.O'C. Drury and H. Völk, *A&A* **291** (1994) 613.
[69] P. Duffy, J.G. Kirk, Y.A. Gallant and R.O. Dendy, *A&A* **302** (1995) L21.
[70] J.P. Edminston, C.F. Kennel and D. Eichler, *Geoph. Res. Lett.* **9** (1982) 531.
[71] D. Eichler, *ApJ* **229** (1979) 419.
[72] D. Eichler, *ApJ* **277** (1984) 429.
[73] D. Eichler, M. Livio, T. Piran and D.N. Schramm, *Nature* **340** (1989) 126.
[74] D.C. Ellison, F.C. Jones and S.P. Reynolds, *ApJ* **360** (1990) 702.
[75] D.C. Ellison, J. Giacalone, D. Burgess and S.J. Schwartz, *J. Geoph. Res.* **98** (1993) 21 085.
[76] D.C. Ellison, M.G. Baring and F.C. Jones, *ApJ* **453** (1995) 873.
[77] D.C. Ellison, L.O'C. Drury and J.-P. Meyer, *ApJ* **487** (1997) 197.
[78] R. Enomoto *et al.*, *Nature* **416** (2002) 823.

[79] R.I. Epstein, *MNRAS* **193** (1980) 723.
[80] S.A.E.G. Falle and J.R. Giddings, *MNRAS* **225** (1987) 399.
[81] G.R. Farrar and T. Piran, *Phys. Rev. Lett.* **84** (2000) 3257.
[82] E. Fermi, *Phys. Rev.* **75** (1949) 1169.
[83] M. Forman and G. Morfill, *Proc. 16th Int. Cosmic Ray Conf., Kyoto* **5** (1979) 328.
[84] J.W. Fowler, L.F. Fortson, C.C.H. Jui, D.B. Kieda, R.A. Ong, C.L. Pryke and P. Sommers, *Astrop. Phys.* **15** (2001) 49.
[85] D.A. Frail *et al.*, *ApJ* **572** (2001) L55.
[86] M.S. Fulbright and S.P. Reynolds, *ApJ* **357** (1990) 591.
[87] T.K. Gaiser, Cosmic Rays and Particle Physics (Cambridge Univ. Press, 1990).
[88] T. Galema, Ph.D. Thesis, University of Amsterdam (The Netherlands, 2001).
[89] Y.A. Gallant and J. Arons, *ApJ* **435** (1994) 230.
[90] Y.A. Gallant and A. Achterberg, *MNRAS* **305** (1999) L6.
[91] U.D.J. Gieseler, J.G. Kirk, Y.A. Gallant and A. Achterberg, *A&A* **345** (1999) 298.
[92] U.D.J. Gieseler, T.W. Jones and H. Kang, *A&A* **364** (2000) 911.
[93] H. Goldstein, Classical Mechanics, 2nd Ed. (Addison-Wesley Publ. Co., 1980).
[94] J. Goodman, *ApJ* **308** (1986) L47.
[95] C.C. Goodricht and J.D. Scudder, *J. Geoph. Res.* **89** (1984) 6654.
[96] P.W. Guilbert, A.C. Fabian and M.J. Rees, *MNRAS* **205** (1983) 593.
[97] D.E. Hall and P.A. Sturrock, *Phys. Fl.* **10** (1967a) 1593.
[98] D.E. Hall and P.A. Sturrock, *Phys. Fl.* **12** (1967b) 2620.
[99] N. Hayashida *et al.*, *Phys. Rev. Lett.* **77** (1996) 1000.
[100] A.F. Heavens and L.O'C. Drury, *MNRAS* **235** (1988) 997.
[101] A.M. Hillas, *ARA&A* **22** (1984) 425.
[102] M. Hoshino and N. Shimada, *ApJ* **572** (2002) 880.
[103] P.D. Hudson, *MNRAS* **131** (1965) 23.
[104] M.B. Isichenko, *Plasma Phys. Contr. Fus.* **33** (1991a) 795.
[105] M.B. Isichenko, *Plasma Phys. Contr. Fus.* **33** (1991b) 809.
[106] J.R. Jokipii, *ApJ* **255** (1982) 716.
[107] J.R. Jokipii, *ApJ* **313** (1987) 842.
[108] J.R. Jokipii, *Space. Sci. Rev.* **86** (1998) 161.
[109] J.R. Jokipii and E.N. Parker, *ApJ* **155** (1969) 777.
[110] F.C. Jones and D.C. Ellison, *J. Geoph. Res.* **92** (1987) 11205.
[111] F.C. Jones and D.C. Ellison, *Space Sci. Rev.* **58** (1991) 259.
[112] B.B. Kadomtsev and O.P. Pogutse, *Nucl. Fus. Suppl.* **1** (1979) 649.
[113] H. Kang and T.W. Jones, *ApJ* **353** (1990) 149.
[114] H. Kang and T.W. Jones, *MNRAS* **249** (1991) 439.
[115] H. Kang, T.W. Jones and U.D.J. Gieseler, *ApJ* **579** (2002) 337.
[116] D. Kazanas and D.C. Ellison, *ApJ* **304** (1986) 178.
[117] C.F. Kennel and H.E. Petchek, *J. Geoph. Res.* **71** (1966) 1.
[118] C.F. Kennel and F.V. Coroniti, *ApJ* **283** (1984) 694.
[119] J.G. Kirk, in Plasma Astrophysics, Saas-Fee Advanced Course 24, edited by A.O. Benz and T.J-L. Couvoisier (Springer Verlag, Berlin, 1994) 262.
[120] J.G. Kirk and P. Schneider, *ApJ* **315** (1987a) 425.
[121] J.G. Kirk and P. Schneider, *ApJ* **322** (1987b) 256.

[122] J.G. Kirk and P. Schneider, *A&A* **201** (1988) 177.

[123] J.G. Kirk and A.F. Heavens, *MNRAS* **239** (1989) 995.

[124] J.G. Kirk, P. Duffy and Y.A. Gallant, *A&A* **314** (1996) 1010.

[125] K. Koyama, R. Petre, E.V. Gotthelf *et al.*, *Nature* **378** (1995) 255.

[126] G.E. Krimsky, *Dokladay Acad. Nauk. SSR* **242** (1977) 1306.

[127] P.P. Kronberg, *Space Sci. Rev.* **75** (1996) 387.

[128] P.P. Kronberg, in High Energy Gamma-Ray Astronomy, edited by F.A. Aharonian and H.J. Völk (2001) 451.

[129] J. Kuijpers, *Publ. Astron. Soc. Aust.* **18** (2001) 407.

[130] L.D. Landau and E.M. Lifshitz, Fluid Mechanics (Pergamonn Press, Oxford, 1959).

[131] S. Lee, A.V. Olinto and G. Sigl, *ApJ* **21** (1995) 455.

[132] M. Lemoine, G. Sigl, A.V. Olinto and D.N. Schramm, *ApJ* **486** (1997) L115.

[133] A. Levinson, *ApJ* **401** (1992) 73.

[134] A. Levinson and E. Boldt (2002), preprint [`astro-ph 0012314v2`]

[135] M.S. Longair, High Energy Astrophysics, 2nd Ed., Vol. I (Cambridge University Press, 1992a).

[136] M.S. Longair, High Energy Astrophysics, 2nd Ed., Vol. II (Cambridge University Press, 1992b).

[137] S.G. Lucek and A.R. Bell, *MNRAS* **268** (1994) 581.

[138] Y. Lyubarsky and J.G. Kirk, *ApJ* **547** (2001) 437.

[139] A. MacFadyen and S. Woosley, *ApJ* **550** (1999) 372.

[140] M.A. Malkov, *ApJ* **485** (1997) 638.

[141] M.A. Malkov, *Phys. Rev. E* **58** (1998) 4911.

[142] M.A. Malkov and H.J. Völk, *A&A* **300** (1995) 605.

[143] M.A. Malkov and H.J. Völk, *ApJ* **473** (1996) 347.

[144] M.A. Malkov and L.O'C. Drury, *Rep. Prog. Phys.* **64** (2001) 429.

[145] M.A. Malkov, P.H. Diamond and T.W. Jones, *ApJ* **571** (2002) 856.

[146] K. Meisenheimer, H.-J. Röser, P.R. Hiltner *et al.*, *A&A* **219** (1989) 63.

[147] G.A. Medina Tanco, E.M. de Gouveia Dal Pino and J.E. Horvath, *Astrop. Phys.* **6** (1997) 337.

[148] M.V. Medvedev and A. Loeb, *ApJ* **526** (1999) 697.

[149] A. Melatos and D.B. Melrose, *MNRAS* **279** (1996) 1168.

[150] D.B. Melrose and M.H. Pope, *Proc. A.S.A.* **10** (1993) 222.

[151] D.B. Melrose and A. Crouch, *Publ. Astron. Soc. Aust.* **14** (1997) 251.

[152] P. Mészáros, *Science* **291** (2001) 79.

[153] P. Mészáros, *ARA&A* **40** (2002).

[154] J.-P. Meyer, L.O'C. Drury and D.C. Ellison, *ApJ* **487** (1997) 182.

[155] F.C. Michel, Theory of Neutron Star Magnetospheres (Univ. of Chicago Press, 1991).

[156] I.F. Mirabel and L.F. Rodríguez, *ARA&A* **37** (1999) 409.

[157] J. Miralda-Escudé and J. Waxman, *ApJ* **462** (1996) L59.

[158] M. Mond and L.O'C. Drury, *A&A* **332** (1998) 385.

[159] M. Nagano and A.A. Watson, *Rev. Mod. Phys.* **72** (2000) 689.

[160] T. Naito and F. Takahara, *J. Phys. G* **20** (1994) 477.

[161] T. Naito and F. Takahara, *MNRAS* **275** (1995) 1077.

[162] R. Narayan, T. Piran and A. Shemi, *ApJ* **379** L1.

[163] R. Narayan, B. Pacyński and T. Piran, *ApJ* **395** (1992) L83.
[164] C.A. Norman, D.B. Melrose and A. Achterberg, *ApJ* **454** (1995) 60.
[165] A.V. Olinto, *Phys. Rep.* **333-334** (2000) 329.
[166] M. Ostrowski, *MNRAS* **249** (1991) 551.
[167] B. Pacyński, *ApJ* **308** (1986) L43.
[168] B. Pacyński, *ApJ* **494** (1998) L45.
[169] E. Parizot, *A&A* **362** (2000) 786.
[170] E. Parizot and L.O'C. Drury, *A&A* **346** (1999a) 329.
[171] E. Parizot and L.O'C. Drury, *A&A* **346** (1999b) 686.
[172] E. Parizot and L.O'C. Drury, *A&A* **349** (1999c) 673.
[173] E.N. Parker, *Phys. Rev.* **99** (1955) 241.
[174] E.N. Parker, *Planet. Sp. Sci.* **13** (1965) 9.
[175] J.A. Peacock, *MNRAS* **196** (1981) 135.
[176] G. Pelletier, *A&A* **350** (1999) 705.
[177] G. Pelletier and M. Karsalé, *A&A* **361** (2000) 788.
[178] E.S. Phinney, *ApJ* **380** (1991) L17.
[179] T. Piran, *Phys. Rep.* **314** (1999) 575.
[180] T. Piran, A. Shemi and R. Narayan, *MNRAS* **263** (1993) 861.
[181] M.H. Pope and D.B. Melrose, *Proc. A.S.A.* **11** (1994) 175.
[182] V.S. Ptuskin and A. Soutoul, *Space Sci. Rev.* **86** (1998) 225.
[183] R.J. Protheroe and T. Stanev, *Astrop. Phys.* **10** (1999) 185.
[184] J.R. Rachen and P.L. Biermann, *A&A* **272** (1993) 161.
[185] A.B. Rechester and M.N. Rosenbluth, *Phys. Rev. Lett.* **40** (1978) 38.
[186] M.J. Rees and P. Mézáros, *MNRAS* **258** (1992) 41P.
[187] S.P. Reynolds, *ApJ* **90** (1994) 845.
[188] S.P. Reynolds, *ApJ* **459** (1996) L13.
[189] S.P. Reynolds, *ApJ* **493** (1998) 375.
[190] G.B. Rybicki and A.P. Lightman, Radiative Processes in Astrophysics (John Wiley & Sons, New York, 1979).
[191] R. Schlickeiser, *A&A* **136** (1984) 227.
[192] R. Schlickeiser, Cosmic Ray Astrophysics (Springer Verlag, 2002).
[193] M. Schmidt, *ApJ* **523** (1999) L117.
[194] H. Schmitz, S.C. Chapman and R.O. Dendy, *ApJ* **570** (2002) 673.
[195] H. Schmitz, S.C. Chapman and R.O. Dendy, *ApJ* **579** (2002) 327.
[196] P. Schneider, *A&A* **378** (1993) 315.
[197] P. Schneider and J.G. Kirk, *A&A* **217** (1989) 344.
[198] M. Scholer, H. Kucharek and K.J. Trattner, *Adv. Space Res.* **21** (1997) 547.
[199] M. Scholer, H. Kucharek and K.J. Trattner, *Adv. Space Res.* **21** (1998) 533.
[200] S.T. Scully and F.W. Stecker, *Astrop. Phys.* **16** (2002) 271.
[201] E. Shatzman, *Ann Astr.* **26** (1967) 234.
[202] N. Shimada and M. Hoshino, *ApJ* **543** (2000) L67.
[203] Y. Shirasaki *et al.*, *Astrop. Phys.* **15** (2001) 357.
[204] G. Sigl, Lecture given at the Summer School on Astroparticle Physics and Cosmology, Trieste (2000), `sigl@iap.fr`
[205] G. Sigl, *Science* **291** (2001) 73.

[206] G. Sigl, M. Lemoine and A.V. Olinto, *Phys. Rev. D* **56** (1997) 4470.
[207] G. Sigl, M. Lemoine and P. Biermann, *Astrop. Phys.* **10** (1999) 141.
[208] M. Sikora, Rudak, J.G. Kirk, M.C. Begelman, and P. Schneider, *ApJ* **341** (1987) L33.
[209] H.C. Spruit, *A&A* **194** (1988) 319.
[210] J. Skilling, *MNRAS* **172** (1975a) 557.
[211] J. Skilling, *MNRAS* **172** (1975b) 567.
[212] T. Stanev, P.L. Biermann, J. Loyd-Evans and A.A. Watson, *Phys. Rev. Lett.* **75** (1995) 3056.
[213] F.W. Stecker, *Phys. Rev. Lett.* **80** (1998) 1816.
[214] F.W. Stecker, *Astrop. Phys.* **14** (2000) 207.
[215] J. Szabelski, T. Wibig and A.W. Wolfendale, *Astrop. Phys.* **17** (2002) 125.
[216] M. Takeda *et al.*, *Phys. Rev. Lett.* **81** (1998) 1163.
[217] J.H. Taylor, R.N. Manchester and A.G. Lyne, *ApJS* **88** (1993) 529.
[218] I.N. Toptygin, Cosmic Rays in Interplanetary Magnetic Fields (D. Reidel Publ. Co., Dordrecht, The Netherlands, 1985), Ch. 18.
[219] D. Ucer and V.D. Shapiro, *Phys. Rev. Lett.* **87** (2001) 075001-1.
[220] Y. Uchihori *et al.*, *Astrop. Phys.* **13** (2000) 151.
[221] S. Usami, H. Hasegawa and Y. Oshawa, *Phys. Plasmas* **8** (2001) 2666.
[222] S. Usami and Y. Oshawa, *Phys. Plasmas* **9** (2002) 1069.
[223] E. van der Swaluw, A. Achterberg, Y.A. Gallant and G. Tóth, *A&A* **380** (2001), 309
[224] J. van Paradijs *et al.*, *Nature* **386** (1997) 686.
[225] M. Vietri, *ApJ* **453** (1995) 883.
[226] H.-D. Wang, M. Vlad, E. Vanden Eijnden, *et al.*, *Phys. Rev. E* **51** (1995) 4844.
[227] A.A. Watson, *Phys. Rep.* **333** (2000) 309.
[228] E. Waxman, *Phys. Rev. Lett.* **75** (1995a) 386.
[229] E. Waxman, *ApJ* **452** (1995b) L1.
[230] E. Waxman, Lecture given at the Summer School on Astroparticle Physics and Cosmology, Trieste (2000), waxmann@wicc.weizmann.ac.il
[231] E. Waxman and J. Miralda-Escudé, *ApJ* **472** (1996) L89.
[232] E. Waxman and J.N. Bahcall, *ApJ* **541** (2000) 707.
[233] G.M. Webb, *ApJ* **296** (1985) 319.
[234] G.M. Webb and L.J. Gleeson, *Astrophys. Sp. Sci.* **60** (1978) 335.
[235] G.M. Webb, W.I. Axford, T. Terasawa, *ApJ* **270** (1983) 537.
[236] G.M. Webb, L.O'C. Drury and P. Biermann, *A&A* **137** (1984) 185.
[237] D.G. Wentzel, *ARA&A* **12** (1974) 71.
[238] A.J. Westphal, P. Buford Price, B.A. Weaver and V.G. Afanasiev, *Nature* **396** (1998) 50.
[239] R.A.J.M. Wijers, J.S. Bloom, J.S. Bagla and P. Natarajan, *MNRAS* **294** (1998) L13.
[240] E.C. Whipple, T.G. Northrop and T.J. Birmingham, *J. Geoph. Res.* **91** (1986) 4149.
[241] S. Woosley, *ApJ* **405** (1993) 273.
[242] S. Yoshida *et al.*, *Astrop. Phys.* **3** (1995) 105.
[243] G.P. Zank, W.I. Axford and J.F. McKenzie, *A&A* **233** (1990) 275.
[244] A.A. Zdziarski, *ApJ* **305** (1986) 45.

Section III: Black holes and AGN's

COURSE 8

THE BLACK HOLE ENVIRONMENTS

M. CAMENZIND

Landessternwarte Königstuhl,
69117 Heidelberg,
Germany

Contents

THE BLACK HOLE ENVIRONMENTS

M. Camenzind[1]

Abstract

Compact objects are astrophysical entities with radii smaller than about 10 Schwarzschild radii. The mass spectrum of these objects ranges from the Chandrasekhar mass of neutron stars to a few solar masses for stellar Black Holes to hundreds of millions of solar masses for supermassive Black Holes in the centers of galaxies. Besides mass, all these objects also carry angular momentum with known spin periods for neutron stars, but unknown spin periods for Black Holes. Black Holes represent exact solutions to Einstein's equations having a horizon, they are characterized by only two parameters, the mass M and the angular momentum J which is restricted to be $J \leq GM^2/c$. Noncollapsed objects dispose however of at least three parameters; in addition to mass and angular momentum, the mass quadrupole moment Q_2 is not completely arbitrary, but essentially determined by rotation and the equation of state. In the last years, new analytical solutions for rapidly rotating neutron stars have been worked out.

Black Holes in galactic nuclei form now a huge class of compact objects. Jet sources represent an interesting sub–sample of active galaxies. According to the spin paradigm, the rotational energy of the central compact object is the ultimate source of energy for the jet launch. There is some progress in the understanding of large–scale structure of jets in the last years, in particular for sources propagating in a dense cluster environment. In distinction to these macro–jets, the formation of the micro–jets on the parsec–scale is still largely unknown. A magnetic coupling between the rotation of the central object and the outflowing disk plasma is the most probable process. Due to the complexity of modelling this process, we are still not able to present time–dependent simulations proving the evidence for this process. This certainly one of the major challenge of future research in relativistic Astrophysics.

This work is partially supported by the Sonderforschungsbereich 439 in Heidelberg, Germany.

1 Introduction

Einstein's general theory of relativity describes gravity as curvature of space-time due to the presence of matter. If the curvature is fairly weak, Newton's laws of gravity can explain most of what is observed. For example, the regular motions of the planets. Very massive or dense objects generate much stronger gravity. The most compact objects imaginable are predicted by General Relativity to have such strong gravity that nothing, not even light, can escape their grip. Scientists today call such an object a Black Hole. Though the history of the term is interesting, the main reason is that no light can escape from inside a Black Hole: it has, in effect, disappeared from the visible universe.

Do Black Holes actually exist? Most physicists believe they do, basing their views on a growing body of observations. In fact, present theories of how the cosmos began rest in part on Einstein's work and predict the existence of both singularities and the Black Holes that contain them. Yet Einstein himself vigorously denied their reality, believing, as did most of his contemporaries, that Black Holes were a mere mathematical curiosity. He died in 1955, before the term Black Hole was coined or understood and observational evidence for Black Holes began to mount. The actual search for Black Holes started with the advent of X–ray astronomy with the launch of the UHURU satellite in 1971. At that time it has been realized that Black Holes can only become visible through accretion processes. Matter falling in disklike structure towards the horizon should be heated to X–ray temperatures. Before this epoch, Black Holes had been postulated to solve the energy problem for Quasars – cosmological objects already detected in 1963. The archetypical Quasar 3C 273 is still one of the most frequently observed object in the Universe. Emission from this object is broadband similar to the emission from the Crab nebula, extending from the low–frequency radio regime to gamma–rays.

Stellar mass Black Holes are not very abundant in our Galaxy, but galaxies are much more frequent. In the visible Universe we find about 10 billion bright galaxies. In the last years it has become evident that all galaxies built around a spheroidal bulge component contain a central dark object whose mass is a certain fraction of the bulge stellar mass. Our Galaxy is no exception to this rule. There is even a better correlation between this dark mass and the stellar dispersion velocity. These dark masses are identified with Black Holes, though the proof of this fact is extremely painful. The Black Holes in the centers of galaxies are therefore no longer hypothetical entities. The concrete search for these objects can be done in nearby galaxies of the local Universe or by means of the investigations of the emission from Quasars and Radio Galaxies in the distant Universe. This requires a definite cosmological model, since absolute luminosities *e.g.* depend very much on the particular cosmological model. Accretion is the main energy reservoir for most of these objects. This accretion is driven by large gas accumulations in the parsec–scale cores of the underlying galaxies. While cores of nearby galaxies usually only dispose of small amounts of gas, this gas mass was about a factor 100 higher in the early Universe, largely due to enhanced gas injection from stellar evolution. Time–dependent accretion from this reservoir is one of the major future topics of research in this area, though there is some progress on much more moderate scale. New methods have to be invented for this vehicle, in particular cooling of the accreting gas has to be included in the simulations. In the present simulations, most of the mass is lost to outgoing winds and no accretion into the horizon is left over – a somewhat problematic prediction from numerical astrophysics.

2 Black Holes and beyond

"The Black Holes of nature are the most perfect macroscopic objects there are in the universe: the only elements in their construction are our concepts of space and time. And since the general theory of relativity provides only a single unique family of solutions for their descriptions, they are the simplest objects as well". (Chandrasekhar in [9].)

2.1 Astrophysically important aspects of Black Holes

From this simplicity in the structure of these objects, a few general items can be derived:

- **No hair theorem**: spacetime only depends on the mass and spin of the hole.
- **Event horizon**: this is the defining property of a Black Hole. It is a one way surface, you can go in, but you can not come out, and it has a radius
$$R_{\rm H} = \frac{2GM_{\rm H}}{c^2} = 3\,{\rm km}\,\frac{M_{\rm H}}{M_\odot}. \tag{2.1}$$
For rotating holes, the horizon shrinks.
- **Gravitational redshift:** from the point of view of distant observers, time appears to slow down near a hole. Photons escaping from near the hole lose energy and are redshifted to lower frequencies and longer wavelengths.
- **Frame dragging:** space is dragged around close to a rotating Black Hole. Even objects with absolutely no angular momentum revolve around the hole!

- **Second law of Black Hole dynamics:** the horizon area of astrophysical Black Holes can never decrease. In the cosmic evolution of the Universe, the mass of Black Holes can only grow, never decrease.

- **Deep potential well:** the most important property that Black Holes have for an astronomer is their very DEEP potential well. This makes them the objects with the highest efficiency for conversion of gravitational energy to radiation.

2.2 *Classes of compact objects*

Compact objects essentially come in five classes, depending on their mass:

- **Neutron stars** (and quark stars) with masses $M < M_{\rm crit} \simeq 1.6\, M_\odot$;

- **Stellar Black Holes** with masses $M_{\rm crit} < M < 100\, M_\odot$;

- Very massive Black Holes with masses in the range of a few hundred to a few thousand solar masses (formed only in the early Universe);

- **Supermassive Black Holes** with masses $10^6\, M_\odot < M < 10^{10}\, M_\odot$ nowadays found in the centers of nearby galaxies;

- **Primordial Black Holes** with masses $M \leq M_\odot$ formed either in the very early Universe or as remnants in the quark–hadron phase transition, when the Universe had a temperature of about 200 MeV.

Ordinary Black Holes are thought to form from stars or other massive objects if and when they collapse from their own gravity to form an object whose density is infinite: in other words, a singularity. During most of a star's lifetime, nuclear fusion in the core generates electromagnetic radiation, including photons, the particles of light. This radiation exerts an outward pressure that exactly balances the inward pull of gravity caused by the star's mass.

As the nuclear fuel is exhausted, the outward forces of radiation diminish, allowing the gravitation to compress the star inward. The contraction of the core causes its temperature to rise and allows remaining nuclear material to be used as fuel. The star is saved from further collapse – but only for a while.

Eventually, all possible nuclear fuel is used up and the core collapses. How far it collapses, into what kind of object, and at what rate, is determined by the star's final mass and the remaining outward pressure that the burnt-up nuclear residue (largely iron) can muster. If the star is sufficiently massive or compressible, it may collapse to a Black Hole. If it is less massive or made of stiffer material, its fate is different: it may become

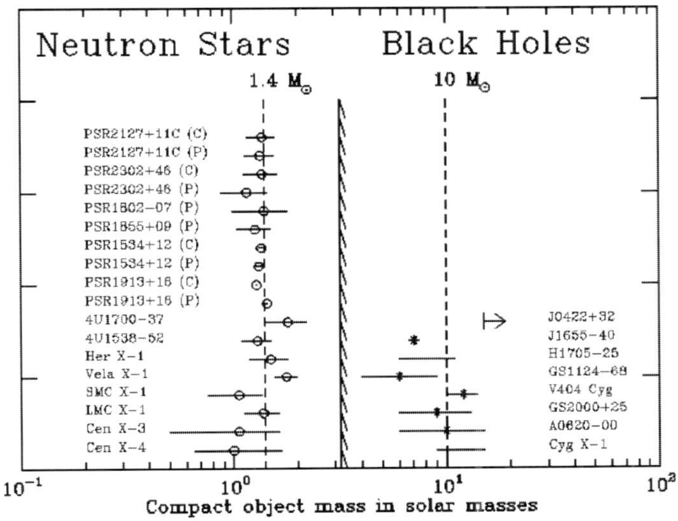

Fig. 1. The masses of neutron stars and Black Holes as derived from binary systems. It seems that all neutron stars have a universal mass of 1.4 solar mass and the Black Hole candidates typically a mass of 10 solar masses.

a white dwarf or a neutron star. When small stars (up to 8 times the size of the Sun) exhaust their nuclear fuel, they typically shed large amounts of matter, leaving a core that eventually cools and contracts gravitationally to about the size of the Earth. The result is a white dwarf: the more massive it is, the greater its inward gravitational pull, and the smaller it becomes. Chandrasekhar derived a relationship between the star's mass and its radius which sets an upper limit to the mass a white dwarf can have, beyond which it will collapse to a neutron star or, if sufficiently massive, to a Black Hole. Calculations put the Chandrasekhar limit at 1.4 solar masses. Decades later Chandrasekhar's fundamental contributions were recognized when he won the 1983 Nobel Prize in Physics.

More massive stars tend to burn hotter and faster. Once all the nuclear fuel has been exhausted, such stars quickly collapse, shedding much of their mass in dramatic explosions called supernovae. The most recent event of this kind was observed in 1987 when a star weighting the equivalent of 20 suns blew up in a neighboring galaxy 160 000 light years away. As heavy as neutron stars are, if they are less than 2 solar masses, they too can only contract so far and no further. That's because, as crushed as they are, the neutrons also resist the inward pull of gravity, just as a white dwarf's electrons do. However, if after a star collapses, the remaining core exceeds approximately 2 solar masses, the outcome is thought to be very different. The precise mass limit is uncertain and depends on the nuclear physics going on within the core, a topic of much debate within the physics community. If the star's final mass exceeds much beyond 2 solar masses, there is no outward force that can resist gravity. The core continues to collapse to a critical size or circumference beyond which there is only one fate: to form a Black Hole.

What happens to a Black Hole after it forms? Does it vibrate? Radiate? Lose mass? Grow? Shrink? Partial solutions of the Einstein equations point to two possible outcomes:

- A non-rotating, spherically symmetric Black Hole, first postulated by Schwarzschild.
- A rotating, spherical Black Hole, predicted in 1963 by the New Zealand mathematician Roy Kerr [28].

These two types of Black Holes have become known as Schwarzschild and Kerr Black Holes, respectively. Both types of Black Holes are stationary in that they do not change in time, unless they are disturbed in some way. As such, they are among the simplest objects known in General Relativity. They can be completely described in terms of just two parameters: their mass M and their angular momentum $J = Ma$, a is called the Kerr–parameter. Theoretically, Black Holes may also possess electric charge, Q, but it would quickly attract enough charge of the opposite sign. The net result is that any realistic or astrophysical Black Hole would tend to exhibit zero charge. This simplicity of Black Holes is summed up in the saying *Black Holes have no hair*, meaning that, apart from its mass and angular momentum, there is no other characteristic (or hair) that a Black Hole can exhibit.

However, both the Schwarzschild and Kerr Black Holes represent end states. Their formation may result from various processes, all of them quite complicated. When a real Black Hole forms from, the collapse of a very massive star, or when a Black Hole is disturbed by, say, another Black Hole spiralling into it, the resulting dynamics cause disturbances in spacetime that should lead to the generation of gravitational waves. By numerically solving the Einstein equations on powerful computers, scientists have been able to simulate the gravitational waves emitted by perturbed or interacting Black Holes. When visualized in movies generated by advanced computer graphics, the unfolding wave patterns are not only intriguing but strikingly beautiful. By emitting gravitational waves, non-stationary Black Holes lose energy, eventually become stationary and cease to radiate in this manner. In other words, they decay into stationary Black Holes, namely holes that are perfectly spherical or whose **rotation is perfectly uniform**. According to Einstein's Theory of General Relativity, such objects cannot emit gravitational waves, they are stable – except for quantum effects, which have no relevance for macroscopic objects.

Supermassive Black Holes: besides stellar Black Holes, the class of supermassive Black Holes with masses from a few million solar masses to a few billion solar masses seems to be a well established class of objects. It is now believed that at the center of each (bulgy) galaxy there is a supermassive Black Hole that is millions to billions of times heavier than our Sun. The massive Black Hole captures nearby stars and gas and drags them into a swirling accretion disk. A torus in the inner accretion shields the Black Hole in those systems that are viewed edge on. In many of these systems (which are called AGN = active galactic nuclei), a jet is ejected perpendicular to the disk and is seen in the optical and radio wavebands. In the very central regions, the disk becomes so hot (tens of millions of degrees) that the emission is in the X-ray and Gamma-ray bands.

2.3 Accretion flows are not well understood

If gas is placed in orbit around a Black Hole, any source of dissipation will cause the gas to spread out to form a disk. Material closer to the hole will orbit faster than material which is further out, very much like the planets in our solar system: Mercury revolves around the Sun in only 88 days, whereas Pluto takes 249 years. Friction will cause the rapidly rotating inner parts to slow down, while the slowly moving outer parts will speed up. The inner parts will therefore spiral inwards, releasing gravitational potential energy which can be converted into heat, radiation, and other forms of energy.

Before it flows across the Black Hole's event horizon, material can in principle convert an equivalent of up to 32% of its rest mass energy into radiation. Compare with nuclear fusion of hydrogen into helium inside the Sun: here, only 0.7% of the rest mass energy of the four protons gets converted into energy.

Black Holes are tiny and very far away. It is currently impossible to image the central accretion flow directly, so we cannot be certain of the flow geometry. Is it really disk-like in the inner parts where all the action is? The only hope is therefore to model the spectrum, polarization, and variability of the radiation we believe comes from the central accretion flow. Unfortunately, the physics is enormously complex (but extremely interesting), involving **turbulent radiation magnetohydrodynamics**, in addition to the physics of being in a relativistic spacetime which is trivial by comparison.

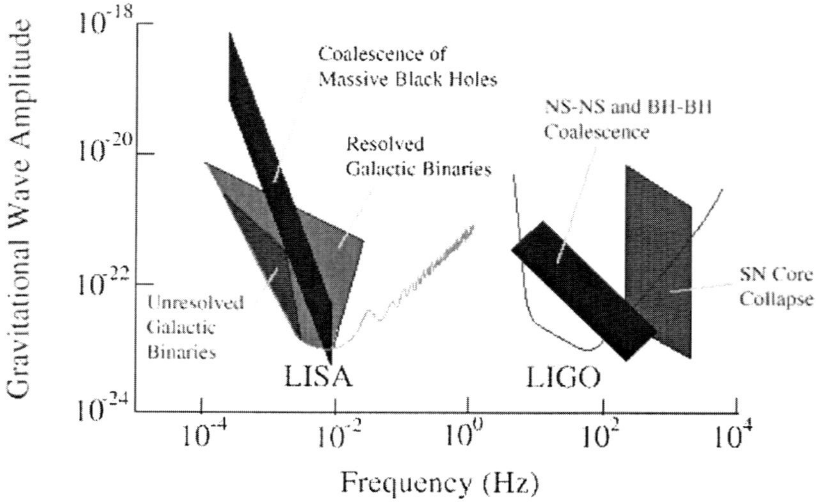

Fig. 2. Gravitational wave spectrum.

2.4 Direct detection of Black Holes

The ultimate proof for the formation of Black Holes will be the detection of gravitational waves emitted in the collapse process (Fig. 2). Favorable processes are the merging of two neutron stars forming from progenitor systems such as PSR 1913+16. In this process, high frequency gravitational waves are emitted with frequencies $f > 100$ Hz. On the other hand, low frequency gravitational waves are copiously produced in the early Universe by merging of supermassive Black Holes in the cores of galaxies; the corresponding frequencies are in the range of mHz. These waves are only detectable in space–born interferometers, such as LISA.

2.5 Present status

- There is very good evidence from astronomical observations that the universe is full of Black Holes with sizes ranging from five to a few billion solar masses in size.

- Black Hole accretion power is responsible for some of the most spectacular phenomena in the universe. These phenomena are not well understood, however, largely because of the complexity of the physics of the central accretion flow. There is little doubt, though, that Black

Holes exist at the heart of (active) galactic nuclei, quasars, and certain X-ray binaries.

- Accretion power is an important contributor to the overall evolution and ecology of the universe (UV–background *e.g.*).
- How can one prove the existence of Black Holes, short of a suicidal leap across an event horizon? Detection of gravitational waves is perhaps the only way.

3 The black hole paradigm for galactic nuclei

Most models of the energy production in quasars and active galactic nuclei invoke the presence of a nuclear Black Hole with a mass $M_{\rm H}$ in the range of 10^6 to 10^{10} $M_\odot$ [5]. The number of quasars at high redshift indicates that many normal galaxies we observe today must have gone through an active phase, if the lifetime of the activity is short in comparison to the Hubble time. This would indicate that supermassive Black Holes could be common in normal galactic nuclei. The high luminosity of quasars at high redshifts also requires that those objects populate the more massive end of the mass distribution. This excludes *e.g.* normal spiral galaxies as hosts of bright high redshift quasars.

3.1 Two aspects of the black hole paradigm

Since the pioneering works by Salpeter (1964), Lynden–Bell (1969), the standard model for the nuclear activity in galaxies has envisaged a supermassive compact object, usually a Black Hole with mass 10^5–$10^{10}\,M_\odot$, which acts as a central engine in the production of optical–UV and radio emission. In this picture, the radio power *e.g.* is derived from twin collimated beams of relativistic material which are ejected along the Black Hole spin axis.

The accretion paradigm: the **accretion paradigm** states that most, and perhaps all, AGN are powered by accretion onto a supermassive Black Hole. Within this scenario, the parameter $\dot{m} \equiv \dot{M}/\dot{M}_{\rm Edd}$ plays the most important role, determining the emission properties, and therefore the appearance of the central source. Objects with high accretion rate ($\dot{m} > 0.01$) appear as an *optical* quasar (of course, equally bright, if not brighter, in X-rays as well), while low sub-Eddington accretion ($\dot{m} \leq 10^{-2}$) produces a weak *radio* core with substantially less optical emission. A zero accretion rate produces a *dead* quasar – a Black Hole detectable only through its gravitational influence on the galactic nucleus. For a given level of activity, the Black Hole mass determines mainly the luminosity scaling.

The total energy output is related to both the mass M_H of the central object and the accretion rate $\dot{m}$. The maximum source luminosity is set by the Eddington limit,

$$L_\mathrm{Ed} = 1.3 \times 10^{47}\,\mathrm{erg\,s^{-1}}\,\frac{M_\mathrm{H}}{10^9\,M_\odot} \tag{3.1}$$

and the required accretion rate

$$\dot{M}_\mathrm{Ed} \simeq 20\,M_\odot\,\mathrm{yr}^{-1}\,\frac{0.1}{\epsilon_\mathrm{H}}\,\frac{L}{10^{47}\,\mathrm{erg\,s^{-1}}}. \tag{3.2}$$

ϵ_H is the efficiency of mass–energy conversion. To produce powerful 3C radio galaxies and quasars with luminosities $L \simeq 10^{46}–10^{47}$ erg s^{-1} requires minimum Black Hole masses of $10^9\,M_\odot$ and accretion rates of a few $M_\odot\,\mathrm{yr}^{-1}$ of infalling gas. The total amount of gas accreted in the life–time of the quasar is similar to the Black Hole mass itself. This amount of gas is indeed available on the parsec–scale of the host galaxies. The total amount of gas assembled in the core of these giant elliptical galaxies can be an appreciable fraction of the core mass itself.

The mass of a dead quasar can also be estimated from the total dissipation generated in the accretion process (thereby only a small fraction of the accreted mass can be lost from the system)

$$M_\mathrm{H} = \frac{L_\mathrm{Q}\tau}{\epsilon c^2} = 7 \times 10^8\,M_\odot\,\frac{L_\mathrm{Q}}{10^{13}\,L_\odot}\,\frac{\tau}{10^8\,\mathrm{yr}}\,\frac{0.1}{\epsilon_\mathrm{H}}, \tag{3.3}$$

where L_Q is the typical quasar luminosity and τ its lifetime. 10^8 years is an upper limit to the lifetime. The Black Hole mass in a typical quasar would be 100 Mio solar masses, while bright quasars with luminosities upto $10^{14}\,L_\odot$ would require 10 times higher masses.

The density of such remnants follows from the integrated comoving energy density in quasar light

$$u \simeq 1.3 \times 10^{-15}\,\mathrm{erg\,cm^{-3}}\,. \tag{3.4}$$

The corresponding mass density is then a question of efficiency

$$\rho_\mathrm{H} = \frac{u}{\epsilon_\mathrm{H} c^2} = 2.2 \times 10^5\,M_\odot\,\mathrm{Mpc}^{-3}\,\frac{0.1}{\epsilon_\mathrm{H}}. \tag{3.5}$$

On the other hand, the luminosity density of galaxies is $j \simeq 1.5 \times 10^8\,L_\odot\,\mathrm{Mpc}^{-3}$, and a typical bright hot component contributes $8.5 \times 10^9\,L_\odot$, depending somewhat on the Hubble constant. This provides us with a number density of 2×10^{-2} hot galaxies Mpc^{-3}. The mean Black Hole mass required per hot galaxy is only $10^8\,M_\odot$, which is in accordance with the observations. Here, one has to be aware of the fact that bright quasars are hosted by giant elliptical galaxies and not by spiral galaxies. This is not the case for low luminosity quasars which are essentially just Seyfert type objects.

The spin paradigm: rotating objects have in general an additional source of energy. The rotational energy of neutron stars is behind the powerful pair winds which are injected into surrounding nebulae (the Crab nebula *e.g.*). For Black Holes, energies (and therefore masses) add quadratically (see Sect. 4.6)

$$M_{\rm H} = \sqrt{M_{\rm irr}^2 + \left(\frac{J_{\rm H}}{2GM_{\rm irr}/c}\right)^2}\,. \tag{3.6}$$

The irreducible mass $M_{\rm irr}$ is a kind of rest mass for a rotating Black Hole that cannot be further reduced by any physical processes, except Hawking radiation – which is however unimportant for macroscopic Black Holes. The second term is due to the rotational energy of Black Holes and is given in terms of the angular momentum $J_{\rm H}$ of the Black Hole. Therefore, each Black Hole contains a rotational energy

$$E_{\rm rot} = (M_{\rm H} - M_{\rm irr})c^2 = M_{\rm H}c^2\left(1 - \sqrt{\frac{1}{2}(1 + \sqrt{1 - (a_{\rm H}/M_{\rm H})^2})}\right), \tag{3.7}$$

which could be extracted by means of some electrodynamic processes. $a_{\rm H}$ denotes the specific angular momentum (Kerr parameter). Since $E_{\rm rot} \leq 0.29\,M_{\rm H}c^2$, the rotational energy is a considerable amount of energy

$$E_{\rm rot} < 5 \times 10^{55}\,{\rm Watt\,s}\,\frac{M_{\rm H}}{10^9\,M_\odot}, \tag{3.8}$$

and no other object can approach this upper limit for the rotational energy. The mass $M_{\rm H}$ of Black Holes residing in giant elliptical galaxies can easily exceed $10^9\,M_\odot$. If this rotational energy could be dissipated in a kind of pulsar process, this would represent a considerable luminosity

$$L_{\rm rot} \simeq \frac{E_{\rm rot}}{t_{\rm diss}} \simeq 1.6 \times 10^{40}\,{\rm Watt}\,\frac{M_{\rm H}}{10^9\,M_\odot}\,\frac{10^9\,{\rm yr}}{t_{\rm diss}}, \tag{3.9}$$

essentially comparable to the mean luminosity of radio quasars at redshift of 2 – provided the system is able to dissipate the energy on a time–scale of a few billion years. This process could be behind the energetization of bright quasars, such as 3C 273 and 3C 345. In fact, the spin–paradigm assumes that the non–thermal (jet) power of radio–loud objects (quasars and radio galaxies) could be accounted for by this rotational energy.

This energy loss rate is an interesting number which can be compared to the bulk kinetic power Q_j in jets estimated from the by–products of the jets, the large–scale lobes

$$Q_\mathrm{j} = \frac{k\,U}{\tau_\mathrm{j}}. \tag{3.10}$$

Here U is the energy stored in the lobes, taken from equipartition energy U_eq, $k \simeq 2$ allows for $P\,\mathrm{d}V$ work expended by the jet on pushing back and warming up the extended medium. τ_j is the age of the jet, estimated *e.g.* from the spectral age of the radio lobes. This amounts to a maximal jet power $\simeq 10^{40}$ Watt for bright radio quasars and narrow–line radio galaxies for $0.5 < z < 1.0$. On the other hand, material radiatively excited by an AGN cools by line emission. Radio galaxies are usually only narrow–line emitters and one can therefore easily estimate the total narrow–line luminosity L_NLR in all narrow–lines. Rawlings & Saunders [57] obtained for an unbiased sample of FR II radio galaxies and low–power FR Is a correlation between the jet power and L_NLR

- $Q_\mathrm{j} \propto L_\mathrm{NLR}$ for FR I and FR II radio galaxies, as well as for radio quasars;
- $Q_\mathrm{j} \simeq 100\,L_\mathrm{NLR} \simeq 10^{36}–10^{40}$ Watt;
- Radio–quiet quasars do not satisfy the $Q_\mathrm{j} - L_\mathrm{NLR}$ correlation.

The last point strongly indicates that the jet power is some extra power provided *e.g.* by the rotational energy of the central source. But also radio–loud objects do have photoionizers which by virtue of $Q_\mathrm{j} \propto L_\mathrm{NLR}$ are controlled by the jet driving mechanism. It is also interesting that jet sources with given Q_j have a higher low–frequency luminosity when the sources are in a dense cluster environment. This probably indicates that the narrow line emission is due to the interaction of the jets with the ambient medium. Since the above correlation extends over more than 4 orders of magnitude, this could reflect the scaling of the central mass from $\simeq 10^6\,M_\odot$ in faint ellipticals to $\simeq 10^{10}\,M_\odot$ in giant ellipticals. If jet power were related to rotational energy of the Black Hole, then its essentially only the mass and the angular momentum that dictate the jet power. This is known as the **spin paradigm** of Quasar physics.

3.2 The Black Holes in nuclei of galaxies

If Black Holes are the drivers of the activity seen at high redshifts, these objects should be still present at low redshifts, though the fuel has declined

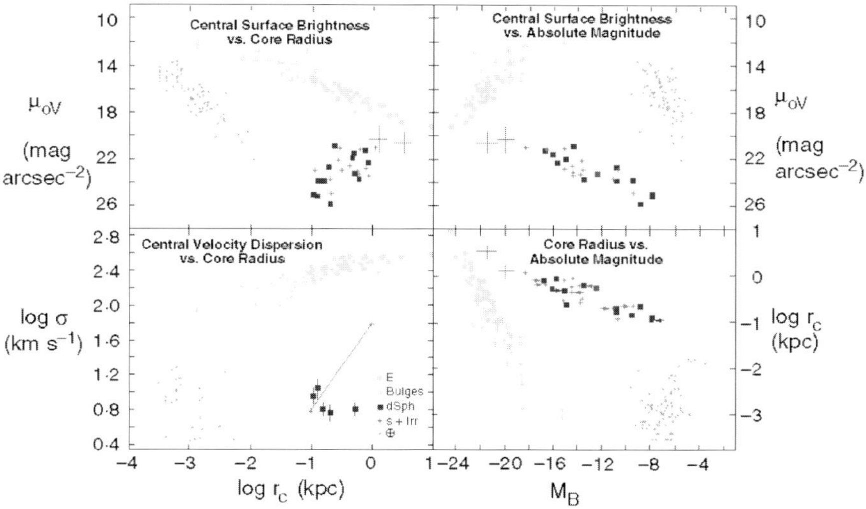

Fig. 3. Fundamental plane of galaxies as relation between the observed parameters: central surface brightness μ_c (in mag/arcsec^2), the stellar core radius R_c, absolute magnitude M_B of the core and the stellar velocity dispersion σ_* (adapted from Kormendy).

by many orders of magnitude. The search for these remnants in the centers of nearby galaxies is a key project in astronomy and can be done in various ways.

A massive central Black Hole significantly influences the motion of the surrounding stars and gas out to a radius

$$R_{\rm BH} = \frac{GM_{\rm H}}{\sigma_*^2} \simeq 10^6\, R_{\rm g} \simeq 100\,{\rm pc}\, \frac{M_{\rm H}}{2\times 10^9\, M_\odot} \left(\frac{\sigma_*}{300\,{\rm km\ s^{-1}}}\right)^{-2}, \qquad (3.11)$$

where σ_* denotes the characteristic velocity dispersion of stars in the center of the host galaxy. σ_* is observed to be $\simeq$100 to 350 km s^{-1} for elliptical and S0 galaxies. This amounts to radii of about 100 parsecs in giant ellipticals such as M 87 and therefore to a scale of about one arcsecond in nearby galaxies. This scale can be resolved with modern HST observations. In this inner part, the velocity dispersion and circular velocities should diverge as $1/\sqrt{r}$ when $r \to 0$. As a result, high resolution photometric and kinematic observations are required to find direct dynamical evidence for the existence of massive central Black Holes [35].

3.2.1 Surface brightness profiles

Surface brightness cusps have been observed in many elliptical galaxies. The ground–based measurements have now been superseded by HST data (for a recent review of these data [13, 34, 56]). These cusps are consistent with the presence of central Black Holes, but cannot prove their existence. HST observations show that the observed surface brightness $I(R)$ can be well fitted with a two–power law model (for the centers of ellipticals and spiral bulges that are called *hot galaxies*)

$$I(R) = 2^{(\beta_1-\gamma_1)/\alpha_1}\, I_{\rm b} \left(\frac{R_{\rm b}}{r}\right)^{\gamma_1} \left[1 + \left(\frac{r}{R_{\rm b}}\right)^{\alpha_1}\right]^{(\gamma_1-\beta_1)/\alpha_1} . \tag{3.12}$$

The radius $R_{\rm b}$ represents a kind of core radius, inside $R_{\rm b}$ the asymptotic logarithmic slope is $-\gamma_1$, the asymptotic outer slope is $-\beta_1$ and the parameter α_1 parametrizes the sharpness of the break. $I_{\rm b}$ is the surface brightness at the radius $R_{\rm b}$. Two types of hot galaxies are found [13]:

- *Core galaxies* have a broken power–law surface brightness profile that changes the slope considerably at the break radius $R_{\rm b}$. The slope in the inner region (the core) is found in the region $0 \leq \gamma_1 < 0.3$. This indicates the existence of some cusp in the density distribution. Core galaxies are luminous objects and are often found in clusters. Black holes are found in this class of galaxies.

- *Power–law galaxies* have still fairly steep surface–brightness profiles in the center and show no significant break. The slope γ_1 is typically 0.8 and these galaxies are generally fainter than core galaxies. The small Virgo ellipticals and the bulges of disk galaxies are of this type.

This surface profile can be deprojected to give a stellar density profile, nowadays called Zhao–profile [68]

$$\rho_*(r) = \frac{M_*}{4\pi R_{\rm b}^3} \frac{C(\alpha,\beta,\gamma)}{r^\gamma\,(1+r^\alpha)^{(\beta-\gamma)/\alpha}} . \tag{3.13}$$

C is a normalisation constant which depends on the slopes of the profile, $r = r/R_B$. For $\beta \leq 3$, the total mass M_* in general diverges. The set of parameters $(\alpha_1, \beta_1, \gamma_1)$ depends in a complicated way on the density parameters (α, β, γ) [68] – except for the asymptotic behavior where $\beta = \beta_1 + 1$. Special examples are:

- Plummer profile: $\alpha = 2$, $\gamma = 0$ and $\beta = 5$ (too steep in the asymptotic region);

- Hernquist profile (1992): $\alpha = 1$, $\gamma = 1$ and $\beta = 4$;

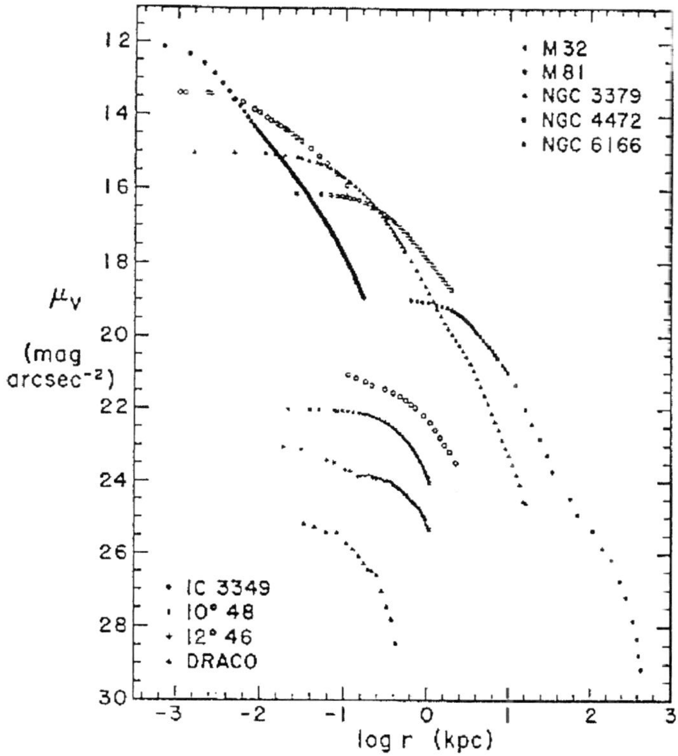

Fig. 4. Surface brightness profiles for various galaxies, ranging from very compact ellipticals (M 32) to cD galaxies in clusters NGC 6166 (adapted from Kormendy).

- Jaffe profile (1983): $\alpha = 1$, $\gamma = 2$ and $\beta = 4$;
- Dehnen profile (1992): $\alpha = 1$, γ free and $\beta = 4$;
- NFW profile (Navarro *et al.* 1997): $\alpha = 1$, $\gamma = 1$, $\beta = 3$. This profile adequately describes the density distribution of collisionless dark matter, as found in cosmological numerical simulations.

3.2.2 Stellar and gas spectroscopic observations

Spectroscopic observations have provided also hints for a central dark mass in the nuclei of dozen nearby galaxies. The evidence rests on an inwards increase of the line–of–sight velocity dispersion σ_* and on a steep central gradient in the mean light–of–sight velocity $\langle v_{\rm los} \rangle$. Such a gradient is expected near a Black Hole, but it could also be the signature of a rapidly rotating nuclear stellar disk.

The compact E3 galaxy M 32 has been carefully analyzed for the presence of a nuclear Black Hole. Though it shows no sign of nuclear activity or emission–line gas, it harbours a Black Hole of $3 \times 10^6\ M_\odot$ [34].

M 87 (E0 galaxy) is probably the archetypical object for an old quasar residing in a big cluster of galaxies. The stellar velocity dispersion increases from 270 $\mathrm{km\,s^{-1}}$ at $15''$ to 400 $\mathrm{km\,s^{-1}}$ at $0\farcs5$ from the nucleus. The emission–line gas in the nucleus of M 87 shows evidence for bulk motion and for a strong inwards increase of the gas velocity [40]. HST measurements have now shown that this gas resides in a small nuclear disk rotating at a speed of 550 $\mathrm{km\,s^{-1}}$ at a distance of $0\farcs26$ from the nucleus (corresponding to about 20 pc from the nucleus). This makes the case for a $3 \times 10^9\ M_\odot$ Black Hole in M 87 very compelling.

The best Black Hole candidate is the S0 galaxy NGC 3115 that shows evidence for a mass of $2 \times 10^9\ M_\odot$ [35]. Again, there is no obvious activity associated with this mass.

3.2.3 Maser disks

In a very few cases, it has been possible to infer the mass and volume of the central mass concentration within an AGN by observing the orbital dynamics of the mega-masers within the galaxy. This is of great scientific interest as the quantities which are obtained can only be explained by invoking the presence of a Black Hole, thus providing the most definite evidence to date of these exotic objects [47]. Aside from the possibility of Black Hole detections, with accurate knowledge of the masers' velocities, we can measure their proper motion as they pass in front of the nucleus (*i.e.* when their radial velocity is equal to the systemic velocity of the galaxy). By combining the measured Doppler velocities of the masers with the time to transverse an angular distance, we obtain the linear distance travelled. This angular to linear conversion factor allows a direct measurement of the distance to the galaxy. For example, in the case of NGC 4258, this method has reduced the uncertainty in the galaxy's distance from $\simeq$50% to 4%, thus having major implications in the determination of the cosmic distance scale.

On this scale of 0.1 pc, the physical conditions are favourable ($T \geq 100$ K, $\rho \simeq 10^{10}$ $\mathrm{cm^{-3}}$) for the molecules to mase. Since the amplification factor varies exponentially with the path length, masers are seen where the path has large velocity coherence, *i.e.* where the velocity varies only slightly across the molecular disk, upon which the masers are located. This occurs at three points; directly in front of radio continuum source, where the radial velocity of the disk is equal to that of the galaxy, and at the two tangent points on the disk, where the projected velocity gradient across the

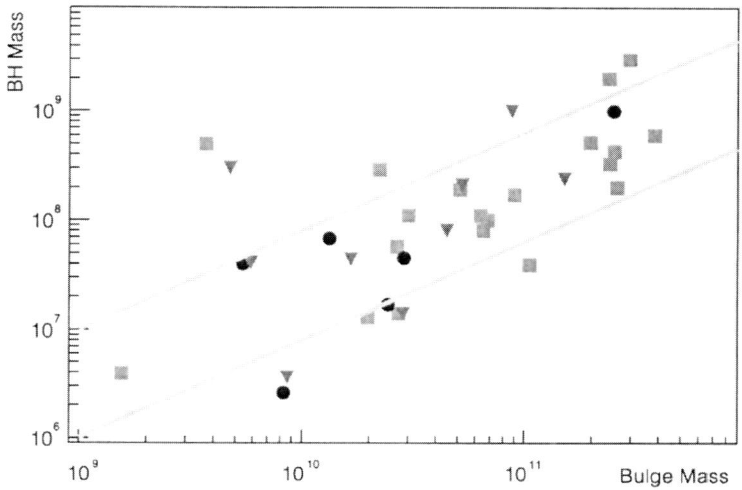

Fig. 5. BH masses (in solar masses) as a function of the bulge mass in nearby galaxies. Filled squares are dark masses in ellipticals, filled circles in spirals and triangles in S0s. Data are from [35].

disk is a minimum. Not lying directly between us and the radio continuum source, the tangent point masers appear much weaker than those directly in the path since they are the result of maser amplification of the molecules' own spontaneous emissions and not those of the central source. When the velocities of the masers are measured, unlike the molecular ring structures on larger scales, whose enclosed mass is comprised of stars and dust, the masers are seen to follow Keplerian orbits, implying that the dynamics are dominated by a compact object. The enclosed mass measured within the maser orbits then constrains the mass of the central object.

3.2.4 Mass correlations

In 2002, data are available for about 45 galaxies [35] (Fig. 5). There is a loose correlation between the Black Hole mass and the mass contained in the galactic nucleus (bulge). The scatter is appreciable indicating that the dark mass is in a way related to the bulge, but not to the entire mass of a galaxy. In particular, the mass contained in the disk has nothing to do with the Black Hole mass. It has recently been found that there is a tighter correlation between the central Black Hole mass and the velocity dispersion σ_e of the bulge (Fig. 6) [14,15,65]. This sheds some light on the evolutionary history and the formation of bulges and Black Holes. There is some evidence

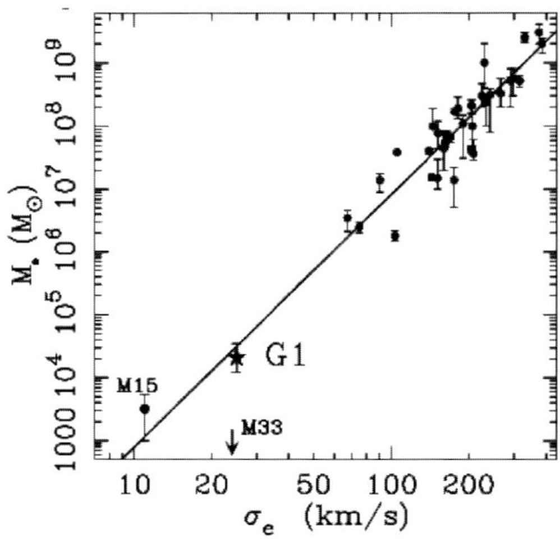

Fig. 6. BH masses as a function of the velocity dispersion in galactic nuclei, adapted from [65]. Included are the upper limit for M 33, the Black Hole mass estimate for the globular cluster M 15, and the mass for the cluster G1 in M 31. The solid line is the linear fit given by [65], not including G1 and M 15, $M_{\rm H} \propto \sigma_{\rm e}^{4.02}$.

that at least some globular clusters may be nuclei of accreted galaxies, and thus may contain Black Holes if all galaxies contain them. The clusters G1 in M 31 and M 15 are excellent candidates for such a search. They both have high central densities and short relaxation times. M 15 likely contains a Black Hole of a few thousand solar masses.

3.3 *The Fermion ball hypothesis*

Self–gravitating degenerate neutrino matter has been suggested as a model for quasars, with neutrino masses in the 0.2 keV $\leq m_\nu \leq$ 0.5 MeV range, at the same time when the Black Hole hypothesis of the quasars was deceived [44]. More recently, supermassive compact dark objects consisting of weakly interacting degenerate fermionic matter, with Fermion masses in the 10 keV $\leq m_\nu \leq$ 20 keV range have been proposed by many authors as an alternative to the supermassive Black Holes [4]. Since the most massive dark objects ever observed is in M 87 in the Virgo cluster with a mass of $3 \times 10^9\, M_\odot$, this has been identified with a degenerate Fermionic ball at the Oppenheimer–Volkoff limit

$$M_{\rm OV} = 0.54\, \frac{M_{\rm Pl}^3}{m_\nu^2}\, \frac{1}{\sqrt{g}} \simeq 3 \times 10^9\, M_\odot \tag{3.14}$$

for $m_\nu \simeq 15$ keV and a spin–degeneracy factor $g = 2$. $M_{\rm Pl} = \sqrt{\hbar c/G}$ denotes the Planck mass. Such an object would have a radius $R_{\rm OV} = 4.45 R_{\rm S} \simeq 1.5$ light days. It would be virtually indistinguishable from a Black Hole of the same mass.

Lower mass objects are then related to much more extended objects [4]. The dark mass in the Galactic center would have a radius of 7×10^4 Schwarzschild radii, or 21 light days. Such a non–relativistic object is then far from a Black Hole. The observed motions of stars within a projected distance of 5 to 50 light days from Sag A* yields an upper limit for the radius of the fermion ball of 22 light days. This alternative is probably now ruled out by the most recent observations of stellar orbits in the Galactic center [59].

3.4 Messier 87 as a dead quasar

M 87 is a well observed active galaxy [58] and you could deliver a whole lecture only on this object. M 87 is the central galaxy in the Virgo cluster and as such is surrounded by hot gas emitting in X–rays. The profile of this gas density has been well determined from Chandra observations. The core of M 87 is not only filled up with stars of a total mass of $M_* \simeq 10^{11}\, M_\odot$, but also with a hot interstellar medium of temperature $kT_{\rm X} \simeq 0.8$ keV and a central density of $n_{\rm e} \simeq 0.2\ {\rm cm}^{-3}$. This gas distribution also has a core radius of about one kpc, so that the total gas mass in the core is $M_{\rm c} \simeq 2 \times 10^7\, M_\odot$. This is only a small fraction of the stellar mass, in view of a mass–injection rate in the core of presently $\dot{M}_{\rm inj} \simeq 0.1\, M_\odot\ {\rm yr}^{-1}$, depending somewhat on the initial mass function in the core. The core gas mass is however exactly $M_{\rm c} \simeq \dot{M}_{\rm inj}\, t_{\rm cool}$ with a cooling time $t_{\rm cool} \simeq 10^8$ yrs. This would mean that the gas in the core only stays hot for the length of the cooling time and then settles down into cold gas with temperature below 10^4 K. This cool gas has been detected in M 87 in form of the cool disk surrounding the central part with a diameter of $\simeq$2 arcsecs corresponding to 150 parsecs. The cool gas is also visible in many filaments emitting in optical emission lines [58].

The present mass in this disk is smaller than the gas mass in the core, probably due to enhanced accretion phases towards the central Black Hole. The gas settles into a disk [40], since the stars have some low angular momentum with rotational velocities of $\simeq$10 km s^{-1}, corresponding to a specific angular momentum $j_* \simeq R_{\rm D} V_{\rm rot} \simeq 10^{27}\ {\rm cm}^2\ {\rm s}^{-1}$. This is in fact exactly the specific angular momentum of a Keplerian accretion disk around the central mass of $M_{\rm H} = 3 \times 10^9\, M_\odot$, $j_{\rm K} = \sqrt{G M_{\rm H} R_{\rm D}} = c R_{\rm D} \sqrt{R_{\rm g}/R_{\rm D}} \simeq 3 \times 10^{27}\ {\rm cm}^2\ {\rm s}^{-1}\ \sqrt{R_{\rm D}/100\,{\rm pc}}$ [7]. Accretion onto the central Black Hole now occurs from this disk.

Many people have argued that the present accretion rate in M 87 is given by Bondi accretion [10, 11, 54] with accretion radius

$$R_{\rm A} = b\,\frac{GM_{\rm H}}{2c_{\rm S}^2} \simeq 25\,{\rm pc}\,T_{0.8}^{-1}\,M_{\rm H,9} \simeq 5\times 10^5\,R_{\rm g} \tag{3.15}$$

well within the radius of influence $R_{\rm BH}$. The accretion radius is taken to be the sonic radius in Bondi accretion, and the sound speed $c_{\rm S} \simeq 0.1$ km s^{-1} $\sqrt{T}$. The parameter b depends on the detailed physics near the accretion radius. The accretion rate is then related to density and sound speed at the sonic radius

$$\dot{M}_{\rm Bondi} = 4\pi R_{\rm A}^2 \rho(R_{\rm A}) c_{\rm S}(R_{\rm A}). \tag{3.16}$$

The physical conditions in the core of M 87 would correspond to a Bondi accretion rate [11]

$$\dot{M}_{\rm Bondi} \simeq 6\times 10^{24}\,M_{\rm H,9.5}^2\,T_{0.8}^{-3/2}\,n_{0.17}\,{\rm g\,s^{-1}} \simeq 0.1\,M_\odot\,{\rm yr}^{-1} \simeq 10^{-3}\,\dot{M}_{\rm Edd}. \tag{3.17}$$

At $\dot{m} \simeq 10^{-3}$ we would not expect a standard accretion disk near the horizon. Bondi accretion is however largely inhibited by the outflow, so that accretion from the disk is the main source of mass provided to the Black Hole. This accretion rate is driven by turbulence in the parsec–scale disk or by angular momentum extraction via magnetic fields. This inflow rate is however in agreement with an outflow rate of $10^{-3}\,M_\odot$ yr^{-1} [7].

The X–ray luminosity of the active nucleus in M 87 has been measured by Chandra [11] to $L_{0.5-7\,{\rm keV}} \simeq 7\times 10^{40}$ erg/s with a steep photon index $\Gamma_{\rm X} = 2.2$. This would be about 4 orders of magnitude less than the luminosity expected from standard accretion

$$L_{\rm Bondi} = \eta \dot{M}_{\rm Bondi} c^2 \simeq 5\times 10^{44}\,{\rm erg\,s^{-1}}. \tag{3.18}$$

The core spectrum of M 87 is probably determined by a hot magnetized ion torus around the Black Hole which cools by the emission of cyclotron photons in the submm–range and inverse Compton emission in the X–rays. The typical frequency in the cyclotron spectrum is at high harmonics of the cyclotron frequency $\nu_{\rm B} = 3\times 10^6\,B$ Hz $\simeq 10^{10}$ Hz for $B \simeq 1000$ Gauss. The density can be estimated at the marginal stable orbit

$$n \simeq \frac{2\dot{M}}{\pi R_{\rm ms}^2 c m_{\rm p}} \simeq 10^8\,{\rm cm}^{-3}\,\dot{M}_{24}. \tag{3.19}$$

This corresponds to an ion pressure near the horizon of $P_{\rm ion} \simeq 10^4$ cgs and therefore to an equipartition field strength of $\simeq$1000 Gauss. This high magnetic field would however lead to an instantaneous cyclotron cooling of the electrons with a time–scale, $\Theta_\mathrm{e} \equiv kT_\mathrm{e}/m_\mathrm{e}c^2$,

$$t_\mathrm{cyc} = \frac{\pi m_\mathrm{e} c^2}{\sigma_\mathrm{T} c B^2 \Theta_\mathrm{e}} \simeq 1000\,\mathrm{s}\,\frac{1}{B_2^2 \Theta_\mathrm{e}} . \tag{3.20}$$

Compared to the inflow time–scale this is by far too short to keep the electrons very hot, *i.e.* at $\Theta_\mathrm{e} \simeq 1$. This short cooling would in addition lead to a too high cyclotron emissivity given by,

$$\epsilon_\mathrm{c} = 16\Theta_\mathrm{e}^2\, n_\mathrm{e}\sigma_\mathrm{T} c\,\frac{B^2}{8\pi} \simeq 0.1\,\mathrm{erg\,cm^{-3}\,s^{-1}}\,\Theta_\mathrm{e}^2 B_2^2 \tag{3.21}$$

with a corresponding cyclotron luminosity from the ion torus

$$L_\mathrm{cyc} = V_\mathrm{T}\epsilon_\mathrm{c} \simeq 10^{44}\,\mathrm{erg\,s^{-1}}\,\Theta_\mathrm{e}^2 B_2^2 , \tag{3.22}$$

with the volume $V_\mathrm{T} \simeq \pi R_\mathrm{ms}^3 \simeq 10^{45}\ \mathrm{cm}^3$. The typical field strength near the horizon is therefore probably in the range of 10–100 Gauss. In view of the observed sub–mm–luminosities, the sub–mm–IR emission is therefore not produced by cyclotron cooling from the ion torus, or inner ADAF (as proposed by [11]). It is more probable from cyclotron cooling at the shock produced in the outflow on the scale of the light cylinder, where the typical field strength has now decayed to $B_\mathrm{J} \simeq 1$ Gauss and the electrons are heated by shock heating to a temperature $\Theta_\mathrm{e} \simeq 1000$, corresponding now to a cyclotron frequency of about 10^{13} Hz when a slight beaming is also involved. X–ray emission is then probably IC emission from this region as proposed by [67] for the Galactic center emission.

Nuclear disks have now been detected in many nearby ellipticals. They always contain a mass $M_\mathrm{D} \simeq 10^6\, M_\odot$ and a lot of dust. This mass is always a fraction of the total interstellar mass in the core of the elliptical. Since only a fraction of the ellipticals is active, the accumulation of cold gas in the center of the core must be an intermittent process.

3.5 Cosmology for quasars

Quantitative analysis of quasars at redshift $z > 1$ cannot be done without invoking a particular cosmological model. Besides the redshift, which is an expression for the expansion of the Universe, $1 + z = R_0/R(t)$ for given expansion factor $R(t)$, the particular form of $R(t)$ enters essentially into

- **luminosity distance** $d_\mathrm{L}(z)$ which is defined by the observed flux f_B in some wavelength filter and the corresponding intrinsic luminosity L_B

$$f_B = \frac{L_B}{4\pi d_\mathrm{L}^2(z)} ; \tag{3.23}$$

- **angular width** Θ of an object of dimension D (a core or jet *e.g.*)

$$\Theta = \frac{D}{d_{\rm A}(z)} = \frac{D(1+z)^2}{d_{\rm L}(z)} \tag{3.24}$$

  for given angular distance $d_{\rm A}(z)$;

- **number counts** of objects in a redshift interval $\mathrm{d}N/\mathrm{d}z\,\mathrm{d}\Omega$ for given solid angle $\mathrm{d}\Omega$ (galaxies, quasars etc.).

The expansion factor $R(t)$ is determined by the Friedmann equation for the normalized expansion factor $a(t) = R(t)/R_0 = 1/(1+z)$

$$\frac{\mathrm{d}a}{\mathrm{d}t} = H_0 \sqrt{\frac{\Omega_{\rm M}}{a} + \Omega_\Lambda\, a^2 + 1 - \Omega_{\rm M} - \Omega_\Lambda}. \tag{3.25}$$

The present Universe is determined by the Hubble constant $H_0 = (68 \pm 10)$ km/Mpc/s and the three density parameters

- **Matter density parameter:**

$$\Omega_{\rm M} \equiv \frac{8\pi G}{3H_0^2}\,\rho_{\rm M,0} = \frac{\rho_{\rm M,0}}{\rho_{\rm crit}} \tag{3.26}$$

- **Vacuum energy parameter:**

$$\Omega_\Lambda = \frac{\Lambda c^2}{3H_0^2} \tag{3.27}$$

- **Curvature parameter:**

$$\Omega_k = -\frac{kc^2}{R_0^2 H_0^2}. \tag{3.28}$$

On behalf of the Friedmann equation we find

$$\Omega_{\rm tot} = \Omega_{\rm M} + \Omega_\Lambda + \Omega_k = 1. \tag{3.29}$$

There is abundant evidence for the dominance of dark matter and dark energy on the largest distance scales [50]. The cosmological microwave background (CMB) radiation tells us that the total energy density of the Universe, $\Omega_{\rm tot}$, is very close to the critical value marking the boundary between open and closed universes. This information is provided, in particular, by the value of the multipole $l \simeq 210$ at which the first acoustic peak appears in the CMB. This tells us, in effect, the relative sizes of the Universe today and when the nuclei and free electrons in the primordial plasma combined to form neutral atoms. There are now indications for a second and even a third acoustic peak in the CMB at higher l. The magnitudes of the fluctuations $\Delta T/T$ at these larger values of l already tell us that the overall baryon density $\Omega_{\rm b} \ll 1$, agreeing to within 50% with the value estimated on the basis of Big Bang nucleosynthesis calculations. Other information about the large-scale geometry of the Universe for redshifts $z \leq 1$ is provided by data on high-z supernovae, which constrain a combination of the matter density $\Omega_{\rm m}$ and the vacuum energy density Ω_Λ. Combining the CMB and high-z supernova data, one finds fairly accurate values for the cosmological density parameters (Fig. 7):

$$\Omega_{\rm tot} = 1.02 \pm 0.06, \quad \Omega_{\rm m} h^2 = 0.13 \pm 0.05 \tag{3.30}$$

$$\Omega_\Lambda = 0.5 \pm 0.2, \quad \Omega_{\rm b} h^2 = 0.022 \pm 0.004. \tag{3.31}$$

The standard paradigm is that large-scale structures in the Universe are formed by gravitational instabilities, building on the primordial density perturbations observed in the CMB, with baryons falling into the "holes" that are amplified by cold dark matter. Galaxy formation is considered to be more complex than cluster formation, with nonlinear astrophysical processes coming into play. The general belief is that clusters formed before galaxies, which were formed by mergers of smaller structures.

The distance module is a fundamental quantity for Quasars

$$DM \equiv m(z) - M = 5\,\log d_{\rm L}(z; \Omega_{\rm M}, \Omega_\Lambda, H_0) + 25, \tag{3.32}$$

which relates the apparent magnitude $m(z)$ with the absolute magnitude M and the distance $d_{\rm L}$[Mpc]. This is calculated from $d_{\rm L}(z) = r_1 R_0 (1+z)$, where r_1 is the solution of the equation

$$\begin{aligned} R_0 \int_0^{r_1} \frac{{\rm d}r}{\sqrt{1-kr^2}} &= R_0 \int_{t_1}^{t_0} \frac{c\,{\rm d}t}{R(t)} \\ &= \frac{c}{H_0} \int_{t_1}^{t_0} \frac{{\rm d}a}{a\sqrt{\Omega_{\rm M}/a + \Omega_\Lambda a^2 + 1 - \Omega_{\rm M} - \Omega_\lambda}}. \end{aligned} \tag{3.33}$$

Since ${\rm d}a = -{\rm d}z/(1+z)^2$, one obtains

$$3.34 \frac{R_0 {\rm d}r}{\sqrt{1-kr^2}} = -\frac{c}{H_0} \frac{{\rm d}z}{E(z)} \tag{3.34}$$

with

$$E(z) \equiv \sqrt{\Omega_k (1+z)^2 + \Omega_{\rm M} (1+z)^3 + \Omega_\Lambda} \tag{3.35}$$

and $E(0) = 1$. This provides us the solution for the luminosity distance

$$d_{\rm L}(z; \Omega_{\rm M}, \Omega_\Lambda, H_0) = \frac{c(1+z)}{H_0 \sqrt{|\Omega_k|}}\, \mathcal{S}(x[z]), \tag{3.36}$$

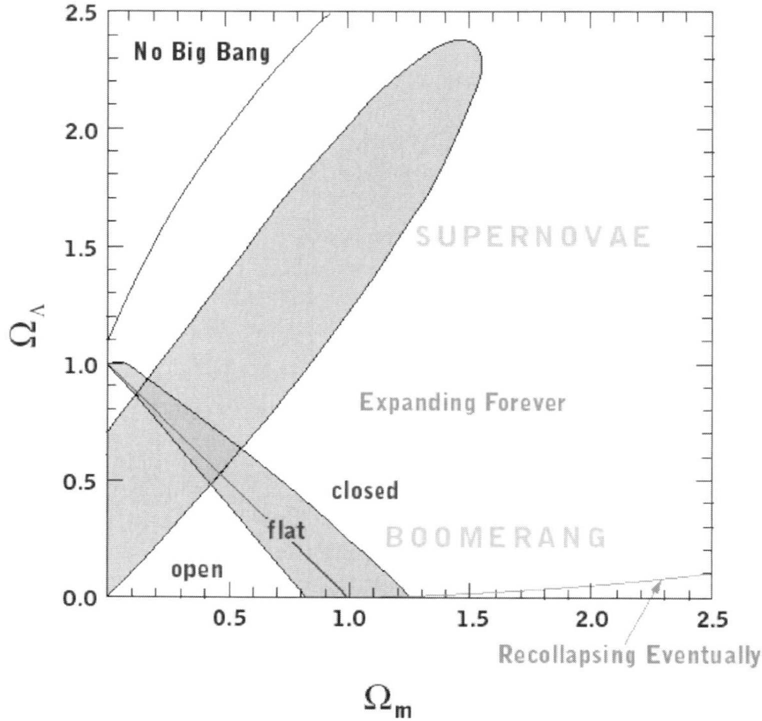

Fig. 7. The fundamental plane of cosmology.

where

$$x[z] = \sqrt{|\Omega_k|} \int_0^z \frac{\mathrm{d}z'}{\sqrt{(1+z')^2(1+\Omega_\mathrm{M} z') - z'(2+z')\Omega_\Lambda}} . \tag{3.37}$$

$\mathcal{S}(x)$ denotes the functions $\sinh(x)$, x oder $\sin(x)$ depending on the curvature of space, *i.e.* for $\Omega_k > 0$, $\Omega_k = 0$ or $\Omega_k < 0$, respectively. For a flat Universe (ΛCDM), we obtain simply

$$d_\mathrm{L}(z; \Omega_\mathrm{M}, \Omega_\Lambda, H_0) = \frac{c(1+z)}{H_0} \int_0^z \frac{\mathrm{d}z'}{\sqrt{\Omega_\mathrm{M}(1+z')^3 + 1 - \Omega_\mathrm{M}}} . \tag{3.38}$$

This has as a limit the famous luminosity distance in a flat classical Friedmann Universe, $\Omega_\Lambda = 0$,

$$d_\mathrm{L}(z; H_0) = \frac{2c}{H_0} \left[1 + z - \sqrt{1+z}\right] . \tag{3.39}$$

3.6 Cosmology with quasars

Quasars at high redshifts can be used to study the universe early in its history, and absorption features from intervening objects can yield information about conditions subsequent to the time when the quasar emitted its light in our direction [38, 39]. Furthermore the spatial and luminosity distribution functions of these high-z quasars can shed light on the formation of structure in the early universe. The Sloan Digital Sky Survey (SDSS) has the potential to identify more of these objects than any survey to date, and thus provide an unprecedented sample for followup study.

Because high-redshift quasars are receeding at relativistic velocities, the Lyman alpha emission line is shifted from the ultraviolet into the red region of the electromagnetic spectrum. The Next Generation Space Telescope (JWST) will be optimized in the infrared in order to detect very distant objects at redshifts greater than $z = 5$. The properties of the intergalactic medium can be studied back to the time of galaxy formation using distant quasars. The number density of absorption lines as a function of epoch will provide an important cosmological test. Observations of correlated absorption, both in redshift and in nearby quasars on the plane of the sky, are leading to important information on the evolution of the large-scale distribution of galaxies in space. The Lyman alpha clouds offer the opportunity to investigate the evolution of the intergalactic medium and the meta-galactic ionizing radiation flux and to provide information on primordial abundances of the elements.

4 Gravity of compact objects

For non–rotating objects, the Schwarzschild metric is the correct description of the gravitational field. In general, this has only two free functions

$$\mathrm{d}s^2 = -\alpha^2\,\mathrm{d}t^2 + g_{rr}\mathrm{d}r^2 + r^2\,(\mathrm{d}\theta^2 + \sin^2\theta\,\mathrm{d}\phi^2). \tag{4.1}$$

where α is the redshift factor. The external gravitational field is then only given by the mass M of the object

$$\alpha^2 = 1 - \frac{2GM}{c^2 r} \tag{4.2}$$

and $g_{rr} = 1/\alpha^2$. For collapsed objects, this metric has a horizon with vanishing redshift factor $\alpha(r_\mathrm{H}) = 0$, $r_\mathrm{H} = 2R_\mathrm{g}$, where $R_\mathrm{g} \equiv GM/c^2$ denotes the gravitational radius of mass M, which is numerically given by $R_\mathrm{g} = 1.48$ km $(M/M_\odot)$. So a supermassive Black Hole of $10^9\,M_\odot$ has a gravitational radius of 10 AU. A Black Hole of this mass represents a true hole in space with a radius out to Jupiter's orbit. The Newtonian concept of a gravitational force is then given by the expression

$$\boldsymbol{g} = -\nabla \ln \alpha. \tag{4.3}$$

Most of the objects in the Universe are, however, rotating so that the Schwarzschild solution is no longer a correct representation of their gravitational fields.

4.1 Relativistic gravity of rotating neutron stars

The most general form of a stationary and axisymmetric metric has been given by Papapetrou [49]

$$\mathrm{d}s^2 = -f(\mathrm{d}t - w\,\mathrm{d}\phi)^2 + f^{-1}[\exp(2\gamma)(\mathrm{d}\rho^2 + \mathrm{d}z^2) + \rho^2\mathrm{d}\phi^2], \tag{4.4}$$

where ρ and z are quasi–cylindrical Weyl–Papapetrou coordinates. f, γ and w are functions independent of t and ϕ. Solutions to Einstein's equations can be given then in terms of two complex potentials (the so–called Ernst potentials) $\mathcal{E}$ and Φ [12]

$$\Phi = A_{\mathrm{t}} + i\phi' \tag{4.5}$$

$$\mathcal{E} = (f - |\Phi|^2) + i\phi \tag{4.6}$$

where the potentials ϕ and ϕ' are defined as

$$\rho^{-1} f^2 \nabla w - 2\boldsymbol{n} \times Im(\Phi^* \nabla \Phi) = \boldsymbol{n} \times \nabla\phi \tag{4.7}$$

$$\rho^{-1} f(\nabla A_\phi - w \nabla A_{\mathrm{t}}) = \boldsymbol{n} \times \nabla\phi'. \tag{4.8}$$

$\boldsymbol{n}$ is the unit vector in azimuthal direction. The complex potentials satisfy the coupled Ernst–equations

$$(Re\mathcal{E} + |\Phi|^2)\nabla^2\mathcal{E} = (\nabla\mathcal{E} + 2\Phi^*\nabla\Phi)\cdot\nabla\mathcal{E} \tag{4.9}$$

$$(Re\mathcal{E} + |\Phi|^2)\nabla^2\Phi = (\nabla\mathcal{E} + 2\Phi^*\nabla\Phi)\cdot\nabla\Phi\,. \tag{4.10}$$

The metric functions are then recovered from the following relations

$$f = Re\mathcal{E} + \Phi\Phi^* \tag{4.11}$$

$$w_{,\rho} = -\rho f^{-2}\,(\Phi_{,z} + Im(\Phi^*\Phi_{,z})) \tag{4.12}$$

$$w_{,z} = \rho f^{-2}\,(\Phi_{,\rho} + Im(\Phi^*\Phi_{,\rho})) \tag{4.13}$$

$$\begin{aligned}\gamma_{,\rho} = {} & \frac{1}{4}\rho f^{-2}\,[(\mathcal{E}_{,\rho} + 2Im(\Phi^*\Phi_{,\rho}))(\mathcal{E}^*_{,\rho} + 2Im(\Phi\Phi^*_{,\rho})) \\ & -(\mathcal{E}_{,z} + 2Im(\Phi^*\Phi_{,z})(\mathcal{E}^*_{,z} + 2Im(\Phi\Phi^*_{,z})] \\ & -\rho f^{-1}(\Phi_{,\rho}\Phi^*_{,\rho} - \Phi_{,z}\Phi^*_{,z})\end{aligned} \tag{4.14}$$

$$\begin{aligned}\gamma_{,z} = {} & \frac{1}{2}\rho f^{-2}\,Re[(\mathcal{E}_{,\rho} + 2\Phi^*\Phi_{,\rho})(\mathcal{E}^*_{,z} + 2\Phi\Phi^*_{,z})] \\ & -2\rho f^{-1} Re(\Phi^*_{,\rho}\Phi_{,z}).\end{aligned} \tag{4.15}$$

The two Ernst equations can then formally be solved by means of integral equations [60] which introduce the axis data of the Ernst potentials, $e(z) = \mathcal{E}(\rho = 0, z)$ and $f(z) = \Phi(\rho = 0, z)$.

To derive a new exact analytical solution of Einstein's equations one starts with an ansatz for $e(z)$ and $f(z)$, checks whether this solution has suitable properties (multipole moments) and derives the metric functions f, w and γ. In the last ten years new efforts have been done to find new solutions for rotating neutron stars, which would involve the essential parameters mass, mass quadrupole moment, angular momentum and a magnetic dipole moment [61]. Finally Manko *et al.* [41, 42] have found a suitable parametrisation of such a solution by using the ansatz

$$e(z) = \frac{(z - M - ia)(z + ib) + d - \delta - ab}{(z + M - ia)(z + ib) + d - \delta - ab} \tag{4.16}$$

$$f(z) = \frac{Qz + i\mu}{(z + M - ia)(z + ib) + d - \delta - ab} \tag{4.17}$$

with

$$\delta = \frac{\mu^2 - M^2 b^2}{M^2 - (a - b)^2 - Q^2} \tag{4.18}$$

$$d = \frac{1}{4}\,[M^2 - (a - b)^2 - Q^2]. \tag{4.19}$$

The parameter M represents the total mass, a the specific angular momentum, Q the total charge (= 0 for astrophysical bodies), μ is the magnetic dipole moment, and b is related to the mass quadrupole moment. In distinction to a Black Hole solution, the mass quadrupole moment Q_2 is not fixed, but is related to the rotation of the star and its internal structure. The Schwarzschild solution is contained in this family with

$$e(z) = \frac{z - M}{z + M}, \quad f(z) = 0, \quad Q = 0 = a = \mu, \quad b = iM. \tag{4.20}$$

The metric functions proposed by Manko *et al.* can be expressed as rational functions in terms of spheroidal coordinates

$$x = \frac{1}{2\kappa}\,(r_+ + r_-), \quad y = \frac{1}{2\kappa}\,(r_+ - r_-) \tag{4.21}$$

with $r_\pm = \sqrt{\rho^2 + (z \pm \kappa)^2}$ and $\kappa = \sqrt{d + \delta}$. In terms of spheroidal coordinates, the metric assumes the form

$$\begin{aligned} \mathrm{d}s^2 = {} & f(\mathrm{d}t - w\,\mathrm{d}\phi)^2 \\ & - \frac{\kappa^2}{f}\left[\exp(2\gamma)(x^2 - y^2)\left(\frac{\mathrm{d}x^2}{x^2 - 1} + \frac{\mathrm{d}y^2}{1 - y^2}\right) + (x^2 - 1)(1 - y^2)\,\mathrm{d}\phi^2\right]. \end{aligned} \tag{4.22}$$

The solutions are given in terms of the following representation

$$f = \frac{E}{D}, \quad \exp(2\gamma) = \frac{E}{16\kappa^8(x^2-y^2)^4}, \quad w = -\frac{(1-y^2)F}{E}. \tag{4.23}$$

The three functions D, E and F are now lengthy polynomials in the variables x and y, as well as in the parameters M, a, b and μ (for the expressions, see [63]).

Quasi–spherical Boyer–Lindquist coordinates: for astrophysical applications, it is more appropriate to transform the metric into Boyer–Lindquist type quasi–spherical coordinates [63]

$$\mathrm{d}s^2 = -\alpha^2\,\mathrm{d}t^2 + h_{\phi\phi}(\mathrm{d}\phi - \omega\,\mathrm{d}t)^2 + h_{rr}\mathrm{d}r^2 + h_{\theta\theta}\mathrm{d}\theta^2 \tag{4.24}$$

with the following coordinate tranformations

$$r(\rho,z) = M + \frac{1}{2}\left(\sqrt{\rho^2+(z+\kappa)^2} + \sqrt{\rho^2+(z-\kappa)^2}\right) \tag{4.25}$$

$$\theta(\rho,z) = \arccos\frac{1}{2\kappa}\left(\sqrt{\rho^2+(z+\kappa)^2} - \sqrt{\rho^2+(z-\kappa)^2}\right). \tag{4.26}$$

α is the redshift factor, $\sqrt{h_{\phi\phi}}$ relates to the cylindrical radius and ω is the frame–dragging frequency. These functions are given by

$$\alpha^2 = \frac{f[(r-M)^2-\kappa^2]\sin^2\theta}{[(r-M)^2-\kappa^2]\sin^2\theta - f^2w^2} \tag{4.27}$$

$$\omega = \frac{f^2 w}{f^2w^2 - [(r-M)^2-\kappa^2]\sin^2\theta} \tag{4.28}$$

$$h_{rr} = \frac{\exp(2\gamma)[(r-M)^2-\kappa^2\cos^2\theta]}{f[(r-M)^2-\kappa^2} \tag{4.29}$$

$$h_{\theta\theta} = \frac{\exp(2\gamma)[(r-M)^2-\kappa^2\cos^2\theta]}{f} \tag{4.30}$$

$$h_{\phi\phi} = \frac{[(r-M)^2-\kappa^2]\sin^2\theta - f^2w^2}{f}. \tag{4.31}$$

In these quasi–spherical coordinates, the exterior analytic solution can be fitted to interior numerical solutions of Einstein's equations for given rotation and EOS. In [63] we have determined the mass M of a rotating neutron star

$$M(j) = M_0 + M_1 j^2 + M_2 j^4 + M_3 j^6 \tag{4.32}$$

for given specific angular momentum $j = J/M^2 = a/M$. The parameter b can then suitably be fitted by the ansatz

$$b = \sqrt{\Delta_1 a^2 - M^2 + \Delta_2\frac{a^4}{M^2} + \Delta_3\frac{a^6}{M^4}}. \tag{4.33}$$

This ansatz ensures $b(j) = b(-j)$ and includes the Schwarzschild solution for $b = iM$. The parameters $M_{\rm A}$ and $\Delta_{\rm A}$ are given for various EOS and masses in [63].

In the asymptotic region, $r \gg R_{\rm g}$, the redshift factor is determined in lowest order by the gravitational mass M and the quadrupole moment Q_2

$$\alpha = 1 - \frac{GM}{r} + \frac{GQ_2}{r^3} P_2(\cos\theta) + O(1/r^4). \tag{4.34}$$

Similarly, the frame–dragging frequency ω is given in lowest order of approximation by the angular momentum J_* of the star

$$\omega = \frac{2GJ_*}{r^3} + O(R_{\rm g}/r)^4. \tag{4.35}$$

This potential drops very rapidly, as one moves away from the stellar surface. For neutron stars the angular momentum is suitably given in terms of mass, radius and angular frequency $\Omega_* = 2\pi/P_*$

$$J_* = k^2 M R_*^2 \Omega_* \tag{4.36}$$

with $k^2 \simeq 0.4$ depending on the EOS. With this, we can relate the frame–dragging frequency to the spin frequency Ω_*

$$\omega(r,\theta) \simeq 2k^2 \frac{GM}{c^2 r} \left(\frac{R_*}{r}\right)^2 \Omega_* = \omega_* \left(\frac{R_*}{r}\right)^3 \tag{4.37}$$

with

$$\omega_* = 2k^2 \frac{2GM}{c^2 R_*} \Omega_* \simeq 0.4\Omega_* \tag{4.38}$$

for a standard neutron star, where $R_* \simeq 4R_{\rm g}$. For a quark star, this could be a factor 2 higher. For a neutron star rotating with 2 ms in a LMXB the corresponding frame–dragging frequency $p_* = 2\pi/\omega_*$ would be about 5 ms.

In the following we discuss two important effects related to relativistic gravity.

Gravitational redshift: emission form the surface of a neutron star or from the accretion disk near the surface will be redshifted by gravitational effects. Photons emitted at the equator in forward (backward) direction with respect to the rotational direction have a 4–momentum

$$k^\mu = \text{const} \times \left(\xi_{\rm t}^\mu + \frac{f}{fw \pm \rho} \xi_\phi^\mu\right) \tag{4.39}$$

where $\xi_{\rm t}^\mu$ and ξ_ϕ^μ are the two Killing vectors defined by stationarity and axisymmetry. The frequency of the emitter is then

$$\nu_{\rm E} = k^\mu U_\mu \tag{4.40}$$

where U^μ is the emitter's 4–velocity

$$U^\mu = \frac{1}{\sqrt{f}\,\sqrt{1-v^2}}\,(\xi^\mu_{\rm t} + \Omega\xi^\mu_\phi) \tag{4.41}$$

v represents the fluid velocity measured by a ZAMO

$$v = \sqrt{2w\Omega - \Omega^2(w^2 - \rho^2/f^2)}. \tag{4.42}$$

The redshift measured at infinity is then

$$z = \frac{k^\mu U_\mu}{k^\mu \xi^\mu_{\rm t}} - 1. \tag{4.43}$$

With this relation, the redshift of photons emitted from the equator can be calculated for a given neutron star structure [63].

Existence of an innermost stable orbit: a particle moving in the background metric is characterized by three constants of motion: the mass $m_{\rm p}$ of the particle, the total energy E and the total angular momentum L. For motions in the equatorial plane the equations of motion are therefore fully integrable. The motion is given by an effective potential

$$V(\rho) \equiv \frac{\exp(2\gamma)}{f}\,\dot{\rho}^2 = \frac{E^2}{f} - \frac{f}{\rho^2}(L - Ew)^2 - 1. \tag{4.44}$$

For circular orbits we have two conditions

$$V(\rho) = 0 \tag{4.45}$$

$$\begin{aligned}\frac{\mathrm{d}V(\rho)}{\mathrm{d}\rho} &= \frac{2f(L-Ew)^2}{\rho^2} - \frac{(L-Ew)^2 f_{,\rho}}{\rho^2} \\ &\quad - \frac{E^2 f_{,\rho}}{f^2} + \frac{2Ef(L-Ew)w_{,\rho}}{\rho^2} = 0. \end{aligned} \tag{4.46}$$

This fixes the energy E and the angular momentum L for given radius ρ

$$E = \frac{\sqrt{f}}{\sqrt{1 - f^2\chi^2/\rho^2}} \tag{4.47}$$

$$L = E(w + \chi) \tag{4.48}$$

$$\chi = \frac{\rho[-f^2 w_{,\rho} - \sqrt{f^4 w^2_{,\rho} + \rho f_{,\rho}(2f - \rho f_{,\rho})}]}{f(2f - \rho f_{,\rho})}. \tag{4.49}$$

The condition for a marginally stable orbit is then $\mathrm{d}^2V(\rho)/\mathrm{d}\rho^2 = 0$. For slow rotation, $j < 0.1$, the position of the marginal stable orbit coincides with the expressions for the Kerr metric, while for faster rotation slightly higher values are obtained [63].

4.2 Rotating Black Holes

A Black Hole is a very special solution of Einstein's vacuum equations which has a global horizon, *i.e.* a surface of no return. This horizon acts as a surface which divides the space into an exterior region which is asymptotically flat (Minkowski space) and an interior region which is also vacuum, except for a ring singularity. The solution for a rotating Black Hole is usually also expressed in quasi–spherical Boyer–Lindquist coordinates [9, 28]

$$ds^2 = -\alpha^2 \,dt^2 + e^{2\Psi}(d\phi - \omega \,dt)^2 + e^{2\mu_2} \,dr^2 + e^{2\mu_3} \,d\theta^2 \tag{4.50}$$

with the following parametrisation

$$\alpha^2 = \frac{\rho^2 \Delta}{\Sigma^2} \tag{4.51}$$

$$\omega = \frac{2aMr}{\Sigma^2}, \quad e^{\Psi} = \frac{\Sigma}{\rho} \sin\theta \tag{4.52}$$

$$e^{\mu_2} = \frac{\rho}{\sqrt{\Delta}}, \quad e^{\mu_3} = \rho. \tag{4.53}$$

Thereby, we use the canonical functions

$$\Delta = r^2 - 2Mr + a^2, \quad \rho^2 = r^2 + a^2 \cos^2\theta \tag{4.54}$$

$$\Sigma^2 = (r^2 + a^2)^2 - a^2 \Delta \sin^2\theta. \tag{4.55}$$

With respect to the previous discussion, the quadrupole moment is not a free parameter, but is fixed by the rotation. Without charge, a Black Hole has only two parameters (**no hair theorem**): the mass M and the specific angular momentum a. This is in contrast to rotating neutron stars, which have at least the mass–quadrupole moment as a third parameter.

The horizon is located at a radius $r_{\rm H}$ where $\Delta(r_{\rm H}) = 0$, *i.e.*

$$r_{\rm H} = M + \sqrt{M^2 - a^2} \tag{4.56}$$

and the ergoregion is bounded by the surface, $r_{\rm H} \le r \le r_{\rm E}(\theta)$,

$$r_{\rm E}(\theta) = M^2 + \sqrt{M^2 - a^2 \cos^2\theta}. \tag{4.57}$$

The redshift now diverges at the horizon, so that spots rotating on the horizon will be never observable. The frame–dragging frequency can be written as

$$\omega(r,\theta) = \Omega_{\rm H} \left(\frac{r_{\rm H}}{r}\right)^3 \frac{4M^2}{r_{\rm H}^2} \frac{r^4}{(r^2 + a^2)^2 - a^2\Delta \sin^2\theta}. \tag{4.58}$$

$\Omega_{\rm H}$ is the spin frequency of the horizon

$$\Omega_{\rm H} = \frac{ca}{r_{\rm H}^2 + a^2} = \frac{1}{2}\frac{a}{M}\frac{c}{r_{\rm H}}. \tag{4.59}$$

For a stellar BH of 10 $M_\odot$ the corresponding period is

$$P_{\rm H} \simeq 0.8\,{\rm ms}\,\frac{M}{a}\,\frac{M_{\rm H}}{10\,M_\odot}, \tag{4.60}$$

for the BH in M 87 we find

$$P_{\rm H} \simeq 2\,{\rm day}\,\frac{M}{a}\,\frac{M_{\rm H}}{3\times 10^9\,M_\odot}. \tag{4.61}$$

4.3 The source of gravity for Black Holes – or what is inside?

The source of gravity for a rotating BH is a ring singularity located in the equatorial plane where the curvature tensor diverges. This ring collapses to a point for the non–rotating limit (Schwarzschild). Physically, this means that the source of gravity for a rotating BH is a mass–current located in the equatorial plane. This also means that this source of gravity is a mere mathematical construct; physically, the ring will be spread out at least by quantum effects. One could imagine that the total mass of the BH is generated by a collection of Planck particles with mass $M_{\rm Pl}$. A true understanding of this fact is still missing in BH physics, though many attempts have been undertaken in the last years. The existence of this singularity is also one of the reasons why the numerical simulations of merging BHs is so difficult – a computer does not allow to handle singularities, since a computer is a physical design! The singularity is a key feature of the Kerr solution; when you cut it out, the new space is no longer the Kerr solution.

Besides the existence of the singularity in the interior, the internal structure of a Black Hole even soon after its formation depends on the conditions on the event horizon at very distant future of the external observer. This would mean that its fate depends on collisions with other Black Holes, or on the fate of the Universe itself. This is somewhat uncomfortable!

Some new twist on this question appeared in the last years. As discussed previously, on the cosmological level vacuum energy is probably an essential ingredient for modelling the expansion of the Universe. If so, then vacuum energy could also play some major role in the interior of BHs – the interior region of a BH should then be given by a solution of Einstein's equations including vacuum energy. Mazur & Mottola [46] have in fact constructed a solution to this question for non–rotating objects, now called grava(c)stars (gravitational vacuum stars). They propose new solution for the endpoint of gravitational collapse. By extending the concept of Bose-Einstein condensation to gravitational systems, a cold, compact object with an interior de Sitter condensate phase and an exterior Schwarzschild geometry of arbitrary total mass M is constructed. These are separated by a phase boundary with a small but finite thickness of fluid (of the order of the Planck scale) with equation of state $P = \rho$, replacing the Schwarzschild horizon. The new solution has no singularities, no event horizons, and has a global time. Its entropy is maximized under small fluctuations and is given by the standard hydrodynamic entropy of the thin shell, instead of the Bekenstein-Hawking entropy. Unlike Black Holes, a collapsed star of this kind is thermodynamically stable and has no information paradox. The exterior solution is not affected by this constructiuon, except for the horizon which is no longer a surface with infinite redshift – **BHs are now gray**, the redshift is still huge, but finite.

4.4 Particle motion in Kerr

As for neutron stars, orbits in the equatorial plane are specified by three integrals of motion, the energy E, angular momentum L and mass μ. In the case of the Kerr solution, Carter has found in 1968 even a fourth constant of motion (based on the separability of the Hamilton Jacobi equations) so that the geodesic equations are reduced to a set of four first order differential equations. This is a very special property of the Kerr solution, not available in the more general solutions for neutron stars discussed in the previous section.

Particle motion can be derived from the Lagrangian $\mathcal{L} = g(\dot{x}, \dot{x})/2$, which reads in Boyer–Lindquist coordinates [9]

$$\begin{aligned} 2\mathcal{L} = \left(1 - \frac{2M_{\rm H}r}{\rho^2}\right) \dot{t}^2 + \frac{4a_{\rm H}M_{\rm H}r\sin^2\theta}{\rho^2}\,\dot{t}\dot{\phi} \\ - \frac{\rho^2}{\Delta}\,\dot{r}^2 - \rho^2\,\dot{\theta}^2 - \left(r^2 + a_{\rm H}^2 + \frac{2a_{\rm H}^2M_{\rm H}r\sin^2\theta}{\rho^2}\right)\sin^2\theta\,\dot{\phi}^2 \end{aligned} \tag{4.62}$$

Due to stationarity and axisymmetry the total energy E and angular momentum L are conserved. Together with mass conservation, these are three constants of motion. Thanks to the existence of a fourth constant of motion for the Kerr solution, the equations for geodesics are reduced to a set of four first order differential equations (Carter 1968, see *e.g.* [9])

$$\rho^2\,\dot{r} = \sqrt{\mathcal{R}(r)} \tag{4.63}$$

$$\rho^2\,\dot{\theta} = \sqrt{\Theta(\theta)} \tag{4.64}$$

$$\Delta\,\dot{\phi} = \frac{L}{\sin^2\theta}\left(1 - \frac{2M_{\rm H}r}{\rho^2}\right) + \frac{2M_{\rm H}r}{\rho^2}\,a_{\rm H}E \tag{4.65}$$

$$\alpha^2\,\dot{t} = E - \omega L \tag{4.66}$$

with the following potentials written as polynomials in r and $\cos\theta$, respectively,

$$\mathcal{R} = (E^2 - \mu^2)r^4 + 2M\mu r^3 + 2(a^2[E^2 - \mu^2] - L^2 - Q)r^2 + 2M([aE - L]^2 + Q)r - a^2 Q \tag{4.67}$$

$$\Theta = Q - \cos^2\theta\left(\frac{L^2}{\sin^2\theta} - (E^2 - \mu^2)a^2\right). \tag{4.68}$$

E is the total conserved energy, L the total angular momentum, μ the rest mass of the particle and Q Carter's constant of motion. This reduced set of geodesic equations can now be treated with some suitable direct integration method (Runge–Kutta), or by means of elliptic integrals [55].

The geodesics for photons follow from these relations in the limit $\mu \to 0$. The last two equations determine the angular velocity $\Omega = {\rm d}\phi/{\rm d}t$ for test particles, using the specific angular momentum $\lambda = L/E$,

$$\Delta(1 - \omega\lambda)\Omega = \frac{2\alpha^2 a_{\rm H} M_{\rm H} r}{\rho^2} + +\alpha^2\left(1 - \frac{2M_{\rm H}r}{\rho^2}\right)\frac{\lambda}{\sin^2\theta}. \tag{4.69}$$

Using the expression for α^2 in the first term, this reduces to

$$\Omega = \omega + \frac{\alpha^2}{R^2}\,\frac{\lambda}{1 - \omega\lambda}. \tag{4.70}$$

As for any matter, also test particles are dragged along by frame–dragging near the horizon. In particular photon trajectories are heavily affected by frame dragging near the horizon.

Innermost Stable Circular Orbit (ISCO): for motion in the equatorial plane, $\cos\theta = 0$, Carter's constant vanishes, $Q = 0$, and motion in r can be written in terms of an effective potential

$$\dot{r}^2 + W(r) = 0. \tag{4.71}$$

Circular orbits, *i.e.* relativistic Keplerian orbits are then given by the two conditions $W(r) = 0$ and $W'(r) = 0$. This determines the total energy per unit mass and the total angular momentum per unit mass of circular orbits

$$\frac{E}{\mu} = \frac{r^2 - 2M_{\rm H}r \pm a_{\rm H}\sqrt{M_{\rm H}r}}{r\sqrt{r^2 - 3M_{\rm H}r \pm 2a_{\rm H}\sqrt{M_{\rm H}r}}} \tag{4.72}$$

$$\frac{L}{\mu} = \pm\sqrt{M_{\rm H}r}\,\frac{(r^2 \mp 2a_{\rm H}\sqrt{M_{\rm H}r} + a_{\rm H}^2)}{r\sqrt{r^2 - 3M_{\rm H}r \pm 2a_{\rm H}\sqrt{M_{\rm H}r}}}. \tag{4.73}$$

For large radii we recover the well–known relations of Newtonian orbits

$$\frac{E}{\mu} = 1 - \frac{1}{2}\frac{M_{\rm H}}{r}, \quad \frac{L}{\mu} = \pm\sqrt{M_{\rm H} r}. \tag{4.74}$$

For the marginally stable orbit we have an additional condition, $\mathrm{d}^2 V/\mathrm{d}r^2 = 0$. This yields a condition for the radius of the form

$$r_{\rm ms}^2 - 6M_{\rm H} r_{\rm ms} \pm 8a_{\rm H}\sqrt{M_{\rm H} r_{\rm ms}} - 3a_{\rm H}^2 = 0. \tag{4.75}$$

For a non–rotating BH, $a_{\rm H} = 0$, this gives the famous result $r_{\rm ms} = 6M_{\rm H}$, *i.e.* the innermost stable orbit is located at 3 Schwarzschild radii. Inside this radius no Keplerian orbits can occur. In the general case, the solution can be given in parametrized form

$$r_{\rm ms} = M_{\rm H}\left(3 + Z_2 - \sqrt{(3 - Z_1)(3 + Z_1 + 2Z_2)}\right), \tag{4.76}$$

with the following abbreviations

$$\begin{aligned} Z_1 &= 1 + (1 - a_{\rm H}^2/M_{\rm H}^2)^{1/3}\left((1 + a_{\rm H}/M_{\rm H})^{1/3} + (1 - a_{\rm H}/M_{\rm H})^{1/3}\right) \\ Z_2 &= \sqrt{3a_{\rm H}^2/M_{\rm H}^2 + Z_1^2}. \end{aligned} \tag{4.77}$$

4.5 The boundary layer around rotating Black Holes

Since there are no stable circular orbits available inside the innermost stable circular orbit (ISCO), a kind of boundary layer is expected to be formed there. Angular frequency as measured by distant observers (fixed stars) is given by (with time t) $\Omega = U^\phi/U^{\rm t}$, where U^μ is the 4–velocity of matter. This is however not the basic quantity which enters into hydrodynamic equations. Due to axisymmetry, hydro and MHD provide an evolutionary equation for the specific angular momentum $j = U_\phi$ which is a scalar quantity, or for a related quantity $\lambda = U_\phi/U_{\rm t}$. Independent of the poloidal motion of matter, angular frequency and specific angular momentum are related over a kinematic correspondence

$$\Omega = \omega + \frac{\alpha^2}{R^2}\frac{\lambda}{1 - \omega\lambda}. \tag{4.78}$$

$R \equiv \sqrt{h_{\phi\phi}}$ is the cylindrical radius. In the Newtonian world, $\alpha = 1$ and $\omega = 0$, this degenerates to the well known expression $\Omega \simeq \lambda/R^2$. The above expression shows that even for vanishing specific angular momentum, $\lambda = 0$, matter is dragged along by space, $\Omega = \omega$. It also demonstrates that near the horizon of a Black Hole, where $\alpha \simeq 0$, matter is also dragged along by mere rotation of space. In particular, at the horizon itself, all types of matter have to corotate with the horizon, $\Omega = \Omega_{\rm H} \equiv \omega(r_{\rm H})$.

This has the consequence that in Schwarzschild space the classic relation between specific angular momentum and angular frequency is broken down, and all matter is forced to **no–rotation near the horizon**! As a consequence, in accretion onto a Schwarzschild Black Hole we always form a boundary layer with a rotation law which has a maximum at a few Schwarzschild radii (Fig. 8, [52]). For rapid rotation, this boundary layer is reduced, depending on the distribution of the specific angular momentum. In general, angular momentum transport is reduced near a horizon, since the accretion drift is high and viscosity has no chance to redistribute the angular momentum. For this case, $\lambda \simeq$ const, the above relation can be easily investigated. Since in a Keplerian disk, $\lambda \propto \sqrt{r}$, at large distances, the rotational frequency would decay as $\Omega \propto 1/r^{3/2}$.

4.6 The mass-formula for Black Holes

Most of the properties of a Black Hole are hidden behind its surface $A_{\rm H}$, given by

$$A_{\rm H} = 4\pi(r_{\rm H}^2 + a_{\rm H}^2) = 8\pi M_{\rm H} r_{\rm H}. \tag{4.79}$$

Due to this equality, the surface can also be written as

$$A_{\rm H} = \frac{A_{\rm H}^2}{16\pi M_{\rm H}^2} + 4\pi \,\frac{J_{\rm H}^2}{M_{\rm H}^2}. \tag{4.80}$$

This provides the mass as a function of the surface and the angular momentum

$$M_{\rm H} = M_{\rm H}(A_{\rm H}, J_{\rm H}) = \sqrt{\frac{A_{\rm H}}{16\pi} + 4\pi \,\frac{J_{\rm H}^2}{A_{\rm H}}}. \tag{4.81}$$

The total energy of a rotating Black Hole consists of an irreducible part given by the surface of the hole, $M_{\rm irr}^2 = A_{\rm H}/16\pi$, and the contribution from the angular momentum, summed up quadratically, as usual in Relativity.

A change of the total energy can therefore be written in the following form

$$\mathrm{d}M_{\rm H} = \frac{g_{\rm H}}{8\pi}\,\mathrm{d}A_{\rm H} + \Omega_{\rm H}\,\mathrm{d}J_{\rm H}, \tag{4.82}$$

where the intensive variables are

$$\frac{g_{\rm H}}{8\pi} = \left(\frac{\partial M_{\rm H}}{\partial A_{\rm H}}\right)_{J_{\rm H}}, \quad \Omega_{\rm H} = \left(\frac{\partial M_{\rm H}}{\partial J_{\rm H}}\right)_{A_{\rm H}}. \tag{4.83}$$

This is the analog of the first law of thermodynamics, called the **first law of Black Hole dynamics**. In a merging process *e.g.*, the area of a Black Hole can only grow, *i.e.* $\mathrm{d}A_{\rm H} > 0$ (**second law of Black Hole dynamics**).

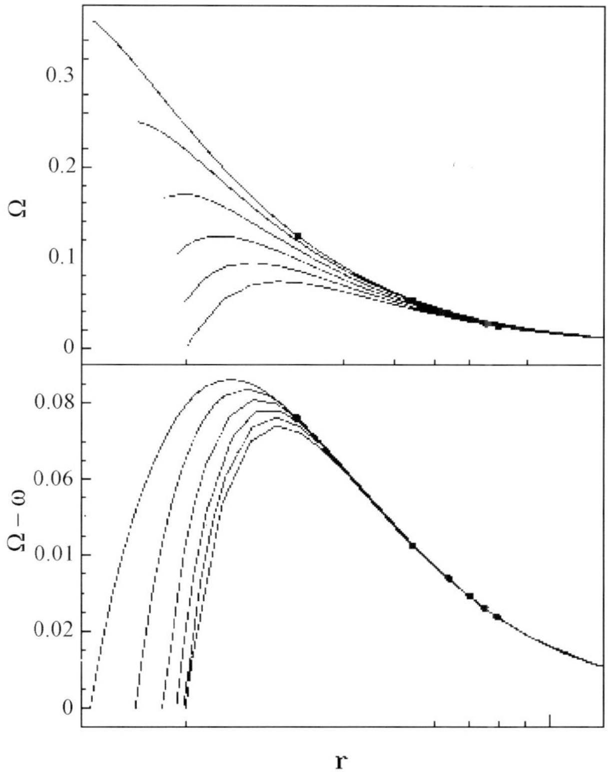

Fig. 8. Formation of a boundary layer near the horizon of a Kerr Black Hole from slow rotation to rapid rotation [52]. The dots denote the location of the sonic point in the radial accretion.

The analogy with thermodynamics tells us that $g_{\rm H}$ plays the role of a temperature and $A_{\rm H}$ the role of entropy. The surface gravity $g_{\rm H}$ can explicitly be calculated

$$g_{\rm H} = c^2 \, \frac{r_{\rm H} - M_{\rm H}}{2M_{\rm H} r_{\rm H}} \tag{4.84}$$

when $M_{\rm H}$ is taken in units of a length, $M_{\rm H} = GM/c^2$. This expression shows that the surface gravity vanishes in the case of extreme Kerr, $a_{\rm H} = M_{\rm H}$, where $r_{\rm H} = M_{\rm H}$. As a consequence, the extreme Kerr solution is a limiting case which cannot be realized in a finite number of processes operating on Black Holes (**third law of Black Hole dynamics**).

As a consequence of these considerations, a Black Hole accreting from the innermost stable orbit will grow in angular momentum, but never reach the extreme Kerr solution, since the innermost stable orbit moves towards the horizon, as the angular momentum increases.

4.7 Spin evolution of Black Holes

When BHs are accreting for a long time, the angular momentum of the ISCO is advected into the hole, thereby increasing mass and angular momentum. The question is therefore, how the specific angular momentum $a_{\rm H}$ evolves under accretion processes.

The time–evolution of a BH is described by the set of equations [48]

$$c^2\,\frac{{\rm d}M_{\rm H}}{{\rm d}t} = e_{\rm in}\dot{M}_{\rm acc} - \mathcal{P} \tag{4.85}$$

$$\frac{{\rm d}J_{\rm H}}{{\rm d}t} = j_{\rm in}\dot{M}_{\rm acc} - \frac{\mathcal{P}}{\Omega_{\rm F}} \tag{4.86}$$

where $\dot{M}_{\rm acc}$ is the ultimate accretion rate through the horizon, $e_{\rm in}$ and $j_{\rm in}$ are the specific energy and angular momentum of matter at the inner edge of the accretion disk (at ISCO *e.g.*). We also allow for some electromagnetic power extraction $\mathcal{P}$, as will be discussed later on. Blandford & Znajek (1976) have proposed a mechanism based on force–free electrodynamics

$$\mathcal{P}_{\rm BZ} \simeq \frac{1}{8}\,\frac{B_{\rm H}^2 r_{\rm H}^4}{c}\,\Omega_{\rm F}(\Omega_{\rm H} - \Omega_{\rm F}) \tag{4.87}$$

with a magnetic filed $B_{\rm H}$ threading the horizon and rotating with angular frequency $\Omega_{\rm F} < \Omega_{\rm H}$. The magnetic filed strength is usually limited by the pressure in the accretion disk

$$\frac{B_{\rm H}^2}{8\pi} \leq P_{\rm Disk} \simeq \frac{\dot{M}_{\rm acc}c^2}{4\pi\alpha_{\rm T} r_{\rm H}^2 c}\left(\frac{H}{r}\right)^{-1} . \tag{4.88}$$

Neglecting power extraction, $\mathcal{P} = 0$, the evolution equations can be written as

$$\frac{{\rm d}\ln M_{\rm H}}{{\rm d}t} = \frac{\dot{M}_{\rm acc}}{M_{\rm H}}\,\frac{e_{\rm in}}{c^2} \tag{4.89}$$

$$\frac{{\rm d}a}{{\rm d}t} = \frac{\dot{M}_{\rm acc}}{M_{\rm H}}\,[j_0(a) - 2ae_0(a)] \tag{4.90}$$

when $j_0 = cj_{\rm in}/GM_{\rm H}$, $e_0 = e_{\rm in}/c^2$ and a is the dimensionless Kerr–parameter. This shows that both mass–evolution and angular momentum evolution are given by the accretion time–scale

$$t_{\rm a} = \frac{M_{\rm H}}{\dot{M}_{\rm acc}} = \frac{1}{\dot{m}} \frac{M_{\rm H}}{\dot{M}_{\rm Edd}} = 4 \times 10^7 \,{\rm yrs}\, \frac{\epsilon_{\rm H}}{0.1\,\dot{m}} \cdot \tag{4.91}$$

Accretion from the marginally stable orbit satisfies the third law of BH dynamics, since $j_0(a) - 2ae_0(a) \to 0$ for $a \to 1$ (*i.e.* in the extreme Kerr limit). Initially non–rotating BHs with mass M_0 will be spun–up to $a \simeq \Delta m/M_0$ after accreting a mass Δm from the surrounding. Since BHs in galactic nuclei accrete for long times, they are expected to always rotate very rapidly, $a \geq 0.9$ – except if the accretion process is alternating between prograde and retrograde accretion.

4.8 Accretion beyond the innermost stable circular orbit

Accretion onto a Kerr BH should be modelled within relativistic hydrodynamics and magnetohydrodynamics.

SAD–ADAFs: stationary standard accretion disk theory (SADs) will not be covered here (for this see the lecture given by O. Blaes). The flux emitted by a SAD and observed at a frequency $\nu_0 = \nu_{\rm e}/(1+z)$ is given by, not including photon transport in the background gravitational field,

$$\nu_0 F_{\nu_0} = \frac{4\pi|\cos i|\, h\nu_{\rm e}^4}{c^2 {\rm d}_{\rm L}^2(z)} \int_{R_{\rm in}}^{R_{\rm out}} \frac{R\,{\rm d}R}{\exp(h\nu_{\rm e}/kT[R]) - 1}, \tag{4.92}$$

where $\nu_{\rm e}$ is the frequency at emission and $d_{\rm L}$ is the luminosity distance. For $h\nu \ll kT_{\rm max}$, this yields the famous disk spectra $\nu F_\nu \propto \nu^{4/3}$. The observed Quasar spectra (Fig. 9) are somewhat steeper in the optical regime. For a Kerr BH the gravitational effects are important, as the spectrum is formed closer to the horizon and is more influenced by gravitational redshift and lens effects. For quasars this mainly concerns the non–observable part of the spectrum (EUV), while the optical spectrum is formed much beyond the horizon. In a thin disk with local energy release, the viscosity prescription does not influence the temperature distribution. The effective temperature at distance R is then

$$\sigma T^4(R) = \frac{3GM_{\rm H}\dot{M}}{8\pi R^3}\, f(R,a), \tag{4.93}$$

where the non–dimensional factor $f(R,a)$ accounts for the inner boundary conditions depending on radius and angular momentum a of the BH. Setting the inner bc $f = 1$ for an infinite disk, one obtains

$$\nu_0 F_{\nu_0} \propto R_{\rm g}^2 T_*^{8/3} \propto (M_{\rm H}\dot{M})^{2/3}. \tag{4.94}$$

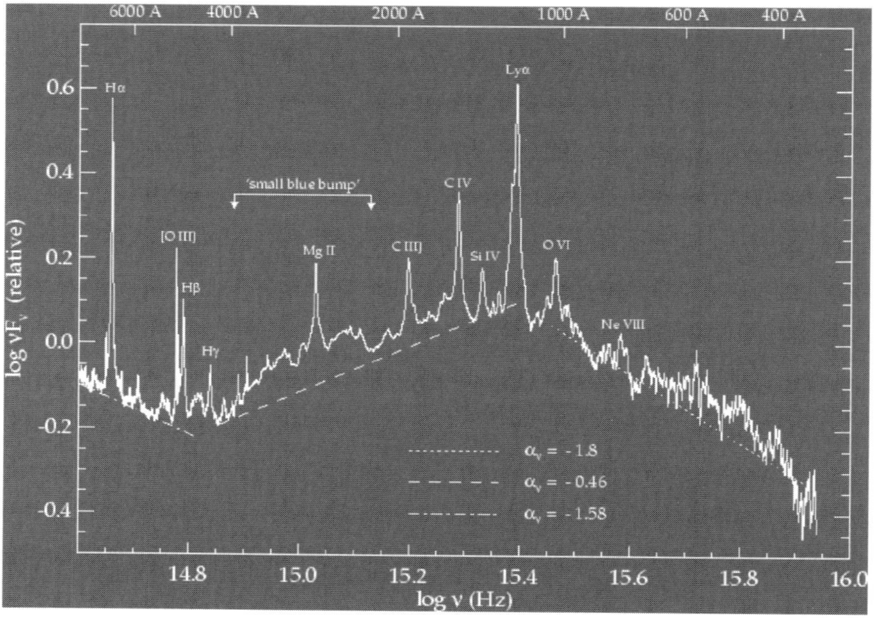

Fig. 9. Optical–UV composite spectrum of quasars is due to emission from optically thick disks around rotating Black Holes. Longward of Lyα, data is compiled from 2200 quasars from the Sloan Digital Sky Survey [66]. The change in the slope at 5000 Å is partly due to the host galaxy contamination. Shortward of Lyα, data has been taken from HST/FOS observations of radio quiet quasars, corrected for Galactic and intergalactic absorption by [69].

From these relations, you can construct a kind of Hertzsprung–Russell diagram for optically thick accretion disks, by identifying the effective temperature with the maximum temperature T_* at the inner edge of the disk

$$T_* = \left(\frac{3GM_{\rm H}\dot{M}}{8\pi R_{\rm in}^3 \sigma} \right)^{1/4} \simeq 1.4 \times 10^5 \, K \, M_{\rm H,8}^{-1/4} \, \dot{m}^{1/4} . \qquad (4.95)$$

There are large deviations to a standard disk spectrum, in particular in the EUV–band and the soft X–ray part, which is emitted by inner most hottest region in the disk where non–LTE effects and Comptonisation become important [19, 33]. Relativistic effects have been included in the modelling of standard accretion disks [19].

SADs are only valid in the range of $0.3\dot{M}_{\rm Edd} \geq \dot{M} \geq 0.01\dot{M}_{\rm Edd}$. For low accretion rates, the plasma becomes optically thin and is no longer able to cool down. Most of the energy dissipated away will be advected by the flow (these are the ADAF solutions).

Relativistic accretion near the horizon: for causality reasons, plasma has to pass the horizon supersonically and at the speed of light. This requires a sound speed close to the speed of light. For this reason, there are only two realisations for this:

- When plasma is optically thick and radiation pressure dominated, the sound speed is essentially given by the speed of light, $c_S^2 \simeq c/3$. This turns out to be in fact one possible solution near the horizon known under the name of **slim disk accretion**, since the high radiation pressure bloats up the disk thickness [3]. These branches are in fact stable.

- When plasma is optically thin near the horizon, then the plasma itself must be hot, *i.e.* at least the ions acquire temperatures of the order of 10^{12} K. The electrons need not stay hot, since for low densities the Coulomb coupling between ions and electrons is weak. So electrons can rapidly cool down by cyclotron emission, Bremsstrahlung and Comptonisation. This is probably the origin of the hard power–laws observed in the X–ray emission of Seyfert galaxies and Galactic Black Hole candidates. The transition between the optically thick solution and this optically thin solution is observable in Galactic Black Hole sources known as X–ray novae. Here the outer accretion rate is driven *e.g.* by eccentric binary orbits. In general, a **hot boundary layer** is formed between the horizon and the inner edge of the standard disk, the extension of this boundary layer is a question of the accretion rate. The inner edge of SAD occurs at the position where the optical depth is of order unity [16, 17, 43]. There is a critical accretion rate, $\dot{m}_{\rm crit} \leq 0.3$, below which a boundary layer always occurs.

The description of accretion near the horizon requires a complete relativistic formulation. The velocity field U of the plasma is given in Boyer–Lindquist coordinates

$$U = (U^t, \Omega U^t, U^r, 0) \tag{4.96}$$

$u \equiv U^r = {\rm d}r/{\rm d}\tau$ is the accretion drift and $\Omega = U^\phi/U^t$ the angular velocity of the plasma as measured by distant observers. Equations of motion for the plasma follow from the dissipative energy–momentum tensor, including a stress tensor Π and heat flux q [53, 62]

$$T = (\epsilon + p)\, U \otimes U + p\, g + \Pi + U \otimes q + q \otimes U. \tag{4.97}$$

ϵ is the total energy density (with rest mass density ρ_0), p pressure, q energy flux (radiative and heat). The basic assumption is here that stress is essentially generated by differential rotation in the disk (by MRI *e.g.*)

$$\Pi_{\alpha\beta} = \Pi_{\alpha\beta}(\sigma) \simeq -2\nu\rho_0\mu\,\sigma_{\alpha\beta}, \tag{4.98}$$

which is determined by the symmetric shear tensor σ

$$2\sigma_{\alpha\beta} \equiv U_{\alpha;\beta} + U_{\beta;\alpha} + a_\alpha U_\beta + a_\beta U_\alpha - \frac{2}{3}\Theta\, h_{\alpha\beta} \tag{4.99}$$

with the 4–acceleration

$$a_\alpha = U^\beta U_{\alpha;\beta}. \tag{4.100}$$

We also use the projector into the rest system of the plasma

$$h_{\alpha\beta} = g_{\alpha\beta} + U_\alpha U_\beta\,, \quad U^\alpha h_{\alpha\beta} = 0 \tag{4.101}$$

and the volume expansion Θ

$$\Theta = \nabla \cdot U. \tag{4.102}$$

Heat flux is given by heat conductivity K and temperature gradients

$$q^\alpha = -K\,h^{\alpha\beta}\,(T_{,\beta} + Ta_\beta). \tag{4.103}$$

Stress tensor and energy flux are real 3D objects

$$U^\alpha\,\Pi_{\alpha\beta} = 0 = U^\alpha\,q_\alpha. \tag{4.104}$$

The first law of thermodynamics implies

$$\mathrm{d}\epsilon = \frac{\epsilon + p}{\rho_0}\,\mathrm{d}\rho_0 + \rho_0 T\,\mathrm{d}s. \tag{4.105}$$

The equations of motion are then a consequence of Einstein's equations

$$h_{\gamma\alpha}\,T^{\alpha\beta}{}_{;\beta} = 0. \tag{4.106}$$

Besides mass conservation

$$M^\alpha_{;\alpha} = 0, \quad M^\alpha = \rho_0 U^\alpha \tag{4.107}$$

stationarity and rotation symmetry of space–time can now be used to derive some general consequences for disk accretion [62]. Based on the Killing equation, we find two conserved currents, the energy current

$$P^\alpha \equiv -T^\alpha_\beta\,\xi^\beta, \quad P^\alpha_{;\alpha} = 0 \tag{4.108}$$

and the angular momentum current

$$J^\alpha \equiv T^\alpha_\beta\, m^\beta, \quad J^\alpha_{;\alpha} = 0. \tag{4.109}$$

These currents have the following form, $E \equiv -U_\alpha \xi^\alpha = -U_\mathrm{t}$ is the energy measured at infinity,

$$P^\alpha = M^\alpha\,(\mu E - q_\mathrm{t}/\rho_0) - p\,\xi^\alpha + q^\alpha E - \Pi^\alpha_\mathrm{t} \tag{4.110}$$

and, using the specific angular momentum $j \equiv U_\alpha m^\alpha = U_\phi$,

$$J^\alpha = M^\alpha\,(\mu j + q_\phi/\rho_0) + p\,m^\alpha + q^\alpha j + \Pi^\alpha_\phi. \tag{4.111}$$

Π^α_ϕ and Π^α_t are real three–vectors, $U_\alpha \Pi^\alpha_\phi = 0$. They have essential poloidal components Π^A_ϕ and Π^A_t. We also use the relativistic specific enthalpy

$$\mu = 1 + \frac{\epsilon + p}{\rho_0}\cdot \tag{4.112}$$

$\mathcal{B} \equiv \mu E - q_\mathrm{t}/\rho_0$ is the generalized Bernoulli number. Its Newtonian expression is $\mathcal{B} \simeq 1 + h + \Phi + V^2/2$.

In the **stationary limit**, the above equations can be reduced to [2, 51, 52], using $\sqrt{-g} = r$ in the equatorial plane of the Kerr geometry, and integrating over latitudinal directions, $U^\theta = 0$:

- the vertical integration leads to a surface mass density Σ_0, surface energy density U and integrated pressure P, as in the Newtonian case. The specific enthalpy is therefore given by

$$\mu = \frac{U + P}{\Sigma_0 c^2} = 1 + \frac{\mathcal{E} + P}{\Sigma_0 c^2}\cdot \tag{4.113}$$

- mass conservation

$$-\frac{\dot M}{2\pi} = r\,\Sigma_0\,u = \sqrt{\Delta}\,\Sigma_0\,\frac{V}{\sqrt{1 - V^2}} \tag{4.114}$$

  with V as the radial velocity with respect to ZAMOs

$$u^2 = \frac{(-g^{rr})V^2}{1 - V^2}\cdot \tag{4.115}$$

- Angular momentum and energy conservation, neglecting q^r and q_t,

$$\frac{\mathrm{d}}{\mathrm{d}r}\left[\mu\left(\frac{\dot M j}{2\pi} + 2\nu\Sigma_0 r\,\sigma^r_\phi\right)\right] = \frac{Q^-}{c^2}\,rj \tag{4.116}$$

$$\frac{\mathrm{d}}{\mathrm{d}r}\left[\mu\left(\frac{\dot M E}{2\pi} - 2\nu\Sigma_0 r\,\sigma^r_\mathrm{t}\right)\right] = \frac{Q^-}{c^2}\,rE. \tag{4.117}$$

$Q^- = 2q^z(H)$ is the vertical energy flux per unit area. Angular momentum conservation is completely analogous to its Newtonian counterpart, except for the last term which denotes the angular momentum carried away by radiation from the disk. Neglecting the angular momentum lost by photons, angular momentum can be integrated to give

$$\frac{\dot{M}}{2\pi}\left[\mu j(r) - L_0\right] = -2\nu\Sigma_0\, r\sigma^r_\phi. \tag{4.118}$$

L_0 is a constant of integration given at the position $r = r_0$, where $\sigma^r_\phi(r_0) = 0$, or $\Pi^r_\phi(r_0) = 0$. For slowly rotating Black Holes, the rotation law $\Omega(r)$ always has a maximum at some radius $r_0 \simeq r_{\rm ms}$. This radius moves inwards for rapid rotation.

- **Accretion efficiency**: in analogy to the angular momentum conservation, we also can integrate the energy conservation (4.117) (Bernoulli theorem). The term on the right hand side describes the radiation carried away by photons. With the relation $\sigma_{\rm At} = -\Omega\sigma_{\rm A\phi}$, this equation determines the total energy lost from the disk

$$L_{\rm acc} = 2\pi\int_{r_0}^{\infty} E\,Q^-\,r\,{\rm d}r = \dot{M}c^2\left[1 - \mu(r_0)E(r_0)\right], \tag{4.119}$$

provided $\sigma_{\rm r\phi}(r_0) = 0 = \Pi^r_{\rm t}(\infty)$ and $\mu(\infty) = 1 = E(\infty)$. This relation determines the radiation efficiency for disk accretion

$$\boxed{\eta_{\rm acc} = 1 - \mu(r_0)\,E(r_0).} \tag{4.120}$$

For a not too hot plasma, $T < 10^{11}$ K, we have $\mu \simeq 1$ and the efficiency of disk accretion $\eta_{\rm K} = 1 - E(r_0)$ is only determined by the binding energy at the special radius r_0. For bounded orbits we always have $E < 1$. In SAD E is determined by the binding energy of ISCO, in Schwarzschild $E(r_{\rm ms}) = 2\sqrt{2}/3 = 0.94$, *i.e.* the efficiency is 6%. In a maximally rotating Kerr solution this amounts to 42%.

- For the **radial Euler equation** we need the radial acceleration [52]

$$a_{\rm r} = \frac{1}{2}\frac{\rm d}{{\rm d}r}(u_{\rm r}u^r) + \frac{1}{2}\frac{\partial g_{\phi\phi}}{\partial r}g^{tt}\,\gamma^2(\Omega - \Omega^+_{\rm K})(\Omega - \Omega^-). \tag{4.121}$$

The radial Euler equation can then be written as [52], $\mathcal{U} \equiv \Delta/r^2 + u^r u^r$,

$$\frac{u^2}{\mathcal{U}^2}\frac{{\rm d}\ln u}{{\rm d}r} + \frac{u^2}{\mathcal{U}^2}\mathcal{H} = -\mathcal{G} - \frac{1}{\mu\Sigma_0}\frac{{\rm d}P}{{\rm d}r} \tag{4.122}$$

which can be solved by some suitable relaxation technique. The gravitational force, $A(r) \equiv \Sigma^2(r, \theta = \pi/2)$,

$$\mathcal{G} = \frac{\gamma^2 A(r) M}{r^4 \Delta} \frac{(\Omega - \Omega_K^+)(\Omega - \Omega_K^-)}{\Omega_K^+ \Omega_K^-} \tag{4.123}$$

is given by the deviation from Keplerian rotation. $\mathcal{H} = \partial_r \ln \sqrt{g_{rr}}$.

Advective–dominated solutions have been discussed in [52] and radiation pressure dominated slim disk solutions in [3]. In both types of solutions, the sonic point occurs near the ISCO. The sonic point is the critical point of equation (4.122) written as

$$\frac{du}{dr} = \frac{u\mathcal{N}}{\mathcal{D}}. \tag{4.124}$$

Ion tori and truncated accretion disks: the physics of advection–dominated flows (ADAF) is, however, much more complicated. In particular, the heat flux should not be neglected, when huge gradients in the temperature occur. Since ADAFs bloat up in vertical direction, it is however necessary to tackle the problem at least in 2.5D [22–24]. In addition, the inner edge in SAD is never at the marginal stable orbit, as is assumed in the 1D calculations. Near the horizon, always a kind of hot ion torus will be formed, which grows by lowering the accretion rate for a given mass. In Figure 10 we show the solution obtained in 2.5D accretion (2D axisymmetry) after 5000 revolution times measured at the marginally stable orbit. The calculations are based on the 2.5D time–implicit diffusive pseudo–Newtonian MHD–code IRMHD [20, 25]. In this inner part, heat conduction is a very important process, since the ion temperature is so high in the inner region. Truncation is a generic feature when conduction and the possibility of a two temperature plasma are included in the calculations. A considerable heat flux is carried by conduction outwards from the inner hot disk which will heat up the ions in the innermost part of the standard disk. Since turbulence increases with increasing ion temperature, matter starts to accelerate, thereby reducing the density in this region. This weakens the radiative cooling as well as the Coulomb interaction ($\Lambda_{\rm i-e} \propto \rho^2/\sqrt{T_{\rm e}}$). Therefore, the ion torus expands and the truncation radius settles at around $9R_{\rm S}$ for an accretion rate $\dot{M} = 0.01\,\dot{M}_{\rm Edd}$ around a $10\,M_\odot$ BH. This **truncation radius** will depend on the overall accretion rate and is near the marginal stable orbit for higher accretion rates, but recedes to larger radii for lower accretion rates. Truncation has also been discussed within 1D models including a turbulent convective energy flux [16, 17, 43].

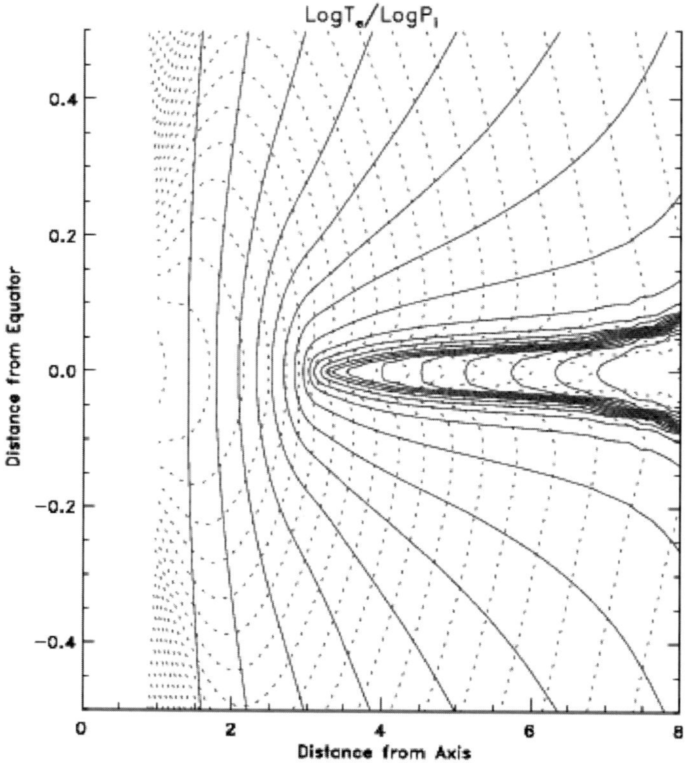

Fig. 10. Formation of an ion torus and truncation of a standard disk around a Schwarzschild BH of 10 $M_\odot$ with an accretion rate of 10^{17} g s^{-1} from the right boundary [22]. Radii are given in units of the radius of the marginally stable orbit, $3R_S$; note that the scale is different in vertical direction (left axis). The dotted lines show contourlines for the ion pressure, solid lines for the electron temperature. The optically thick disk gets truncated at $R_{tr} \simeq 9R_S$, and a hot ion pressure dominated torus forms inside. This snapshot occurs after 5000 orbital periods, measured at the marginal stable orbit. The standard disk is extremely thin which requires extreme spatial resolutions in the computations.

4.9 *Magnetic energy production by rotating compact objects*

4.9.1 Physical fields

The 3+1 formalism is of special importance for the discussion of electromagnetic fields in General Relativity. This technique is based on a slicing of the spacetime such that the slices are spacelike. This defines a vector field

$\partial_{\rm t}$ which can be decomposed into normal and parallel components relative to the slicing

$$\partial_{\rm t} = \alpha \boldsymbol{n} + \boldsymbol{\beta}. \tag{4.125}$$

Here $\boldsymbol{n}$ is the unit normal field and $\boldsymbol{\beta}$ is tangent to the slices. α is called the lapse function (or redshift factor) and $\boldsymbol{\beta}$ the shift vector field. The line elements of rapidly rotating neutron stars and of the Kerr geometry are a special decompositions of the general expression

$$\mathrm{d}s^2 = -\alpha^2\,\mathrm{d}t^2 + h_{ij}(\mathrm{d}x^i + \beta^i \mathrm{d}t)(\mathrm{d}x^j + \beta^j \mathrm{d}t). \tag{4.126}$$

The four one–forms $\Theta^0 = \alpha\,\mathrm{d}t$ and $\Theta^i = \theta^i + \beta^i \mathrm{d}t$ $(i = 1, 2, 3)$ are therefore orthonormal and form a natural basis for one–forms. θ^i are normalized one–forms in the spatial slice. In Kerr space, α is explicitly known, and due to axisymmetry the shift vector has only one nonvanishing component $\beta^\phi = -\omega$.

The derivation of Maxwell's equations adapted to the slicing is quite complicated when usual techniques are involved. There is a very elegant derivation based on the technique of differential forms, the electromagnetic potential A is essentially a one–form and the Faraday tensor F a two–form (for a derivation, see [6]).

In Astrophysics, we will not work with forms, but with the corresponding vector fields. $\boldsymbol{E}$ and $\boldsymbol{B}$ denote the electric and magnetic fields, respectively, as measured by ZAMOs, which are special observers having a 4–velocity U perpendicular to the absolute space

$$U = \frac{1}{\alpha}\,(\partial_{\rm t} - \beta^i\,\boldsymbol{e}_i). \tag{4.127}$$

Together with the orthonormal tetrad in Boyer–Lindquist coordinates $\boldsymbol{e}_{\rm r}$, $\boldsymbol{e}_\theta$ and $\boldsymbol{e}_\phi = (1/\bar{\omega})\,\partial_\phi$ this forms a physical basis in the tangential space. This is the basis that is dual to the above natural one–form basis. The circumference of a circle around the rotational axis is measured by $\tilde{\omega}$. The 4–velocity of the plasma u is then expressed as

$$u = \gamma(U + \boldsymbol{v}), \tag{4.128}$$

where $\boldsymbol{v}$ is now the 3–velocity of the plasma with respect to ZAMOs.

Instead of using the forms, Maxwell's equations are expressed for the vector fields $\boldsymbol{E}$ and $\boldsymbol{B}$

$$\nabla \cdot \boldsymbol{E} = 4\pi\rho_{\rm e}, \quad \nabla \cdot \boldsymbol{B} = 0 \tag{4.129}$$

$$\nabla \times (\alpha \boldsymbol{E}) = -(\partial_{\rm t} - \mathcal{L}_\beta)\boldsymbol{B} \tag{4.130}$$

$$\nabla \times (\alpha \boldsymbol{B}) = (\partial_{\rm t} - \mathcal{L}_\beta)\boldsymbol{E} + 4\pi\alpha\,\boldsymbol{j}. \tag{4.131}$$

The current density $\boldsymbol{j}$ is given by Ohm's law,

$$\boldsymbol{j} = \sigma\gamma\,(\boldsymbol{E} + \boldsymbol{v}\times\boldsymbol{B}) + \rho'_{\rm e}\,\gamma\boldsymbol{v}, \tag{4.132}$$

where σ denotes the conductivity, $\boldsymbol{v}$ the bulk velocity of the plasma, γ the corresponding Lorentz factor, and $\rho'_{\rm e}$ the charge density in the rest frame of the plasma. As compared to flat spacetimes, there are two important additional terms related to the shift vector of the slicing (or the frame–dragging effect in Kerr space, given by the Lie–derivative–term).

For axisymmetric fields, Maxwell's equations assume the simple form

$$\nabla\times(\alpha\boldsymbol{E}) = -\frac{\partial\boldsymbol{B}}{\partial t} + (\boldsymbol{B}\cdot\nabla\omega)\,\tilde{\omega}\boldsymbol{e}_\phi \tag{4.133}$$

$$\nabla\times(\alpha\boldsymbol{B}) = \frac{\partial\boldsymbol{E}}{\partial t} - (\boldsymbol{E}\cdot\nabla\omega)\,\tilde{\omega}\boldsymbol{e}_\phi + 4\pi\alpha\boldsymbol{j}. \tag{4.134}$$

On an axisymmetric spacetime, it is now useful to split all the vector fields into poloidal and toroidal components, $\boldsymbol{B} = \boldsymbol{B}_{\rm p} + \boldsymbol{B}_{\rm T}$ with $\boldsymbol{B}_{\rm T} = B^{\hat{\phi}}\,\boldsymbol{e}_\phi$ and $\boldsymbol{B}_{\rm p} = B^{\hat{r}}\boldsymbol{e}_{\rm r} + B^{\hat{\theta}}\boldsymbol{e}_\theta$. $\boldsymbol{B}_{\rm p} = \nabla\times(A^{\hat{\phi}}\boldsymbol{e}_\phi)$. The induction and Ampère's equations give the following relations, when decomposed into poloidal and toroidal components

$$\frac{\partial A^{\hat{\phi}}}{\partial t} = -\alpha\,E^{\hat{\phi}} \tag{4.135}$$

$$\frac{\partial B^{\hat{\phi}}}{\partial t} = \tilde{\omega}\,\boldsymbol{B}_{\rm p}\cdot\nabla\omega - \boldsymbol{e}_\phi\cdot(\nabla\times\alpha\boldsymbol{E}_{\rm p}) \tag{4.136}$$

$$\frac{\partial E^{\hat{\phi}}}{\partial t} = \tilde{\omega}\boldsymbol{E}_{\rm p}\cdot\nabla\omega - \mathcal{G}_2[A^{\hat{\phi}}] - 4\pi\alpha j^{\hat{\phi}} \tag{4.137}$$

$$\frac{\partial\boldsymbol{E}_{\rm p}}{\partial t} = \nabla\times(\alpha B^{\hat{\phi}}) - 4\pi\alpha\boldsymbol{j}_{\rm p}. \tag{4.138}$$

$\mathcal{G}_2[A] \equiv -\boldsymbol{e}_\phi\cdot[\nabla\times\alpha\boldsymbol{B}_{\rm p}]$ is the Grad–Shafranov operator for the poloidal flux function $\Psi \equiv \tilde{\omega}A^{\hat{\phi}}$

$$\mathcal{G}_2[\Psi] \equiv \tilde{\omega}\nabla\cdot\left[\frac{\alpha}{\tilde{\omega}^2}\,\nabla\Psi\right]. \tag{4.139}$$

In terms of this flux function Ψ, the poloidal magnetic field is given in the standard form

$$\boldsymbol{B}_{\rm p} = \frac{1}{\tilde{\omega}}\,\nabla\Psi\times\boldsymbol{e}_\phi\,. \tag{4.140}$$

In addition, Ohm's law has the two components

$$\boldsymbol{j}_{\rm p} = \sigma\gamma\left[\boldsymbol{E}_{\rm p} + \boldsymbol{v}_{\rm T}\times\boldsymbol{B}_{\rm p} + \boldsymbol{v}_{\rm p}\times\boldsymbol{B}_{\rm T}\right] \tag{4.141}$$

$$j^{\hat{\phi}} = \sigma\gamma\left[E^{\hat{\phi}} + \boldsymbol{e}_\phi\cdot(\boldsymbol{v}_{\rm p}\times\boldsymbol{B}_{\rm p})\right], \tag{4.142}$$

provided the charge density in the plasma frame vanishes.

4.9.2 Time evolution of magnetic flux and current

The above formulation can now be used to investigate the time evolution of magnetic flux in accretion disks around rapidly rotating objects.

The Grad–Shafranov equation: when we combine equation (4.137) with the first equation (4.135) of Maxwell's equations, this provides us a kind of wave equation for the poloidal flux Ψ

$$\frac{\partial^2 \Psi}{\partial t^2} - \alpha\tilde{\omega}\,\mathcal{G}_2[\Psi] = -\alpha\tilde{\omega}^2 \boldsymbol{E}_\mathrm{p} \cdot \nabla\omega + 4\pi\tilde{\omega} j^{\hat{\phi}}. \tag{4.143}$$

This shows explicitly that the equation is hyperbolic, as required by Maxwell's theory. Using Ohm's law, this can be rewritten as

$$\frac{\partial^2 \Psi}{\partial t^2} + 4\pi\gamma\sigma\,\alpha\frac{\partial \Psi}{\partial t} - \alpha\tilde{\omega}\,\mathcal{G}_2[\Psi] = - \tilde{\omega}^2\alpha\boldsymbol{E}_\mathrm{p} \cdot \nabla\omega + 4\pi\alpha^2\gamma\sigma\tilde{\omega}\boldsymbol{e}_\phi \cdot (\boldsymbol{v}_\mathrm{p} \times \boldsymbol{B}_\mathrm{p}). \tag{4.144}$$

Using the expression for the poloidal magnetic field and the definition of the magnetic diffusivity $\eta = c^2/4\pi\sigma$, we obtain the equation

$$\eta\,\frac{\partial^2 \Psi}{c^2\partial t^2} + \alpha\gamma\,\frac{\partial \Psi}{\partial t} + \alpha\gamma\,(\alpha\boldsymbol{v}_\mathrm{p} \cdot \nabla)\Psi - \eta\,\alpha\tilde{\omega}\,\mathcal{G}_2[\Psi] = -\eta\,\tilde{\omega}^2\,\alpha\boldsymbol{E}_\mathrm{p} \cdot \nabla\omega. \tag{4.145}$$

Causality requires that the second derivative is present in this transport equation for the poloidal magnetic flux. This term regulates the relaxation. For long term evolution we may neglect this part and end up with a parabolic diffusion type equation used in some simulations of the time evolution of magnetic fields [29]

$$\gamma\,\frac{\partial \Psi}{\partial t} + \gamma\,(\alpha\boldsymbol{v}_\mathrm{p} \cdot \nabla)\Psi - \eta\,\tilde{\omega}\,\mathcal{G}_2[\Psi] = -\eta\,\tilde{\omega}^2\,\boldsymbol{E}_\mathrm{p} \cdot \nabla\omega. \tag{4.146}$$

The current flux equation: the second equation determines the time–evolution of the current function $T(t,r,\theta) = 2\int \alpha\boldsymbol{j}_\mathrm{p} \cdot \mathrm{d}\boldsymbol{A} = \alpha\tilde{\omega}B^{\hat{\phi}}(t,r,\theta)$. From equation (4.136) we get

$$\frac{\partial T}{\partial t} = \alpha\tilde{\omega}^2\,\boldsymbol{B}_\mathrm{p} \cdot \nabla\omega - \alpha\tilde{\omega}\boldsymbol{e}_\phi \cdot \left[\nabla \times \alpha\left(\frac{1}{\sigma\gamma}\boldsymbol{j}_\mathrm{p} - \boldsymbol{v}_\mathrm{T} \times \boldsymbol{B}_\mathrm{p} - \boldsymbol{v}_\mathrm{p} \times \boldsymbol{B}_\mathrm{T}\right)\right], \tag{4.147}$$

when Ohm's law is used. Together with the poloidal component of Ampère's law this can be written as

$$\frac{\partial T}{\partial t} - \alpha\tilde{\omega}\boldsymbol{e}_\phi \cdot \left[\nabla \times \frac{\eta}{\gamma}(\nabla \times \alpha \boldsymbol{B}_{\rm T})\right] = \alpha\tilde{\omega}^2\, \boldsymbol{B}_{\rm p} \cdot \nabla\omega$$
$$+ \alpha\tilde{\omega}\boldsymbol{e}_\phi \cdot \nabla \times \left[\frac{\eta}{\gamma}\frac{\partial \boldsymbol{E}_{\rm p}}{\partial t} - \boldsymbol{e}_\phi \cdot \Big(\boldsymbol{v}_{\rm T} \times \boldsymbol{B}_{\rm p} - \boldsymbol{v}_{\rm p} \times \boldsymbol{B}_{\rm T}\Big)\right]. \quad (4.148)$$

This finally leads to a diffusion type equation for the current function

$$\frac{\partial T}{\partial t} + \alpha(\boldsymbol{v}_{\rm p} \cdot \nabla)T - \alpha\tilde{\omega}^2\, \nabla \cdot \left(\frac{T}{\tilde{\omega}^2}\boldsymbol{v}_{\rm p}\right) - \alpha\tilde{\omega}^2\, \nabla \cdot \left(\frac{\eta}{\gamma\tilde{\omega}^2}\nabla T\right) =$$
$$\alpha\tilde{\omega}^2\, \boldsymbol{B}_{\rm p} \cdot \nabla\Omega + \alpha\tilde{\omega}\, \boldsymbol{e}_\phi \cdot \nabla \times \left(\frac{\eta}{\gamma}\frac{\partial \boldsymbol{E}_{\rm p}}{\partial t}\right). \quad (4.149)$$

This equation is exactly similar to the Newtonian equation, $T = RB_\phi$, in the limit $\alpha = 1$, except for the appearance of the displacement current (the last term on the right hand side). Frame–dragging enters merely over the form of the angular frequency Ω which is given in general by equation (4.78). Near the horizon, $\Omega \simeq \omega$ and frame–dragging is dominating the shearing of the poloidal field. Even for vanishing specific angular momentum of the plasma, $\lambda = 0$, a current flux is generated by differential rotation of absolute space around **any rapidly rotating compact object**. This condition is in particular satisfied for the coronal plasma above the accretion disk. The above property is not a specific feature of Black Holes, this mechanism would also operate around rapidly rotating neutron stars and quark stars. We have performed simulations of this effect by neglecting the displacement current and the second order time–derivative in the GS equation, by starting with a homogeneous poloidal magnetic field threading an accretion disk around a rapidly rotating BH [8, 29]. A modern version of this effect is discussed in [30]. Differential rotation of absolute space, given by the ω–potential, always generates toroidal magnetic fields which must escape from the vicinity of the horizon. How this magnetic energy (in the form of Poynting flux) is converted into kinetic energy is only understood in the stationary approximation. Some aspects of time–dependent processes have been worked out in the force–free limit [31, 32], which is, however, never satisfied near the horizon of a rotating Black Hole [8].

On the validity of Cowling's theorem: cowling's theorem states that stationary axisymmetric magnetic fields cannot be maintained by purely axisymmetric plasma motions. The gravitomagnetic effect in the induction equation changes this conclusion of classical electrodynamics [8, 29]. The

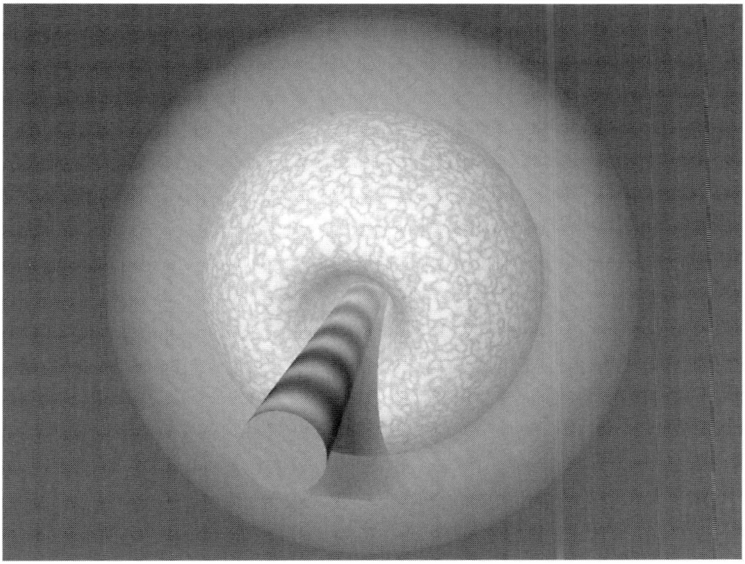

Fig. 11. Micro–jets are generated in the center of galactic nuclei near the horizon of the rotating Black Hole. They propagate into the interstellar medium of the galactic core on the scale of a few hundred parsecs (red). In Quasars, the core is filled with a torus of dusty material confined to clouds.

reason is that in Kerr space the coupling between the gravitomagnetic shear $\nabla\omega$ and the poloidal electric field $\boldsymbol{E}_{\rm p}$ is a potential source for the poloidal magnetic flux Ψ. This term is absent in nonrotating space and leads to the usual formulation of Cowling's theorem. In order to see this effect, one has to simulate the two equations, there are no analytical results known.

4.10 Jets as collimated outflows from near the horizon

The jets of quasars and radio galaxies are certainly the most spectacular collimated plasma flows in the Universe [36]. Their plasma is probably very exotic and is only visible in non–thermal synchrotron and inverse Compton emission. In contrast to these objects, jets of normal stellar objects (young stellar objects, protostars and White Dwarfs) can be detected in line emission of a normal gas [64]. The plasma in the jets of active galactic nuclei moves close to the speed of light, as derived from VLBI observations. The speeds of the superluminally moving components in these jets range from the speed of light to about 20 times the speed of light, indicating that the bulk speed of the beam–plasma is very near to the speed of light [45]. Simulations of this phenomenon require relativistic hydrodynamical and magneto–hydrodynamical solvers. As these jets penetrate into the kiloparsec–scale, they come into close contact with the ambient cluster medium. The overall structure of the jets (cocoon and shocked external medium) depends very much on the density contrast [36].

5 Conclusions and future prospects

Relativistic gravity is essential only in the nearest environment of rotating neutron stars and Black Holes. The Black Hole paradigm for the explanation of the activity seen in galaxy centers has been confirmed in the last years through measurements of the corresponding masses. But these objects live in close contact with the gas reservoir on the parsec–scale, and this connection is still far from understood. In particular, mass transfer from millions of Schwarzschild radii down to the horizon is one of the challenging problems to be solved in the future.

In this respect, Galactic Black Holes offer some advantage. The time–scales are at least a million times shorter. It is however by far not understood whether a scaling from the stellar Black Holes to supermassive ones is allowed or not. In this respect, one has to recognize that gravity around a rotating Black Hole is not much different from gravity around a rotating neutron or quark star. It is therefore very helpful to have analytic exterior solutions for rotating neutron stars, *e.g.* in order to work out differences between accretion towards neutron stars and Black Holes.

The future in understanding Black Hole physics will be mainly based on heavy numerical simulations of accretion processes, magnetic coupling near rotating Black Holes and time–dependent formation of outflows.

Another topic for future research is the understanding of the formation of Black Holes in the centers of young galaxies. This process starts very early on in the young Universe, at redshifts of about 25. In the most probable scenario, massive Black Holes with masses in the range of a few hundred solar masses are formed in the first generation of stars. These objects must grow very rapidly and achieve a fraction of their present mass already at redshift 5. The bright Quasars now found at redshift 6 must then be located in the cores of the anchestors of present galaxy clusters.

It is a great pleasure to thank my research partners over the years in this field of relativistic astrophysics. Among them are Stefan Appl, Ramon Khanna, Christian Fendt, Jochen Peitz, Ahmed Hujeirat, Jose Gracia, Martin Krause and Andreas Mueller. Many projects have been funded by our Sonderforschungsbereich 439 of the Deutsche Forschungsgemeinschaft on the topics of "Galaxies in the Early Universe". Finally, I enjoyed very much to stay in this marvelous area of Les Houches.

References

[1] M.A. Abramowicz, G. Björnsson and J.E. Pringle, "Theory of Black Hole Accretion Discs" (Cambridge University Press, Cambridge, 1999).
[2] A.M. Beloborodov, M.A. Abramowicz and I.D. Novikov, *ApJ* **491** (1997) 267.
[3] A.M. Beloborodov, *MNRAS* **297** (1998) 739.
[4] N. Bilic, F. Munyaneza, G.B. Tupper and R.D. Viollier, *Progr. Part. Nucl. Phys.* **48** (2002) 291.
[5] M. Camenzind and A. Boucher, "Les noyaux actifs de galaxies", *Lect. Notes Phys.* **46** (Springer–Verlag, Heidelberg, 1997).
[6] M. Camenzind, "Magnetohydrodynamics of Rotating Black Holes", in Relativistic Astrophysics, edited by H. Riffert, H. Ruder, H.-P. Nollert and F.W. Hehl, Vieweg, *Astrophysics* (1998), 82.
[7] M. Camenzind, *Lect. Notes Phys. LNP* **530** (1999) 252.
[8] M. Camenzind and R. Khanna, *Il Nuovo Cimento* **115B** (2000) 815.
[9] Chandrasekhar, "The Mathematical Theory of Black Holes" (Oxford Univ. Press, 1983).
[10] T. di Matteo, A.c. Fabian *et al.*, *MNRAS* **305** (1999) 492.
[11] T. di Matteo *et al.*, *Lighthouses of the Universe* (2002) 443.
[12] F.J. Ernst, *Phys. Rev.* **167** (1968) 1175; **168** (1968) 1415.
[13] S. Faber *et al.*, *AJ* **114** (1997) 1771.
[14] L. Ferrarese and D. Merritt, *ApJ* **539** (2000) L9.
[15] K. Gebhardt *et al.*, *ApJ* **539** (2000) L13.
[16] J. Gracia, "Time–dependent accretion flows onto Black Holes", Ph.D. Thesis (University of Heidelberg, 2002).
[17] J. Gracia, J. Peitz, Ch. Keller and M. Camenzind, *MNRAS* (2002) submitted.
[18] M. Haas, S.A.H. Müller, R. Chini *et al.*, *A&A* **354** (2000) 453.
[19] Hubeny *et al.*, *ApJ* (2001).
[20] A. Hujeirat, A., *MNRAS* **298** (1998) 310.
[21] A. Hujeirat and M. Camenzind, *A&A* **360** (2000a) L17.
[22] A. Hujeirat and M. Camenzind, *A&A* **361** (2000b) L53.
[23] A. Hujeirat and M. Camenzind, *A&A* **362** (2000c) L41.
[24] A. Hujeirat, M. Camenzind and A. Burkert, *A&A* (2002).
[25] A. Hujeirat and R. Rannacher, *New Astron. Rev.* **45** (2001) 425.
[26] A. Hujeirat, M. Camenzind and M. Livio, *A&A* **394** (2002) L9.
[27] A. Hujeirat, M. Camenzind and A. Burkert, *A&A* **386** (2002) 757.
[28] R. Kerr, *Phys. Rev. Lett.* **11** (1965) 237.
[29] R. Khanna and M. Camenzind, *A&A* **307** (1996) 665.
[30] S. Koide, K. Shibata, T. Kudoh and D.L. Meier, *Science* **295** (2002) 1688.
[31] S.S. Komissarov, *MNRAS* **326** (2000) L41.
[32] S.S. Komissarov, *MNRAS* **336** (2002) 759.
[33] A. Koratkar and O. Blaes, *PASP* **111** (1999) 1.
[34] J. Kormendy and D. Richstone, *ARA&A* **33** (1995) 581.
[35] J. Kormendy and K. Gebhardt, in The 20th Texas Symposium on Relativistic Astrophysics, edited by H. Martel and J.C. Wheeler (AIP, 2001) 363
[36] M. Krause, "On the Interaction of Jets with the Dense Medium in the Early Universe", Ph.D. Thesis (University of Heidelberg, 2002).

[37] M. Krause, "A 3D Hydrodynamic Simulation for the Cygnus A Jet as a Prototype for Hygh Redshift Radio Galaxies", in: High Performance Computing in Science and Engeneering '01, edited by E. Krause and W. Jäger (Springer–Verlag, Heidelberg).

[38] A. Loeb, *Phys. Rev. D* **65** (2002).

[39] A. Loeb and R. Barkana, *ARA&A* **39** (2001) 19.

[40] D. Machetto, *Lect. Notes Phys. LNP* **530** (1999) 291.

[41] V.S. Manko, E.W. Mielke and J.D. Sanabri–Gomez, *Phys. Rev. D* **61** (2000) issue 8.

[42] V.S. Manko *et al.*, *Phys. Rev. D* **62** (2001) issue 4.

[43] T. Manmoto and S. Kato, *ApJ* **538** (2000) 295.

[44] M.A. Markov, *Ann. Phys.* **155** (1984) 333.

[45] A.P. Marscher S. Jorstadt, J.-L. Gomez *et al.*, *Nature* **417** (2002) 625.

[46] P.O. Mazur and E. Mottola, "Gravitational Condensate Stars" [gr-qc/0109035] (2001).

[47] M. Miyoshi *et al.*, *Nature* **373** (1995) 127.

[48] R. Moderski and M. Sikora, *MNRAS* **283** (1996) 854.

[49] Papapetrou, *Ann. Phys.* **12** (1953) 309.

[50] P.J.E. Peebles and B. Ratra, "The Cosmological Constant and Dark Energy", [astro-ph/0207347] (2002).

[51] J. Peitz, "On the structure of relativistic hydrodynamic accretion discs around rotating Black Holes", Ph.D. Thesis (University Heidelberg, 1997).

[52] J. Peitz and S. Appl, *MNRAS* **286** (1997) 681.

[53] J. Peitz and S. Appl, *MNRAS* **296** (1998) 231.

[54] E. Quataert and R. Narayan, *ApJ* **528** (2000) 236.

[55] K.P. Rauch and R.D. Blandford, *ApJ* **421** (1994) 46.

[56] S. Ravindranath *et al.*, *AJ* **122** (2002) 653.

[57] S. Rawlings and R. Saunders, *Nature* **349** (1991) 138.

[58] H.-J. Röser and K. Meisenheimer, "The Radio Galaxy Messier 87", *Lect. Notes Phys.* (1999), *LNP* **530** (Springer–Verlag, Heidelberg).

[59] R. Schödel, T. Ott, R. Genzel *et al.*, *Nature* **419** (2002) 694.

[60] N.R. Sibgatullin and N.M. Queen, "Oscillations and Waves in Strong Gravitational and Electromagnetic Fields" (Springer–Verlag, Berlin, 1991) 362.

[61] N.R. Sibgatullin and R.A. Sunyaev, *Astron. Lett.* **24** (1998) 774.

[62] N. Straumann, "General Relativity and Relativistic Astrophysics" (Springer–Verlag, Heidelberg, 1984).

[63] M. Stute and M. Camenzind, *MNRAS* **336** (2002) 831.

[64] M. Thiele and M. Camenzind, *A&A* **381** (2002) L53.

[65] S. Tremaine *et al.*, *ApJ* **574** (2002) 740.

[66] D.E. Vanden Berk *et al.*, *AJ* **122** (2001) 549.

[67] F. Yuan, S. Markoff and H. Falcke, *A&A* **383** (2002) 854.

[68] Zhao, HongShen, *MNRAS* **278** (1996) 488.

[69] Zheng, Wei *et al.*, *ApJ* **475** (1997) 469.

COURSE 9

ACCRETION AROUND ACTIVE GALACTIC NUCLEI

B. CZERNY

Copernicus Astronomical Center,
Bartycka 18, Warsaw, Poland

Contents

ACCRETION AROUND ACTIVE GALACTIC NUCLEI

B. Czerny[1]

Abstract

In bright AGN, at distances of several Schwarzschild radii, the accretion onto a black hole proceeds predominantly in a form of a cool accretion disk. However, very close to a black hole the accreting material forms a multi-phase medium (coexisting cold optically thick gas, very hot optically thin plasma and an intermediate warm partially ionized material), and the geometrical arrangement is not clear. Several different geometries satisfy the current observational constraints. In faint AGN the observationally estimated Bondi accretion rate is orders of magnitudes higher than expected at the basis of the observed luminosity so again the accretion pattern is not understood, with angular momentum problem, outflow of material and/or non-stationarity possibly playing a role. Better theoretical models are being developed but still a number of important physical processes are to be introduced. Such better models, and extensive monitoring of the variability are needed to understand such a complex phenomenon as the accretion onto a supermassive black hole.

1 Introduction

Active Galactic Nuclei (hereafter AGN) are the most powerful long lasting energy sources in the Universe. Bolometric luminosities of the brightest quasars reach values as high as 10^{48} erg s^{-1} cm^{-2}. Their luminosity is only overpassed by gamma bursts which can be still brighter by a few orders of magnitude but last for seconds only. Nuclear activity on the other hand is a long lasting stage in the evolution of most, probably all, galaxies. Its importance for galactic evolution only now becomes apparent. The most distant presently known AGN is a quasar located at the redshift $z = 6.28$, and another candidate, at 6.4, is currently considered. One of the nearest galaxies customarily included into AGN class is the Seyfert galaxy NGC 4945

[1] Copernicus Astronomical Center.

(z = 0.00187, corresponding to a distance of ~3.7 Mpc). However, the transition between so called active and non-active galaxies is smooth and the signatures of some level of activity are now discovered in more and more galaxies with the progress of the sensitivity of the instruments. Some activity is going on also at the center of our Milky Way. Therefore, the division of the galaxies into active and non-active is now arbitrary and we can say that we count a galaxy as active if the bolometric luminosity of an active nucleus is above 10^{41} erg s^{-1} cm^{-2}.

1.1 Short history of the subject

The history of AGN research is much longer than the name of the subject itself. Observational data on some of AGN cover more than a hundred years while the term "active galactic nucleus" is in broad use only for the last 10 years. Objects which now belong to this class were not immediately recognized as "active galaxies" and originally belonged to different research areas. Many nearby AGN were first classified as nebulae and included in the famous Messier catalog and NGC catalog of Dreyer (*e.g.* M 77 ≡ NGC 1068). Prototype of so called BL Lac class was classified originally as a variable star. First systematic research aimed indeed at active galaxies was performed by Carl Seyfert who studied in detail a group of galaxies with exceptionally bright point like nuclei. His work published in 1943 included now famous AGN like NGC 4151.

Better hint of the nature of nuclear activity came with the onset of radio and X-ray astronomy in the second half of XX century. First some of the well known galaxies were recognized as radio galaxies, but some of the radio sources had no apparent optical counterpart. First optical identification was done for the source 3C 48 but the optical counterpart was a point like source – like a star – with the optical spectrum very much different from normal stars. In such way quasars – quasi-stellar sources – were born. The true understanding of this class came with the discovery by Marteen Schmidt that broad intense emission lines seen in the spectra of another source – 3C 273 – are just hydrogen lines but strongly redshifted (z = 0.158). Follow-up optical searches showed that such point-like sources with high redshifts are quite numerous and most of them actually are not strong radio emitters so they were called QSO – quasi-stellar objects. Observations in X-ray band which were started in sixties showed that radio galaxies are also good X-ray emitters. The next major step came with the launch of EXOSAT. Long uninterrupted observations of Seyfert galaxies performed due to the high elliptical orbit of the satellite showed that their X-ray emission is strongly variable in a timescale of hours.

For years there was an on-going discussion whether quasars and QSO are the same as radio galaxies and Seyfert galaxies since the first two classes did not show the presence of a host galaxy while in the last two classes the presence of the galaxy was obvious. However, the observations performed with the use of Hubble Space Telescope allowed to detect host galaxies around AGN and even to analyze their morphological properties since the lack of atmospheric interference allowed for much better spatial resolution of the image and its subsequent division into a point-like source (active nucleus) and an extended source (a host).

It is now clear that all historically introduced various classes of AGN have more or less the same basic properties. However, the division is still frequently used since it emphasizes the relative strength of a given property although the transition between various classes is smooth. So, an object is usually called a QSO if its absolute magnitude in B band M_B is below -23 and a Seyfert galaxy otherwise. An object is also qualified as radio-loud if the value of the parameter $R_{\rm L} = \log F(5\,{\rm GHz})/F(B)$ is larger than 1, where $F(5\,{\rm GHz})$ is the observed radio flux of the core at 6 cm and $F(B)$ is the observed optical flux in B band (*i.e.* 4400 Å). Seyfert galaxies themselves are classified according to the properties of their emission lines into Sy1 and Sy2 (with intermediate types Sy1.2, Sy1.5, Sy1.8 and Sy1.9) basically according to the ratio of the broad component of H_β line to [OIII] λ5007 being above 5 and 0 (and 2–5, 0.333–2, below 0.333, and traces of broad component only in H_α in the intermediate cases, correspondingly). A separate important class is formed by Narrow Line Seyfert 1 galaxies which are type 1, but their broad lines are relatively narrow, below 2000 $\rm km\,s^{-1}$. Radio galaxies in turn are classified as Fanaroff-Riley Class I and Fanaroff-Riley Class II, according to their value of the parameter $R_{\rm FR}$ being smaller or larger than 0.5, where $R_{\rm FR}$ is the ratio of the distance between the highest surface brightness regions on the opposite side of the central galaxy to the total extent of the radio source.

1.2 Recent observational developments

Our understanding of the details of nuclear activity is rapidly increasing. There seems to be no alternative to the basic picture: nuclear activity is powered by accretion onto a massive black hole and the material surrounding a black hole consists of two principal phases: relatively cold optically thick gas and a hot optically thin plasma (see *e.g.* [1]). However, the details of the accretion process are still under study. Major progress is being done and is expected due to the new observational material. I will emphasize three aspects which seem to me particularly important from the point of view of understanding the behavior of the material close to a black hole.

1.2.1 X-ray spectroscopy

Two excellent X-ray satellites are currently providing us with the new data: CHANDRA and XMM. The energy resolution of the first one can be as high as 0.03–0.05 Å with the use of ACIS-S/HETG instrument, at energy smaller than 1 keV [2]. The XMM instead has larger surface area and therefore better photon statistics which is important for relatively faint sources like AGN. Together, they take us into the field of true X-ray spectroscopy, making all traditional spectroscopic tools now available. It supplements the previous studies being done in optical and UV band and at the same time surpasses them since the spectroscopic features characteristic of the material very close to a black hole are expected to be seen, and they are seen, in X-ray band. In the future, important data are expected from ASTRO E-2 expected to be launched in 2005 and next major qualitative improvement will come with the Constellation X project, which may be in operation within several years.

1.2.2 Broad band AGN monitoring

Active Galactic Nuclei radiate in broad energy band: roughly comparable luminosity is emitted at each band from IR through optical, UV and X-rays, sometimes extending towards gamma ray band as well. Broad band observation is needed in order to asses the bolometric luminosity of a source. They are also strongly variable in all energy bands. Therefore, simultaneous observations in various energy bands are of much higher value than just random collection of measurements. However, even simultaneous measurements do not provide a perfect information on the state of an active nucleus since the source, although frequently point-like on a plate, is in reality extended and various more distant parts of the source respond to the changes in the innermost part with some delay. Observations of such delays, in turn, are crucial to understanding the complex structure of the nucleus, including black hole mass measurement. The program of broad band monitoring was first introduced extensively by AGN Watch team and now increasing amount of data is collected in such a mode, combining satellite and are crucial to ground-based measurements. Papers reporting such measurements take astronomers close to high energy physics people, with a hundred of names on a single paper. However, the value of such data collection is enormous even if not exploited fully yet due to the lack of appropriate broad band models of a nucleus.

1.2.3 Large surveys

Most of our believes on what is typical and what is not for an AGN relies on extensive observations of a few dozens of objects. All fantastic results about the accretion flow pattern close to the horizon of a black hole are based practically on a single object – MCG-6-30-15! This is by no means enough so systematic searches for new objects are under way. Particularly valuable will be the Sloan Digital Sky Survey (hereafter SDSS), in progress. It already contains more than two thousands of AGN and brought the discovery of the currently known most distant AGN. Such a large sample will allow both to determine much better the typical properties of AGN and to find extreme cases which are always a challenge to the models.

2 General overview of AGN

AGN cover a broad range of bolometric luminosities but when viewed closely they seem to form quite a uniform family of objects. Most unification schemes conclude that the activity is characterized by only a few basic global parameters. They are: mass of the black hole, M, accretion rate, $\dot{M}$, inclination angle, $\imath$, and a parameter describing radio loudness. This last parameter is frequently thought to be related to the angular momentum parameter of a black hole, a, although this hypothesis remains to be checked. For the moment it seems that a set of such four basic parameters should be able to explain the observed diversity. It does not mean, however, that indeed, adopting the values of these parameters, we are able at present to reconstruct an active nucleus.

The applicable range of masses and accretion rates can be easily judged from the observed range of bolometric luminosities. Astronomical sources usually do not exceed significantly the Eddington luminosity determined by the mass M of an object

$$L_{\mathrm{Edd}} = \frac{4\pi G M m_{\mathrm{p}}}{c\sigma_{\mathrm{T}}} \sim 1.3 \times 10^{38} \frac{M}{M_{\odot}} \ [\mathrm{erg\,s^{-1}\,cm^{2}}], \tag{2.1}$$

where G is the gravitational constant and c is the light speed, σ_{T} is the Thomson cross-section and m_{p} is the proton mass. Therefore, the brightest quasars contain black holes with masses up to $10^{10}\ M_{\odot}$. As for the faintest objects, there are arguments based on independent determination of the black hole mass that they still rather contain considerably massive black holes (10^5–$10^6\ M_{\odot}$) but their luminosities are much lower than the Eddington luminosity. The typical accretion rate needed to support AGN activity can be nicely estimated at the basis of statistical considerations. Comparison of the integrated light of all QSO with the activity remnants in the form of black hole masses of in nearby galaxies yields rest mass to energy conversion between 0.04 (D. Merritt, private communication) and 0.15 [3], on average about 0.1 as already determined in 1982 by [4]. Adopting efficiency 0.1, an accretion rate about 100 $M_{\odot}$ is needed for the brightest sources and $10^{-5}\ M_{\odot}$ for the faintest ones.

2.1 Spatial resolution

The direct cause of the difficulties with detailed understanding of the nuclear activity is the relatively low spatial resolution. Characteristic radius of an active nucleus is provided by the radius of a non-rotating Schwarzschild black hole horizon, $R_{\rm Schw}$

$$R_{\rm Schw} = \frac{2GM}{c^2} \approx 3 \times 10^5 \frac{M}{M_\odot} \ [\rm cm]. \tag{2.2}$$

Most of the energy, without much doubts, is released within the distance $10-100\ R_{\rm Schw}$. Nearest AGN are close to the center of the Virgo Cluster, at a distance 20 Mpc. Typical angular resolution of observations is of order of $1''$, and at such a distance it corresponds to a spatial resolution of $\sim$100 pc. For a standard black hole mass of $10^8\ M_\odot$, 100 pc corresponds to $10^6\ R_{\rm Schw}$. Special technics like VLBI in radio band or speckle photometry in the optical band allow to do better but still we map only the outer region of an active nucleus. Clues to the processes going on close to a black hole come only from the analysis of the radiation spectra and their variability.

2.1.1 Jets

It is a paradox that while the activity of AGN is powered by accretion what we actually see directly is always an outflow of material, with jets being the most spectacular form. Jets – narrow streams of material outflowing from the innermost region – are well resolved since they form an extended feature which sometimes goes far beyond the optical image of the host galaxy. From radio VLBI images we can see the formation of a jet at a distance as small as $100\ R_{\rm Schw}$ in the case of the nearest strong radio galaxy M 87 ([5]). Jets are also seen in the optical band – an excellent example being the most famous quasar 3C 273. CHANDRA satellite allowed also to resolve jet emission in X-ray band in a number of objects [6] (and the references therein).

2.1.2 Water masers

In a number of relatively faint but nearby Seyfert galaxies the observations made at 22 GHz allow to map an active nucleus at distances of a fraction of a pc. Water maser emission comes from several clumps of material and the measured Doppler shift of the observed maser line allow to determine the kinematics of the emitting gas. In all but one source (NGC 1068) the emitting gas clumps are parts of an edge-on disk in Keplerian motion around the gravity center.

2.1.3 True nucleus; starlight contamination (excess) and absorption in host galaxy or by dusty/molecular torus (type II objects)

True nucleus is unresolved and our understanding of accretion is based on radiation spectra. Therefore, precise determination of the intrinsic spectrum of the nucleus is of vital importance. However, this is by no mean easy. There are four problems.

The first one is the extinction by the material in our own galaxy. This is an important effect in soft X-ray and optical/UV band. However, the properties of the interstellar medium are rather well known and the extinction in various directions on the sky is mapped (*e.g.* [7]), therefore correcting any AGN spectrum (*i.e.* dereddening) is relatively easy.

The second one is the extinction by the interstellar material of the host galaxy. The importance of this effect is more difficult to asses since the result depends both on the orientation of the host galaxy and the luminosity of an active nucleus. The extinction is unessential for most of the bright quasars since they clear more easily the line of sight across the host galaxy. However, it is very important for many Seyfert galaxies and some quasars. It is quite possible that the effect is frequently confused with the effect of the presence of dust/molecular torus considered below.

The third one is the contribution of the inner stellar cluster to the observed spectrum in the optical band. This effect is unimportant for QSO but for Seyfert galaxies the starlight component frequently dominates. The effect depends on the aperture of the instrument which complicates the comparison of the spectra obtained with various instruments. Stellar subtraction is needed but both the amount of the starlight and its spectral shape are not easy to determine (*e.g.* [8]).

The fourth one is the presence of the dusty/molecular torus. Its presence and role was demonstrated by [9] who analyzed polarized spectra of several AGN. Those AGN classified as type 2 did not show strong UV emission and the presence of broad emission lines typical for type 1 (normal) AGN. However, when viewed in polarized light, they had typical continuum with broad lines although relatively faint and therefore indistinguishable from the starlight, some possible extended emission and contribution from the Narrow Line Region. It is now widely believed that such a torus exists in all (at least in all bright) AGN. It obscures the view to the nucleus at high inclination angle with respect to its equatorial plane and its emission is directly seen in the IR. The chemical composition of the dust is still under discussion but it rather seems to be different from the dust of interstellar medium of the Galaxy (*e.g.* [11]). Object heavily obscured by its dusty/molecular torus is just type 2 AGN and the emission from its nucleus is only seen indirectly, due to the scattering of photons by electrons or distant dust, or sometimes directly in very hard X-rays if the column density of the torus $N_{\rm H}$ does not exceed considerably 10^{24} cm^{-2} [12]. However, moderate extinction may well be either due to the dusty torus or due to the host galaxy and therefore may be not all objects classified as type 2 at the basis of their spectra are indeed type 2 as defined above which complicates statistical studies.

Obscuration due to the dusty/molecular torus is the basis of the unification scheme, with an inclination being the only parameter responsible for the apparent existence of type 2 objects. Therefore, if we are interested in the accretion flow pattern close to a black hole we should concentrate solely on type 1 unobscured objects. We will adopt this approach although the reality may be more complicated: there are some arguments in favor of certain **intrinsic** differences between the Broad Line Region itself in type 2 and type 1 objects (*e.g.* see [15] for scattering efficiency and [16] on variation of BLR size during type 1/type 2 transitions in a single object).

2.2 Broad band spectra

2.2.1 Bright radio loud AGN

Broad band spectra of radio quiet objects usually consist of several (up to four) major components. An example of well studied object is show in Figure 2 of [17]. Such a strong emission in gamma range and very small contribution from the disk is characteristic for blazars. We see those sources almost along the jet and with the bulk motion of the jet characterized by the Lorentz factor Γ of order of 10 the Doppler boosting is very strong. In some sources (most BL Lac type objects) only the jet emission is visible. It is now frequently accepted that those sources are basically the same as other radio galaxies apart from the special orientation towards the observer but the issue is still under discussion – some systematic intrinsic differences like higher Lorentz factor are also not excluded.

2.2.2 Bright radio quiet AGN

The broad band spectra of radio quiet objects are not so well known due to the problems discussed in Section 2.1.3 but collecting the data for various classes of objects and combining it allows to see a basic pattern.

The total spectrum is dominated by a Big Blue Bump component which extends through optical and UV band to (sometimes) soft X-rays. Separate broad hump is seen in the IR, and a basically power low component is seen in the X-ray band. Big Blue Bump and IR hump are best seen in bright quasars but they are present also in Seyfert galaxies although the subtraction of the starlight and correction for extinction make its exact determination difficult for many objects, including the famous Seyfert galaxy

MCG-6-15-30. Particularly well determined are the composite quasar spectra made from several objects since by combining the high and low redshift quasars we narrow the unobserved gap between the far UV and soft X-ray band (*e.g.* [28]). On the other hand X-ray spectra are better studied for Seyfert galaxies since those objects are relatively nearby and have higher count rate. The X-ray photon index is usually around 1.9, although with significant dispersion and the power law extends up to ~100 keV, as determined from Beppo-SAX and OSSE (*e.g.* [18]).

Combining knowledge from sources of different luminosity to the Eddington luminosity ratio has to be done carefully. Studies of galactic accreting black holes indicate significant spectral trends and similar trends seem to be present in AGN, with brighter objects having stronger Big Blue Bump and weaker X-ray emission. Also the extension of the X-ray spectrum may not be the same in all objects: galactic sources with strong soft component (*i.e.* in soft/high state) have an X-ray emission extending far beyond 1 MeV while those with week/absent soft component (*i.e.* sources in their low/hard state) have an X-ray power law extending only up to 100 keV. In the case of bright radio quiet quasars there are no actual measurements of the extension of the X-ray power law.

2.2.3 Faint AGN/normal galaxies

Determination of the broad band spectra of faint AGN is extremely difficult. Errors of the measured X-ray slopes are large, and optical/radio measurements usually provide only upper limits to the nuclear emission.

2.3 Models of the principal components

2.3.1 Jet models

The formation and the dynamics of jets is not well studied yet but if the jet kinematics (bulk motion, electron distribution) is assumed the jet spectra can be computed. The principal components are formed as a result of synchrotron emission, Comptonized synchrotron (SSC), and Comptonization of external photons, including those from the cosmic microwave background.

2.3.2 X-ray component in radio quiet objects

X-ray power law component in AGN is basically modeled as Comptonization by the thermal plasma. Assuming plasma electron temperature, optical depth, geometrical factors and the source of soft photons the spectrum can be successfully reproduced. The external photons are provided by the Big Blue Bump but intrinsic synchrotron photons may be also important for sources with strong magnetic field and/or fainter Big Blue Bump component. Very crude estimate of the spectrum of the thermal plasma can be done in the following way: cut-off energy is of order of the electron temperature and the slope is determined by the Compton parameter representing the amplification factor in the Compton process (see *e.g.* [19]):

$$F_{\rm X} = F_{\rm soft} A(\tau, T_{\rm e}) f_{\rm geom}; \qquad A(\tau, T_{\rm e}) \approx y = \tau \frac{4kT_{\rm e}}{m_{\rm e}c^2} \tag{2.3}$$

$$\alpha_{\rm E} = \sqrt{\frac{4}{y} + \frac{9}{4}} - \frac{3}{2}. \tag{2.4}$$

Since only the slope is measured accurately there is significant degeneracy in parameters of such a family of models so the geometry of the hot plasma cannot be uniquely determined.

2.3.3 Big Blue Bump component

There is no unquestionable observational proof in favor of the view that accretion in active galactic nuclei (AGN) proceeds predominantly in the form of an accretion disks. Constraints on the compactness of the region emitting the optical light in an Einstein cross (Q2237+0305) only suggests a very compact region, in a marginal agreement with the irradiated accretion disk [20]. Microlensing studies may in principle enhance significantly the accessible spatial resolution and provide us with the radial temperature distribution very close to the center if the suitable caustic passage is observed [21, 22].

However, there are three very strong arguments for such disks.

The first argument comes from the obvious axial symmetry of the flow which is most spectacularly reflected in the unification scheme of AGN (*e.g.* [9]).

The second argument is the striking similarity of the spectra and variability properties of AGN and X-ray transients, apart from obvious scaling of some properties with mass of the central black hole. Those galactic sources have low mass companions and accrete through the inner Lagrange point, so the accretion disk in those sources must form and, indeed, it is directly observed (*e.g.* [10]).

The third one is the interpretation of broad iron K_α line as originating in the innermost part of the disk. More such lines are now being detected in CHANDRA data. We will discuss this issue in some detail in Section 2.4.2.

Modeling an accretion disk spectrum is simple if we can assume that the disk is Keplerian and stationary, and radiates locally as a black body. The energy flux emitted at a given radius is in this case given by the formula

$$F(r) = \frac{3GM\dot{M}}{8\pi r^3} f(r) = \sigma T_{\rm eff}^4, \tag{2.5}$$

and the integrated disk spectrum is

$$F_\nu = \frac{\cos i}{D^2} \int 2\pi r B_\nu(T_{\rm eff}(r)) {\rm d}r, \qquad (2.6)$$

which at low frequencies leads to a power law behavior

$$F_\nu \propto \nu^{1/3} M^{2/3} \dot{M}^{2/3}, \qquad (2.7)$$

if neither the presence of the outer edge nor the inner edge of the disk modifies the spectrum.

Any departure from this simple model requires the knowledge of the disk structure.

2.3.4 Soft X-ray excess

A number of sources like NLS1 and radio quiet quasars show a significant emission in soft X-ray band of roughly a power law shape with a steep slope (photon index $\sim$3 or even more) seen above an extrapolation of a flatter hard X-ray power low. This emission is well modeled as the Comptonized disk component and it contains noticeable fraction of the total energy, up to $\sim$30% (*e.g.* [23]). In many other sources there is only some small excess above a hard X-ray power law and it may be due to the reflection by a partially ionized colder medium, being thus soft X-ray extension of the Compton reflection component (*e.g.* [24]; see also Sect. 2.4.2).

2.3.5 Dusty/molecular torus (IR component)

IR component of radio quiet objects is thought to be due to the dust emission. It is well modeled as a number of components with various temperatures. More advanced models based on radiative transfer do not reproduce well yet the observed spectra since they are based on too simplified geometry. The considerable geometrical thickness of the torus is most frequently explained by postulating a warped thin disk or by identifying torus with dusty wind from much thinner disk.

2.4 Spectral features

2.4.1 BAL, NLR, WA, BELR

On the top of the broad band spectral components there is a number of characteristic spectral features which are keys to understand the structure of an active nucleus, including black hole mass determination. In particular, line energy, net shift and profile allow to determine the kinematics of the gas and its thermodynamical state.

The outermost features are broad absorption lines (BAL) seen in the UV spectra of many radio quiet quasars. The velocity of the absorbing material, as determined from the line width, is up to 0.1 c. Lines have usually several components and the absorption is not complete, they originate in an outflowing material partially covering the central source, located at a distance of a parsec or more from the black hole.

Narrow Line Region (NLR) is best studied in the nearby Seyfert galaxies. The relatively narrow lines (below 1000 $\mathrm{km\,s^{-1}}$) are due to the interstellar material of the host galaxy irradiated by the nucleus also some contribution from the shock heating is not excluded.

Warm absorber (WA) seen in the soft X-ray band of many Seyfert galaxies as well as in the UV spectra (in from of narrow absorption features superimposed on much broader emission lines) come from a highly ionized medium, outflowing from the nucleus with velocity of order of 2000 $\mathrm{km\,s^{-1}}$. The medium has low velocity dispersion and frequently several velocity components are identified, making this outflow somewhat similar to much faster BAL seen in quasars. This flow is located several light weeks from the nucleus.

Broad Emission Lines originating from Broad Emission Line Region (BELR) are characterized by the velocity dispersion of order of 5000 $\mathrm{km\,s^{-1}}$ and they come from the material predominantly in Keplerian motion, located at a distance of several light days from the black hole, as suggested by the detailed studies of the nearest Seyfert galaxies. They are usually thought to come from the gas near the disk surface, possibly from a slow wind (velocity below 1000 $\mathrm{km\,s^{-1}}$) from the disk surface.

2.4.2 X-ray disk lines and Compton reflection

The reflection of the X-rays by the colder material close to the black hole leads to formation of the so called "reflection" component which arises both due to electron scattering (*i.e.* true Compton reflection) and due to absorption and reemission of the radiation due to atomic processes. The shape of the component has characteristic turn-off at energies above 30 keV (due to Klein-Nishina effect) and below 10 keV (due to strong absorption), and the change in absorption coefficient at frequencies characteristic for each of abundant elements leads to the presence of absorption edges. Superimposed, are emission lines: predominantly strong K_α line at $\sim$6.4 keV, much fainter K_β line. Many other much fainter lines should appear in soft X-ray band. The exact energy of the K_α line depends on the ionization stage of the material: it varies from 6.4 keV from cold medium to 6.9 keV for highly ionized hydrogen-like iron. Also the shape of the reflection component at lower energies depends strongly on the ionization level of the material: with increasing ionization, the absorption effect decreases and the reflected spectrum approaches the shape of the incident spectrum.

The shape of the observed reflection component and the emission lines depends both on the local state of the irradiated material and its kinematics. The basic model assumes disk-type geometry for a reprocessor, the radial dependence of the irradiation flux (in a form of a "primary" power law), inner and outer radius and inclination angle of an observer.

The two types of processes (scattering and reprocessing) are taking place in the same medium and the efficiency of both is related but frequently, for convenience, they are treated separately in data analysis and in that case Compton reflection and iron K_α line are fitted independently.

The K_α line and the Compton reflection component are well seen in all Seyfert galaxies. In most of them the line is not very broad and its position is consistent with 6.4 keV. The interpretation complicates the possibility that a fraction of the line emission and Compton reflection may come from the distant region (*e.g.* dusty torus; see [12]) so the broad disk line may be sometimes difficult to detect. ASCA results nevertheless indicated generally quite broad line while recent CHANDRA data not always support this view (*e.g.* in NGC 5548 the line is very narrow, coming from the BLR, according to CHANDRA [13] and XMM-BeppoSAX data [14]). In some sources, with most famous being MCG-6-15-30, the line is extremely broad, having a red wing extending down to 4 keV or even lower when the source is fainter. The parameters obtained from the data for this object are extraordinary: the inner radius is very small (below the marginally stable orbit), the outer radius is not much larger – indicating a ring – and a power law index in radial distribution of irradiating flux is unexpectedly high. It is sometimes concluded that only an extraction of rotational energy from the black hole may lead to such a line profile. On the other hand, more detailed study of the entire reflection component (including partial covering by warm absorber) and a possibility of an intrinsic curvature of the "primary" may be needed in order to asses reliably the extension of the line red wing.

2.5 The basic model of accretion flow pattern in bright AGN

2.5.1 Outermost region (0.1 pc or 10^5 $R_{\rm Schw}$)

This outer region is the least understood part of the accretion flow. Only a few papers speculated about the processes there, including ionizational instability, gravitational instability and possible star formation within the disk itself (see [25]). The nature of viscosity leading to accretion is difficult to asses: simple adoption of α viscosity leads to very long timescales at the region of order of a few parsecs. As for MRI instability, it may not operate there efficiently anyway due to low ionization of material. A mechanism of angular momentum transfer based on gravitational instabilities may be a more promising hypothesis. Water maser observations in rather faint objects are the only observational constraints for this region.

2.5.2 Intermediate region (100 $R_{\rm Schw}$)

Excellent review of [26] discusses most of the issues which are important from the point of view of this region so we only summarize it here shortly.

Optical/UV spectra are rather well reproduced by a standard optically thick stationary Keplerian accretion disk, if the spectra of AGN are taken with an aperture small enough to diminish strongly the contamination by the starlight.

The extension of the region described by the disk actually strongly depends on the accretion rate. In bright quasars disk radiation strongly dominates the bolometric luminosity, with a marginal contribution from the hot plasma. Perhaps, in analogy with the soft state of Cyg X-1, the X-ray emission is non-thermal (*e.g.* [27]), coming from the scarce flares, and the standard disk most possibly extends down to the marginally stable orbit. In fainter quasars and many Narrow Line Seyfert 1 galaxies the disk again dominates but there is a considerable contribution of Comptonization by a moderately hot plasma (up to $\sim$30%) shaping the soft X-ray band (*e.g.* [28] for a composite spectrum, [29] for PG 1211+143), so the disk structure close to the black hole is already somehow modified. In Seyfert galaxies the hard X-ray part of the spectrum contains a considerable fraction of the bolometric luminosity so the standard disk part starts at rather larger radii and the disk is strongly irradiated.

2.5.3 Innermost region (10 $R_{\rm Schw}$) (inflow/outflow)

The need to explain strong X-ray production in many AGN and the observed variability leads to conclusion that the flow close to a black hole is more complex and a number of different geometrical arrangements of various flow components: cold emitting/reflecting material, very hot Comptonizing medium and moderately hot emitting/reflecting/Comptonizing medium were suggested.

We can organize those models roughly in four major classes (see Fig. 1).

♠ Lamp post/shock

In a number of papers an X-ray source irradiating the disk was envisioned as a point source located at the symmetry axis above a black hole, at a distance of a few – a few hundred $R_{\rm Schw}$. The physically plausible scenario for such a model was proposed by [30]. Hot material somehow evaporated from the disk surface forms a turbulent magnetohydrodynamical outflowing stream.

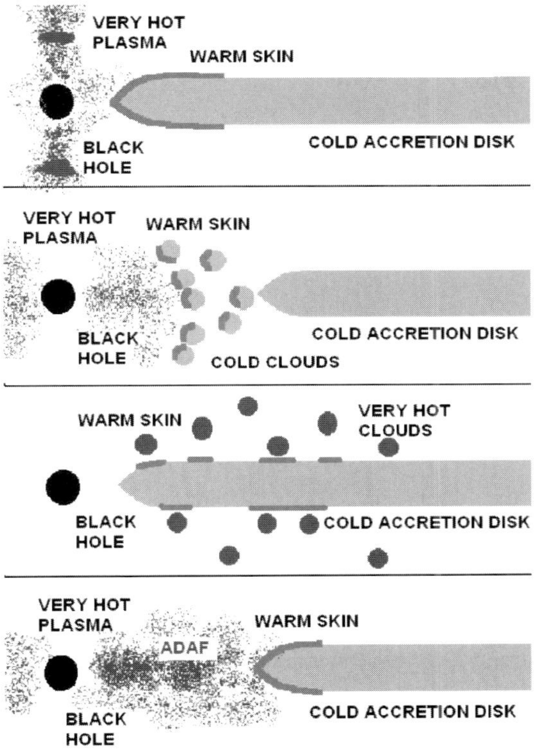

Fig. 1. The schematic picture of the three types of models of the innermost part of accretion flow into a black hole: **a)** lamp post **b)** clouds **c)** flares and **d)** inner hot flow.

This stream may form a well collimated jet further out, as seen in radio loud objects. At the basis of this stream, considerable energy dissipation takes place, thus accounting for the observed X-ray power law component.

♠ Clouds
The inner disk may be disrupted into large cold fragments due to some violent instabilities. Such clouds will subsequently collide and loose mass leading to formation of the very hot medium very close to the black hole. Cold clumps may survive and carry a significant fraction of the mass into black hole or they may evaporate, and close to the black hole the flow may be totally dominated by a very hot plasma. We refer to such a quasi-spherical flow of clouds surrounding an inner source of hard X-ray emission as "clouds". A model based on this motivation was discussed by [31] and in many subsequent papers. Other type of cloud models based on strong

magnetic fields was introduces by [32]. Clouds forming spontaneously within a hot medium were discussed by [33] and [34] but those clouds appeared to be optically too thin to provide a good model.

♠ **Flares**

Another approach is based on assumption that the cold disk is not disrupted and extends down to the marginally stable orbit. Magnetic field is amplified in its interior at the expense of the rotational energy and finally field lines leave the body of the disk due to buoyancy and lead to formation of strong short-lasting flares above the disk surface. Accelerated hot plasma cloud reach considerable outflow velocity. Disk surface below the flares becomes considerably ionized due to the X-ray radiation. Such a model, known from the solar corona, was already considered by [35] and [36] and it was recently developed by many authors, although mostly in the context of galactic sources (*e.g.* [37,38]) since it accounts in a natural way for stochastic nature of variability [39] and red noise character of the power spectra.

♠ **Hot inner flow/ADAF**

Finally, it is also possible that the formation of the initially faint corona above the outer parts of the disk leads to subsequent disk evaporation and the complete replacement of the cold disk flow with a very hot optically thin flow. Such an optically thin two-temperature inner flow was considered by [40] and particularly well studied in case of advection dominated flows (ADAF; [41,42]). Warm medium in this picture forms in the transition zone where the hot and cold flows overlap.

The simplest spectra models usually applied to the data cannot differentiate between these geometries. Such spectra are frequently characterized by the inner radius of the (cold) disk, its temperature, the hot plasma parameters (electron temperature and the optical depth) and the ratio of the disk to hot plasma luminosity. Sometimes disk black body emission is replaced with the Comptonized disk emission to account for the soft X-ray excess. Reflection component from the cold phase is also included, and its normalization determined. In Seyfert galaxies only the hot plasma parameters and the reflection component are relatively well determined (electron temperature typically 50–500 keV, optical depth 0.3–1.5 or 0.1–0.6 if spherical or slab geometry is adopted, reflection amplitude $\sim$0.75; [43]).

Such a description is appropriate for all geometrical models discussed above and derived parameters simply translate into model parameters. In the case of flare models inner disk radius translates into the radius where ionization becomes very high so the reflection looses its characteristic cold reflection spectral properties (strong absorption below a few keV, iron line and edge). An example of the model degeneracy is shown below.

An observational relation was found between the photons index Γ of the hard X-ray power law and the normalization of the reflected component [44].

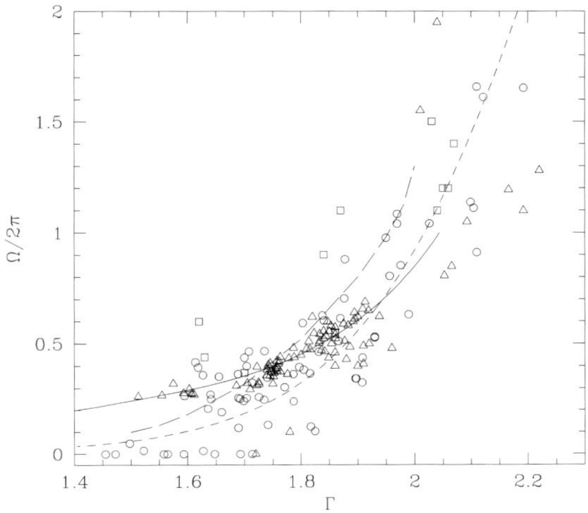

Fig. 2. The correlation between the slope of the hard X-ray power law and the amplitude of the reflection component. Points represent the data for a number of AGN and GBHs [43, 44]. Continuous line represents the flare model (after [46], short dashed line the inner hot flow/ADAF [43, 44]), and long dashed line the cloud model (after [45]).

However, in Figure 2 we plot the data from [44] and the relations expected from all three models (cloud model using results of [45], flare model after [46], and inner hot flow after [44]). We see that the theoretical curves roughly coincide with the data since all models give the coupling between the amount of soft photons, the cooling efficiency of the hot plasma and amount of reflection combined with hard X-ray thermalisation.

2.6 *The accretion flow in faint sources*

Accretion flow in faint sources may have a different character. A cold stream of matter may not be present. In some nearby elliptical galaxies and in the case of the Milky Way center, due to the high resolution X-ray images, it is possible to estimate directly the temperature and the density of the available gas. This temperature is high, of order of 1 keV, and the density is of order of a few (*e.g.* [47,48]). This material may be a subject of spherically symmetric Bondi accretion onto a central black hole. The accretion rate of such a flow is easy to estimate

$$\dot{M}_{\rm B} = 4\pi R_{\rm B}^2 \rho v_{\rm s}; \qquad R_{\rm B} = \frac{GM}{v_{\rm s}^2}; \qquad v_{\rm s}^2 = \frac{kT}{m_{\rm H}}. \tag{2.8}$$

The expected luminosity from such an accretion flow, $L_{\rm bol} = \eta \dot{M} c^2$, assuming typical efficiency $\eta \sim 0.1$ is many order of magnitude higher than the observed luminosity. It means that either the flow is extremely inefficient, or the Bondi accretion flow does not represent the flow close to a black hole. The second possibility may be due to the presence of considerable angular momentum halting inflow or due to strong outflow. There is also a possibility that such accretion is strongly non-stationary and periods of high accretion are separated by periods of mass accumulation. Recent measurements of polarization in Milky Way center seems to indicate that there is indeed much less material closer to a black hole than at Bondi radius.

3 Recent progress in stationary models of bright radio quiet objects

Very simple phenomenological models do not allow for much of an insight into the accretion pattern at the innermost 10 $R_{\rm Schw}$. With a continuous accumulation of observational data of increasingly higher accuracy/complexity a comparable progress in theory is expected to bring better understanding and this theory is needed for the interpretation. In this section we will discuss the current progress in modeling of the mean spectra while in the next section we will concentrate on various aspects of the variability.

The current theoretical progress is extremely non-uniform. We have quite detailed knowledge of some aspects while others, equally important, are almost untouched. Therefore, if we try to go any further than the simple facts overviewed above we do not have a complete self-consistent picture any more. Instead, we have a list of problems which are partially solved, and a list of unsolved issues. Generally, there was considerable progress in the description of the radiative transfer in a hot and a cold(disk) plasma which is directly related to the spectra formation. However, we have poor understanding of the plasma dynamics and, consequently, the energy dissipation, and any spectra computations need this information as an input.

Therefore, in AGN study we are at a similar stage as stellar structure people before the "invention" of nuclear processes. However, we are in much worse position. In stars the region of energy generation is deeply inside most of the time, hidden from our sight, and the description of the energy generation is unessential from the point of view of the stellar atmosphere (although it is essential for the stellar evolution). In AGN the energy generation region is exposed and its structure is clearly much more complex that that of any spherically symmetric star.

The lack of dynamical information and energy dissipation results in the development of a number of specific "models" or "scenarios".

The basic step forwards dates back to the papers which provided the major step in accretion description, namely [49] and [50]. The complete set of equations for an accretion disk required the description of the viscosity – without a viscous torque accretion would not proceed. In those papers an α viscosity prescription was proposed: viscous torque is proportional to the pressure. The idea was invented at the basis of simple considerations of the probable properties of either turbulence or random self-generated magnetic field and originally applied to the description of vertically averaged disk structure.

For a specific case of a stationary Keplerian disk model emitting locally as a black body the disk spectrum actually did not depend on the adopted assumption. However, any departure of the model from those assumptions leads to the explicit dependence on the viscosity, and any model predicting the X-ray spectra must do just that. In order to make a progress, two extreme approaches were adopted.

The first one was the search for the true physical viscosity mechanism and development of its description. This approach, best implemented by Balbus, Hawley and their collaborators (for a review, see [51]) resulted in rediscovery of the magneto-rotational instability, hereafter MRI) which develops spontaneously in any ionized differentially rotating plasma. Current status of progress shows that the α mechanism is actually not a bad description. In 3-D simulations the value of α computed from the model is fluctuating more or less around 0.1, with some increase close to the marginally stable orbit [52]. Considerable fraction of energy (up to 30%) is dissipated in large magnetic loops emerging from the disk interior [53]. However, these computations are done for a plasma without cooling, not mentioning two-temperature plasma which may be present. Recently [54] included cooling and radiation pressure but in that case also the computations never went longer than a single thermal timescale while in principle it should cover several viscous timescales. The 3-D dynamical computations are so computer time consuming that the progress is not easy.

The second approach was to adopt the scaling, apply it even locally (*i.e.* to the description of the disk vertical structure as well) and look at possible observational consequences in search for a support of the adopted scaling or the need to its modification. The biggest discovery made along this line was the explanation of the nature of dwarf novae outbursts [55]. The same kind of instability is most probably responsible for the transient phenomena in many X-ray novae [56]. This kind of description was applied to AGN as well, either in the context of a single phase disk accretion or to the description of two-phase disk/coronal flow (*e.g.* [57,58]). Prescription for a viscous torque allowed to study classical Shakura-Sunyaev disk structure in detail and to find three other branches of solutions applicable either for very high accretion rates (slim disks, [59]) or for very low accretion rates (radiatively efficient disks – SLE, [40]; advection dominated accretion flows – ADAF, [41, 42]).

The third approach is to further develope phenomenological models, including new phenomena, and eventually trying to constrain the parameters involved at the basis of comparison with observational data.

3.1 Hot plasma cooling – modeling the "primary" continuum

As was shown in [60], plane-parallel geometry of disk/corona leads naturally to a spectrum slope of order of 0.9 (photon index 1.9) if all gravitational energy is dissipated in the corona and half of it later reprocessed by a disk providing soft photon input for Comptonization in the corona. Any dissipation in th disk will lead to steeper (softer) spectra so relatively hard spectra in most objects must result from somewhat modified geometry: clumpy corona, partial overlapping, very high ionization of disk surface and/or outflow.

The Comptonization of soft photons by a hot thermal distribution of electrons leads to formation of a spectrum which is not a strict power law so the process was considered in more detail by several authors. Various assumptions about the exact shape of the region were considered (*e.g.* [61]) but the effect is predominantly in the modification of the power law slope and therefore difficult to distinguish from the change in the optical depth. Slight departures from a strict power law dependence, even if predicted, are difficult to find in the present data.

Interesting direction of the study is considering the mixture of thermal/nonthermal plasma since in realistic situations (like solar corona) such mixed distribution is expected. This will be particularly important when more constraints of the gamma-ray spectrum for radio quiet objects will be available due to INTEGRAL and GLAST.

In case of some models, like flares, which must be characterized by strong magnetic fields, interesting constraints are obtained if synchrotron photons are considered as seed photons for Comptonization. [63] concluded that actually in flare models plasma will tend to cool much too fast, or in other words, in some objects it is impossible to reconcile the required magnetic field with the observed level of X-ray emission.

Most detailed predictions taking into account several processes (synchrotron, bremsstrahlung and Comptonization of those photons) were made for ADAF models, with computations performed at each radius which was possible due to assumption about the dynamics of the flow (*e.g.* [62]).

Computational methods used to predict the spectrum can be divided into three groups. First group form semi-analytical methods. They are simple and convenient for X-ray data analysis. Two such models, COMPTT and COMPS are included in a standard package XSPEC for data reduction.

In many applications like computing the disk hot skin a diffusion approximation is used in a form of Kompaneets equation.

However, if geometry is complex and the plasma optically thin, the Monte Carlo computations must be performed since they include all the effects, including angular dependence.

3.2 Hot plasma heating

The progress in this field is quite poor although there are only two basic possibilities: either the gravitational energy is first used to heat ions, and later on the energy is transmitted to electrons, or a significant fraction (most?) of energy is used directly to heat electrons. The issue is discussed in a few papers only, without a firm conclusion [64]. The general believe is that in case of hot accretion flow ions are heated while in the case of magnetic flares electrons are heated directly. If ions are the prime collectors of energy, they will transfer the energy to electrons (at least) at the rate given by the Coulomb interaction. Since Coulomb interaction is not efficient this leads to development of the ADAF type of flow, with ions retaining most of its energy and carrying it down to the black hole.

Classical ADAF solutions are based on the assumption that the hot flow viscosity is parameterized by α and ion-electron energy transfer is provided by Coulomb interaction. This allow to compute the radial dependence of the flow parameters. The characteristic property is that ion temperature is close to the virial temperature and ADAF solution exists only at limited range of radii and accretion rates. Modified solutions of this type take into account the possibility of significant outflow (ADIOS – Advection Dominated Inflow Outflow Solutions proposed by [65]) or the effect of convection (CDAF – convection dominated advection flow; [66]).

Similar solutions are constructed for coronal accretion flow in two-phase accretion flow where the accretion flow is naturally sandwiched into hot and cold flow due to differences in the cooling mechanism. Specific models must assume either the radial distribution of the corona strength (*e.g.* [62]) or to consider a physical mechanism for disk/corona mass exchange [67,68].

3.3 Radiative transfer in X-ray irradiated medium in hydrostatic equilibrium

Any modification of the disk emission from the simplest black body description depends on the assumption of the disk structure. One of the reasons for this departure is the irradiation of the disk by the hard X-ray flux.

Irradiation leads to a thermal instability in an atmosphere in hydrostatic equilibrium, as can be indirectly concluded from the studies of [69] and [70]. Since several cooling and heating mechanisms operate (Compton heating/cooling, bremsstrahlung and atomic processes), there is a range of pressure values when three solutions for the temperature and density are allowed, with an intermediate one being unstable. The transition from hot branch (almost at Inverse Compton temperature) to a cold branch is possible at any pressure as a discontinuous passage. Unique solution for the temperature profile can be achieved only if the electron conduction is included [71]. It shows that the transition is steep and happens close to the pressure of maximum extension of the hot branch.

However, solving the radiative transfer with the electron conduction is a difficult task and is was not performed. Either the heating/cooling is described approximately and the radiation spectra cannot be computed (*e.g.* [72]) or the instability strip is treated approximately. In computations of radiative transfer done by [73] for galactic sources and by [74] for AGN the transition was done numerically, by subsequent iteration between the density determined from hydrostatic equilibrium and the temperature determined from radiative transfer which apparently does not display a discontinuity (see the discussion in [74]). In solutions obtained by [75] and [76] the sharp transition was adopted at a position fixed arbitrarily during the first iteration.

The actual transfer equation is solved including the effect of Comptonization and all atomic processes. Various technics are applied by various groups but the outcome is similar: sharp transition in the disk vertical structure caused by the thermal instability leads to a presence of a Compton reflection component coming from relatively cold material if the irradiating spectrum is hard. If the irradiating spectrum is soft the contribution the upper disk skin is not completely ionized and the contribution from ionized reflector shows up. Apart from the iron K_α line, a number of soft X-ray line are predicted with the equivalent width of order of 10 eV. Such lines are possibly just marginally detected in NGC 5548 [77] and Ton S180 [78]. The postulated detection of broad soft X-ray disk lines by [79] does not seem probable since the lines are an order of magnitude more intense than predicted by any of the models.

3.4 Electron conduction; disk evaporation/condensation

A global model should predict the formation of a hot X-ray emitting plasma. One of the possibilities is the formation of a hot accreting corona due to the disk evaporation. Such a model was first considered for the case of cataclysmic variables by [67] and adopted to an AGN environment by [68]. The estimate of the evaporation rate is based on the consideration of the radiative heating/cooling as well as electron conduction close to the disk surface. A stationary model is determined uniquely if the viscosity in the accreting coronal phase is adopted. Such a stationary model predicts significant evaporation in the outer part and condensation tendency in the inner part of the flow, and the importance of evaporation increases with the decrease of accretion rate, in agreement with the observed trend. However, the description of the interaction of the two-temperature accreting corona with the disk is not quite satisfactory yet since ions have large mean free path, as stressed by [80].

4 Recent progress in variability studies of bright radio quiet objects

Variability is one of the basic properties of active galaxies distinguishing them from their non-active relatives. Variability was a key in establishing several of the distance scales in an active nucleus and remains a promising tool for a better understanding of the character of the innermost part of the accretion flow although the results with respect to the last point are not quite satisfactory yet.

4.1 General outline of variability studies

Observations suggest that there are some systematic trends in the spectral properties with the luminosity. On the other hand the light curves of AGN do not show a specific pattern but look "noisy". Therefore, in variability studies either trends are explored, or a standard approach to a noise is adopted, that is the power spectrum is constructed from the light curve in a given energy band, and studies of correlations/delays are performed to shed some light on the nature of interactions. Some less standard technics are being used as well.

The variability of radio loud objects is the most spectacular in all energy bands. However, radio quiet objects are also significantly variable. In short timescales – minutes/hours – the most significant variations are seen in the X-ray band, frequently by a factor of 2, sometimes more, with short periods of significant decrease in the level of X-ray emission. While the regular variability is best explained as intrinsic (*i.e.* related to accretion rate in the innermost part) the dim periods may be either intrinsic or caused by obscuration effect. An argument in favor of the intrinsic character is the change of the size of the BLR after such events.

In the optical/UV band the variability seems to be much more smooth but if viewed for a longer periods (years or more) the amplitude is possibly even larger than in X-ray band.

4.2 Time delays

Time delays between the continuum emission in different wavelengths are short but they were recently measured in some Seyfert galaxies. The variations seem to originate somewhere in far/UV – soft X-ray band and propagate both towards higher and lower energies. Hard X-ray emission shows a delay of order of a thousand of seconds [82] and a few thousand seconds delays were measured in the optical band [81]. The interpretation of these delays is not quite clear and the origin of the variability is not understood.

There was also an extensive search of the delay between the iron K_α line and the continuum and between the Compton reflection and the continuum. However, even after long ASCA and SAX monitoring there is no clear answer to this issue.

Much more clear situation is in the case of the delays of broad emission lines with respect to the continuum. Those delays, of order of a few days up to one month (depending on the source and the studied emission line), measured in many objects simply represent the distance of the BLR from a black hole. Determination of the delay, together with determination of the line width is the best way to determine the mass of the black hole. Accuracy of this method is quite high – systematic errors are smaller than a factor 2–3 – and results were obtained for several objects (*e.g.* [83, 84]).

4.3 Power spectra

Long and well covered light curves allow also to construct power spectra although errors are large. Power spectra in X-ray band were determined by several authors. Those spectra are broad band, generally similar to broad band power spectra of galactic sources although the frequencies are correspondingly lower due to much larger black hole mass. If long enough light curve is available, the power spectra show significant departure from a single power law shape and can be modeled as a power law with a break (more complex models are also allowed due to large errors).

There seems to be a rather good correspondence between the power spectra of Sy1 galaxies and GBHs in their low/hard state, with the frequency scaling strictly with the black hole mass [85] although errors in AGN power spectra are large and Monte Carlo simulations are needed to determine those errors reliably [86]. The situation is less clear in the case of NLS1 galaxies – their power spectra seem to extend relatively more towards high frequencies than expected but again, errors are large and further monitoring is clearly needed.

No quasi-periodic features were reliably detected so far in AGN power spectra.

Power spectra in the optical band are more difficult to obtain and the errors are even larger. The results obtained so far for NGC 5548 and NGC 4151 [87, 88] suggest that they are significantly different from the

X-ray power spectra: much stronger variability is seen at longer timescales (years) while the variability at short timescales (days) is reduced.

Power spectra contain information about the plasma dynamics but at present we have no clear understanding of the meaning even of the breaking frequencies (for a discussion in the context of Cyg X-1, see [89]). Optical power spectra may suggest an agreement with the predictions of α viscosity approach to the disk structure although no quantitative comparison of the disk evolution and observations was performed.

4.4 Stationarity issue and new technics

More sophisticated approach to the analysis of the AGN variability is clearly needed since it offers an additional tool for selecting viable models of the activity. For example, a stationarity issue raised by [90] and [91] allows in principle to differentiate between the uncorrelated coronal flares (they give linear signal) and jet basis model (rather non-linear evolution). At present, the data quality is still not good enough to give firm results but some indications may come from the analysis [91].

Other methods include the structure function (*e.g.* [92]), coherence function [93] and phase delays [94]. Fourier resolved spectroscopy was successfully applied to the data for galactic sources (see [95]) and used later to models [96, 97] supporting hot inner flow against flare model.

The work was partially supported by grant No. 2P03D00322 of the Polish State Committee for Scientific Research, and by Jumelage/CNRS No. 16 "Astronomie France/Pologne".

References

[1] J.H. Krolik, "Active galactic nuclei: from the central black hole to the galactic environment" (Princeton U Press, 1999).

[2] Chandra Proposer's Observatory Guide, v.4, `http://asc.harvard.edu/edocs/docs.html`

[3] M. Elvis, G. Risaliti and G.M. Zamorani, *ApJ* **565** (2002) L75.

[4] A. Sołtan, *MNRAS* **200** (1982) 115.

[5] J.A. Biretta, W. Junor and M. Livio, *New Astron. Rev.* **46** (2002) 239.

[6] A. Siemiginowska, J. Bechtold, T.L. Aldcroft, M. Elvis, D.E. Harris and A. Dobrzycki, *ApJ* **570** (2002) 543.

[7] O. Malkov and E. Kilpio, *Ap&SS* **280** (2002) 115.

[8] J.K. Kotilainen and M.J. Ward, *MNRAS* **266** (1994) 953.

[9] R. Antonucci, *ARA&A* **31** (1993) 473.

[10] J. Frank, A.R. King and D. Raine, "Accretion power in astrophysics", 3rd Ed. (Cambridge, UK: Cambridge University Press, 2002).

[11] B. Czerny, Z. Loska, R. Szczerba, J. Cukierska and G. Madejski, *Acta Astron.* **45** (1995) 623.

[12] J.H. Krolik, P. Madau and P.T. Życki, *ApJ* **420** (1994) L57.
[13] T. Yaqoob, I.M. George, K. Nandra, T.J. Turner, P.J. Serlemitsos and R.F. Mushotzky, *ApJ* **546** (2001) 759.
[14] K.A. Pounds, J.N. Reeves, K.L. Page, R. Edelson, G. Matt and G.C. Perola, *MNRAS* (2002), submitted.
[15] S.L. Lumsen and D.M. Alexander, *MNRAS* **328** (2001) L32.
[16] B.M. Peterson *et al.*, *ApJ* **511** (1999) 185.
[17] L. Ballo *et al.*, *ApJ* **567** (2002) 50.
[18] A.A. Zdziarski, J. Poutanen and W.N. Johnson, *ApJ* **542** (2000) 703.
[19] G.B. Rybicki and A.L. Lightman (John Wiley Sons, Inc., New York, 1979).
[20] B. Czerny, M. Czerny and J. Jaroszyński, *MNRAS* **268** (1994) 135.
[21] P.Woźniak, A. Udalski, M. Szymański *et al.*, *ApJ* **540** (2000) L65.
[22] J.S.B. Wyithe, R.L. Webster and E.L. Turner, *MNRAS* **318** (2000) 762.
[23] B. Czerny and M. Elvis, *ApJ* **321** (1987) 305.
[24] B. Czerny and P.T. Życki, *ApJ* **431** (1994) L5.
[25] S. Collin and J.-P. Zahn, *A&A* **344** (1999) 433.
[26] A. Koratkar and O. Blaes, *PASP* **111** (1999) 1.
[27] M. Gierliński, A.A. Zdziarski, J. Poutanen, P.S. Coppi, K. Ebisawa and W.N. Johnson, *MNRAS* **309** (1999) 496.
[28] A. Laor, F. Fiore, M. Elvis, B.J. Wilkes and J.C. McDowell, *ApJ* **477** (1997) 93.
[29] A. Janiuk, B. Czerny and G.M. Madejski, *ApJ* **557** (2001) 408.
[30] G. Henri and G. Pelletier, *ApJ* **383** (1991) L7.
[31] S. Collin-Souffrin, B. Czerny, A.-M. Dumont and P.T. Życki, *A&A* **314** (1996) 393.
[32] A. Celotti, A.C. Fabian and M.J. Rees, *MNRAS* **255** (1992) 419.
[33] J.H. Krolik, *ApJ* **498** (1998) L13.
[34] G. Torricelli-Ciamponi and T.J.-L. Courvoisier, *A&A* **325** (1998) 881.
[35] A.A. Galeev, R. Rosner and G.S. Vaiana, *ApJ* **229** (1979) 318.
[36] G.S. Bisnovatyi-Kogan and S.I. Blinnikov, *A&A* **59** (1977) 111.
[37] J. Poutanen and A.C. Fabian, *MNRAS* **306** (1999) L31.
[38] J. Malzac and E. Jourdain, *A&A* **359** (2000) 843.
[39] B. Czerny and H. Lehto, *MNRAS* **286** (1997) 365.
[40] S.L. Shapiro, A.P. Lightman and D.M. Eardley, *ApJ* **204** (1976) 187.
[41] S. Ichimaru, *ApJ* **214** (1977) 840.
[42] R. Narayan and I. Yi, *ApJ* **428** (1994) L13.
[43] A.A. Zdziarski, J. Poutanen and W.N. Johnson, *ApJ* **542** (2000) 703.
[44] A.A. Zdziarski, P. Lubiński and D.A. Smith, *MNRAS* **303** (1999) L11.
[45] B. Czerny and A.-M. Dumont, *A&A* **338** (1998) 386.
[46] A. Beloborodov, *ApJ* **510** (1999) L123.
[47] F.K. Baganoff *et al.*, "Chandra X-ray Spectroscopic Imaging of Sgr A* and the Central Parsec of the Galaxy" (2001) [astro-ph/0102151].
[48] T. Di Matteo, S.W. Allen, A.C. Fabian, A.S. Wilson and A.J. Young, "Accretion onto the Supermassive Black Hole in M 87" (2002) [astro-ph/0202238].
[49] N.I. Shakura, *AZh* **49** (1972) 921.
[50] N.I. Shakura and R.A. Sunyaev, *A&A* **24** (1973) 337.
[51] S.A. Balbus, "Accretion disk turbulence", in The Physics of Cataclysmic Variables and Related Objects, ASP Conf. Proc. 261, edited by B.T. Gänsicke, K. Beuermann and K. Reinsch (San Francisco: Astronomical Society of the Pacific, 2002) 356.

[52] J.F. Hawley and J.H. Krolik, *ApJ* **548** (2001) 348.

[53] K.A. Miller and J.M. Stone, *ApJ* **534** (2000) 398.

[54] E. Agol, J.H. Krolik, J.N. Turner and J.M. Stone, *ApJ* **558** (2001) 543.

[55] F. Meyer and E. Meyer-Hofmeister, *A&A* **104** (1981) L10.

[56] J.K. Cannizzo, J.C. Wheeler and P. Ghosh, in Pulsations in Classical and Cataclysmic Variable Stars, edited by J.P. Cox and J. Patterson (Dordrecht, Reidel, 1982) 13.

[57] A. Siemiginowska, B. Czerny and V. Kostyunin, *ApJ* **458** (1996) 491.

[58] H.J. Witt, B. Czerny and P.T. Życki, *MNRAS* **286** (1997) 848.

[59] M.A. Abramowicz, B. Czerny, J.-P. Lasota and E. Szuszkiewicz, *ApJ* **332** (1988) 646.

[60] F. Haardt and L. Maraschi, *ApJ* **380** (1991) L51.

[61] B.E. Stern, J. Poutanen, R. Svensson, M. Sikora and M.C. Begelman, *ApJ* **449** (1995) L13.

[62] A.A. Esin, J.E. McClintock and R. Narayan, *ApJ* **498** (1997) 865.

[63] G. Wardziński and A.A. Zdziarski, *MNRAS* **314** (2001) 325.

[64] E. Quataert, "Low Radiative-Efficiency Accretion Flows", in Probing the Physics of Active Galactic Nuclei, ASP Conf. Proc. 224, edited by B.M. Peterson, R.W. Pogge and R.S. Polidan (San Francisco: Astronomical Society of the Pacific) 71.

[65] R.D. Blandford and M.C. Begelman, *MNRAS* **303** (1999) L1.

[66] R. Narayan, I.V. Igumenshchev and M.A. Abramowicz, *ApJ* **539** (2000) 798.

[67] F. Meyer and E. Meyer-Hofmeister, *A&A* **288** (1994) 175.

[68] A. Różańska and B. Czerny, *A&A* **360** (2000) 1055.

[69] G.B. Field, *ApJ* **142** (1965) 531.

[70] J.H. Krolik, C.F. McKee and C.B. Tarter, *ApJ* **249** (1981) 422.

[71] M.C. Begelman and C.F. McKee, *ApJ* **358** (1990) 375.

[72] A. Różańska, *MNRAS* **308** (1999) 751.

[73] J.C. Raymond, *ApJ* **412** (1993) 267.

[74] A. Różańska, A.-M. Dumont, B. Czerny and S. Collin, *MNRAS* **332** (2002) 799.

[75] S. Nayakshin, D. Kazanas and T.R. Kallman, *ApJ* **537** (2000) 833.

[76] D.R. Ballantyne, R.R. Ross and A.C. Fabian, *MNRAS* **327** (2001) 10.

[77] J.S. Kaastra *et al.*, *A&A* **386** (2002) 427.

[78] A. Różańska, B. Czerny and A. Siemiginowska, *ApJ*, submitted.

[79] G. Branduardi-Raymont, M. Sako, S.M. Kahn, A.C. Brinkman, J.S. Kaastra and M.J. Page, *A&A* **365** (2001) L140.

[80] H.C. Spruit and B. Deufel, *A&A* **387** (2002) 918.

[81] S. Collier *et al.*, *ApJ* **561** (2001) 146.

[82] O. Shemmer *et al.*, *ApJ* **561** (2001) 162.

[83] S. Kaspi, P.S. Smith, H. Netzer, D. Maoz, B.T. Januzzi and U. Giveon, *ApJ* **533** (2000) 631.

[84] A. Wandel, B.M. Peterson and M.A. Malkan, *ApJ* **526** (1999) 579.

[85] B. Czerny, M. Nikołajuk, M. Piasecki and J. Kuraszkiewicz, *MNRAS* **325** (2001) 865.

[86] P. Uttley, I.M. McHardy and I.E. Papadakis, *MNRAS* **332** (2002) 231.

[87] B. Czerny, A. Schwarzenberg-Czerny and Z. Loska, *MNRAS* **303** (1999) 148.

[88] B. Czerny, V.T. Doroshenko, M. Nikołajuk *et al.*, "Variability of accretion flow in the core of the Seyfert galaxy NGC 4151", *MNRAS*, submitted.

[89] K. Pottschmidt *et al.*, *A&A* (2002), in press.
[90] K. Leighly, *ApJS* **125** (1999) 297.
[91] M. Gliozzi, W. Brinkmann, C. Räth *et al.*, *A&A* **391** (2002) 875.
[92] S. Paltani, T.J.-P. Courvoisier and R. Walter, *A&A* **340** (1998) 47.
[93] M.A. Nowak and B.A. Vaughan, *MNRAS* **280** (1995) 227.
[94] S. Miyamoto, S. Kitamoto, K. Mitsuda and T. Otani, *Nature* **336** (1988) 450.
[95] M. Revnivtsev, M. Gilfanov and E. Churazov, *A&A* **347** (1999) L23.
[96] P.T. Życki, *MNRAS* **333** (2002) 800.
[97] P.T. Życki, "Simulations of X-ray spectral/timing properties in a propagation model of variability of accreting black holes" [astro-ph/0209496].

COURSE 10

HIGH ENERGY EMISSION OF AGN

A. CELOTTI

S.I.S.S.A., Trieste, Italy

Contents

HIGH ENERGY EMISSION OF AGN

A. Celotti[1]

Abstract

We present the most commonly accepted scenarios to account for the properties of the high energy emission in Active Galactic Nuclei (AGN). Because of the significant overlapping with other lectures (specifically on the high energy emission associated with the gas accreting onto the black hole) here we will concentrate mostly on non-thermal emission from radio–loud AGN.

1 Introduction

AGN have been studied for about 40 years, but many fundamental questions on how they work and their relation with the host galaxy and larger scale environment are still open.

They emit over all of the detectable electromagnetic spectrum, ranging from several tens of MHz to TeV energies, and phenomena related to nuclear activity take place from scales comparable with the Schwarzschild radius of the supermassive black hole ($\sim 3\times 10^{13} M_8$ cm, where the black hole mass has been expressed as $M_{\rm bh} = 10^8 M_8 M_\odot$) – as inferred from the fast variability timescales at high energies – to visible structures up to Mpc scales, well into the intergalactic medium. This immediately suggest a huge range of physical processes and coexisting different conditions, presumably out of thermodynamical equilibrium.

Activity in the nuclei of galaxies is believed to be ultimately powered by the gravitational energy of matter into the potential well of a supermassive black hole. Indications of (non-stellar) activity are however various in nature and origin, ranging from high energy emission (X– and γ-ray radiation), non-thermal radio emission, broad and narrow emission lines, variable flux, polarized light, images of large scale collimated outflows (jets), evidence of superluminal motion, etc. Because of this variety, many classes of AGN

[1]S.I.S.S.A. (celotti@sissa.it).

have been defined on the basis on one or more of these properties. The first (oversimplifying) choice here is to avoid a classification of the AGN family. For our purposes, we will refer to a few (basic) definitions, and basically focus on two populations. The (hope and) believe is that part of the sometimes complex taxonomy is only due to the selection band and criteria (emission line widths and ratios, variability, radio power, point-likeness, etc.), and that there are only a few key fundamental ingredients responsible for the functioning of any AGN. Furthermore and perhaps more importantly the degree of activity ranges from powers of order 10^{47-48} erg s^{-1} to levels barely detectable in contrast with the host galaxy emission. In the following we will focus on powerful and "bona fide" AGN.

As also described in other lectures, the current phenomenological "standard" picture for an active nucleus involves the presence of (see Fig. 1): a supermassive black hole ($10^{6-10}M_{\odot}$), an accretion flow – different possible physical configurations have been studied (see Blaes' lecture) – a large scale (>pc) obscuring structure, presumably geometrically flattened ("torus"), the presence (in some systems) of two symmetric well collimated jets of plasma, relatively cold clumps of gas responsible for the reprocessing of the primary continuum into broad and narrow line emission.

Within this general (simplified) scenario, we will discuss the possible origin of different high energy spectral components and the implications on the physical conditions in AGN. Given the wide range of expertise of the audience at the School and the several good specific books and reviews already available (books and articles on high energy emission in Active Galactic Nuclei include [6, 20, 29, 33]) this contribution will try to only provide a broad overview and sketch a general picture, while refer the reader interested in specific aspects to the available material. We apologize from the beginning with colleagues for the omissions, simplifications and paucity of references to important work in the field.

2 Basic properties

In order to describe the basic properties of AGN, we will simply refer to two main classes, namely radio-quiet and radio-loud sources. In fact, several studies have shown that typically only a fraction of order ten per cent of AGN are powerful radio emitters (the definition of radio loud/quiet is based on the value of the radio to optical flux ratio or radio power). The character of the radio-loud/radio-quiet distinction, and in particular the possibility that the radio loudness present a dichotomic behavior, is still under debate as recently constructed large quasar samples appear to show a smooth, although rapid, transition between the radio-loud and the radio-quiet levels of emission (*e.g.* [46]) and deep radio observations of objects traditionally

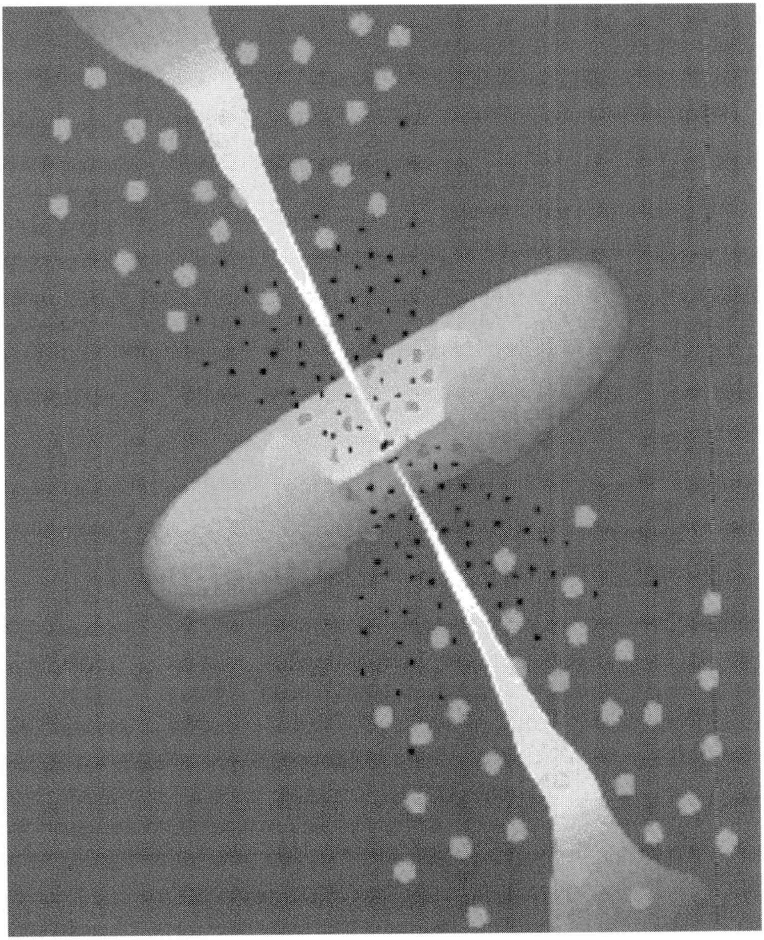

Fig. 1. Schematic cartoon (not in scale) of the main components believed to be present in an active nucleus (from [42]): a central supermassive black hole, accreting matter in the form of a disc, a larger scale dusty toroidal structure, well collimated jets (in some sources), clouds of gas responsible for the emission of broad and narrow lines (the former at smaller distances, *i.e.* higher orbital velocity).

belonging to the radio-quiet class are showing (presumably) non-thermal radio nuclei (*e.g.* [41]). Nevertheless for the purpose of these lectures we will maintain such a classification: this in fact characterizes reasonably well the different spectral energy distributions (SED) of these sources and the corresponding nature of the dominant emission processes. In fact, a part from the radio structure/jets which can be imaged on scales larger than ~0.1 pc (indicatively corresponding to milliarcseconds resolution of radio interferometers), inferences on the functioning of the nuclei have to be inferred from the spectral, variability and polarization properties of the observed radiation.

2.1 The spectral energy distributions

Average broad band spectral energy distributions of radio-quiet and radio-loud AGN (see Fig. 2) show that basically the mm to EUV continuum is similar, but that radio-loud sources emit typically a radio luminosity two or three orders of magnitude larger than the radio-quiet ones. Furthermore radio-loud objects show a flatter X–ray spectral index (defined hereafter as $F(\nu) \propto \nu^{-\alpha}$), *i.e.* harder spectrum, and the high energy emission can extend up to GeV or even TeV energies. The high energy emission of radio-quiet sources is instead typically characterized by a steeper X–ray spectrum which appears to cut off at few hundreds keV (at least in the brightest Seyfert galaxies which have been observed). Let us then look at the possible origin of such spectral components.

For sake of completeness, before focusing on the high energy emission, let us briefly recall the believed origin of the two peaks characterizing the mm to extreme ultraviolet SED of most AGN. The higher energy one, which often dominates the total energetics (although the lack of a complete observational coverage does not allow to actually measure the intensity at the peak) is believed to arise as quasi-thermal emission from the dissipation of energy of the gas accreting onto the black hole (as detailed in Blaes' lectures). The lower energy one is instead thought to arise as emission from dust at different temperatures – presumably in the form of the obscuring torus – intercepting and reprocessing part of the high energy continuum, as supported by the typical minimum in the SED around ~1 μm which well corresponds to the dust sublimation temperature.

3 Radio-quiet AGN

Typically this class comprises high redshift, powerful quasars and nearby, fainter Seyfert galaxies. These are believed to be of similar nature, Seyfert being the lower power counterparts of quasars (although Seyfert are hosted in spiral galaxies while powerful quasars are also and typically hosted in ellipticals). Seyferts and quasars are also sub-classified as type 1 and 2, depending on whether broad emission lines are present in their optical-UV spectra (absent in type 2 sources). This difference is thought to be only due to an orientation effect: the dusty torus would "hide" because of its

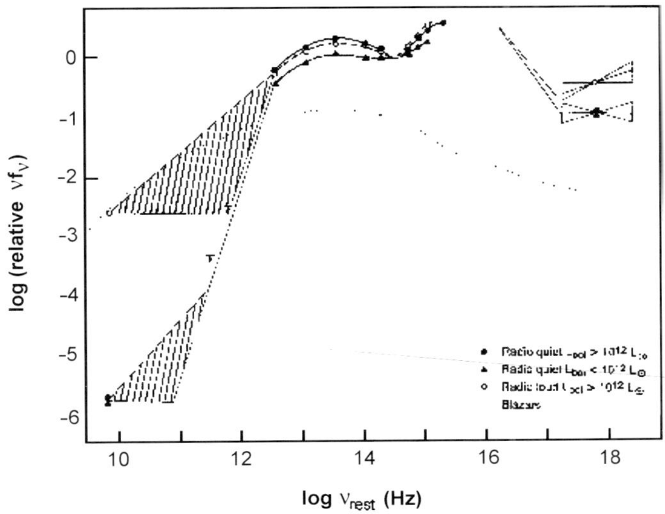

Fig. 2. Average spectral energy distributions of radio-quiet and radio-loud AGN, from the radio to the X-ray band (from [34]). Blazars are also represented: these are radio-loud sources where most of the emission is dominated by non-thermal radiation (see Sect. 5). Above the frequency range reproduced in the figure, the spectrum of radio-quiet (Seyfert galaxies) presents a cut-off at a few hundreds keV, while in radio-loud objects the harder X–ray spectrum extends approximately as a power-law at higher frequencies, peaking and turning over at GeV or even TeV energies. The vertical axis is in the $\log \nu F(\nu)$ representation.

geometrical thickness the presence of broad line emitting gas in sources where the line of sight intercepts it. The torus would also partly absorb the optical-UV continuum and (*via* photoelectric absorption) the soft X-rays photons, depending on the dust and gas column densities. Strongly supporting such view, the presence of these hidden components is instead seen through the scattering of part of the radiation toward our line of sight by material located "above" the torus [3].

The X-ray spectrum of Seyfert 1 (*i.e.* unobscured objects) is quite complex and several components and effects have been identified (see Czerny's lectures). These include a "primary" continuum (typically a power-law with a spectral index $\sim$1) produced in a relatively tenuous corona (*e.g.* above the accretion disc), which partly impinges onto optically thick gas (possibly the accretion disc itself). In the interaction with the thick material the primary photons are thus both (down) scattered and photoelectrically absorbed and reprocessed (at the lower energies). The resulting continuum "reflected back" by the optically thick matter is thus peaked at a few tens keV (where the two processes have similar importance) and contributes to the overall observed spectrum, as a broad bump at such energies. The reprocessing also gives rise to atomic line emission and in particular highly ionized Fe has the largest probability of emitting line photons *via* fluorescence. These lines (around 6–7 keV) together with absorption edges have been indeed detected. In some cases their width is so large that highly spinning holes are required (see *e.g.* [31]) as the emission has to originate in material closer to the black hole than the marginal stable orbit of a Schwarzschild hole (see *e.g.* [31]). In the soft X–ray band, excess absorption is also detected, produced by warm gas located at distance of a fraction of a pc.

What is the origin of the primary continuum? In the nearest objects this extends to few hundreds keV where a cut-off is required by the data. Such a feature, together with its power-law nature (with a rather typical spectral index) and the presence of a large density of accretion photons in the coronal environment, makes thermal Comptonization of soft ultraviolet disc photons by hot electrons the most likely raditive process. The electrons- with a temperature of order hundreds keV and a Thomson optical depth of tenths of unity – are assumed to be located and energized in such a corona. The actual geometry of the system and the nature of the corona is still unclear. The variability properties of the X–ray emission and the varying ratio of the ultraviolet and X-ray components in different objects, suggest that it might take place in "flaring" dissipation regions localized above the disc. The energy might be provided by the disc itself, being possibly magnetic energy in the accreting flow which would reach the corona *via* buoyancy and dissipating through reconnection.

As mentioned this scenario is discussed in relation to powerful AGN, whose high energy emission appears also to be responsible for the energetics and spectrum of the diffuse X–ray background (*e.g.* [18]). Much discussed in the last few years is however the paucity and nature of emission in low power or even "inactive" black holes, expected to arise from the accretion of the estimated interstellar gas onto the black hole [12]. This problem prompted a large amount of theoretical work, as discussed in Blaes' lectures, and deep observations of nearby galaxies to detect or tightly constrain the actual level of emission. A satisfactory scenario has not been however established: accretion might be occurring in a low radiative efficiency regime, matter could be expelled by the accreting system, *e.g.* in the form of a wind, or accretion could be episodic.

4 Radio-loud AGN

Radio-loud AGN can be thought as comprising the physical components described above but showing also the presence of well collimated jets (Fig. 3). In powerful sources, the energy collimated and transported by the can be compared to that of the whole of the accreting system, making jets an important ingredient in the energetics of the nucleus.

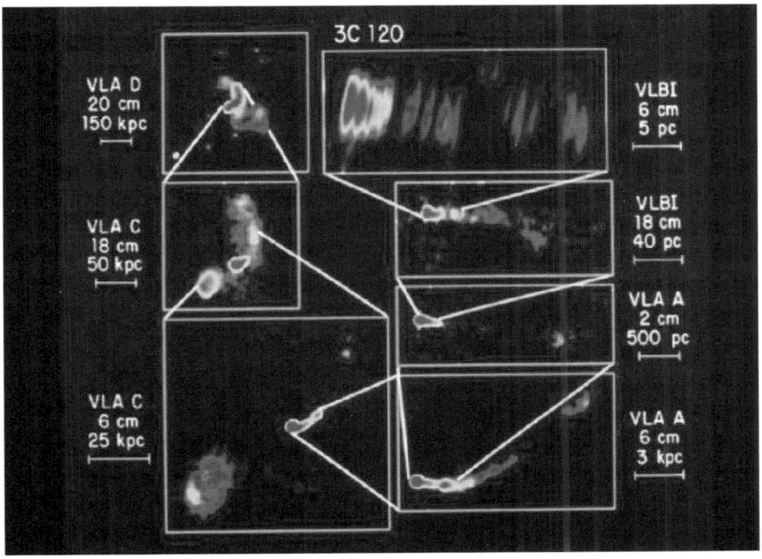

Fig. 3. Radio maps of the jet associated with the source 3C 120 on different physical scales, spanning more than four decades in size. The jet is continuous and quite aligned on all of its structure (from [44]). The radio telescopes used and the observational frequencies are also indicated.

Several independent pieces of evidence indicate that the plasma in the jet is moving with ultrarelativistic bulk speeds, corresponding to Lorentz factors of order of a few to tens (see Sect. 7). This is a crucial element for understanding, not only obviously the physics of jets, but also the radiative properties. In fact, due to the relativistic motion the radiation will be affected by significant beaming. This accounts for several characteristics of the observed phenomenology and gives raise to strong anisotropy in the observed emission with respect to the direction of motion of the plasma. The SED of sources observed at small angles with respect to the jet axis result thus dominated by the emission from the jet itself ("blazars"), while, when viewed at larger angles the observed properties are believed to be those typical of radio–galaxies. More precisely, the so called unification scenario accounting for the orientation effect, identifies low power blazars (BL Lac objects) with low power radio galaxies (FR I, [13]) and similarly, the high power blazars (radio-loud quasars) with FR II radio galaxies. The SED of the radio-loud AGN shown in Figure 2 would thus arise from the quasi–thermal contribution from the accreting system and the emission from jets, dominating in blazars.

4.1 Blazar spectral energy distribution

Let us therefore concentrate on the high energy properties of the emission from jets. The corresponding SED is characterized by two broad components (see Fig. 4), the low energy one peaking at IR-UV energies, and the high energy one in the MeV-TeV band. The strong linear polarization of the radio (up to $\sim$60%) and optical emission, together with the non-thermal character of the spectral components, indicate as the emission process responsible for the first component synchrotron by a non-thermal distribution of particles. Their Lorentz factor in some cases reach $\gamma \sim 10^5$, requiring an efficient mechanism to accelerate electrons on a timescale shorter than their cooling one.

4.2 The high energy component

The discovery by the Compton Gamma-Ray Observatory of intense γ–ray emission from radio-loud AGN and in particular from blazars has been an important step forward in the field. It first revealed that the bulk of the radiative dissipation was occurring (at least in radio-loud quasars) in this band, where – although highly variable and often not detectable – more than 10–100 times the luminosity of the synchrotron component can be radiated. But this is not only important for the energetics of the dissipation.

The detection itself of intense and rapidly variable γ-ray emission (see Fig. 5) above MeV energies provided a strong clue for the γ–ray emitting plasma to be moving at ultrarelativistic speeds. As mentioned in Section 7, in fact, the compactness parameter inferred from the observed luminosities and variability timescales would imply strong absorption of the γ–ray photons *via* electron-positron pair production and reprocessing of this energy into (a steep spectrum) X-ray emission, contrary to observations. The effects of relativistic beaming of the luminosity and contraction of observed variability timescale can thus "solve" the problem.

Furthermore, the same argument on pair opacity applies with regard to target photons outside the emitting region. The requirement that the γ-rays can propagate in the intense nuclear radiation field external to the jet, turns into a constraint on the distance (from the central object) where

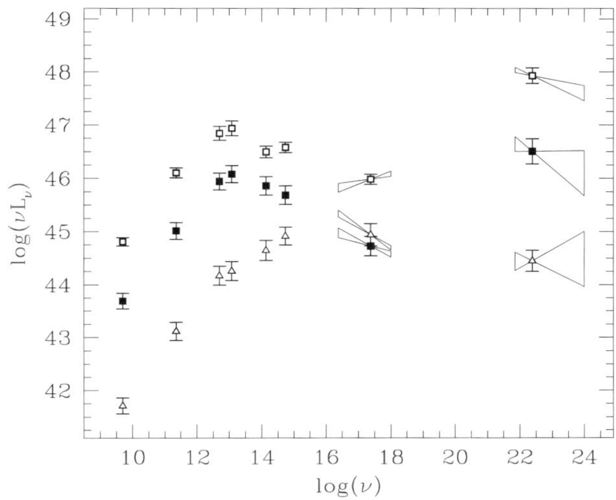

Fig. 4. The average spectral energy distributions of blazars: the most powerful sources are flat spectrum radio–loud quasars while the lowest power ones are low luminosity BL Lac objects. It can be noticed a suggestive trend of the position of the synchrotron and high energy peaks to decrease and their relative power to increase with the increase of the source power. (From [15].)

the high energy photons can be produced in order not to violate the observational constraints on the lack/paucity of pair reprocessing. On the other side, variability arguments provide an indication on the maximum distance where this occurs (if the distance is related to the size of the emitting source). These two constraints suggest that the dissipation region is located at distances of order 10^{2-3} the Schwarzschild radius.

Let us consider the origin of this high energy component. One process occurring in the source which has been proposed to account for the γ–ray emission is synchrotron self-Compton (see Sect. 7, [24]) through which the synchrotron photons (say at the peak of this component) are upscattered by a factor $\langle\gamma^2\rangle$ (where the average is over the non–thermal particle distribution). The large amplification of the luminosity required to account for the ratio of the high energy vs synchrotron components in powerful blazars is however not straightforward to be accounted by the model, and a process more likely to dominate in these sources has been proposed [10,35]. In fact, the nuclear region of quasars is believed to be permeated by an intense radiation field produced by the accretion flow. This photon field is partly also isotropized *via* either scattering by ambient electrons or reprocessing in broad line emitting clouds or even in larger scale dust (re-emitting in

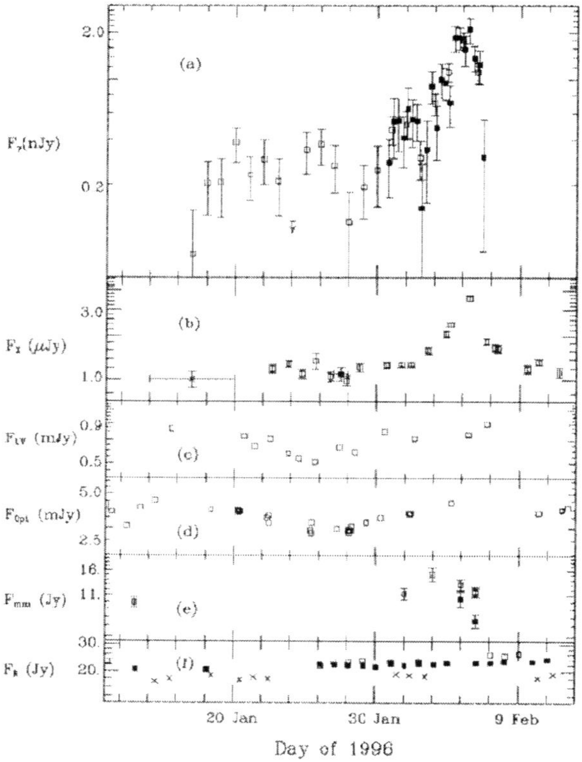

Fig. 5. Light curves in different bands for the powerful γ-ray emitting blazar 3C 279 (from [45]): in the top panel a strong and rapid variation of the γ-ray flux is reported. The lower frequency emission (at least in X–rays and possibly in the optical) appears to vary nearly simultaneously with the γ–rays.

the near-IR) (Fig. 6). As the plasma in the jet is moving at relativistic speeds, the actual energy density of the radiation field produced externally to the jet $U_{\rm ext}$ results to be relativistically boosted in the plasma frame as $\sim\Gamma^2\, U_{\rm ext}$. Thanks to such an amplification, for typical parameters this well exceed the energy density in synchrotron radiation. The external inverse Compton mechanism (external as the seed field is produced outside the emitting region) well accounts for the spectral properties of most powerful blazars (*e.g.* Fig. 7), while low power blazars (BL Lacs) for which there is indication that the nuclear external radiation field is in fact weak, synchrotron self-Compton can be responsible for a significant or all of the high energy spectral component.

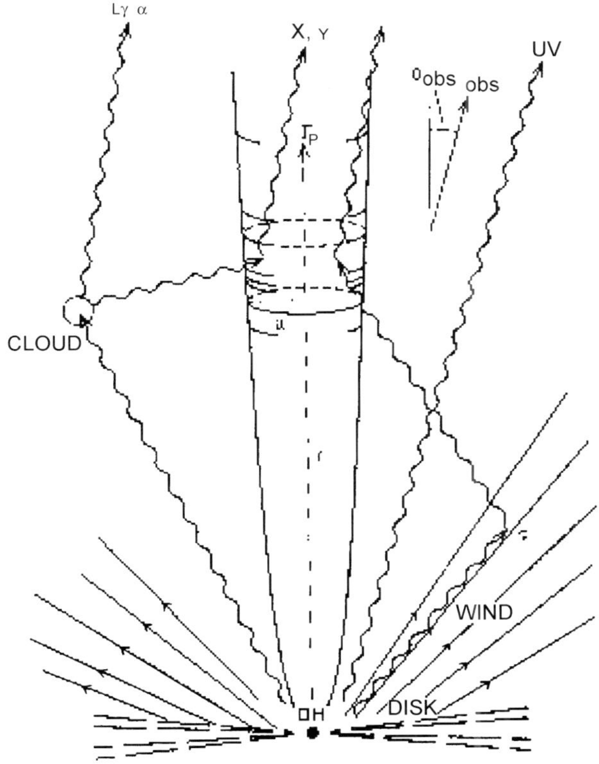

Fig. 6. A schematic cartoon (from [35]) representing the emitting region in a jet, *i.e.* the location where part of the jet energy is dissipated (*e.g.* *via* reconnection, shocks) into particle internal energy. In this cartoon the electrons are supposed to radiate *via* synchrotron, synchrotron self-Compton but also, as the jet propagates in an intense external radiation field, by inverse Compton scattering of photons produced by the disc, scattered and/or reprocessed in broad line emitting gas.

An alternative class of models proposed to explain the origin of the high energy component postulates that also hadrons (and not only electrons/leptons) are efficiently accelerated during the dissipation. Different scenarios have been suggested on the most efficient way for the protons to loose energy. This can be *via* proton-proton or proton-photon interactions, giving raise to the formation of pions – which then decay in a cascade - and the photons and leptons thus created are responsible for the observed high energy emission [23, 26]. It has been also proposed that the radiation could originate directly as synchrotron emission from highly energetic

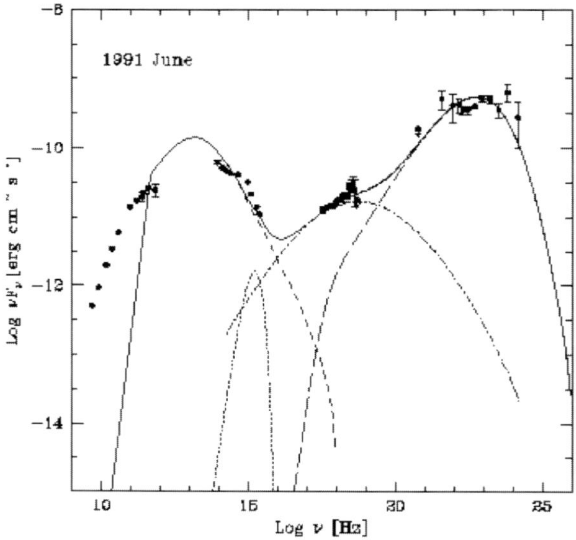

Fig. 7. The simultaneous broad band spectral energy distribution of the powerful blazar 3C 279, compared to a model involving a homogeneous region where a non-thermal electron distribution emits predominantly *via* synchrotron (short dashed line), and inverse Compton emission on the synchrotron (long dashed line) and an isotropic radiation field external to the jet (dotted line). (From [4].)

protons [2]. While the proton-proton and proton synchrotron scenarios appear to require an implausibly high proton density and cannot simply account for fast variability, respectively, the proton-photons scenario appears viable, although might have a low efficiency.

Unfortunately spectral information alone is not sufficient to select a specific model [43]. In principle a tool to distinguish among the different possibilities would be to test the model predictions on the variability in different spectral bands, which in the various cases are expected to correlate differently (*e.g.* could be expected to occur simultaneously at frequencies produced by the same electron *via* two mechanisms). Much effort has been therefore put into the design and realization of large multifrequency campaigns, from radio to γ-rays. Unfortunately, although in some cases the findings have allowed to set constraints on the models, the complexity of the behaviors have not uniquely selected a radiation mechanism or the origin of the variable flux.

Variability studies have also been performed through intense and long monitoring in specific spectral bands. While again no conclusive result has been reached, indication of typical variability timescales (derived *e.g.* *via* the estimate of power density spectra and structure functions) and correlations and lags at different energies are leading to promising constraints on the source dimension, cooling and acceleration timescales.

We note that most of the proposed scenarios to account for the γ–ray radiation involve a single dissipation region in the jet dominating the emission. Earlier work (before the Compton Gamma-Ray Observatory discovery) developed instead frames where the emission was due to the superposition of radiation from different zones of an inhomogeneous jet structure [7]. This different approach follows in part results of multifrequency campaigns which showed variability consistent with one emitting zone (*e.g.* Fig. 5) and from the condition that but even more crucially the condition on the location of the production of γ–ray is distant (and thus large) enough not to require extreme values of the Doppler factor to account for the observed radiation (as it was required by having a single region at a few Schwarzschild radii). Inhomogeneous jets are of course expected and the flat spectra of blazars in the radio band indeed are interpreted (and in some cases observed) to originate as superposition of emission from regions of different compactness (*i.e.* different synchrotron self-absorption frequencies).

4.3 Large scale jet emission

High energy emission is now detectable also from the large scale structures in some jets. In particular the Hubble Space Telescope and recently the X-ray satellite Chandra have sufficient sensitivity and spatial resolution to detect structures on scales of hundreds of kpc. The very same detection of high energy emission revealed the necessity of having extremely energetic particles up to large distances.

The comparison of images in the different spectral bands shows (often) similar features. However a closer analysis reveals differences in the extension of individual emitting regions (knots) as well as different relative brightness of knots at different frequencies as a function of the distance from the nucleus (*e.g.* Fig. 8).

While the optical emission appears to be consistent with being produced (as the radio) *via* synchrotron emission, the origin of the X–rays is less clear. Thermal bremsstrahlung is disfavored by the required high particle density, inconsistent with upper limits on the density imposed by Faraday rotation in some sources. Analogously to what occurs on smaller scales, likely non-thermal processes which can be at work are synchrotron and inverse Compton (SSC and/or external). The SED of some individual knots, especially in the powerful sources, is not consistent with the X–rays being

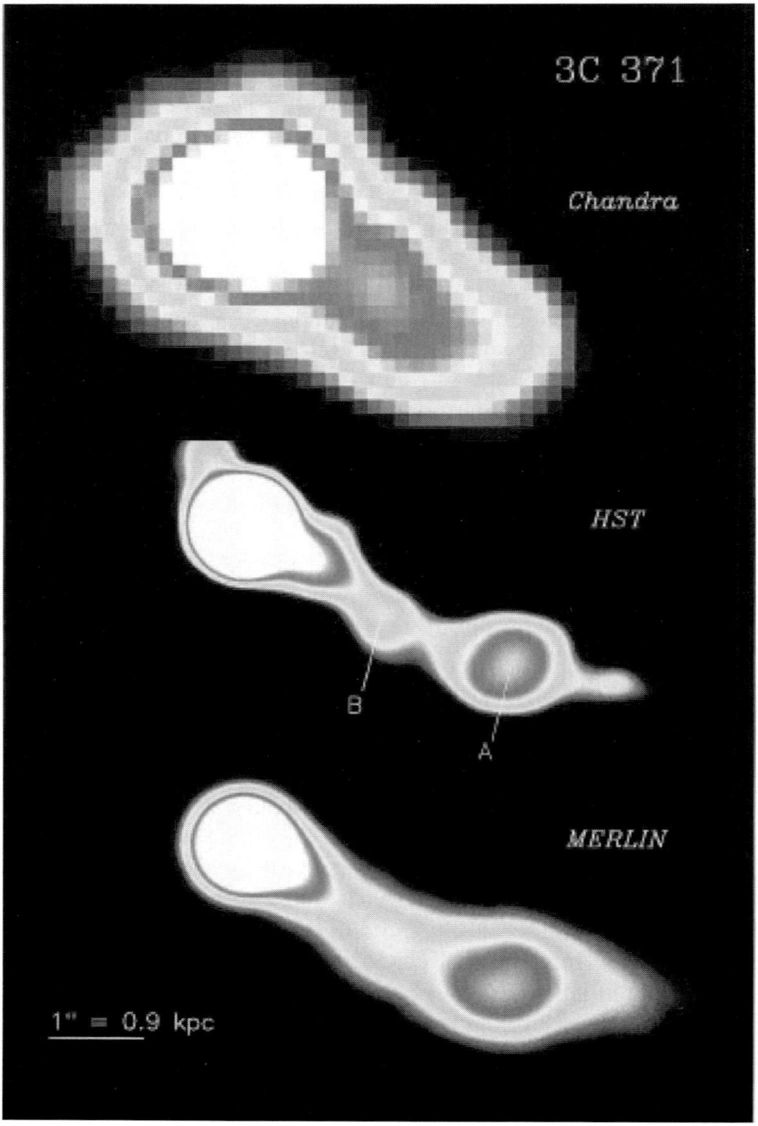

Fig. 8. Images of a jet (source 3C 371) on arcsec scales, in the X–rays (Chandra), optical (HST) and radio (MERLIN) bands. The radio and optical images appear to be rather similar, while the X–rays are more concentrated toward the nucleus. The relative differences of brightness in different bands provide important clues on the emission conditions. From [28].

part of the lower energy synchrotron component from a single population of electrons, favoring an inverse Compton origin. The external radiation fields which might provide seed photons for the scattering include the cosmic microwave background and the (relativistically beamed) emission from the inner nuclear jet.

Although the issue is still open, and more data is being gathered, one plausible possibility (avoiding some of the difficulties of alternative scenarios) is that the plasma is still moving highly relativistically on these scales and thus the cosmic microwave background field, amplified by relativistic boosting, provides most of the seed photons for inverse Compton emission [47].

4.4 Jet properties

The understanding of the dissipation and emission processes at work provides then information on the energetics of jets and on the form in which such energy is transported. Estimates of a minimum energy are possible *via* the synchrotron emission of extended (lobe) structures, assuming a condition of equipartition between relativistic emitting particles and magnetic field [9]. However it is still not clear how this energy is transported, namely the composition of jets is still unknown (whether they are dominated by ordinary electron-proton plasma, electron-positron pairs or electromagnetic fields [5]).

The high energy emission in these sources provide some constraints, although clearly model dependent. In fact a broad band spectral "fit" of the SED gives information on the intrinsic radiative power dissipated (*i.e.* a minimum to the carried power of course), on a minimum power in those particles which are emitting, on the energy in the magnetic field responsible for the synchrotron radiation, etc. Furthermore, limits on the content of leptons flowing in the jet come from the lack of observable signature (expected in the soft X-rays) of the interaction of particles moving with the jet bulk motion with the external photon field (bulk Compton process). The latest claims suggest that at least powerful blazars cannot be dynamically dominated by an electron-positron pair plasma, and either a proton or an electromagnetic component have to dominate the energetics [16, 36]. Note that these considerations do not exclude the presence of a numerically (but not dynamically) relevant electron–positron population partly responsible for the high energy emission, as postulated in some scenarios [37]. The role and energy of protons will be possibly probed by future neutrino experiments.

Another powerful tool to address these issues is provided by dynamical constraints on the jet propagation. Chandra imaging of radio sources in the core of clusters of galaxies have revealed clear signs of the interaction

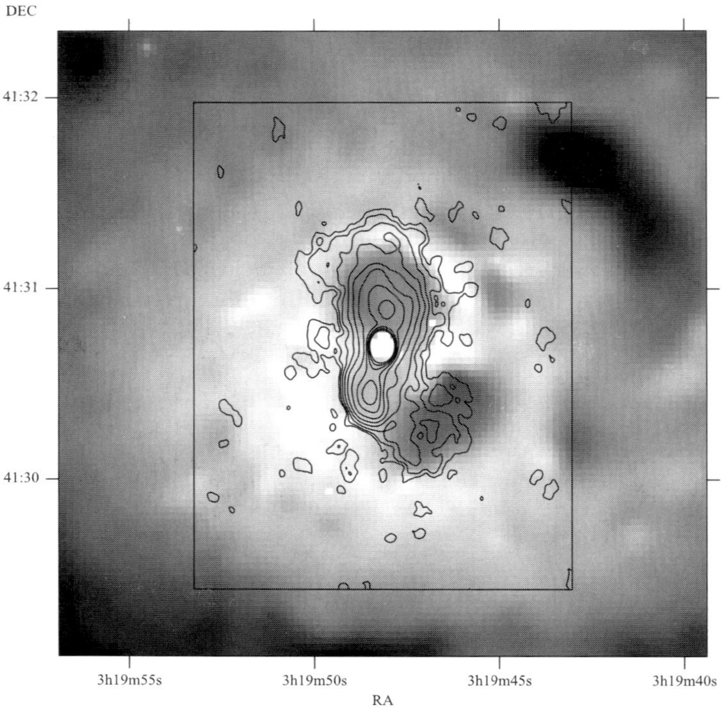

Fig. 9. Overlay of the radio contours (at 330 MHz) and X–ray (Chandra) image of the radio source in the center of the Perseus cluster (from [11]). The morphology (and spectra) reveal clear signs of interaction between the radio emitting plasma (in the form of two lobes at the center) and the thermal X-ray gas in the cluster. Information from the emitted radiation as well as the modeling of the dynamics of the system allows to constrain the energetics of the radio source.

of the radio plasma and the X–ray emitting cluster gas [11, 47] (Fig. 9). The former appears to be displacing the gas during its propagation and expansion, possibly *via* buoyancy, without formation of strong shocks. These findings, by setting constraints on the dynamics of the relativistic plasma, offer an important piece of information on the content and energetics of jets (*e.g.* [11]).

4.5 Cosmological backgrounds and intergalactic medium

Not only the γ–ray emission of blazars is crucial to understand the jet physics and dissipation processes, but also provides important information on the cosmological backgrounds. It is in fact possible that most of the contribution to the diffuse γ–ray background is indeed provided by the radio–loud AGN population [38], although uncertainties – especially on the duty cycle of the variable γ-ray activity in each source – do not allow to tightly constraint such contribution.

Furthermore, TeV measures provide us with constraints on the uncertain diffuse IR background. In fact, TeV photons can interact with such background field to produce electron-positron pairs. The detection of TeV photons from distant sources can thus set a limit on the IR photon intensity. The major difficulty here is to estimate the degree of attenuation of the TeV photons, which requires a knowledge of the intrinsically emitted spectrum. Recent results suggest that indeed such attenuation has been detected [1].

A final note is on the importance of high energy emission from blazars for the study of the intergalactic medium. In fact a warm/hot component of the IGM has been revealed from absorption features in the soft X–ray spectrum of bright blazars, thanks to the increased spectral resolution and sensitivity of the current generation of X-ray telescopes (*e.g.* [18]).

5 Analogous systems?

In the spirit of this School, let me just mention two astrophysical systems where not only accretion and jet formation occur, but where the physics of the flows mostly resemble that of radio-loud AGN.

The first class of sources comprises the so-called microquasars, *i.e.* binary systems which show at times the presence of relativistic jets [27]. The constraints on their velocity are uncertain (because of the uncertainty on the distance of these systems) and the lower limits estimated so far are lower than those typical of AGN. Of course, if these are a scaled down version (with respect to the black hole mass) of radio-loud AGN, they provide a powerful tool to study these systems being much brighter and evolving on much shorter timescales. One important clue which microquasars seem to provide is related to the presence of jets only during some of the spectral (accretion) states, and in particular jet components appear to follow a minimum in the X–ray emission presumably produced in the inner part of the disc (*e.g.* [14]). Intriguingly, the search for such a behavior in one radio-loud AGN has given promising results [25].

Gamma-ray Bursts constitute a further astrophysical context where – which much more uncertainty – it is thought that accretion onto a black hole could be the engine responsible for the production of an extremely relativistic (and possibly collimated) flow. Although more extreme in terms of power, the (magneto-)hydrodynamical evolution of the flow, the dissipation mechanism and the radiation process(es) could be well analogous to the case of AGN.

6 Non-thermal radiation processes and relativistic beaming

Here we simply recall very basic information on the non-thermal emission processes and the role of relativistic beaming mentioned in the previous Sections. For proper treatments and descriptions see *e.g.* [8, 17, 19, 21, 22, 30, 32, 39, 40].

The emitting particles are relativistic and are typically assumed to have a non-thermal distribution described by a power-law, $N(\gamma) = K\gamma^{-p}$, between a minimum and a maximum Lorentz factors, where γmc^2 is the particle energy (in the following only leptons will be considered).

6.1 Synchrotron emission

A relativistic charged particle in the presence a static magnetic field, of intensity B, is subject to the Lorentz force and thus follows a helicoidal trajectory along a field line, with an angular frequency $\nu_{\rm g} = eB/2\pi\gamma mc$. Because of the corresponding acceleration the particle radiates.

The dipole expression for the emitted power, generalized to the relativistic case, reads

$$P = \frac{2e^2\gamma^6}{3c}\left(\dot{\beta}^2 - (\beta\times\dot{\beta})^2\right) = 2\sigma_{\rm T}c\gamma^2\beta_\perp^2\,\frac{B^2}{8\pi} = \frac{4}{3}\sigma_{\rm T}c\beta^2\gamma^2 U_{\rm B}$$

where $\beta = v/c$, $\sigma_{\rm T} = 8\pi r_{\rm e}^2/3$ is the Thomson cross section, $r_{\rm e}$ is the electron radius, $U_{\rm B} = B^2/8\pi$ is the magnetic energy density and it has been assumed an isotropic distribution of particles.

While a non-relativistic electron would emit a narrow band (cyclotron) spectrum, in the ultrarelativistic relativistic case ($\gamma >> 1$), the superposition of high order harmonics can be approximated by a broad band continuum, $P(\nu) \propto (\nu/\nu_{\rm c})^{1/3}\exp^{-\nu/\nu_{\rm c}}$, where $\nu_{\rm c}$ corresponds to the Fourier transformation of the duration of each pulse detected by the observer (*i.e.* the time it takes for an emission cone to sweep the line of sight of the observer). Numerically, the spectrum is therefore centered at a frequency $\nu_{\rm c} \sim 10^6\gamma^2 B$ Hz (where B is measured in Gauss). This implies an almost one-to-one correspondence between the electron energy (squared) and the radiated frequency.

The corresponding synchrotron cooling timescale is thus of order

$$t_{\rm syn} \sim 6\times 10^8 B^{-3/2}\nu_{\rm MHz}^{-1/2}\ {\rm s}.$$

Because of the symmetry of the motion and the beaming of the radiation emitted by a relativistic particle, the synchrotron radiation can be highly linearly polarized, up to a degree of polarization of 70 per cent. (Circular polarization is also expected and indeed several cases of circularly polarized radio emission from radio-loud AGN at a few per cent level have been detected, although this appears to be difficult to be accounted for as intrinsic synchrotron polarization.)

In the case of a non-thermal particle distribution, the overall emitted spectrum will be the convolution of the peaked single particle spectrum – often approximated as a delta-function for simplicity – with the particle power-law distribution. The spectral emissivity per unit volume results

$$\epsilon_{\rm syn}(\nu) \propto \int P(\nu)N(\gamma){\rm d}\gamma \propto KB^{1+\alpha}\nu^{-\alpha},$$

where the resulting spectral index $\alpha = (p-1)/2$. At the lower and upper ends, the spectrum will resemble that of a single particle at the corresponding energies. Note that the two limits of the particle distribution critically determine the total energy, synchrotron emissivity, particle number density (which limit is crucial of course depends on the slope of the power-law, p).

If the source is compact enough, part of the emitted flux can be re-absorbed within the source itself. The effect of self-absorption is often quantified using the so-called brightness temperature, defined as the temperature of a black body emitting the observed (measured) intensity $I(\nu)$ in its Rayleigh-Jeans part, *i.e.*

$$T_{\rm B} = \frac{c^2}{2k}\frac{I(\nu)}{\nu^2}$$

where k is the Boltzmann constant. From a thermodynamical point of view, the radiation temperature cannot be higher than the kinetic temperature of the electrons emitting it and therefore in such a case the source would become self-absorbed. In other words, the brightness temperature is limited to $kT_{\rm B} < \gamma m_{\rm e}c^2/3$, *i.e.* of order $T_{\rm B} \sim 10^9 \nu_{\rm MHz}^{1/2} B^{-1/2}$ K. In a homogeneous source, the corresponding flux density is $\propto \nu^2 T_{\rm B} \propto \nu^{5/2}$ in the self-absorbed regime, peaks at the self-absorption frequency and steepens to $\nu^{-\alpha}$ in the optically thin regime.

6.2 *Inverse Compton emission*

Compton scattering, *i.e.* the interaction of an electromagnetic wave with an electron which is thus accelerated, is called "inverse" when the electron is relativistic. Again, *via* the dipole formula for the total emitted power by a particle accelerated by a wave, one can obtain

$$P = \frac{4}{3}\sigma_{\rm T} c\gamma^2 U_{\rm rad},$$

where it has been assumed that the radiation is isotropic, with an energy density $U_{\rm rad}$, and that the particle is highly relativistic $\gamma >> 1$. If the process is interpreted as the interaction of electrons and photons, the scattering rate (per unit volume) will be $\tau_{\rm r} = \sigma_{\rm T} c n$ (again for a isotropic distribution). It results from these two expressions that the average energy of a scattered photon is

$$\nu \sim \frac{4}{3}\gamma^2 \nu_0,$$

where ν_0 is the typical original energy of the photon. Again the spectrum results peaked and there is a correspondence between photon and electron energy. The γ^2 dependence can be seen as resulting from two Lorentz transformations (to the particle rest frame and back to the radiation frame after the scattering).

The similarity of the expressions of the powers emitted *via* synchrotron and inverse Compton and the dependence of the energy of the emitted radiation on γ^2 not only allows to interpret synchrotron emission as scattering of virtual photons, but it also implies a similar dependence of the resulting spectrum from the shape of the (power-law) distribution of relativistic particles, *i.e.* $\epsilon_{\rm iC}(\nu) \propto \nu^{-\alpha}$.

The radiation field which is scattered can be of course of any origin. In particular it can be synchrotron radiation produced by the same electron population which then scatters it (Synchrotron-Self Compton, SSC). This is typically effective if the source is compact and the resulting SSC spectrum has a similar slope as the synchrotron one. Furthermore, from the above relations, the ratio of the luminosity emitted *via* the two processes will scale as

$$\frac{L_{\rm c}}{L_{\rm s}} \sim \frac{U_{\rm rad}}{U_{\rm B}},$$

while the ratio of the monochromatic luminosities (at same frequency) is proportional to the optical depth, *i.e.*

$$\frac{L_{\rm c}(\nu)}{L_{\rm s}(\nu)} \propto \sigma_{\rm T} K R,$$

where R is a typical source size.

If the scattered radiation exceeds the magnetic field one, then the (second order) scattering will be more effective than the synchrotron emission and so on. A limit to this cascade (which would cause dramatic cooling) is reached when the energy of the (increasingly) incident photons is so high that the electron recoil cannot be ignored (as implicitly assumed above). Above this limit ($h\nu_0\gamma << m_{\rm e}c^2$) the Klein-Nishina scattering cross section decreases and the process saturates. The condition that the SSC radiation does not exceed the synchrotron one – so called inverse Compton limit –

can be re-expressed as a limit on the brightness temperature (proportional to the synchrotron radiation intensity) as $T_{\rm B} < 10^{12}$ K (for typical spectral parameters).

6.3 Pair production

The presence of γ–ray photons with energies above $m_{\rm e}c^2$ in sources with high radiation densities of target (lower energy) photons implies a high efficiency in the production of electron–positron pairs *via* two photons interactions. The threshold energy for the process is given by

$$x_\gamma x_{\rm t} > \frac{2}{(1-\cos\theta)},$$

where $x \equiv h\nu/m_{\rm e}c^2$ is the photon energy in unit of electron rest mass and θ is the angle of incidence. The optical depth for the process for a γ–ray photon of energy ~MeV is $\tau_{\gamma\gamma} \sim \frac{\ell}{60}$ where the (adimensional) compactness

$$\ell \equiv \frac{\sigma_{\rm T}}{m_{\rm e}c^3}\frac{L}{R}$$

has been introduced. For high values of compactness, which can be reached in active nuclei, the γ–ray photons can thus not escape from the emitting region. The process can of course occur also outside the region responsible for the production of the γ-rays if enough target photons are present along the photon path.

The process is of course in principle non-linear, as the newly produced electron-positron pairs can emit themselves above threshold and thus induce more pair production. The evolution of a source where the phenomenon takes place therefore strongly depends on the source conditions and the distribution of particles (thermal or non-thermal). An equilibrium can be reached when the pair creation rate can be balanced against the pair annihilation one. For relatively cold, non–relativistic particles, the annihilation proceeds at a rate of $\sim 0.4\sigma_{\rm T}cn^2$ (here n is the total particle number density), and decrease with increasing particle energy. A typical resulting optical depth in pairs (assumed to rapidly cool) is of order $0.3\ell^{1/2}x_\gamma^{-1/2}$.

For a thermal distribution of particles which is continuously energized the production of pairs naturally leads to a thermostat effect: when the temperature of the particles is such that pairs can be produced, the increase in energy can result in an increase in the particle density and not in the temperature. This has been thought to be a powerful mechanism to limit the typical temperature in an accretion disc corona. However the found cut-off in the high energy spectrum of radio-quiet sources (Seyferts) and its spectral modeling appear to indicate that pair production is not a dominant effect in such a region.

In the case of non-thermal relativistic plasma, in conditions similar to those of radio-loud jets of AGN, there is copious production of photons above threshold. It is expected that the pairs themselves emit (*via* synchrotron and inverse Compton) above threshold and that more than one generation of pairs can be produced. In the limiting case of many pair generations a full cascade develops, and saturation can be reached. The net effect with respect to the initially emitted spectrum will corresponds to the luminosity absorbed above threshold being reprocessed and emitted by pairs to lower energies. This produces a steepening of the X–ray spectrum, with a limiting slope $\alpha \sim 1$ (corresponding to a conserved flow of energy in the particle distribution space).

6.4 Relativistic beaming

When a source, such as plasma in AGN jets, is moving relativistically, transformations between different frames have to be taken into account. In the case of AGN, several observed phenomena, which otherwise could not be understood, have been ascribed to the highly relativistic speed of the emitting sources. In particular, because of observed brightness temperatures in excess of 10^{12} K, it was proposed [30] that the radio emitting plasma was moving at speeds close to c and at the same time predicted that apparent superluminal motion would have been detectable (!). Indeed, the high brightness temperature (note that the recent findings appear to suggest that, at least part of the variability detected in intraday variable sources, is due to interstellar scintillation, reducing the corresponding inferred temperature to $\sim 10^{14}$ K), the observation of jet components moving at an apparent speed in excess of c, the detection of copious and variable γ–ray emission, the apparent one-sidedness of jets, ..., all argue in favor of such interpretation.

In fact, what is the effect of the relativistic motion? Note that in the astrophysics situation, information is collected through the detection of photons simultaneously reaching the observer. This implies that, not only the classical Lorentz transformations have to be taken into account (which are based on *e.g.* transformations of length at simultaneous times, of time intervals at the same location in space etc.), but also the effects due to the propagation of photons and in particular the Doppler effect, altering *e.g.* the time interval of detection of photons leaving the source at different times (*e.g.* [17]).

The net effect is that energies, time intervals (and dimensions inferred from them), and angular distributions of the emitted radiation are affected.

The transformations between the observer and source frame can thus be expressed in terms of the relativistic Doppler factor,

$$\delta \equiv \frac{1}{\Gamma(1-\beta\cos\theta)},$$

where β and Γ are the velocity and Lorentz factor of the bulk motion of the emitting plasma, and θ is the angle between the direction of motion and the observer (supposing a uniform component with a velocity field parallel to the jet axis).

Because of the transformation of energy ($\propto \delta$), time ($\propto \delta^{-1}$) and solid angle ($\propto \delta^2$), the observed (integrated) luminosity results δ^3 (δ^4) times the radiated one. (The different exponents depend on whether the emitting source can be better approximated *e.g.* as individual "blobs" of plasma or as a continuous jet, see [21].) As typically δ ($\sim\Gamma$ at small angles of sight) is of order of tens, the resulting amplification at small viewing angles with respect to the jet axis can reach three/four orders of magnitude.

This means that once these effects are taken into account, the intrinsic luminosity can dramatically decrease and the actual source size is intrinsically larger (if estimated *via* variability timescale arguments). Thus appears clear why the hypothesis of bulk relativistic motion of the plasma accounts for an overestimate of the photon energy density and compactness of the source, giving raise to the interpretational problems mentioned above. The aberration of the emitted photons also leads to a high degree of anisotropy with respect to the plasma direction of motion.

I wish to thank the Organizers for inviting me to such a stimulating School, interestingly organized in a cross field topic and in a beautiful environment. Francois Menard is also thanked for his (infinite!) patience with my delaying the completion of this manuscript. The participants to the School are acknowledged for their active role, curiosity and friendliness, which provided a scientifically stimulating and warm atmosphere.

References

[1] F. Aharonian *et al.*, *A&A* **384** (2002) L23.
[2] F. Aharonian, *New Astron.* **5** (2000) 377.
[3] R. Antonucci, *ARA&A* **31** (1993) 473.
[4] L. Ballo *et al.*, *ApJ* **567** (2002) 50.
[5] M.C. Begelman, R.D. Blandford and M.J. Rees, *Rev. Mod. Phys.* **56** (1984) 255.
[6] R.D. Blandford, in "Active Galactic Nuclei", edited by T.J.-L. Courvoisier and M. Mayor (Springer Verlag, Berlin, 1990) 161.
[7] R.D. Blandford and A. Königl, *ApJ* **232** (1979) 34.

[8] R.D. Blandford and M.J. Rees, in "Pittsburgh Conference on BL Lac Objects" (University of Pittsburgh, 1978) 328.

[9] G.R. Burbidge, *ApJ* **129** (1959) 849.

[10] C.D. Dermer and R. Schlikeiser, *ApJ* **416** (1993) 458.

[11] A.C. Fabian, A. Celotti, K.M. Blundell, N.E. Kassim and R.A. Perley, *MNRAS* **331** (2002) 369.

[12] A.C. Fabian and M.J. Rees, *MNRAS* **277** (1995) L55-L58.

[13] B.L. Fanaroff and J.M. Riley, *MNRAS* **167** (1974) 31.

[14] R. Fender, in "Relativistic Flows in Astrophysics", A.W. Guthmann, M. Georganopoulos, edited by A. Marcowith and K. Manolakou (Springer Verlag, Berlin) **589** (2002) 101.

[15] G. Fossati, L. Maraschi, A. Celotti, A. Comastri and G. Ghisellini, *MNRAS* **299** (1998) 433.

[16] G. Ghisellini, in "Stellar Endpoints, AGN and the Diffuse Background", edited by N.E. White, G. Malaguti and G.G.C. Palumbo, *AIP* **599** (2001) 120.

[17] G. Ghisellini, in "Recent developments in General Relativity", edited by B. Casciaro, D. Fortunato, M. Francaviglia and A. Masiello (Springer-Verlag, Milano), 5 [astro-ph/9905181].

[18] G. Hasinger, in "High Energy Processes and Phenomena in Astrophysics", IAU Symp. 214, edited by X. Li, Z. Wang and V. Trimble (2003), in press [astro-ph/0301040].

[19] J.D. Jackson, "Classical Electrodynamics" (John Wiley and Sons, New York, 1975).

[20] J.H. Krolik, "Active Galactic Nuclei: from the Central Black Hole to the Galactic Environment" (Princeton University Press, Princeton, 1999).

[21] K.R. Lind and R.D. Blandford, *ApJ* **295** (1985) 358.

[22] M. Longair, "High Energy Astrophysics" (Cambridge University Press, Cambridge, 1992, 1997), revised edition.

[23] K. Mannheim, *A&A* **269** (1993) 67.

[24] L. Maraschi, G. Ghisellini and A. Celotti, *ApJ* **397** (1992) L5-L9.

[25] A.P. Marscher, S.G. Jorstad, J.-L. Gomez *et al.*, *Nature* **417** (2002) 625.

[26] A. Mastichiadis and R.J. Protheroe, *MNRAS* **246** (1990) 279.

[27] I.F. Mirabel and L.F. Rodriguez, *Nature* **371** (1994) 46.

[28] J.E. Pesce, R.M. Sambruna, F. Tavecchio *et al.*, *ApJ* **556** (2001) L79-L82.

[29] B.M. Peterson, "An Introduction to Active Galactic Nuclei" (Cambridge University Press, Cambridge, 1997).

[30] M.J. Rees, *Nature* **211** (1966) 468.

[31] C.S. Reynolds and M.A. Nowak, *Phys. Rep.* (2003) in press.

[32] G. Rybicki and A.P. Lightman, "Radiative Processes in Astrophysics" (John Wiley and Sons, New York, 1979).

[33] I. Robson, "Active galactic nuclei" (Praxis Publishing, Chichester, 1996).

[34] D.B. Sanders, E.S. Phinney, G. Neugebauer, B.T. Soifer and K. Matthews, *ApJ* **347** (1989) 29.

[35] M. Sikora, M.C. Begelman and M.J. Rees, *ApJ* **421** (1994) 153.

[36] M. Sikora and G. Madejski, *ApJ* **534** (2000) 109.

[37] H. Sol, G. Pelletier and E. Asseo, *MNRAS* **237** (1989) 411.

[38] P. Sreekumar *et al.*, *ApJ* **494** (1998) 523.

[39] R. Svensson, *MNRAS* **209** (1984) 175.

[40] R. Svensson, *MNRAS* **227** (1987) 403.

[41] J.S. Ulvestad, in "Radio Astronomy at the Fringe", ASP Conf. Ser. (2003) in press [astro-ph/0310057].

[42] C.M. Urry and P. Padovani, *PASP* **107** (1995) 803.

[43] C. von Montigny *et al.*, *ApJ* **483** (1997) 161.

[44] R.C. Walker, J.M. Benson and S.C. Unwin, *ApJ* **316** (1987) 546.

[45] A.E. Wehrle *et al.*, *ApJ* **497** (1998) 178.

[46] R.L. White *et al.*, *ApJS* **126** (2000) 133.

[47] A.S. Wilson, in "The Physics of Relativistic Jets in the Chandra and XMM Era", *New Astron. Rev.*, in press.

Section IV: Pre-main sequence objects

COURSE 11

ACCRETION SIGNATURES IN YOUNG STELLAR OBJECTS

N. CALVET

Smithsonian Astrophysical Observatory, Cambridge, U.S.A.

Contents

ACCRETION SIGNATURES IN YOUNG STELLAR OBJECTS

N. Calvet

Abstract

Mass accretion onto the star is taken place in young stellar objects, providing the energy needed to power the excess emission observed at all wavelength bands. Signatures that allow us to identify and quantify accretion processes are reviewed in these lectures.

1 Introduction

In the process of star formation, matter is transfered from molecular clouds to stars. This accretion of matter leaves distinctive signatures, for instance, redshifted absorptions in spectral lines indicating that matter is falling in. In addition, some energy losses observed in young stellar objects (YSO) need to be compensated by external sources. Gravitational potential energy is the best candidate to supply this need. Present models based on this hypothesis can consistently explain most of the emission properties of YSO. We will review the evidence in these lectures.

2 YSO

Samples of visible YSO are located in the HR diagram in Figure 1, superimposed to the theoretical isochrones of [1]. The well known T Tauri stars (TTS) in the Taurus cloud have typical masses of a few tenths of $M_\odot$ [2]. However, the type extends to masses $\sim$2–3 $M_\odot$, in the high mass side, to objects in the sub-stellar limits, the brown dwarfs [3]. The Herbig Ae/Be stars (HAeBe) have masses $>$1.5 $M_\odot$ [4]. Typical ages are $\sim$1–10 Myr.

The emerging flux of the visible YSO is characterized by excesses over photospheric fluxes appearing in emission lines and continua (see [5] for complete list of references). The emission lines show a wide range of conditions of formation. The forbidden lines indicate densities of $n_{\rm H} \sim 10^5$ cm^{-3}

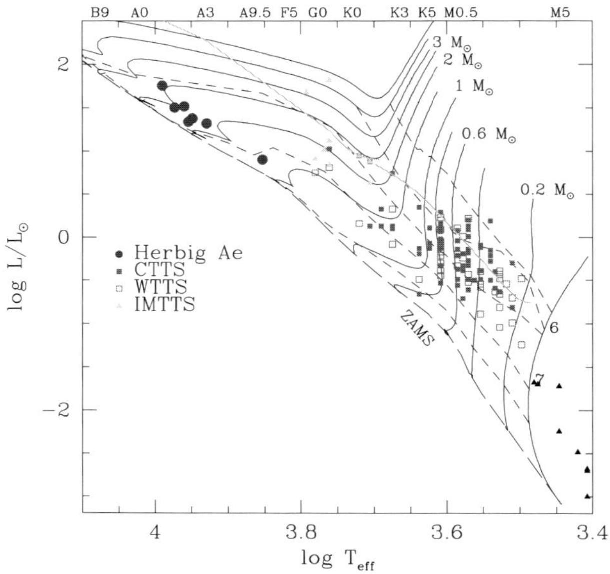

Fig. 1. HR diagram showing the location of visible YSO: Herbig Ae/Be stars, Classical T Tauri stars, Weak T Tauri stars, and intermediate mass T Tauri stars. The luminosity refers to the purely stellar luminosity. Evolutionary tracks for masses between 0.1 and 4 $\dot{M}$ (light solid lines) and isochrones between log $t = 5.5$ and 8 (dashed lines) are indicated. The heavy solid line is the birthline.

while the permitted lines require densities of up to $n_{\rm H} \sim 10^{13}$ cm^{-3}. Forbidden and optical/infrared (IR) permitted lines indicate temperatures in the formation region of $T \sim 10^4$ K, while lines of highly ionized species seen in the ultraviolet (UV) indicate $T \sim 10^{5-6}$ K. Typical velocities range from ~0 to a few hundred km s^{-1}. The continuum excess appears in the optical as a flux veiling the photospheric absorption lines, and in fluxes orders of magnitude higher than expected photospheric fluxes in the UV, IR, millimeter (mm) and radio emission.

TTS can be classified en two types: Classical T Tauri stars (CTTS) and weak T Tauri stars (WTTS). Initially, the classification was based on the equivalent width of Hα, with EW(Hα) < 10 Å in WTTS [6]. It was soon recognized that the classification represented clear physical differences. CTTS have large IR excesses, which the WTTS lack. While WTTS have emission lines, these lines are not only generally weaker but much *narrower* than CTTS. WTTS show no indication of veiling in their absorption lines, while CTTS show a wide range of veiling that correlates with near IR colors. Similarly, WTTS show *no* forbidden emission lines, while these lines are always present in CTTS [7]. The strength of the UV lines in WTTS, although higher or comparable to other type of very active stars, are order of magnitude lower than in CTTS [8]. The X-ray luminosity, on the other hand, appears to be similar in both types [9].

The total luminosity of a given TTS, including the excess emission if present, can be compared to the purely stellar luminosity, that is, the luminosity the stellar photosphere would have by itself. This can be estimated from the J band magnitude, where the contribution of the excess flux seem to be minimal [2] (see however, measurements of veiling at J by [10]). The observed luminosity of the WTTS agrees with the stellar luminosity. In contrast, the CTTS show luminosities in average 10% higher, and in some cases much higher than that due to the star alone [5] (Fig. 6.3). *An additional source of energy is required to compensate for this extra energy.* The most likely source is accretion energy, that is *gravitational potential energy* [11]. This source must also be responsible for other peculiarities that stars with excess energy show, namely, the broad emission lines, the veiling, the IR excess, and the forbidden line emission. In the next sections, we will review models of mass accretion for the CTTS which can account for these properties.

TTS share properties with other active stars in the HR diagram. Strong magnetic fields, of the order of a few KG, have been detected in both types of TTS, by measuring the Zeeman splitting in magnetic sensitive absorption lines [12, 13]. Similar magnetic field strengths are indicated by circular spectropolarimetry of emission lines [14]. These magnetic fields will induce surface activity as it does in other active stars. The general level of activity observed in WTTS, comparable or higher than active stars of other types, can be explained by these type of phenomena. The generally high rotational velocities observed in WTTS would support this hypothesis [15, 16], but cannot explain the energy output of the CTTS, which are rotating more slowly.

Some YSO are also found embedded in molecular clouds, so heavily extincted that only appear in the infrared and beyond. The characteristics of these objects can be understood if they are *protostars*, that is, they are still actively receiving mass from the cloud. Besides the close association with the densest cores, direct evidence of infall motions in molecular line profiles has been obtained [17, 18]. These objects are also called Class I, in which the flux increases with wavelength in the near-mid infrared. In contrast, the slope is negative in the visible objects, classified as Class II [19]. Typical ages for these objects can be estimated from the relative numbers of embedded and visible objects in a given cloud. For Taurus, there are $\sim$10 times as many visible stars as protostars. With a typical age of 1–2 Myr for the visible TTS, the protostellar phase is $\sim$0.1–0.2 Myr.

It is difficult to asses the *stellar* properties of protostars, since they are so embedded. We can estimate the luminosities of the objects by measuring the total energy emitted in the infrared; this energy has been absorbed by the dust in the circumstellar material and re-radiated at longer wavelengths. However, this luminosity can be either stellar luminosity, that is, due to the release of thermal energy as the star contracts, or energy released as matter is accreted onto the star. As we will see later, present evidence indicates that the observed luminosity corresponds to stellar luminosity for most protostars.

3 Formation mechanisms

Accretion energy is the most likely source for the extra emission observed in YSO, and it is naturally expected to be released in the process of star formation. Stars are formed by gravitational contraction of cores, the densest part of molecules clouds. Observations indicate that these cores are slowly rotating [20]. Collapse under conservation of energy and angular momentum results in the formation of a disk [21,22]. The material with the highest angular is that at larger radius, where most of the mass of the core is for reasonable density laws. This means that most of the mass of the core lands on the disk, since it has too much angular momentum to go directly to the star. Thus, accretion occurs schematically in two stages: cloud to disk, and disk to star, which we refer to in these lectures as *infall* and *accretion* proper, respectively.

Relevant luminosities in these processes are the potential energy released in the (radial) infall to the star, $L_{\rm inf}(\rm star)$, the energy released in the infall to the disk, $L_{\rm inf}(\rm disk)$, the accretion luminosity released as matter is accreted from the disk onto the star, $L_{\rm acc}$, the stellar luminosity, L_*, and the observed luminosity $L_{\rm obs}$. We will estimate and compare these luminosities next.

The infall luminosities can be estimated by $L_{\rm inf} \sim G\dot{M}_{\rm inf}M_*/R$, where $\dot{M}_{\rm inf}$ is the infall mass accretion rate, M_* is the stellar mass and R is the radius where material lands. R is equal to the stellar radius R_* for infall to the star and to the disk radius for infall to the disk. Objects in this phase are expected to be on the *birthline*, that is, the locus in the HR diagram of objects which are accreting mass [23, 24]. Birthline calculations yield an estimate of $M_*/R_* \sim 0.5$, and $R_* \sim 2\ R_\odot$ [23, 24], while disk radii are of the order of 10–100 AU.

The mass infall rate can be estimated from the density of the infalling envelope model that produces a spectral energy distribution (SED) similar to that observed, since by mass conservation $\dot{M}_{\rm inf} \sim 4\pi r^2 \rho v$, assuming spherical infall at large distances, with v the free-fall velocity, $v = (GM_*/r)^{1/2}$. Radiative transfer calculations for infalling envelope models have been carried out for Class I objects in the Taurus cloud [25], yielding an average $\dot{M}_{\rm inf} \sim 4 \times 10^{-6}\ M_\odot\ {\rm yr}^{-1}$.

With this mass infall rate, $L_{\rm inf}({\rm star}) \sim 15\text{--}30\ L_\odot$, higher than the observed luminosities, $L_{\rm obs} \leq 1\ L_\odot$. This has been named the "luminosity problem" [26], but actually is not a problem. For typical disk radii, $L_{\rm inf}({\rm disk}) \sim 0.002\text{--}0.03\ L_\odot$, so if infall occurs onto the disk, as expected from angular momentum conservation, then the infall luminosities are much lower than the observed luminosities.

The stellar luminosity can be estimated from the birthline, see Figure 1. It can be seen that for low mass objects and typical mass infall rates, protostars luminosities are expected to be of the same order of magnitude as visible objects. In fact, in the Taurus cloud, the observed distribution of luminosities of protostars and visible stars are similar, which seems to suggest that the luminosities of protostars are actually stellar luminosities.

It still remains to quantify the accretion luminosity, that is, the energy released in the transfer of matter from the disk onto the star, to completely identify the source of luminosity in Class I objects, and thus determine if the luminosity is an accretion signature in these objects.

4 Accretion luminosity and mass accretion rate

Disk emission in protostars tends to be hidden by the absorption and emission of the infalling envelope, which has sizes $\sim$ thousands of AU, surrounding the star and the 10–100 AU disk. The best determination of the accretion luminosity can be obtained in the visible objects, in which the envelope has already dissipated.

In CTTS, the excess flux veiling the absorption lines in the optical and dominating in the UV can be extracted by subtraction of the photospheric fluxes. The luminosity in this excess flux, if identified with accretion energy, gives an estimate of the accretion luminosity. With M_* and R_* from the position in the HR diagram of a given star, the accretion luminosity $L_{\rm acc} = G\dot{M}M_*/R_*$ yields the mass accretion rate $\dot{M}$. Figure 2 shows the mass accretion rate measurements for CTTS in Taurus and Chameleon [27, 28]. The mean accretion luminosity for CTTS in these clouds is $\sim$0.1 $L_\odot$, and the corresponding mean mass accretion rate is $10^{-8}\ M_\odot\ {\rm yr}^{-1}$.

5 Magnetospheric accretion

We can now begin to identify the mechanisms by means of which accretion onto the star takes place. The star is surrounded by a disk through which mass is flowing onto the star at a rate $\dot{M}$. The star, in turn, has a magnetic field of a few KG at its surface. For spherical accretion, and in the case of a

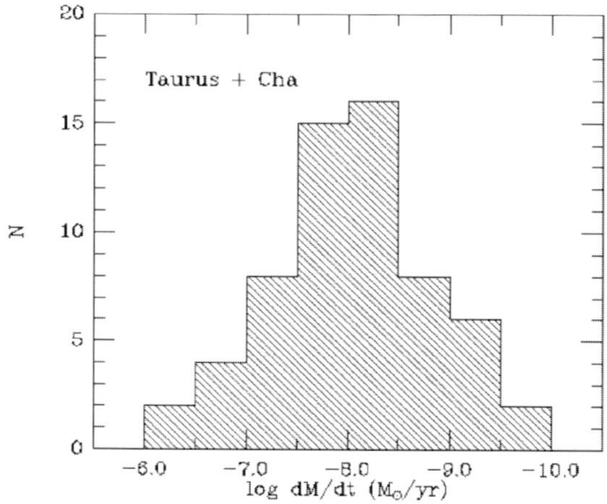

Fig. 2. Distribution of mass accretion rates for CTTS in the associations of Taurus and Chameleon [28].

dipolar magnetic field, accretion is halted by the magnetic field at a radius

$$r_{\mathrm{t}}=7R_*\left(\frac{B}{1\ \mathrm{KG}}\right)^{4/7}\left(\frac{\dot{M}}{10^{-8}\ M_\odot\ \mathrm{yr}^{-1}}\right)^{-2/7}\left(\frac{M_*}{0.5\ M_\odot}\right)^{-1/7}\left(\frac{R_*}{2\ R_\odot}\right)^{5/7} \tag{5.1}$$

[29]. For disk accretion, $R_{\mathrm{t}} \sim \gamma r_{\mathrm{t}}$, with $\gamma < 1$, 1/3–2/3 [30–32]. So, for the typical parameters characterizing CTTS (see Fig. 1), the inner disk is truncated at a few stellar radii by the stellar magnetic field. Matter falls onto the star along field lines with essentially the free-fall velocity (the thermal pressure is expected to be much smaller than the ram pressure).

Before the role of magnetic fields was recognized, disks were expected to join the star through a boundary layer. In this layer, material in the innermost disk annulus, rotating at the Keplerian rotation at the stellar surface, $v_{\mathrm{K}} \sim 217(M_*/0.5\ M_\odot)^{1/2}(R_*/2\ R_\odot)^{-1/2}\ \mathrm{km\,s^{-1}}$, would slow down to the low surface rotational velocities observed, $<15\ \mathrm{km\,s^{-1}}$. Magnetospheric accretion, in addition to being consistent with the measured values of magnetic fields and mass accretion rates, implies a much larger emitting volume and larger infall velocities than the boundary layer model; both properties predict emission indicators in much better agreement with observations.

Magnetospheric accretion has several important predictions which can be compared to the observations. Lines formed in the magnetospheric flow have characteristic profiles; in addition, matter joins the star through an accretion shock formed at the stellar surface, which also produces characteristic emission. We review them in turn.

5.1 Line formation in the magnetospheric flow

The emission profiles of permitted lines in CTTS are characterized by peaks at nearly zero velocities, slightly blueshifted in a few cases. Superimposed to the emission, they show a blue-shifted absorption component and in some cases a *red-shifted* absorption component (*cf.* [34]). These profiles cannot be due to a stellar wind, which produces the typical P-Cygni profile, with a peak red-shifted to velocities of the order of the terminal velocity, a few hundred $\mathrm{km\,s^{-1}}$, and a blue-shifted absorption component that goes below the continuum [35]. They cannot be produced in a boundary layer wind either, because profiles would be too broadened by the rotation of the emitting region [36]. In contrast, the maximum contribution to the lines in magnetospheric infall comes from the region where matter is getting lifted from the disk, which has the largest volume; the velocity of the matter in this region is nearly zero. The wings, on the other hand, are formed where the flow approaches the stellar surface, resulting in a line width comparable to the free-fall velocity at the stellar surface. In addition, if the line of sight intersects the infalling material in front of the accretion shock, which is emitting as we will see, at much higher temperatures than the stellar photosphere, a red-shifted absorption component is formed (Fig. 3). Thus, magnetospheric infall can naturally explain the main features of observed emission line profiles in CTTS.

Detailed calculations for dipolar fields and axially-symmetric flows have confirmed these expectations for CTTS [37]. Observed line profiles of H and Na lines have been calculated, producing excellent agreement with observations of CTTS covering the observed range of mass accretion rate [37]. The wind only has a major contribution to the Hα emission in stars with the highest $\dot{M}$, in agreement with the characteristic P Cygni profiles of these lines. But even in these stars, lines with lower opacity like the Paschen and Bracket lines or the Na lines, show typical magnetospheric profiles [37]. Therefore, *broad, nearly centered profiles in emission lines of low mass TTS represent one of the most important accretion signatures.* The presence of red-shifted absorption components, of course, reinforces the interpretation of accretion.

Magnetospheric accretion has also been identified in HAeBe. In this case, however, the typical magnetospheric profile is characterized by a broad central or slightly red-shifted absorption. In contrast to CTTS, where the infalling material can only absorb radiation from the accretion shock, in HAeBe, the flow can absorb radiation from the much hotter stellar disk, giving rise to the broad absorption. As illustration, Balmer line profiles of the HAeBe star UX Ori are shown in Figure 4, compared to the predictions of a magnetospheric model [38].

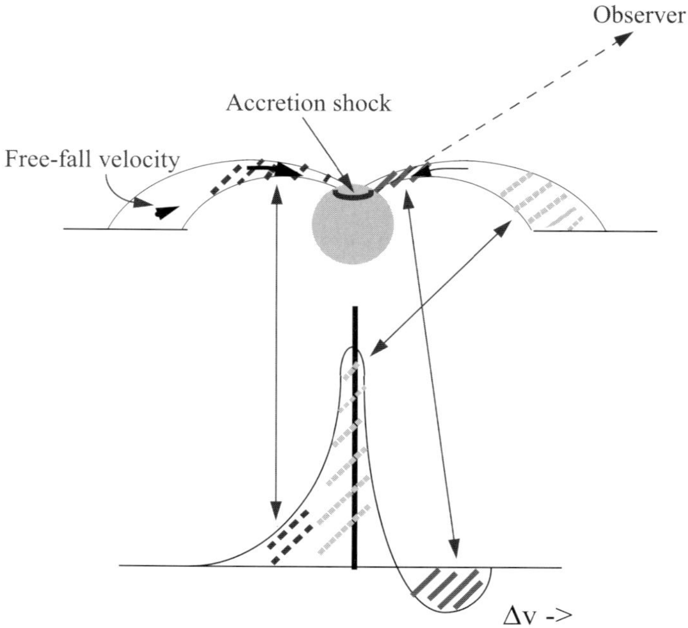

Fig. 3. Schematic of line formation in the magnetospheric accretion flow. Most flux at line center comes from the regions where matter is lifted from the disk, at nearly zero velocity. Blue-shifted emission comes from flow at the back of the star, falling in. Red-shifted emission is formed if the line of sight crosses the infalling material in front of the hot accretion shock.

Broad line profiles characteristic of magnetospheric accretion have been identified in stars below the substellar limit, the brown dwarfs [39]. Because their low accretion luminosity yields little continuum flux excess, line profiles have been crucial in finding the accreting stars in the 10 Myr association TW Hya, other than in TW Hya itself [40, 41]. Finally, magnetospheric accretion as indicated by broad and nearly symmetric line profiles has been identified in protostars. Figure 5 shows the observed Brγ profile of a protostar, PV Cep, together with the excellent fit with a magnetospheric model [42].

Empirical correlations between line luminosities and accretion luminosities for a sample of CTTS with well determined $\dot{M}$ from veiling studies have been found [43, 44]. These correlations can be used to determine the mass accretion rates in objects which are too extincted to be observed in the optical or in the UV. Using this method, accretion luminosities have been estimated for Class I objects [43]; it was found that $L_{\rm acc}$ is lower than the

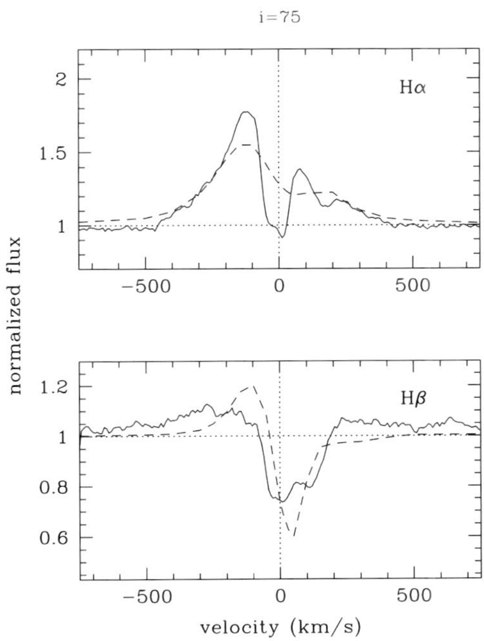

Fig. 4. Line profiles of Hα and Hβ of the Herbig Ae/Be star UX Ori (solid lines), compared to predictions of magnetospheric accretion models (dotted lines). Model profiles are calculated for an inclination of 75 degrees [38].

observed luminosity by at least by a factor of 10. Two consequences follow: (1) the observed luminosity in Class I must be due to stellar luminosity, and thus is not an accretion indicator, and (2) the typical mass accretion rate in the disks of protostars, $\sim 10^{-7}\ M_{\odot}\ \mathrm{yr}^{-1}$, estimated from L_{acc}, is lower than the mass infall rate by roughly a factor of 10 (in Taurus). These facts have profound implications for the evolution of YSO (see Hartmann lectures).

5.2 The accretion shock

Material falling along magnetic field lines will reach the stellar surface at free-fall velocities. It has to slow down through a shock to merge with the photosphere below. For typical parameters, the shock forms just above the stellar surface [45]. The total luminosity emitted by the accretion columns is $L_{\mathrm{shock}} = \zeta L_{\mathrm{acc}}$, where $\zeta = 1 - R_*/R_{\mathrm{t}}$. So, for typical values of R_{t}/R_*, most of the accretion luminosity is emitted in the shock [45].

Soft X-rays from the shock heat the photosphere below it and the pre-shock region just above it along the accretion column. Roughly, 1/2 of

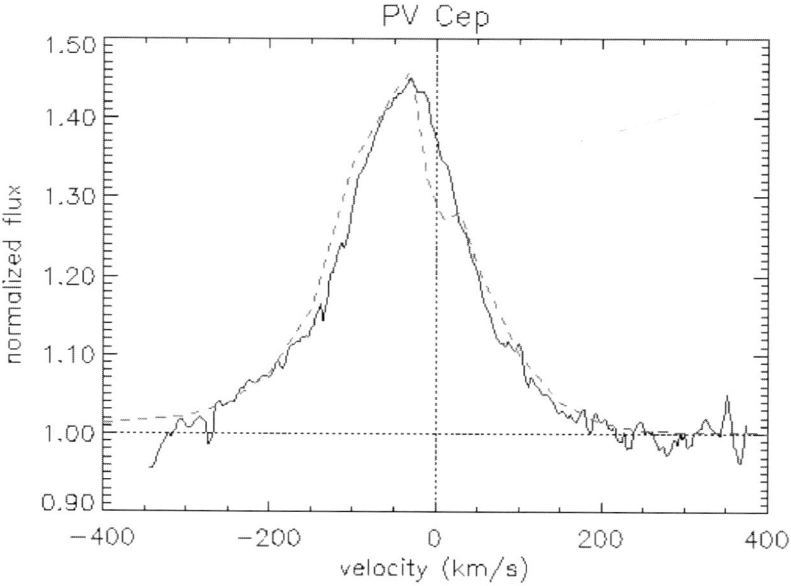

Fig. 5. Line profiles of Brγ in the Class I object PV Cep (solid line), compared to predictions of magnetospheric accretion models (dotted line) [42].

$L_{\rm shock}$ is emitted in each direction. In turn, half of the energy entering the pre-shock is re-emitted down towards the heated photosphere, while the rest leaves the shock region. As a first approximation, the heated photosphere emits as a black body, with $\sigma T_{\rm hp} \sim 3/4 L_{\rm shock}/A \sim 3/4\mathcal{F}$, where $\mathcal{F}$ is the energy flux carried by each column and A is total surface area covered by accretion columns. The pre-shock region, on the other hand is optically thin with a temperature $\sim$10 000$-$20 000 K.

Accretion shock models based in these simple principles can explain very well the observed optical and UV excesses of CTTS [45, 46]. Typical energy fluxes carried by the accretion columns are found to be $\sim 1-3 \times 10^{11}$ erg s^{-1} cm^{-2}, with corresponding temperatures of the heated photosphere of the order of 7000–9000 K. The surface filling factor of the shock is found to be very small, $<1\%$. The so-called continuum stars, those so heavily veiled that absorption lines are difficult to detect, carry similar energy fluxes in their accretion columns, but the filling factors are much higher, $\sim 10\%$ [45, 46]. Line fluxes formed in the pre-shock region have been calculated for typical parameters, and found to compare well with observations [47].

Shock emission has also been calculated for HAeBe. However, the *effects* of shock emission are much more difficult to discern in this case. The reason is that if the energy carried by accretion columns in HAeBe is similar to that of the lower mass CTTS, then the emission from the heated photosphere, which really stands out from the cool photospheres of CTTS, is now similar to that of the stellar photosphere. Standard de-veiling procedures applicable to CTTS cannot be implemented in HAeBe because of the differential veiling introduced by line formation in the heated photosphere [38]. One conspicuous indicator of accretion seems to be the Balmer jump, which tends to go from absorption to emission as the mass accretion rate increases. Comparison with observed strengths of Balmer jumps indicate mass accretion rates of the order of 10^{-8}–10^{-7} $M_\odot$ yr^{-1} for the disks of HAeBe [38].

In conclusion, the observed continuum excess in the optical and the UV in YSO can be explained fairly well with emission from the accretion shock. This justifies the measurement of the accretion luminosity from this excess, and thus the estimates of $\dot{M}$ on which the magnetospheric model is based.

6 Observations of disks in YSO

While the optical and UV flux excesses in CTTS arise in the region of interaction between the disk and the star, the disk itself is responsible for the flux excess at long wavelengths. The lack of correlation between the extinction A_V and the IR excess was interpreted long ago as due to a flat geometry [48], but it was not until the 1990's that disks were imaged directly with HST. Disk images seen dark against the nebular background in the so-called proplyds in the Orion Nebula Cluster [49] or in light scattered in the surface of edge-on disk [50] indicate the disks are flared rather than flat, with typical sizes 100–500 AU.

Disks around CTTS have also been imaged in the mm wavelength range, both in dust continuum emission [51, 52] and in molecular lines [53–55]. Velocity gradients measured in molecular observations are consistent with Keplerian velocity, which would be expected if disks were geometrically thin and moved in the central potential of the star. Disk masses have been estimated from the dust continuum emission, assuming that the disk is optically thin and that the disk is isothermal vertically. With these assumptions, the disk flux can be written as the sum of the emission of all annuli,

$$F_\nu = \frac{2\pi}{d^2} \int_{R_1}^{R_2} B_\nu(T(R)) \Sigma(R) \kappa_\nu R \, \mathrm{d}R \tag{6.1}$$

where it has been used that the specific intensity at each annulus is $I_\nu = B_\nu(1 - \exp(-\tau_\nu/\cos i) \sim B_\nu \tau_\nu/\cos i$, with B_ν the Planck function. The optical depth is given by $\tau_\nu = \kappa_\nu \Sigma$, where κ_ν is the dust opacity. Assuming that the dependence of temperature and surface density on radius is given by power laws, $T = T_0(R/R_0)^{-q}$ and $\Sigma = \Sigma(R/R_0)^{-p}$, the observed flux is related to the disk mass $M_\mathrm{d} = \int_{R_1}^{R_2} 2\pi\Sigma R\,\mathrm{d}R$ as

$$\nu F_\nu = \frac{4k\nu^3}{c^2}\kappa_\nu M_\mathrm{d} T(R_\mathrm{d})\left(\frac{2-p}{2-p-q}\right) \tag{6.2}$$

where $T(R_\mathrm{d})$ the temperature at the disk radius [5, 51]. With estimates for p, q, and the dust opacity, the measured flux gives the mass of the disk. Assuming $\kappa_\nu = \kappa_0(\nu/\nu_0)^\beta = \kappa_0(\lambda_0/\lambda)^\beta$,

$$\nu F_\nu \propto \nu^{3+\beta} \tag{6.3}$$

that is, the slope of the SED in the mm gives the frequency dependence of the opacity, under this approximations. The dust opacity frequently used gives $\kappa_0 = 0.1$ cm^2 gr^{-1} at $\lambda_0 = 250$ μm [58]. With this value, $M_\mathrm{d} \sim 0.01$–$0.1\ M_\odot$ [51], which indicates that indeed disk is less massive than the star.

In summary, to estimate the disk mass, we require knowledge of the dependences of T and Σ with radius, in addition to the dust opacity. Simplifying assumptions as those described are useful to get order of magnitude estimates, but more detailed treatments are needed to get more accurate estimates and to determine how accretion properties can be obtained from observations. Appropriate solutions of the disk structure equations are required to calculate the physical quantities. In addition, the dust opacity depends on the compounds, the grain size distribution, and the degree of dust settling in the disk. Results from more detailed treatments will be discussed below.

7 Accretion disks

Since disks in YSO are accreting, we attempt to describe them using the theories of accretion disks. Many excellent reviews and books on accretion disks have been published (*cf.* [29]) even specifically to the case of YSO [5]. I refer the reader to those for more details. In here, I will only list the basic properties of steady accretion disks.

Assume a disk with a differential rotational velocity, moving in the potential of a central star. If the disk is geometrically thin, the rotational velocity will be the Keplerian velocity, $\Omega_\mathrm{K} = \sqrt{GM_*/R_*^3}$. Friction between nearby radii will release heat. As a result, the material loses energy and sinks in the potential well, drifting towards the star. At the same time, some material must move outwards to conserve angular momentum. So, matter drifts inwards and angular momentum outwards.

A geometrically thin accretion disk with a steady mass accretion rate $\dot{M}$ has a surface density distribution given by

$$\Sigma = \frac{\dot{M}}{4\pi\nu}\left[1 - \left(\frac{R_*}{R}\right)^{1/2}\right] \tag{7.1}$$

where ν is the viscosity, and a temperature distribution given by

$$T_{\rm vis} = \left(\frac{3GM_*\dot{M}}{8\pi\sigma R^3}\left[1 - \left(\frac{R_*}{R}\right)^{1/2}\right]\right)^{1/4} \tag{7.2}$$

so, $T_{\rm vis} \propto R^{-3/4}$, $R >> R_*$. If the disk is optically thick, the SED resulting from this temperature distribution is given by

$$\lambda F_\lambda \propto \lambda^{-4/3}. \tag{7.3}$$

8 FU Ori objects

The FU Ori objects constitute a perfect example of the steady accretion disk. The first objects of these class were discovered by an increase in flux by 2–3 orders of magnitude over time scales of months, reaching luminosities of $100-1000\ L_\odot$. In only one case a pre-outburst optical spectrum exists, and it was similar to that of a CTTS After the outburst, the spectrum changed to that of a ~G-F star. FU Ori objects are found in close association with molecular clouds in star-forming regions.

It is now accepted that FU Ori objects are high mass accretion rate disks surrounding a low mass YSO (see review [56, 57] and references therein). This interpretation naturally explains the observed features of the objects. Spectra at different wavelength ranges indicate that the spectral type becomes later with increasing wavelength. So, while the optical spectrum is similar to that of an F star, the spectrum at 2 μm appears similar to that of a K supergiant. This property can be explained naturally by a disk, in which a range of temperatures is present (cf. Eq. (7.2)). An accretion disk can also naturally explain the double-line profile of the lines, expected from a ring-like geometry, and moreover, it can explain the decrease of rotational velocity between the optical and the infrared, consistent with Keplerian velocity. Finally, the SED of the least embedded of these objects, FU Ori itself, is consistent with the expected emission from steady disks, equation (7.3). In others, the disk emission is more contaminated by envelope emission and it is difficult to extract.

The properties of the disk can be inferred from the luminosity, obtained by integration of the observed fluxes, the measured rotational velocities, and the maximum temperature observed. These three conditions can be used to infer the mass accretion rate, and the stellar properties (with certain uncertainty for the inclination). The mass accretion rates inferred are of the order of 10^{-5}–10^{-4} $M_\odot$ yr^{-1}, much higher than the mean $\dot{M}$ of CTTS, and consistent with accretion luminosity dominating the emission over stellar luminosities.

Disk atmospheres with effective temperature equal to equation (7.2) with these mass accretion rates and gravities corresponding to a geometrically thin disk in a central potential, can explain well the observed features of the spectrum. In particular, deep and broad near infrared CO bands at 2.2 μm are a characteristic feature. As mentioned, the SED of all known FU Ori objects shows a contribution from surrounding envelope material. It is smaller in the first discovered visible objects, but other, heavily embedded FU Ori objects have only been discovered by the presence of deep CO absorption.

In the case of FU Ori objects, then, the total luminosity measures the accretion. However, independent confirmation of the existence of a disk with high mass accretion rate is required to assign an object to this class.

9 Irradiated accretion disks

In contrast to FU Ori objects, the SEDs of CTTS are much flatter than expected from equation (7.3) [2,59], which first raised questions about their accretion disk nature. We have seen that disks in CTTS are accreting mass at a rate given by $\dot{M} \sim 10^{-8}$ $M_\odot$ yr^{-1}, which results in an accretion luminosity of <0.1 $L_\odot$, mostly released at the accretion shock on the surface. With $L_* \sim 1$ $L_\odot$, $L_* \gg L_{\rm acc}$, which means that stellar irradiation must play an important role in the heating of the disk [59]. In fact, they are irradiated accretion disks.

For a flat disk, the irradiation flux can be estimated as

$$F_{\rm irr} \sim I_* \Omega_* \cos\theta \sim \frac{I_* \pi R_*^2}{R^2} \frac{h}{R} \propto I_* \left(\frac{R_*}{R}\right)^3 \tag{9.1}$$

where θ is the angle between the line from a representative point on the stellar surface to the point R in the disk and the line along the midplane, approximating the stellar radiation as coming in a single beam, and $h \sim R_*$ is the height of the representative point above the midplane. With $F_{\rm irr} \sim \sigma T^4$, $T \propto R^{-3/4}$, similar to an accretion disk, and thus resulting in SEDs that cannot explain the observations either.

However, disks of CTTS are not flat, they are flared, which means that they can capture more stellar radiation than flat disks. In the case of a vertically isothermal, geometrically thin disk affected by the central potential of the star, the density can be found from the equation of hydrostatic equilibrium in the vertical direction to be

$$\rho(z,R) = \rho_\mathrm{c}\mathrm{e}^{-\frac{z^2}{2H^2}} \tag{9.2}$$

where $\rho_\mathrm{c}(R)$ is the density at the midplane and H is the scale height, given by

$$H = \frac{c_\mathrm{s}}{(GM_*/R^3)^{1/2}} = \frac{c_\mathrm{s}}{\Omega_\mathrm{K}} \propto T^{1/2} \tag{9.3}$$

where $c_\mathrm{s} = (2kT/m)^1/2$ is the sound speed and m the mass of the particle. If T increases, the cross section to capture of stellar photons increases, so the disk gets heated. A stable solution for the disk surface exists, corresponding to a flared surface [59, 60]. In these solutions, $T(R) \to R^{-1/2}$, and the corresponding SEDs are much flatter than those of the purely viscous disk.

Disks are not vertically isothermal. Stellar radiation enters the disk at an angle θ_0 to the local normal, (again approximating the stellar radiation as coming in a single beam), so the energy captured by the disk is $\sim(\sigma T_*^4/\pi)(R_*/R)^2\mu_0$, with $\mu_0 = \cos\theta_0$ [61–65]. As stellar flux enters the disk, a fraction $\mathrm{d}\tau_*/\mu_0$ is absorbed at each z, where τ_* is the optical depth at the wavelength range characterizing the stellar energy (the "stellar" range). This energy re-emerges at the wavelength characterizing the local temperature (the "disk" range). So, the stellar flux decreases with height. The main opacity source in CTTS disks is due to dust grains, because temperatures are low enough that dust is not sublimated. Dust opacity increases as λ decreases, and stellar radiation is emitted at a shorter wavelength than the radiation emitted by the disk, so the hotter the star, the larger the opacity at the stellar range and thus the heating.

The vertical temperature profile can be obtained from the equation of conservation of energy

$$\int \kappa_\nu B_\nu(T)\,\mathrm{d}\nu - \int \kappa_\nu J_\mathrm{d}\,\mathrm{d}\nu = \frac{\mathrm{d}}{\mathrm{d}z}F_\mathrm{d} \tag{9.4}$$

where κ_ν is the monochromatic opacity, J_d the mean intensity of the local radiation at the disk wavelength range, and F_d the local flux. Neglecting viscous heating, the local flux only changes by deposition of stellar energy, so

$$\frac{\mathrm{d}}{\mathrm{d}z}F_\mathrm{d} \sim 4\pi\kappa^*\rho J_* = 4\pi\kappa^*\rho J_{*,0}\,\mathrm{e}^{-\tau_*/\mu_0} \tag{9.5}$$

where κ^* is the opacity at the stellar wavelength, and $J_{*,0}$ is the mean intensity entering the disk. With $J_{*,0} = 1/4\pi \int I\,\mathrm{d}\Omega \sim I_*\Omega_*/4\pi$,

$$\kappa_\mathrm{P}(T)\frac{\sigma T^4(z)}{\pi} = \kappa_\mathrm{P}(T)J_\mathrm{d}(z) + \kappa_\mathrm{P}(T_*)\frac{\sigma T_*^4}{\pi}\left(\frac{R_*}{R}\right)^2 \mathrm{e}^{-\tau_*/\mu_0} \tag{9.6}$$

with $\kappa^* \sim \kappa_\mathrm{P}(T_*)$, and κ_P the Planck mean opacity [65, 66].

In the surface, where the local field is much smaller than the stellar field, $J_{\rm d} \ll J_{*,0}$, and $\tau_*/\mu_0 \ll 1$,

$$\kappa_{\rm P}(T_0)T_0^4(z) \sim \kappa_{\rm P}(T_*)T_*^4 \left(\frac{R_*}{R}\right)^2 . \tag{9.7}$$

This is the expression for the surface temperature T_0, given by the optically thin limit. Note that this is the "hot layer" temperature in the 2-level approximation [64].

Stellar heating and thus the temperature decrease as radiation penetrates the disk, because the optical depth increases (*cf.* Eq. (9.6)). The actual T profile depends on μ_0, that is, on the actual shape of the *surface* of the disk, defined as the surface where $\tau_*/\mu_0 \sim 1$, that is, where most of the stellar energy is deposited. The size of the upper optically thin region above the surface is given by $\Delta z \sim \mu_0/\kappa^*$; the more inclined the surface, the smaller the optically thin region.

Results of the detailed solution of the set of equations of vertical structure including viscous dissipation for a disk with typical CTTS parameters are shown in Figure 6 [66]. The figure on the right shows the height of the surface, $z_{\rm s}$, the scale height H, and the photospheric height $z_{\rm phot}$, where the Rosseland mean optical depth $\tau_{\rm Ross} \sim 1$. Since the opacity at the wavelength where the stellar radiation is absorbed ($\lambda_* \sim 1$ μm for $T_* \sim 4000$ K), is large, the surface is flared out to a few hundred AU, even though the disk becomes optically thin to its own radiation ($\tau_{\rm Ross} < 1$) at ~20 AU.

The left side of Figure 6 shows characteristic temperatures in the disk. It can be seen that the midplane temperature $T_{\rm m}$ is higher than the photospheric temperature $T_{\rm phot} = T(z_{\rm phot})$ in the inner disk, where it is optically thick to its own radiation. This can be understood using the diffusion approximation

$$\frac{\rm d}{{\rm d}z} J_{\rm d} = \frac{\rm d}{{\rm d}z}\sigma T^4 = -\frac{3}{4\pi}\chi_{\rm R}\rho F_{\rm d} \tag{9.8}$$

from which we can write approximately

$$\Delta(\sigma T^4) \sim \frac{3}{4\pi}\tau_{\rm R} F_{\rm d}. \tag{9.9}$$

In the inner annuli, $\tau_{\rm R} \gg 1$, $T_{\rm m} > T_{\rm phot}$; the temperature gradient allows the flux (viscous plus local radiation) to emerge from the disk. In contrast, when the disk becomes optically thin to its own radiation, $\tau_{\rm R} \ll 1$, the disk becomes nearly isothermal in the regions near the midplane. On the other hand, the surface temperature T_0 is higher than $T_{\rm phot}$ in the inner regions and than $T_{\rm m}$ when the disk becomes optically thin in the outer regions, as predicted by equation (9.6). By comparison of $T_{\rm m}$ and $T_{\rm vis}$ in Figure 6, it can be seen that viscous heating is only important near the midplane of the

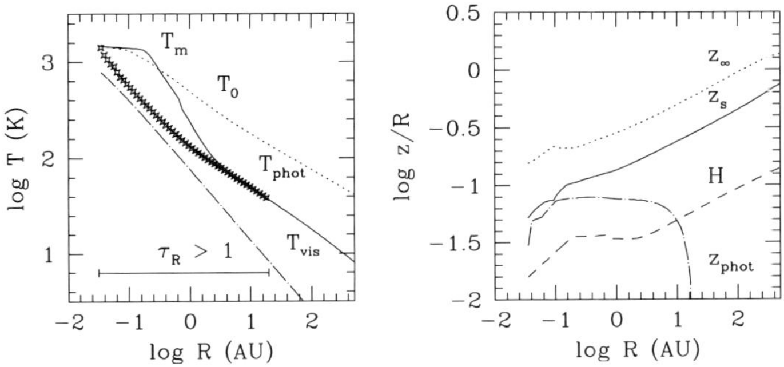

Fig. 6. Left: characteristic temperatures in the disk: midplane temperature ($T_{\rm m}$), surface temperature (T_0), photospheric temperature ($T_{\rm phot}$), and viscous temperature ($T_{\rm vis}$). The disk is optically thick (to its own radiation, $\tau_{\rm R} > 1$) inside ∼6 AU. Right: characteristic heights in the disk: surface ($z_{\rm s}$), scale height (H), photospheric height $z_{\rm phot}$ (defined in regions where the disk is thick to its own radiation). $z_{\rm inf}$ is the top height of the disk. The disk model parameters are $M_* = 0.7\ M_\odot$, $R_* = 2.5\ R_\odot$, $T_* = 4000$ K and $\dot{M} = 10^{-8}\ M_\odot\ {\rm yr}^{-1}$. Adapted from [66].

regions inside 1 AU, given the low $\dot{M}$ characteristic of the the typical CTTS. Finally, it can be seen that temperatures behave as $1/R^{1/2}$ for $R \gg R_*$.

This particular shape of the temperature profile has important observational implications. For one thing, features formed in the optically thin upper regions will appear in emission, even if the disk is optically thick, because the local temperature is so much higher than that of deeper regions where the continuum forms. These includes strong features like the CO near infrared lines or the silicate features [61, 62, 67]. The higher temperatures of the upper layers also imply that molecules can exist in the gas phase in the disk, even when the midplane temperatures are so low that molecules are settled onto grains surfaces [69–71]. For example, if the disk shown in Figure 6 was isothermal at $T_{\rm m}$, the CO molecules would be on grains surfaces for $R > 60$ AU, for which $T < 20$ K. However, the hot upper layers are kept at a temperature high enough for molecules to be in the gas phase out to ∼400 AU, in agreement with mm molecular observations [53–55].

The surface density of the disk can be self-consistently calculated from the disk equations. The limiting behavior can be obtained from equation (7.1). In the parametric α prescription [72], the viscosity can be written as $\nu = \alpha c_{\rm s} H = \alpha c_{\rm s}^2/\Omega_{\rm K}$, using equation (9.3). With

$c_s \propto T^{1/2} \propto R^{-1/4}$ and $\Omega_K \propto R^{-3/2}$, we obtain at large radii

$$\Sigma \sim 4 \left(\frac{\dot{M}}{10^{-8} \frac{M_\odot}{\mathrm{yr}}} \right) \left(\frac{\alpha}{0.01} \right)^{-1} \left(\frac{T_{100\ \mathrm{AU}}}{10\ \mathrm{K}} \right)^{-1} \left(\frac{R}{100\ \mathrm{AU}} \right)^{-1} \mathrm{gr\ cm}^{-2} \quad (9.10)$$

using values of α found in modeling CTTS disks and expected from theories [73], and typical temperatures at 100 AU. The surface density dependence on radius of irradiated accretion disks, $\propto 1/R$, is much flatter then the usually assumed dependence $p = 3/4$, and it has been confirmed by observations [74].

Supposing this dependence holds at all radii, the disk mass is given by

$$M_\mathrm{d} = 0.03 \, \frac{\left(\frac{\dot{M}}{10^{-8} \frac{M_\odot}{\mathrm{yr}}} \right) \left(\frac{R_\mathrm{d}}{200\ \mathrm{AU}} \right)^{1/2}}{\left(\frac{\alpha}{0.01} \right) \left(\frac{M_*}{0.5\ M_\odot} \right)^{1/2} \left(\frac{T_{\mathrm{Rd}}}{10\ \mathrm{K}} \right)} \, M_\odot \quad (9.11)$$

in agreement with values determined from dust mm emission. Therefore, the mass accretion rates determined from the inner disk properties are consistent with large-scale properties like the disk mass. Note that if $\dot{M}$ is known for a given star, then M_d and α are complementary parameters, since the temperature at the outer disk radii, T_{Rd} is fixed by stellar irradiation and sizes can be estimated from observations.

10 Effects of dust properties

The expression normally used to calculate the disk flux is equation (6.1), with $I_\nu = B_\nu[1 - \exp(-\tau_\nu / \cos i)]$ to account for optically thick emission. However, this expression is a simplification of the actual situation. As we have seen, the temperature and surface density profiles in the disk are not power laws, and disks are not vertically isothermal. In addition, since actual disks are flared, emission from the innermost disk annuli may be attenuated by the outer disk for certain ranges of inclination [66].

Nonetheless, equation (6.1) shows the main effects. In particular, it indicates that the SED is highly dependent on the properties of the dust, because it depends explicitly on κ_ν and on the temperature structure, which as we have seen is determined by the dust opacity. It is then important to understand the effects of dust.

The dust opacity depends on the shape, size distribution, and constitution of the dust grains. Schematically, a spherical grain of size a will have a cross section to radiation of wavelength λ of the order of the geometrical cross section πa^2, if $\lambda \ll 2\pi a$, but the cross section will decrease as $\sim\lambda^{-2}$ for $\lambda \gg 2\pi a$. The dust opacity, in $\mathrm{cm}^2\ \mathrm{gr}^{-1}$, will be given by the cross section

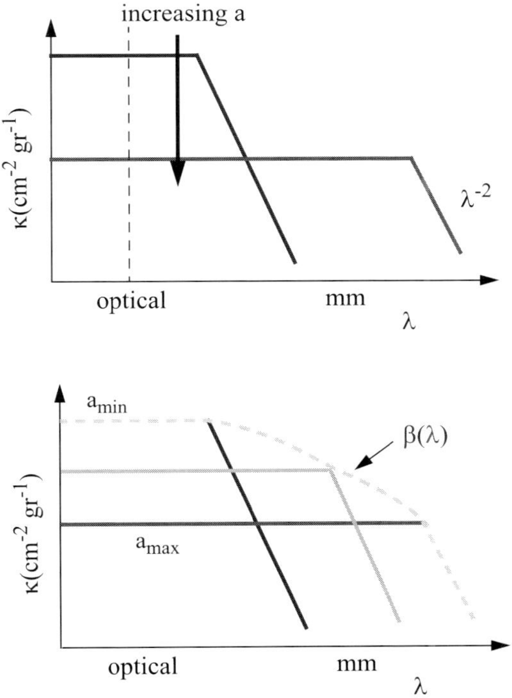

Fig. 7. Dependence of opacity on grain size. Upper: opacity for single grains. As the grain size a increases, κ decreases at short λ, and increases at long λ. Lower: opacity for a grain size distribution. The slope β of κ *vs.* λ depends on λ.

over the mass of the grain, $\rho_g 4/3\pi a^3$, where ρ_g is the density of the grain, so it will be proportional to $1/a$. So, the bigger the grain, the lower the opacity at short wavelengths and the higher at long wavelengths, $\lambda \leq 2\pi a$, because the transition to the λ^{-2} regime occurs at longer wavelengths. This is shown as shown schematically in Figure 7 for grains of sizes a_1 and a_2, $a_1 < a_2$.

A simple characterization of the dust can be done by assuming that the dust particles are homogeneous spheres with a size distribution, $n(a)\mathrm{d}a = a^{-p}\mathrm{d}a$, where a is the grain size and the exponent p is usually taken as 3.5, describing the properties of the interstellar medium dust [75], or 2.5, if there has been some degree of coagulation [76]. Additional parameters are the minimum and maximum sizes, $a_{\rm min}$ and $a_{\rm max}$. If $a_{\rm min}$ and the amount of dust are fixed, the larger the $a_{\rm max}$ the less the smaller particles, because the larger particles take more mass. Therefore, the lower the opacity at short wavelengths and the higher at long wavelengths. For a mixture of sizes, the transition from the "flat" to the $\propto \lambda^{-2}$ regime occurs over a large range of wavelengths (see Fig. 7), so that the local slope of the function $\kappa(\lambda)$ changes slowly from $\sim$0 to $\sim$−2. Therefore, the form usually assumed to represent the dust opacity $\kappa \propto \lambda^{-\beta}$ is not actually valid, since β depends on λ [68,76], although it may be applicable over for a sufficiently small λ interval.

The surface of the disk, that is, the height where the stellar radiation is absorbed, depends on the dust opacity. For instance, a dust mixture characterized by $a_{\rm max} \leq 1$ μm will result in more absorption of stellar radiation, at $\lambda \sim 1$ μm, that one characterized by $a_{\rm max} \sim 1$ mm. Therefore, the resultant surface will be higher for the lower $a_{\rm max}$ case. Since the flux absorbed by the disk depends on the inclination of the surface through $\mu_{\rm o}$, more stellar radiation is absorbed as the disk gets more flared and the hotter the disk will get, resulting in higher infrared fluxes for smaller $a_{\rm max}$. On the other hand, the mm emission of the $a_{\rm max} \leq 1$ μm disk will be lower than that of the $a_{\rm max} \sim 1$ mm disk, because larger grains are more efficient emitters at those wavelengths. Therefore, the SED gives direct information of the dust properties. In particular, the median SED of the CTTS in Taurus can be much better explained by $a_{\rm max} \sim 1$ mm dust than by interstellar medium dust [68].

11 SEDs of HAeBe

The $\dot{M}$ inferred from magnetospheric and shock analysis of (visible) HAeBe are similar to those in CTTS. However, the disks are expected to be hotter since the effective temperatures and luminosities of the central stars are higher. SEDs of HAeBe surrounded by disks including these effects can explain well the observations [67]. However, they cannot explain the "bump" at 2–3 μm, which has been a puzzle for almost 10 years.

Early interpretations of the SEDs of HAeBe required high mass accretion rates $\dot{M} \sim 10^6$ $M_\odot$ yr^{-1} to explain the "bump", that is, the high fluxes at 2–3 μm. On the other hand, they also required holes in the disk of the order of 0.1–1 AU to explain the lack of emission at shorter wavelengths [4]. The hole was generally located at the radius where the dust would be destroyed, $R_{\rm dust}$. However, it was soon recognized that with such high $\dot{M}$ the inner gas disk would be optically thick and thus still have large emission at J [77].

A breakthrough came with the realization that if the dust disk had some vertical size, then the disk would show a wall facing the star at the dust destruction radius. In this wall, radiation from the star would enter frontally, and thus it would be much hotter than the surface of the disk. Radiation from this wall can explain very well the extra emission [67, 78]. Even with the lower $\dot{M}$ we now expect for the HAeBe, the inner gaseous disk will be marginally thick, but the scale height in those regions is low enough for radiation to reach the dust wall [38].

12 Winds as accretion diagnostics

YSO are losing matter. Powerful outflows in molecular and atomic material are observed in objects still surrounded by infalling envelopes. In the visible CTTS, they show through blue-shifted absorption components in the permitted emission lines and in forbidden line emission.

Many models have been presented for the origin of the winds in YSO, as discussed by other lectures in these classes. They all have in common the hypothesis that winds are powered by accretion. This hypothesis is confirmed by observations. In particular, the equivalent width of the forbidden line [OI], expected to form in the low density regions of the wind, correlates well with accretion indicators as veiling and with K-L excesses [5, 7]. Moreover, the luminosity of [OI] correlates very well with the accretion luminosity [81]. In addition, the strength of the blueshifted absorption components in the emission lines increases with $\dot{M}$, such that the emission disappears in the objects with the highest $\dot{M}$, the FU Ori objects. In these objects, the mass accretion is so high that the magnetosphere, where the emission is formed, is quenched (*cf.* Eq. (5.1)).

Models differ in the location where the wind forms. In broad terms, the wind either arises in the so-called X-region, located at the corotation radius [32] or it originates in a wider region in the disk [80]. It is difficult to find observational diagnostics that help discriminate between models. I will not discuss them here; rather, I will refer to a case in which the origin of the wind can be proven to be in the disk, the FU Ori objects.

As I mentioned, absorption lines formed in a disk have well defined double-peaked profiles. These profiles have been observed in mm molecular lines in CTTS, which are formed in the outermost regions, where dust opacity is very low. However, the hot inner gaseous annuli of FU Ori objects produce spectrum rich in metallic absorption lines, in which the characteristic disk profile shows very well. Nonetheless, not all lines are double in FU Ori. The strongest lines, generally located in the blue region of the spectrum, are single and blueshifted [82]. This is a direct consequence of the presence of a disk wind. If material is leaving the disk surface, it is expected to accelerate until it gets to a terminal velocity at several scale heights above the midplane. Weak lines will form on the disk surface and will be broadened by the rotational velocity of the disk, so they will be double. Strong lines will form much higher, where the expansion velocity dominates over rotation, so they will be single and blueshifted [83]. Therefore, the observed behavior of the lines gives clear indication of the presence

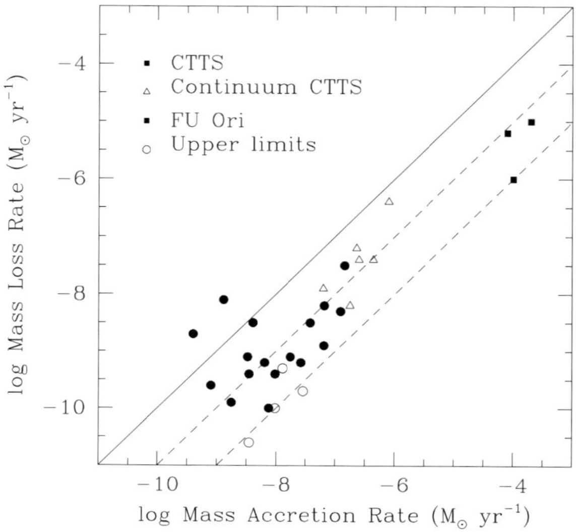

Fig. 8. Mass loss rate *vs.* mass accretion rate for YSOs. Different type of objects are indicated in the figure.

of a disk wind. Modeling of the lines has yield the mass loss rate of this wind [83].

In CTTS, mass loss rates have been estimated from the measured fluxes and velocities of the forbidden lines, with various assumptions for the temperature and geometry [7]. Figure 8 shows the mass loss rate *vs.* the mass accretion rate for several types of YSO. It includes the CTTS, the more active, heavily veiled "continuum" CTTS, and the FU Ori objects. A remarkable correlation shows holding over 4–5 orders of magnitude of the mass accretion and loss rates, and it suggests a similar origin for the winds in all objects. In any event, this correlation indicates that the presence and strength of a wind is an important accretion diagnostic. Objects with strong winds must be powered by accretion disks where matter is accreted at a rate ∼10 higher than the mass at which matter is being lost.

13 Validity of the models

We have seen that a model based on accretion of matter onto the star can explain the observed properties of YSO. The model includes an irradiated accretion disk, from which ∼10% of the matter is lost in the wind, and the rest falls onto the star along magnetic field lines, merging with the photosphere through an accretion shock.

The models are axially symmetric, uniform, and assume the simplest geometry for the magnetic field. Reality is much more complicated than this.

First of all, a large degree of variability is observed in both lines and continua. This variability is nearly periodic in visual and red optical bands, as is interpreted as rotational modulation of spots in the surface [85–87], in a configuration where the magnetic field axis is inclined relative to the rotation axis.

In addition, the magnetospheric flow probably is not axially symmetric. The observed correlation between line luminosities and accretion luminosities cannot be explained by axially symmetric models, which predict higher fluxes than observed at the lowest values of $\dot{M}$. However, it can be understood if the area covered by accreting columns decreases with $\dot{M}$, as suggested by the comparison of shock emission between normal CTTS and the continuum stars [37]. To support this, accretion by sectors is required to explain the line profile of very low mass objects with extremely low values of $\dot{M}$ [39].

The magnetospheric sectors are probably not uniform in density or temperature either. Broad velocity wings are observed in lines requiring high temperatures as the C IV lines [88], indicating that they form in the extended flow. So, it appears that high and low temperature coexist in the flow.

Similarly, disks may not be axisymmetric. Absorption episodes, as observed in the UX Ori objects [67], and in some CTTS [89] may be due to warps in the disk [90]. Nor is dust homogeneously distributed in the disk, radially or vertically.

Given all these facts, it is remarkable that models so simple can explain such a large range of observations, from the ultraviolet to the mm range. This may indicate the models contain the essential physics of the problem, namely, accretion as the main mechanism driving the activity.

14 Summary

To briefly summarize these lectures, accretion is taking place in YSO. In protostars, it occurs in two stages, from cloud to disk, and from disk to star. The observed luminosity of protostars is not due to energy released in the infall from the cloud, because matter lands on the disk too far from the star. In general, it is due to the stellar luminosity. There are cases of high $\dot{M}$ disks in which $L_{\rm obs} \sim L_{\rm acc}$, but independent confirmation of the existence of such disks, as deep near IR CO absorption bands, is required to confirm the assumption. In the accreting TTS, the CTTS, magnetospheric accretion transfers mass from the disk onto the star. This is indicated by broad emission lines formed in the accretion flow, and by excess continuum emission in the optical and UV, due to accretion shock emission. Measurements of this excess yield $L_{\rm acc}$. Correlations of luminosities in the U band and of optical and IR lines with $L_{\rm acc}$ are useful to obtain $L_{\rm acc}$ for large samples and/or heavily extincted stars. Although CTTS are surrounded by accretion disks, the emission properties of these disks do not provide a direct measurement of the mass accretion rate, because for the average $\dot{M} \sim 10^{-8}\ M_\odot\ {\rm yr}^{-1}$ of CTTS, stellar irradiation dominates the heating.

I would like to thank the organizers for the invitation to participate in the Summer School and the European Union for sponsoring the visit. I also would like to thank Bruno Merin for valuable comments to the manuscript. Work reported in these lectures was supported by NASA grant NAG5-4282.

References

[1] L. Siess, E. Dufour and M. Forestini, *AA* **358** (2000) 593.
[2] S.J. Kenyon and L. Hartmann, *ApJS* **101** (1995) 117.
[3] C. Briceño and K. Luhman.
[4] L.A. Hillenbrand, St.E. Strom, F.J. Vrba and J. Keene, *ApJ* **397** (1992) 613.
[5] L. Hartmann, "Accretion processes in star formation" (New York: Cambridge University Press, 1998).
[6] G.H. Herbig and K.R. Bell, "Catalog of emission line stars of the orion population" (Lick Observatory Bulletin, Santa Cruz, 1988).
[7] P. Hartigan, S. Edwards and L. Ghandour, *ApJ* **452** (1995) 736.
[8] V.M. Costa, M.T.V.T. Lago, L. Norci and E.J.A. Meurs, *AA* **354** (2000) 621.
[9] R. Neuhaeuser, M.F. Sterzik, J.H.M.M. Schmitt, R. Wichmann and J. Krautter, *AA* **297** (1995) 391.
[10] D.F.M. Folha and J.P. Emerson, *AA* **352** (1999) 517.
[11] D. Lynden-Bell and J.E. Pringle, *MNRAS* **168** (1973) 603.
[12] C.M. Johns-Krull, J.A. Valenti and C. Koresko, *ApJ* **516** (1999) 900.
[13] E.W. Guenther, H. Lehmann, J.P. Emerson and J. Staude, *AA* **341** (1999) 768.
[14] C.M. Johns-Krull, J.A. Valenti, A.P. Hatzes and A. Kanaan, *ApJ* **510** (1999) L41.
[15] L. Hartmann, R. Hewett, S. Stahler and R.D. Mathieu, *ApJ* **309** (1986) 275.
[16] J. Bouvier, C. Bertout, W. Benz and M. Mayor, *AA* **165** (1986) 110.
[17] S. Zhou, N.J.II. Evans, Y. Wang, R. Peng and K.Y. Lo, *ApJ* **433** (1994) 131.
[18] D. Mardones, "Signatures of Infall in Regions of Low Mass Star Formation", Ph.D. Thesis (Harvard University, 1998).
[19] B.A. Wilking and C.J. Lada, *ApJ* **274** (1983) 698.
[20] A.A. Goodman, P.J. Benson, G.A. Fuller and P.C. Myers, *ApJ* **406** (1993) 528.
[21] P. Cassen and A. Moosman, *Icarus* **48** (1981) 353.
[22] S. Terebey, F.H. Shu and P. Cassen, *ApJ* **286** (1984) 529.
[23] S.W. Stahler, *ApJ* **332** (1988) 804.
[24] L. Hartmann, P. Cassen and S.J. Kenyon, *ApJ* **475** (1997) 770.
[25] S.J. Kenyon, N. Calvet and L. Hartmann, *ApJ* **414** (1993) 676.

[26] S.J. Kenyon, L.W. Hartmann, K.M. Strom and S.E. Strom, *AJ* **99** (1990) 869.

[27] E. Gullbring, L. Hartmann, C. Briceño and N. Calvet, *ApJ* **492** (1998) 323.

[28] L. Hartmann, N. Calvet, E. Gullbring and P. D'Alessio, *ApJ* **495** (1998) 385.

[29] J. Frank, A. King and D. Raine, "Accretion power in astrophysics" (UK: Cambridge University Press, 2002).

[30] P. Ghosh and F.K. Lamb, *ApJ* **234** (1979) 296.

[31] M. Camenzind, *Rev. Mod. Astron.* **3** (1990) 234.

[32] F. Shu, J. Najita, E. Ostriker, F. Wilkin, S. Ruden and S. Lizano, *ApJ* **429** (1994) 781.

[33] C. Bertout, G. Basri and J. Bouvier, *ApJ* **330** (1988) 350.

[34] J. Muzerolle, L. Hartmann and N. Calvet, *AJ* **116** (1998) 455.

[35] L. Hartmann, E.H. Avrett, R. Loeser and N. Calvet, *ApJ* **349** (1990) 168.

[36] N. Calvet, L. Hartmann and R. Hewett, *ApJ* **386** (1992) 229.

[37] J. Muzerolle, N. Calvet and L. Hartmann, *ApJ* **550** (2001) 944.

[38] J. Muzerolle, P. D'Alessio, N. Calvet and L. Hartmann, in preparation.

[39] J. Muzerolle, C. Briceño, N. Calvet *et al.*, *ApJ* **545** (2000) L141.

[40] J. Muzerolle, N. Calvet, C. Briceño, L. Hartmann and L. Hillenbrand, *ApJ* **535** (2000) L47.

[41] J. Muzerolle, L. Hillenbrand, N. Calvet, L. Hartmann, C. Briceño, C., "Disk Accretion at 10 Myr: Results from the TW Hydrae Association", Young Stars Near Earth: Progress and Prospects, ASP Conf. Ser. 244, edited by Ray Jayawardhana and T. Greene (San Francisco: Astronomical Society of the Pacific).

[42] J. Muzerolle, N. Calvet and L. Hartmann, in preparation.

[43] J. Muzerolle, L. Hartmann and N. Calvet, *AJ* **116** (1998) 2965.

[44] N. Calvet, L. Hartmann and S.E. Strom, "Evolution of Disk Accretion", Protostars and Planets IV, edited by V. Mannings, A.P. Boss and S.S. Russell (Tucson: University of Arizona Press, 2000) 37.

[45] N. Calvet and E. Gullbring, *ApJ* **509** (1998) 802.

[46] E. Gullbring, N. Calvet, J. Muzerolle and L. Hartmann, *ApJ* **544** (2000) 927.

[47] A.I. Gomez de Castro and S.A. Lamzin, *MNRAS* **304** (1999) L41.

[48] M. Cohen and L.V. Kuhi, *ApJS* **41** (1979) 743.

[49] C.R. O'dell, *AJ* **115** (1998) 263.

[50] K.R. Stapelfeldt, J.E. Krist, F. Menard *et al.*, *ApJ* **502** (1998) L65.

[51] S.V.W. Beckwith, A.I. Sargent, R.S. Chini and R. Guesten, *AJ* (99) 924.

[52] A. Dutrey, S. Guilloteau, G. Duvert *et al.*, *AA* **309** (1996) 493.

[53] A. Dutrey, S. Guilloteau, L. Prato *et al.*, *AA* **338** (1998) L63.

[54] S. Guilloteau and A. Dutrey, *AA* **339** (1998) 467.

[55] C. Qi, "Aperture synthesis studies of the chemical composition of protoplanetary disks and comets", Ph.D. Thesis (California Institute of Thechnology, 2001).

[56] L. Hartmann and S.J. Kenyon, *ApJ* **299** (1985) 462.

[57] L. Hartmann and S.J. Kenyon, *ARA&A* **34** (1996) 207.

[58] S.V.W. Beckwith and A.I. Sargent, *ApJ* (381) 250.

[59] S.J. Kenyon and L. Hartmann, *ApJ* **323** (1987) 714.

[60] P. D'Alessio, J. Cantó, L. Hartmann, N. Calvet and S. Lizano, *ApJ* **511** (1999) 896.

[61] N. Calvet, A. Patiño, G.C. Magris and P. D'Alessio, *ApJ* **380** (1991) 617.

[62] N. Calvet, G.C. Magris, A. Patiño and P. D'Alessio, *Rev. Mex. Astron. Astrofis.* **24** (1992) 27.

[63] F. Malbet and C. Bertout, *ApJ* **383** (1991) 814.

[64] E.I. Chiang and P. Goldreich, *ApJ* **490** (1997) 368.

[65] P. D'Alessio, J. Canto, N. Calvet and S. Lizano, *ApJ* **500** (1998) 411.

[66] P. D'Alessio, N. Calvet, L. Hartmann, S. Lizano and J. Cantó, *ApJ* **527** (1999) 893.

[67] A. Natta, T. Prusti, R. Neri *et al.*, *AA* **371** (2001) 186.

[68] P. D'Alessio, N. Calvet and L. Hartmann, *ApJ* **321** (2001) 321.

[69] K. Willacy and W.D. Langer, *ApJ* **544** (2000) 903.

[70] G.-J. van Zadelhoff, E.F. van Dishoeck, W.-F. Thi and G.A. Blake, *AA* **377** (2001) 566.

[71] Y. Aikawa, G.J. van Zadelhoff, E.F. van Dishoeck and E. Herbst, *AA* **386** (2002) 622.

[72] N.I. Shakura and R.A. Sunyaev, *AA* **24** (1973) 337.

[73] J.M. Stone, C.F. Gammie, S.A. Balbus and J.F. Hawley, "Transport Processes in Protostellar Disks", Protostars and Planets IV, edited by V. Mannings, A.P. Boss and S. Russell (Tucson: University of Arizona Press, 2000) 589.

[74] D.J. Wilner, P.T.P. Ho, J.H. Kastner and L.F. Rodríguez, *ApJ* **534** (2000) L101.

[75] B.T. Draine and H.M. Lee, *ApJ* **285** (1984) 89.

[76] K. Miyake and Y. Nakagawa, *Icarus* **106** (1993) 20.

[77] L. Hartmann, S.J. Kenyon and N. Calvet, *ApJ* **407** (1993) 219.

[78] C.P. Dullemond, C. Dominik and A. Natta, *ApJ* **560** (2001) 957.

[79] P. D'Alessio, N. Calvet and L. Hartmann, *ApJ* **474** (1997) 397.

[80] A. Konigl and R.E. Pudritz, "Disk Winds and the Accretion-Outflow Connection", Protostars and Planets IV, edited by V. Mannings, A.P. Boss and S. Russell (Tucson: University of Arizona Press, 2000) 759.

[81] N. Calvet, "Properties of Winds of T Tauri stars", Proc. IAU Symp. 182, edited by B. Reipurth and C. Bertout (Kluwer, 1997) 417.

[82] P.P. Petrov and G.H. Herbig, *ApJ* **392** (1992) 209.

[83] N. Calvet, L. Hartmann and S.J. Kenyon, *ApJ* **402** (1993) 623.

[84] K. Croswell, L. Hartmann and E.H. Avrett, *ApJ* **312** (1987) 227.

[85] J. Bouvier, S. Cabrit, M. Fernandez, E.L. Martin and J.M. Matthews, *AA* **176** (1993).

[86] C. Eiroa, R.D. Oudmaijer and J.K. Davies, *AA* **384** (2002) 1038.

[87] W. Herbst, "The Rotation and Variability of T Tauri Stars: Results of Two Decades of Monitoring at Van Vleck Observatory", in Small Telescope Astronomy on Global Scales, ASP Conf. Ser. 246, IAU Coll. 183, edited by B. Paczynski, W.-P. Chen and C. Lemme (San Francisco: Astronomical Society of the Pacific, 2001) 177.

[88] N. Calvet, L. Hartmann, R. Hewett, J.A. Valenti, G. Basri and F. Walter, *Astron. Soc. Pac. Conf. Ser.* **109** (1996) 419.

[89] J. Bouvier, A. Chelli, S. Allain *et al.*, *AA* **349** (1999) 619.

[90] C. Terquem and C. Bertout, *MNRAS* **279** (1996) 415.

COURSE 12

EVOLUTION OF YSO DISKS

L. HARTMANN

Smithsonian Astrophysical Observatory, Cambridge, MA 02138, U.S.A.

Contents

EVOLUTION OF YSO DISKS

L. Hartmann[1]

Abstract

As stars and planets form from accretion processes in disks, it is important to understand these processes and their evolutionary implications. While much remains unclear, it seems likely that the accretion of material through protoplanetary disks is driven by some combination of the magnetorotational instability and gravitational torques. Angular momentum transport by disk winds can be important, especially in inner disks, but this is unlikely to dominate the overall disk evolution. A key uncertainty in both disk physics and the interpretation of observations is the rate and manner of dust particle growth. Finally, the ultimate fate of the gas component in protoplanetary disks remains unclear for systems not in proximity to hot stars with strong photoionizing radiation.

1 Introduction

Because protostellar cores are many orders of magnitude larger than the stars they produce, and because they collapse so rapidly that it is difficult for substantial amounts of angular momentum to be transferred away, the principal result of such core collapse must be a flattened, rotationally-supported structure – a disk. The implication is that most of the mass that initially ends up in stars must pass through an extended disk. After the main phase of stellar accretion, some of the remaining disk material is thought to form planets. An understanding of the physical processes of disk evolution is therefore an essential component of theories of star and planet formation.

The observational basis for determining disk accretion rates has been discussed by Nuria Calvet in her lectures; here I summarize their time dependence. Figure 1 outlines our present understanding of the time evolution

[1]Smithsonian Astrophysical Observatory.

of disk accretion in early stellar evolution. It is now thought that the infall rate of the protostellar core to the disk is strongly time-dependent, with the highest rates occurring early on in the collapse (see Sect. 6). Because angular momentum is thought to be nearly conserved during the (essentially free-fall) collapse of the core, most of the mass falls onto a disk, with subsequent accretion through the disk onto the central star(s). There is evidence that, at least in some phases, the rate of accretion through the disk does not necessarily keep up with the rate of infall onto the disk, which would not be surprising since the mechanisms of mass transport in the two processes are distinct. In this case, material may pile up in the disk and then be released in relatively short-lived rapid accretion events called FU Ori outbursts. It is not clear why FU Ori events occur, or whether all systems undergo FU Ori events, but at least some stars must undergo these outbursts multiple times [31].

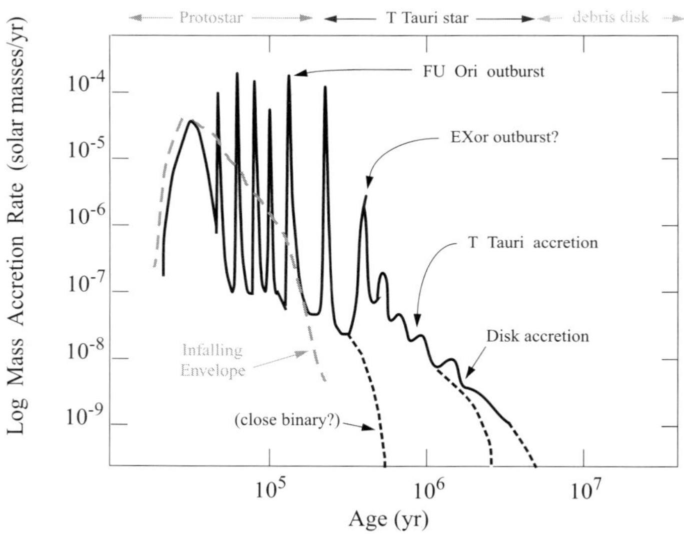

Fig. 1. Schematic behavior of accretion during early stellar evolution (see text).

After infall to the disk ceases, the protostellar phase ends and the T Tauri phase begins (in low-mass stars, loosely defined as stars with masses $\lesssim 1.5\ M_\odot$; intermediate-mass objects $\sim 1.5-3\ M_\odot$ are generally called Herbig Ae/Be stars). T Tauri stars are divided into two groups, "Classical" T Tauri stars (CTTS), which are accreting from their circumstellar disks, and "Weak" T Tauri stars (WTTS), which are not accreting (at a detectable level). The fraction of pre-main sequence low-mass stars which exhibit disk accretion decreases steadily with increasing age; at ages of $\sim$1–2 Myr, about half of all such stars are accreting [39], while at ages of $\sim$10 Myr it appears that only a small fraction $\lesssim$10% of low-mass stars are still CTTS (Muzerolle *et al.* 2000). For the half of the $\sim$1 Myr population accreting at a typical rate $\sim 10^{-8}\, M_\odot\, \mathrm{yr}^{-1}$, the observations imply that of order 0.01 $M_\odot$, or 10 Jupiter masses, is accreted during the typical T Tauri phase.

Because it is likely that all stars are formed with accretion disks, as discussed above, the increasing fraction of WTTS implies that disk evolutionary timescales are not a simple function of stellar mass and age but must depend on other properties as well. One factor which clearly affects disk lifetimes is the presence of a binary stellar companion, which can restrict the regions where disk material can stably orbit, potentially shutting off accretion onto the central star.

At ages $\gtrsim$10 Myr, there is little evidence for disk accretion or much infrared emission from dusty circumstellar material. Detailed observations of a number of bright, nearby stars indicate that many systems of ages 10–100 Myr exhibit so-called “debris disks”. The infrared excess emission of these stars is much smaller than that typical of T Tauri disks, indicating a dramatic reduction in the amount of small dust particles. In addition, these systems typically show evidence of evacuation in inner disk regions. Because radiation pressure and the Poynting-Robertson effect either blow away or drag in (relatively small) particles, it has been suggested that the observed dust is not simply material left behind from the T Tauri phase; instead, these particles may be continually produced by the collisional erosion of larger bodies.

The physical processes producing this complex picture of disk evolution are not very well understood at present. However, substantial progress has been made in recent years, and more should follow in the next decade with improved computational and observational tools. In the following I discuss some aspects of disk evolution, starting with considerations of the T Tauri phase, then working backward in time to consider protostellar disks, finishing with a discussion of the transition from T Tauri to debris disks.

2 Viscous disk evolution

To begin, it is useful to develop standard results for viscous accretion disks. While it is not clear that real disks behave as simple viscous disks, this development helps place other possibilities into perspective.

For a disk to accrete material inward, angular momentum must be transferred outward. In the viscous disk picture, “friction” between material at adjacent radial positions in the differentially rotating disk results in slowing down inner, faster-rotating material and spinning up outer, slower-rotating material, producing the needed angular momentum transfer. The dissipation of energy produced by the "frictional" process enables most of the material to fall in deeper into the gravitational potential well of the central star.

Here I make the usual simplifying assumptions that the disk is thin and in Keplerian rotation. I further simplify by making the assumption that variations in the vertical coordinate can be neglected, so that I consider only the integrated surface density of the disk Σ as a function of time t and cylindrical radius R.

Mass conservation yields

$$R\frac{\partial \Sigma}{\partial t} + \frac{\partial}{\partial R}(R\Sigma v_{\rm R}) = 0, \tag{2.1}$$

where $v_{\rm R}$ is the *net* radial velocity. In a viscous disk, the angular momentum flux outward across R is

$$g = -2\pi R\Sigma\nu_{\rm v}R^2{\rm d}\Omega/{\rm d}R, \tag{2.2}$$

where $\nu_{\rm v}$ is the viscosity, with units $[L^2/t]$. Then conservation of angular momentum in Keplerian rotation yields

$$\frac{\partial}{\partial t}(\Sigma R^2\Omega) + \frac{1}{R}\frac{\partial}{\partial R}(\Sigma R^3\Omega v_{\rm R}) = \frac{1}{R}\frac{\partial}{\partial R}\left(\nu_{\rm v}\Sigma R^3\frac{{\rm d}\Omega}{{\rm d}R}\right), \tag{2.3}$$

where $\Omega = (GM_*/R^3)^{1/2}$.

Multiplying the mass conservation equation by ΩR and substituting into the angular momentum equation yields the mass flux

$$\dot{M} = 2\pi R\Sigma v_{\rm R} = -6\pi R^{1/2}\frac{\partial}{\partial R}\nu_{\rm v}\Sigma R^{1/2}. \tag{2.4}$$

Finally, using the mass conservation equation again one finds the non-linear diffusion equation for the surface density,

$$\frac{\partial \Sigma}{\partial t} = \frac{3}{R}\frac{\partial}{\partial R}\left[R^{1/2}\frac{\partial}{\partial R}(\nu_{\rm v}\Sigma R^{1/2})\right]. \tag{2.5}$$

Thus, the evolution of the viscous disk is controlled, not surprisingly, by the viscosity.

It is instructive to consider a case where $\nu_{\rm v}$ = constant and with the mass confined initially to a ring. The formal solution can be expressed analytically with a complicated modified Bessel function [44]. For our purposes, it suffices simply to look at the result in Figure 2. The ring spreads out, with most of the material moving inward but a fraction moving to larger radii to take up the angular momentum. The timescale for this evolution

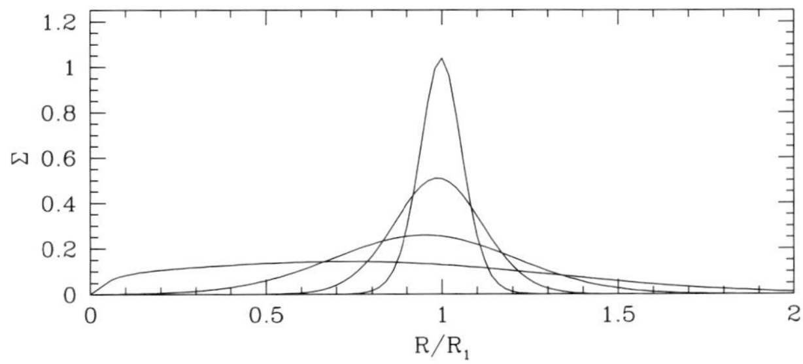

Fig. 2. Diffusion of a ring initially at $R = R_1$ for ν_v = constant. Results are shown for multiples of 0.004, 0.016, 0.064, and 0.256 of the time measured in units of $R_1^2/(6\nu_v)$. After [44].

is of order R_1^2/ν_v, where R_1 is the initial radius of the ring, as expected on dimensional grounds.

To take this a bit further, it is useful to develop a simple similarity solution for the viscous disk [28, 44] (Sect. 5.2). Here I assume that the viscosity is linearly proportional to the radial distance, $\nu_v = \nu_c(R/R_\circ)$. Then, as long as we consider regions far from the inner radius of the star, one can find a similarity solution, where the disk retains its basic shape as a function of time. It is of some interest to note that, with this assumed dependence of the viscosity, even if the *initial* distribution of mass in the disk does not follow the similarity form, the disk arranges itself asymptotically with increasing time to the similarity solution.

If R_1 is an initial length scale, then the corresponding scaling for the time is in units of the viscous time at that radius,

$$t_s \equiv \frac{1}{3} R_1^2/\nu_1, \tag{2.6}$$

and the similarity variable is

$$\tau = t/t_s + 1. \tag{2.7}$$

Finding the similarity solution is complicated (*e.g.*, LBP). One can verify that the following is a solution of (2.5) by substitution:

$$\Sigma(R,\tau) = \frac{M_d(0)}{2\pi R_1^2} \frac{R_1}{R} \tau^{-3/2} \exp\left(-R/(R_1\tau)\right), \tag{2.8}$$

$$M_d(R,\tau) = \frac{M_d(0)}{\tau^{1/2}} \left[1 - \exp\{-R/(R_1\tau)\}\right], \tag{2.9}$$

$$\frac{dM_d}{dt} = \frac{M_d(0)}{2t_s} \frac{1}{\tau^{3/2}} \exp\{-R/(R_1\tau)\}.\left[1 - 2R/(R_1\tau)\right]. \tag{2.10}$$

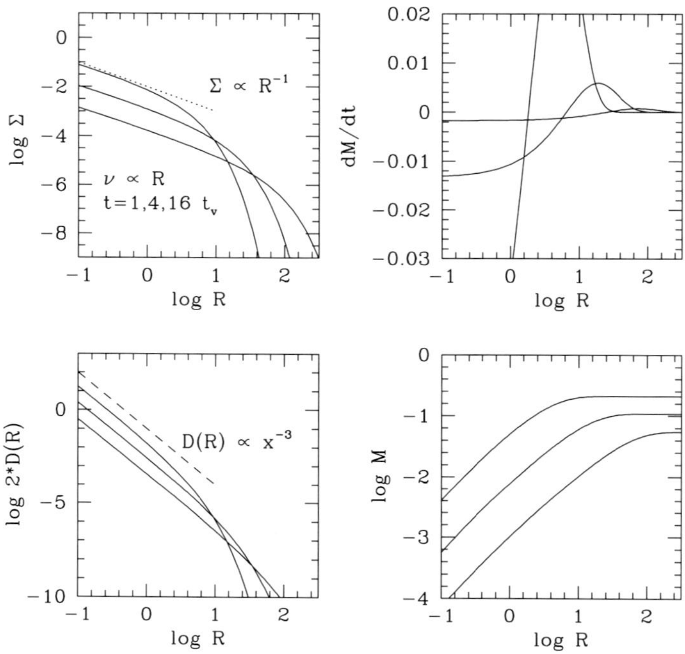

Fig. 3. Similarity solution. The surface density (upper left panel) exhibits the same scaled shape at all times. The mass flux (upper right) is negative at small radii (accretion) and positive at large radii (expansion). The viscous dissipation rate is shown in the lower left panel, while the disk mass shown in the lower right panel exhibits a $\tau^{-1/2}$ scaling. The dashed lines show the $\Sigma \propto R^{-1}$ and $D(R) \propto R^{-3/4}$ scaling expected for steady accretion.

This solution is shown in Figure 3. The disk expands with time as it empties out; the "transition" radius, where the mass flux changes sign, is

$$R_{\rm t} = \frac{R_1}{2}\,\tau. \tag{2.11}$$

Understanding the basic time dependence of this solution is straightforward. The viscous disk spreads out with time, asymptotically as $R_{\rm t} \propto t$, because the viscous time varies as R. Now almost all of the angular momentum of the system remains in the disk; very little of the angular momentum is accreted onto the central star if the stellar radius is much smaller than the disk outer radius. Thus to first order (and exactly in the similarity solution, which has no inner radius) one can assume that the angular momentum of the disk remains constant. As the disk retains the same scaled surface density distribution, the result can be given in terms of R_t,

$$J_d \propto M_d \Omega(R_t) R_t^2 \sim \text{constant}. \tag{2.12}$$

But $\Omega = (GM_*/R^3)^{1/2}$, so

$$M_d \propto R_t^{-1/2} \propto t^{-1/2}. \tag{2.13}$$

Now consider an application to observations. In her lectures, Calvet has described how most T Tauri disks have temperature distributions that are dominated by irradiation by the central star, not by local accretion energy release. Furthermore, at large radii the disks tend to be flared, *i.e.*, they have surfaces which intercept radiation from the star that are curved away from the midplane. Such disks tend to have temperature distributions which scale approximately as $T \propto R^{-1/2}$, and I make this simplifying assumption here.

I next make the standard assumption that the viscosity can be written as

$$\nu_v = \alpha c_s H = c_s^2/\Omega, \tag{2.14}$$

where α is a constant, c_s is the disk temperature (assumed vertically isothermal in this approximation), and H is the disk scale height. The final equality results from applying vertical hydrostatic equilibrium in the disk, which implies $c_s/v_K = H/R$, v_K being the local Keplerian velocity.

Because $c_s^2 \propto R^{-1/2}$ and $\Omega \propto R^{-3/2}$, the further assumption of constant α means that the viscosity is linearly proportional to R and so we can use the similarity solution derived earlier. Defining $t_d = 1 + t/t_s$, where

$$t_s \sim 8 \times 10^4\,\text{yr} \left(\frac{R_1}{10\ \text{AU}}\right) \left(\frac{\alpha}{10^{-2}}\right)^{-1} \left(\frac{M_*}{0.5\,M_\odot}\right)^{1/2} \left(\frac{T_{100}}{10\ \text{K}}\right)^{-1}, \tag{2.15}$$

where T_{100} is the disk temperature at 100 AU, the solution is

$$M_d = M_d(0)\, t_d^{-1/2}, \tag{2.16}$$

$$\dot{M} \sim 6 \times 10^{-7}\, M_\odot\,\text{yr}^{-1} \frac{e^{-R/(R_1 t_d)}}{t_d^{3/2}} \left(1 - \frac{2R}{R_1\, t_d}\right) \left(\frac{M_d(0)}{0.1\, M_\odot}\right)$$
$$\times \left(\frac{R_1}{10\ \text{AU}}\right)^{-1} \left(\frac{\alpha}{10^{-2}}\right) \left(\frac{M_*}{0.5\,M_\odot}\right)^{-1/2} \left(\frac{T_{100}}{10\ \text{K}}\right). \tag{2.17}$$

The radius which includes half the disk mass is then

$$R_{1/2} = 2(\ln 2)^{1/2} R_t \sim 80\,\text{AU}\, t_d \left(\frac{R_1}{100\ \text{AU}}\right). \tag{2.18}$$

For example, suppose we take $\alpha = 0.01$, a rough value justified to some extent in the following section. We further take an initial disk mass of $M_\mathrm{d}(0) = 0.1\,M_\odot$, which for typical T Tauri stars is roughly the limit at which gravitational instabilities would become important (see below). Then for an initial characteristic disk radius $R_1 \sim 10$ AU, and a typical T Tauri star age of 1 Myr, $M_\mathrm{d} \sim 0.027\,M_\odot$, $\dot{M} \sim 1.2 \times 10^{-8}\,M_\odot\,\mathrm{yr}^{-1}$, and $R_{1/2} \sim 110$ AU. These are all typical parameters of T Tauri disks, suggesting a rough consistency.

While this "toy" model is likely to be incorrect in detail – for example, the temperature distributions in real disks are not power laws – it provides a very useful means by which to explore parameter space. Consider, for example, changing α. In Figure 4 I have compared models with two values of α to observations of Taurus objects, as summarized in [30]. The figure shows that both $\alpha = 10^{-2}$ and $\alpha = 10^{-3}$ do moderately well in explaining the range of observed disk masses and mass accretion rates, with the $\alpha = 10^{-2}$ models perhaps explaining the range of disk masses slightly better. Although there are no good observational results for the disk radii at the present, the curves for $R_{1/2}$ help explain what is happening in the models. As α increases, the radius from which the disk can accrete at a typical age of 1 Myr also increases. To maintain a certain accretion rate for a specified age, there must be enough mass within the disk viscous radius given by $R = (t\nu(R))^{1/2}$. For smaller α, there needs to be more mass at a smaller radius to maintain the same accretion rate. The $\alpha = 10^{-3}$ models deplete the disk mass more slowly than the $\alpha = 10^{-2}$ models, helping to maintain a higher accretion rate at 1 Myr.

If the disk masses shown were reasonably accurate, then it would difficult to explain a combination of low disk masses and high accretion rates with a low α, unless the disks were very small. Unfortunately, there are difficulties with measuring disk masses. Mass estimates are derived almost universally from mm-wave dust emission; therefore such estimates depend upon the magnitude of the mm-wave opacity. The problem is that the dust in pre-main sequence disks is very unlikely to be comparable to that of the interstellar medium; one expects substantial growth in dust sizes over T Tauri lifetimes (see Sect. 8). Standard calculations indicate that the dust opacity is quite sensitive to the size distribution; if most of the mass is in meter-sized bodies or larger, there is very little mm-wave opacity, and disk masses could therefore be underestimated by large factors. In general, one can make an argument that the usual opacities used for disk masses are likely to be near the upper range of likely values (see [18]), indicating that disk masses may be systematically underestimated. On the other hand, it seems unlikely that the dust opacities are in general very far off; if they were much higher, the low disk masses would be inadequate to supply T Tauri

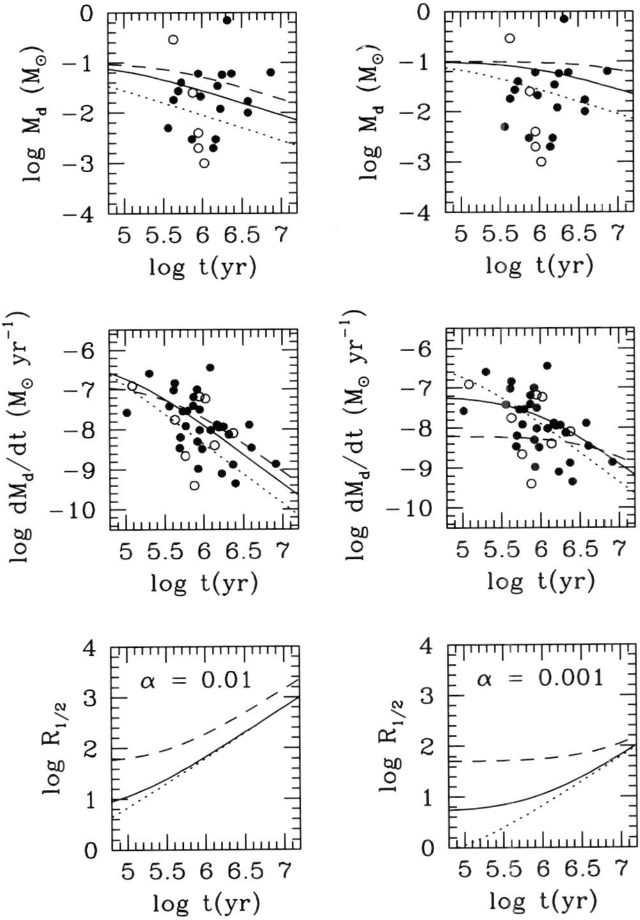

Fig. 4. Similarity solutions compared to observations of Taurus stars. All models have an initial disk mass $M_{\rm d}(0) = 0.1\, M_\odot$, with $R_1 = 1, 10, 100$ AU (dashed, solid, dotted lines). Left sequence: $\alpha = 10^{-2}$; right sequence: $\alpha = 10^{-3}$. Filled circles: single or unknown multiplicity stars; open circles: binaries.

accretion rates for T Tauri lifetimes; if much smaller, the disks would be highly unstable to gravitational perturbations. In any event, it is difficult to make detailed tests of disk evolutionary models with our current understanding of disk masses.

With the above caveats, very high or very low values of α seem unlikely given current observations. If disk viscosities were characterized globally by

$\alpha \gtrsim 0.1$, then they would expand so rapidly that they all would be much larger than 100 AU at 1 Myr, and it would be difficult to explain the largest disk masses observed. On the other hand, if $\alpha \lesssim 10^{-4}$, then disks would have to be extremely small, with all the mass concentrated in regions of $\sim$1 AU or so, to explain observed accretion rates; in more extended disks, with most of the mass outside of the viscous radius at 1 Myr, the supply of material which can be accreted would be too low.

Thus, crudely speaking, the observations suggest that T Tauri disks have $\alpha \sim 10^{-2}$–10^{-3} – *assuming* that they can be modelled as viscous disks, and be characterized by a roughly constant value of α. Neither of these assumptions is certain. Perhaps a more general way of describing the observational constraints would be to say that it appears that T Tauri stars must be able to accrete material from distance scales of 10–100 AU on a timescale of 1 Myr, if the observed accretion rates and disk masses are to be reconciled with disk lifetimes.

3 What is α (if any)?

One of the major developments of the last decade in the study of astrophysical disks has been the emergence of the magnetorotational instability, or MRI, as a prime candidate mechanism for producing a turbulent viscosity (see [3], and references therein). The MRI has been discussed in previous lectures in this school; here I wish only to make a few general remarks, and then discuss in more detail its possible applications to protoplanetary and protostellar disks.

For some time it was thought that hydrodynamic turbulence, such as might be driven by convection, could provide the anomalous viscosity. There is a problem with convection in T Tauri disks; specifically, for typical accretion rates and dusty disk structure, the irradiation heating of the central star dominates the central energy production due to viscous dissipation; this means that the disk is generally hotter on its surfaces than at its midplane (see Calvet's lectures), stabilizing against convection. There is also a generic problem with hydrodynamic turbulence. Consider the situation shown in the left-most panel of Figure 5, where two blobs originating at different radii are exchanging radii (for some reason). In Cartesian shear flow, such an exchange could lead to destabilization and turbulence. However, in the rotating disk things are different. The inner blob, moving outward, carries *less* angular momentum; the outer blob, moving inward, carries *more* angular momentum. This is the wrong direction for accretion. Detailed analytic studies [52] and numerical simulations [55] indicate that hydrodynamic turbulence in a rotating disk results in little if any angular momentum transfer, and that which does occur tends to be inward, not outward.

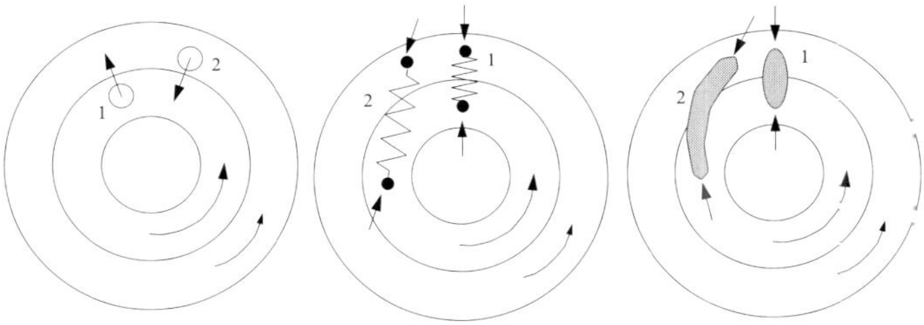

Fig. 5. Angular momentum transport. Pure hydrodynamic turbulence (left) tends to mix angular momentum inward, and so appears to be ineffective in driving accretion. In the MRI (center), the magnetic fields couple outer and inner material, transferring angular momentum outward. If the disk is self-gravitating (right), gravity can create non-axisymmetric disturbances which then act to concentrate material; this disturbance tends to generate trailing waves, and gravity can then act to provide a torque to accelerate outer material and decelerate inner disk regions.

In the MRI, the magnetic coupling of inner and outer disk regions causes outer material to be accelerated forward, adding angular momentum, while inner regions are decelerated and thus lose angular momentum (see Fig. 5). The magnetic fields can be stretched only so far, so that they break up and result in a turbulent flow. The instability appears to be robust; even if there is no initial imposed magnetic field, any small perturbations tend to grow; as the field is stretched, it increases; and so the MRI can generate its own (turbulent) magnetic field [12, 17, 32].

Furthermore, the MRI can be subsumed within an α disk formalism [4], with the viscosity written in terms of a magnetic stress tensor composed of correlated fluctuating velocity and magnetic fields. Numerical studies suggest that the MRI produces values of $\alpha \sim 10^{-1}$–10^{-2}, with larger values resulting in situations with an imposed external magnetic field [12, 56] (see also [33]). Given the observational estimates for T Tauri disks outlined in the previous section, it would appear that we are approaching a coherent physical explanation of T Tauri disk accretion.

However, a major problem remains. The MRI requires some level of ionization to be maintained; the magnetic field couples directly only to the ions and electrons; the neutral material responds only to the extent that collisions between charged particles and neutrals allow. Because protostellar disks are so dense, the minimal fractional ionization levels are extremely low, of order 10^{-13}; but even these low fractional ionizations are difficult to maintain in the extremely cold interiors of protoplanetary disks. Thermal ionization is adequate only for temperatures $\gtrsim$1000 K, which generally occur only at $R \lesssim 10^{-1}$ AU [51]. Something else is required if the MRI is to operate in T Tauri disks.

Gammie [24] addressed this problem by supposing that cosmic rays were able to ionize at least the outer layers of YSO disks. In this picture, accretion occurs in active layers exterior to a central "dead zone" where the magnetic field does not couple effectively to the neutral gas. In Gammie's original picture, cosmic rays can penetrate and ionize effectively down to a relatively constant surface density of order 100 $\mathrm{g\,cm^{-1}}$. Now, inserting a constant Σ into equation (2.4) results in $\dot{M} \propto R^{1/2}(\partial/\partial R)\nu_{\mathrm{v}} R^{1/2}$. Unless ν_{v} = constant, $\dot{M}$ cannot be constant with radius. For the specific case treated above, $\nu_{\mathrm{v}} \propto R$, $\dot{M} \propto R$. Such a behavior of increasing inward accretion as a function of increasing radius implies that material must pile up. Gammie suggested that mass would pile up until the disk becomes gravitationally unstable, at which point the gravitational torques of non-axisymmetric perturbations would create angular momentum transport in the dead zone, allowing this material to accrete rapidly; this might be the origin of FU Ori outbursts.

A concern with this picture is that the low-energy cosmic rays most important for ionization might be scattered away by the powerful outflows of YSOs, in an analogous manner to their scattering out of the solar system by the solar wind. An alternative suggestion is that X-rays from the central YSO are responsible for the ionization [26,35]. X-rays are certainly produced at high luminosities and with energies greater than 1 KeV (needed for substantial disk penetration) by young low-mass stars. Whether intermediate-mass (late B, A type) stars also produce the needed X-rays is not clear; there are reasons to doubt this [23].

Igea & Glassgold [35] suggested that X-rays from a typical T Tauri star might penetrate to surface densities of order 10 $\mathrm{g\,cm^{-2}}$ at distances of 1 AU. In terms of the similarity solutions, this suggests that disks with low accretion rates, high α, and low masses might remain magnetically active, and thus the α formalism would apply to the entire disk (*e.g.*, [22]). However, this depends upon crucial assumptions regarding the dust size and spatial distributions, as well as the amount of heavy element depletion in the disk. The Igea & Glassgold conclusions hold when the dust grains have entirely condensed at the disk midplane (Sect. 8), and no metal ions (only molecular species) are present. If such is not the case, the ionization is strongly reduced, because small grains [53] and metal ions [22] are sinks for electrons. Sano *et al.* [53] show that for the dust size distribution and abundance typical of the interstellar medium, the active regions are extremely small; reductions of the small grain population by factors of 10^4 or more are needed to achieve substantial active regions.

As discussed by Calvet, there is evidence for grain growth in T Tauri disks; but just exactly how much growth, and the spatial distribution of the larger grains, is uncertain. Moreover, there must be some moderately small dust $\lesssim 1$ μm in upper disk layers to explain spectral energy distributions and scattered light images. About all that can be said at the present time is that there may be enough dust processing and metal depletion to allow enough ionization in T Tauri disks to allow the MRI to operate. Beyond this, numerical simulations are needed to understand just how a layered disk might really accrete.

4 Angular momentum loss by magnetically-coupled winds

The discovery of powerful bipolar outflows and jets from young stellar objects (and also from AGN) led to theoretical suggestions that magnetic fields coupled to accretion disks could fling out material efficiently, accounting for these outflows [7, 41, 48–50]. This mechanism is discussed in much more detail in the lectures of Pudritz and Heyvaerts. Here I confine myself to a few considerations relevant to young stellar objects.

It seems very likely that the winds and jets of YSOs are indeed driven by magnetic fields rooted in disks. As described in Calvet's contribution, the pieces of evidence in favor of this mechanism are: (1) massive winds and jets are observed only in systems with accreting disks; (2) the mass ejection rates, as best we can determine, scale with accretion rates; (3) in the case of FU Ori, we can detect the ejection of the wind directly from the rotating disk surface [29]; and (4) radiation and thermal pressure are clearly insufficient to accelerate material from T Tauri stars [28].

In extreme versions of the theory (*e.g.*, [41, 48]), the wind carries away all of the angular momentum needed for accretion. In such a case it would seem that there is little need for an MRI. However, both mechanisms share a similar requirement for sufficient ionization to couple the magnetic field ultimately to the neutral gas. If there is a sizeable dead zone in the middle of the disk, a disk wind is not going to do anything to help this material accrete, because its magnetic field is not coupled to the dead zone.

Another problem with the extreme version of wind angular momentum loss is a sort of "chicken and egg" problem; is it really necessary to have a wind before material can accrete? In the case of FU Ori outbursts, does this mean that one requires an outburst of wind to produce the outburst of accretion?

Some numerical simulations (*e.g.*, [60]) exhibit very strong mass ejection with collapse on dynamical timescales. However, it is not clear whether such simulations exhibit this behavior primarily as a transient result to a situation far out of dynamical equilibrium to begin with.

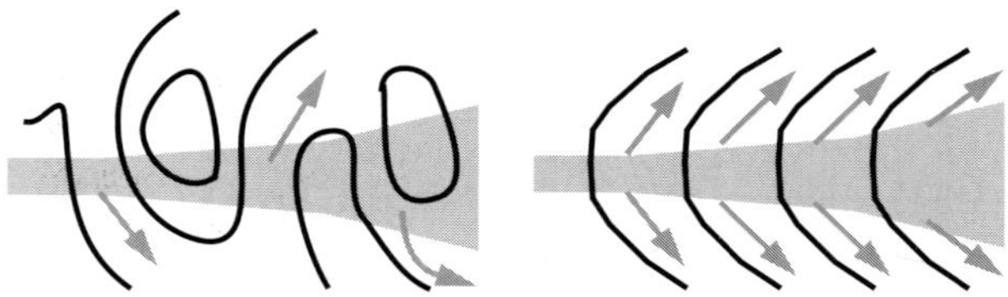

Fig. 6. Two possible (schematic) disk magnetic topologies: (left) complex, as might arise from the MRI; (right) highly ordered, producing maximum open field geometry for mass ejection (arrows).

A crucial issue in this problem is the topology of the magnetic field. Standard simple models presume (mostly for simplicity) a regular open field which passes through the disk, as in the right panel of Figure 6. On the other hand, Blandford & Payne [7] suggested a much more complicated structure for disk fields (left panel), in which only a fraction of the magnetic field lines open to infinity; this could reduce the effectiveness of magnetic braking by the wind. The BP field is suggestive of the complicated field structure that the MRI can produce naturally from very small initial fields; the open structure is more suggestive of a magnetic field that might be dragged in from the interstellar medium.

We have little idea of the protostellar field that remains trapped in the infalling gas, for exactly the same reason that a dead zone might exist. The low ionization state likely in protostellar envelopes probably means that the a large fraction of the initial magnetic field diffuses out of the collapsing material [61]. There are additionally the possibilities of diffusion and magnetic field reconnection in the disk. One thing is clear; very little of the initial protostellar magnetic flux gets put into the central star. If the magnetic flux passing through typical protostellar cloud cores were preserved in the final star, stellar surface fields would be of the order of 10^6 Gauss, whereas observational estimates suggest stellar surface field strengths of order 10^3 Gauss [36]. Moreover, it is quite possible that these magnetic fields are actually produced by a stellar dynamo than by any remnant interstellar field, especially as the starspots in these systems change with time (unlike, for example, magnetic Ap stars, which have extremely stable fields, thought to be remnants of formation).

Given these large theoretical uncertainties, observations must be called upon to provide some constraints. One important prediction of the wind theory is the relation between energy and angular momentum fluxes.

Suppose a magnetic flow tube is connected to an annular area on the disk of $\mathrm{d}R$ at $R_\circ$. In steady state, the flow cannot extract more angular momentum than corresponds to the accretion rate through the disk. Away from the inner disk edge, the angular momentum lost as material accretes through $\mathrm{d}R$ at $R_\circ$ is

$$\frac{\mathrm{d}j_{\rm acc}}{\mathrm{d}t} = \dot{M}\frac{\mathrm{d}}{\mathrm{d}R}\Omega_{\rm K}R^2\mathrm{d}R = \frac{1}{2}\dot{M}\Omega_\circ R_\circ^2\frac{\mathrm{d}R}{R_\circ}. \tag{4.1}$$

If the angular momentum carried away by the wind flux tube spanning $\mathrm{d}R$ is exactly equal to the angular momentum transport needed for accretion, the energy carried away by the wind is

$$\dot{e}_{\rm w} = \Omega_\circ l_{\rm B\circ}\dot{m}_{\rm w} = \Omega_\circ\frac{\mathrm{d}j_{\rm acc}}{\mathrm{d}t} = \frac{1}{2}\frac{GM_*\dot{M}}{R_\circ}\frac{\mathrm{d}R}{R_\circ}. \tag{4.2}$$

Now the right-hand side of the above equation is precisely the energy released locally by material accreting from $R_\circ + \mathrm{d}R$ to $R_\circ$. (A viscous disk has additional energy advected from inner regions by disk torques, but these are absent in the present case by assumption.) Therefore, a wind which removes all of the angular momentum needed for disk accretion carries off all of the accretion energy as well. An accretion disk whose angular momentum transport is entirely the result of magnetocentrifugal mass loss is generally not self-luminous, since all of the accretion energy is put into the wind, leaving nothing to radiate.

We observe accretion energy as radiation. In T Tauri systems, most of this energy is visible as hot continuum emission; in in high accretion rate systems such as FU Ori objects and some strong-emission CTTS, the accretion-powered radiative losses are seen from the disk as well (see Calvet's lectures). This means that the wind does not carry away all of the accretion energy. The question is, what is the relative balance of the wind and radiative energy losses?

It is extremely difficult to estimate mass loss rates; the best we can do currently is basically order-of-magnitude estimates. Currently it appears that $\dot{M}(\mathrm{wind}) \sim 10^{-1}\dot{M}(\mathrm{acc})$, a relation that seems to hold equally well for FU Ori objects and T Tauri stars [13]. The terminal velocities of winds and jets are also uncertain, although less so; a reasonable estimate is that maximum wind/jet velocities are typically $\lesssim$3 times the Keplerian velocity of the inner disk. Then we have that $\dot{M}v_\infty^2 \sim L_{\rm acc}$, where $L_{\rm acc}$ is the *radiative* energy release due to accretion.

In summary, observations suggest that, to order of magnitude, the energy losses due to wind ejection are comparable to the radiative losses powered by accretion. This is a remarkable result; it demonstrates that disks must be extremely effective engines of converting accretion energy to wind power. Nevertheless, it is *not* the case that $\dot{M}v_\infty^2 \gg L_{\rm acc}$, and therefore winds cannot be the only mechanism of outward angular momentum transport. Moreover, to the extent that the observational estimates can be relied upon, the two mechanisms might provide *comparable* levels of angular momentum transport. This would imply modifications of the simple viscous disk picture by modest factors, not complete revision.

Wind angular momentum loss in the X-wind theory of Shu *et al.* [54] is essentially negligible from the point of view of disk evolution, because in this picture all the wind emerges from the inner edge of the disk, where it is truncated by the stellar magnetosphere. In my opinion, this extreme version is also unlikely, because I think the disk must have its own magnetic field to accrete unless gravity does the entire job; and if there is a disk magnetic field, there is no reason why some of the field lines do not open up, a la Blandford & Payne (Fig. 6). More significantly, it is clear that in FU Ori, which does not apparently have a disk truncated by a disk magnetosphere, the wind must emanate from a significant range of stellar radii in order to explain the observed line profiles [15].

My own guess is that the MRI does operate at some level in T Tauri disks, that it creates a field like the left panel in Figure 6 perhaps overlaid with a weak, open, remnant organized primordial field, and that at certain points the field geometry is favorable for the ejection of mass. This picture results in lower but non-negligible angular momentum transport by the wind, which can be accommodated easily within theory (see, for instance, [50]). The MRI is then the fundamental driving mechanism for accretion, except for the case of massive disks, as discussed next.

5 Gravity

If the disk is sufficiently massive, gravitational torques can be set up which can transfer angular momentum outwards and permit accretion. For the moment I ignore the possibility that the disk might simply fragment into multiple self-gravitating bodies, and consider small perturbations.

As shown in Figure 5, suppose a perturbation places an excess of mass at a certain point in the disk. If gravity is important, this excess can cause neighboring material to move toward it, increasing the density of its surroundings. Because the disk is differentially rotating, such perturbations tend to be sheared out into trailing waves. As shown in the figure, this results in a non-radial component of the local gravitational acceleration which tries to speed up outer regions and slow down inner parts; thus, such disturbances can transfer angular momentum outward.

Although non-axisymmetric modes are needed for angular momentum transport, study of axisymmetric disturbances gives results that are roughly applicable to tightly-wound spiral waves. The dispersion relation for symmetric modes in a thin rotating disk is [9]

$$\omega^2 = \kappa^2 + c_s^2 k^2 - 2\pi G \Sigma \mid k \mid . \tag{5.1}$$

This differs from the dispersion relation for a non-rotating sheet by the term involving the epicyclic frequency κ. When the disk rotation is Keplerian, $\kappa = \Omega$, the local angular velocity.

For negative ω^2, perturbations grow exponentially and the disk is unstable. The limiting condition occurs when $\omega = 0$; the condition for axisymmetric instability is then

$$Q \equiv \frac{c_s \kappa}{\pi G \Sigma} < 1. \tag{5.2}$$

To understand this result, rewrite the relation as

$$\Omega < \pi G \Sigma / c_s. \tag{5.3}$$

Using the relation $c_s = H/\Omega$,

$$\Omega^2 < \pi G \rho_o, \tag{5.4}$$

where ρ_o is an average density in the disk. Now the local free-fall time is of order $(G\rho_o)^{-1/2}$, so this "Toomre criterion" is roughly equivalent to requiring the local free-fall time to be shorter than the orbital timescale. This makes sense; otherwise the motion would shear out the perturbation.

For quick comparison with observations (which have their own uncertainties) it is useful to develop a rule of thumb. Multiplying the numerator and denominator of (5.2) by the outer disk radius R_d^2, and making the approximation that the disk mass is $M_d \sim \pi R_d^2 \Sigma$, the condition for gravitational instability becomes

$$c_s \kappa R_d^2 < G M_d. \tag{5.5}$$

Making the further approximation that the disk motion is nearly Keplerian, we have the following rough rule for instability:

$$M_d > \frac{H}{R} M_*. \tag{5.6}$$

Numerical simulations show that $Q \lesssim 1$ disks do indeed generate non-axisymmetric structure and angular momentum transfer. If $Q \ll 1$, the disks are strongly unstable and material moves on a dynamical timescale. Models have been constructed which assume that massive disks tend to maintain $Q \sim 1$ by accreting as necessary, if there is a supply of material being added to the disk, as for example would occur during protostellar collapse [43]. Similar behavior is seen in simulations in which the disk is cooled [58].

One problem is that self-gravitating disks in general [42], and perhaps usually [4] do not act like viscous disks. The reason is that gravity is a long-range force. Unlike viscosity, angular momentum and energy can be transferred by gravity between distant, non-adjacent regions. Under certain circumstances it may be possible to model weakly self-gravitating disks as α disks [25], but there is no guarantee that this is generally possible.

For the present, it seems safe to conclude that, however massive disks evolve, the result is that $Q \ll 1$ must correspond to extremely transient stages, if any, and that T Tauri disks must have $Q \gtrsim 1$. As shown in Figure 4, disk masses for T Tauri stars are typically estimated to mostly lie between $\sim 10^{-3}\,M_\odot$ (which is really an observational rather than a physical lower limit), and $\sim 0.1\,M_\odot$, with most objects at ~ 0.003–$0.03\,M_\odot$. Since the typical masses of the Taurus stars contributing to figure whatever are $\sim 0.7\,M_\odot$, these masses conveniently fall below the rough mass limit for gravitational instability (5.6), assuming a typical value of $H/R \sim 0.1$.

6 From protostars to T Tauri stars

At this point I return to the initial stages of YSO evolution, when the infalling envelope is adding mass to the central stellar core and then to the protostellar disk. It would seem logical that if gravitational instabilities ever play a role in protoplanetary disks, they do so in early phases, when most of the mass of the core is being added to the disk, not directly to the central protostar. The question is, does the disk simply pass on the infalling material to the central star in a roughly steady flow, hovering near $Q \sim 1$, or even above if viscous stresses are large enough? Or does the disk fragment into binary or multiple stellar companions, or even massive planets?

The answer to these questions depends on how fast the disk cools, *i.e.* how fast it can get rid of excess energy. If it cools slowly, then the gravitational spiral waves generated can heat up the disk, reducing the instability (*e.g.*, [58]); this raises the possibility of steady accretion as mass is added to the disk. On the other hand, if the disk cools rapidly, then self-gravitating fragments can form and separate from the flow. On the basis of a local analysis (which may not be adequate; see above), Gammie [25] suggests that the criterion for fragmentation is $t(\mathrm{cool}) < (3/2\pi)P(\mathrm{orbital})$. Unfortunately, we don't know how to calculate $t(\mathrm{cool})$ with any assurance at present.

Although in principle binary and multiple stellar systems can fragment out of disks, there is no clear consensus that they do. My own opinion is that many binaries must form in disks. It is difficult to see how separate protostellar cores, which should be separated by distances of the order of typical core sizes, thousands of AU, can really dissipate enough energy to end up in a bound system at <30 AU (corresponding to more than half of all binaries in the field; [20]). Capture is impossible at the low densities of most nearby star-forming regions, and fission, if it is at all possible, (and this is far from clear; [19]) could only possibly account for the very closest systems. Finally, we observe protostellar binary systems in which the components seem to be lying within disk-like structures, while infall still occurs [47].

The structure of the initial protostellar cloud core must play an important role in this problem. There is increasing evidence that protostellar cores have central density distributions that are relatively flat rather than highly centrally-concentrated [2]. For such models, the infall rate is not constant; instead, there is a very high infall rate for a short period of time, followed by a slower collapse. This behavior tends to deposit a small but significant fraction of the total mass initially in a relatively small region (still larger than a T Tauri star radius), with a slower accretion of material at larger radii as the later material, with more angular momentum, falls in.

There are two possible points of significance here. First, protostellar accretion is likely to be dominated by a rapid, dynamic phase of accreting the first cores; thus the fragmentation process may qualitatively differ from the evolution of perturbations growing in a nearly-Keplerian disk with a central mass. Second, it seems clear that if fragments that do form form early, they will then accrete material later from a circumbinary (or circumobject) disk [6]. Although the fragments will try to push away accreting outer disk material initially, once the outer disk has a comparable mass to the inner fragments, gravitational instabilities drive the circumsystem disk inward, accreting material onto the central protostars until Q moves above unity.

I think this picture also makes it difficult to make planets by gravitational instability, as urged by Boss [11] (et seq.). Studies of fragmentation have always been limited by numerical limitations; it is not possible to follow the evolution of a disk over enough orbits to decide whether individual fragments really remain, merge, or simply shear away. In the protostellar case, because there is so much mass that must be accreted from large radii, it is hard to see how even a small initial fragment will not grow eventually to stellar or at least sub-stellar masses. The Boss picture also does not explain what happens to the CTTS disks, which must disappear in a few Myr; in the more conventional picture of giant planet formation, the disappearance of the disk can be attributed to the accretion of material by the planets.

A long-standing puzzle concerning protostellar luminosities also has some implications for disk structure and evolution. In nearby, well-studied, low-mass star-forming regions such as Taurus, maximum accretion luminosities [38] and other estimates of mass accretion [45] indicate that many protostars are accreting at rates $\lesssim 10^{-7}\,M_\odot\,\mathrm{yr}^{-1}$. At this rate, to form a typical $0.7\,M_\odot$ Taurus star would take 7 Myr. This is time span is larger than the average age ~2 Myr of stars in the region. Moreover, based on population statistics and estimates of free-fall times for cores, the true protostellar lifetimes cannot be much more than a few $\times 10^5$ yr. This problem can be somewhat alleviated by "front-loading" the mass infall, as described above, and as indicated in Figure 1. A significant fraction of the mass can then be added at very high accretion rates, which imply very high luminosities, but only for a short period of time; and if the sample is small enough and the high accretion period short enough, it might be missed observationalliy. However, this is unlikely to be the entire explanation, as there is evidence for higher *current* mass infall rates in envelopes than disk accretion rates [38]. The implication is that material may be piling up in the outer disk. Indeed, Kenyon & Hartmann [40] suggested that FU Ori outbursts might be produced by such an accumulation of mass in outer disks, with periodic short releases of material to the inner star to produce the outburst behavior. In this picture, FU Ori outbursts occur preferentially during the protostellar infall phase. This is consistent with observations, in that all objects appear to be surrounded by close dusty envelopes [40]. This mechanism would also allow for lower time-averaged protostellar accretion, since most of the mass could get dumped in during short bursts with a small duty cycle (*e.g.*, Fig. 1).

The mechanism(s) driving FU Ori events is(are) not clear. Suggestions include thermal instabilities [8], passage of a companion star on an elliptical orbit [10], and periodic release of dead-zone material due to gravitational instabilities [24]. All these mechanisms have difficulties; the dead zone theory might be the most promising. In any event, the FU Ori outbursts (see discussion by Calvet) show that protostellar disk accretion can be very far from smooth and steady.

7 Decay of T Tauri accretion

At some point, pre-main sequence stars no longer exhibit the strong infrared-sub-mm-mm emission disk emission, nor is there evidence for significant accretion onto the central star. Figure 7 shows the most recent data on accretion rates, along with the estimated fraction of accreting stars as a function of age. Most T Tauri stars stop accreting at some age between 1 and 10 Myr, with a corresponding reduction in infrared emission.

The statistics of accreting stars indicate that the previous discussion of accretion rate behavior is incomplete. Most stars must undergo a fairly rapid shutoff in mass accretion. The question is, why?

One possibility is that some (massive) body prevents material from accreting from the outer disk, limiting the mass supply for accretion. This is likely to happen in many systems with binary companions. Consider a case

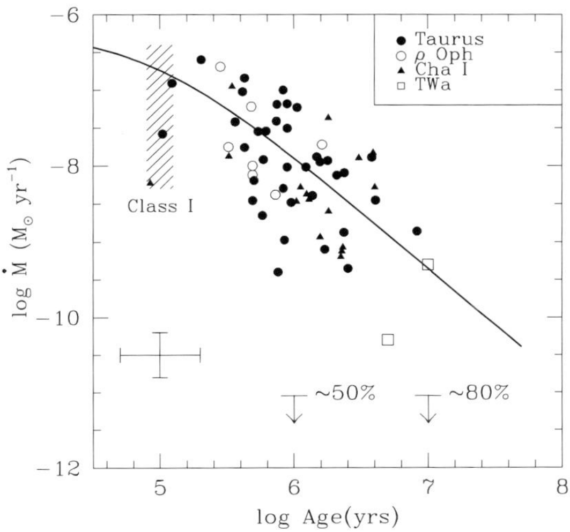

Fig. 7. Observational mass accretion rates as a function of age. The shaded region indicates mass accretion rates estimated for some protostellar (Class I) objects; the open squares are two accreting members of the TW Hya association. Also indicated are the number of accreting (CTTS) stars as a fraction of the total population. (After [46].)

in which the disk is initially small in comparison to the binary companion, both in radius and in mass. The disk tries to expand, but at some point cannot do so because of the companion star; the star exerts a torque on the disk which opposes its expansion. Angular momentum is then transferred from the disk to the star, but the star hardly moves outward if it is sufficiently massive.

To address this in a simple way, I have calculated viscous disk models with $\alpha = 10^{-2}$ as before, but with a boundary condition which does not allow the disk to expand beyond a certain radius, mimicing the effect of a massive outer companion star. As shown in Figure 8, the disk empties out rapidly upon reaching the outer boundary.

The reason for this behavior is straightforward. Earlier, I showed that the behavior of the similarity solution could be understood in terms of a nearly constant disk angular momentum, coupled with the variation of the viscous time with R. However, in this case once the disk reaches the outer boundary, it can no longer expand; the angular momentum must therefore decrease with time. The relevant viscous time also remains constant. Thus,

$$\dot{M} \sim -M_{\rm d}/t_{\rm vis}, \tag{7.1}$$

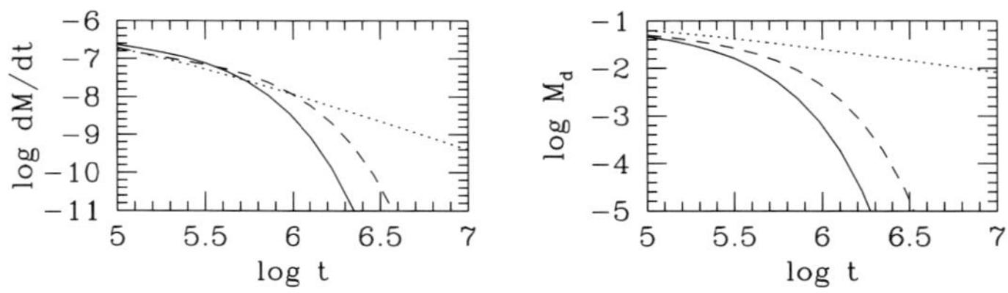

Fig. 8. Viscous disk evolution in a case where the outer boundary is fixed, simulating a case in which a massive companion star prevents the disk from expanding beyond 15 and 30 AU (dotted and dashed curves). The solid curve is the standard viscous solution for reference.

and so the disk mass decreases exponentially on the viscous timescale at the outer radius.

This behavior suggests another check on the value of α; close binary systems may not exhibit disk emission if they are sufficiently old. For example, the GG Tau system is a pair of close pairs; the primary binary is separated in projection by about 35 AU, and appears to have an age $\sim 10^6$ yr, and both are accreting [64]. Given that the circumstellar disks of the inner binary would ordinarily be truncated within about 0.3 of their separation, Figure 8 would suggest a lower value of α than 0.01 to avoid depleting the inner disks too rapidly. Unfortunately, this test is complicated by the possibility of accretion from the outer circumbinary disk, which be enhanced by a large viscosity or by a highly-eccentric binary orbit [1].

The disk accretion rate curves shown in Figure 8 are qualitatively what one would like to explain the statistics of disk accretion; the overall trend is basically an upper envelope of behavior, with individual systems dropping out rapidly at various times, presumably because of differing physical parameters. It seems unlikely that all of the depletion of disks and accretion rates is due to companion stars. Alternatively, massive planetary bodies could also have the same effect. To address this issue, one must first take up the issue of dust growth and evolution.

8 From T Tauri to debris disks

The evolution of dust particles in T Tauri disks must be an extremely complicated function of many parameters and processes which are poorly understood at present. In the following, I outline a few basic considerations and some current numerical simulations.

Figure 9 shows the median spectral energy distribution of T Tauri stars, along with two detailed disk models as discussed in Calvet's lectures. Here I wish to emphasize that the amount of mm-wave flux, as well as the spectral

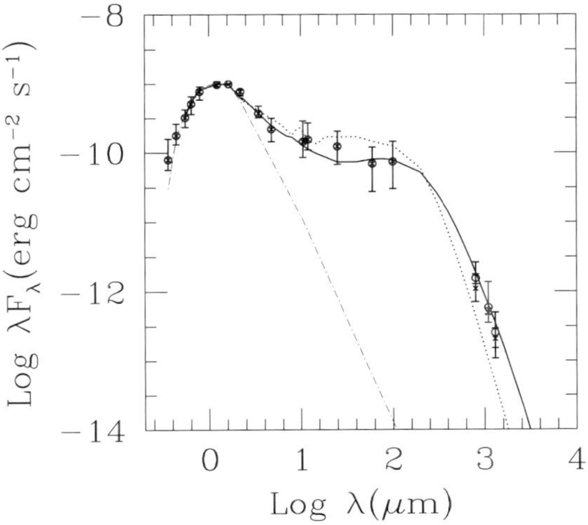

Fig. 9. Median spectral energy distribution of T Tauri stars in Taurus compared with disk models using interstellar medium dust (dotted line) and grain growth in a power law distribution out to a size of 1 mm (solid line). From [18].

slope in that region, is consistent with models which require grain growth from maximum sizes of interstellar medium dust of order 0.1 μm to ~1 mm. It is difficult to be certain of this, as the composition of dust is unknown, but it seems quite likely that the observations can best be explained with larger dust particles than were initially present in the material accumulated from the protostellar cloud.

Dust growth in protoplanetary disks is expected theoretically, although the details are complicated and poorly understood. To begin, aerodynamic forces are important in describing the dust motion. The drag equation for a spherical grain of radius a is

$$m_{\rm d}\frac{{\rm d}v_{\rm d}}{{\rm d}t} = m_{\rm d}v_{\rm d}\frac{{\rm d}v_{\rm d}}{{\rm d}x} = -F_{\rm d} = C\pi a^2 \rho v_{\rm d} c_{\rm s}, \tag{8.1}$$

where it is assumed that the dust velocity $v_{\rm d}$ is small with respect to the thermal velocity of the gas $c_{\rm s}$. This suggests the definition of a response time, the time it takes the grain to be struck with its own mass in gas due to thermal motions,

$$\pi a^2 \rho c_{\rm s} t_{\rm e} = 4/3\pi a^3 \rho_{\rm d} \tag{8.2}$$

or

$$t_{\rm e} = \frac{4}{3}\frac{a}{c_{\rm s}}\frac{\rho_{\rm d}}{\rho}. \tag{8.3}$$

Dust particles are suspended in the gas only in so far as they are supported by the gas. In general, the dust will tend to settle to the disk midplane under the effects of gravity;

$$m_{\rm d} v_{{\rm d}z} \frac{{\rm d}v_{{\rm d}z}}{{\rm d}z} = -\frac{GM}{R^2}\frac{z}{R} m_{\rm d} - C\pi a^2 \rho v_{{\rm d}z} c_{\rm s}. \tag{8.4}$$

The dust then attains a terminal velocity

$$v_{{\rm t}z} = -\frac{GM}{R^2}\frac{z}{R}\frac{4/3\pi a^3 \rho_{\rm d}}{C\pi a^2 \rho c_{\rm s}} = -\Omega^2 z\, t_{\rm e}. \tag{8.5}$$

The settling time is then

$$t_{\rm s} = \frac{z}{v_{{\rm t}z}} = \Omega^{-2} t_{\rm e}^{-1}. \tag{8.6}$$

As the dust particles sink toward the equatorial plane, they can grow if sticking occurs. This growth can be described by

$$\frac{{\rm d}m_{\rm d}}{{\rm d}z} = p_{\rm s}\pi a^2 \rho_{\rm dn}, \tag{8.7}$$

where $p_{\rm s}$ is the sticking probability and $\rho_{\rm dn}$ is the dust density in the nebula. Consider the timescale for sinking from one scale height allowing for dust growth. Assume roughly constant gas density;

$$\frac{\rm d}{{\rm d}z}\frac{4}{3}\pi a^3 \rho_{\rm d} = p_{\rm s}\pi a^2 \frac{\Sigma_{\rm d}}{2H} = 4\pi a^2 \rho_{\rm d}\frac{{\rm d}a}{{\rm d}z}, \tag{8.8}$$

so that

$$\frac{{\rm d}a}{{\rm d}z} = \frac{p_{\rm s}}{4\rho_{\rm d}}\frac{\Sigma_{\rm d}}{2H}. \tag{8.9}$$

Integrating this equation from an initial dust radius $a_{\rm o}$,

$$a = a_{\rm o} = \int_{\rm o}^{H} {\rm d}z \frac{p_{\rm s}}{4\rho_{\rm d}}\frac{\Sigma_{\rm d}}{2H} = a_{\rm o} + (1 - z/H) a_{\rm max} \tag{8.10}$$

where

$$a_{\rm max} = \frac{p_{\rm s}}{8\rho_{\rm d}}\Sigma_{\rm d}. \tag{8.11}$$

One can then solve for the sedimentation time,

$$\int {\rm d}t = \int \frac{{\rm d}z}{v_z} = \int \frac{{\rm d}z}{(\rho_{\rm d}/\rho)(a/c_{\rm s})\Omega^2 z}. \tag{8.12}$$

After a certain amount of manipulation one finds

$$t = \frac{\rho_{\rm dn}}{\rho}\frac{4}{p_{\rm s}}\Omega^{-1}(1 + a_{\rm o}/a_{\rm max})^{-1} \ln\left(\frac{a}{a_{\rm o}}\frac{H}{z}\right). \tag{8.13}$$

Note that the timescale is generally proportional to the orbital period $P = 2\pi/\Omega$. Much more sophisticated numerical simulations (*e.g.*, [63]) tend to confirm the scaling with orbital period. However, it is not possible to use the above considerations in detail for two reasons. First, theory and experiment indicate that real dust particles are highly unlikely to be spherical in early phases (*e.g.*, [65]). Instead, they exhibit a spidery, open structure. The net effect is that the cross-sectional area of the particle increases more rapidly with increasing mass than is the case for a sphere. In turn, this means that the dust particles are much more highly coupled to the gas than would be assumed in the spherical grain model. Only when the particles grow to sizes of order a cm are they likely to compact and become more spherical.

The other problem with the above equations is that they ignore the radial motion of dust particles. This is because (sufficiently) massive particles try to move at Keplerian velocities, whereas the gas in the disk rotates at slightly *sub Keplerian* velocities. The equation of motion in the plane of the disk for the gas is

$$v \frac{\mathrm{d}v}{\mathrm{d}R} - \frac{v_\phi^2}{R} = -\frac{1}{\rho}\frac{\mathrm{d}P}{\mathrm{d}R} - \frac{GM}{R^2} . \tag{8.14}$$

Ignore the radial component of the motion. In gerneral $\mathrm{d}P/\mathrm{d}R < 0$ for almost all disk models, so this provides support against gravity. To first order,

$$v_\phi^2 \approx GM/R - c_\mathrm{s}^2 , \tag{8.15}$$

$$v_\mathrm{phi} \approx \left(\frac{GM}{R}\right)^{1/2} \left(1 - \frac{1}{2}\frac{c_\mathrm{s}^2}{v_\mathrm{K}^2}\right) \approx v_\mathrm{K}\left(1 - \frac{H^2}{2R^2}\right) . \tag{8.16}$$

Thus, for typical values of H/R, the gas rotates just slightly below Keplerian velocities by an amount of 10^{-2}–$10^{-3} v_\mathrm{K}$. Though this is a negligible correction for understanding disk energetics, it can easily be many orders of magnitude larger than the vertical settling velocity. Thus, if the dust is not perfectly coupled to the gas, collisions occur, because different size dust particles have differing velocities.

The most sophisticated calculations following this evolution are those of Weidenschilling (1997). In this particular case, evolution of a solar nebula was considered at a radius $R = 30$ AU. Weidenschilling found that the dust particles became strongly depleted exterior to one scale height above the midplane in only 1–2 $\times$ 10^5 yr, and the disk became optically thin (in the Rosseland mean) in a similar timescale. This seems to be about an order of magnitude too fast, compared with observations of T Tauri stars; as explained in Calvet's lectures, explaining the median SED at 1 Myr seems to require the suspension of considerable amounts of dust up to 3–4 scale heights above the midplane.

Two factors might mitigate this problem. As emphasized by Weidenschilling (1997), disk turbulence is not taken into account in these models; but turbulence will be present if the disk is accreting, and particularly if the MRI is present, although the MRI is not particularly effective in producing vertical motions (*e.g.*, Fromang *et al.* 2002). Secondly, there may be more destruction than in the Weidenschilling models, with consequent production of small particles as well as dispersal. Nevertheless, it seems likely that some particle growth has occurred in T Tauri disks, even if not quite as rapidly as might be expected in a non-turbulent disk. The mm-wave emission, which suggests the presence of larger particles, is generally dominated by contributions at large radii, $\sim$100 AU or more. If growth to mm-sized particles has already occurred at 100 AU, then it would seem inevitable, given the orbital period scaling discussed above, that there must be much larger bodies orbiting at smaller radii, hidden from our view by clouds of much smaller dust particles.

Following this line of thought further, if large bodies are already present in T Tauri disks at radii of a few AU or more at 1 Myr, then it is quite plausible for runaway growth to occur and produce large planets within the next few Myr. Runaway growth results when bodies reach a threshhold of mass where gravitational focussing allows them to capture material from much wider regions. Thus there is a natural way in which a threshhold can be reached, at which point disk evolution becomes rapid.

In this general picture, it is to be expected that disks will end up with inner "holes". The reason: large bodies should form first at smaller radii, given the faster evolution there. The inner regions can then be cleared relatively rapidly through a combination of planetary sweeping and accretion onto the central star. This schematic behavior is in principle capable of explaining the so-called "debris disk" or "Vega-like" systems, which all have inner disk regions that are evacuated of small (in this case, few micron-sized) grains.

A particularly interesting system for study in this regard is the nearest known classical T Tauri star, TW Hya [37]. As shown in Figure 10, there is evidence of a clearing in the inner disk at about 4 AU [14]. However, the far-infrared and submm to mm fluxes of TW Hya are high in comparison with many T Tauri stars. The modelling of Calvet *et al.* [14] suggests that grains may have grown in this 10 Myr-old system to sizes of order 10 cm in the outer disk; but the disk itself remains optically thick to far-infrared wavelengths. The suggestion is that this system is a transition object, on a path to making giant planets in the few AU region, since there is still accretion onto the central star [46]. Clearly this object will be studied carefully in the near future, both for reasons of its interesting disk structure and possible evolutionary state, as well as its near distance (50 pc).

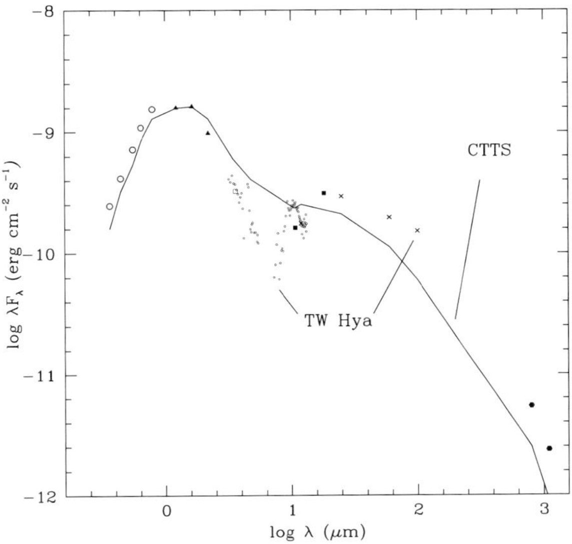

Fig. 10. Spectral energy distribution of TW Hya compared with the median T Tauri SED. There is a clear deficit of near- infrared excess, even though the disk emission of TW Hya exceeds that of the typical T Tauri star [14].

9 Gas dispersal

It is not clear whether the disk gas disappears when the dust emission vanishes. There is some reason to believe that the two are related; certainly when there is little near-infrared radiation from hot dust close to the star, there is no evidence for accretion [27].

Beyond this observational question, theories of planet formation often invoke some kind of disk gas dispersal at a convenient time, for example to stop type II migration which might otherwise carry giant planets all the way into the star [59].

The problem is, how is the gas dispersed? The old idea was that the strong T Tauri wind would disperse outer disk gas on a relevant timescale. However, now we know that T Tauri winds are accretion-disk-driven winds (see Calvet's discussion). This means (a) that the wind emerges more nearly perpendicular than parallel to the disk, and (b) that the wind/jet shuts off when the inner disk accretion shuts off. So this possibility seems highly unlikely.

Another possibility is photoevaporation or photoionization. Gaseous mass loss from the disk is very strong, for example, for stars in the Orion

Nebula Cluster close to the central O7 star θ_1 C Ori (see [34], and references therein). Unfortunately (or fortunately) most stars are not situated near O stars, so this can hardly be a universal mechanism. Nevertheless, it is worth developing estimates to see the scope of the problem.

The basic idea is that photoionization or photoevaporation leads to a thermal wind emanating from regions with escape velocity < thermal velocity. Recall that the critical point of a spherical thermal wind is characterized by

$$v = c_{\rm s}, \quad 2c_{\rm s}^2 = GM_*/r. \tag{9.1}$$

Here we concentrate on the possibility of photoionization; the effect of heating by FUV (~1000–2000 Å) photons is likely to be unimportant for T Tauri stars.

Photoionized gas typically has temperatures of $T \sim 10^4$ K, so $c_{\rm s} \sim 10\,{\rm km\,s^{-1}}$; then the escape radius is approximately $r \sim 10$ AU. Assume rough spherical symmetry. Then the ionized region is constrained by a Strömgren sphere estimate. Take $v \propto r$ in the region near the critical point, so that the number density $n \propto r^{-3}$ assuming steady spherical flow. Then the emission measure is

$$EM = \int_{r_{\rm o}}^{\infty} 4\pi r^2 {\rm d}r n_{\rm o}^2 (r/r_{\rm o})^6 = 4\pi r_{\rm o}^3 n_{\rm o}^2/3. \tag{9.2}$$

For a given hydrogen recombination rate α_B, this implies a total recombination rate in the region which must be balanced in steady state by an ionizing flux of photons $\Phi_{\rm i}$,

$$\Phi_{\rm i} \frac{{\rm d}\omega}{4\pi} = 4\pi r_{\rm o}^3 n_{\rm o}^2 \alpha_{\rm B}/3, \tag{9.3}$$

where ${\rm d}\omega$ represents the solid angle that the disk material subtends at the star (assuming the ionizing flux is isotropic). The mass loss rate is then

$$\dot{M} \sim 4\pi r_{\rm o}^2 c_{\rm s} \rho_{\rm o} = (12\pi r_{\rm o})^{1/2} c_{\rm s} \mu m_{\rm H} \alpha_{\rm B}^{-1/2} \Phi_{\rm i}^{1/2} ({\rm d}\omega/4\pi)^{1/2}. \tag{9.4}$$

Adopting $\alpha_{\rm B} = 2.6 \times 10^{-13}\ {\rm cm^6\,s^{-1}}$ and $({\rm d}\omega/4\pi) \sim 0.1$,

$$\dot{M} \sim 6 \times 10^{-10}\, M_{\odot}\,{\rm yr^{-1}} \left(\frac{r_{\rm o}}{10\ {\rm AU}}\right)^{1/2} \left(\frac{c_{\rm s}}{10\,{\rm km\,s^{-1}}}\right) \left(\frac{\Phi_{\rm i}}{10^{41}\ {\rm s^{-1}}}\right)^{1/2} . \tag{9.5}$$

which agrees within factors of order unity of the more detailed theory [34].

This mass loss would get rid of a $0.06\,M_{\odot}$ disk ($r > 10$ AU) in 10^8 yr. Moreover, unlike accretion, it gets rid of the outer disk gas. However, it is not at all clear that T Tauri stars have such high UV fluxes; the fiducial value is about 10^3 times solar. It is likely, of course, that young stars have much more active chromospheres than the Sun, but this seems to be a large factor for chromospheric emission. Moreover, even this rate would be insufficient to explain the loss of disk gas on timescales $\lesssim 10^7$ yr over which disk emission and mass accretion subsides, and required by some theories of planet formation and migration (see below). This remains a serious problem.

The FUV photons can heat the gas only to $\sim 10^3$ K and thus can promote escape from disk radii $\gtrsim 100$ AU. This mechanism works at distances [34]

$$d \lesssim 10^{15}\,{\rm cm} \left(\frac{\Phi_{\rm FUV}}{10^{43}\,{\rm s^{-1}}}\right)^{1/2} , \tag{9.6}$$

where $\Phi_{\rm FUV}$ is the number of FUV ($\sim$10 eV) photons emitted per second. I have made an estimate of the FUV photons emitted by the 10 Myr-old Classical T Tauri star TW Hya, based on observations with IUE and HST; the result is $\Phi_{\rm FUV} \sim 2 \times 10^{39}\,{\rm s^{-1}}$. Thus it would seem very unlikely that this mechanism is important in dispersing protoplanetary disk gas.

In general, it seems unlikely that photoevaporative fluxes are responsible for eliminating disk gas on timescales of 10 Myr (although they could be quite effective over much longer periods of time). This is in a way reassuring; TW Hya (see previous section), with an age of 10 Myr, seems to have quite a substantial dust disk, with a good complement of gas; otherwise the dust grains would not be suspended to sufficient heights from the disk midplane to explain the far-infrared emission, and of course some source of gas is needed to explain the observed accretion onto the central star.

This conclusion, however, leads to a difficulty; if outer disk gas is not dispersed, then as it accretes it can carry in giant planets in type II migration [62], possibly on an uncomfortably short timescale. Of course, the discovery of close giant planets has led to the idea of substantial radial migration, moving these planets from disk outer regions where there is enough mass to form them and moving them inward due to some mechanism. Viscous disk evolution could easily accomplish this; the problem is that it is easy to have too much of a good thing; an $\alpha \sim 10^{-2}$ would result in giant planets zipping through the inner 10 AU in 0.1 Myr or less. Unless the outer disk pushing is turned off, what stops the conveyor belt? Possibly the dead zone plays a role here, in producing enough mass and angular momentum in the planetary zone to help hold off the inward forcing of the outer disk material until it can be dissipated on longer timescales by stellar winds and photoevaporation.

Finally, what does become of the disk gas, if it is not ejected on 10 Myr timescales? My guess is that much of the gas is incorporated as CO and water ice into larger bodies, and thus disappears from detection. What happens to the H_2, which is relatively inert? Is it still there, and we just can't see it [57]? I don't know; my guess is that a substantial amount is still there, and we just haven't found it, due to the difficulty in detecting cold H_2.

This work was supported in part by NASA grant NAG5-9670.

References

[1] P. Artymowicz and S.H. Lubow, *ApJ* **467** (1996) L77.
[2] A. Bacmann, P. André, J.-L. Puget, A. Abergel, S. Bontemps and D. Ward-Thompson, *A&Ap* **361** (2000) 555.
[3] S.A. Balbus and J.F. Hawley, *Rev. Mod. Phys.* **70** (1998) 1.
[4] S.A. Balbus and J.C.B. Papaloizou, *ApJ* **521** (1999) 650.
[5] J. Bally, R.S. Sutherland, D. Devine and D. Johnstone, *AJ* **116** (1998) 293.
[6] M.R. Bate, *MNRAS* **314** (2000) 33.
[7] R.D. Blandford and D.G. Payne, *MNRAS* **199** (1982) 883.
[8] K.R. Bell and D.N.C. Lin, *ApJ* **427** (1994) 987.
[9] J. Binney and S. Tremaine, Galactic Dynamics (Princeton University, Press, 1987).
[10] I. Bonnell and P. Bastien, *ApJ* **401** (1992) 654.
[11] A.P. Boss, *ApJ* **563** (2001) 367.
[12] A. Brandenburg, A. Nordlund, R.F. Stein and U. Torkelsson, *ApJ* **458** (1996) L45.
[13] N. Calvet, in Accretion Processes in Astrophysical Systems: Some Like it Hot!, edited by S.S. Holt and T.R. Kallman, AIP Conf. Proc. **431** (1998) 495.
[14] N. Calvet, P. D'Alessio, L. Hartmann, D. Wilner, A. Walsh and M. Sitko, *ApJ* **568** (2002) 1008.
[15] N. Calvet, L. Hartmann and S.J. Kenyon, *ApJ* **402** (1993) 623.
[16] P. Cassen and A. Moosman, *Icarus* **48** (1981) 353.
[17] C. Curry, R.E. Pudritz and P.G. Sutherland, *ApJ* **434** (1994) 206.
[18] P. D'Alessio, N. Calvet and L. Hartmann, *ApJ* **553** (2001) 321.
[19] R.H. Durisen, R.A. Gingold, J.E. Tohline and A.P. Boss, *ApJ* **305** (1986) 281.
[20] A. Duquennoy and M. Mayor, *A&Ap* **248** (1991) 485.

[21] P.N. Foster and R.A. Chevalier, *ApJ* **416** (1993) 303.
[22] S. Fromang, C. Terquem and S.A. Balbus, *MNRAS* **329** (2002) 18.
[23] M. Gagne, J. Caillault and J.R. Stauffer, *ApJ* **445** (1995) 280.
[24] C.F. Gammie, *ApJ* **457** (1996) 355.
[25] C.F. Gammie, *ApJ* **553** (2001) 174.
[26] A.E. Glassgold, J. Najita and J. Igea, *ApJ* **480** (1997) 344.
[27] P. Hartigan, L. Hartmann, S.J. Kenyon, S.E. Strom and M.F. Skrutskie, *ApJ* **354** (1990) L25.
[28] L. Hartmann, Accretion Processes in Star Formation (Cambridge University Press, 1998).
[29] L. Hartmann and N. Calvet, *AJ* **109** (1995) 1846.
[30] L. Hartmann, N. Calvet, E. Gullbring and P. D'Alessio, *ApJ* **495** (1998) 385.
[31] L. Hartmann and S.J. Kenyon, *ARA&A* **34** (1996) 207.
[32] J.F. Hawley, C.F. Gammie and S.A. Balbus, *ApJ* **464** (1996) 690.
[33] J. Heyvaerts, E.R. Priest and A. Bardou, *ApJ* **473** (1996) 403.
[34] D.J. Hollenbach, H.W. Yorke and D. Johnstone, in Protostars and Planets IV, edited by V. Mannings, A.P. Boss and S.S. Russell (Tucson: University of Arizona Press, 2000) 401.
[35] J. Igea and A.E. Glassgold, *ApJ* **518** (1999) 848.
[36] C.M. Johns-Krull, J.A. Valenti and C. Koresko, *ApJ* **516** (1999) 900.
[37] J.H. Kastner, B. Zuckerman, D.A. Weintraub and T. Forveille, *Science* **277** (1997) 67.
[38] S.J. Kenyon, N. Calvet and L. Hartmann, *ApJ* **414** (1993) 676.
[39] S.J. Kenyon and L. Hartmann, *ApJ* **101** (1995) 117.
[40] S.J. Kenyon and L. Hartmann, *ApJ* **383** (1991) 664.
[41] A. Königl, *ApJ* **342** (1989) 208.
[42] G. Laughlin and M. Rozyczka, *ApJ* **456** (1996) 279.
[43] D.N.C. Lin and J.E. Pringle, *ApJ* **358** (1990) 515.
[44] D. Lynden-Bell and J.E. Pringle, *MNRAS* **168** (1974) 603.
[45] J. Muzerolle, L. Hartmann and N. Calvet, *AJ* **116** (1998) 2965.
[46] J. Muzerolle, N. Calvet, C. Briceño, L. Hartmann and L. Hillenbrand, *ApJ* **535** (2000) L47.
[47] D.L. Padgett, W. Brandner, K.R. Stapelfeldt *et al.*, *AJ* **117** (1999) 1490.
[48] R.E. Pudritz and C.A. Norman, *ApJ* **274** (1983) 677.
[49] R.E. Pudritz and C.A. Norman, *ApJ* **301** (1986) 571.
[50] G. Pelletier and R.E. Pudritz, *ApJ* **394** (1992) 117.
[51] M. Reyes-Ruiz and T.F. Stepinski, *ApJ* **438** (1995) 750.
[52] D. Ryu and J. Goodman, *ApJ* **338** (1992) 438.
[53] T. Sano, S.M. Miyama, T. Umebayashi and T. Nakano, *ApJ* **543** (2000) 486.
[54] F. Shu, J. Najita, E. Ostriker, F. Wilkin, S. Ruden and S. Lizano, *ApJ* **429** (1994) 781.
[55] J.M. Stone and S.A. Balbus, *ApJ* **464** (1996) 364.
[56] J.M. Stone, J.F. Hawley, C.F. Gammie and S.A. Balbus, *ApJ* **463** (1996) 656.
[57] W.F. Thi *et al.*, *ApJ* **561** (2001) 1074.
[58] L. Tomley, T.Y. Steiman-Cameron and P. Cassen, *ApJ* **422** (1994) 850.
[59] D.E. Trilling, W. Benz, T. Guillot *et al.*, *ApJ* **500** (1998) 428.

[60] Y. Uchida and K. Shibata, *PASJ* **37** (1985) 515.
[61] T. Umebeyashi and T. Nakano, *Prog. Theor. Phys. Suppl.* **96** (1988) 151.
[62] W.R. Ward, *ApJ* **482** (1997) L211.
[63] S.J. Weidenschilling, *Icarus* **127** (1997) 290.
[64] R.J. White, A.M. Ghez, I.N. Reid and G. Schultz, *ApJ* **520** (1999) 811.
[65] G. Wurm and J. Blum, *Icarus* **132** (1998) 125.

COURSE 13

X-RAYS FROM YSOS: DESPERATELY SEEKING THE CENTRAL ENGINE

T. MONTMERLE

*Service d'Astrophysique,
Centre d'Études de Saclay,
91191 Gif-sur-Yvette, France
and Laboratoire d'Astrophysique
de Grenoble, BP. 53,
38041 Grenoble Cedex, France*

Contents

X-RAYS FROM YSOS: DESPERATELY SEEKING THE CENTRAL ENGINE

T. Montmerle

Abstract

The earliest stages of the evolution of low-mass stars are characterized by the presence of powerful jets, likely originating from accretion via circumstellar disks. In this review, we examine if, and how, the ubiquitous X-ray emission from young stars, taken as a proxy for magnetic reconnection events, can constrain the "central engine", *i.e.*, the magnetic structures widely thought to couple accretion and ejection.

1 Background: YSOs and magnetic fields

1.1 Pre-main sequence evolution in a nutshell

At their earliest (pre-main sequence) evolutionary stages, "young stellar objects" ("YSOs": ~solar-mass young forming stars) are characterized by the presence of circumstellar material out to large distances (for a review, see, *e.g.*, André *et al.* 2000). At the youngest protostellar stage ("Class 0" protostars, age $\sim 10^4$ yrs), the central, forming object is surrounded by a dense, extended envelope (radius ~10 000 AU). As collapse (and accretion, building the central object into a star) proceeds, the envelope while still being large becomes tenuous ("Class I" protostars, age $\sim 10^5$ yrs). Because of rotation, accretion takes place *via* a circumstellar disk, observed to have a typical radius ~500–1 000 AU, *i.e.*, ~10–20 times the size of the present-day solar system. When the envelope material is exhausted, only the disk remains around the nascent star, which becomes visible in the optical: this is the "classical T Tauri" stage (CTTS; or equivalently "Class II" stage when observed in the IR range). Eventually, this circumstellar disk itself disappears, for reasons which are still unclear (complete accretion onto the star, and/or dust grain growth followed by planet formation, or some other mechanism), leaving a diskless T Tauri star, sometimes called a "naked", or "weak (emission)-line T Tauri star (WTTS; "Class III" in the IR). The

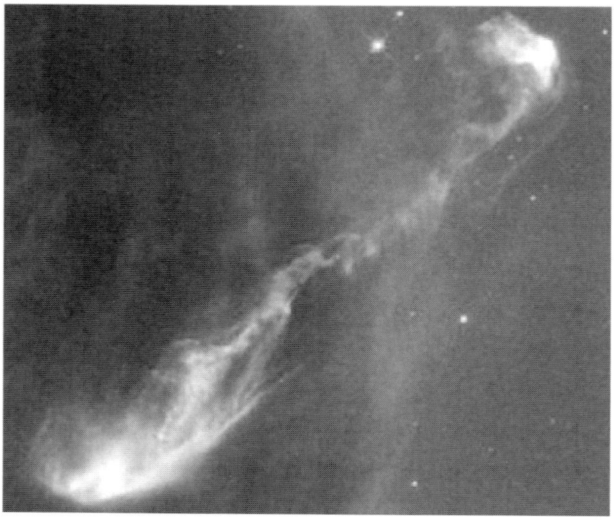

Fig. 1. A bipolar jet from a young, low-mass star: HH47. The exciting star is a faint dot almost excatly at the center of the picture. (HST WFPC2 image; J. Morse, NASA, 1995.)

transition between the Class II and Class III stages appears to take place at an ill-defined moment, between $\sim 10^6$ and $\sim 10^7$ yrs.

While the collapse of a dense envelope can be more or less intuitively understood to lead to star formation at its center, this is not true if rotation is present, unless angular momentum is lost in the process. Mother Nature seems to have found a solution, in the form of powerful jets and outflows observed to emanate from the central regions of star-disk-envelope system, out to very large distances (sometimes up to a fraction of a pc or more), as illustrated in Figure 1. (See, *e.g.*, Dougados *et al.* 2000, and Reipurth *et al.* 2002 for recent references.) Millimeter observations have convincingly shown that accretion and ejection of material near the central, forming star, are intimately connected: $\approx 10\%$ of the accreted mass is found to be ejected. According to various estimates, the accretion rate declines roughly from $\dot{M} \sim 10^{-6} M_\odot$ during the protostar stages, down to $\dot{M} \sim 10^{-8} M_\odot$ at the Class II T Tauri stage (*e.g.*, Bontemps *et al.* 1996).

Understanding how a significant part of the accreted material is ejected from the star-disk-envelope system, and carries away angular momentum, remains one of the major challenges of the astrophysics of star formation. As I will discuss in this review, magnetic fields and X-ray emission hold the key to unraveling the mysteries of this "central engine" powering accretion and ejection, although much debate is going on about the physics involved.

1.2 X-rays as tracers of YSO magnetic activity

The X-ray variability of YSOs is ubiquitous, and was discovered as soon as imaging satellites became available (the *Einstein Observatory* in the 1980's). A large number of YSOs (mainly TTS, see below) were already known to be X-ray emitters in the nineties thanks to the wide field-of-view of previous satellites (*ROSAT*, *ASCA* = $30'$–$1°$ in diameter), as reviewed by Feigelson & Montmerle (1999). X-ray luminosities spanned the range $L_X \sim 10^{29}$–10^{31} erg s^{-1} (or up to $\sim 10^4$ times the X-ray luminosity of the active Sun), *i.e.*, a typical X-ray luminosity to bolometric luminosity ratio $L_X/L_{bol} \sim 10^{-4}$. At present, the improved sensitivity, as well as the extension of the energy range towards hard X-rays, of the new generation of satellites (*Chandra* and *XMM-Newton*) more than compensate for the reduction in field-of-view ($17' \times 17'$ for *Chandra*, $30'$ diameter for *XMM-Newton*)[1]. Star-forming regions like Orion now typically contain several hundreds of X-ray sources, almost all identified with YSOs (>1000 in the Orion Nebula Cluster), for which a continuous coverage of several 10 ksec is available (Garmire *et al.* 2000; Feigelson *et al.* 2002).

1.3 The "solar paradigm" and the solar-stellar connection

Such a continuous time coverage is crucial. *ROSAT* and *ASCA*, owing to their low Earth orbit, could only provide coverage interrupted by Earth occultations, except around the ecliptic pole. In addition, with a few exceptions the comparatively modest sensitivity led to a mediocre signal-to-noise ratio for YSOs, so that analytical fits to the light curves were at best a crude representation of the real light curves. Prior to *Chandra* and *XMM-Newton*, large variations in the X-ray light curves were usually fit by a steep, almost vertical rise, followed by an exponential decrease (with a typical e-folding time of a few hrs). Simplified in this way, this behaviour became synonymous with "X-ray flares", characterized by a short, energetic heating phase, followed by a long cooling phase, the cooling being attributed to some combination of radiation and conduction, in analogy with solar X-ray flares (Van den Oord & Mewe 1989). Based on this "solar paradigm", *i.e.*, the identity in nature between YSO and solar X-ray flares, the observed X-ray variability in YSOs was taken as a proxy for *magnetic activity*, expected

[1] For general descriptions of the *Chandra* and *XMM-Newton* missions, see Weisskopf *et al.* (2002) and Jansen *et al.* (2001) respectively.

as a result of convection-induced dynamo in late-type stars. The basic interpretation of such light curves is then that a high-temperature plasma ($T_X \sim 10^6$–10^8 K, $n_e \sim 10^{10-12}$ cm^{-3}) is suddenly heated by *magnetic reconnection processes*, while being confined within large magnetic loops ($B_\star \approx 0.1$–1 kG, $h \sim 1$–2 $R_\star$), and cools by bresstrahlung X-ray emission.

With the advent of *Chandra* and *XMM-Newton*, the continuous coverage (up to ~140 ksec or more) resulting from their highly eccentric orbit, and their much better sensitivity, have led to a different, and unfortunately much less simple, picture.

- The assumption of a pure cooling of the plasma after the heating phase is likely incorrect: as shown by *Yohkoh* observations of solar flares, some heating appears to be present even during the cooling phase. The result is that the flare plasma volume, which is usually derived from the observed X-ray parameters (temperature, luminosity, e-folding time for luminosity decrease) is actually smaller than computed when assuming pure radiative cooling (characteristic loop sizes 2 to 5 times smaller, see Favata *et al.* 2001, and Refs. therein);
- Only a small fraction of the nearly 1000 light curves obtained by *Chandra* in the ONC (Garmire *et al.* 2000), in two separate (continuous) observations lasting ~40 ksec each, display explicit, "simple" flarelike exponential decays. As described by Feigelson *et al.* (2002), the vast majority of the sources display apparently disordered light curves. Perhaps several flares occur more or less simultaneously on the same stars, or perhaps the heating and cooling of the plasma in a given volume takes place on long and short timescales, respectively, or something else happens.

In spite of these complications, there can be little doubt that the basic origin of the X-ray emission is magnetic reconnection resulting from large-scale differential movements of the surface magnetic field. In the Sun and in ordinary, late-type main-sequence stars, and also in Class III T Tauri stars (WTTS), such movements take place in the convective photosphere in which the magnetic loops are anchored. In fact, magnetic fields have been measured in TTS by Zeeman broadening of optical spectral lines (see, *e.g.*, Johns-Krull & Gafford 2002), with strengths consistent with X-ray-derived values.

However, in Class II T Tauri stars (CTTS) and in protostars, the presence of a circumstellar disk also offers the possibility of *star-disk magnetic interactions*. Since this is the configuration in which jets and outflows are thought to be produced, as discussed in the next section, it is extremely important to search for specific X-ray characteristics proving that in YSOs the central stars and their surrounding disks indeed mutually interact.

2 Challenging the solar paradigm

2.1 X-ray emission at high latitudes

While the solar paradigm would imply that, as in the Sun, the X-ray activity comes from reconnecting loops in the equatorial region, other observational evidences suggest instead that the X-ray emission may take place at high stellar latitudes, even in the polar regions.

In young stars, this evidence is so far limited, but would be extremely significant if confirmed. It concerns two Class I protostars belonging to the ρ Oph cloud, abundantly observed by X-ray satellites, in particular in the course of long exposures with *ASCA* (200 ksec; Tsuboi *et al.* 2000) and *Chandra* (100 ksec; Imanishi *et al.* 2000).

The Class I protostar YLW15 in ρ Oph, detected several times with *ASCA*, showed on one occasion a "triple flare", which could be well fitted by three successive exponential decays separated by ~20 hrs (Tsuboi *et al.* 2000). According to a scenario suggested by Montmerle *et al.* (2000) and sketched in Figure 2, this quasi-periodic behavior could be explained by successive magnetic reconnections between a fast rotating (near break-up) central star and the inner boundary of the circumstellar disk. Although it can be argued that this picture is fully consistent with the X-ray data and flare modelling, the situation is likely more complex: IR spectroscopy has recently shown that YLW15, although rotating much more rapidly than

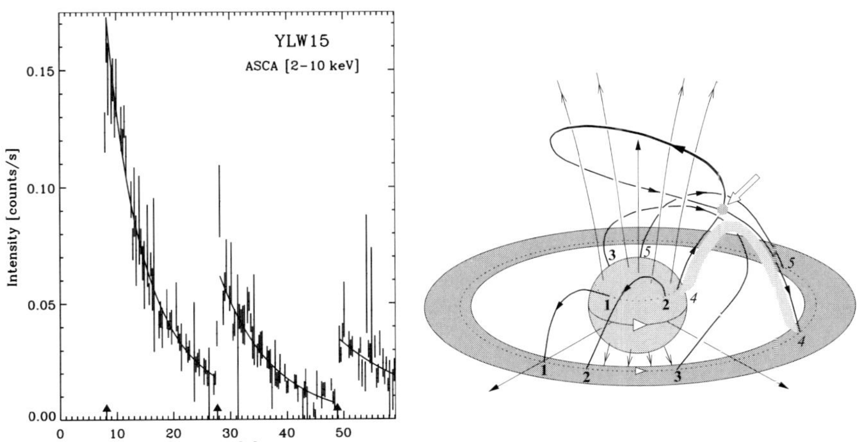

Fig. 2. Triple, quasi-periodic flare in the Class I protostar YLW15 observed with *ASCA* (left: Tsuboi *et al.* 2000), and the rotating star-disk reconnection scenario proposed by Montmerle *et al.* (2000; right).

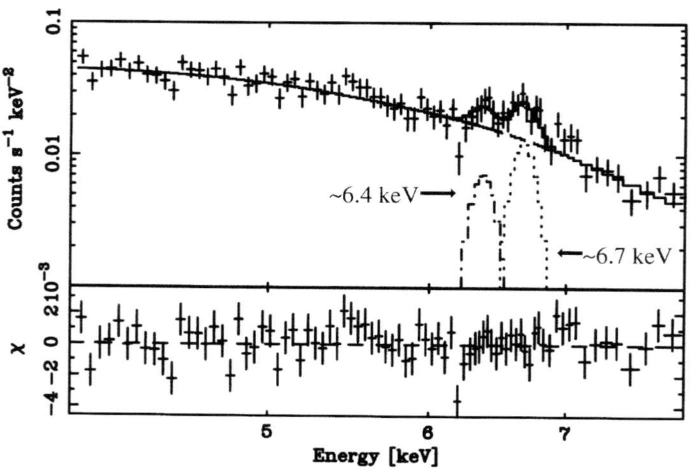

Fig. 3. *Chandra* ACIS spectroscopy of the Class I protostar YLW16A. The spectrum reveals a double Fe line: the standard K line at 6.7 keV, and the fluorescence line at 6.4 keV (Imanishi *et al.* 2001).

TTS, does it less than break-up (Greene & Lada 2002); perhaps more importantly, VLA observations have shown that YLW15 is a double, perhaps even a triple, system (Girart *et al.* 2000). Further multiwavelength observations are obviously needed to clarify the status of this object. If confirmed, this case would be the first of a direct star-disk magnetic interaction. So far, however, existing *Chandra* and *XMM-Newton* observations of Class I protostars in various star-forming regions (including new ones of YLW15 and its vicinity) have revealed no new case for periodicity.

In the close vicinity of YLW15, another Class I protostar, YLW16A, displayed a powerful flare during an observation with *Chandra* (Imanishi *et al.* 2000). The luminosity peak ($L_X \sim 10^{31}$ erg^{-1}) was bright enough that a reasonably good low-resolution spectrum could be taken with the ACIS detector, as shown in Figure 3. In general, hot plasmas emit a strong line at 6.7 keV, corresponding to almost fully ionized, He-like Fe atoms. This line can be detected if the signal-to-noise ratio is large enough. In the case of YLW16A, this line was indeed detected, but another line, at 6.4 keV, was clearly resolved from the 6.7 keV line. This line is interpreted as fluorescence from cold Fe atoms in the disk irradiated by the X-rays from the flaring plasma. The most likely scenario is inspired from AGN physics: the X-rays are somehow emitted at high latitudes, in a "lamp post" fashion-otherwise they would be directly absorbed by the disk.

2.2 Polar X-rays?

Whereas the most commonly invoked magnetic configuration calls for an equatorial, dipolar magnetosphere, some evidence may even favor instead *polar* configurations. One possibility to map active, X-ray emitting regions is to observe an eclipsing binary system, for which the orbit of the companion lies close to a plane perpendicular to the plane of the sky. The best documented case is Algol, a B8V + K2IV system with a 2.87 orbital period. This system was observed during 240 ksec up to very hard X-ray energies (~20 keV) using various detectors aboard *Beppo-SAX*. A powerful flare erupted during the observation, but the light curve displayed a deep gap, consistent with zero intensity (Favata & Schmitt 2000). A look at the system's ephemeris showed that the only explanation was a total X-ray eclipse event produced by the smaller, late-type star. Knowing the system's orientation, the authors showed that the volume eclipsed by the companion had to be located at the south pole. Much the same picture emerges from radio cm observations (Favata *et al.* 2000). Other examples include V284 Ara, a PMS binary for which optical Doppler imaging of active regions showed that the activity (spots) was concentrated near the pole facing the observer (Hatzes & Kürster 1999). In fact, polar configurations may be more frequent than previously recognized: according to Schrijver & Title (2000), out of 53 solar-like stars for which Doppler imaging exists, 51 show high-latitude or polar spots, and 32 have spots covering the rotational poles, and all are rapid rotators.

How is this somewhat counter-intuitive (and non-solar) configuration possible? Clearly, *rapid rotation* plays an important role. Numerical simulations (Schrijver & Title 2000) of rotation-induced dynamo in solar-like stars suggest two reasons: (i) in a rapidly rotating, solar-like star, the Coriolis force may pull convective cells towards the poles; (ii) the magnetic tension may pull entire magnetic loops towards high latitudes. Since T Tauri stars are not, on average, fast rotators, this result would preferably apply only to the younger TTS, or to protostars, before they have had time to spin down by way on star-disk magnetic interactions (Königl 1991).

2.3 A new look a stellar dynamos

Now these simulations are done in the framework of the so-called "$\alpha\omega$" dynamo, in which magnetic field generation (and by inference magnetic activity indicators) is proportional to the star's differential rotation, itself more or less proportional to the global rotation velocity as measured, for instance, by starspot photometry. This picture seems basically correct for main sequence stars, in which the X-ray luminosity (taken as a proxy for the magnetic activity) is inversely proportional to the rotation period, at least for slow rotators (Randich 2000). For optically visible T Tauri stars, this was thought to be also valid, based on relatively scarce statistics (mainly of reliable rotation periods -as opposed to $v$sin$i$) (Stelzer & Neuhaüser 2001). But these statistics have been vastly increased in the Orion nebula, where nearly 400 rotation periods are now known (Herbst *et al.* 2002), along with the spectral types and X-ray emission (Feigelson *et al.* 1999, 2001, 2003). This allows to identify the stellar masses, and thus to retain sufficient statistics in relatively small mass ranges and consider only the effect of rotation.

The result is extremely puzzling: a sample of 41 stars with masses between 0.8 and 1.2 $M_\odot$ and well-determined rotation periods between $\sim$1 and 30 days, shows no correlation at all, but a large dispersion of points scattered over nearly three orders of magnitude in L_X.

The conclusion reached by Feigelson *et al.* (2003) is that the so-called "$\alpha\omega$" dynamo, usually invoked to explain the generation of magnetic fields in late-type, convective stars, is probably not the right mechanism in the case of T Tauri stars. This is not completely unexpected, since T Tauri stars are *fully* convective: in the "$\alpha\omega$" dynamo, the field is generated mainly at the base of the convective zone -which does not exist if the star is fully convective. An alternative mechanism may be invoked in this case, the "α^2" dynamo, aka the "turbulent" dynamo. Is this dynamo more likely than the "$\alpha\omega$" dynamo to give rise to polar structures? This is not known, but since apparently rotation effects are wiped out by other effects like turbulence, the dynamo should be predominantly spherically symmetric, and thus so should be active regions.

3 Starting the central engine: X-rays from protostars

As discussed in the preceding sections, young stellar objects are ubiquitous X-ray emitters. The question is: when does the X-ray activity start? Since jet/outflow activity starts as soon as gravitational collapse begins, *i.e.*, at the Class 0 protostar stage, we want now to adress the issue of the X-ray emission of protostars in general: when does the "magnetic central engine" start?

We have mentioned above that many "Class I" protostars have been detected in X-rays. What are the differences with the "Class 0" protostars just mentioned? As summarized in Sect. 1, Class 0 protostars are the youngest stage of stellar evolution ($\sim 10^4$ yrs); the envelope is still significantly opaque, although the opacity decreases with time (equivalent $A_V \sim$ 100–1000, or $N_H \sim 10^{23}$–10^{24} cm^{-2}). When the mass of the central object becomes comparable to the (remnant) mass of the envelope, the protostar evolves into a "Class I" protostar; in the course of the next $\sim 10^5$ yrs, the central object, which already has stellar characteristics, acquires its final mass, the envelope becomes more tenuous ($N_{\rm H} \approx 10^{22}$ cm^{-2}). In other words, although the transition is certainly continuous, there is a very significant difference in structure between a Class 0 protostar (dominated by its dense envelope) and a Class I protostar (dominated by the central star and its circumstellar disk).

In view of the X-ray/magnetic properties of the Class I protostars described above, one would expect that, since they are active jet sources, the X-rays would give strong observational constraints on their central engine. However, the case of Class 0 protostars is now clear: *they are not detected in X-rays.* Using *Chandra* and *XMM-Newton*, we have observed nearly a dozen Class 0 protostars, in various star-forming regions, having both high and low bolometric luminosities ($L_{\rm bol} \sim 0.5$–$10\ L_\odot$), large and small distances from the Sun ($d \sim 100$–300 pc) (Montmerle *et al.*, in prep.). We have also re-analyzed the *Chandra* data of one X-ray source in the OMC-3 cloud, claimed by Tsuboi *et al.* (2001) to be a "candidate" Class 0 protostar, discovered in the mm range. We find that this X-ray source is actually marginally resolved by *Chandra*, and associated with two Subaru IR sources (from Tsujimoto *et al.* 2002). Our preliminary interpretation is thus that the *Chandra* source is associated not with a Class 0-type object, but rather with two marginally resolved Class I sources.

How can we explain the non-detection of Class 0 sources in X-rays? Either or both of the following explanations are possible:

- The extinction is so high that the X-rays are absorbed. This puts upper limits on the intrinsic (absorption corrected) $L_{\rm X}$ (or $(L_{\rm bol}/L_{\rm X})$, as a function of $T_{\rm X}$: the higher the assumed $T_{\rm X}$, the lower the upper limit on $L_{\rm X}$. For most Class 0 protostars in our sample, these upper limits are comparable to, or higher than, values typical for T Tauri stars ($L_{\rm bol}/L_{\rm X} \sim 10^{-4}$). This conclusion is therefore not very contraining. But for a few Class 0s in our sample, the upper limits are actually significantly lower than for TTS ($L_{\rm bol}/L_{\rm X} < 10^{-5}$ for $T_{\rm X} > 5$ keV);

- This opens an alternative, exciting possibility, *i.e.*, that Class 0 protostars simply do not emit X-rays. A possible explanation may be that, since the stellar embryo is still accreting dynamically, no dynamo-generating convection, and/or no reconnection configuration exists yet, so that neither purely stellar, nor star-disk, X-ray activity may be triggered.

Combined with the fairly common X-ray detection of Class I protostars, this could mean that *"an X-ray birth cry" of stars occurs at the transition between the Class 0 and Class I stages.* Such a statement, clearly crucial for star formation theories, unfortunately may be hard to demonstrate. A time signature in the form of a quasi-periodic signal, linked with rotation as perhaps for YLW15, could be the best evidence.

4 Magnetic topology of YSOs: Confronting theory and observation

To summarize, apart from the very earliest, short-lived Class 0 phases for which no X-rays are observed, YSOs are ubiquitous flaring X-ray sources. On the other hand, theories explain ejection of material accreted from a circumstellar disk by coupling with magnetic fields. Can X-ray observations help constrain these magnetic fields?

First, it must be emphasized again that X-rays can at best be a tracer of *reconnecting* magnetic fields, *i.e.*, when field lines of opposite parity move and get in contact. If the magnetic field topology moves but is (or is assumed to be) kept rigid, then no reconnection occurs and no X-rays are emitted. As recently reviewed by Montmerle (2003), while they qualitatively use the same framework, most theories do not incorporate reconnection. The framework is that of a complex, but basically dipolar, magnetic star-disk topology, inspired from solar loops. Figure 4 illustrates a number of configurations studied in the literature. Rotation is present, but more or less implicitly through the presence of the circumstellar disk, which is in keplerian rotation. But circular symmetry must be broken for accretion to occur across the disk -otherwise the disk would be made of rings. Turbulent viscosity is thought to be the main mechanism through which matter can be transferred radially, but the microphysics are still unclear; the "magnetorotational instability" proposed by Balbus & Hawley (1991) is a strong contender, but other efficient, non-magnetic instabilities may exist. Most theories more or less also use the fact that there must be a point (rather, a ring) in corotation with the star, in such a way that a rigid magnetosphere links the star and the inner region of the disk. At this so-called "X-ring", the material is in an unstable gravitational equilibrium: it can either be accreted radially onto the star, or be ejected centrifugally outwards. The key idea here is that *is both accreted and ejected, from the vicinity of the X-ring.* The most popular scheme is due to Shu *et al.* (*e.g.*, 1997), but variants exist depending on the details of how the material and the magnetic fields are coupled, and where these magnetic fields come from: either from the star itself only (by some convection-induced dynamo), or also tied to the disk (coming from the original interstellar magnetic field, more or less frozen during gravitational collapse).

At any rate, such structures must be twisted at least on large scales; any amount of differential rotation, whether on the disk itself, or between the star and the disk, leads to a deformation of the field lines and therefore to

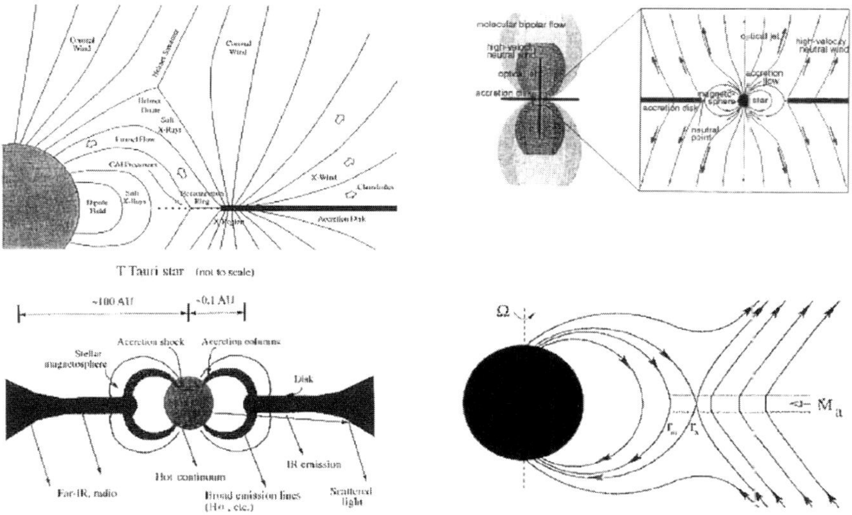

Fig. 4. YSO magnetic configurations taken from the literature. Clockwise from top left: Shu *et al.* (1997), Hirose *et al.* (1997), Ferreira *et al.* (2000), Hartmann (1998).

a potentially reconnecting situation. However, only one series of theoretical models explicitly includes reconnection, hence predictions for X-ray emission: those of the "Japanese school" (*e.g.*, Shibata & Yokoyama 2002, and Refs. therein). This group has made extensive numerical simulations of solar X-ray flares imaged by *Yohkoh*, being able in this fashion to parametrize the reconnection parameters, and adapted the results to other stars and YSOs keeping the same parameters. In particular, they found scaling laws between the various parameters involved (X-ray parameters like T_X, L_X; magnetic field parameters like $B_\star$, loop size, etc.), and applied them to star-disk interactions in protostars and T Tauri stars (Hayashi *et al.* 1996). These scaling laws are in fact very similar to the ones originally found for the Sun (Rosner *et al.* 1978). The "triple X-ray flare" event on the YLW15 protostar, described above, could be explained in this framework, giving support to the general connection between accretion-ejection and magnetic fields. Unfortunately, in view of the fact that such a (possibly) rotationnally-induced event is so far unique, this support remains limited. More generally, even though the theoretical scaling laws appear to be verified over a large range of flaring events, from solar microflares to "superflares" of YSOs, the correlations span a wide range of parameters with a large dispersion (for instance, two orders of magnitude in loop size for a given plasma temperature), and furthermore there is no specific signature of star-disk events, as opposed to stellar events. It is clear that solar-like activity (if enhanced) dominates in YSO, and a detailed model of a rotating star-disk configuration is still lacking.

An additional complication comes from the way "magnetospheric accretion" takes place. In the idealized theoretical models, accretion takes place radially along field lines. In reality, accretion may be taking place in a more complicated manner, though distinct "funnels", perhaps connecting the disk to active regions (see the example of AA Tau, Bouvier *et al.* 1999). Optical spectra show Doppler-broadened emission lines consistent with free fall, but transfer models suggest that the geometry is rather in the form of accretion "columns" (*e.g.*, Muzerolle *et al.* 2001), covering only a fraction (∼10–20%) of the total stellar area. One must not forget that even in the presence of star-disk magnetic fields the standard convection-induced, X-ray emitting magnetic activity is always present, with its associated loops anchored on the star. As a result, the real magnetic field topology must be extremely complex, with closed loops corresponding to the stellar activity, intertwined with star-disk accretion tubes, and outgoing centrifugal ejection field lines rooted near the X-ring!

5 Conclusions

There is a large consensus that magnetic fields are the essential ingredient to explain accretion and ejection phenomena in YSOs, and how they are mutually coupled. However, most theoretical models are in steady-state, with no reconnection explicitly included, thus are unable to make X-ray predictions which one could compare with the large number of X-ray detections now available at all YSO stages except the youngest, Class 0 stage. One class of models however stands out, because it uses reconnection "normalized" to solar flares observed by *Yohkoh*: X-ray observations of YSOs are consistent with such models, but within a large range of X-ray and magnetic parameters, so that in practice X-ray observations are not extremely contraining. There is however one parameter which X-rays give a reliable access to, which is the magnetic field strength at the top of the loops, obtained to a good approximation by equating the plasma and magnetic field pressures: typical equipatition values are ≈100–200 G, or $B_\star \approx 1$ kG at the stellar surface for dipolar loops. Such a value is basically confirmed by recent Zeeman broadening measurements in optical spectra.

Progress in now possible on two grounds:

- *Theory.* Realistic 3D numerical models for star-disk interactions, accretion and ejection, *including energy release by magnetic reconnections*, are needed, *i.e.*, extending for instance the Japanese models.

Only such theoretical models may allow to better understand magnetic braking and angular momentum transfer between the star and the disk. Attention should be paid to observational predictions, in particular X-rays.

- *Observations.* Since X-ray emission is the signature of reconnecting events, but not of how magnetic fields reconnect, and in view of the importance of the rotation history of the star-disk system, it is crucial to look for *time signatures*, in particular periodic ones, with typical time scales of <1 to several days (break-up rotation periods, depending on the stellar mass). To get a better insight into the topology of the magnetic field, evidence for X-ray irradiation from high stellar latitudes must be sought. The best indication is possibly the presence of a spectral line at 6.4 keV, corresponding to fluorescence of neutral Fe atoms. Ideally, this should lead to the "reverberation mapping" method used in AGN spectra. X-ray eclipse studies, similar to Algol (eclipse by a companion) or AA Tau (eclipse by a warped disk) are also important to determine the location of the dominant active regions.

It remains however unclear how standard jet models, which so far assume only equatorial, dipolar structures, can tie in with observations suggesting polar, or at least high-latitude, magnetic fields.

References

[1] P. André, D. Ward-Thompson and M. Barsony, in Protostars & Planets IV, edited by V. Mannings *et al.* (Tucson: U. Arizona, 2000), 59.

[2] S.A. Balbus and J.F. Hawley, *ApJ* **376** (1991) 223.

[3] S. Bontemps, P. André, S. Terebey and S. Cabrit, *A&A* **311** (1996) 858.

[4] J. Bouvier *et al.*, *A&A* **349** (1999) 619.

[5] C. Dougados, S. Cabrit, C. Lavalley and F. Ménard, *A&A* **357** (2000) L61

[6] F. Favata *et al.*, *A&A* **362** (2000) 628.

[7] F. Favata, G. Micela and F. Reale, *A&A* **375** (2001) 485.

[8] E.D. Feigelson *et al.*, *ApJ* **574** (2002) 258.

[9] E.D. Feigelson *et al.*, *ApJ* **584** (2003) 911.

[10] E.D. Feigelson and T. Montmerle, *ARA&A* **37** (1999) 363.

[11] J. Ferreira, G. Pelletier and S. Appl, *MNRAS* **312** (2000) 387.

[12] G. Garmire *et al.*, *AJ* **120** (2000) 1426.

[13] T. Greene and C.J. Lada, *AJ* **124** (2002) 2185.

[14] N. Grosso, T. Montmerle, E.D. Feigelson *et al.*, *Nature* **387** (1997) 56.

[15] L. Hartmann, Accretion processes in astrophysics (Cambridge U. Press, 1999).

[16] A.P. Hatzes and M. Kürstner, *A&A* **346** (1999) 432.

[17] M.R. Hayashi, K. Shibata and R. Matsumoto, *ApJ* **468** (1996) L37.

[18] W. Herbst *et al.*, *A&A* **396** (2002) 513.

[19] S. Hirose, Y. Uchida, K. Shibata and R. Matsumoto, *PASJ* **49** (1997) 193.

[20] K. Imanishi, K. Koyama and Y. Tsuboi, *ApJ* **557** (2001), 747 (Erratum in *ApJ* **579** (2002) 920).

[21] F. Jansen *et al.*, *A&A* **365** (2001) L1.

[22] C.M. Johns-Krull and A.D. Gafford, *ApJ* **573** (2002) 685.

[23] A. Königl, *ApJ* **370** (1991) L39.

[24] T. Montmerle, *Adv. Sp. Res. (COSPAR)* (2003), in press.

[25] T. Montmerle, N. Grosso, Y. Tsuboi and K. Koyama, *ApJ* **532** (2000) 1097.

[26] J. Muzerolle, N. Calvet and L. Hartmann, *ApJ* **550** (2001) 944.

[27] S. Randich, *ASP Conf. Ser.* **198** (2000) 401.

[28] F. Reale, G. Peres and S. Orlando, *ApJ* **557** (2001) 906.

[29] B. Reipurth, L.F. Rodríguez, G. Anglada and J. Bally, *AJ* **124** (2002) 1045.

[30] R. Rosner, W.H. Tucker and G.S. Vaiana, *ApJ* **220** (1978) 643.

[31] C.J. Schrijver and A.M. Title, *ApJ* **551** (2001) 1099.

[32] K. Shibata and T. Yokoyama, *ApJ* **577** (2002) 422.

[33] F.H. Shu, H. Shang, A.E. Glassgold and T. Lee, *Science* **277** (1997) 1475.

[34] B. Stelzer and R. Neuhaüser, *A&A* **377** (2001) 538.

[35] Y. Tsuboi, K. Imanishi, K. Koyama, N. Grosso and T. Montmerle, *ApJ* **532** (2000) 1089.

[36] Y. Tsuboi *et al.*, *ApJ* **554** (2001) 734.

[37] M. Tsujimoto *et al.*, *ApJ* **573** (2002) 270.

[38] M. Weisskopf *et al.*, *PASP* **114** (2002) 1.

[39] G.H.J. van den Oord and R. Mewe, *A&A* **213** (1989) 245.

Section V: New instrumentation

COURSE 14

PIERRE AUGER OBSERVATORY: STATUS AND PROSPECTS

C. LACHAUD

Laboratoire de Physique Corpusculaire et de Cosmologie, Collège de France, Paris, France

Contents

PIERRE AUGER OBSERVATORY: STATUS AND PROSPECTS

C. Lachaud[1] (for the Auger collaboration)

Abstract

The Pierre Auger Observatory is the next generation ground based cosmic ray detector fully devoted to the study of the ultra-high energy cosmic rays (UHECR) with energies above 10^{19} eV. The Auger Observatory, with its unique features (unprecented statistical power, hybrid detection possibilities, full sky aperture, excellent sensitivity for all zenith angles...), will bring in a decisive step, if not definitive answers, in the understanding of one of the most puzzling phenomena of modern astrophysics.

1 Introduction

For its designers and builders, the Auger Observatory is the best (and for the time being, unique) tool to unveil (conclusively?) the mystery of the origin and/or nature of the UHECR. Let us give a few arguments in favour of such a presumptious statement. When completed, *i.e.* with a site in each hemisphere, the Observatory can be considered as a single detector having a full sky coverage, an almost compulsory feature to be able to study weak anisotropies (*e.g.* for sources accumulating in the galactic halo). The detector is "hybrid", meaning that some 10% of the events will be observed both with a ground array and a fluorescence telescope[2]. This unique feature improves the detector's performance in many aspects: cross-calibration of the two components, improving the energy and angular resolution; two independent methods for the reconstruction of the air-shower (and hence a weaker dependence on models used for simulation); an increased number of observables related to the incident cosmic ray's nature, therefore a stronger

[1]LPCC, Paris, France; e-mail: `lachaud@cdf.in2p3.fr`

[2]This article will be restricted to the parameters and performances of the ground array.

lever to study the chemical composition (*e.g.* by a multi-dimensional analysis), etc. The use of Cherenkov tanks rather than scintillators for the ground array increases the angular acceptance to values much larger than the usual 60°. Actually, showers with zenith angles up to 90° can be detected. The specific case of tau neutrinos and angles even larger than 90° is mentioned in Section 3.4 [1]. This allows the opening of a new window for UHE neutrino astronomy, in an energy range where the Earth becomes opaque to neutrinos and makes life harder for underwater or under-ice neutrino telescopes. Finally, maybe the most important performance needed for this physics is the necessary statistical power. Large statistics are needed for good spectrum reconstruction (a necessary indicator of the operating production mechanism), anisotropy studies and chemical composition studies. The complete Auger observatory, with its two sites, will have an aperture of more than 14 000 km^2 sr, nearly two orders of magnitude above the largest operating ground array (AGASA) and 15 to 40 times that of the most powerful existing fluorescence detector (HiRes).

2 Main parameters of the ground array

The full Auger Observatory will consist in two sites of 3000 km^2 each. The ground array stations are to be distributed with a regular spacing of 1.5 km over sites whose shape has to be adapted to the undulations and access conditions of the field. The two sites were chosen on the basis of a list of specifications of which the most important were the size (and flatness of the landscape for easier wireless communication), the latitude (between 35 and 40° North and South for optimum sky coverage), the altitude (around 1400 m, an altitude close to the shower maximum for a vertical shower to minimize statistical fluctuations), dry atmosphere, clear skies and low light pollution for the optical component (and partly for the solar power). After two years of prospecting, the choice was the following. Northern site: Millard County, Utah, U.S.A. (39.1° N and 112.6° W); southern site: Malargüe, Mendoza, Argentina (35.2° N and 69.2° W). A prototype observatory called the engineering array (EA) is already built on the southern site (see below).

Detector stations. Each surface detector is a 3.6 m diameter, 1.2 m high roto-molded plastic tank, filled with filtered and de-ionized water. The internal walls of the tank are lined with a highly reflecting and diffusing material (Tyvek). The charged particles (mainly electrons – including those from Compton scattering and photon conversion – and muons) produce Cherenkov light in the water that is detected by three phototubes installed on the top of the tank.

The electrical power of the station is provided by solar panels.

The electronics and gains of the phototubes are designed so that one can collect signals as close and as far as possible from the shower core (*e.g.* from 1 to 10^5 photoelectrons in a 25 ns time slot) without requiring a too sophisticated system. The linear gain range of the photo-multipliers (PMTs) is required to be 10^5–10^6 (with an operational value of 2×10^5). Each of the three PMTs has two outputs, one (with a low gain amplification) from the anode, another (with a high gain amplification) from the last dynode. The outputs are read with six 10-bit flash-ADCs cycling with a frequency of 40 MHz.

This allows data at distances from the core of the expected highest energy showers in the range of 500 m to several kilometers to be recorded.

Communications. Data from the detector stations are transferred to the CDAS (Central Data Acquisition System) in a two-step process. The first step is the communication between the local stations and base station units (BSU) installed on collector towers situated near the fluorescence detector (FD) buildings (in principle, four of them). The BSUs then transfer the data through a microwave link to the CDAS.

3 Global performance of the ground array

3.1 Aperture

Based on a total area of 3000 km^2, the aperture of the Auger ground array is (for one site) 7350 km^2 sr, when we take 60° as the maximum zenith angle. Since, as was stated in the introduction, a Cherenkov tank is sensitive to particles penetrating its volume at any angle, the value given here can be considered as a lower limit. However, one should keep in mind that showers with very large zenith angles will be difficult to fully analyze in a model independent way (see below).

3.2 Energy resolution

Below 60°, the energy resolution is weakly dependent on the energy and zenith angle. As the energy increases from 10 to 100 EeV or more, its rms value is expected to slightly improve from 12 to 10% or less[1]. This result is almost independent on the nature of the primary particle, with slightly better resolution for showers generated by heavy nuclei because of smaller shower-to-shower fluctuations. At larger zenith angles, the electromagnetic component of the shower is more or less totally absorbed and for the energy

[1] To be strict, this error does not include part of the systematic errors coming *e.g.* from the shower model adopted.

measurement one has to rely mainly on the lateral distribution function for penetrating muons. Therefore, in this range we expect the systematic errors due to hadronic modelling to dominate.

3.3 Angular resolution

The measurement of the incident cosmic ray's angle is made by fitting an adapted surface (paraboloid) to the shower front reconstructed from the arrival time of the particles on the array stations. This of course is a totally model-independent measure, and the precision on the angles depends on a few parameters: the number of stations hit by the shower front particles *i.e.* the projection of the front on the ground (which depend on the zenith angle and the primary energy); the fluctuations on the arrival time of the particles (therefore the thickness of the shower front); the precision on the *relative* synchronization of the detector stations (the interpolated and corrected GPS time). The average angular resolution[2] for a hadronic shower at 30° for all energies in the detector's optimum energy range (>10 EeV) is about 1°. This is obtained with a timing precision of about 15 ns (GPS time interpolated with a 100 MHz clock). This resolution improves with the energy ($\simeq$0.5° at the same angle above 100 EeV), and with the incident angle ($\simeq$0.3° at $\theta = 80°$, all energies).

3.4 Composition analysis

The number of known stable particles that can propagate over large distances and then be detected on earth are quite limited in numbers: nuclei (heavy or light, with special emphasis on protons), photons and neutrinos. In the search for a solution to the UHECR puzzle, several authors invoke exotic particles or interactions (meaning not yet established with experimental evidence). Such models have called for mechanisms which would violate the GZK cutoff. Examples are: neutrinos whose cross-sections on nuclei are substantially increased by graviton exchanging interactions [4]; vortons [5] (stable, superconducting cosmic string loops); magnetic monopoles [6], and so on. Here we shall limit our comments to standard particles and interactions.

Different particles leave different fingerprints on the EAS parameters. Two such parameters are the position of the shower maximum $X_{\max}$ and the relative muon content of the shower at ground level (which of course depends on the slant-depth $X = X_0/\cos\theta$ where X_0 is the vertical atmospheric depth at the altitude of the detector). The shower maximum can be

[2] All the values given here for the angular resolution mean the uncertainty which contains 68% of all events.

measured directly by the fluorescence detector. Only the ground array can measure (to some extent and at large distances from the shower core) the muon content by the analysis of the the flash-ADC traces. Other indicators such as the shower front curvature, rise time of the signal detected by the array stations, steepness of the lateral distribution functions (LDF) are parameters to use in a multi-dimensional analysis of the nature of the primary cosmic ray. However, all simulation and reconstruction studies show that the discrimination between heavy and light nuclei (the most likely ones to be found in the incident samples being iron nuclei and protons) will be possible only on a statistical basis. This is due to physics: the depth at which the shower maximum occurs (and the parameters which derive from it and are measured by the ground array) is a strongly fluctuating parameter and its values, even for extreme cases such as protons and iron nuclei, have such an overlap that only their average over large numbers will be distinguishable.

The situation improves dramatically for gammas and neutrinos, strong signatures of exotic processes (top-down mechanisms) if found in large proportions at extreme energies.

The identification of UHE gammas is based on two physical phenomena: photon conversion in interactions with the geomagnetic field and the LPM effect. Gamma rays with energies E_γ propagating through a magnetic field with a transverse component $B_\perp$ have a large probability of converting into electron-positron pairs if the product $E_\gamma B_\perp$ is large. This happens in particular for $E_\gamma > 10$ EeV for typical values of the geomagnetic field. The conversion occurs at altitudes of several thousands kilometers (therefore well above the atmosphere). The e^+e^- pairs then undergo energy loss by magnetic bremsstrahlung and so on. By the time the secondary electrons and photons reach the atmosphere, they all have energies less than a few EeV. The development of this pre-shower in the atmosphere is like that of a superposition of several low energy electromagnetic showers.

However, whenever the incident gamma arrives with its direction parallel to the field vector, the conversion becomes negligible and the gamma penetrates the atmosphere with its full energy. Then a second phenomenon, the Landau–Pomeranchuk–Migdal (LPM) effect, takes over. This effect consists of a decreasing of the cross-sections of electromagnetic processes with the energy of photons/electrons and the density of the propagation medium. It becomes dominant over the Bethe–Heitler processes for particles with typical energies of around 100 EeV even in the rarefied layers of the upper atmosphere.

Figure 1 shows how these effects operate [8] at a given place, namely at the site of the southern Auger Observatory. One can see how the gamma conversion probability depends on the energy of the photons and their direction in the earth's reference frame. The dashed circles show the directions

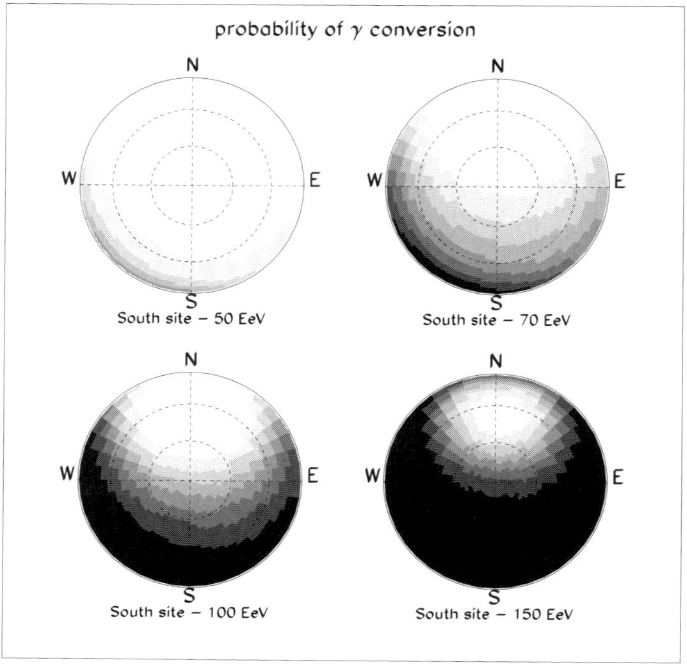

Fig. 1. Photon conversion probabilities as a function of the direction in the earth's frame of reference as seen from the southern Auger site [8]. The conversion probabilities range from 100% (black areas) to 0% (white zones).

with zenith angles of 30 and 60 degrees. One can see that for all values of the gamma ray energy, the conversion probability is zero in a direction close to 50° North. UHE gammas coming from this direction will deeply penetrate the atmosphere before starting to shower. Therefore their geometry compared to those coming from the other directions will be quite distinctive and detectable.

With the use of the curvature, as well as the steepness of the LDF or the muon content of the EAS, it is expected that a 5–10% contamination by UHE photons of a sample of protons will be detectable.

The detection of neutrinos in the same energy range with a ground array is based on the fact that a particle penetrating the atmosphere tangentially to the earth will encounter an enormous amount of matter [10]: more than 800 radiation lengths, about 350 nuclear interaction lengths. For horizontal airshowers (HAS) the atmosphere absorbs all particles except neutrinos and muons. Using cross-section values extrapolated from lower energy data [9], the probability that UHE (anti)neutrinos interact in such a thickness of air is larger than 10^{-4} above 0.1 EeV, which is far from being negligible. Moreover, this energy range is just above the limit accessible to the neutrino telescopes for up-going neutrinos, since above 10^{16} eV the earth becomes opaque to them. It was therefore interesting to see what a giant ground array such as Auger can do with the detection of UHE neutrinos. The studies are based on EAS incident with a zenith angle above 75 degrees and energies between 0.1 and 100 EeV. For neutrinos above 1 EeV, the acceptance of the ground array is more than 10 km^3 water equivalent. The background to the expectedly weak neutrino signal is of course the numerous hadronic showers incoming at large angles. However, making use of the same methods as in the case of UHE gammas (in particular the shape of the shower front and the signal risetime) it is expected that even a few neutrino events per year should be detectable.

An equally interesting case is that of the tau-neutrinos if neutrino oscillations as observed by Super-K and SNO are confirmed. With a maximum mixing, and given the large propagation distances, it is expected to have equal proportions of the three neutrino species incident on earth, if the ν_μ/ν_e ratio at the source is 2/1. In this case, it becomes very interesting to look for tau neutrinos through their interaction in the ground (*i.e.* with zenith angles larger than 90°) producing a τ. In our energy ranges, the taus can propagate over tens of km before decaying (or interact with the earth and start an iterative process). If the decay occurs in the volume of air above the ground array, the direction of the showers could, at least statistically, give indications to discriminate them from the "standard" HAS.

Detailed studies were made [1, 11] to estimate the performance of the Auger ground array with respect to neutrino detection. Figure 2 is a summary of recent studies where the sensitivity for one event per year per decade of energy is shown for ν_μ/ν_e (top thick line) and ν_τ (shaded area). The curves limiting the shaded zone on this figure are for deep-inelastic scattering with strong energy loss (top) and no such loss (bottom). The expected event rate depends of course on the models of UHE neutrino production. Whenever the sensitivity curve remains below the flux curves for more than one decade of neutrino energy, more than one event per year is expected to be observed. For models considered as being solid [12], such as the so-called GZK neutrinos (those produced as secondaries of the interactions of UHECR with the 2.7 K microwave background, full line on the figure) the detection rate appears to be rather low. For more speculative estimates (GRBs, AGN etc.) several events per year should be detected. The horizontal line shown as the "ν_τ limit" is the 90% confidence level limit (background-free detection) achievable with five years of data taking with Auger for an E^{-2} flux between 0.3 and 3 EeV.

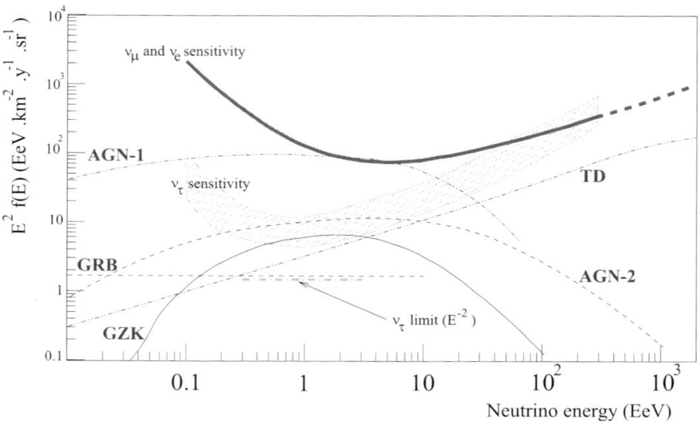

Fig. 2. Sensitivity of Auger to neutrino detection [1]. The ν_μ/ν_τ flux estimates are from [12] and are divided by a factor 2 to take into account the full mixing hypothesis. See text for comments.

4 Status and prospects

The Collaboration decided to phase the construction of the southern site in two steps. During 2001, the prototype hybrid system called the Engineering Array (EA) was built. It consists of 40 tanks covering an area of about 50 km^2 and two elements of fluorescence telescopes overlooking the area equipped with the tanks. The EA aims to test all the technical issues (including those where the final choice is still open) before the launching of the final production. The deployment of the full observatory is expected to be completed by the end of 2005.

The data presented here come from unnumerable sources within the Auger collaboration. Those who provided this information will know that my deep gratitude goes to them. A special thanks to Murat Boratav who helped me in the writing of this lecture.

References

[1] X. Bertou *et al.*: Tau neutrinos in the Auger Observatory: a new window to UHECR sources. *Astropart. Phys.*, in press [astro-ph/0104452].

[2] P. Auger and R. Maze, *Comptes rendus, Acad. Sci.* **207** (1938) 228.

[3] Cosmic Rays above 10^{19} eV. *Nucl. Phys. B (Proc. Suppl.)* **28B** (1992); edited by M. Boratav, J.W. Cronin and A.A. Watson.

[4] P. Jain *et al.*, *Phys. Lett. B* **484** (2000) 267.

[5] R.L. Davis and E.P.S. Shellard, *Phys. Rev. D* **38** (1988) 4722. S. Bonazzola and P. Peter, *Astropart. Phys.* **7** (1997) 161.

[6] T.W. Kephart and T.J. Weiler, *Astropart. Phys.* **4** (1996) 271.

[7] B. McBreen and C.J. Lambert, *Phys. Rev. D* **24** (1981) 2536.

[8] X. Bertou, P. Billoir and S. Dagoret-Campagne, *Astropart. Phys.* **14** (2000) 121.

[9] R. Gandhi *et al.*, *Astropart. Phys.* **5** (1996) 81.

[10] See *e.g.* K.S. Capelle *et al.*, *Astropart. Phys.* **8** (1998) 321; D. Fargion, A. Aiello and R. Conversano [astro-ph/9906450] etc. among the abundant literature on this issue.

[11] M. Ave *et al.*, *Astropart. Phys.* **14** (2000) 109.

[12] R.J. Protheroe [astro-ph/9809144].

COURSE 15

OBSERVATIONS OF THE HIGH ENERGY γ-RAY UNIVERSE

C. MASTERSON

Max-Planck-Institute for Nuclear Physics, Saupfercheckweg 1, 69117 Heidelberg, Germany

Contents

OBSERVATIONS OF THE HIGH ENERGY γ-RAY UNIVERSE

C. Masterson

Abstract

This article describes the nature and current status of observational programs studying astrophysical high energy γ-rays. It begins by describing ground based γ-ray observatories, using the H.E.S.S. project as an example. Major satellite based γ-ray detectors, past and present are discussed. A brief overview of interesting astrophysical and cosmological phenomena which may be studied using these classes of detectors is also provided.

1 Experimental techniques in γ-ray astronomy

The most successful technique for observing astrophysical γ-rays at energies higher than 100 GeV is the imaging air-Čerenkov technique (IAČT), which is based on their indirect detection through Čerenkov light emitted in the atmosphere. A γ-ray impingent on the atmosphere pair produces, typically at an altitude of 20 km, and a an electromagnetic cascade of photons, electrons and positrons results. The charged particles emit Čerenkov light over an opening angle of roughly $1°$, which is detectable over a radius of typically 150 m on the ground.

This Čerenkov light is detected by a telescope with fast light detectors sensitive to the very rapid pulses of light produced (a few nanoseconds duration). Very large mirrors are required to generate images of the shower in the focal plane of the telescope camera. Based on shape and intensity of these images it is possible to distinguish γ-ray showers from the much larger number of showers from charged particles such as protons and helium nuclei, which make up the bulk of the isotropic flux of cosmic rays which strike the earth.

By reconstructing the arrival directions of the γ-rays and their energies it is possible to study the sources producing this radiation. The major advantage of this detection technique is the large collection area possible due to the use of the atmosphere as a detection medium, of the order of 10^5 sq m. This is vital for the detection of the low fluxes involved. There are a number of major new ground-based telescopes under construction at the moment, including the VERITAS MAGIC and CANGAROO projects, however this paper will concentrate on the H.E.S.S. project as an example of this type of detector.

At lower energies (below 100 GeV), the flux of γ-rays from astrophysical sources is sufficiently high for direct detection in space using satellite based telescopes. These have the advantage of allowing easy discrimination of background charged particles but suffer from low collection area and large angular resolution. The duty cycle of sensitive field of view of a satellite is also better than ground-based telescopes.

1.1 The stereo imaging technique

The majority of major ground based IAČT projects currently under construction are based on the Stereo technique, pioneered by the HEGRA collaboration [1]. This technique uses a group of identical telescopes to record multiple images of the same air shower. Some of the main advantages of this method are the better measurement of shower parameters, due to multiple images of each shower, the reduction in random triggers due to the coincidence demanded between a number of telescopes and the fact that local muons passing near to a single telescope will not trigger the system.

In a stereo system the impact position of the shower can be geometrically reconstructed, thus allowing the primary particle energy to be more precisely measured. Angular resolution is also improved as the shower direction is more precisely reconstructed on the sky.

1.2 The H.E.S.S. telescope array

The H.E.S.S. telescope array is situated in the Khomas Highlands in Namibia (long. 23° 16′ S, 16° 30′ E, at 1800 m asl.). It consists of four identical telescopes, as shown in Figure 1, with a 120 m separation in a square configuration. Each telescope structure consists of an azimuthal mount supporting a dish of 12 m in diameter (in the Davies – Cotton configuration), with a 15 m focal length. The dish carries 382 front-aluminised glass mirrors which focus light onto the camera focal plane. The camera (shown in Fig. 2) consists of 960 photomultiplier tubes (PMTs) in a hexagonal arrangement, giving a 5° field of view. The camera electronics are designed using a modular construction, whereby a group of 16 tubes are mounted together with the necessary power supply and readout electronics in a single unit (called a Drawer).

Fig. 1. The first operational telescope of the H.E.S.S. array.

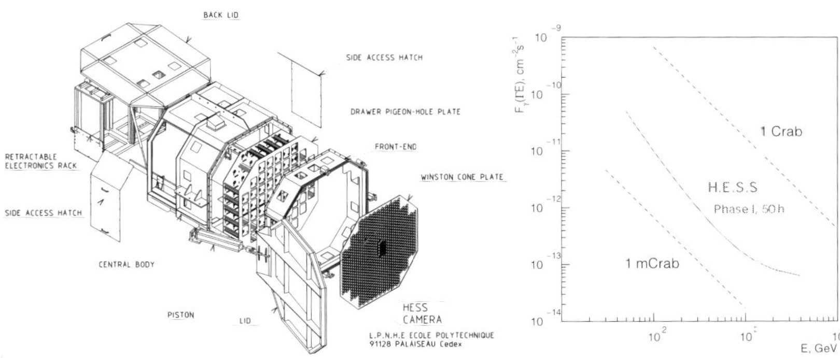

Fig. 2. A diagram of the H.E.S.S. camera, showing the modular structure. Expected performance of the H.E.S.S. array compared to the flux from the Crab Nebula [2].

As H.E.S.S. is based on a self-triggering data acquisition scheme, it is necessary to store the signals received by each channel while the trigger decision is made for an event. This is done using solid state analog ring sampler (ARS) circuits which sample the analog signal once per nanosecond and store it for a maximum of 128 ns. The trigger is formed from the coincidence of signals above a certain level in a number of PMTs. The delay chips are then read out and multiplexed into ADCs to record the shower images. Two channels are used with different gain for each PMT in order to extend the dynamic range of the system. A camera control CPU reads out the Drawers and transmits the data to the central data acquisition system.

1.3 H.E.S.S. system performance

It is expected that the combination of the greater light collection capability of H.E.S.S. compared with current instruments, together with the finer pixellation and the faster electronics will allow for detailed imaging of γ-ray air showers in the atmosphere. This will allow better rejection of background showers at lower energies, improved angular resolution and better energy reconstruction.

Figure 2 compares the expected sensitivity of the H.E.S.S. experiment with the flux measured from the Crab Nebula, the standard reference of TeV astrophysics. H.E.S.S. will have a sensitivity improved by up to an order of magnitude over current instruments and a greater sensitive energy range.

Observations commenced with the first telescope of the H.E.S.S. system in July 2002. Calibration and engineering data have been taken and the current focus is on understanding the individual telescope and camera system before the planned completion of the second telescope in early 2003, when stereo observations will begin. The system will be completed in 2004.

1.4 Satellite based γ-ray astrophysics

The EGRET detector, shown in Figure 3, on the Compton γ-ray observatory combined a gas filled spark chamber detector and a Sodium Iodide (NaI) scintillator calorimeter to detect γ-rays in the range from 30 MeV to 30 GeV from astrophysical targets.

EGRET made a large number of important discoveries in its lifespan during its service life of nine years. [5] describes the third, and probably final, EGRET catalogue of high energy γ-ray sources, which includes a complete all-sky survey of point sources of γ-ray emission in this energy range. Of the more then 200 discovered sources, most are unidentified, mainly due to the poor angular resolution of the instrument.

The natural successor to EGRET is the GLAST experiment, shown also in Figure 3, a planned satellite based γ-ray detector, scheduled to be launched in 2007. It will use silicon-strip technology in the tracking detector, which has a large surface area, providing a large (1 sq m) collection area for γ-rays and a higher angular resolution. A NaI calorimeter mounted in series with the tracker will allow for good energy resolution also. It is expected that

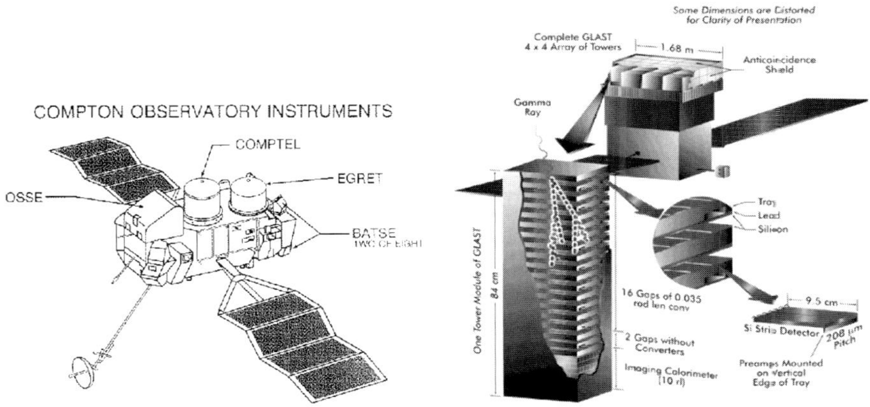

Fig. 3. The Compton γ-ray observatory, containing the EGRET experiment [3]; the planned GLAST γ-ray telescope [4].

GLAST will have significant sensitivity up to 300 GeV, overlapping with ground-based experiments.

2 Astrophysics at VHE energies

The EGRET all-sky survey raised a large number of questions about the nature of emission from both point sources and from diffuse emission regions of the Galactic Plane itself. Complementary observations by current ground based detectors at higher energies have only detected a small number of these sources, as well as some not seen by EGRET. Further observations with higher sensitivity and over wider energy ranges are vital to the understanding of these objects.

2.1 Active Galactic Nuclei

Some of the most important results in VHE γ-ray astronomy have come from the Bl Lac category of Active Galactic Nuclei (AGN). Many observations of this type of source have been made at TeV energies and a number detected; some spectral energy distributions of AGN are shown in Figure 4. The high sensitivity and low energy thresholds of the new class of ground based detectors will allow searches for baseline emission from these sources, which is currently undetected. It will also allow more accurate measurement of the spectral characteristics of the emission and study of the physical processes involved. Very high energy γ-rays have a significant cross-section for pair production with infrared photons. As a consequence absorption of γ-rays

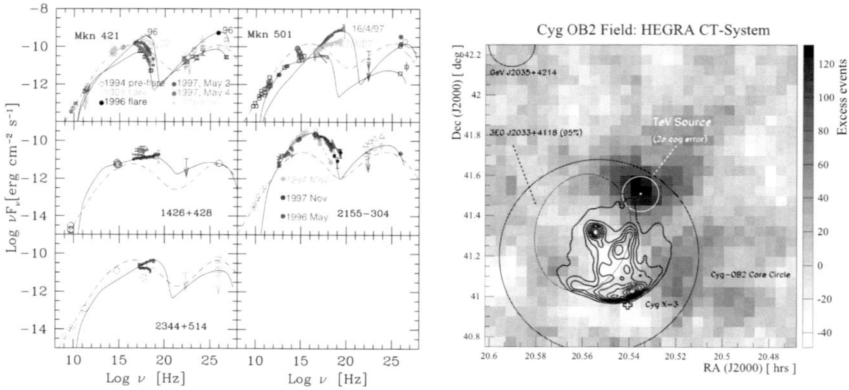

Fig. 4. Spectral energy distributions for a number of AGN detected at TeV energies [6]; a reconstructed map of γ-ray intensity on the sky for the unidentified source in the Cygnus region [7]. The X-ray emission is shown as contours.

from extragalactic sources modifies the observed energy spectrum. The detection of more distant, AGN will allow more accurate measurement of this infrared absorption and give better insight into the density of infrared light outside our Galaxy.

2.2 Supernova remnants

Another major question in the field of high energy astrophysics is the origin of hadronic cosmic rays. It is thought that this radiation may originate in Supernova remnants, the left over shock front from supernova explosions. Due to the effect of Galactic magnetic fields these particles impinge on the earth isotropically. The detection of γ-ray emission from these objects could provide conclusive proof of this hypothesis by indicating the existence of high energy protons at the source.

The CANGAROO telescope in Australia has observed two Southern SNRs (SN 1006 [8] and RX J1713 [9]) at TeV energies but it is thought that emission from the first is possibly due to Inverse Compton processes, which does not require Cosmic Ray acceleration. Possible evidence for hadronic emission is claimed for the second.

2.3 Pulsars

Current models of charged particle acceleration in pulsars differ strongly in their predictions of emission at energies above 1 GeV. The Polar Cap model [10], which predicts acceleration close to the polar region of the pulsar places a strong upper limit on maximum γ-ray energy. The Outer Gap model [11], allows for the possibility of detectable γ-ray emission up to hundreds of GeV. Thus detections of γ-rays would be very important for understanding the processes involved. EGRET detected seven pulsars, including Geminga which was previously only seen in X-rays, however no evidence of pulsed emission has been seen at higher energies. Models are most strongly constrained by CELESTE observations [12] of the Crab Nebula at an energy threshold below 100 GeV. Improved sensitivity of the new generation of detectors will allow much better understanding of this area.

2.4 Unidentified EGRET sources

A major legacy of the EGRET experiment has been the large number of unidentified sources discovered. Many of these lie along the Galactic Plane and are impossible to identify with known sources at other wavelengths due the crowded nature of this region. New ground based experiments, such as H.E.S.S., should have the angular resolution, sensitivity and energy threshold to improve the understanding of a large number of these objects. An example of this is the recent serendipitous discovery near the Cyg OB2 star forming region, found during a search of HEGRA archive data for unidentified EGRET sources. Full Galactic Plane surveys are also planned with these new instruments. An eventual all sky survey is possible, although the larger field of view of GLAST will make it an instrument better suited to this work.

2.5 Fundamental physics

It is thought that some γ-rays may be produced by exotic particle annihilation in the Galactic Centre or Galactic Clusters. Detections of this type of emission, or upper limits on the flux of these particles, would be very useful as limits on models of non-standard physics and could provide insights into the nature of non-baryonic dark matter.

3 Conclusions

Currently the field of γ-ray astrophysics is exciting, as more sensitive new ground based detectors, such as H.E.S.S., are coming into operation. New satellites are also planned in the near term to replace the very successful EGRET experiment. Complementary observations will soon be possible with these two technologies.

With the order of magnitude improvement in sensitivity, there are a large number of interesting astrophysical questions which will be addressed.

These include the question of the origin of cosmic rays, acceleration processes in AGN and pulsars, as well as some fundamental particle physics processes.

References

[1] A. Daum *et al.*, *Astropart. Phys.* **8** (1997) 1.
[2] K. Bernlöhr *et al.*, in Proc. of the Rencontres de Moriond 2001: Very High Energy Phenomena in the Universe, Les Arcs (2002).
[3] G. Kanbach *et al.*, *Space Sci. Rev.* **49** (1988) 69.
[4] D.A. Kniffen, D.L. Bertsch and N. Gehrels, in GeV-TeV Gamma Ray Astrophysics Workshop: towards a major atmospheric Cherenkov detector (2000) 492.
[5] R.C. Hartman *et al.*, *ApJS* **123** (1999) 79.
[6] L. Costamante and G. Ghisellini, *A&A* **384** (2002) 56.
[7] F. Aharonian *et al.*, *A&A* **393** (2002) L37.
[8] T. Tanimori *et al.*, *ApJ* **497** (1998) L25.
[9] H. Muraishi *et al.*, in *Proc. 26th Int. Cosm. Ray Conf. (Salt Lake City)* **3** (1999) 500.
[10] A.K. Harding, TeV Gamma-Ray Astrophysics (Kluwer Acedemic Press, Dordrecht, 1996) 257.
[11] K. Hirotani and S. Shibata, *ApJ* **558** (2001) 216.
[12] M. de Naurois *et al.*, *ApJ* **566** (2002) 343.

SESSIONS PRÉCÉDENTES

1951–2002

I	1951	Quantum mechanics. Quantum field theory
II	1952	Quantum mechanics. Statistical mechanics. Nuclear physics
III	1953	Quantum mechanics. Solid state physics. Statistical mechanics. Elementary particle physics
IV	1954	Quantum mechanics. Collision theory. Nucleon-nucleon interaction. Quantum electrodynamics
V	1955	Quantum mechanics. Non-equilibrium phenomena. Nuclear reactions. Interaction of a nucleus with atomic and molecular fields
VI	1956	Quantum perturbation theory. Low temperature physics. Quantum theory of solids; dislocations and plastic properties. Magnesium; ferromagnetism
VII	1957	Scattering theory. Recent developments in field theory. Nuclear interaction; strong interactions. High energy electrons. Experiments in high energy nuclear physics
VIII	1958	The many body problem
IX	1959	The theory of neutral and ionized gases*
X	1960	Elementary particles and dispersion relations*
XI	1961	Low temperature physics*
XII	1962	Geophysics; the earth's environment*
XIII	1963	Relativity groups and topology*
XIV	1964	Quantum optics and electronics*
XV	1965	High energy physics*
XVI	1966	High energy astrophysics*
XVII	1967	Many body physics*
XVIII	1968	Nuclear physics*
XIX	1969	Physical problems in biological systems*
XX	1970	Statistical mechanics and quantum field theory*
XXI	1971	Particle physics*
XXII	1972	Plasma physics*
XXIII	1972	Black holes*
XXIV	1973	Fluids dynamics*

* Sessions ayant reçu l'appui de Comité Scientifique de l'OTAN.

XXV	1973	Molecular fluids*
XXVI	1974	Atomic and molecular physics and the interstellar matter*
June Inst.	1975	Structural analysis of collision amplitudes
XXVII	1975	Frontiers in laser spectroscopy*
XXVIII	1975	Methods in field theory*
XXIX	1976	Weak and electromagnetic interactions at high energy*
XXX	1977	Nuclear physics with heavy ions and mesons*
XXXI	1978	Ill-condensed matter*
XXXII	1979	Membranes and intercellular communication*
XXXIII	1979	Physical cosmology*
XXXIV	1980	Laser plasma interaction
XXXV	1980	Physics of defects*
XXXVI	1981	Chaotic behaviour of deterministic systems*
XXXVII	1981	Gauge theories in high energy physics*
XXXVIII	1982	New trends in atomic physics*
XXXIX	1982	Recent advances in field theory and statistical mechanics*
XL	1983	Relativity, groups and topology*
XLI	1983	Birth and infancy of stars*
XLII	1984	Cellular and molecular aspects of developmental biology*
XLIII	1984	Critical phenomena, random systems, gauge theories*
XLIV	1985	Architecture of fundamental interactions at short distances*
XLV	1985	Signal processing*
XLVI	1986	Chance and matter*
XLVII	1986	Astrophysical fluid dynamics*
XLVIII	1988	Liquids at interfaces*
XLIX	1988	Fields, strings and critical phenomena*
L	1988	Oceanographic and geophysical tomography*
LI	1989	Liquids, freezing and glass transition*
LII	1989	Chaos and quantum physics*
LIII	1990	Fundamental systems in quantum optics*
LIV	1990	Supernovae*
LV	1991	Particles in the nineties*
LVI	1991	Strongly interacting fermions and high T_c superconductivity*
LVII	1992	Gravitation and quantizations*
LVIII	1992	Progress in picture processing*

* Sessions ayant reçu l'appui de Comité Scientifique de l'OTAN.

LIX	1993	Computational fluid dynamics*
LX	1993	Cosmology and large scale structure*
LXI	1994	Mesoscopic quantum physics*
LXII	1994	Fluctuating geometries in statistical mechanics and quantum field theory*
LXIII	1995	Quantum fluctuations*
LXIV	1995	Quantum symmetries*
LXV	1996	From cell to brain*
LXVI	1996	Trends in nuclear physics, 100 years later
LXVII	1997	Modélisation du climat de la terre et de sa variabilité*
LXVIII	1997	Particules et interactions : le modèle standard mis à l'épreuve*
LXIX	1998	Topological aspects of low dimensional systems*
LXX	1998	Infrared space astronomy, today and tomorrow*
LXXI	1999	The primordial universe
LXXII	1999	Coherent atomic matter waves*
LXXIII	2000	Atomic clusters and nanoparticles*
LXXIV	2000	New trends in turbulence*
LXXV	2001	Physics of bio-molecules and cells*
LXXVI	2001	Unity from duality: Gravity, gauge theory and strings*
LXXVII	2002	Slow relaxations and nonequilibrium dynamics in condensed matter*

* Sessions ayant reçu l'appui de Comité Scientifique de l'OTAN.
Publishers: Session VIII: Dunod, Wiley, Methuen; Sessions IX & X: Herman, Wiley – Session XI: Gordon and Breach, Presses Universitaires – Sessions XII-XXV: Gordon and Breach – Sessions XXVI-LXVIII: North-Holland – Session LXIX-LXXVII: EDP Sciences, Springer.

Achevé d'imprimer par Corlet, Imprimeur, S.A.
14110 Condé-sur-Noireau
N° d'Imprimeur : 74811 - Dépôt légal : novembre 2003
Imprimé en France